EMBASE

List of Journals Indexed 2006

ELSEVIER

Bibliographic Databases

http://www.info.embase.com

Global Customer Support and Sales Offices

Americas (North/South/Central)
US Customers
Toll Free: +1 (800) 545-2522
Fax: +1 (800) 535-9935

Customers Outside US
Toll Free: +1 (800) 460-3110
Tel: +1 (314) 453-7010
Fax: +1(314) 453-7095
E-mail: usbkinfo@elsevier.com

Asia Pacific
Tel: +65 6349 0222
Fax: +65 6733 1817
E-mail: asiabkinfo@elsevier.com

Australia, New Zealand & Fiji
Tel: +61 (02) 9517 8999
Fax: +61 (02) 9517 2249
E-mail: service@elsevier.com.au

Europe, Middle East and Africa
Tel: +44 (0) 1865 474000
Fax: +44 (0) 1865 474001
E-mail: eurobkinfo@elsevier.com

Japan
Tel: +81 3 3589 6370
Fax: +81 3 3589 6371
E-mail: jp.bkinfo@elsevier.com

ELSEVIER

Bookmark our site at:
www.elsevier.com

ISBN: 0-444-52939-X

RESEARCH
Within your reach

TABLE OF CONTENTS

INTRODUCTION

This guide gives a complete list of all journals which are screened for inclusion in EMBASE, as well as for other EMBASE information products, including EMBASE Alert.

The EMBASE journal collection includes over 5,000 biomedical journals[1] that are covered comprehensively.

List of Current Titles

In the **List of Current Titles** journals are arranged in alphabetical order by full journal title.

The full journal title is followed by the International Standard Serial Number (ISSN), the abbreviated journal title, the name, address, telephone and fax numbers (if available) of the publisher, and the publication frequency (issues/yr).

Title abbreviations follow the guidelines established by the International Council of Scientific Unions Abstracting Board (ICSU-AB) and the International Center for the Registration of Serials (ICRS).

The journals marked with ● are the priority journals within the EMBASE collection. These are the journals which are considered to be the most important in biomedicine. In EMBASE, priority journals can be retrieved using the index term 'priority journal'. There are currently 1835 priority journals[1].

Additional Lists

Changes in the EMBASE journal collection are noted in **New Titles**, **Changed Titles**, and **Discontinued Titles**. Following on from last year, this edition also includes a list of Discontinued Titles (1995-2004) containing the ISSN, full journal title, and the last volume/issue/publication year. Other lists include: **List of Priority Journals**, **Number of Journals Per Classification**, **Subject Classification List**, **Number of Journals by Country of Publication**, and **List of Journals by Country of Publication**.

Journal Collection Update Procedure

The EMBASE journal collection is reviewed by an editorial committee and is updated at the beginning of each calendar year. Throughout the year new journals are being evaluated for coverage. Important criteria for journal selection are:

- **Subject scope:** emphasis on drug research, pharmacology, pharmaceutics and toxicology; human medicine; clinical and experimental, including human biology; health policy and management, including pharmacoeconomics; public, occupational and environmental health, and pollution; substance dependence and abuse; psychiatry; forensic medicine; biotechnology, biomedical engineering and medical devices.
- **Scientific and editorial quality:** the most important criterion is peer review.
- **Language requirements:** although EMBASE covers journals in all languages, English-language abstracts and article titles are generally required for journals with articles written in languages other than English.

---

1. Please note that the number of journals fluctuates during the year.

NEW TITLES

This list contains journal titles that have been added to the EMBASE journal collection at the beginning of 2006.

Academic Journal of Second Military Medical University
Acta Academiae Medicinae Sinicae
Acta Anatomica Sinica
Acta Biomaterialia
Acta Cirurgica Brasileira
Actualites Pharmaceutiques Hospitalieres
Adipocytes
Advances in Breast Cancer
Advances in Gastrointestinal Cancers
African Journal of Traditional, Complementary and Alternative Medicines
Aging Health
AIDS Research and Therapy
Aktualnosci Neurologiczne
Alzheimer's and Dementia
American Journal of Oncology Review
American Journal of Psychiatric Rehabilitation
Anadolu Psikiyatri Dergisi
Annali Degli Ospedali San Camillo e Forlanini
Annals of Hepatology
Annual Review of Clinical Psychology
Annual Review of Pathology
Archives of Medical Science
Archivos de la Facultad de Medicina de Zaragoza
Archivos de Medicina
ASGE Clinical Update
Asia-Pacific Journal of Clinical Oncology
Asia-Pacific Journal of Public Health
Australia and New Zealand Health Policy
Autophagy
Behavioral and Brain Functions
Biochemia Medica
Biomedical Digital Libraries
Biomedical Materials
Biomedical Signal Processing and Control
BMC Evolutionary Biology
Breast Cancer Online
Breast Care
British Journal of Hospital Medicine
Cancer Informatics
Case Reports and Clinical Practice Review
Cell Communication and Signaling
Cell Metabolism
Cerebrospinal Fluid Research
Chinese Critical Care Medicine
Chinese Journal of Andrology
Chinese Journal of Cerebrovascular Diseases
Chinese Journal of Clinical Nutrition
Chinese Journal of Contemporary Neurology and Neurosurgery
Chinese Journal of Emergency Medicine
Chinese Journal of Infection and Chemotherapy
Chinese Journal of Interventional Imaging and Therapy
Chinese Journal of Medical Imaging Technology
Chinese Journal of Ophthalmology
Chinese Journal of Pathology
Chinese Journal of Radiology
Chinese Journal of Rehabilitation Medicine
Chinese Traditional and Herbal Drugs
Chiropractic and Osteopathy
Chronic Illness
Clinical and Translational Oncology
Clinical Dermatology: Retinoids and other Treatments
Clinical Medicine and Research
Clinical Neuropsychiatry
Clinical Practice and Epidemiology in Mental Health
Clinical Radiology Extra
Clinical Teacher
Cochrane Database of Systematic Reviews
Colombia Medica
Comparative Biochemistry and Physiology - Part D: Genomics and Proteomics
Comprehensive Ophthalmology Update
Congenital Heart Disease
Contagion: Reports, Cases, and Commentaries in HIV and Infectious Disease Research
CONTINUUM Lifelong Learning in Neurology
Cor et Vasa
Core Evidence
Current Cancer Therapy Reviews
Current Cardiology Reviews
Current Computer-Aided Drug Design
Current Diabetes Reviews
Current Enzyme Inhibition
Current Heart Failure Reports
Current Hematology Reports
Current Hepatitis Reports
Current HIV/AIDS Reports
Current Hypertension Reviews
Current Immunology Reviews
Current Medical Imaging Reviews
Current Nanoscience
Current Nutrition and Food Science
Current Osteoporosis Reports
Current Pediatric Reviews
Current Pharmaceutical Analysis
Current Prostate Reports
Current Psychiatry Reviews
Current Psychosis and Therapeutics Reports
Current Respiratory Medicine Reviews
Current Rheumatology Reports
Current Rheumatology Reviews
Current Sexual Health Reports
Current Sports Medicine Reports
Current Topics in Neurochemistry
Current Topics in Peptide and Protein Research
Current Topics in Pharmacology
Current Topics in Toxicology
Current Trends in Immunology
Current Urology Reports
Current Women's Health Reviews
Deutsches Arzteblatt
Drugs in Context
Drugs in Context (US)
Electrolyte and Blood Pressure
European Journal of Emergency Medicine
Experimental and Clinical Transplantation
Expert Opinion on Drug Metabolism and Toxicology
Expert Review of Clinical Immunology
Explore: The Journal of Science and Healing
Family Medicine and Primary Care Review
Frontiers in Drug Design and Discovery
Frontiers in Medicinal Chemistry
Future Cardiology
Future Oncology
Gastroenterology and Hepatology
Ginekologia Onkologiczna
Ginekologia Praktyczna

New

Globalization and Health
Heart and Metabolism
Heart Failure Clinics
Hippokratia
Human Movement
Human Resources for Health
Human Vaccines
Immunity and Ageing
Insulin
International Journal of Bioinformatics Research and Applications
International Journal of Biological Sciences
International Journal of Biostatistics
International Journal of Cancer Research
International Journal of Health Geographics
International Journal of Low Radiation
International Journal of Medical Robotics and Computer Assisted Surgery
International Journal of Men's Health
International Journal of Neuroprotection and Neurogeneration
International Journal of Pharmaceutical Compounding
International Journal of Pharmacology
International Journal of Prisoner Health
International Journal of Respiratory Care
International Journal of Stroke
International Journal of Virology
Iranian Journal of Allergy, Asthma and Immunology
Iranian Journal of Immunology
Iranian Journal of Nuclear Medicine
Iranian Journal of Pharmacology and Therapeutics
Iranian Journal of Radiation Research
Jordan Medical Journal
Journal of Biological Sciences
Journal of Cardiothoracic-Renal Research
Journal of Chinese Integrative Medicine
Journal of Clinical Dermatology
Journal of Clinical Sleep Medicine
Journal of Complementary and Integrative Medicine
Journal of Cosmetic Dermatology
Journal of Geriatric Cardiology
Journal of Geriatric Physical Therapy
Journal of HIV Therapy
Journal of Indian Association for Child and Adolescent Mental Health
Journal of Interventional Radiology
Journal of Liaquat University of Medical and Health Sciences
Journal of Medical Sciences
Journal of Organ Dysfunction
Journal of Pediatric Neurosciences
Journal of Pediatric Urology
Journal of Pharmacy of Istanbul University
Journal of Plastic Dermatology
Journal of Public Health Management and Practice
Journal of the American Pharmacists Association
Journal of the Canadian Academy of Child and Adolescent Psychiatry
Journal of the Royal Society Interface
Journal of Veterinary Cardiology
Journal of Water and Health
Kathmandu University Medical Journal
Klinik Psikiyatri Dergisi
Koomesh
Korean Circulation Journal
Medicinal Chemistry
Medicinski Glasnik
Middle East Journal of Emergency Medicine
Middle East Paediatrics
Molecular Pain
Molecular Systems Biology
Morphologie
Nano Today
Nanomedicine: Nanotechnology, Biology, and Medicine
Nanotoxicology
Nature Chemical Biology
Nature Clinical Practice Endocrinology and Metabolism
Nature Clinical Practice Nephrology
Nature Clinical Practice Neurology
Nature Clinical Practice Rheumatology
Neuropsychiatric Disease and Treatment
Nobel Medicus
Obesity in Practice
Obesity Management
Ophthalmology in China
Oral Biosciences and Medicine
Pakistan Journal of Nutrition
Pan Arab Journal of Neurosurgery
Particle and Fibre Toxicology
PAT
Pathology Case Reviews
Patologia Del Aparato Locomotor
Pediatria i Medycyna Rodzinna
Personalized Medicine
PLoS Computational Biology
PLoS Genetics
PLoS Pathogens
Practical Cardiovascular Risk Management
Prevention and Control
Proktologia
Psychiatria i Psychologia Kliniczna
Psychiatry Investigation
Psychogeriatria Polska
Respiratory Medicine Extra
Respiratory Medicine: COPD Update
Reumatologia Clinica
Reviews in Analgesia
Revista da Sociedade Brasileira de Cancerologia
Revista de Psiquiatria do Rio Grande do Sul
Revista Medica de Rosario
Ricerche in Psichiatria
Salud (i) Sciencia
Seguimiento Farmacoterapeutico
Seleccion
Sexual Health
Supportive Cancer Therapy
Surgery for Obesity and Related Diseases
Techniques in Foot and Ankle Surgery
Techniques in Knee Surgery
Techniques in Ophthalmology
Tecniche Chirurgiche in Ortopedia e Traumatologia
Therapeutics and Clinical Risk Management
TheScientificWorldJournal
Topics in Clinical Nutrition
Topics in Emergency Medicine
Tuberkuloz ve Toraks
Turk Geriatri Dergisi
Turk Pediatri Arsivi
Turkish Journal of Pharmaceutical Sciences
Vascular Disease Management
Vascular Health and Risk Management
Virologica Sinica
Visual Impairment Research
Women's Health
Work Based Learning in Primary Care
Wounds UK
Yeni Symposium

CHANGED TITLES

This list contains journal titles that were changed in 2005.

AAPS PharmSci
cont. as : AAPS Journal

ACC Current Journal Review
cont. as : ACC Cardiosource Review Journal

Acta Orthopaedica Scandinavica
cont. as : Acta Orthopaedica

Acta Orthopaedica Scandinavica, Supplement
cont. as : Acta Orthopaedica, Supplement

Acta Oto-Rhino-Laryngologica Belgica
cont. as : B-ENT

Acta Physiologica Scandinavica
cont. as : Acta Physiologica

Acta Scientiarum - Biological and Health Sciences
cont. as : (1) Acta Scientiarum - Biological Sciences, and (2) Acta Scientiarum - Health Sciences

Anaesthesiologie und Reanimation
incorp. in : Anasthesiologie Intensivmedizin Notfallmedizin Schmerztherapie

Annales de Genetique
cont. as : European Journal of Medical Genetics

Annals of General Hospital Psychiatry
cont. as : Annals of General Psychiatry

Annals of the College of Surgeons of Hong Kong
cont. as : Surgical Practice

Archives des Maladies Professionnelles et de Medecine du Travail
cont. as : Archives des Maladies Professionnelles et de l'Environnement

Artificial Cells, Blood Substitutes, and Immobilization Biotechnology
cont. as : Artificial Cells, Blood Substitutes, and Biotechnology

Arztezeitschrift fur Naturheilverfahren
cont. as : Arztezeitschrift fur Naturheilverfahren und Regulationsmedizin

CPD Journal Acute Medicine
cont. as : Acute Medicine

Canadian Pharmaceutical Journal
cont. as : Canadian Pharmacists Journal

Cardiovascular Radiation Medicine
cont. as : Cardiovascular Revascularization Medicine

Child Psychology and Psychiatry Review
cont. as : Child and Adolescent Mental Health

Clinical Lymphoma
cont. as : Clinical Lymphoma and Myeloma

Clinical Otolaryngology and Allied Sciences
cont. as : Clinical Otolaryngology

Clinical Prostate Cancer
cont. as : Clinical Genitourinary Cancer

Clinical Psychology
cont. as : Clinical Psychology Forum

Clinical and Diagnostic Laboratory Immunology
cont. as : Clinical and Vaccine Immunology

Comments on Toxicology
incorp. in : (1) Journal of Toxicology and Environmental Health - Part A, and (2) Journal of Toxicology and Environmental Health - Part B: Critical Reviews

Controlled Clinical Trials
cont. as : Contemporary Clinical Trials

Current Drug Targets - Cardiovascular and Haematological Disorders
cont. as : Cardiovascular and Hematological Disorders - Drug Targets

Current Drug Targets - Infectious Disorders
cont. as : Infectious Disorders - Drug Targets

Current Drug Targets: Immune, Endocrine and Metabolic Disorders
cont. as : Endocrine, Metabolic and Immune Disorders - Drug Targets

Current Drug Targets: Inflammation and Allergy
cont. as : Inflammation and Allergy - Drug Targets

Current Medicinal Chemistry - Anti-Cancer Agents
cont. as : Anti-Cancer Agents in Medicinal Chemistry

Current Medicinal Chemistry - Central Nervous System Agents
cont. as : Central Nervous System Agents in Medicinal Chemistry

Current Medicinal Chemistry: Anti-Infective Agents
cont. as : Anti-Infective Agents in Medicinal Chemistry

Current Medicinal Chemistry: Anti-Inflammatory and Anti-Allergy Agents
cont. as : Anti-Inflammatory and Anti-Allergy Agents in Medicinal Chemistry

Current Medicinal Chemistry: Cardiovascular and Hematological Agents
cont. as : Cardiovascular and Hematological Agents in Medicinal Chemistry

Current Medicinal Chemistry: Immunology, Endocrine and Metabolic Agents
cont. as : Immunology, Endocrine and Metabolic Agents in Medicinal Chemistry

Cytokines, Cellular and Molecular Therapy
incorp. in : Cytotherapy

Electronic Journal of Oncology
incorp. in : Bulletin du Cancer

Ethical Human Sciences and Services
cont. as : Ethical Human Psychology and Psychiatry

European Journal of Surgery, Acta Chirurgica
incorp. in : British Journal of Surgery

European Journal of Surgery, Acta Chirurgica, Supplement
incorp. in : British Journal of Surgery

Evidence-Based Eye Care
cont. as : Evidence-Based Ophthalmology

Forschende Komplementarmedizin und Klassische Naturheilkunde
cont. as : Forschende Komplementarmedizin

HeartDrug
incorp. in : Cardiology

Changed

Hybridoma and Hybridomics
cont. as : Hybridoma

International Journal of Cardiovascular Interventions
cont. as : Acute Cardiac Care

International Journal of Integrative Medicine
cont. as : Integrative Medicine

Japanese Heart Journal
cont. as : International Heart Journal

Journal d'Echographie et de Medecine par Ultrasons
incorp. in : Journal de Radiologie

Journal of Exposure Analysis and Environmental Epidemiology
cont. as : Journal of Exposure Science and Environmental Epidemiology

Journal of Intravenous Nursing
cont. as : Journal of Infusion Nursing

Journal of Learning Disabilities
cont. as : Journal of Intellectual Disabilities

Journal of Neuro-AIDS
cont. as : Journal of LGBT Health Research

Journal of Peptide Research
cont. as : Chemical Biology and Drug Design

Journal of Toxicology - Clinical Toxicology
cont. as : Clinical Toxicology

Journal of Toxicology - Cutaneous and Ocular Toxicology
cont. as : Cutaneous and Ocular Toxicology

Journal of Toxicology - Toxin Reviews
cont. as : Toxin Reviews

Journal of Vascular Technology
cont. as : Journal for Vascular Ultrasound

Journal of the American Academy of Psychoanalysis
cont. as : Journal of the American Academy of Psychoanalysis and Dynamic Psychiatry

Journal of the American Association of Gynecologic Laparoscopists
cont. as : Journal of Minimally Invasive Gynecology

Letters in Peptide Science
cont. as : International Journal of Peptide Research and Therapeutics

Medicina Oral
cont. as : Medicina Oral, Patologia Oral y Cirugia Bucal

Nephrology, Hemodialysis and Transplantation
cont. as : Nephrology, Dialysis and Transplantation

Neurology in Practice
incorp. in : Practical Neurology

New Zealand Public Health Report
cont. as : New Zealand Public Health Surveillance Report

Pediatric Pathology and Molecular Medicine
cont. as : Fetal and Pediatric Pathology

Pediatrik Cerrahi Dergisi
cont. as : Cocuk Cerrahisi Dergisi

Pharmacogenetics
cont. as : Pharmacogenetics and Genomics

Pharmacology and Toxicology, Supplement
cont. as : Basic and Clinical Pharmacology and Toxicology, Supplement

Polish Journal of Pharmacology
cont. as : Pharmacological Reports

Primary Care Psychiatry
cont. as : Primary Care and Community Psychiatry

Proceedings of the Royal Society - Biological Sciences (Series B)
cont. as : Biology Letters

Recent Progress in Hormone Research
incorp. in : Endocrine Reviews

Romanian Journal of Gastroenterology
cont. as : Journal of Gastrointestinal and Liver Diseases

Scandinavian Journal of Nutrition/Naringsforskning
cont. as : Scandinavian Journal of Food and Nutrition

Seminars in Laparoscopic Surgery
cont. as : Surgical Innovation

South Pacific Underwater Medicine Society Journal
cont. as : Diving and Hyperbaric Medicine

THOD - Turk Hematoloji-Onkoloji Dergisi
cont. as : UHOD - Uluslararasi Hematoloji-Onkoloji Dergisi

Twin Research
cont. as : Twin Research and Human Genetics

DISCONTINUED TITLES

This list contains journal titles that have been deleted from the EMBASE journal collection at the beginning of 2006.

Acta Dermato-Venereologica
Acta Dermato-Venereologica, Supplement
Acta Neurologica Scandinavica, Supplement
Acta Technologiae et Legis Medicamenti
AIDS, Supplement
Alcologia
Alimentary Pharmacology and Therapeutics, Supplement
BJU International, Supplement
British Journal of Dermatology, Supplement
Cahiers d'Anesthesiologie
Cardiovascular Engineering
Connective Tissue
Contemporary Clinical Gynecology and Obstetrics
Continuing Education in Radiologic Technology
Current Drug Discovery
Current Interventional Cardiology Reports
Current Treatment Options in Infectious Diseases
Dermatology and Psychosomatics
Drug Design and Discovery
European Journal of Gastroenterology and Hepatology, Supplement
Farmaco
Gastrointestinal Oncology
Gynaecological Endoscopy
Heart Disease
Hematologia - Citocinas, Inmunoterapia y Terapia Celular
Herzschrittmacher
In Vitro and Molecular Toxicology: Journal of Basic and Applied Research
Informatik Biometrie und Epidemiologie in Medizin und Biologie
Journal of Drug Evaluation
Journal of Institute of Postgraduate Medicine and Research
Journal of Perinatal Medicine, Supplement
Journal of Trace Elements in Experimental Medicine
Klinische Pharmakologie Aktuell
Lekar a Technika
Medicinal Chemistry Reviews - Online
MEDIFAM - Revista de Medicina Familiar y Comunitaria
Mental Health Reforms
NeuroScience News
Neuroscience Research Communications
New Trends in Experimental and Clinical Psychiatry
Onkologie Service Aktuell
Operative Techniques in Neurosurgery
Oral Biosciences and Medicine
Papillomavirus Report
Pediatric Allergy and Immunology, Supplement
Pharmaceutical Development and Regulation
Pratiques Medicales et Therapeutiques
Problems in Anesthesia
Problems in General Surgery
Progress in Anesthesiology
Receptors and Channels
Revisiones en Ginecologia y Obstetricia
Revista de Psicogeriatria
Revue Medicale Libanaise
Techniques in Neurosurgery
Toko-Ginecologia Practica
UroOncology
Zeitschrift fur Kardiovaskulare Medizin

DISCONTINUED TITLES (1995-2004)

This list contains journal titles that have been deleted from the EMBASE journal collection from 1995-2004.

ISSN: 8910162
AAOHN Journal
PUB-YEAR: 1999 VOLUME: 47 ISSUE:11

ISSN: 15466
Acta Chirurgica Italica
PUB-YEAR: 1999 VOLUME: 55 ISSUE:1

ISSN: 8674140
Acta Endoscopica Polona 2000
PUB-YEAR: 10 VOLUME: 4 ISSUE:

ISSN: 15938
Acta Leprologica
PUB-YEAR: 1999 VOLUME: 11 ISSUE:3

ISSN: 3794857
Acta Medica Dominicana
PUB-YEAR: 1996 VOLUME: 18 ISSUE:3

ISSN: 0004654X
Acta Naturalia de l'Ateneo Parmense
PUB-YEAR: 1998 VOLUME: 34 ISSUE:4

ISSN: 290386
Acta Neonatologica Japonica
PUB-YEAR: 1999 VOLUME: 35 ISSUE:4

ISSN: 3009165
Acta Obstetrica et Gynaecologica Japonica
PUB-YEAR: 1999 VOLUME: 51 ISSUE:11

ISSN: 3008835
Acta Obstetricia et Gynecologica Scandinavica, Supplement
PUB-YEAR: 1998 VOLUME: 77 ISSUE:168

ISSN: 3502309
Acta Orthopaedica Iugoslavica
PUB-YEAR: 2003 VOLUME: 34 ISSUE:2

ISSN: 14137852
Acta Ortopedica Brasileira
PUB-YEAR: VOLUME: ISSUE:

ISSN: 0392100X
Acta Otorhinolaryngologica Italica
PUB-YEAR: 1997 VOLUME: 17 ISSUE:5

ISSN: 16519
Acta Otorrinolaringologica Espanola
PUB-YEAR: 2000 VOLUME: 51 ISSUE:8

ISSN: 3239950
Acta Physiologica et Pharmacologica Bulgarica
PUB-YEAR: 2000 VOLUME: 25 ISSUE:2

ISSN: 3276309
Acta Physiologica Pharmacologica et Therapeutica Latinoamericana
PUB-YEAR: 1999 VOLUME: 49 ISSUE:1

ISSN: 3710874
Acta Physiologica Sinica
PUB-YEAR: 2003 VOLUME: 55 ISSUE:6

ISSN: 16896
Acta Psiquiatrica y Psicologica de America Latina
PUB-YEAR: 1996 VOLUME: 42 ISSUE:2

ISSN: 0303464X
Acta Reumatologica Portuguesa
PUB-YEAR: 1996 VOLUME: 21 ISSUE:77

ISSN: 3780619
Acta Therapeutica
PUB-YEAR: 1997 VOLUME: 23 ISSUE:1

ISSN: 17183
Acta Urologica Belgica
PUB-YEAR: 2000 VOLUME: 68 ISSUE:2

ISSN: 2104806
Actas Urologicas Espanolas
PUB-YEAR: 2003 VOLUME: 27 ISSUE:10

ISSN: 10185275
Adli Tip Dergisi
PUB-YEAR: 1997 VOLUME: 13 ISSUE:1-4

ISSN: 13866206
Advances in Cardiac Echo-Contrast
PUB-YEAR: 1998 VOLUME: 5 ISSUE:3

ISSN: 2674874
Advances in Contraception
PUB-YEAR: 1999 VOLUME: 15 ISSUE:4

ISSN: 1436112
Advances in Contraceptive Delivery Systems
PUB-YEAR: 1997 VOLUME: 13 ISSUE:1

ISSN: 652490
Advances in Drug Research
PUB-YEAR: 1997 VOLUME: 30 ISSUE:

ISSN: 0065275X
Advances in Human Genetics
PUB-YEAR: 1994 VOLUME: 22 ISSUE:

ISSN: 2765063
Advances in Modern Environmental Toxicology
PUB-YEAR: 1992 VOLUME: 21 ISSUE:

ISSN: 9259848
Advances in MRI Contrast
PUB-YEAR: 1998 VOLUME: 5 ISSUE:3

ISSN: 10158618
African Journal of Neurological Sciences
PUB-YEAR: 1998 VOLUME: 17 ISSUE:1

ISSN: 2540517
Afro-Asian Journal of Ophthalmology
PUB-YEAR: 1995 VOLUME: 14 ISSUE:2

ISSN: 8928762
Aging: Immunology and Infectious Disease
PUB-YEAR: 1996 VOLUME: 6 ISSUE:3

ISSN: 1793098
AIDS-Forschung
PUB-YEAR: 1996 VOLUME: 11 ISSUE:5

ISSN: 11381795
AINE
PUB-YEAR: VOLUME: ISSUE:

ISSN: 3009092
Akusherstvo i Ginekologiya
PUB-YEAR: 1996 VOLUME: 72 ISSUE:6

ISSN: 10885412
Allergy and Asthma Proceedings
PUB-YEAR: 1997 VOLUME: 18 ISSUE:1

ISSN: 13608495
Alzheimer Insights
PUB-YEAR: 1999 VOLUME: 5 ISSUE:4

ISSN: 1905279
American Druggist
PUB-YEAR: 1999 VOLUME: 216 ISSUE:12

ISSN: 913960
American Journal of Acupuncture
PUB-YEAR: 1999 VOLUME: 27 ISSUE:3-4

ISSN: 10784500
American Journal of Anesthesiology
PUB-YEAR: 1996 VOLUME: 23 ISSUE:3

ISSN: 8877971
American Journal of Cardiac Imaging
PUB-YEAR: 1996 VOLUME: 10 ISSUE:4

ISSN: 10767460
American Journal of Geriatric Cardiology
PUB-YEAR: 2003 VOLUME: 12 ISSUE:5

ISSN: 10784519
American Journal of Orthopedics
PUB-YEAR: 1996 VOLUME: 25 ISSUE:5

ISSN: 7307780
American Journal of Pharmacy and the Sciences Supporting Public Health
PUB-YEAR: 1995 VOLUME: 167 ISSUE:

ISSN: 13622013
Anaesthetic Pharmacology and Physiology Review
PUB-YEAR: 1996 VOLUME: 4 ISSUE:1

ISSN: 2130599
Anales de Psiquiatria
PUB-YEAR: 1999 VOLUME: 15 ISSUE:8

ISSN: 3024342
Anales Espanoles de Pediatria
PUB-YEAR: 1999 VOLUME: 51 ISSUE:3

ISSN: 3038874
Anales Otorrinolaringologicos Ibero-Americanos
PUB-YEAR: 1999 VOLUME: 26 ISSUE:3

ISSN: 2017563
Anesteziologiya i Reanimatologiya
PUB-YEAR: 1996 VOLUME: ISSUE:1

ISSN: 33006
Anesthesia Progress
PUB-YEAR: 1995 VOLUME: 42 ISSUE:3

ISSN: 7219318
Angio
PUB-YEAR: 1994 VOLUME: 16 ISSUE:6

ISSN: 33898
Annales de Biologie Clinique
PUB-YEAR: 1999 VOLUME: 57 ISSUE:6

ISSN: 662070
Annales de Gastroenterologie et d'Hepatologie
PUB-YEAR: 1998 VOLUME: 34 ISSUE:5

ISSN: 9244204
Annales de l'Institut Pasteur/Actualites
PUB-YEAR: 2003 VOLUME: - ISSUE:18

ISSN: 34185
Annales de Radiologie
PUB-YEAR: 1997 VOLUME: 40 ISSUE:4

ISSN: 34657
Annali di Ostetricia Ginecologia Medicina Perinatale
PUB-YEAR: 1996 VOLUME: 117 ISSUE:3

ISSN: 906964
Annals of Biomedical Engineering
PUB-YEAR: 1998 VOLUME: 26 ISSUE:2

ISSN: 10139087
Annals of Dermatology
PUB-YEAR: 1996 VOLUME: 8 ISSUE:1

ISSN: 2352990
Antibiotiki i Khimioterapiya
PUB-YEAR: 1995 VOLUME: 40 ISSUE:10

ISSN: 9464832
Antiinfective Drugs and Chemotherapy
PUB-YEAR: 2000 VOLUME: 17 ISSUE:1

ISSN: 1069417X
Antimicrobics and Infectious Diseases Newsletter
PUB-YEAR: 2000 VOLUME: 18 ISSUE:12

ISSN: 3930726
Antologia Medica Italiana
PUB-YEAR: VOLUME: ISSUE:

ISSN: 9368671
Arbeiten aus dem Paul-Ehrlich-Institut (Bundesamt fur Sera und Impfstoffe) Langen
PUB-YEAR: 1997 VOLUME: 91 ISSUE:

ISSN: 39152
Archiv fur Japanische Chirurgie
PUB-YEAR: 1998 VOLUME: 67 ISSUE:2

ISSN: 0395501X
Archives d'Anatomie et de Cytologie Pathologiques
PUB-YEAR: 1999 VOLUME: 47 ISSUE:6

ISSN: 2536269
Archives of Pharmacal Research
PUB-YEAR: 1996 VOLUME: 19 ISSUE:2

ISSN: 10710906
Archives of STD/HIV Research
PUB-YEAR: 1997 VOLUME: 11 ISSUE:2

ISSN: 13811118
Archives of Suicide Research
PUB-YEAR: 1999 VOLUME: 5 ISSUE:4

ISSN: 0004010X
Archivio di Medicina Interna
PUB-YEAR: 1998 VOLUME: 50 ISSUE:3-4

ISSN: 40436
Archivo Espanol de Morfologia
PUB-YEAR: 1998 VOLUME: 3 ISSUE:2

ISSN: 3656691
Archivos de la Sociedad Espanola de Oftalmologia
PUB-YEAR: 1997 VOLUME: 72 ISSUE:8

ISSN: 40614
Archivos Espanoles de Urologia
PUB-YEAR: 2003 VOLUME: 56 ISSUE:10

ISSN: 1045652
Arquivos Brasileiros de Pediatria
PUB-YEAR: 1997 VOLUME: 4 ISSUE:6

ISSN: 8714592
Arquivos da Maternidade Dr. Alfredo da Costa
PUB-YEAR: 1995 VOLUME: ISSUE:12

ISSN: 8723265
Arquivos de Fisiatria e Doencas Osteo-Articulares
PUB-YEAR: 1999 VOLUME: 6 ISSUE:3

ISSN: 42897
Ars Medici
PUB-YEAR: 1997 VOLUME: 87 ISSUE:18

ISSN: 42927
Ars Pharmaceutica
PUB-YEAR: 1998 VOLUME: 39 ISSUE:2

ISSN: 986127
Artery
PUB-YEAR: 1998 VOLUME: 23 ISSUE:1

ISSN: 13000594
Artroplasti Artroskopik Cerrahi
PUB-YEAR: 2003 VOLUME: 14 ISSUE:4

ISSN: 9390448
Arztliche Praxis Dermatologie
PUB-YEAR: 2001 VOLUME: ISSUE:5

ISSN: 14357739
Arztliche Praxis Gynakologie
PUB-YEAR: 2002 VOLUME: 6 ISSUE:

ISSN: 9353224
Arztliche Praxis Neurologie Psychiatrie
PUB-YEAR: 2002 VOLUME: 6 ISSUE:

ISSN: 14362651
Arztliche Praxis Padiatrie
PUB-YEAR: 2002 VOLUME: 6 ISSUE:

ISSN: 14351080
Arztliche Praxis Urologie Nephrologie
PUB-YEAR: 2001 VOLUME: ISSUE:6

ISSN: 10948465
Assisted Reproductive Technology/Andrology
PUB-YEAR: 1997 VOLUME: 9 ISSUE:1

ISSN: 1023618X
Auditory Neuroscience
PUB-YEAR: 1997 VOLUME: 3 ISSUE:4

ISSN: 14400243
Australian Journal of Psychopharmacology
PUB-YEAR: 1999 VOLUME: 9 ISSUE:

ISSN: 950963
Automedica
PUB-YEAR: 2001 VOLUME: 19 ISSUE:3-4

ISSN: 9610421
Bailliere's Clinical Neurology
PUB-YEAR: 1997 VOLUME: 6 ISSUE:3

ISSN: 10234853
Bangladesh Journal of Neuroscience
PUB-YEAR: 1994 VOLUME: 10 ISSUE:1

ISSN: 0210895X
Biologia y Clinica Hematologica
PUB-YEAR: 1995 VOLUME: 17 ISSUE:4

ISSN: 14309858
Biomedical Ethics
PUB-YEAR: 2000 VOLUME: 5 ISSUE:3

ISSN: 1415492
Biotechnology Letters
PUB-YEAR: 1999 VOLUME: 21 ISSUE:10

ISSN: 8982848
Biotechnology Therapeutics
PUB-YEAR: 1994 VOLUME: 5 ISSUE:3

ISSN: 0921299X
Biotherapy
PUB-YEAR: 1998 VOLUME: 11 ISSUE:4

ISSN: 8708363
Boletim do Hospital de Pulido Valente
PUB-YEAR: 2002 VOLUME: 15 ISSUE:4

ISSN: 5396115
Boletin Medico del Hospital Infantil de Mexico
PUB-YEAR: 1998 VOLUME: 55 ISSUE:6

ISSN: 13891987
Brain and Mind
PUB-YEAR: 2003 VOLUME: 4 ISSUE:3

ISSN: 14621665
British Journal of Homecare
PUB-YEAR: VOLUME: ISSUE:

ISSN: 7537417
Bulletin de la Societe Francaise de Cancerologie Privee
PUB-YEAR: 1997 VOLUME: 16 ISSUE:51

ISSN: 0391481X
Bulletin of Molecular Biology and Medicine
PUB-YEAR: 1995 VOLUME: 20 ISSUE:3

ISSN: 1203844X
Canadian Journal of Allergy and Clinical Immunology
PUB-YEAR: 2002 VOLUME: 7 ISSUE:5-6

ISSN: 8280827
Canadian Journal of Rehabilitation
PUB-YEAR: 1997 VOLUME: 10 ISSUE:3

ISSN: 0740820X
Cancer Bulletin
PUB-YEAR: 1995 VOLUME: 47 ISSUE:6

ISSN: 10907491
Cancer Case Presentations: The Tumor Board
PUB-YEAR: 2001 VOLUME: 5 ISSUE:6

ISSN: 10654704
Cancer Practice
PUB-YEAR: 2002 VOLUME: 10 ISSUE:6

ISSN: 10640525
Cancer Research, Therapy and Control
PUB-YEAR: 2002 VOLUME: 11 ISSUE:3-4

Discontinued

ISSN: 2612429
Cancer Surveys
PUB-YEAR: 1998 VOLUME: 33 ISSUE:

ISSN: 11200421
Cardiovascular Imaging
PUB-YEAR: 1999 VOLUME: 11 ISSUE:4

ISSN: 1973118
Cardiovascular Reviews and Reports
PUB-YEAR: 2003 VOLUME: 24 ISSUE:11

ISSN: 11307501
Cardiovascular Risk Factors
PUB-YEAR: 1996 VOLUME: 6 ISSUE:6

ISSN: 9688773
Cellular and Molecular Biology Research
PUB-YEAR: 1995 VOLUME: 41 ISSUE:6

ISSN: 89176
Central African Journal of Medicine
PUB-YEAR: 1999 VOLUME: 45 ISSUE:4

ISSN: 13005227
Cerrahpasa Tip Dergisi
PUB-YEAR: 1999 VOLUME: 30 ISSUE:2

ISSN: 3933512
Cervix and the Lower Female Genital Tract
PUB-YEAR: 1995 VOLUME: 13 ISSUE:4

ISSN: 12107875
Cesko-Slovenska Patologie a Soudni Lekarstvi
PUB-YEAR: 1999 VOLUME: 35 ISSUE:2

ISSN: 2983850
Champ Psychosomatique: Medecine Psychoanalyse Anthropologie
PUB-YEAR: 1997 VOLUME: 38 ISSUE:11

ISSN: 8995869
Children's Hospital Quarterly
PUB-YEAR: 2000 VOLUME: 11 ISSUE:4

ISSN: 2533758
Chinese Journal of Cardiology
PUB-YEAR: 1999 VOLUME: 27 ISSUE:5

ISSN: 10008179
Chinese Journal of Clinical Oncology
PUB-YEAR: 1999 VOLUME: 26 ISSUE:10

ISSN: 2533766
Chinese Journal of Oncology
PUB-YEAR: 1999 VOLUME: 21 ISSUE:4

ISSN: 2541793
Chinese Journal of Pharmaceutical Analysis
PUB-YEAR: 1994 VOLUME: 14 ISSUE:1

ISSN: 10067884
Chinese Journal of Psychiatry
PUB-YEAR: 1999 VOLUME: 32 ISSUE:3

ISSN: 8945853
Choices in Cardiology
PUB-YEAR: 1995 VOLUME: 9 ISSUE:1

ISSN: 111759
Chronica Dermatologica
PUB-YEAR: 1997 VOLUME: 7 ISSUE:6

ISSN: 11266929
Cibus
PUB-YEAR: VOLUME: ISSUE:

ISSN: 2646900
Circulation et Metabolisme du Cerveau
PUB-YEAR: 1995 VOLUME: 12 ISSUE:4

ISSN: 2121808
Clinica Cardiovascular
PUB-YEAR: 2001 VOLUME: 19 ISSUE:2

ISSN: 3921344
Clinica e Terapia Cardiovascolare
PUB-YEAR: 1997 VOLUME: 16 ISSUE:1

ISSN: 3918998
Clinica Oculistica e Patologia Oculare
PUB-YEAR: 1997 VOLUME: 18 ISSUE:3

ISSN: 14611449
Clinical Acupuncture and Oriental Medicine
PUB-YEAR: 2003 VOLUME: 4 ISSUE:4

ISSN: 9602178
Clinical and Experimental Allergy, Supplement
PUB-YEAR: 2000 VOLUME: 30 ISSUE:1

ISSN: 9642536
Clinical and Experimental Immunology, Supplement
PUB-YEAR: 1999 VOLUME: 118 ISSUE:1

ISSN: 10894195
Clinical Bulletin of Myofascial Therapy
PUB-YEAR: 1998 VOLUME: 3 ISSUE:1

ISSN: 10430660
Clinical Consultations in Obstetrics and Gynecology
PUB-YEAR: 1996 VOLUME: 8 ISSUE:2

ISSN: 8887950
Clinical Laboratory Management Review
PUB-YEAR: 1996 VOLUME: 10 ISSUE:4

ISSN: 0894959X
Clinical Laboratory Science
PUB-YEAR: 1996 VOLUME: 9 ISSUE:1

ISSN: 13813390
Clinical Neonatology
PUB-YEAR: 1998 VOLUME: 5 ISSUE:2

ISSN: 10656766
Clinical Neuroscience
PUB-YEAR: 1998 VOLUME: 5 ISSUE:2

ISSN: 13670123
CME Bulletin Dermatology
PUB-YEAR: 2001 VOLUME: 3 ISSUE:1

ISSN: 14628783
CME Bulletin Gynaecology
PUB-YEAR: 2001 VOLUME: 2 ISSUE:3

ISSN: 13679007
CME Bulletin Haematology
PUB-YEAR: 2001 VOLUME: 3 ISSUE:3

ISSN: 13678965
CME Bulletin Respiratory Medicine
PUB-YEAR: 2001 VOLUME: 3 ISSUE:2

ISSN: 1367014X
CME Bulletin Sexually Transmitted Infections and HIV
PUB-YEAR: 2001 VOLUME: 5 ISSUE:2

ISSN: 14630141
CME Urology
PUB-YEAR: 2001 VOLUME: 3 ISSUE:1

ISSN: 1053749X
Complications in Surgery
PUB-YEAR: 1996 VOLUME: 15 ISSUE:1

ISSN: 2939908
Comptes Rendus de Therapeutique et de Pharmacologie Clinique
PUB-YEAR: 1996 VOLUME: 13 ISSUE:132

ISSN: 2101424
Comunicacion Psiquiatrica
PUB-YEAR: 1993 VOLUME: 18 ISSUE:

ISSN: 15275299
Congestive Heart Failure
PUB-YEAR: 2003 VOLUME: 9 ISSUE:3

ISSN: 10853324
Contemporary Treatments in Cardiovascular Disease
PUB-YEAR: 1998 VOLUME: 3 ISSUE:1

ISSN: 108650
Cor et Vasa
PUB-YEAR: 1994 VOLUME: 36 ISSUE:5

ISSN: 9398147
Cor Europaeum - European Journal of Cardiac Interventions
PUB-YEAR: 2000 VOLUME: 8 ISSUE:4

ISSN: 13623265
Coronary Health Care
PUB-YEAR: 2001 VOLUME: 5 ISSUE:4

ISSN: 1463046X
CPD Bulletin Psychiatry
PUB-YEAR: 2000 VOLUME: 2 ISSUE:3

ISSN: 13678922
CPD Rheumatology
PUB-YEAR: 2002 VOLUME: 3 ISSUE:2

ISSN: 9562257
Critical Ischaemia
PUB-YEAR: 2001 VOLUME: 11 ISSUE:3

ISSN: 8939675
Critical Reviews in Oncogenesis
PUB-YEAR: 2000 VOLUME: 11 ISSUE:3-4

ISSN: 10482687
CRNA: Clinical Forum for Nurse Anesthetists
PUB-YEAR: 2001 VOLUME: 12 ISSUE:1

ISSN: 11347872
Cuadernos de Artroscopia
PUB-YEAR: 1999 VOLUME: 6 ISSUE:2

ISSN: 11227176
Cuore
PUB-YEAR: 1996 VOLUME: 13 ISSUE:5

ISSN: 9698868
Current Opinion in Surgical Infections
PUB-YEAR: 1996 VOLUME: 4 ISSUE:4

ISSN: 9520627
Current Practice in Surgery
PUB-YEAR: 1996 VOLUME: 8 ISSUE:2

ISSN: 10400486
Current Problems in Dermatology
PUB-YEAR: 2003 VOLUME: 15 ISSUE:6

ISSN: 15345874
Current Women's Health Reports
PUB-YEAR: 2003 VOLUME: 3 ISSUE:6

ISSN: 114162
Cutis
PUB-YEAR: 1997 VOLUME: 59 ISSUE:4

ISSN: 10954155
CVD Prevention
PUB-YEAR: 2000 VOLUME: 3 ISSUE:4

ISSN: 10916326
Depression and Stress
PUB-YEAR: 1996 VOLUME: 2 ISSUE:2

ISSN: 10195815
Developmental Brain Dysfunction
PUB-YEAR: 1997 VOLUME: 10 ISSUE:6

ISSN: 10669442
Diabetes Reviews
PUB-YEAR: 1999 VOLUME: 7 ISSUE:4

ISSN: 10703608
Diagnostic and Therapeutic Endoscopy
PUB-YEAR: 2000 VOLUME: 7 ISSUE:1

ISSN: 10883371
Disease Management and Clinical Outcomes
PUB-YEAR: 2001 VOLUME: 3 ISSUE:2

ISSN: 126683
Drugs Made in Germany
PUB-YEAR: 2002 VOLUME: 45 ISSUE:4

ISSN: 13555618
Drug Stability
PUB-YEAR: 1997 VOLUME: 1 ISSUE:4

ISSN: 14677733
Ear
PUB-YEAR: 2000 VOLUME: 1 ISSUE:3

ISSN: 0012835X
East African Medical Journal
PUB-YEAR: 1999 VOLUME: 76 ISSUE:1

ISSN: 15248909
e-biomed
PUB-YEAR: 2003 VOLUME: 4 ISSUE:

ISSN: 9480382
Electronic Journal of Pathology and Histology
PUB-YEAR: 2003 VOLUME: 9 ISSUE:2

ISSN: 2506882
Emirates Medical Journal
PUB-YEAR: 1997 VOLUME: 15 ISSUE:1

ISSN: 7490003
Employee Assistance Quarterly
PUB-YEAR: 1999 VOLUME: 14 ISSUE:4

ISSN: 10242430
Enantiomer
PUB-YEAR: 2002 VOLUME: 7 ISSUE:6

ISSN: 0933811X
Endoskopie Heute
PUB-YEAR: 1999 VOLUME: 12 ISSUE:2

ISSN: 15347621
Endo Trends
PUB-YEAR: 2001 VOLUME: 8 ISSUE:2

ISSN: 15227987
Environmental Epidemiology and Toxicology
PUB-YEAR: 2000 VOLUME: 2 ISSUE:4

ISSN: 10196773
Enzyme and Protein
PUB-YEAR: 1996 VOLUME: 49 ISSUE:5

ISSN: 9409912
Epithelial Cell Biology
PUB-YEAR: 1995 VOLUME: 4 ISSUE:3

ISSN: 9699252
Epstein-Barr Virus Report
PUB-YEAR: 2002 VOLUME: 9 ISSUE:6

ISSN: 9566309
Erythropoiesis: New Dimensions in the Treatment of Anaemia
PUB-YEAR: 2001 VOLUME: 11 ISSUE:2

ISSN: 141755
Ethiopian Medical Journal
PUB-YEAR: 1998 VOLUME: 36 ISSUE:4

ISSN: 10995129
Europace, Supplement
PUB-YEAR: VOLUME: ISSUE:

ISSN: 9396780
European Journal of Cardiac Pacing and Electrophysiology
PUB-YEAR: 1998 VOLUME: 8 ISSUE:1

ISSN: 8035288
European Journal of Experimental Musculoskeletal Research
PUB-YEAR: 1995 VOLUME: 4 ISSUE:3

ISSN: 14391147
European Journal of Geriatrics
PUB-YEAR: 2003 VOLUME: 5 ISSUE:4

ISSN: 11218967
European Journal of Internal Medicine, Supplement
PUB-YEAR: VOLUME: ISSUE:

ISSN: 10176721
European Journal of Physical Medicine and Rehabilitation
PUB-YEAR: 1998 VOLUME: 8 ISSUE:6

ISSN: 1401610
European Journal of Rheumatology and Inflammation
PUB-YEAR: 1995 VOLUME: 15 ISSUE:2

ISSN: 15923894
European Journal of Sports Traumatology and Related Research
PUB-YEAR: 2001 VOLUME: 23 ISSUE:4

ISSN: 9298266
European Journal of Ultrasound
PUB-YEAR: 2002 VOLUME: 16 ISSUE:3

ISSN: 1025448X
European Respiratory Monograph
PUB-YEAR: 2001 VOLUME: 6 ISSUE:16

ISSN: 2549670
Experimental and Clinical Immunogenetics
PUB-YEAR: 2001 VOLUME: 18 ISSUE:4

ISSN: 8692092
Experimental and Clinical Pharmacology
PUB-YEAR: 1995 VOLUME: 58 ISSUE:3

ISSN: 12824038
Facts, Research and Intervention in Geriatrics
PUB-YEAR: VOLUME: ISSUE:

ISSN: 2126583
Farmacia Clinica
PUB-YEAR: 1998 VOLUME: 15 ISSUE:7

ISSN: 11231254
Farmacia e Clinica
PUB-YEAR: 1994 VOLUME: 33 ISSUE:4

ISSN: 3673057
Farmatsevtichnii Zhurnal
PUB-YEAR: 1993 VOLUME: 48 ISSUE:4

ISSN: 3673014
Farmatsiya
PUB-YEAR: 1994 VOLUME: 43 ISSUE:6

ISSN: 8882401
Female Patient - OB/GYN Edition
PUB-YEAR: 1996 VOLUME: 21 ISSUE:9

ISSN: 4282779
Feuillets de Biologie
PUB-YEAR: 1999 VOLUME: 40 ISSUE:230

ISSN: 13690191
Fibrinolysis and Proteolysis
PUB-YEAR: 2001 VOLUME: 15 ISSUE:1

ISSN: 13006614
Fizik Tedavi Rehabilitasyon Dergisi
PUB-YEAR: 1996 VOLUME: 20 ISSUE:3

ISSN: 8616485
Fizikalna Kurortna i Rekhabilitatsionna Meditsina
PUB-YEAR: 2001 VOLUME: 40 ISSUE:2

ISSN: 8698139
Fiziologicheskii Zhurnal Imeni I.M. Sechenova
PUB-YEAR: 1995 VOLUME: 81 ISSUE:6

ISSN: 155454
Folha Medica
PUB-YEAR: 1998 VOLUME: 116 ISSUE:1

ISSN: 2112558
Folia Neuropsiquiatrica
PUB-YEAR: 1996 VOLUME: 31 ISSUE:2

ISSN: 10116788
Formosan Journal of Surgery
PUB-YEAR: 2003 VOLUME: 36 ISSUE:5

ISSN: 1703331
Fortschritt und Fortbildung in der Medizin
PUB-YEAR: 1999 VOLUME: 23 ISSUE:

ISSN: 9213775
Frontiers of Medical and Biological Engineering
PUB-YEAR: VOLUME: ISSUE:

ISSN: 0016254X
Fukuoka Acta Medica
PUB-YEAR: 2000 VOLUME: 91 ISSUE:9

ISSN: 162582
Fukushima Medical Journal
PUB-YEAR: 1999 VOLUME: 49 ISSUE:4

ISSN: 3905845
Gaslini
PUB-YEAR: 2002 VOLUME: 34 ISSUE:2

ISSN: 10137696
Gastroenterological Journal of Taiwan
PUB-YEAR: 2000 VOLUME: 17 ISSUE:3

ISSN: 2345730
Gematologiya i Transfuziologiya
PUB-YEAR: 1995 VOLUME: 40 ISSUE:4

ISSN: 163503
GEN
PUB-YEAR: 1996 VOLUME: 50 ISSUE:1

ISSN: 0394249X
Gene Geography
PUB-YEAR: 1996 VOLUME: 10 ISSUE:3

ISSN: 0947823X
General and Diagnostic Pathology
PUB-YEAR: 1998 VOLUME: 143 ISSUE:5

ISSN: 13607413
Genes and Function
PUB-YEAR: VOLUME: ISSUE:

ISSN: 1466920X
GeneScreen
PUB-YEAR: VOLUME: ISSUE:

ISSN: 166758
Genetika
PUB-YEAR: 1995 VOLUME: 31 ISSUE:6

ISSN: 11225807
Geriatria
PUB-YEAR: 1997 VOLUME: 9 ISSUE:2

ISSN: 170224
Giornale della Arteriosclerosi
PUB-YEAR: 1997 VOLUME: 22 ISSUE:2

ISSN: 11204818
Giornale di Gastroenterologia
PUB-YEAR: VOLUME: ISSUE:

ISSN: 170305
Giornale di Gerontologia
PUB-YEAR: 2000 VOLUME: 48 ISSUE:4

ISSN: 170437
Giornale Italiano della Malattie del Torace
PUB-YEAR: 1994 VOLUME: 48 ISSUE:5

ISSN: 3921387
Giornale Italiano di Angiologia
PUB-YEAR: 1995 VOLUME: 15 ISSUE:4

ISSN: 3922227
Giornale Italiano di Chimica Clinica
PUB-YEAR: 1997 VOLUME: 21 ISSUE:1

ISSN: 11269952
Giornale Italiano di Malattie Infettive
PUB-YEAR: 2001 VOLUME: 7 ISSUE:4

ISSN: 3935590
Giornale Italiano di Nefrologia
PUB-YEAR: 1999 VOLUME: 16 ISSUE:2

ISSN: 0392128X
Giornale Italiano di Oncologia
PUB-YEAR: 1998 VOLUME: 18 ISSUE:1

ISSN: 3935957
Giornale Italiano di Ricerche Cliniche e Terapeutiche
PUB-YEAR: 1995 VOLUME: 16 ISSUE:3

ISSN: 11203560
Giornale Italiano di Riflessoterapia ed Agopuntura
PUB-YEAR: 2000 VOLUME: 12 ISSUE:2

ISSN: 11203846
Giornale Italiano di Ultrasonologia
PUB-YEAR: 1997 VOLUME: 8 ISSUE:4

ISSN: 0956523X
Growth Regulation
PUB-YEAR: 1996 VOLUME: 6 ISSUE:4

ISSN: 12503347
Gynecologie - Revue du Gynecologue
PUB-YEAR: 1995 VOLUME: 3 ISSUE:8

ISSN: 176036
Gyogyszereszet
PUB-YEAR: 2000 VOLUME: 44 ISSUE:10

ISSN: 176559
Haematologia
PUB-YEAR: 2002 VOLUME: 32 ISSUE:4

ISSN: 7350732
Healthcare Financial Management
PUB-YEAR: 2001 VOLUME: 55 ISSUE:11

ISSN: 87557673
Heart Failure
PUB-YEAR: 1997 VOLUME: 13 ISSUE:1

ISSN: 9273859
Helix
PUB-YEAR: VOLUME: ISSUE:

ISSN: 467324
Herz Kreislauf
PUB-YEAR: 2000 VOLUME: 32 ISSUE:12

ISSN: 181935
Hindustan Antibiotics Bulletin
PUB-YEAR: 1994 VOLUME: 36 ISSUE:3

ISSN: 10278230
Hong Kong Journal of Ophthalmology
PUB-YEAR: VOLUME: ISSUE:

ISSN: 10282637
Hong Kong Journal of Orthopaedic Surgery
PUB-YEAR: VOLUME: ISSUE:

ISSN: 1922262
Hospital Materiel Management Quarterly
PUB-YEAR: 2000 VOLUME: 22 ISSUE:3

ISSN: 87502836
Hospital Practice
PUB-YEAR: 2001 VOLUME: 36 ISSUE:9

ISSN: 8948569
HPB Surgery
PUB-YEAR: 1999 VOLUME: 11 ISSUE:6

ISSN: 3939375
Human Evolution
PUB-YEAR: 1998 VOLUME: 13 ISSUE:3-4

ISSN: 9706054
IDMA Bulletin
PUB-YEAR: 1999 VOLUME: 30 ISSUE:32

ISSN: 191655
Igiene Moderna
PUB-YEAR: 2001 VOLUME: 116 ISSUE:4

ISSN: 13528513
Illustrated Case Reports in Gastroenterology
PUB-YEAR: 1997 VOLUME: 4 ISSUE:3

ISSN: 3401162
Immunitat und Infektion
PUB-YEAR: 1995 VOLUME: 23 ISSUE:6

ISSN: 8935300
Immunodeficiency
PUB-YEAR: 1995 VOLUME: 5 ISSUE:3

ISSN: 9701915
Indian Journal of Clinical Biochemistry
PUB-YEAR: 1996 VOLUME: 11 ISSUE:2

ISSN: 3786323
Indian Journal of Dermatology, Venereology and Leprology
PUB-YEAR: 1995 VOLUME: 61 ISSUE:6

ISSN: 195189
Indian Journal of Experimental Biology
PUB YEAR. 1999 VOLUME: 37 ISSUE:11

ISSN: 0250474X
Indian Journal of Pharmaceutical Sciences
PUB-YEAR: 2000 VOLUME: 62 ISSUE:6

ISSN: 9701591
Indian Journal of Urology
PUB-YEAR: 2003 VOLUME: 19 ISSUE:2

ISSN: 10535586
Infant-Toddler Intervention
PUB-YEAR: 2001 VOLUME: 11 ISSUE:3-4

ISSN: 10562044
Infectious Agents and Disease
PUB-YEAR: 1996 VOLUME: 5 ISSUE:4

ISSN: 3274438
Infectologia y Microbiologia Clinica
PUB-YEAR: VOLUME: ISSUE:

ISSN: 10479422
Infertility and Reproductive Medicine Clinics of North America
PUB-YEAR: 2003 VOLUME: 14 ISSUE:4

ISSN: 11350946
Inflamacion y Regulacion del Crecimiento Celular
PUB-YEAR: 1998 VOLUME: 9 ISSUE:2

ISSN: 2139626
Inmunologia
PUB-YEAR: 1996 VOLUME: 15 ISSUE:1

ISSN: 16602250
Innovation in Medicine
PUB-YEAR: VOLUME: ISSUE:

ISSN: 10962190
Integrative Medicine
PUB-YEAR: 1999 VOLUME: 2 ISSUE:2-3

ISSN: 7353847
Integrative Psychiatry
PUB-YEAR: 1994 VOLUME: 10 ISSUE:4

ISSN: 9565272
Interferons and Cytokines
PUB-YEAR: 1995 VOLUME: ISSUE:27

ISSN: 5062772
Internal Diseases
PUB-YEAR: 2001 VOLUME: 33 ISSUE:2-3

ISSN: 9652310
International Antiviral News
PUB-YEAR: 2001 VOLUME: 9 ISSUE:12

ISSN: 9330445
International Journal of Feto-Maternal Medicine
PUB-YEAR: 1993 VOLUME: 6 ISSUE:1

ISSN: 13648233
International Journal of Geriatric Psychopharmacology
PUB-YEAR: 2000 VOLUME: 2 ISSUE:3

ISSN: 9255710
International Journal of Hematology
PUB-YEAR: 1998 VOLUME: 68 ISSUE:4

ISSN: 10967028
International Journal of Intensive Short-Term Dynamic Psychotherapy
PUB-YEAR: VOLUME: ISSUE:

ISSN: 14331128
International Journal of Medical Microbiology, Supplement
PUB-YEAR: 2004 VOLUME: 293 ISSUE:37

ISSN: 10798110
International Journal of Neuroradiology
PUB-YEAR: VOLUME: ISSUE:

ISSN: 10803548
International Journal of Occupational Safety and Ergonomics
PUB-YEAR: 1996 VOLUME: 2 ISSUE:4

ISSN: 10196439
International Journal of Oncology
PUB-YEAR: 1998 VOLUME: 13 ISSUE:6

ISSN: 8848297
International Journal of Psychosomatics
PUB-YEAR: 1995 VOLUME: 42 ISSUE:1

ISSN: 9431675
International Journal of Thymology
PUB-YEAR: 1999 VOLUME: 7 ISSUE:12-13

ISSN: 8696683
International Medical Reviews
PUB-YEAR: 1994 VOLUME: 2 ISSUE:5

ISSN: 3936066
International Journal of Sports Cardiology
PUB-YEAR: 1999 VOLUME: 8 ISSUE:3

ISSN: 208868
International Surgery
PUB-YEAR: 1997 VOLUME: 82 ISSUE:4

ISSN: 2511789
Invasion and Metastasis
PUB-YEAR: 1998 VOLUME: 18 ISSUE:5-6

ISSN: 14204002
Ippocrate
PUB-YEAR: VOLUME: ISSUE:

ISSN: 3017362
Istanbul Tip Fakultesi Mecmuasi
PUB-YEAR: 1997 VOLUME: 60 ISSUE:1

ISSN: 11211709
Italian Journal of Mineral and Electrolyte Metabolism
PUB-YEAR: 2001 VOLUME: 15 ISSUE:1-4

ISSN: 2898055
Japan Journal of Medical Informatics
PUB-YEAR: 1999 VOLUME: 19 ISSUE:2

ISSN: 214809
Japanese Archives of Internal Medicine
PUB-YEAR: 1996 VOLUME: 43 ISSUE:6

ISSN: 13418963
Japanese Journal of Alcohol Studies and Drug Dependence
PUB-YEAR: 1999 VOLUME: 34 ISSUE:5

ISSN: 3000818
Japanese Journal of Artificial Organs
PUB-YEAR: 1999 VOLUME: 28 ISSUE:3

ISSN: 3850684
Japanese Journal of Cancer and Chemotherapy
PUB-YEAR: 1997 VOLUME: 24 ISSUE:15

ISSN: 9110836
Japanese Journal of Cardiac Pacing and Electrophysiology
PUB-YEAR: 1997 VOLUME: 13 ISSUE:3

ISSN: 3009173
Japanese Journal of Geriatrics
PUB-YEAR: 1999 VOLUME: 36 ISSUE:10

ISSN: 215082
Japanese Journal of Hygiene
PUB-YEAR: 1999 VOLUME: 54 ISSUE:3

ISSN: 3850307
Japanese Journal of Psychosomatic Medicine
PUB-YEAR: 1999 VOLUME: 39 ISSUE:8

ISSN: 1691163
Japanese Journal of Rheumatology
PUB-YEAR: 1999 VOLUME: 9 ISSUE:4

ISSN: 3011542
Japanese Journal of Thoracic Diseases
PUB-YEAR: 1997 VOLUME: 35 ISSUE:

ISSN: 216968
Jikeikai Medical Journal
PUB-YEAR: 2000 VOLUME: 47 ISSUE:3

ISSN: 3681416
Jornal Brasileiro de Ginecologia
PUB-YEAR: 1997 VOLUME: 107 ISSUE:11

ISSN: 217689
Journal de Chimie Physique et de Physico-Chimie Biologique
PUB-YEAR: 1999 VOLUME: 96 ISSUE:9-10

ISSN: 217697
Journal de Chirurgie
PUB-YEAR: 2000 VOLUME: 137 ISSUE:6

ISSN: 217905
Journal de Medecine de Strasbourg
PUB-YEAR: 1997 VOLUME: 28 ISSUE:7

ISSN: 2480018
Journal d'Urologie
PUB-YEAR: 1997 VOLUME: 103 ISSUE:1

ISSN: 3989771
Journal Francais d'Oto-Rhino-Laryngologie
PUB-YEAR: 1996 VOLUME: 45 ISSUE:3

ISSN: 7448481
Journal of American College Health
PUB-YEAR: 1999 VOLUME: 48 ISSUE:3

ISSN: 9701842
Journal of Anatomical Sciences
PUB-YEAR: 1996 VOLUME: 15 ISSUE:1

ISSN: 3855201
Journal of Antibacterial and Antifungal Agents, Japan
PUB-YEAR: 1999 VOLUME: 27 ISSUE:11

ISSN: 218960
Journal of Applied Nutrition
PUB-YEAR: 1997 VOLUME: 49 ISSUE:4

ISSN: 15211401
Journal of Applied Psychoanalytic Studies
PUB-YEAR: 2003 VOLUME: 5 ISSUE:4

ISSN: 10292659
Journal of Applied Therapeutic Research
PUB-YEAR: 2000 VOLUME: 3 ISSUE:1

ISSN: 10258140
Journal of Biochemistry, Molecular Biology and Biophysics
PUB-YEAR: 2002 VOLUME: 6 ISSUE:6

ISSN: 10944540
Journal of Brachytherapy International
PUB-YEAR: 2001 VOLUME: 17 ISSUE:4

ISSN: 14766647
Journal of Cancer Epidemiology and Prevention
PUB-YEAR: 2002 VOLUME: 7 ISSUE:4

ISSN: 15299775
Journal of Cannabis Therapeutics
PUB-YEAR: 2004 VOLUME: 4 ISSUE:1

ISSN: 10737774
Journal of Cardiovascular Diagnosis and Procedures
PUB-YEAR: 1999 VOLUME: 16 ISSUE:2

ISSN: 1412760
Journal of Clinical and Laboratory Immunology
PUB-YEAR: 1992 VOLUME: 38 ISSUE:4

ISSN: 8958831
Journal of Clinical Dentistry
PUB-YEAR: 1997 VOLUME: 8 ISSUE:5

ISSN: 3638855
Journal of Clinical Engineering
PUB-YEAR: 1995 VOLUME: 20 ISSUE:6

ISSN: 15246175
Journal of Clinical Hypertension
PUB-YEAR: 2003 VOLUME: 5 ISSUE:4

ISSN: 14683768
Journal of Congestive Heart Failure and Circulatory Support
PUB-YEAR: 2001 VOLUME: 2 ISSUE:1

ISSN: 2704145
Journal of Craniofacial Genetics and Developmental Biology
PUB-YEAR: 2000 VOLUME: 20 ISSUE:4

ISSN: 10743219
Journal of Cranio-Maxillofacial Trauma
PUB-YEAR: 2001 VOLUME: 7 ISSUE:1

ISSN: 13579215
Journal of Drug Development and Clinical Practice
PUB-YEAR: 1996 VOLUME: 8 ISSUE:3

ISSN: 10951539
Journal of Environmental Medicine
PUB-YEAR: VOLUME: ISSUE:

ISSN: 15910989
Journal of Functional syndromes
PUB-YEAR: 2002 VOLUME: 2 ISSUE:2

ISSN: 15237036
Journal of Gender-Specific Medicine
PUB-YEAR: 2003 VOLUME: 6 ISSUE:4

ISSN: 221414
Journal of Geriatric Psychiatry
PUB-YEAR: 2002 VOLUME: 35 ISSUE:2

ISSN: 10692673
Journal of Gynecologic Techniques
PUB-YEAR: 2000 VOLUME: 6 ISSUE:3

ISSN: 10787852
Journal of Inflammation
PUB-YEAR: 1998 VOLUME: 48 ISSUE:2

ISSN: 9557873
Journal of Internal Medicine, Supplement
PUB-YEAR: 1997 VOLUME: 242 ISSUE:740

ISSN: 2680882
Journal of Interventional Radiology
PUB-YEAR: 1999 VOLUME: 14 ISSUE:4

ISSN: 214671
Journal of Japan Society for Cancer Therapy
PUB-YEAR: 1997 VOLUME: 32 ISSUE:4

ISSN: 290343
Journal of Japanese Association of Physical Medicine Balneology and Climatology
PUB-YEAR: 1999 VOLUME: 63 ISSUE:1

ISSN: 290203
Journal of Japanese Ophthalmological Society
PUB-YEAR: 1999 VOLUME: 103 ISSUE:11

ISSN: 9396322
Journal of Maternal-Fetal Investigation
PUB-YEAR: 1998 VOLUME: 8 ISSUE:4

ISSN: 10457984
Journal of Myocardial Ischemia
PUB-YEAR: 1996 VOLUME: 8 ISSUE:3

ISSN: 10945512
Journal of Noninvasive Cardiology
PUB-YEAR: 2000 VOLUME: 4 ISSUE:1

ISSN: 10954082
Journal of Occupational Hearing Loss
PUB-YEAR: VOLUME: ISSUE:

ISSN: 10612130
Journal of Optimal Nutrition
PUB-YEAR: 1995 VOLUME: 4 ISSUE:1

ISSN: 9519580
Journal of Orthopaedic Rheumatology
PUB-YEAR: 1996 VOLUME: 9 ISSUE:4

ISSN: 10225536
Journal of Orthopaedic Surgery
PUB-YEAR: 2003 VOLUME: 11 ISSUE:1

ISSN: 3340236
Journal of Orthopaedic Surgical Techniques
PUB-YEAR: 1992 VOLUME: 7 ISSUE:4

ISSN: 10567437
Journal of Orthopaedic Techniques
PUB-YEAR: 1996 VOLUME: 4 ISSUE:1

ISSN: 11212233
Journal of Preventive Medicine and Hygiene
PUB-YEAR: 1995 VOLUME: 36 ISSUE:3

ISSN: 15229580
Journal of Psychotherapy in Independent Practice
PUB-YEAR: 2001 VOLUME: 2 ISSUE:4

ISSN: 1055050X
Journal of Psychotherapy Practice and Research
PUB-YEAR: 2001 VOLUME: 10 ISSUE:4

ISSN: 13205455
Journal of Quality in Clinical Practice
PUB-YEAR: 2001 VOLUME: 21 ISSUE:4

ISSN: 0194259X
Journal of Respiratory Diseases - For Pediatricians
PUB-YEAR: 2003 VOLUME: 5 ISSUE:6

ISSN: 3855074
Journal of Saitama Medical School
PUB-YEAR: 1999 VOLUME: 26 ISSUE:4

ISSN: 13507702
Journal of Serotonin Research
PUB-YEAR: 1998 VOLUME: 4 ISSUE:4

ISSN: 2810662
Journal of Social and Administrative Pharmacy
PUB-YEAR: 2003 VOLUME: 20 ISSUE:6

ISSN: 10841288
Journal of Sports Chiropractic and Rehabilitation
PUB-YEAR: 2001 VOLUME: 15 ISSUE:2

ISSN: 8993289
Journal of Substance Abuse
PUB-YEAR: 2001 VOLUME: 13 ISSUE:4

ISSN: 10871306
Journal of Sudden Infant Death Syndrome and Infant Mortality
PUB-YEAR: 1997 VOLUME: 2 ISSUE:4

ISSN: 10871306
Journal of Sudden Infant Death Syndrome and Infant Mortality
PUB-YEAR: 1998 VOLUME: 2 ISSUE:3

ISSN: 3010902
Journal of the Aichi Medical University Association
PUB-YEAR: 2000 VOLUME: 28 ISSUE:4

ISSN: 28223
Journal of the American Dietetic Association
PUB-YEAR: 2000 VOLUME: 100 ISSUE:1

ISSN: 10865802
Journal of the American Pharmaceutical Association
PUB-YEAR: 1997 VOLUME: 37 ISSUE:6

ISSN: 30651
Journal of the American Psychoanalytic Association
PUB-YEAR: 1999 VOLUME: 47 ISSUE:3

ISSN: 13639579
Journal of the British Fertility Society
PUB-YEAR: 1997 VOLUME: 2 ISSUE:2

ISSN: 154148
Journal of the Florida Medical Association
PUB-YEAR: 1997 VOLUME: 84 ISSUE:4

ISSN: 10907173
Journal of the Gay and Lesbian Medical Association
PUB-YEAR: 2002 VOLUME: 6 ISSUE:3-4

ISSN: 10512306
Journal of the International Federation of Clinical Chemistry
PUB-YEAR: 1995 VOLUME: 7 ISSUE:2

ISSN: 471801
Journal of the Japan Society of Colo-Proctology
PUB-YEAR: 1999 VOLUME: 52 ISSUE:10

ISSN: 215325
Journal of the Japanese Orthopaedic Association
PUB-YEAR: 1996 VOLUME: 70 ISSUE:4

ISSN: 319341
Journal of the Physiological Society of Japan
PUB-YEAR: 1995 VOLUME: 57 ISSUE:7

ISSN: 409022
Journal of Tokyo Women's Medical College
PUB-YEAR: 2001 VOLUME: 71 ISSUE:9-10

ISSN: 225274
Journal of Transportation Medicine
PUB-YEAR: 2000 VOLUME: 54 ISSUE:2-3

ISSN: 13538012
Journal of Vascular Investigation
PUB-YEAR: 1998 VOLUME: 4 ISSUE:4

ISSN: 1047930
Journal of Venomous Animals and Toxins
PUB-YEAR: 1998 VOLUME: 4 ISSUE:1

ISSN: 2580659
Journal of Xi'an Medical University, Chinese Edition
PUB-YEAR: 1999 VOLUME: 20 ISSUE:3

ISSN: 3543447
Jugoslovenska Medicinska Biokemija
PUB-YEAR: 1999 VOLUME: 18 ISSUE:3

ISSN: 3948218
Jura Medica
PUB-YEAR: 1999 VOLUME: 12 ISSUE:1

ISSN: 3850234
Kawasaki Medical Journal
PUB-YEAR: 2001 VOLUME: 27 ISSUE:1

ISSN: 229040
Kardiologiya
PUB-YEAR: 1995 VOLUME: 35 ISSUE:11

ISSN: 188247
Khigiena i Zdraveopazvane
PUB-YEAR: 1998 VOLUME: 41 ISSUE:2

ISSN: 231207
Khirurgiya
PUB-YEAR: 1995 VOLUME: 71 ISSUE:6

ISSN: 232149
Klinicheskaya Meditsina
PUB-YEAR: 1995 VOLUME: 73 ISSUE:6

ISSN: 13000675
Klinik Gelisim
PUB-YEAR: 1996 VOLUME: 9 ISSUE:3

ISSN: 234516
Krankenhaus Arzt
PUB-YEAR: 1997 VOLUME: 70 ISSUE:12

ISSN: 8860467
Lasers in the Life Sciences
PUB-YEAR: 2001 VOLUME: 9 ISSUE:4

ISSN: 3008622
Leber Magen Darm
PUB-YEAR: 2000 VOLUME: 30 ISSUE:6

ISSN: 10686991
Lower Extremity
PUB-YEAR: 1998 VOLUME: 5 ISSUE:2

ISSN: 0399032X
Lyon Mediterranee Medical Medecine du Sud-Est
PUB-YEAR: 1999 VOLUME: 35 ISSUE:3-4

ISSN: 247804
Lyon Pharmaceutique
PUB-YEAR: 1998 VOLUME: 49 ISSUE:8

ISSN: 252514
Manuelle Medizin
PUB-YEAR: 1998 VOLUME: 36 ISSUE:6

ISSN: 2624230
Manufacturing Chemist
PUB-YEAR: 2001 VOLUME: 72 ISSUE:12

ISSN: 255246
Materia Medica Polona
PUB-YEAR: 1996 VOLUME: 28 ISSUE:4

ISSN: 12798479
Medecine de Catastrophe Urgences Collectives
PUB-YEAR: 2000 VOLUME: 3 ISSUE:4

ISSN: 476412
Medecine et Chirurgie Digestives
PUB-YEAR: 1999 VOLUME: 28 ISSUE:7-8

ISSN: 3987604
Medecine et Nutrition
PUB-YEAR: 1998 VOLUME: 34 ISSUE:1

ISSN: 10270256
Medical Biochemistry
PUB-YEAR: 2001 VOLUME: 1 ISSUE:4

ISSN: 3786358
Medical Bulletin of Istanbul Medical Faculty
PUB-YEAR: 1997 VOLUME: 30 ISSUE:1

ISSN: 0037380X
Medical Consultation and New Remedies
PUB-YEAR: VOLUME: ISSUE:

ISSN: 3858367
Medical Journal of Kinki University
PUB-YEAR: 1995 VOLUME: 20 ISSUE:2

ISSN: 10161430
Medical Journal of the Islamic Republic of Iran
PUB-YEAR: 1998 VOLUME: 12 ISSUE:1

ISSN: 9668454
Medical Mycology, Supplement
PUB-YEAR: 2003 VOLUME: 39 ISSUE:1

ISSN: 476552
Medical Progress through Technology
PUB-YEAR: 1997 VOLUME: 21 ISSUE:

ISSN: 10827579
Medical Update for Psychiatrists
PUB-YEAR: 1998 VOLUME: 3 ISSUE:6

ISSN: 11239395
Medicina Clinica e Termale
PUB-YEAR: 1998 VOLUME: 12 ISSUE:43

ISSN: 3924548
Medicina Oggi
PUB-YEAR: 1995 VOLUME: 15 ISSUE:3-4

ISSN: 2128292
Medicina Veterinaria
PUB-YEAR: 1999 VOLUME: 16 ISSUE:11

ISSN: 15313069
Medicine of the Americas
PUB-YEAR: 2001 VOLUME: 2 ISSUE:2

ISSN: 258024
Medicine, Science and the Law
PUB-YEAR: 1996 VOLUME: 36 ISSUE:

ISSN: 0971720X
Medico-Legal Update
PUB-YEAR: 1998 VOLUME: 3 ISSUE:1-2

ISSN: 0394025X
Mediterranean Journal of Infectious and Parasitic Diseases
PUB-YEAR: 1999 VOLUME: 14 ISSUE:

ISSN: 258334
Meditsinskaya Radiologiya i Radiatsionnaya Bezopasnost
PUB-YEAR: 1995 VOLUME: 40 ISSUE:3

ISSN: 15212076
Medscape Women's Health eJournal
PUB-YEAR: VOLUME: ISSUE:

ISSN: 11352000
Medula Espinal
PUB-YEAR: 1996 VOLUME: 2 ISSUE:2

ISSN: 258601
Medycyna Doswiadczalna i Mikrobiologia
PUB-YEAR: 2003 VOLUME: 55 ISSUE:4

ISSN: 10236597
Membrane and Cell Biology
PUB-YEAR: 2001 VOLUME: 14 ISSUE:6

ISSN: 476765
Mental Retardation
PUB-YEAR: 1998 VOLUME: 36 ISSUE:4

ISSN: 7930291
Metal-Based Drugs
PUB-YEAR: 2002 VOLUME: 9 ISSUE:1-2

ISSN: 8987750
Methods in Molecular and Cellular Biology
PUB-YEAR: 1995 VOLUME: 6 ISSUE:2

ISSN: 11200146
Microbiologia Medica
PUB-YEAR: 1997 VOLUME: 12 ISSUE:4

ISSN: 3749096
Mikrobiyoloji Bulteni
PUB-YEAR: 2003 VOLUME: 37 ISSUE:4

ISSN: 3780392
Mineral and Electrolyte Metabolism
PUB-YEAR: 1999 VOLUME: 25 ISSUE:4-6

ISSN: 10998209
Modern Drug Discovery
PUB-YEAR: 2001 VOLUME: 4 ISSUE:12

ISSN: 10447393
Molecular and Chemical Neuropathology
PUB-YEAR: 1999 VOLUME: 35 ISSUE:1

ISSN: 14685698
Molecular Biology Today
PUB-YEAR: 2002 VOLUME: 3 ISSUE:3

ISSN: 15224724
Molecular Cell Biology Research Communications
PUB-YEAR: 2001 VOLUME: 4 ISSUE:6

ISSN: 10915362
Molecular Urology
PUB-YEAR: 2001 VOLUME: 5 ISSUE:4

ISSN: 268984
Molekulyarnaya Biologiya
PUB-YEAR: 1995 VOLUME: 29 ISSUE:3

ISSN: 2144735
Monografias de Dermatologia
PUB-YEAR: 1997 VOLUME: 10 ISSUE:6

ISSN: 1473964
Multivariate Experimental Clinical Research
PUB-YEAR: 1997 VOLUME: 11 ISSUE:3

ISSN: 13835726
Mutation Research - Mutation Research Genomics
PUB-YEAR: 2001 VOLUME: 458 ISSUE:3-4

ISSN: 1657992
Mutation Research - Mutation Research Letters
PUB-YEAR: 1995 VOLUME: 348 ISSUE:4

ISSN: 1694294
Nederlands Tijdschrift voor Intensive Care
PUB-YEAR: 2001 VOLUME: 16 ISSUE:DEC.

ISSN: 2116995
Nefrologia
PUB-YEAR: 1999 VOLUME: 19 ISSUE:5

ISSN: 2129787
Neoplasia
PUB-YEAR: 1998 VOLUME: 15 ISSUE:6

ISSN: 12168068
Neurobiology
PUB-YEAR: 2001 VOLUME: 9 ISSUE:3-4

ISSN: 11202254
Neurologia Psichiatria Scienze Umane
PUB-YEAR: 1997 VOLUME: 17 ISSUE:6

ISSN: 10844759
Neurological Infections and Epidemiology
PUB-YEAR: 1997 VOLUME: 2 ISSUE:4

ISSN: 1777955
Neuro-Orthopedics
PUB-YEAR: 2000 VOLUME: 28 ISSUE:1

ISSN: 15294145
Neuro-Psychoanalysis
PUB-YEAR: 2001 VOLUME: 3 ISSUE:2

ISSN: 3935302
New Trends in Arrhythmias
PUB-YEAR: 1994 VOLUME: 10 ISSUE:1

ISSN: 3935345
New Trends in Clinical Neuropharmacology
PUB-YEAR: 1994 VOLUME: 8 ISSUE:2

ISSN: 3943755
New Trends in Ophthalmology
PUB-YEAR: 1997 VOLUME: 12 ISSUE:3

ISSN: 5460352
Nihon University Journal of Medicine
PUB-YEAR: 2001 VOLUME: 43 ISSUE:4

ISSN: 13000667
Noropsikiyatri Arsivi
PUB-YEAR: 1995 VOLUME: 32 ISSUE:4

ISSN: 295604
Nuclear Safety
PUB-YEAR: 1997 VOLUME: 38 ISSUE:1

ISSN: 1122035X
Nuova Rivista di Neurologia
PUB-YEAR: 2003 VOLUME: 13 ISSUE:6

ISSN: 11207434
Occhio Clinico
PUB-YEAR: 2002 VOLUME: 8 ISSUE:6

ISSN: 15921093
Occhio Clinico Pediatria
PUB-YEAR: 2002 VOLUME: 6 ISSUE:8

ISSN: 15269396
Office and Emergency Pediatrics
PUB-YEAR: 2001 VOLUME: 14 ISSUE:1

ISSN: 3784835
Oncologia
PUB-YEAR: 1999 VOLUME: 22 ISSUE:7

ISSN: 13900110
Oncologia
PUB-YEAR: 1998 VOLUME: 8 ISSUE:2

ISSN: 1021335X
Oncology Reports
PUB-YEAR: 1998 VOLUME: 5 ISSUE:6

ISSN: 9650407
Oncology Research
PUB-YEAR: 1996 VOLUME: 8 ISSUE:12

ISSN: 13862618
Opportunistic Pathogens
PUB-YEAR: 1997 VOLUME: 9 ISSUE:1

ISSN: 10591516
Orthopaedic Physical Therapy Clinics of North America
PUB-YEAR: 2002 VOLUME: 11 ISSUE:2

ISSN: 3921417
Ortopedia e Traumatologia Oggi
PUB-YEAR: 1997 VOLUME: 17 ISSUE:3

ISSN: 4734378
Ortopediya i Travmatologiya
PUB-YEAR: 2003 VOLUME: 39 ISSUE:1-2

ISSN: 298786
Osterreichische Arztezeitung
PUB-YEAR: 1997 VOLUME: 52 ISSUE:21

ISSN: 309338
Padiatrie und Padologie
PUB-YEAR: 1997 VOLUME: 32 ISSUE:6

ISSN: 9681302
Pain Reviews
PUB-YEAR: 2002 VOLUME: 9 ISSUE:2

ISSN: 10245359
Pathogenesis
PUB-YEAR: 2000 VOLUME: 1 ISSUE:4

ISSN: 312983
Pathologica
PUB-YEAR: 1998 VOLUME: 90 ISSUE:1

ISSN: 312991
Patologicheskaya Fiziologiya i Eksperimentalnaya Terapiya
PUB-YEAR: 1995 VOLUME: 39 ISSUE:4

ISSN: 3979180
Pediatre
PUB-YEAR: 1995 VOLUME: 31 ISSUE:149

ISSN: 8829225
Pediatric Reviews and Communications
PUB-YEAR: 1996 VOLUME: 8 ISSUE:4

ISSN: 10640517
Perspectives on Developmental Neurobiology
PUB-YEAR: 1998 VOLUME: 5 ISSUE:4

ISSN: 9399488
Pharmaceutical and Pharmacological Letters
PUB-YEAR: 2001 VOLUME: 11 ISSUE:2

ISSN: 1071894X
Pharmaceutical News
PUB-YEAR: 2002 VOLUME: 9 ISSUE:6

ISSN: 3699579
Pharmacie Hospitaliere Francaise
PUB-YEAR: 1999 VOLUME: ISSUE:126

ISSN: 14608081
Pharmacy and Pharmacology Communications
PUB-YEAR: 2000 VOLUME: 6 ISSUE:12

ISSN: 5560071
Philippine Journal of Internal Medicine
PUB-YEAR: 1995 VOLUME: 33 ISSUE:2

ISSN: 1090378X
Phonoscope
PUB-YEAR: 1999 VOLUME: 2 ISSUE:4

ISSN: 3000508
Physiotherapy Canada
PUB-YEAR: 1994 VOLUME: 46 ISSUE:1

ISSN: 12107182
Pohybove Ustroji
PUB-YEAR: 1998 VOLUME: 5 ISSUE:3-4

ISSN: 1377183
Polski Przeglad Radiologii
PUB-YEAR: 1999 VOLUME: 64 ISSUE:4

ISSN: 14329190
PRAXIS med.
PUB-YEAR: 1997 VOLUME: 6 ISSUE:9

ISSN: 327611
Prescribers' Journal
PUB-YEAR: 2000 VOLUME: 40 ISSUE:2

ISSN: 1520037X
Preventive Cardiology
PUB-YEAR: 2003 VOLUME: 6 ISSUE:4

ISSN: 3635104
Primary Cardiology
PUB-YEAR: 1995 VOLUME: 21 ISSUE:12

ISSN: 1068607X
Primary Care Update for Ob/Gyns
PUB-YEAR: 2003 VOLUME: 10 ISSUE:6

ISSN: 1355185X
Probe Microscopy
PUB-YEAR: 2001 VOLUME: 2 ISSUE:2

ISSN: 3239179
Problemi na Khigienata
PUB-YEAR: 1997 VOLUME: 22 ISSUE:

ISSN: 3759660
Problemy Endokrinologii
PUB-YEAR: 1995 VOLUME: 41 ISSUE:5

ISSN: 329533
Problemy Tuberkuleza
PUB-YEAR: 1996 VOLUME: 74 ISSUE:1

ISSN: 1081650X
Proceedings of the Association of American Physicians
PUB-YEAR: 1999 VOLUME: 111 ISSUE:6

ISSN: 3016331
Proceedings of the University of Otago Medical School
PUB-YEAR: 1995 VOLUME: 73 ISSUE:3

ISSN: 3045013
Progresos en Obstetricia y Ginecologia
PUB-YEAR: 1999 VOLUME: 42 ISSUE:9

ISSN: 1060913X
Progress in Liver Diseases
PUB-YEAR: 1997 VOLUME: 15 ISSUE:

ISSN: 9349545
Progress in Pharmacology and Clinical Pharmacology
PUB-YEAR: 1995 VOLUME: 10 ISSUE:3

ISSN: 10955100
Prostate Journal
PUB-YEAR: 2001 VOLUME: 3 ISSUE:2

ISSN: 3535053
Psychiatria Danubina
PUB-YEAR: 1996 VOLUME: 8 ISSUE:1

ISSN: 10683178
Psychiatric Clinics of North America: Annual of Drug Therapy
PUB-YEAR: 2001 VOLUME: ISSUE:

ISSN: 8896313
Psychobiology
PUB-YEAR: 2000 VOLUME: 28 ISSUE:4

ISSN: 485756
Psychologie Medicale
PUB-YEAR: 1995 VOLUME: 27 ISSUE:1

ISSN: 485764
Psychopharmacology Bulletin
PUB-YEAR: 1998 VOLUME: 34 ISSUE:4

ISSN: 7317158
Psychotherapy in Private Practice
PUB-YEAR: 1998 VOLUME: 17 ISSUE:4

ISSN: 14154315
Pulmao RJ
PUB-YEAR: VOLUME: ISSUE:

ISSN: 3935930
Quaderni di Medicina e Chirurgia
PUB-YEAR: 2001 VOLUME: 17 ISSUE:2-3

ISSN: 13883593
Quantitative Microbiology
PUB-YEAR: 2000 VOLUME: 2 ISSUE:3

ISSN: 8698031
Radiatsionnaya Biologiya Radioekologiya
PUB-YEAR: 1995 VOLUME: 35 ISSUE:3

ISSN: 338354
Radiologia Diagnostica
PUB-YEAR: 1994 VOLUME: 35 ISSUE:6

ISSN: 7203322
Radiologie - Journal du CEPUR
PUB-YEAR: 2001 VOLUME: 21 ISSUE:5

ISSN: 10691286
Radiologist
PUB-YEAR: 2003 VOLUME: 10 ISSUE:6

ISSN: 339490
Rassegna di Dermatologia e Sifilografia
PUB-YEAR: 1995 VOLUME: 48 ISSUE:1

ISSN: 339636
Rassegna di Studi Psichiatrici
PUB-YEAR: 1996 VOLUME: 85 ISSUE:1

ISSN: 339695
Rassegna Internazionale di Clinica e Terapia
PUB-YEAR: 1995 VOLUME: 75 ISSUE:6

ISSN: 7655290
Reanimation Soins Intensifs Medecine d'Urgence
PUB-YEAR: 1997 VOLUME: 13 ISSUE:4

ISSN: 9350411
Regional Cancer Treatment
PUB-YEAR: 1996 VOLUME: 9 ISSUE:4

ISSN: 343501
Rehabilitation in South Africa
PUB-YEAR: 1995 VOLUME: 39 ISSUE:

ISSN: 3920607
Relata Technica
PUB-YEAR: 1996 VOLUME: 28 ISSUE:2

ISSN: 1050964X
Report on Pediatric Infectious Diseases
PUB-YEAR: 1997 VOLUME: 7 ISSUE:10

ISSN: 15284840
Reproductive Technologies
PUB-YEAR: 2001 VOLUME: 10 ISSUE:6

ISSN: 2145987
Research in Surgery
PUB-YEAR: 1996 VOLUME: 8 ISSUE:1

ISSN: 13981714
Reviews in Immunogenetics
PUB-YEAR: 2000 VOLUME: 2 ISSUE:4

ISSN: 11387165
Revisiones en Psicofarmacologia
PUB-YEAR: 2000 VOLUME: 4 ISSUE:3

ISSN: 25151
Revista Alergia Mexico
PUB-YEAR: 2002 VOLUME: 49 ISSUE:3

ISSN: 0370372X
Revista Brasileira de Farmacia
PUB-YEAR: 2000 VOLUME: 81 ISSUE:1-2

ISSN: 347280
Revista Brasileira de Oftalmologia
PUB-YEAR: 2003 VOLUME: 62 ISSUE:9

ISSN: 1023616
Revista Brasileira de Ortopedia
PUB-YEAR: 2000 VOLUME: 35 ISSUE:7

ISSN: 347434
Revista Colombiana de Obstetricia y Ginecologia
PUB-YEAR: 1996 VOLUME: 47 ISSUE:2

ISSN: 3750906
Revista de Gastroenterologia de Mexico
PUB-YEAR: 1997 VOLUME: 62 ISSUE:2

ISSN: 10186212
Revista de la Sanidad de la Policia Nacional del Peru
PUB-YEAR: 1995 VOLUME: 56 ISSUE:1

ISSN: 13714
Revista de Microbiologia
PUB-YEAR: 1999 VOLUME: 30 ISSUE:2

ISSN: 767131
Revista del Instituto Nacional de Cancerologia
PUB-YEAR: 1999 VOLUME: 45 ISSUE:3

ISSN: 349536
Revista Espanola de Anestesiologia y Reanimacion
PUB-YEAR: 2001 VOLUME: 48 ISSUE:10

ISSN: 3774732
Revista Espanola de Medicina Legal
PUB-YEAR: 1999 VOLUME: 23 ISSUE:86/87

ISSN: 2126982
Revista Espanola de Medicina Nuclear
PUB-YEAR: 1999 VOLUME: 18 ISSUE:5

ISSN: 2134233
Revista Espanola de Neurologia
PUB-YEAR: 1995 VOLUME: 10 ISSUE:5

ISSN: 3044815
Revista Espanola de Reumatologia
PUB-YEAR: 1999 VOLUME: 26 ISSUE:8

ISSN: 1405101X
Revista Iberolatinoamericana de Cuidados Intensivos
PUB-YEAR: 1997 VOLUME: 6 ISSUE:2

ISSN: 8715718
Revista Oncologia
PUB-YEAR: 1993 VOLUME: 3 ISSUE:2

ISSN: 5577721
Revue de Medecine de Tours
PUB-YEAR: 2000 VOLUME: 34 ISSUE:5-6

ISSN: 9892737
Revue du Praticien - Medecine Generale
PUB-YEAR: 2002 VOLUME: 16 ISSUE:596/597

ISSN: 352888
Revue Francaise de Gastro-Enterologie
PUB-YEAR: 1998 VOLUME: 34 ISSUE:341-342

ISSN: 353655
Revue Medicale de la Suisse Romande
PUB-YEAR: 1999 VOLUME: 119 ISSUE:11

ISSN: 2497581
Rhumatologie
PUB-YEAR: 1999 VOLUME: 51 ISSUE:8

ISSN: 5579430
Riabilitazione
PUB-YEAR: 1999 VOLUME: 32 ISSUE:3

ISSN: 3906019
Rivista di Tossicologia Sperimentale e Clinica
PUB-YEAR: 1995 VOLUME: 25 ISSUE:3

ISSN: 3931137
Rivista Italiana di Biologia e Medicina
PUB-YEAR: 2002 VOLUME: 22 ISSUE:3-4

ISSN: 2542307
Romanian Journal of Gerontology and Geriatrics
PUB-YEAR: 2001 VOLUME: 23 ISSUE:2

ISSN: 353973
Romanian Journal of Internal Medicine
PUB-YEAR: 1995 VOLUME: 33 ISSUE:3

ISSN: 3009157
Ryumachi
PUB-YEAR: 2003 VOLUME: 43 ISSUE:4

ISSN: 1018077X
Saudi Heart Journal
PUB-YEAR: 1995 VOLUME: 6 ISSUE:2

ISSN: 8917035
Scanning Microscopy
PUB-YEAR: 1996 VOLUME: 10 ISSUE:4

ISSN: 367508
Schweizer Apothekerzeitung
PUB-YEAR: 1997 VOLUME: 135 ISSUE:22

ISSN: 371777
Semaines des Hopitaux
PUB-YEAR: 1999 VOLUME: 75 ISSUE:35-36

ISSN: 10843612
Seminars in Clinical Neuropsychiatry
PUB-YEAR: 2003 VOLUME: 8 ISSUE:4

ISSN: 10495118
Seminars in Gastrointestinal Disease
PUB-YEAR: 2003 VOLUME: 14 ISSUE:4

ISSN: 15318192
Seminars in Infection Control
PUB-YEAR: VOLUME: ISSUE:

ISSN: 10445765
Seminars in Neuroscience
PUB-YEAR: 1998 VOLUME: 9 ISSUE:5-6

ISSN: 8820546
Seminars in Respiratory Infections
PUB-YEAR: 2003 VOLUME: 18 ISSUE:4

ISSN: 5826802
Seoul Journal of Medicine
PUB-YEAR: 1995 VOLUME: 36 ISSUE:4

ISSN: 13850229
Sepsis
PUB-YEAR: 2001 VOLUME: 4 ISSUE:3

ISSN: 373699
Shikoku Acta Medica
PUB-YEAR: 2000 VOLUME: 56 ISSUE:5

ISSN: 373826
Shinshu Medical Journal
PUB-YEAR: 2000 VOLUME: 48 ISSUE:3

ISSN: 14385163
Single Molecules
PUB-YEAR: 2002 VOLUME: 3 ISSUE:5-6

ISSN: 11370351
S.N.
PUB-YEAR: VOLUME: ISSUE:

ISSN: 3796175
South African Journal of Physiotherapy
PUB-YEAR: 1998 VOLUME: 54 ISSUE:2

ISSN: 9457712
Sozialpadiatrie und Kinderarztliche Praxis
PUB-YEAR: 1996 VOLUME: 18 ISSUE:11

ISSN: 15911195
Sports Cardiology
PUB-YEAR: VOLUME: ISSUE:

ISSN: 13510029
Sports Exercise and Injury
PUB-YEAR: 1998 VOLUME: 4 ISSUE:4

ISSN: 492116
Star
PUB-YEAR: 1999 VOLUME: 58 ISSUE:2

ISSN: 5362180
Stomach and Intestine
PUB-YEAR: 1999 VOLUME: 34 ISSUE:11

ISSN: 391735
Stomatologiya
PUB-YEAR: 1995 VOLUME: 74 ISSUE:6

ISSN: 394521
Sudebno-Meditsinskaya Ekspertiza
PUB-YEAR: 1995 VOLUME: 38 ISSUE:4

ISSN: 8829233
Surgical Research Communications
PUB-YEAR: 1998 VOLUME: 19 ISSUE:2

ISSN: 10239332
Swiss Surgery
PUB-YEAR: VOLUME: ISSUE:

ISSN: 10793259
Techniques in Urology
PUB-YEAR: 2001 VOLUME: 7 ISSUE:4

ISSN: 403660
Terapevticheskii Arkhiv
PUB-YEAR: 1995 VOLUME: 67 ISSUE:11

ISSN: 9670149
Therapeutic Immunology
PUB-YEAR: 1995 VOLUME: 2 ISSUE:4

ISSN: 1433111X
Therapie und Erfolg
PUB-YEAR: 1998 VOLUME: 2 ISSUE:3

ISSN: 405973
Therapiewoche
PUB-YEAR: 1996 VOLUME: 46 ISSUE:36

ISSN: 405973
Therapiewoche
PUB-YEAR: 2002 VOLUME: 18 ISSUE:12

ISSN: 1656090
Thymus
PUB-YEAR: 1997 VOLUME: 24 ISSUE:4

ISSN: 3782778
Tijdschrift voor Alcohol, Drugs en Andere Psychotrope Stoffen
PUB-YEAR: 1999 VOLUME: 24 ISSUE:3

ISSN: 9605592
Transfusion Medicine, Supplement
PUB-YEAR: VOLUME: ISSUE:

ISSN: 10236171
Transgenics
PUB-YEAR: 2001 VOLUME: 3 ISSUE:2-4

ISSN: 1134315X
Transplantology: Journal of Cell and Organ Transplantation
PUB-YEAR: 1999 VOLUME: 10 ISSUE:2

ISSN: 7436637
Trauma Quarterly
PUB-YEAR: 2002 VOLUME: 15 ISSUE:1

ISSN: 3855643
Tropical Medicine
PUB-YEAR: 2001 VOLUME: 43 ISSUE:3-4

ISSN: 5643783
Tsitologiya i Genetika
PUB-YEAR: 1995 VOLUME: 29 ISSUE:3

ISSN: 13000659
Turk Oftalmoloji Gazetesi
PUB-YEAR: 1995 VOLUME: 25 ISSUE:1

ISSN: 13004301
Turkish Journal of Dermatopathology
PUB-YEAR: 1998 VOLUME: 7 ISSUE:2

ISSN: 11266937
UCP News
PUB-YEAR: VOLUME: ISSUE:

ISSN: 1484818
U.S. Pharmacist
PUB-YEAR: 1996 VOLUME: 21 ISSUE:3

ISSN: 10567909
Vaccine Research
PUB-YEAR: 1997 VOLUME: 6 ISSUE:3

ISSN: 10321012
Venereology
PUB-YEAR: 2001 VOLUME: 14 ISSUE:4

ISSN: 424609
Vestnik Dermatologii i Venerologii
PUB-YEAR: 1995 VOLUME: 71 ISSUE:6

ISSN: 0042465X
Vestnik Oftalmologii
PUB-YEAR: 1995 VOLUME: 111 ISSUE:4

ISSN: 424676
Vestnik Rentgenologii i Radiologii
PUB-YEAR: 1995 VOLUME: 70 ISSUE:3

ISSN: 3009858
Veterinary Pathology
PUB-YEAR: 2001 VOLUME: 38 ISSUE:6

ISSN: 10817050
Viral Hepatitis Reviews
PUB-YEAR: 2000 VOLUME: 6 ISSUE:1-4

ISSN: 428450
Vojnosanitetski Pregled
PUB-YEAR: 1998 VOLUME: 55 ISSUE:2

ISSN: 428809
Voprosy Meditsinskoj Khimii
PUB-YEAR: 1995 VOLUME: 41 ISSUE:6

ISSN: 5073758
Voprosy Onkologii
PUB-YEAR: 1995 VOLUME: 41 ISSUE:1

ISSN: 5074088
Voprosy Virusologii
PUB-YEAR: 1995 VOLUME: 40 ISSUE:6

ISSN: 9212639
Water Treatment
PUB-YEAR: 1995 VOLUME: 10 ISSUE:4

ISSN: 930415
Western Journal of Medicine
PUB-YEAR: 2002 VOLUME: 176 ISSUE:4

ISSN: 10129375
Wiener Zeitschrift fur Suchtforschung
PUB-YEAR: 1999 VOLUME: 22 ISSUE:2

ISSN: 3782255
World Health Organization Regional Publications - European Series
PUB-YEAR: 2002 VOLUME: - ISSUE:97

ISSN: 5131731
Yamaguchi Medical Journal
PUB-YEAR: 2000 VOLUME: 49 ISSUE:5

ISSN: 440531
Yokohama Medical Bulletin
PUB-YEAR: 1994 VOLUME: 45 ISSUE:3

ISSN: 441570
Zacchia
PUB-YEAR: 1999 VOLUME: 72 ISSUE:1

ISSN: 9409599
Zentralblatt fur Kinderchirurgie
PUB-YEAR: 2002 VOLUME: 11 ISSUE:4

ISSN: 3729311
Zhurnal Mikrobiologii Epidemiologii i Immunobiologii
PUB-YEAR: 1995 VOLUME: 72 ISSUE:4

ISSN: 444650
Zhurnal Ushnykh Nosovykh i Gorlovykh Boleznei
PUB-YEAR: 1999 VOLUME: ISSUE:5

ISSN: 428817
Zhurnal Voprosy Nejrokhirurgii Imeni N.N. Burdenko
PUB-YEAR: 1995 VOLUME: 59 ISSUE:3

ISSN: 444677
Zhurnal Vysshei Nervnoi Deyatelnosti Imeni I.P. Pavlova
PUB-YEAR: 1995 VOLUME: 45 ISSUE:3

LIST OF CURRENT TITLES

AANA Journal
ISSN: 0094-6354
AANA J.
AANA Publishing Inc., 222 South Prospect Road, Park Ridge, IL 60068-4001, United States, Tel: +1(708)6927050 ISSUES/YR: 6

AAPS Journal
ISSN: 1550-7416
AAPS Journal
American Association of Pharmaceutical Scientists, 2107 Wilson Blvd., Suite 700, Arlington, 22201, United States, Tel: +1(703)2432800, Fax: +1(703)2439650 ISSUES/YR: 4

AAPS PharmSciTech
ISSN: 1530-9932
AAPS PharmSciTech
AAPS PharmSci Editorial Office, 2107 Wilson Blvd. Suite 700, Arlington, VA 22201-3046, United States, Tel: +1(703)2432800, Fax: +1(703)2439650 ISSUES/YR: 4

• **Abdominal Imaging**
ISSN: 0942-8925
Abdom. Imaging
Springer New York LLC, 233 Springer Street, New York, 10013-1578, United States, Tel: +1-212 460 1501, Fax: +1-212 460 1595 ISSUES/YR: 6

• **Academic Emergency Medicine**
ISSN: 1069-6563
Acad. Emerg. Med.
Hanley and Belfus Inc., 210 South 13th Street, Philadelphia, PA 19107, United States, Tel: +1(215)5467293 ISSUES/YR: 12

Academic Journal of Second Military Medical University
ISSN: 0258-879X
Acad. J. Second Mil. Med. Univ.
Second Military Medical University Press, 800 Xiangyin Road, 200433, Shanghai, China, Tel: +86(21)25070130, Fax: +86(21)25070125 ISSUES/YR: 12

Academic Journal of Xi'an Jiaotong University
ISSN: 1671-8267
Acad. J. Xi'An Jiaotong Univ.
Xi'an Medical University, 16 Scarlet-Bird Avenue, 710061, Xi'an, China, Tel: +86(29)711621 ISSUES/YR: 2

• **Academic Medicine**
ISSN: 1040-2446
Acad. Med.
Lippincott Williams and Wilkins, 530 Walnut Street, Philadelphia, PA 19106-3621, United States, Tel: +1(215)5218342, Fax: +1(215)5218326 ISSUES/YR: 12

Academic Psychiatry
ISSN: 1042-9670
Acad. Psychiatry
American Psychiatric Publishing Inc., 1000 Wilson Blvd, Suite 1825, VA 22209, United States, Tel: +1(703)907-7322, Fax: +1(703)90-1091 ISSUES/YR: 4

• **Academic Radiology**
ISSN: 1076-6332
Acad. Radiol.
Elsevier USA, 6277 Sea Harbor Drive, Orlando, FL 32862 8239, United States, Tel: +1-407 345 4000, Fax: +1-407 363 9661 ISSUES/YR: 12

ACC Cardiosource Review Journal
ISSN: 1556-8571
ACC Cardiosource Rev. J.
Elsevier USA, 6277 Sea Harbor Drive, Orlando, FL 32862 8239, United States, Tel: +1-407 345 4000, Fax: +1-407 363 9661 ISSUES/YR: 12

Acta Academiae Medicinae Hubei
ISSN: 1000-243X
Acta Acad. Med. Hubei
Editorial Board of the Journal of Hubei Medical College, 39 Donghu Road, 430071, Wuhan, Hubei, China, Tel: +86-7812135385 ISSUES/YR: 4

Acta Academiae Medicinae Sinicae
ISSN: 1000-503X
Acta Acad. Med. Sin.
Chinese Academy of Medical Sciences, Panjiayuan, Chaoyang District, 100021, Beijing, China ISSUES/YR: 6

Acta Anaesthesiologica Belgica
ISSN: 0001-5164
Acta Anaesthesiol. Belg.
ASSMB a.s.b.l., Avenue Circulaire 138A, B-1180, Brussels, Belgium, Tel: +32(2)3745158, Fax: +32(2)3749628 ISSUES/YR: 4

Acta Anaesthesiologica Italica / Anaesthesia and Intensive Care in Italy
ISSN: 1124-8882
Acta Anaesthesiol. Ital. Anaesth. Intensive Care Italy
Editrice La Garangola, via E. dalla Costa, 6, 35129, Padova, Italy, Tel: +39(049)8750550, Fax: +39(049)7806580 ISSUES/YR: 3

• **Acta Anaesthesiologica Scandinavica**
ISSN: 0001-5172
Acta Anaesthesiol. Scand.
Blackwell Munksgaard, 1 Rosenorns Alle, P.O. Box 227, DK-1502, Copenhagen V, Denmark, Tel: +45(77)333333, Fax: +45(77)333377 ISSUES/YR: 10

• **Acta Anaesthesiologica Scandinavica, Supplement**
ISSN: 0515-2720
Acta Anaesthesiol. Scand. Suppl.
Blackwell Munksgaard, 1 Rosenorns Alle, P.O. Box 227, DK-1502, Copenhagen V, Denmark, Tel: +45(77)333333, Fax: +45(77)333377 ISSUES/YR: 1

Acta Anaesthesiologica Taiwanica
ISSN: 0254-1319
Acta Anaesthesiol. Taiwan.
Society of Anesthesiology of Republic of China, Unit 3, 4th Floor, 271 Roosevelt Road, Sec. 3, Taipei, Taiwan, Tel: +886(2)23123456 EXT.5496 ISSUES/YR: 4

Acta Anatomica Sinica
ISSN: 0529-1356
Acta Anat. Sin.
Chinese Medical Association, 20 Ave. Appia, CH-1211, Geneva 27, Switzerland, Tel: +86(22)7912111, Fax: +86(22)7914857 ISSUES/YR: 6

Acta Angiologica
ISSN: 1234-950X
Acta Angiolog.
Via Medica, Ul. Swietokrzyska 73, 80-180, Gdansk, Poland, Tel: +48(58)3209494, Fax: +48(58)3209460 ISSUES/YR: 4

Acta Biologica Szegediensis
ISSN: 1588-385X
Acta Biol. Szegediensis
University of Szeged, 2 Egyetem Street, P.O. Box 659, H-6722, Szeged, Hungary, Tel: +36(62)544048, Fax: +36(62)544049 ISSUES/YR: 4

• **Acta Biomaterialia**
ISSN: 1742-7061
Acta Biomater.
Elsevier Ltd, Langford Lane, Kidlington, OX5 1GB, Oxford, United Kingdom, Tel: +44(1865)843000, Fax: +44(1865)843010 ISSUES/YR: 6

Acta Biomedica de l'Ateneo Parmense
ISSN: 0392-4203
Acta Biomed. Ateneo Parmense
L'Ateneo Parmense, Via Gramsci 14, 43100, Parma, Italy, Tel: +39(521)983364 ISSUES/YR: 3

Acta Cardiologica
ISSN: 0001-5385
Acta Cardiol.
Acta Cardiologica, Weldadigheidsstraat 49, B-3000, Leuven, Belgium, Tel: +32-16 23 20 00, Fax: +32-16 23 18 00 ISSUES/YR: 6

Acta Cardiologica Sinica
ISSN: 1011-6842
Acta Cardiol. Sin.
Republic of China Society of Cardiology, 7F, No.27, Min-Chuan W.Road, 104, Taipei, Taiwan, Tel: +886(2)25976177, Fax: +886(2)25976180 ISSUES/YR: 4

Acta Chirurgiae Orthopaedicae et Traumatologiae Cechoslovaca
ISSN: 0001-5415
Acta Chir. Orthop. Traumatol. Cech.
Galen s.r.o., Nadrazni 116, 150 00, Prague 5, Czech Republic, Tel: +420(2)57326178, Fax: +420(2)57326170 ISSUES/YR: 6

Acta Chirurgiae Plasticae
ISSN: 0001-5423
Acta Chir. Plast.
Czech Medical Association J.E. Purkyne, Sokolska 31, P.O. Box 88, 120 26, Praha 2, Czech Republic, Tel: +420(2)24266201-4, Fax: +420(2)24266212 ISSUES/YR: 4

Acta Chirurgica Austriaca
ISSN: 0001-544X
Acta Chir. Austriaca
Blackwell Verlag Wien, Firmiangasse 7, A-1130, Vienna, Austria, Tel: +43(1)8779351, Fax: +43(1)8779351-24 ISSUES/YR: 6

Acta Cientifica Venezolana
ISSN: 0001-5504
Acta Cient. Venez.
Edificio FundaVAC - AsoVAC, Avenida Neveri, Colinas de Bello Monte, Apartado de Correos 47286, Caracas, Venezuela, Tel: +58(2)7531002, Fax: +58(2)7511420 ISSUES/YR: 4

Acta Cirurgica Brasileira
ISSN: 0102-8650
Acta Cir. Bras.
Sociedade Brasileira para o Desenvolvimento de Pesquisa em Cirurgia, Al.Rio Claro 179, 14 and., 01332-010, Sao Paulo, Brazil, Tel: +55-1132878814, Fax: +55-1132878814 ISSUES/YR: 6

Acta Clinica Belgica
ISSN: 0001-5512
Acta Clin. Belg.
Acta Clinica Belgica, De Pintelaan 185, 9000, UZ Gent, Belgium, Tel: +32(9)2404402, Fax: +32(9)2404403 ISSUES/YR: 6

Acta Clinica Croatica
ISSN: 0353-9466
Acta Clin. Croat.
Klinicka Bolnica Sestre Milosrdnice, Vinogradska c. 29, 41000, Zagreb, Croatia, Tel: +385(41)174666 ISSUES/YR: 4

Acta Clinica Croatica, Supplement
ISSN: 0353-9474
Acta Clin. Croat. Suppl.
Klinicka Bolnica Sestre Milosrdnice, Vinogradska c. 29, 41000, Zagreb, Croatia, Tel: +385(41)174666 ISSUES/YR: 1

Acta Criminologiae et Medicinae Legalis Japonica
ISSN: 0302-0029
Acta Criminol. Med. Leg. Jpn.
Japanese Association of Criminology, 3-10, 2-chome, Kandasurugadai, Chiyoda-ku, 101, Tokyo, Japan, Tel: +81(3)2919577 ISSUES/YR: 6

• **Acta Cytologica**
ISSN: 0001-5547
Acta Cytol.
Science Printers and Publishers Inc., 8342 Olive Blvd, St. Louis, MO 63132, United States, Tel: +1(314)9914588, Fax: +1(314)9914654 ISSUES/YR: 6

Acta Dermatovenerologica Alpina, Pannonica et Adriatica
ISSN: 1318-4458
Acta Dermatovenerol. Alp. Pannonica Adriat.
Slovenian Medical Society, Komenskega 4, 61000, Ljubljana, Slovenia ISSUES/YR: 4

Acta Dermatovenerologica Croatica
ISSN: 1330-027X
Acta Dermatovenerol. Croat.
Croatian Dermatovenerological Society, Salata 4, 10 000, Zagreb, Croatia, Tel: +385(1)271085, Fax: +385(1)271085 ISSUES/YR: 4

• **Acta Diabetologica**
ISSN: 0940-5429
Acta Diabetol.
Springer Verlag, Tiergartenstrasse 17, D-69121, Heidelberg, Germany, Tel: +49(6221)4870, Fax: +49(6221)487366 ISSUES/YR: 4

Acta Endoscopica
ISSN: 0240-642X
Acta Endosc.
Springer-Verlag France, 1 rue Paul Cezanne, 75008, Paris, France, Tel: +33(1)53933628, Fax: +33(1)53933683 ISSUES/YR: 5

Acta Facultatis Medicae Fluminensis
ISSN: 0065-1206
Acta Fac. Med. Flum.
Medicinski Fakultet Sveucilista u Rijeci, Brace Branchetta 20/22, HR-51000, Rijeka, Croatia, Tel: +385(51)513222 ISSUES/YR: 2

Acta Farmaceutica Bonaerense
ISSN: 0326-2383
Acta Farm. Bonaer.
Colegio de Farmaceuticos de la Provincia de Buenos Aires, Calle 5 No. 966, 1900, La Plata, Argentina, Tel: +54(21)258603 ISSUES/YR: 4

Acta Gastro-Enterologica Belgica
ISSN: 0001-5644
Acta Gastro-Enterol. Belg.
ASSMB a.s.b.l., Avenue Circulaire 138A, B-1180, Brussels, Belgium, Tel: +32(2)3745158, Fax: +32(2)3749628 ISSUES/YR: 4

Acta Genetica Sinica
ISSN: 0379-4172
Acta Genet. Sin.
Institute of Genetics and Developmental Biology, Datun Road, Andingmenwai, 100101, Beijing, China, Tel: +86-10 648 89354, Fax: +86-10 648 89348 ISSUES/YR: 12

Acta Geneticae Medicae et Gemellologiae
ISSN: 0001-5660
Acta Genet. Med. Gemellol.
The Mendel Institute, Piazza Galeno 5, 00162, Roma, Italy, Tel: +39(06)8554658 ISSUES/YR: 4

Acta Ginecologica
ISSN: 0001-5776
Acta Ginecol.
Editores Medicos S.A. (EDIMSA), Gabriela Mistral 2, 28035, Madrid, Spain, Tel: +34(91)3860033, Fax: +34(91)3860366 ISSUES/YR: 5

• **Acta Haematologica**
ISSN: 0001-5792
Acta Haematol.
S. Karger AG, Allschwilerstrasse 10, P.O. Box, CH-4009, Basel, Switzerland, Tel: +41(61)3061111, Fax: +41(61)3061234 ISSUES/YR: 8

Acta Haematologica Polonica
ISSN: 0001-5814
Acta Haematol. Pol.
Redakcja Acta Haematologica Polonica, Pabianicka str 62, 93-513, Lodz, Poland, Tel: +48(42)6846890 ISSUES/YR: 4

Acta Hepatologica Japonica
ISSN: 0451-4203
Acta Hepatol. Jpn.
Japan Society of Hepatology, 28-10 Hongo 3-chome, Bunkyo-ku, 113, Tokyo, Japan, Tel: +81(3)38121567, Fax: +81(3)38126620 ISSUES/YR: 12

Acta Histochemica
ISSN: 0065-1281
Acta Histochem.
Urban und Fischer Verlag Jena, P.O. Box 100537, 07705, Jena, Germany, Tel: +49(3641)6263, Fax: +49(3641)626500 ISSUES/YR: 6

Current

Acta Histochemica et Cytochemica
ISSN: 0044-5991
Acta Histochem. Cytochem.
Japan Society of Histochemistry and Cytochemistry, Shimotachiuri-Ogawa, Kamikyo-ku, 602-8048, Kyoto, Japan, Tel: +81(75)4153661, Fax: +81(75)4153662 ISSUES/YR: 6

Acta Hospitalia
ISSN: 0044-6009
Acta Hosp.
Katholieke Universiteit, Kapucijnenvoer 35, B-3000, Leuven, Belgium, Tel: +32(16)336973 ISSUES/YR: 4

Acta Medica Austriaca
ISSN: 0303-8173
Acta Med. Austriaca
Springer-Verlag Wien, Sachsenplatz 4-6, P.O. Box 89, A-1201, Vienna, Austria, Tel: +43(1)3302415, Fax: +43(1)3302426 ISSUES/YR: 5

Acta Medica Auxologica
ISSN: 0001-6004
Acta Med. Auxol.
Instituto Auxologica Italiano, Via Spagnoletto, 3, 20149, Milano, Italy, Tel: +39-2 619 111 ISSUES/YR: 3

Acta Medica Bulgarica
ISSN: 0324-1750
Acta Med. Bulg.
Medical Information Center, 1 Sv. G. Sofijski str., 1431, Sofia, Bulgaria, Tel: +359-522342, Fax: +359-522393 ISSUES/YR: 1

Acta Medica Croatica
ISSN: 1330-0164
Acta Med. Croat.
Croatian Academy of Medical Sciences, Subiceva ul. 29, HR-41000, Zagreb, Croatia, Tel: +385(41)419446 ISSUES/YR: 5

Acta Medica et Biologica
ISSN: 0567-7734
Acta Med. Biol.
Niigata University School of Medicine, Asahimachi-dori 1-757, 951, Niigata, Japan, Tel: +81(25)2236161 EXT.2381, Fax: +81(25)2295539 ISSUES/YR: 4

Acta Medica Iranica
ISSN: 0044-6025
Acta Med. Iran.
Medical Sciences University of Teheran, Enghelab Avenue, 14-174, Teheran, Iran, Tel: +98(21)8962510, Fax: +98(21)8962510 ISSUES/YR: 6

Acta Medica Mediterranea
ISSN: 0393-6384
Acta Med. Mediterr.
A. CARBONE Editore, Via Q. Sella 68, 90139, Palermo, Italy, Tel: +39(91)321273, Fax: +39(91)321782VIA ISSUES/YR: 3

Acta Medica Nagasakiensia
ISSN: 0001-6055
Acta Med. Nagasaki.
Nagasaki University School of Medicine, 12-4 Sakamotomachi, 852, Nagasaki, Japan, Tel: +81(95)819-2193, Fax: +81(95)819-2196 ISSUES/YR: 4

Acta Medica Portuguesa
ISSN: 0870-399X
Acta Med. Port.
CELOM, Avenida Almirante Reis 242 Esq, 100-057, Lisbon, Portugal, Tel: +351-21 843 7750, Fax: +351-21 843 7751 ISSUES/YR: 6

Acta Medica Romana
ISSN: 0001-6098
Acta Med. Rom.
Vita e Pensiero, Largo A. Gemelli 1, 20123, Milan, Italy, Tel: +39(02)72341 ISSUES/YR: 4

Acta Microbiologica Hellenica
ISSN: 0438-9573
Acta Microbiol. Hell.
Greek Society of Microbiology Ascent Ltd, 77 Vas. Sofias, GR-115 21, Athens, Greece, Tel: +30(210)7246180, Fax: +30(210)7246180 ISSUES/YR: 6

Acta Myologica
ISSN: 1128-2460
Acta Myologica
Gaetano Conte Academy, Viale dei Pini 101, 80131, Napoli, Italy, Tel: +39(081)7419545, Fax: +39(081)5665100 ISSUES/YR: 3

Acta Neurobiologiae Experimentalis
ISSN: 0065-1400
Acta Neurobiol. Exp.
Nencki Institute of Experimental Biology, ul. Pasteura 3, 02-093, Warsaw, Poland, Tel: +48(22)6598571, Fax: +48(22)8225342 ISSUES/YR: 4

• **Acta Neurochirurgica**
ISSN: 0001-6268
Acta Neurochir.
Springer-Verlag Wien, Sachsenplatz 4-6, P.O. Box 89, A-1201, Vienna, Austria, Tel: +43(1)3302415, Fax: +43(1)3302426 ISSUES/YR: 12

Acta Neurochirurgica, Supplement
ISSN: 0065-1419
Acta Neurochir. Suppl.
Springer-Verlag Wien, Sachsenplatz 4-6, P.O. Box 89, A-1201, Vienna, Austria, Tel: +43(1)3302415, Fax: +43(1)3302426 ISSUES/YR: 1

Acta Neurologica Belgica
ISSN: 0300-9009
Acta Neurol. Belg.
ASSMB a.s.b.l., Avenue Circulaire 138A, B-1180, Brussels, Belgium, Tel: +32(2)3745158, Fax: +32(2)3749628 ISSUES/YR: 4

Acta Neurologica Scandinavica
ISSN: 0001-6314
Acta Neurol. Scand.
Blackwell Publishing Ltd, 9600 Garsington Road, OX4 2XG, Oxford, United Kingdom, Tel: +44(1865)776868, Fax: +44(1865)714591 ISSUES/YR: 12

Acta Neurologica Taiwanica
ISSN: 1019-6099
Acta Neurol. Taiwan.
Neurological Society R.O.C (Taiwan), 7 Chung-Shan S. Road, 100, Taipei, Taiwan, Tel: +886(2)23123456 EXT.2335, Fax: +886(2)23214321 ISSUES/YR: 4

• **Acta Neuropathologica**
ISSN: 0001-6322
Acta Neuropathol.
Springer Verlag, Tiergartenstrasse 17, D-69121, Heidelberg, Germany, Tel: +49(6221)4870, Fax: +49(6221)487366 ISSUES/YR: 12

Acta Neuropsychiatrica
ISSN: 0924-2708
Acta Neuropsychiatr.
Blackwell Publishing Ltd, 9600 Garsington Road, OX4 2XG, Oxford, United Kingdom, Tel: +44(1865)776868, Fax: +44(1865)714591 ISSUES/YR: 6

Acta Neuropsychologica
ISSN: 1730-7503
Acta Neuropsychol.
Agencja Wydawnicza Medsportpress, ul. Marymoncka 34 (skr. 23), 01-813, Warsaw, Poland, Tel: +48(22)8346772, Fax: +48(22)8346772 ISSUES/YR: 4

• **Acta Obstetricia et Gynecologica Scandinavica**
ISSN: 0001-6349
Acta Obstet. Gynecol. Scand.
Taylor and Francis Ltd., 4 Park Square, Milton Park, OX14 4RN, Abingdon, Oxfordshire, United Kingdom, Tel: +44(20)70175541, Fax: +44(20)70175198 ISSUES/YR: 12

Acta Odontologica Scandinavica
ISSN: 0001-6357
Acta Odontol. Scand.
Taylor and Francis AS, Hollandargatan 20, SE-10365, P.O. Box 3255, Stockholm, Sweden, Tel: +46-8 440 8040, Fax: +46-8 440 8050 ISSUES/YR: 6

Acta of Bioengineering and Biomechanics
ISSN: 1509-409X
Acta Bioeng. Biomech.
Oficyna Wydawnicza Politecniki Wroclawskiej, Wybrzeze Wyspianskiego 27, 50-370, Wroclaw, Poland, Tel: +48(71)222940, Fax: +48(71)223664 ISSUES/YR: 2

- **Acta Oncologica**
  ISSN: 0284-186X
  Acta Oncol.
  Taylor and Francis AS, Hollandargatan 20, SE-10365, P.O. Box 3255, Stockholm, Sweden, Tel: +47-8 440 8040, Fax: +47-8 440 8050 ISSUES/YR: 8
- **Acta Oncologica, Supplement**
  ISSN: 1100-1704
  Acta Oncol. Suppl.
  Taylor and Francis A.S., P.O. Box 2562 Solli, 0202, Oslo, Norway, Tel: +47(22)129880, Fax: +47(22)129890 ISSUES/YR: 1
- **Acta Ophthalmologica Scandinavica**
  ISSN: 1395-3907
  Acta Ophthalmol. Scand.
  Blackwell Publishing Ltd, 9600 Garsington Road, OX4 2XG, Oxford, United Kingdom, Tel: +44(1865)776868, Fax: +44(1865)714591 ISSUES/YR: 6
- **Acta Ophthalmologica Scandinavica, Supplement**
  ISSN: 1395-3931
  Acta Ophthalmol. Scand. Suppl.
  Blackwell Munksgaard, 1 Rosenorns Alle, P.O. Box 227, DK-1502, Copenhagen V, Denmark, Tel: +45(77)333333, Fax: +45(77)333377 ISSUES/YR: 1

Acta Orthopaedica
ISSN: 1745-3674
Acta Orthop.
Taylor and Francis Ltd., 4 Park Square, Milton Park, OX14 4RN, Abingdon, Oxfordshire, United Kingdom, Tel: +44(20)70175541, Fax: +44(20)70175198 ISSUES/YR: 6

Acta Orthopaedica Belgica
ISSN: 0001-6462
Acta Orthop. Belg.
ASSMB a.s.b.l., Avenue Circulaire 138A, B-1180, Brussels, Belgium, Tel: +32(2)3745158, Fax: +32(2)3749628 ISSUES/YR: 5

Acta Orthopaedica, Supplement
ISSN: 1745-3690
Acta Orthop. Suppl.
Taylor and Francis Ltd., 4 Park Square, Milton Park, OX14 4RN, Abingdon, Oxfordshire, United Kingdom, Tel: +44(20)70175541, Fax: +44(20)70175198 ISSUES/YR: 1

- **Acta Oto-Laryngologica**
  ISSN: 0001-6489
  Acta Oto-Laryngol.
  Taylor and Francis AS, Hollandargatan 20, SE-10365, P.O. Box 3255, Stockholm, Sweden, Tel: +46-8 440 8040, Fax: +46-8 440 8050 ISSUES/YR: 10
- **Acta Oto-Laryngologica, Supplement**
  ISSN: 0365-5237
  Acta Oto-Laryngol. Suppl.
  Taylor and Francis A.S., P.O. Box 2562 Solli, 0202, Oslo, Norway, Tel: +47(22)129880, Fax: +47(22)129890 ISSUES/YR: 1

Acta Oto-Rhino-Laryngologica Belgica
ISSN: 0001-6497
Acta Oto-Rhino-Laryngol. Belg.
ASSMB a.s.b.l., Avenue Circulaire 138A, B-1180, Brussels, Belgium, Tel: +32(2)3745158, Fax: +32(2)3749628 ISSUES/YR: 4

Acta Paediatrica Taiwanica
ISSN: 1608-8115
Acta Paediatr. Taiwan.
Taiwan Pediatric Association, 10F-1, No. 69, Sec. 1, Hangchow South Road, 100, Taipei, Taiwan, Tel: +886(2)23516446, Fax: +886(2)23516448 ISSUES/YR: 6

- **Acta Paediatrica, International Journal of Paediatrics**
  ISSN: 0803-5253
  Acta Paediatr. Int. J. Paediatr.
  Taylor and Francis A.S., P.O. Box 2562 Solli, 0202, Oslo, Norway, Tel: +47(22)129880, Fax: +47(22)129890 ISSUES/YR: 12
- **Acta Paediatrica, International Journal of Paediatrics, Supplement**
  ISSN: 0803-5326
  Acta Paediatr. Int. J. Paediatr. Suppl.
  Taylor and Francis A.S., P.O. Box 2562 Solli, 0202, Oslo, Norway, Tel: +47(22)129880, Fax: +47(22)129890 ISSUES/YR: 1

Acta Pediatrica Espanola
ISSN: 0001-6640
Acta Pediatr. Esp.
Ediciones Mayo S.A., Aribau 185-187 2e planta, 08021, Barcelona, Spain, Tel: +34(93)2090255, Fax: +34(93)2020643 ISSUES/YR: 11

Acta Pharmaceutica
ISSN: 1330-0075
Acta Pharm.
Croatian Pharmaceutical Society, Masarykova 2, 10 000, Zagreb, Croatia, Tel: +385(41)427944 ISSUES/YR: 4

Acta Pharmaceutica Hungarica
ISSN: 0001-6659
Acta Pharm. Hung.
Magyar Gyogyszereszeti Tarsasag, Zrinyi u. 3, H-1361, Budapest, Hungary, Tel: +36(1)1181573 ISSUES/YR: 4

Acta Pharmaceutica Turcica
ISSN: 1300-638X
Acta Pharm. Turc.
Anadolu Universitesi, Yunusemre Campus, 26470, Eskisehir, Turkey, Tel: +90(222)3350581 ISSUES/YR: 4

Acta Pharmacologica Sinica
ISSN: 1671-4083
Acta Pharmacol. Sin.
Blackwell Publishing Ltd, 9600 Garsington Road, OX4 2XG, Oxford, United Kingdom, Tel: +44(1865)776868, Fax: +44(1865)714591 ISSUES/YR: 12

- **Acta Physiologica**
  ISSN: 1748-1708
  Acta Physiol.
  Blackwell Publishing Ltd, 9600 Garsington Road, OX4 2XG, Oxford, United Kingdom, Tel: +44(1865)776868, Fax: +44(1865)714591 ISSUES/YR: 12

Acta Physiologica Hungarica
ISSN: 0231-424X
Acta Physiol. Hung.
Akademiai Kiado Rt., Prielle Kornelia u. 4, H-1117, Budapest, Hungary, Tel: +36(1)4648282, Fax: +36(1)4648251 ISSUES/YR: 4

Acta Physiologica Scandinavica, Supplement
ISSN: 0302-2994
Acta Physiol. Scand. Suppl.
Blackwell Publishing Ltd, 9600 Garsington Road, OX4 2XG, Oxford, United Kingdom, Tel: +44(1865)776868, Fax: +44(1865)714591 ISSUES/YR: 1

Acta Poloniae Pharmaceutica - Drug Research
ISSN: 0001-6837
Acta Pol. Pharm. Drug Res.
Polish Pharmaceutical Society, ul. Dluga 16, 00-238, Warsaw, Poland, Tel: +48(22)6350970 ISSUES/YR: 6

- **Acta Psychiatrica Scandinavica**
  ISSN: 0001-690X
  Acta Psychiatr. Scand.
  Blackwell Publishing Ltd, 9600 Garsington Road, OX4 2XG, Oxford, United Kingdom, Tel: +44(1865)776868, Fax: +44(1865)714591 ISSUES/YR: 12

Acta Psychiatrica Scandinavica, Supplement
ISSN: 0065-1591
Acta Psychiatr. Scand. Suppl.
Blackwell Munksgaard, 1 Rosenorns Alle, P.O. Box 227, DK-1502, Copenhagen V, Denmark, Tel: +45(77)333333, Fax: +45(77)333377 ISSUES/YR: 1

Acta Scientiarum - Biological Sciences
ISSN: 1679-9283
Acta Sci. Biol. Sci.
Universidade Estadual de Maringa, Avenida Colombo, 5790, 87020-900, Maringa, Parana, Brazil, Tel: +55(44)2614040, Fax: +55(44)2222754 ISSUES/YR: 2

Current

Acta Scientiarum - Health Sciences
ISSN: 1679-9291
Acta Sci. Health Sci.
Universidade Estadual de Maringa, Avenida Colombo, 5790, 87020-900, Maringa, Parana, Brazil, Tel: +55(44)2614040, Fax: +55(44)2222754 ISSUES/YR: 2

Acta Toxicologica
ISSN: 1731-6383
Acta Toxicol.
Polish Society of Toxicology, Sw. Teresy 8, 90-950, Lodz, Poland, Tel: +48(22)8225060, Fax: +48(22)8231487 ISSUES/YR: 2

Acta Tropica
ISSN: 0001-706X
Acta Trop.
Elsevier, P.O. Box 211, 1000 AE, Amsterdam, Netherlands, Tel: +31(20)4853757, Fax: +31(20)4853432 ISSUES/YR: 12

Acta Urologica Japonica
ISSN: 0018-1994
Acta Urol. Jpn.
Editorial Board of Acta Urologica Japonica, 18 Shogoin Sannoumachi, Sakyo-ku, 606, Kyoto, Japan, Tel: +81(75)7520100, Fax: +81(75)7520190 ISSUES/YR: 12

Acta Virologica
ISSN: 0001-723X
Acta Virol.
SAP - Slovak Academic Press, spol. s.r.o., Nam. Slobody 6, P.O. Box 57, 81005, Bratislava, Slovakia, Tel: +421(7)211729 ISSUES/YR: 4

Actas Dermo-Sifiliograficas
ISSN: 0001-7310
Actas Dermo-Sifiliogr.
Ediciones Doyma, S.L., Travesera de Gracia 17-21, 08021, Barcelona, Spain, Tel: +34(93)2000711, Fax: +34(93)2091136 ISSUES/YR: 10

Actualites Pharmaceutiques
ISSN: 0515-3700
Actual. Pharm.
Editions Scientifiques et Medicales Elsevier, 23 rue Linois, 75724, Paris Cedex 15, France, Tel: +33(1)45589826, Fax: +33(1)45589428 ISSUES/YR: 11

Actualites Pharmaceutiques Hospitalieres
ISSN: 1769-7344
Actual. Pharm. Hosp.
Editions Scientifiques et Medicales Elsevier, 23 rue Linois, 75724, Paris Cedex 15, France, Tel: +33(1)45589826, Fax: +33(1)45589428 ISSUES/YR: 4

Actualites Reproduction Humaine
ISSN: 1243-6933
Actual. Reprod. Hum.
Editions ESKA, 12 rue du 4 Septembre, 75002, Paris, France, Tel: +33(01)42865600, Fax: +33(01)42603269 ISSUES/YR: 6

Actualizaciones en Anestesiologia y Reanimacion
ISSN: 1132-0095
Actual. Anestesiol. Reanim.
ARAN Ediciones S.A., Castello 128 - 1o, 28006, Madrid, Spain, Tel: +34(91)7820035, Fax: +34(91)5615787 ISSUES/YR: 4

Acupuncture and Electro-Therapeutics Research
ISSN: 0360-1293
Acupunct. Electro-Ther. Res.
Cognizant Communication Corporation, 3 Hartsdale Road, Elmsford, NY 10523-3701, United States, Tel: +1(914)5927720, Fax: +1(914)5928981 ISSUES/YR: 4

Acupuncture in Medicine
ISSN: 0964-5284
Acupunct. Med.
British Medical Acupuncture Society, 3 Winnington Street, CW8 1AQ, Northwich Chesire, United Kingdom, Tel: +44-1606 786782, Fax: +44-1606 786783 ISSUES/YR: 4

• **Acute Cardiac Care**
ISSN: 1748-2941
Acute Card. Care
Taylor and Francis Ltd., 4 Park Square, Milton Park, OX14 4RN, Abingdon, Oxfordshire, United Kingdom, Tel: +44(20)70175541, Fax: +44(20)70175198 ISSUES/YR: 4

Acute Medicine
ISSN: 1747-4884
Acute Med.
Rila Publications Ltd, 73 Newman Street, W1A 4PG, London, United Kingdom, Tel: +44(207)6311299, Fax: +44(207)5807166 ISSUES/YR: 3

• **Acute Pain**
ISSN: 1366-0071
Acute Pain
Elsevier, P.O. Box 211, 1000 AE, Amsterdam, Netherlands, Tel: +44(20)4853757, Fax: +44(20)4853432 ISSUES/YR: 4

Adapted Physical Activity Quarterly
ISSN: 0736-5829
Adapt. Phys. Act. Q.
Human Kinetics Publishers Inc., 1607 N. Market Street, Champaign, IL 61820-2200, United States, Tel: +1(217)3515076, Fax: +1(217)3511549 ISSUES/YR: 4

Addiction
ISSN: 0965-2140
Addiction
Blackwell Publishing Ltd, 9600 Garsington Road, OX4 2XG, Oxford, United Kingdom, Tel: +44(1865)776868, Fax: +44(1865)714591 ISSUES/YR: 12

• **Addiction Biology**
ISSN: 1355-6215
Addict. Biol.
Blackwell Publishing Ltd, 9600 Garsington Road, OX4 2XG, Oxford, United Kingdom, Tel: +44(1865)776868, Fax: +44(1865)714591 ISSUES/YR: 4

Addictive Behaviors
ISSN: 0306-4603
Addict. Behav.
Elsevier Ltd, Langford Lane, Kidlington, OX5 1GB, Oxford, United Kingdom, Tel: +44(1865)843000, Fax: +44(1865)843010 ISSUES/YR: 9

• **Addictive Disorders and their Treatment**
ISSN: 1531-5754
Addict. Disord. Treat.
Lippincott Williams and Wilkins, 530 Walnut Street, Philadelphia, PA 19106-3621, United States, Tel: +1(215)5218342, Fax: +1(215)5218326 ISSUES/YR: 4

Adicciones
ISSN: 0214-4840
Adicciones
Edita Socidrogalcohol, C/. Rambla 15, 2a 3a, 07003, Palma de Mallorca, Spain, Tel: +34(971)727434, Fax: +34(971)718073 ISSUES/YR: 4

Adipocytes
ISSN: 1550-2082
Adipocytes
New Century Health Publishers, PO BOX 50702, New Orleans, United States, Fax: +1-5042427690 ISSUES/YR: 4

• **Administration and Policy in Mental Health**
ISSN: 0894-587X
Adm. Policy Ment. Health
Kluwer Academic/Human Sciences Press Inc., 233 Spring Street, New York, NY 10013-1578, United States, Tel: +1(212)6208000, Fax: +1(212)4630742 ISSUES/YR: 6

• **Advanced Drug Delivery Reviews**
ISSN: 0169-409X
Adv. Drug Deliv. Rev.
Elsevier, P.O. Box 211, 1000 AE, Amsterdam, Netherlands, Tel: +31(20)4853757, Fax: +31(20)4853432 ISSUES/YR: 12

Advanced Studies in Medicine
ISSN: 1530-3004
Adv. Stud. Med.
Galen Publishing LLC, P.O. Box 340, Somerville, NJ 08876, United States, Tel: +1(908)2539001 ISSUES/YR: 10

• **Advances in Anatomic Pathology**
ISSN: 1072-4109
Adv. Anat. Pathol.
Lippincott Williams and Wilkins, 530 Walnut Street, Philadelphia, PA 19106-3621, United States, Tel: +1(215)5218342, Fax: +1(215)5218326 ISSUES/YR: 6

Advances in Applied Microbiology
ISSN: 0065-2164
Adv. Appl. Microbiol.
Academic Press Inc., 525 B Street, Suite 1900, San Diego, CA 92101-4495, United States ISSUES/YR: 1

- **Advances in Biophysics**
ISSN: 0065-227X
Adv. Biophys.
Japan Scientific Societies Press, 16-9, Honkomagome 5-chome, Bunkyo-ku, 113, Tokyo, Japan, Tel: +81(3)38675801 ISSUES/YR: 1

Advances in Breast Cancer
ISSN: 1742-0946
Adv. Breast Cancer
Mediscript Ltd, 1 Mountview Court,310 Friern Barnet Lande, N20 0LD, London, United Kingdom, Tel: +44-208 446 8898, Fax: +44-208 446 9194 ISSUES/YR: 4

- **Advances in Cancer Research**
ISSN: 0065-230X
Adv. Cancer Res.
Academic Press Inc., 525 B Street, Suite 1900, San Diego, CA 92101-4495, United States ISSUES/YR: 1

- **Advances in Carbohydrate Chemistry and Biochemistry**
ISSN: 0065-2318
Adv. Carbohydr. Chem. Biochem.
Academic Press Inc., 525 B Street, Suite 1900, San Diego, CA 92101-4495, United States ISSUES/YR: 1

- **Advances in Chronic Kidney Disease**
ISSN: 1548-5595
Adv. Chron. Kidney Dis.
W.B. Saunders, Independence Square West, Philadelphia, PA 19106-3399, United States, Tel: +1(215)2387800, Fax: +1(215)2386445 ISSUES/YR: 4

Advances in Clinical and Experimental Medicine
ISSN: 1230-025X
Adv. Clin. Exp. Med.
Wroclaw University of Medicine, Wybrzeze, L. Pasteur 2, 50-367, Wroclaw, Poland, Tel: +48(71)3285154, Fax: +48(71)3285154 ISSUES/YR: 6

- **Advances in Environmental Research**
ISSN: 1093-0191
Adv. Environ. Res.
Elsevier Ltd, Langford Lane, Kidlington, OX5 1GB, Oxford, United Kingdom, Tel: +44(1865)843000, Fax: +44(1865)843010 ISSUES/YR: 4

- **Advances in Experimental Medicine and Biology**
ISSN: 0065-2598
Adv. Exp. Med. Biol.
Springer Science and Business Media Deutschland GmbH, Heidelberger Platz 3, 14197, Berlin, Germany, Tel: +1-30 827875282, Fax: +1-30 827875707 ISSUES/YR: 1

Advances in Gastrointestinal Cancers
ISSN: 1479-9995
Adv. Gastrointest. Cancers
Mediscript Ltd, 1 Mountview Court,310 Friern Barnet Lande, N20 0LD, London, United Kingdom, Tel: +44-208 446 8898, Fax: +44-208 446 9194 ISSUES/YR: 4

- **Advances in Immunology**
ISSN: 0065-2776
Adv. Immunol.
Academic Press Inc., 525 B Street, Suite 1900, San Diego, CA 92101-4495, United States ISSUES/YR: 1

Advances in Microbial Physiology
ISSN: 0065-2911
Adv. Microb. Physiol.
Academic Press, 24-28 Oval Road, NW1 7DX, London, United Kingdom, Tel: +44(207)2674466 ISSUES/YR: 1

Advances in Mind-Body Medicine
ISSN: 0741-9783
Adv. Mind Body Med.
InnoVision Communications, 169 Saxony Road, Suite 103, Encinitas, CA 92024, United States, Tel: +1(760)633 3910, Fax: +1(760)633 3918 ISSUES/YR: 4

- **Advances in Parasitology**
ISSN: 0065-308X
Adv. Parasitol.
Academic Press, 24-28 Oval Road, NW1 7DX, London, United Kingdom, Tel: +44(207)2674466 ISSUES/YR: 1

- **Advances in Pharmacy**
ISSN: 1532-3862
Adv. Pharm.
Facts and Comparisons, 111 W. Port Plaza, Ste. 300, St. Louis, MO 6314603098, United States, Fax: +1(314)8785563 ISSUES/YR: 4

Advances in Physiotherapy
ISSN: 1403-8196
Adv. Physiother.
Taylor and Francis A.S., P.O. Box 2562 Solli, 0202, Oslo, Norway, Tel: +47(22)129880, Fax: +47(22)129890 ISSUES/YR: 4

- **Advances in Protein Chemistry**
ISSN: 0065-3233
Adv. Protein Chem.
Academic Press Inc., 525 B Street, Suite 1900, San Diego, CA 92101-4495, United States ISSUES/YR: 1

Advances in Psychiatric Treatment
ISSN: 1355-5146
Adv. Psychiatr. Treat.
Royal College of Psychiatrists, 17 Belgrave Square, SW1X 8PG, London, United Kingdom, Tel: +44(20)72352351, Fax: +44(20)72451231 ISSUES/YR: 6

Advances in Therapy
ISSN: 0741-238X
Adv. Ther.
Health Communications Inc., 292 Fernwood Avenue, Edison, NJ 08837, United States, Tel: +1(732)346 0027, Fax: +1(732)346 0442 ISSUES/YR: 6

Adverse Drug Reaction Bulletin
ISSN: 0044-6394
Adverse Drug React. Bull.
Lippincott Williams and Wilkins, 241 Borough High Street, SE1 1GB, London, United Kingdom, Tel: +44(207)9407500, Fax: +44(207)9407575 ISSUES/YR: 6

- **Aerobiologia**
ISSN: 0393-5965
Aerobiologia
Kluwer Academic Publishers, Van Godewijckstraat 30, 3311 GZ, Dordrecht, Netherlands, Tel: +31(78)6576000, Fax: +31(78)6576254 ISSUES/YR: 4

- **Aerosol Science and Technology**
ISSN: 0278-6826
Aerosol Sci. Technol.
Taylor and Francis Inc., 325 Chestnut St, Suite 800, Philadelphia PA, PA 19106, United States, Tel: +1(800)3541420, Fax: +1(215)6258914 ISSUES/YR: 12

- **Aesthetic Plastic Surgery**
ISSN: 0364-216X
Aesthet. Plast. Surg.
Springer New York LLC, 233 Springer Street, New York, 10013-1578, United States, Tel: +1-212 460 1501, Fax: +1-212 460 1595 ISSUES/YR: 6

- **Aesthetic Surgery Journal**
ISSN: 1090-820X
Aesthet. Surg. J.
Mosby Inc., 11830 Westline Industrial Drive, St. Louis, MO 63146, United States, Tel: +1(314)8728370, Fax: +1(314)4321380 ISSUES/YR: 6

African Health Sciences
ISSN: 1680-6905
Afr. Health Sci.
Makerere University, Medical School, P.O. Box 7072, Kampala, Uganda, Tel: +256-256 41 530020, Fax: +256-256 41 530022 ISSUES/YR: 3

African Journal of AIDS Research
ISSN: 1608-5906
Afr. J. AIDS Res.
NISC Pty Ltd, 19 Worcester Street, 6140, Grahamstown, South Africa, Tel: +27(46)622 9698, Fax: +27(46)622 9550 ISSUES/YR: 2

African Journal of Biotechnology
ISSN: 1684-5315
Afr. J. Biotechnol.
Academic Journals, P.O. Box 5170-00200, Nairobi, Kenya ISSUES/YR: 12

Current

African Journal of Neurological Sciences
ISSN: 1015-8618
Afr. J. Neurol. Sci.
Pan African Association of Neurological Sciences, Route de Desvres, 62280, Saint Martin, Boulogne, France ISSUES/YR: 1

African Journal of Traditional, Complementary and Alternative Medicines
ISSN: 0189-6016
Afr. J. Trad. Complement. Altern. Med.
African Ethnomedicines Network, Ile-Ife, Nigeria, Tel: +234-08037066089, Fax: +234(44)07092309478 ISSUES/YR: 4

• **Age and Ageing**
ISSN: 0002-0729
Age Ageing
Oxford University Press, Great Clarendon Street, OX2 6DP, Oxford, United Kingdom, Tel: +44(1865)353907, Fax: +44(1865)353486 ISSUES/YR: 6

Ageing Research Reviews
ISSN: 1568-1637
Ageing Res. Rev.
Elsevier Ireland Ltd, P.O. Box 85, Limerick, Ireland, Tel: +353(61)471944, Fax: +353(61)472144 ISSUES/YR: 4

Aggiornamento Pediatrico
ISSN: 0002-0958
Aggiorn. Pediatr.
Verduci Editore, Via Gregorio VII 186, 00165, Roma, Italy, Tel: +39(06)39375224, Fax: +39(06)6385672 ISSUES/YR: 4

• **Aggression and Violent Behavior**
ISSN: 1359-1789
Aggression Violent Behav.
Elsevier Ltd, Langford Lane, Kidlington, OX5 1GB, Oxford, United Kingdom, Tel: +44(1865)843000, Fax: +44(1865)843010 ISSUES/YR: 6

Aging - Clinical and Experimental Research
ISSN: 1594-0667
Aging Clin. Exp. Res.
Editrice Kurtis s.r.l., Via Luigi Zoja 30, 20153, Milan, Italy, Tel: +39(02)48202740, Fax: +39(02)48201219 ISSUES/YR: 6

• **Aging and Mental Health**
ISSN: 1360-7863
Aging Ment. Health
Taylor and Francis Ltd., 4 Park Square, Milton Park, OX14 4RN, Abingdon, Oxfordshire, United Kingdom, Tel: +44(20)70175541, Fax: +44(20)70175198 ISSUES/YR: 6

• **Aging Cell**
ISSN: 1474-9718
Aging Cell
Blackwell Publishing Ltd, 9600 Garsington Road, OX4 2XG, Oxford, United Kingdom, Tel: +44(1865)776868, Fax: +44(1865)714591 ISSUES/YR: 6

Aging Health
ISSN: 1745-509X
Aging Health
Future Medicine Ltd., 2nd Albert Place, Finchley Central, N3 1QB, London, United Kingdom, Tel: +44-208 343 3883, Fax: +44-208 343 2313 ISSUES/YR: 4

• **Aging Male**
ISSN: 1368-5538
Aging Male
Taylor and Francis Ltd., 4 Park Square, Milton Park, OX14 4RN, Abingdon, Oxfordshire, United Kingdom, Tel: +44(20)70175541, Fax: +44(20)70175198 ISSUES/YR: 4

Aging, Neuropsychology, and Cognition
ISSN: 1382-5585
Aging Neuropsychol. Cogn.
Psychology Press Ltd, 4 Park Square, Milton Park, OX14 4RN, Abingdon, Oxfordshire, United Kingdom, Tel: +44(1235)828600, Fax: +44(1235)829000 ISSUES/YR: 4

Agri
ISSN: 1300-0012
Agri
Turkish Society of Algology, Capa Klinikleri, 34390, Istanbul, Turkey, Tel: +90(212)5313147, Fax: +90(212)6310541 ISSUES/YR: 4

• **AIDS**
ISSN: 0269-9370
AIDS
Lippincott Williams and Wilkins, 241 Borough High Street, SE1 1GB, London, United Kingdom, Tel: +44(207)9407500, Fax: +44(207)9407575 ISSUES/YR: 18

AIDS and Behavior
ISSN: 1090-7165
AIDS Behav.
Springer Science and Business Media Deutschland GmbH, Heidelberger Platz 3, 14197, Berlin, Germany, Tel: +1-30 827875282, Fax: +1-30 827875707 ISSUES/YR: 4

AIDS and Public Policy Journal
ISSN: 0887-3852
AIDS Public Policy J.
University Publishing Group Inc., Hagerstown, Maryland, 21740, United States, Tel: +1-240 420 0036, Fax: +1-240 420 0037 ISSUES/YR: 4

• **AIDS Care - Psychological and Socio-Medical Aspects of AIDS/HIV**
ISSN: 0954-0121
AIDS Care Psychol. Socio-Med. Asp. AIDS HIV
Routledge, 4 Park Square, Milton Park, OX14 4RN, Abingdon, Oxfordshire, United Kingdom, Tel: +44(1235)828600, Fax: +44(1235)829000 ISSUES/YR: 8

AIDS Education and Prevention
ISSN: 0899-9546
AIDS Educ. Prev.
Guilford Publications, 72 Spring Street, New York, NY 10012, United States, Tel: +1(212)4319800 ISSUES/YR: 6

AIDS Patient Care and STDs
ISSN: 1087-2914
AIDS Patient Care STDs
Mary Ann Liebert Inc., 140 Huguenot Street, New Rochelle, NY 10801-5215, United States, Tel: +1(914)7402100, Fax: +1(914)7402101 ISSUES/YR: 12

AIDS Reader
ISSN: 1053-0894
AIDS Read
CMP Healthcare Media LLC, 330 Boston Post Road, Darien, CT 06820-4027, United States, Tel: +1(203)6626400, Fax: +1(212)6311600 ISSUES/YR: 12

• **AIDS Research and Human Retroviruses**
ISSN: 0889-2229
AIDS Res. Hum. Retroviruses
Mary Ann Liebert Inc., 140 Huguenot Street, New Rochelle, NY 10801-5215, United States, Tel: +1(914)7402100, Fax: +1(914)7402101 ISSUES/YR: 12

• **AIDS Research and Therapy**
ISSN: 1742-6405
AIDS Res. Ther.
BioMed Central Ltd., 34 - 42 Cleveland Street, W1T 4LB, London, United Kingdom, Tel: +44(207)3230323, Fax: +44(207)6319926 ISSUES/YR: 1

AIDS Reviews
ISSN: 1139-6121
AIDS Rev.
Publicaciones Permanyer, Mallorca 310, E-08037, Barcelona, Spain, Tel: +34(93)2075920, Fax: +34(93)4576642 ISSUES/YR: 4

• **Air Medical Journal**
ISSN: 1067-991X
Air Med. J.
Mosby Inc., 11830 Westline Industrial Drive, St. Louis, MO 63146, United States, Tel: +1(314)8728370, Fax: +1(314)4321380 ISSUES/YR: 6

Aktuality v Nefrologii
ISSN: 1210-955X
Akt. Nefrol.
Tigis s.r.o., Na Habrove 17, 152 00, Prague 5, Czech Republic, Tel: +420-251 813 1924, Fax: +420-251 681 217 ISSUES/YR: 4

Aktualnosci Neurologiczne
ISSN: 1641-9227
Aktual. Neurol.
Medical Communications, Ul.Ojcowska 11, 02-918, Warszawa, Poland, Tel: +48-226519783, Fax: +48-8425363 ISSUES/YR: 4

Aktuelle Dermatologie
ISSN: 0340-2541
Aktuel. Dermatol.
Georg Thieme Verlag, Rudigerstrasse 14, D-70469, Stuttgart, Germany, Tel: +49(711)8931-321, Fax: +49(711)8931-422 ISSUES/YR: 12

Aktuelle Ernahrungsmedizin
ISSN: 0341-0501
Aktuel. Ernahr. Med.
Georg Thieme Verlag, Rudigerstrasse 14, D-70469, Stuttgart, Germany, Tel: +49(711)8931-321, Fax: +49(711)8931-422 ISSUES/YR: 6

Aktuelle Neurologie
ISSN: 0302-4350
Aktuel. Neurol.
Georg Thieme Verlag, Rudigerstrasse 14, D-70469, Stuttgart, Germany, Tel: +49(711)8931-321, Fax: +49(711)8931-422 ISSUES/YR: 10

Aktuelle Neurologie, Supplement
ISSN: 1437-2754
Aktuel. Neurol. Suppl.
Georg Thieme Verlag, Rudigerstrasse 14, D-70469, Stuttgart, Germany, Tel: +49(711)8931-321, Fax: +49(711)8931-422 ISSUES/YR: 1

Aktuelle Rheumatologie
ISSN: 0341-051X
Aktuel. Rheumatol.
Georg Thieme Verlag, Rudigerstrasse 14, D-70469, Stuttgart, Germany, Tel: +49(711)8931-321, Fax: +49(711)8931-422 ISSUES/YR: 6

Aktuelle Rheumatologie, Supplement
ISSN: 1433-7940
Aktuel. Rheumatol. Suppl.
Georg Thieme Verlag, Rudigerstrasse 14, D-70469, Stuttgart, Germany, Tel: +49(711)8931-321, Fax: +49(711)8931-422 ISSUES/YR: 1

Aktuelle Traumatologie
ISSN: 0044-6173
Aktuel. Traumatol.
Georg Thieme Verlag, Rudigerstrasse 14, D-70469, Stuttgart, Germany, Tel: +49(711)8931-321, Fax: +49(711)8931-422 ISSUES/YR: 6

Aktuelle Traumatologie, Supplement
ISSN: 0935-0705
Aktuel. Traumatol. Suppl.
Georg Thieme Verlag, Rudigerstrasse 14, D-70469, Stuttgart, Germany, Tel: +49(711)8931-321, Fax: +49(711)8931-422 ISSUES/YR: 1

Akupunktur und Traditionelle Chinesische Medizin
ISSN: 1614-6891
Akupunkt. Trad. Chin. Med.
Medizinisch Literarische Verlagsgesellschaft mbH, P.O. Box 1151/1152, D-29501, Uelzen, Germany, Tel: +49(581)808150, Fax: +49(581)808158 ISSUES/YR: 4

Alcohol
ISSN: 0741-8329
Alcohol
Elsevier Inc., 360 Park Avenue South, New York, NY 10010, United States, Tel: +1(212)6333950, Fax: +1(212)6333990 ISSUES/YR: 9

• **Alcohol and Alcoholism**
ISSN: 0735-0414
Alcohol Alcohol.
Oxford University Press, Great Clarendon Street, OX2 6DP, Oxford, United Kingdom, Tel: +44(1865)353907, Fax: +44(1865)353486 ISSUES/YR: 6

Alcoholism
ISSN: 0002-502X
Alcoholism
Center for Study and Control of Alcoholism and Addictions, Vinogradska c.29, HR-10000, Zagreb, Croatia, Tel: +385(41)575045 ISSUES/YR: 2

Alcoholism Treatment Quarterly
ISSN: 0734-7324
Alcohol. Treat. Q.
Haworth Press Inc., 10 Alice Street, Binghamton, NY 13904-1580, United States, Tel: +1(607)7225857, Fax: +1(607)7710012 ISSUES/YR: 4

• **Alcoholism: Clinical and Experimental Research**
ISSN: 0145-6008
Alcohol. Clin. Exp. Res.
Blackwell Publishing Ltd, 9600 Garsington Road, OX4 2XG, Oxford, United Kingdom, Tel: +44(1865)776868, Fax: +44(1865)714591 ISSUES/YR: 12

Alergia Astma Immunologia
ISSN: 1427-3101
Alergia Astma Immunol.
University School of Medicine of Lodz, 11 Mazowiecka Street, 92-215, Lodz, Poland, Tel: +48(42)6781536, Fax: +48(42)6782292 ISSUES/YR: 4

Alergie
ISSN: 1212-3536
Alergie
Tigis s.r.o., Na Habrove 17, 152 00, Prague 5, Czech Republic, Tel: +420-251 813 1924, Fax: +420-251 681 217 ISSUES/YR: 4

Alergologia e Inmunologia Clinica
ISSN: 1575-734X
Alergol. Inmunol. Clin.
SANED S.L. (Sanidad y Ediciones), Capitan Haya 60, 28020, Madrid, Spain, Tel: +34(91)7499506, Fax: +34(91)7499507 ISSUES/YR: 6

• **Alimentary Pharmacology and Therapeutics**
ISSN: 0269-2813
Aliment. Pharmacol. Ther.
Blackwell Publishing Ltd, 9600 Garsington Road, OX4 2XG, Oxford, United Kingdom, Tel: +44(1865)776868, Fax: +44(1865)714591 ISSUES/YR: 24

• **Alimentary Pharmacology and Therapeutics Symposium Series**
ISSN: 1746-6334
Aliment. Pharmacol. Ther. Symp. Ser.
Blackwell Publishing Ltd, 9600 Garsington Road, OX4 2XG, Oxford, United Kingdom, Tel: +44(1865)776868, Fax: +44(1865)714591 ISSUES/YR: 1

Allergo Journal
ISSN: 0941-8849
Allergo J.
Urban und Vogel GmbH, Neumarkter Strasse 43, D-81673, Munchen, Germany, Tel: +49(89)43721300, Fax: +49(89)43721399 ISSUES/YR: 8

Allergologia et Immunopathologia
ISSN: 0301-0546
Allergol. Immunopathol.
Ediciones Doyma, S.L., Travesera de Gracia 17-21, 08021, Barcelona, Spain, Tel: +34(93)2000711, Fax: +34(93)2091136 ISSUES/YR: 6

Allergologie
ISSN: 0344-5062
Allergologie
Dustri-Verlag Dr. Karl Feistle, P.O. Box 1351, D-82032, Deisenhofen/Munchen, Germany, Tel: +49(89)6138610, Fax: +49(89)6135412 ISSUES/YR: 12

• **Allergology International**
ISSN: 1323-8930
Allergol. Int.
Japanese Society of Allergology, 1-35-26 Hongo, Bunkyo-ku, 113, Tokyo, Japan, Tel: +81(3)38160280, Fax: +81(3)38160219 ISSUES/YR: 4

Allergy and Clinical Immunology International
ISSN: 0838-1925
Allergy Clin. Immunol. Int.
Hogrefe and Huber Publishers, Rohnsweg 25, D-37085, Gottingen, Germany, Tel: +49(551)49609-0, Fax: +49(551)49609-88 ISSUES/YR: 6

• **Allergy: European Journal of Allergy and Clinical Immunology**
ISSN: 0105-4538
Allergy Eur. J. Allergy Clin. Immunol.
Blackwell Publishing Ltd, 9600 Garsington Road, OX4 2XG, Oxford, United Kingdom, Tel: +44(1865)776868, Fax: +44(1865)714591 ISSUES/YR: 12

- **Allergy: European Journal of Allergy and Clinical Immunology, Supplement**
ISSN: 0108-1675
Allergy Eur. J. Allergy Clin. Immunol. Suppl.
Blackwell Munksgaard, 1 Rosenorns Alle, P.O. Box 227, DK-1502, Copenhagen V, Denmark, Tel: +45(77)333333, Fax: +45(77)333377 ISSUES/YR: 1

Alternative and Complementary Therapies
ISSN: 1076-2809
Altern. Complement. Ther.
Mary Ann Liebert Inc., 140 Huguenot Street, New Rochelle, NY 10801-5215, United States, Tel: +1(914)7402100, Fax: +1(914)7402101 ISSUES/YR: 6

Alternative Medicine Review
ISSN: 1089-5159
Altern. Med. Rev.
Thorne Research Inc., P.O. Box 25, Dover, ID 83825, United States, Tel: +1(208)2631337, Fax: +1(208)2652488 ISSUES/YR: 4

Alternative Therapies in Health and Medicine
ISSN: 1078-6791
Altern. Ther. Health Med.
InnoVision Communications, 169 Saxony Road, Suite 103, Encinitas, CA 92024, United States, Tel: +1(760)633 3910, Fax: +1(760)633 3918 ISSUES/YR: 6

Altex
ISSN: 0946-7785
Altex
Elsevier GmbH, Slevogtstr. 3-5, 69126, Heidelberg, Germany, Tel: +49(6221)9126-200, Fax: +49(6221)9126-370 ISSUES/YR: 4

- **Alzheimer Disease and Associated Disorders**
ISSN: 0893-0341
Alzheimer Dis. Assoc. Disord.
Lippincott Williams and Wilkins, 530 Walnut Street, Philadelphia, PA 19106-3621, United States, Tel: +1(215)5218342, Fax: +1(215)5218326 ISSUES/YR: 4

- **Alzheimer's and Dementia**
ISSN: 1552-5260
Alzheimer's Dementia
Elsevier Inc., 360 Park Avenue South, New York, NY 10010, United States, Tel: +1(212)6333950, Fax: +1(212)6333990 ISSUES/YR: 4

Ambulatory Pediatrics
ISSN: 1530-1567
Ambul. Pediatr.
Thomson Scientific Ltd., 14 Great Queen Street, WC2B 5DF, London, United Kingdom, Tel: +44-20 7344 2800, Fax: +44-20 7344 2900 ISSUES/YR: 6

Ambulatory Surgery
ISSN: 0966-6532
Ambul. Surg.
Elsevier Ltd, Langford Lane, Kidlington, OX5 1GB, Oxford, United Kingdom, Tel: +44(1865)843000 ISSUES/YR: 2

American Family Physician
ISSN: 0002-838X
Am. Fam. Phys.
American Academy of Family Physicians, 11400 Tomahawk Creek Parkway, Suite 440, Leawood, KS 66211, United States, Tel: +1(913)9066000, Fax: +1(913)9066095 ISSUES/YR: 24

- **American Heart Journal**
ISSN: 0002-8703
Am. Heart J.
Mosby Inc., 11830 Westline Industrial Drive, St. Louis, MO 63146, United States, Tel: +1(314)8728370, Fax: +1(314)4321380 ISSUES/YR: 12

- **American Journal Geriatric Pharmacotherapy**
ISSN: 1543-5946
Am. J. Geriatr. Pharmacother.
Excerpta Medica Inc., 105 Raider Blvd, Suite 101, Hillsborough, NJ 08844, United States, Tel: +1(908)8748550, Fax: +1(908)8745633 ISSUES/YR: 4

American Journal of Alzheimer's Disease and other Dementias
ISSN: 1533-3175
Am. J. Alzheimer's Dis. Other Dem.
Prime National Publishing Corp., 470 Boston Post Road, Weston, MA 02193-1529, United States, Tel: +1(781)8992702, Fax: +1(781)8994900 ISSUES/YR: 6

American Journal of Audiology
ISSN: 1059-0889
Am. J. Audiol.
American Speech-Language-Hearing Association, 10801 Rockville Pike, Rockville, MD 20852-3279, United States, Tel: +1(301)8975700 ISSUES/YR: 2

- **American Journal of Cancer**
ISSN: 1175-6357
Am. J. Cancer
Adis International Ltd, 41 Centorian Drive, Private Bag 65901, Mairangi Bay, 1311, Auckland 10, New Zealand, Tel: +64(9)4770700, Fax: +64(9)4770766 ISSUES/YR: 6

- **American Journal of Cardiology**
ISSN: 0002-9149
Am. J. Cardiol.
Elsevier Inc., 360 Park Avenue South, New York, NY 10010, United States, Tel: +1(212)6333950, Fax: +1(212)6333990 ISSUES/YR: 24

- **American Journal of Cardiovascular Drugs**
ISSN: 1175-3277
Am. J. Cardiovasc. Drugs
Adis International Ltd, 41 Centorian Drive, Private Bag 65901, Mairangi Bay, 1311, Auckland 10, New Zealand, Tel: +64(9)4770700, Fax: +64(9)4770766 ISSUES/YR: 6

American Journal of Chinese Medicine
ISSN: 0192-415X
Am. J. Chin. Med.
World Scientific Publishing Co. Pte Ltd, 5 Toh Tuck Link, 596224, Singapore, Singapore, Tel: +65(6466)5775, Fax: +65(6467)7667 ISSUES/YR: 6

- **American Journal of Clinical Dermatology**
ISSN: 1175-0561
Am. J. Clin. Dermatol.
Adis International Ltd, 41 Centorian Drive, Private Bag 65901, Mairangi Bay, 1311, Auckland 10, New Zealand, Tel: +64(9)4770700, Fax: +64(9)4770766 ISSUES/YR: 6

American Journal of Clinical Hypnosis
ISSN: 0002-9157
Am. J. Clin. Hypn.
American Society of Clinical Hypnosis, 140 N; Bloomingdale Rd., Bloomingdale, IL 60108-1017, United States, Tel: +1(630)9804740, Fax: +1(630)3518490 ISSUES/YR: 4

American Journal of Clinical Nutrition
ISSN: 0002-9165
Am. J. Clin. Nutr.
American Society for Clinical Nutrition Inc., 9650 Rockville Pike, Bethesda, MD 20814-3998, United States, Tel: +1(301)5307026, Fax: +1(301)5718303 ISSUES/YR: 12

American Journal of Clinical Oncology: Cancer Clinical Trials
ISSN: 0277-3732
Am. J. Clin. Oncol. Cancer Clin. Trials
Lippincott Williams and Wilkins, 530 Walnut Street, Philadelphia, PA 19106-3621, United States, Tel: +1(215)5218342, Fax: +1(215)5218326 ISSUES/YR: 6

- **American Journal of Clinical Pathology**
ISSN: 0002-9173
Am. J. Clin. Pathol.
American Society of Clinical Pathologists, 2100 West Harrison Street, Chicago, IL 60612-3798, United States, Tel: +1(312)7381336 ISSUES/YR: 12

- **American Journal of Dermatopathology**
ISSN: 0193-1091
Am. J. Dermatopathol.
Lippincott Williams and Wilkins, 530 Walnut Street, Philadelphia, PA 19106-3621, United States, Tel: +1(215)5218342, Fax: +1(215)5218326 ISSUES/YR: 6

American Journal of Drug and Alcohol Abuse
ISSN: 0095-2990
Am. J. Drug Alcohol Abuse
Taylor and Francis Inc., 325 Chestnut St, Suite 800, Philadelphia PA, PA 19106, United States, Tel: +1(800)3541420, Fax: +1(215)6258914 ISSUES/YR: 4

- **American Journal of Drug Delivery**
ISSN: 1175-9038
Am. J. Drug Deliv.
Adis International Ltd, 41 Centorian Drive, Private Bag 65901, Mairangi Bay, 1311, Auckland 10, New Zealand, Tel: +64(9)4770700, Fax: +64(9)4770766 ISSUES/YR: 4

American Journal of Electroneurodiagnostic Technology
ISSN: 1086-508X
Am. J. Electroneurodiagn. Technol.
American Society of Electroneurodiagnostic Tech. Inc., 426 W. 42nd St., Kansas City, MO 64111, United States, Tel: +1-816 931 1120, Fax: +1-816 931 1145 ISSUES/YR: 4

- **American Journal of Emergency Medicine**
ISSN: 0735-6757
Am. J. Emerg. Med.
W.B. Saunders, Independence Square West, Philadelphia, PA 19106-3399, United States, Tel: +1(215)2387800, Fax: +1(215)2386445 ISSUES/YR: 7

American Journal of Epidemiology
ISSN: 0002-9262
Am. J. Epidemiol.
Oxford University Press, 2001 Evans Road, Cary, NC 27513, United States, Tel: +1(919)6770977, Fax: +1(919)6771714 ISSUES/YR: 24

American Journal of Family Therapy
ISSN: 0192-6187
Am. J. Fam. Ther.
Routledge, 4 Park Square, Milton Park, OX14 4RN, Abingdon, Oxfordshire, United Kingdom, Tel: +44(1235)828600, Fax: +44(1235)829000 ISSUES/YR: 5

American Journal of Forensic Medicine and Pathology
ISSN: 0195-7910
Am. J. Forensic Med. Pathol.
Lippincott Williams and Wilkins, 530 Walnut Street, Philadelphia, PA 19106-3621, United States, Tel: +1(215)5218342, Fax: +1(215)5218326 ISSUES/YR: 4

American Journal of Forensic Psychiatry
ISSN: 0163-1942
Am. J. Forensic Psychiatry
American College of Forensic Psychiatry, P.O. Box 5870, Balbao Island, CA 92662, United States, Tel: +1(714)8310236, Fax: +1(714)6737710 ISSUES/YR: 4

American Journal of Forensic Psychology
ISSN: 0733-1290
Am. J. Forensic Psychol.
American College of Forensic Psychology, P.O. Box 5870, Balboa Island, CA 92662, United States, Tel: +1(714)8310236, Fax: +1(714)6737710 ISSUES/YR: 4

- **American Journal of Gastroenterology**
ISSN: 0002-9270
Am. J. Gastroenterol.
Blackwell Publishing Inc., 350 Main Street, Malden, MA 02148, United States, Tel: +1(781)3888200, Fax: +1(781)3888210 ISSUES/YR: 12

American Journal of Geriatric Psychiatry
ISSN: 1064-7481
Am. J. Geriatr. Psychiatry
American Psychiatric Publishing Inc., 1000 Wilson Blvd, Suite 1825, VA 22209, United States, Tel: +1(703)907-7322, Fax: +1(703)90-1091 ISSUES/YR: 12

American Journal of Health Promotion
ISSN: 0890-1171
Am. J. Health Promot.
American Journal of Health Promotion, 1120 Chester Avenue, Suite 470, Cleveland, OH 44114, United States, Tel: +1(248)6820707, Fax: +1(248)6821212 ISSUES/YR: 6

- **American Journal of Health-System Pharmacy**
ISSN: 1079-2082
Am. J. Health-Syst. Pharm.
American Society of Health-Systems Pharmacy, 7272 Wisconsin Ave., Bethesda, MD 20814, United States, Tel: +1(301)6574383, Fax: +1(301)6571251 ISSUES/YR: 24

- **American Journal of Hematology**
ISSN: 0361-8609
Am. J. Hematol.
Wiley-Liss Inc., 111 River Street, Hoboken, NJ 07030-5774, United States, Tel: +1(201)7486645, Fax: +1(201)7486021 ISSUES/YR: 12

- **American Journal of Human Biology**
ISSN: 1042-0533
Am. J. Human Biol.
Wiley-Liss Inc., 111 River Street, Hoboken, NJ 07030-5774, United States, Tel: +1(201)7486645, Fax: +1(201)7486021 ISSUES/YR: 6

- **American Journal of Human Genetics**
ISSN: 0002-9297
Am. J. Hum. Genet.
University of Chicago Press, 1427 E. 60th Street, Chicago, IL 60637-2954, United States, Tel: +1(773)7027600, Fax: +1(773)7020694 ISSUES/YR: 12

- **American Journal of Hypertension**
ISSN: 0895-7061
Am. J. Hypertens.
Elsevier Inc., 360 Park Avenue South, New York, NY 10010, United States, Tel: +1(212)6333950, Fax: +1(212)6333990 ISSUES/YR: 12

American Journal of Industrial Medicine
ISSN: 0271-3586
Am. J. Ind. Med.
Wiley-Liss Inc., 111 River Street, Hoboken, NJ 07030-5774, United States, Tel: +1(201)7486645, Fax: +1(201)7486021 ISSUES/YR: 12

American Journal of Infection Control
ISSN: 0196-6553
Am. J. Infect. Control
Mosby Inc., 11830 Westline Industrial Drive, St. Louis, MO 63146, United States, Tel: +1(314)8728370, Fax: +1(314)4321380 ISSUES/YR: 10

American Journal of Kidney Diseases
ISSN: 0272-6386
Am. J. Kidney Dis.
W.B. Saunders, Independence Square West, Philadelphia, PA 19106-3399, United States, Tel: +1(215)2387800, Fax: +1(215)2386445 ISSUES/YR: 12

American Journal of Law and Medicine
ISSN: 0098-8588
Am. J. Law Med.
American Society of Law, Medicine and Ethics, 765 Commonwealth Avenue, Suite 1634, Boston, MA 02215-1401, United States, Tel: +1(617)2624990, Fax: +1(617)4377596 ISSUES/YR: 4

- **American Journal of Managed Care**
ISSN: 1088-0224
Am. J. Managed Care
American Medical Publishing, LLC, 241 Forsgate Drive, Suite 102, Jamesburg, NJ 08831, United States, Tel: +1(732)6561006, Fax: +1(732)6560818 ISSUES/YR: 12

- **American Journal of Medical Genetics, Part A**
ISSN: 1552-4825
Am. J. Med. Genet. Part A
Wiley-Liss Inc., 111 River Street, Hoboken, NJ 07030-5774, United States, Tel: +1(201)7486645, Fax: +1(201)7486021 ISSUES/YR: 28

- **American Journal of Medical Genetics, Part B: Neuropsychiatric Genetics**
ISSN: 1552-4841
Am. J. Med. Genet. Part B Neuropsychiatr. Genet.
Wiley-Liss Inc., 111 River Street, Hoboken, NJ 07030-5774, United States, Tel: +1(201)7486645, Fax: +1(201)7486021 ISSUES/YR: 8

- **American Journal of Medical Genetics, Part C: Seminars in Medical Genetics**
ISSN: 1552-4868
Am. J. Med. Genet. Part C Semin. Med. Genet.
Wiley-Liss Inc., 111 River Street, Hoboken, NJ 07030-5774, United States, Tel: +1(201)7486645, Fax: +1(201)7486021 ISSUES/YR: 4

- **American Journal of Medicine**
  ISSN: 0002-9343
  Am. J. Med.
  Elsevier Inc., 360 Park Avenue South, New York, NY 10010, United States, Tel: +1(212)6333950, Fax: +1(212)6333990 ISSUES/YR: 24
- **American Journal of Nephrology**
  ISSN: 0250-8095
  Am. J. Nephrol.
  S. Karger AG, Allschwilerstrasse 10, P.O. Box, CH-4009, Basel, Switzerland, Tel: +41(61)3061111, Fax: +41(61)3061234 ISSUES/YR: 6

American Journal of Neuroradiology
ISSN: 0195-6108
Am. J. Neuroradiol.
American Society of Neuroradiology, 2210 Midwest Road, Suite 207, Oak Brook, IL 60521, United States, Tel: +1(630)5740220, Fax: +1-630 786 6251 ISSUES/YR: 10

- **American Journal of Obstetrics and Gynecology**
  ISSN: 0002-9378
  Am. J. Obstet. Gynecol.
  Mosby Inc., 11830 Westline Industrial Drive, St. Louis, MO 63146, United States, Tel: +1(314)8728370, Fax: +1(314)4321380 ISSUES/YR: 12

American Journal of Oncology Review
ISSN: 1542-9520
Am. J. Oncol. Rev.
TyWin Communications, 318 Cleveland Avenue, Unit 1 Highland Park, New Jersey, 08904, United States, Tel: +1(732)2490800, Fax: +1(732)2490071 ISSUES/YR: 12

- **American Journal of Ophthalmology**
  ISSN: 0002-9394
  Am. J. Ophthalmol.
  Elsevier Inc., 360 Park Avenue South, New York, NY 10010, United States, Tel: +1(212)6333950, Fax: +1(212)6333990 ISSUES/YR: 12

American Journal of Orthopsychiatry
ISSN: 0002-9432
Am. J. Orthopsychiatry
American Psychological Association Inc., 750 First Street NE, Washington, DC 20002-4242, United States, Tel: +1(202)3365600, Fax: +1(202)3365568 ISSUES/YR: 4

- **American Journal of Otolaryngology - Head and Neck Medicine and Surgery**
  ISSN: 0196-0709
  Am. J. Otolaryngol. Head Neck Med. Surg.
  W.B. Saunders, Independence Square West, Philadelphia, PA 19106-3399, United States, Tel: +1(215)2387800, Fax: +1(215)2386445 ISSUES/YR: 6
- **American Journal of Pathology**
  ISSN: 0002-9440
  Am. J. Pathol.
  American Society for Investigative Pathology Inc., 9650 Rockville Pike, Bethesda, MD 20814-3993, United States, Tel: +1(301)5710107, Fax: +1(301)5710108 ISSUES/YR: 12
- **American Journal of Perinatology**
  ISSN: 0735-1631
  Am. J. Perinatol.
  Thieme Medical Publishers, Inc., 333 7th Avenue, New York, NY 10001-5004, United States, Tel: +1(212)7600888, Fax: +1(212)7799020 ISSUES/YR: 8

American Journal of Physical Anthropology
ISSN: 0002-9483
Am. J. Phys. Anthropol.
Wiley-Liss Inc., 111 River Street, Hoboken, NJ 07030-5774, United States, Tel: +1(201)7486645, Fax: +1(201)7486021 ISSUES/YR: 12

American Journal of Physical Medicine and Rehabilitation
ISSN: 0894-9115
Am. J. Phys. Med. Rehabil.
Lippincott Williams and Wilkins, 351 West Camden Street, Baltimore, MD 21201-2436, United States, Tel: +1(410)5284000 ISSUES/YR: 12

- **American Journal of Physiology - Cell Physiology**
  ISSN: 0363-6143
  Am. J. Physiol. Cell Physiol.
  American Physiological Society, 9650 Rockville Pike, Bethesda, MD 20814-3991, United States, Tel: +1(301)6347180, Fax: +1(301)6347241 ISSUES/YR: 12
- **American Journal of Physiology - Endocrinology and Metabolism**
  ISSN: 0193-1849
  Am. J. Physiol. Endocrinol. Metab.
  American Physiological Society, 9650 Rockville Pike, Bethesda, MD 20814-3991, United States, Tel: +1(301)6347180, Fax: +1(301)6347241 ISSUES/YR: 12
- **American Journal of Physiology - Gastrointestinal and Liver Physiology**
  ISSN: 0193-1857
  Am. J. Physiol. Gastrointest. Liver Physiol.
  American Physiological Society, 9650 Rockville Pike, Bethesda, MD 20814-3991, United States, Tel: +1(301)6347180, Fax: +1(301)6347241 ISSUES/YR: 12
- **American Journal of Physiology - Heart and Circulatory Physiology**
  ISSN: 0363-6135
  Am. J. Physiol. Heart Circ. Physiol.
  American Physiological Society, 9650 Rockville Pike, Bethesda, MD 20814-3991, United States, Tel: +1(301)6347180, Fax: +1(301)6347241 ISSUES/YR: 12
- **American Journal of Physiology - Lung Cellular and Molecular Physiology**
  ISSN: 1040-0605
  Am. J. Physiol. Lung Cell. Mol. Physiol.
  American Physiological Society, 9650 Rockville Pike, Bethesda, MD 20814-3991, United States, Tel: +1(301)6347180, Fax: +1(301)6347241 ISSUES/YR: 12
- **American Journal of Physiology - Regulatory Integrative and Comparative Physiology**
  ISSN: 0363-6119
  Am. J. Physiol. Regul. Integr. Comp. Physiol.
  American Physiological Society, 9650 Rockville Pike, Bethesda, MD 20814-3991, United States, Tel: +1(301)6347180, Fax: +1(301)6347241 ISSUES/YR: 12
- **American Journal of Physiology - Renal Physiology**
  ISSN: 0363-6127
  Am. J. Physiol. Renal Physiol.
  American Physiological Society, 9650 Rockville Pike, Bethesda, MD 20814-3991, United States, Tel: +1(301)6347180, Fax: +1(301)6347241 ISSUES/YR: 12

American Journal of Preventive Medicine
ISSN: 0749-3797
Am. J. Prev. Med.
Elsevier Inc., 360 Park Avenue South, New York, NY 10010, United States, Tel: +1(212)6333950, Fax: +1(212)6333990 ISSUES/YR: 10

- **American Journal of Psychiatric Rehabilitation**
  ISSN: 1548-7768
  Am. J. Psychiatr. Rehabil.
  Routledge, 325 Chestnut Street, Philadelphia, PA 19106, United States, Tel: +1(215)62508900 ISSUES/YR: 3
- **American Journal of Psychiatry**
  ISSN: 0002-953X
  Am. J. Psychiatry
  American Psychiatric Association, 1000 Wilson Boulevard, Suite 1825, Arlington, VA 22209-3901, United States, Tel: +1(703)9077300 ISSUES/YR: 12

American Journal of Psychoanalysis
ISSN: 0002-9548
Am. J. Psychoanal.
Kluwer Academic Publishers, Van Godewijckstraat 30, 3311 GZ, Dordrecht, Netherlands, Tel: +31(78)6576000, Fax: +31(78)6576254 ISSUES/YR: 4

American Journal of Psychotherapy
ISSN: 0002-9564
Am. J. Psychother.
Association for the Advancement of Psychotherapy Inc., 1300 Morris Park Avenue, Bronx, NY 10461-1602, United States, Tel: +1(718)4303503, Fax: +1(718)4308907 ISSUES/YR: 4

Current

American Journal of Public Health
ISSN: 0090-0036
Am. J. Public Health
American Public Health Association Inc., 800 I Street NW, Washington, DC 20061-3710, United States, Tel: +1(202)7772516, Fax: +1(202)7772531 ISSUES/YR: 12

- **American Journal of Reproductive Immunology**
ISSN: 1046-7408
Am. J. Reprod. Immunol.
Blackwell Publishing Ltd, 9600 Garsington Road, OX4 2XG, Oxford, United Kingdom, Tel: +44(1865)776868, Fax: +44(1865)714591 ISSUES/YR: 12

- **American Journal of Respiratory and Critical Care Medicine**
ISSN: 1073-449X
Am. J. Respir. Crit. Care Med.
American Thoracic Society, 61 Broadway 4th Floor, New York, 10006 - 2755, United States, Tel: +1(212)315 8600, Fax: +1(212)315 8689 ISSUES/YR: 24

American Journal of Respiratory Cell and Molecular Biology
ISSN: 1044-1549
Am. J. Resp. Cell Mol. Biol.
American Thoracic Society, 61 Broadway 4th Floor, New York, 10006 - 2755, United States, Tel: +1(212)315 8600, Fax: +1(212)315 8689 ISSUES/YR: 12

American Journal of Rhinology
ISSN: 1050-6586
Am. J. Rhinol.
OceanSide Publications Inc., 95 Pitman Street, Providence, RI 02906, United States, Tel: +1(401)3312510, Fax: +1(401)3315138 ISSUES/YR: 6

- **American Journal of Roentgenology**
ISSN: 0361-803X
Am. J. Roentgenol.
American Roentgen Ray Society, 44221 Slatestone Court, Leesburg, VA 20176-5109, United States, Tel: +1(703)7293353, Fax: +1(703)7294839 ISSUES/YR: 12

- **American Journal of Sports Medicine**
ISSN: 0363-5465
Am. J. Sports Med.
SAGE Publications Inc., 2455 Teller Road, Thousand Oaks, CA 91320, United States, Tel: +1(805)4990721, Fax: +1(805)4990871 ISSUES/YR: 8

- **American Journal of Surgery**
ISSN: 0002-9610
Am. J. Surg.
Elsevier Inc., 360 Park Avenue South, New York, NY 10010, United States, Tel: +1(212)6333950, Fax: +1(212)6333990 ISSUES/YR: 12

American Journal of Surgical Pathology
ISSN: 0147-5185
Am. J. Surg. Pathol.
Lippincott Williams and Wilkins, 530 Walnut Street, Philadelphia, PA 19106-3621, United States, Tel: +1(215)5218342, Fax: +1(215)5218326 ISSUES/YR: 12

American Journal of the Medical Sciences
ISSN: 0002-9629
Am. J. Med. Sci.
Lippincott Williams and Wilkins, 530 Walnut Street, Philadelphia, PA 19106-3621, United States, Tel: +1(215)5218342, Fax: +1(215)5218326 ISSUES/YR: 12

- **American Journal of Therapeutics**
ISSN: 1075-2765
Am. J. Ther.
Lippincott Williams and Wilkins, 530 Walnut Street, Philadelphia, PA 19106-3621, United States, Tel: +1(215)5218342, Fax: +1(215)5218326 ISSUES/YR: 6

- **American Journal of Transplantation**
ISSN: 1600-6135
Am. J. Transplant.
Blackwell Publishing Ltd, 9600 Garsington Road, OX4 2XG, Oxford, United Kingdom, Tel: +44(1865)776868, Fax: +44(1865)714591 ISSUES/YR: 12

American Journal of Tropical Medicine and Hygiene
ISSN: 0002-9637
Am. J. Trop. Med. Hyg.
American Society of Tropical Medicine and Hygiene, 60 Revere Drive, Suite 500, Northbrook, IL 60062, United States, Tel: +1(847)4809592, Fax: +1(847)4809282 ISSUES/YR: 12

American Journal of Urology Review
ISSN: 1542-8745
Am. J. Urol. Rev.
TyWin Communications, 318 Cleveland Avenue, Unit 1 Highland Park, New Jersey, 08904, United States, Tel: +1(732)2490800, Fax: +1(732)2490071 ISSUES/YR: 12

American Journal of Veterinary Research
ISSN: 0002-9645
Am. J. Vet. Res.
American Veterinary Medical Association, 1931 N. Meacham Road, Suite 100, Schaumburg, IL 60173-4360, United States, Tel: +1(847)9258070, Fax: +1(847)9251329 ISSUES/YR: 12

American Journal on Addictions
ISSN: 1055-0496
Am. J. Addict.
Routledge, 4 Park Square, Milton Park, OX14 4RN, Abingdon, Oxfordshire, United Kingdom, Tel: +44(1235)828600, Fax: +44(1235)829000 ISSUES/YR: 6

American Journal on Mental Retardation
ISSN: 0895-8017
Am. J. Ment. Retard.
American Association on Mental Retardation, 444 North Capitol Street NW, Suite 846, Washington, DC 20001-1512, United States, Tel: +1(202)3871968, Fax: +1(202)3872193 ISSUES/YR: 6

American Pharmaceutical Outsourcing
ISSN: 1529-6318
Am. Pharm. Outsourcing
Russell Publishing LLC, 8888 Keystone Crossing, Suite 1420, Indianapolis, IN 46240-2182, United States, Tel: +1(317)8168787, Fax: +1(317)8168788 ISSUES/YR: 6

American Pharmaceutical Review
ISSN: 1099-8012
Am. Pharm. Rev.
Russell Publishing LLC, 8888 Keystone Crossing, Suite 1420, Indianapolis, IN 46240-2182, United States, Tel: +1(317)8168787, Fax: +1(317)8168788 ISSUES/YR: 6

American Surgeon
ISSN: 0003-1348
Am. Surg.
Southeastern Surgical Congress, 141 West Wieuca Road, Suite B100, Atlanta, GA 30342, United States, Tel: +1(404)2554549, Fax: +1(404)2555442 ISSUES/YR: 12

- **Amino Acids**
ISSN: 0939-4451
Amino Acids
Springer-Verlag Wien, Sachsenplatz 4-6, P.O. Box 89, A-1201, Vienna, Austria, Tel: +43(1)3302415, Fax: +43(1)3302426 ISSUES/YR: 8

- **Amyloid**
ISSN: 1350-6129
Amyloid
Taylor and Francis Ltd., 4 Park Square, Milton Park, OX14 4RN, Abingdon, Oxfordshire, United Kingdom, Tel: +44(20)70175541, Fax: +44(20)70175198 ISSUES/YR: 4

- **Amyotrophic Lateral Sclerosis and Other Motor Neuron Disorders**
ISSN: 1466-0822
Amyotrophic Lateral Scler. Other Mot. Neuron Disord.
Taylor and Francis A.S., P.O. Box 2562 Solli, 0202, Oslo, Norway, Tel: +47(22)129880, Fax: +47(22)129890 ISSUES/YR: 4

Anadolu Kardiyoloji Dergisi
ISSN: 1302-8723
Anadolu Kardiyoloji Derg.
Anadolu Kardiyoloji Dergisi, Inonu Cad. 64/1, 26130, Ilgazhan, Eskisehir, Turkey, Tel: +90(222)2399842, Fax: +90(222)2392817 ISSUES/YR: 4

Current

Anadolu Psikiyatri Dergisi
ISSN: 1302-6631
Anadolu Psikiyatr. Derg.
Cukurova University, Faculty of Medicine, 01330, Balcali/ ADANA, Turkey, Tel: +90-322-3386060, Fax: +90-322-3386505 ISSUES/YR: 4

ANAE - Approche Neuropsychologique des Apprentissages chez l'Enfant
ISSN: 0999-792X
ANAE Approche Neuropsychol. Apprentiss. Enfant
PDG Communication, 30 rue d'Armaille, 75017, Paris, France ISSUES/YR: 4

• **Anaerobe**
ISSN: 1075-9964
Anaerobe
Academic Press, 24-28 Oval Road, NW1 7DX, London, United Kingdom, Tel: +44(207)2674466 ISSUES/YR: 6

Anaesthesia
ISSN: 0003-2409
Anaesthesia
Blackwell Publishing Ltd, 9600 Garsington Road, OX4 2XG, Oxford, United Kingdom, Tel: +44(1865)776868, Fax: +44(1865)714591 ISSUES/YR: 12

Anaesthesia and Intensive Care
ISSN: 0310-057X
Anaesth. Intensive Care
Australian Society of Anaesthetists, P.O. Box 600, NSW 2027, Edgecliff, Australia, Tel: +61(2)3274022 ISSUES/YR: 6

Anaesthesia, Supplement
ISSN: 1461-6904
Anaesth. Suppl.
Blackwell Publishing Ltd, 9600 Garsington Road, OX4 2XG, Oxford, United Kingdom, Tel: +44(1865)776868, Fax: +44(1865)714591 ISSUES/YR: 1

Anaesthesiology and Intensive Care
ISSN: 1310-4284
Anaesthesiol. Intensive Care
Medical Information Center, 1 Sv. G. Sofijski str., 1431, Sofia, Bulgaria, Tel: +359-522342, Fax: +359-522393 ISSUES/YR: 5

Anaesthesist
ISSN: 0003-2417
Anaesthesist
Springer Verlag, Tiergartenstrasse 17, D-69121, Heidelberg, Germany, Tel: +49(6221)4870, Fax: +49(6221)487366 ISSUES/YR: 12

Anais Brasileiros de Dermatologia
ISSN: 0365-0596
An. Bras. Dermatol.
Sociedade Brasileira de Dermatologia, Av. Ipiranga 5311 sala 208, 90610-001, Porto Alegre/RS, Brazil, Tel: +55(51)3339-1811, Fax: +55(51)3339-1811 ISSUES/YR: 6

Anales de Cirugia Cardiaca y Vascular
ISSN: 1135-2760
An. Cir. Cardiaca Vasc.
Nexus Medica, c/Caracas 33, 08030, Barcelona, Spain, Tel: +34(93)4593492, Fax: +34(93)4590283 ISSUES/YR: 5

Anales de la Real Academia Nacional de Farmacia
ISSN: 0034-0634
An. R. Acad. Farm.
Real Academia de Farmacia, Farmacia 11, 28004, Madrid, Spain, Tel: +34(91)5310307 ISSUES/YR: 4

Anales de Medicina Interna
ISSN: 0212-7199
An. Med. Interna
ARAN Ediciones S.A., Castello 128 - 1o, 28006, Madrid, Spain, Tel: +34(91)7820035, Fax: +34(91)5615787 ISSUES/YR: 12

Anales de Pediatria
ISSN: 1695-4033
An. Pediatr.
Ediciones Doyma, S.L., Travesera de Gracia 17-21, 08021, Barcelona, Spain, Tel: +34(93)2000711, Fax: +34(93)2091136 ISSUES/YR: 12

Anales de Psiquiatria
ISSN: 0213-0599
An. Psiquiatr.
ARAN Ediciones S.A., Castello 128 - 1o, 28006, Madrid, Spain, Tel: +34(91)7820035, Fax: +34(91)5615787 ISSUES/YR: 10

Anales del Sistema Sanitario de Navarra
ISSN: 1137-6627
An. Sist. Sanit. Navarra
Anales del Sistema Sanitario de Navarra, C/ Irunlarrea 3, 31008, Pamplona (Navarra), Spain, Tel: +34(948)422008, Fax: +34(948)422009 ISSUES/YR: 3

Analyst
ISSN: 0003-2654
Analyst
Royal Society of Chemistry, Milton Road, CB4 OWF, Cambridge, United Kingdom, Tel: +44(1223)420066, Fax: +44(1223)420247 ISSUES/YR: 12

• **Analytica Chimica Acta**
ISSN: 0003-2670
Anal. Chim. Acta
Elsevier, P.O. Box 211, 1000 AE, Amsterdam, Netherlands, Tel: +31(20)4853757, Fax: +31(20)4853432 ISSUES/YR: 50

• **Analytical and Quantitative Cytology and Histology**
ISSN: 0884-6812
Anal. Quant. Cytol. Histol.
Science Printers and Publishers Inc., 8342 Olive Blvd, St. Louis, MO 63132, United States, Tel: +1(314)9914588, Fax: +1(314)9914654 ISSUES/YR: 6

• **Analytical Biochemistry**
ISSN: 0003-2697
Anal. Biochem.
Academic Press Inc., 6277 Sea Harbor Drive, Orlando, FL 32887-4900, United States, Tel: +1(407)3454040, Fax: +1(407)3639661 ISSUES/YR: 22

Analytical Chemistry
ISSN: 0003-2700
Anal. Chem.
American Chemical Society, 2540 Olentangy River Road, P.O. Box 3337, Columbus, OH 43210-3337, United States, Tel: +1(614)4473776, Fax: +1(614)4473671 ISSUES/YR: 24

Anasthesiologie Intensivmedizin Notfallmedizin Schmerztherapie
ISSN: 0939-2661
Anasthesiol. Intensivmed. Notf.med. Schmerzther.
Georg Thieme Verlag, Rudigerstrasse 14, D-70469, Stuttgart, Germany, Tel: +49(711)8931-321, Fax: +49(711)8931-422 ISSUES/YR: 12

Anasthesiologie Intensivmedizin Notfallmedizin Schmerztherapie, Supplement
ISSN: 1430-7790
Anasthesiol. Intensivmed. Notf.med. Schmerzther. Suppl.
Georg Thieme Verlag, Rudigerstrasse 14, D-70469, Stuttgart, Germany, Tel: +49(711)8931-321, Fax: +49(711)8931-422 ISSUES/YR: 1

Anasthesiologie und Intensivmedizin
ISSN: 0170-5334
Anasthesiol. Intensivmed.
DIOmed Verlags GmbH, Obere Schmiedgasse 11, 90403, Nurnberg, Germany, Tel: +49(911 21440)0, Fax: +49(911 21440)44 ISSUES/YR: 12

• **Anatomical Record - Part A Discoveries in Molecular, Cellular, and Evolutionary Biology**
ISSN: 0003-276X
Anat. Rec. Part A Discov. Mol. Cell. Evol. Biol.
Wiley-Liss Inc., 111 River Street, Hoboken, NJ 07030-5774, United States, Tel: +1(201)7486645, Fax: +1(201)7486021 ISSUES/YR: 12

• **Anatomical Record - Part B New Anatomist**
ISSN: 1552-4906
Anat. Rec. Part B New Anat.
Wiley-Liss Inc., 111 River Street, Hoboken, NJ 07030-5774, United States, Tel: +1(201)7486645, Fax: +1(201)7486021 ISSUES/YR: 6

Current

- **Anatomical Science International**
  ISSN: 1447-6959
  Anat. Sci. Int.
  Blackwell Publishing, 550 Swanston Street, VIC 3053, Carlton South, Australia, Tel: +61(3)8359 1011, Fax: +61(3)8359 1120 ISSUES/YR: 4
- **Anatomy and Embryology**
  ISSN: 0340-2061
  Anat. Embryol.
  Springer Verlag, Tiergartenstrasse 17, D-69121, Heidelberg, Germany, Tel: +49(6221)4870, Fax: +49(6221)487366 ISSUES/YR: 12

Andrologia
ISSN: 0303-4569
Andrologia
Blackwell Publishing Ltd, 9600 Garsington Road, OX4 2XG, Oxford, United Kingdom, Tel: +44(1865)776868, Fax: +44(1865)714591 ISSUES/YR: 6

Andrologie
ISSN: 1166-2654
Andrologie
Editions Vingt et un, 76 rue de la Pompe, 75116, Paris, France, Tel: +33(01)45033196, Fax: +33(01)45047289 ISSUES/YR: 4

Anestesia en Mexico
ISSN: 1405-0056
Anest. Mex.
Comunicaciones Cientificas Mexicanas S.A. de C.V., Calz. Gen. Anaya 330, Col. de Carmen Coyoacan, CP 04100, Mexico D.F., Mexico, Tel: +52(55)5-659-3863, Fax: +52(55)5-554-6054 ISSUES/YR: 6

Anestezi Dergisi
ISSN: 1300-0578
Anestezi Derg.
Anestezi Dergisi, P.O. Box 133, Cebeci, 06591, Ankara, Turkey, Tel: +90(312)2141000 ISSUES/YR: 4

Anesteziologie a Intenzivni Medicina
ISSN: 1214-2158
Anesteziol. Intenzivni Med.
Czech Medical Association J.E. Purkyne, Sokolska 31, P.O. Box 88, 120 26, Praha 2, Czech Republic, Tel: +420(2)24266201-4, Fax: +420(2)24266212 ISSUES/YR: 6

Anestezjologia Intensywna Terapia
ISSN: 0209-1712
Anest. Intens. Ter.
Borgis Ltd, ul. E. Ciolka 11A, 01-445, Warsaw, Poland, Tel: +48(22)8369669, Fax: +48(22)8369669 ISSUES/YR: 4

- **Anesthesia and Analgesia**
  ISSN: 0003-2999
  Anesth. Analg.
  Lippincott Williams and Wilkins, 351 West Camden Street, Baltimore, MD 21201-2436, United States, Tel: +1(410)5284000 ISSUES/YR: 12

Anesthesia and Resuscitation
ISSN: 0385-1664
Anesth. Resusc.
Hiroshima University School of Medicine, Kasumi 1-2-3, Minami-ku, 734, Hiroshima, Japan, Tel: +81(82)2575267 ISSUES/YR: 4

- **Anesthesiology**
  ISSN: 0003-3022
  Anesthesiology
  Lippincott Williams and Wilkins, 530 Walnut Street, Philadelphia, PA 19106-3621, United States, Tel: +1(215)5218342, Fax: +1(215)5218326 ISSUES/YR: 12

Anesthesiology Clinics of North America
ISSN: 0889-8537
Anesthesiol. Clin. North Am.
W.B. Saunders, Independence Square West, Philadelphia, PA 19106-3399, United States, Tel: +1(215)2387800, Fax: +1(215)2386445 ISSUES/YR: 4

Angeiologie
ISSN: 0003-3049
Angeiologie
Editions ESKA, 12 rue du 4 Septembre, 75002, Paris, France, Tel: +33(01)42865600, Fax: +33(01)42603269 ISSUES/ YR: 4

- **Angiogenesis**
  ISSN: 0969-6970
  Angiogenesis
  Kluwer Academic Publishers, Van Godewijckstraat 30, 3311 GZ, Dordrecht, Netherlands, Tel: +31(78)6576000, Fax: +31(78)6576254 ISSUES/YR: 4

Angiologia
ISSN: 0003-3170
Angiologia
Viguera Editores S.L., Place Tetuan 7, 2a planta, 08010, Barcelona, Spain, Tel: +34-932478188, Fax: +34-932317250 ISSUES/YR: 6

Angiology
ISSN: 0003-3197
Angiology
Westminster Publications Inc., 708 Glen Cove Avenue, Glen Head, NY 11545, United States, Tel: +1(516)7590025, Fax: +1(516)7595524 ISSUES/YR: 6

Anil Aggrawal's Internet Journal of Forensic Medicine and Toxicology
ISSN: 0972-8074
Anil Aggrawal's Internet J. Forensic Med. Toxicol.
Anil Aggrawal's Internet Journal of Forensic Medicine and Toxicology, S-299 Greater Kailah-1, 110048, New Delhi, India, Tel: +91-26465460 ISSUES/YR: 2

Animal Welfare
ISSN: 0962-7286
Anim. Welf.
Universities Federation for Animal Welfare, The Old School, Brewhouse Hill, AL4 8AN, Wheathampstead, Herts, United Kingdom, Tel: +44(1582)831818, Fax: +44(1582)831414 ISSUES/YR: 4

Ankara Universitesi Eczacilik Fakultesi Dergisi
ISSN: 1015-3918
Ank. Univ. Eczacilik Fak. Derg.
University of Ankara, Tandogan, 06100, Ankara, Turkey, Tel: +90(4)2135462, Fax: +90(4)2131081 ISSUES/YR: 2

Annales Academiae Medicae Gedanensis
ISSN: 0303-4135
Ann. Acad. Med. Gedanensis
Medical Academy, ul. Debinki 1, 80-952, Gdansk, Poland, Tel: +48(58)323914, Fax: +48(58)323914 ISSUES/YR: 1

Annales d'Endocrinologie
ISSN: 0003-4266
Ann. Endocrinol.
Masson, 21 Rue Camille Desmoulins, 92789, Issy les Moulineaux Cedex 9, France, Tel: +33(1)73281634, Fax: +33(1)73281649 ISSUES/YR: 6

Annales d'Urologie
ISSN: 0003-4401
Ann. Urol.
Editions Scientifiques et Medicales Elsevier, 23 rue Linois, 75724, Paris Cedex 15, France, Tel: +33(1)45589826, Fax: +33(1)45589428 ISSUES/YR: 6

Annales de Cardiologie et d'Angeiologie
ISSN: 0003-3928
Ann. Cardiol. Angeiol.
Editions Scientifiques et Medicales Elsevier, 23 rue Linois, 75724, Paris Cedex 15, France, Tel: +33(1)45589826, Fax: +33(1)45589428 ISSUES/YR: 6

Annales de Chirurgie
ISSN: 0003-3944
Ann. Chir.
Editions Scientifiques et Medicales Elsevier, 23 rue Linois, 75724, Paris Cedex 15, France, Tel: +33(1)45589826, Fax: +33(1)45589428 ISSUES/YR: 10

Annales de Chirurgie Plastique et Esthetique
ISSN: 0294-1260
Ann. Chir. Plast. Esthet.
Editions Scientifiques et Medicales Elsevier, 23 rue Linois, 75724, Paris Cedex 15, France, Tel: +33(1)45589826, Fax: +33(1)45589428 ISSUES/YR: 6

Annales de Dermatologie et de Venereologie
ISSN: 0151-9638
Ann. Dermatol. Venereol.
Masson, 21 Rue Camille Desmoulins, 92789, Issy les Moulineaux Cedex 9, France, Tel: +33(1)73281634, Fax: +33(1)73281649 ISSUES/YR: 12

Current

Annales de Pathologie
ISSN: 0242-6498
Ann. Pathol.
Masson, 21 Rue Camille Desmoulins, 92789, Issy les Moulineaux Cedex 9, France, Tel: +33(1)73281634, Fax: +33(1)73281649 ISSUES/YR: 6

Annales de Pediatrie
ISSN: 0066-2097
Ann. Pediatr.
Societe Edition Assoc. Enseignement Med. Hop. de Paris, 31 Boulevard de Latour-Maubourg, F-75343, Paris Cedex 07, France, Tel: +33(1)40626400, Fax: +33(1)45556920 ISSUES/YR: 10

Annales de Readaptation et de Medecine Physique
ISSN: 0168-6054
Ann. Readapt. Med. Phys.
Editions Scientifiques et Medicales Elsevier, 23 rue Linois, 75724, Paris Cedex 15, France, Tel: +33(1)45589826, Fax: +33(1)45589428 ISSUES/YR: 9

Annales Francaises d'Anesthesie et de Reanimation
ISSN: 0750-7658
Ann. Fr. Anesth. Reanim.
Editions Scientifiques et Medicales Elsevier, 23 rue Linois, 75724, Paris Cedex 15, France, Tel: +33(1)45589826, Fax: +33(1)45589428 ISSUES/YR: 10

Annales Medico-Psychologiques
ISSN: 0003-4487
Ann. Med.-Psychol.
Editions Scientifiques et Medicales Elsevier, 23 rue Linois, 75724, Paris Cedex 15, France, Tel: +33(1)45589826, Fax: +33(1)45589428 ISSUES/YR: 10

Annales Nestle
ISSN: 0517-8606
Ann. Nestle
S. Karger AG, Allschwilerstrasse 10, P.O. Box, CH-4009, Basel, Switzerland, Tel: +41(61)3061111, Fax: +41(61)3061234 ISSUES/YR: 3

Annales Pharmaceutiques Francaises
ISSN: 0003-4509
Ann. Pharm. Fr.
Masson, 21 Rue Camille Desmoulins, 92789, Issy les Moulineaux Cedex 9, France, Tel: +33(1)73281634, Fax: +33(1)73281649 ISSUES/YR: 6

Annali Degli Ospedali San Camillo e Forlanini
ISSN: 1127-4867
Ann. Osp. San Camillo Forlanini
Armando Editore S.r.l., Viale Trastevere 236, 00153, Roma, Italy, Tel: +39-39 06 58.94.525, Fax: +39-39 06 58.18.564 ISSUES/YR: 4

Annali dell'Istituto Superiore di Sanita
ISSN: 0021-2571
Ann. Ist. Super. Sanita
Istituto Superiore di Sanita, Viale Regina Elena 299, 00161, Roma, Italy, Tel: +39(06)4457163 ISSUES/YR: 4

Annali di Ottalmologia e Clinica Oculistica
ISSN: 0003-4665
Ann. Ottalmol. Clin. Ocul.
CEM Casa Editrice Maccari, Via Trento 53, P.O. Box 120, 43100, Parma, Italy, Tel: +39(0521)771268 ISSUES/YR: 12

Annali Italiani di Dermatologia Allergologica Clinica e Sperimentale
ISSN: 1592-6826
Ann. Ital. Dermatol. Allergol. Clin. Sper.
Monte Meru S.r.l., via San Pietro, 06081 PG, Campagna 1000 Assisi, Italy ISSUES/YR: 3

Annali Italiani di Medicina Interna
ISSN: 0393-9340
Ann. Ital. Med. Interna
CEPI s.r.l., Via A. Ristori 38, 00197, Rome, Italy, Tel: +39-+39.06809681, Fax: +39-+39.068072458 ISSUES/YR: 4

Annali Italiani di Medicina Interna, Supplement
ISSN: 1122-0538
Ann. Ital. Med. Interna Suppl.
CEPI s.r.l., Via A. Ristori 38, 00197, Rome, Italy, Tel: +39-+39.06809681, Fax: +39-+39.068072458 ISSUES/YR: 1

Annals of African Medicine
ISSN: 1596-3519
Ann. Afr. Med.
Annals of African Medicine Society, Sokoto, Nigeria ISSUES/YR: 4

Annals of Agricultural and Environmental Medicine
ISSN: 1232-1966
Ann. Agric. Environ. Med.
Institute of Agricultural Medicine, Jaczewskiego 2, P.O. Box 185, 20-950, Lublin, Poland, Tel: +48(81)7478027, Fax: +48(81)7478646 ISSUES/YR: 2

• **Annals of Allergy, Asthma and Immunology**
ISSN: 1081-1206
Ann. Allergy Asthma Immunol.
American College of Allergy, Asthma and Immunology, 7212 Davis Ct., McLean, VA 22101, United States ISSUES/YR: 12

Annals of Anatomy
ISSN: 0940-9602
Ann. Anat.
Urban und Fischer Verlag Jena, P.O. Box 100537, 07705, Jena, Germany, Tel: +49(3641)6263, Fax: +49(3641)626500 ISSUES/YR: 6

Annals of Behavioral Medicine
ISSN: 0883-6612
Ann. Behav. Med.
Lawrence Erlbaum Associates Inc., 10 Industrial Avenue, Mahwah, NJ 07430-2262, United States, Tel: +1(201)2582200, Fax: +1(201)2360072 ISSUES/YR: 6

Annals of Cancer Research and Therapy
ISSN: 1344-6835
Ann. Cancer Res. Ther.
PJD Publications Ltd, P.O. Box 966, Westbury, NY 11590-0966, United States, Tel: +1(516)6260650, Fax: +1(516)6265546 ISSUES/YR: 2

• **Annals of Clinical and Laboratory Science**
ISSN: 0091-7370
Ann. Clin. Lab. Sci.
Association of Clinical Scientists, P.O. Box 1287, Middlebury, VT 05753, United States, Tel: +1(802)4622507, Fax: +1(802)4622673 ISSUES/YR: 4

• **Annals of Clinical Biochemistry**
ISSN: 0004-5632
Ann. Clin. Biochem.
Royal Society of Medicine Press Ltd, P.O. Box 9002, W1A 0ZA, London, United Kingdom, Tel: +44(207)2902927, Fax: +44(207)2902929 ISSUES/YR: 6

Annals of Clinical Microbiology and Antimicrobials
ISSN: 1476-0711
Ann. Clin. Microbiol. Antimicrob.
BioMed Central Ltd., 34 - 42 Cleveland Street, W1T 4LB, London, United Kingdom, Tel: +44(207)3230323, Fax: +44(207)6319926 ISSUES/YR: 1

• **Annals of Clinical Psychiatry**
ISSN: 1040-1237
Ann. Clin. Psychiatry
Taylor and Francis Ltd., 4 Park Square, Milton Park, OX14 4RN, Abingdon, Oxfordshire, United Kingdom, Tel: +44(20)70175541, Fax: +44(20)70175198 ISSUES/YR: 4

• **Annals of Diagnostic Pathology**
ISSN: 1092-9134
Ann. Diagn. Pathol.
W.B. Saunders, Independence Square West, Philadelphia, PA 19106-3399, United States, Tel: +1(215)2387800, Fax: +1(215)2386445 ISSUES/YR: 6

• **Annals of Emergency Medicine**
ISSN: 0196-0644
Ann. Emerg. Med.
Mosby Inc., 11830 Westline Industrial Drive, St. Louis, MO 63146, United States, Tel: +1(314)8728370, Fax: +1(314)4321380 ISSUES/YR: 12

• **Annals of Epidemiology**
ISSN: 1047-2797
Ann. Epidemiol.
Elsevier Inc., 360 Park Avenue South, New York, NY 10010, United States, Tel: +1(212)6333950, Fax: +1(212)6333990 ISSUES/YR: 10

Current

Annals of Family Medicine
ISSN: 1544-1709
Ann. Fam. Med.
Annals of Family Medicine, Inc, 11400 Tomahawk Creek Parkway, Leawood, KS 66211-2672, United States, Tel: +1(800)2742237 ISSUES/YR: 6

Annals of Fundeni Hospital
ISSN: 1224-3450
Ann. Fundeni Hosp.
Fundeni Hospital, Fundeni Street 258, 72432, Bucharest, Romania ISSUES/YR: 4

Annals of General Psychiatry
ISSN: 1744-859X
Ann. Gen. Psychiatry
BioMed Central Ltd., 34 - 42 Cleveland Street, W1T 4LB, London, United Kingdom, Tel: +44(207)3230323, Fax: +44(207)6319926 ISSUES/YR: 1

● **Annals of Hematology**
ISSN: 0939-5555
Ann. Hematol.
Springer Verlag, Tiergartenstrasse 17, D-69121, Heidelberg, Germany, Tel: +49(6221)4870, Fax: +49(6221)487366 ISSUES/YR: 12

● **Annals of Hematology, Supplement**
ISSN: 0945-8077
Ann. Hematol. Suppl.
Springer Verlag, Tiergartenstrasse 17, D-69121, Heidelberg, Germany, Tel: +49(6221)4870, Fax: +49(6221)487366 ISSUES/YR: 1

Annals of Hepatology
ISSN: not available yet
Ann. Hepatol.
Mexican Association of Hepatology, Puente de Piedra 150, 14050, Col.Toriello Guerra, Tlalpan, Mexico, Tel: +52(55606)6222 (EXT 4215), Fax: +52(5)55666-4031 ISSUES/YR: 4

● **Annals of Human Genetics**
ISSN: 0003-4800
Ann. Hum. Genet.
Blackwell Publishing Ltd, 9600 Garsington Road, OX4 2XG, Oxford, United Kingdom, Tel: +44(1865)776868, Fax: +44(1865)714591 ISSUES/YR: 6

● **Annals of Internal Medicine**
ISSN: 0003-4819
Ann. Intern. Med.
American College of Physicians, Sixth Street at Race, Philadelphia, PA 19106-1572, United States, Tel: +1(610)3512400 ISSUES/YR: 24

Annals of Long-Term Care
ISSN: 1524-7929
Ann. Long-Term Care
HMP Communications LLP, 4365 U.S. Highway 1 Suite 250, Princeton, NJ 08540, United States, Tel: +1(609)452 9200, Fax: +1(609)452 8241 ISSUES/YR: 12

Annals of Medical Sciences
ISSN: 1300-0683
Ann. Med. Sci.
University of Cukurova, 01330, Adana, Turkey ISSUES/YR: 3

● **Annals of Medicine**
ISSN: 0785-3890
Ann. Med.
Taylor and Francis A.S., P.O. Box 2562 Solli, 0202, Oslo, Norway, Tel: +47(22)129880, Fax: +47(22)129890 ISSUES/YR: 8

● **Annals of Neurology**
ISSN: 0364-5134
Ann. Neurol.
John Wiley and Sons Inc., P.O.Box 18667, Newark, NJ 07191-8667, United States, Tel: +1(212)8506000, Fax: +1(212)8506012 ISSUES/YR: 12

● **Annals of Noninvasive Electrocardiology**
ISSN: 1082-720X
Ann. Noninvasive Electrocardiol.
Blackwell Publishing Inc., 350 Main Street, Malden, MA 02148, United States, Tel: +1(781)3888200, Fax: +1(781)3888210 ISSUES/YR: 4

● **Annals of Nuclear Medicine**
ISSN: 0914-7187
Ann. Nucl. Med.
Japanese Society of Nuclear Medicine, 28-45 Hon-Komagome 2-chome, Bunkyo-ku, 113-0021, Tokyo, Japan, Tel: +81(3)39470976, Fax: +81(3)39472535 ISSUES/YR: 8

● **Annals of Nutrition and Metabolism**
ISSN: 0250-6807
Ann. Nutr. Metab.
S. Karger AG, Allschwilerstrasse 10, P.O. Box, CH-4009, Basel, Switzerland, Tel: +41(61)3061111, Fax: +41(61)3061234 ISSUES/YR: 6

● **Annals of Occupational Hygiene**
ISSN: 0003-4878
Ann. Occup. Hyg.
Oxford University Press, 2001 Evans Road, Cary, NC 27513, United States, Tel: +1(919)6770977, Fax: +1(919)6771714 ISSUES/YR: 8

● **Annals of Oncology**
ISSN: 0923-7534
Ann. Oncol.
Oxford University Press, Great Clarendon Street, OX2 6DP, Oxford, United Kingdom, Tel: +44(1865)353907, Fax: +44(1865)353486 ISSUES/YR: 12

Annals of Ophthalmology
ISSN: 1530-4086
Ann. Ophthalmol.
Humana Press Inc., 999 Riverview Drive, Suite 208, Totowa, NJ 07512-1165, United States, Tel: +1(973)2561699, Fax: +1(973)2568341 ISSUES/YR: 4

● **Annals of Otology, Rhinology and Laryngology**
ISSN: 0003-4894
Ann. Otol. Rhinol. Laryngol.
Annals Publishing Company, 4507 Laclede Avenue, St. Louis, MO 63108, United States, Tel: +1(314)3674987 ISSUES/YR: 12

● **Annals of Pharmacotherapy**
ISSN: 1060-0280
Ann. Pharmacother.
Harvey Whitney Books Company, 8044 Montgomery Road, Suite 415, Cincinnati, OH 45236, United States, Tel: +1(513)7933555, Fax: +1(513)7933600 ISSUES/YR: 12

● **Annals of Plastic Surgery**
ISSN: 0148-7043
Ann. Plast. Surg.
Lippincott Williams and Wilkins, 530 Walnut Street, Philadelphia, PA 19106-3621, United States, Tel: +1(215)5218342, Fax: +1(215)5218326 ISSUES/YR: 12

● **Annals of Saudi Medicine**
ISSN: 1319-9226
Ann. Saudi Med.
King Faisal Specialist Hospital and Research Centre, P.O. Box 3354, 11211, Riyadh, Saudi Arabia, Tel: +966(1)4647272 EXT.1815 ISSUES/YR: 6

● **Annals of Surgery**
ISSN: 0003-4932
Ann. Surg.
Lippincott Williams and Wilkins, 530 Walnut Street, Philadelphia, PA 19106-3621, United States, Tel: +1(215)5218342, Fax: +1(215)5218326 ISSUES/YR: 12

Annals of Surgical Oncology
ISSN: 1068-9265
Ann. Surg. Oncol.
Springer New York LLC, 233 Springer Street, New York, 10013-1578, United States, Tel: +1-212 460 1501, Fax: +1-212 460 1595 ISSUES/YR: 10

Annals of the Academy of Medicine Singapore
ISSN: 0304-4602
Ann. Acad. Med. Singapore
Academy of Medicine Singapore, Runme Shaw Building, 088871, Singapore, Singapore, Tel: +65-622 45166, Fax: +65-6221 8891 ISSUES/YR: 6

Annals of the ICRP
ISSN: 0146-6453
Ann. ICRP
Elsevier Ltd, Langford Lane, Kidlington, OX5 1GB, Oxford, United Kingdom, Tel: +44(1865)843000, Fax: +44(1865)843010 ISSUES/YR: 4

Annals of the New York Academy of Sciences
ISSN: 0077-8923
Ann. New York Acad. Sci.
Blackwell Publishing Inc., 350 Main Street, Malden, MA 02148, United States, Tel: +1(781)3888200, Fax: +1(781)3888210 ISSUES/YR: 1

- **Annals of the Rheumatic Diseases**
ISSN: 0003-4967
Ann. Rheum. Dis.
BMJ Publishing Group, Tavistock Square, WC1H 9JR, London, United Kingdom, Tel: +44(207)3874499, Fax: +44(207)3836668 ISSUES/YR: 12

Annals of the Royal College of Surgeons of England
ISSN: 0035-8843
Ann. R. Coll. Surg. Engl.
Royal College of Surgeons of England, 35-43 Lincoln's Inn Fields, WC2A 3PN, London, United Kingdom ISSUES/YR: 6

- **Annals of Thoracic Surgery**
ISSN: 0003-4975
Ann. Thorac. Surg.
Elsevier USA, 6277 Sea Harbor Drive, Orlando, FL 32862 8239, United States, Tel: +1-407 345 4000, Fax: +1-407 363 9661 ISSUES/YR: 12

Annals of Tropical Medicine and Parasitology
ISSN: 0003-4983
Ann. Trop. Med. Parasitol.
Maney Publishing, Hudson Road, LS9 7DL, Leeds, United Kingdom, Tel: +44(113)2497481, Fax: +44(113)2486983 ISSUES/YR: 8

Annals of Tropical Paediatrics
ISSN: 0272-4936
Ann. Trop. Paediatr.
Maney Publishing, Hudson Road, LS9 7DL, Leeds, United Kingdom, Tel: +44(113)2497481, Fax: +44(113)2486983 ISSUES/YR: 4

- **Annals of Vascular Surgery**
ISSN: 0890-5096
Ann. Vasc. Surg.
Springer New York LLC, 233 Springer Street, New York, 10013-1578, United States, Tel: +1-212 460 1501, Fax: +1-212 460 1595 ISSUES/YR: 6

Annexins
ISSN: 1547-6294
Annexins
Landes Bioscience, 810 S Church Street, Georgetown, 78626, United States, Tel: +1-512.863.7762, Fax: +1-512.863.0081 ISSUES/YR: 3

Annual Report of Shionogi Research Laboratory
ISSN: 0559-8680
Annu. Rep. Shionogi Res. Lab.
Shionogi Co. Ltd, 12-4 Sagisu, 5-chome, Fukushima-ku, 553, Osaka, Japan, Tel: +81(6)4585861 ISSUES/YR: 1

- **Annual Review of Biochemistry**
ISSN: 0066-4154
Annu. Rev. Biochem.
Annual Reviews Inc., 4139 El Camino Way, P.O. Box 10139, Palo Alto, CA 94303-0139, United States, Tel: +1(415)4934400 ISSUES/YR: 1

Annual Review of Biomedical Engineering
ISSN: 1523-9829
Annu. Rev. Biomed. Eng.
Annual Reviews Inc., 4139 El Camino Way, P.O. Box 10139, Palo Alto, CA 94303-0139, United States, Tel: +1(415)4934400 ISSUES/YR: 1

Annual Review of Biomedical Sciences
ISSN: 1517-3011
Annu. Rev. Biomed. Sci.
Universidade Estadual Paulista - UNESP, Av. Vicente Ferreira 1278, C.P. 71, 17515-901, Marilia, S.P., Brazil ISSUES/YR: 1

- **Annual Review of Biophysics and Biomolecular Structure**
ISSN: 1056-8700
Annu. Rev. Biophys. Biomol. Struct.
Annual Reviews Inc., 4139 El Camino Way, P.O. Box 10139, Palo Alto, CA 94303-0139, United States, Tel: +1(415)4934400 ISSUES/YR: 1

- **Annual Review of Cell and Developmental Biology**
ISSN: 1081-0706
Annu. Rev. Cell Dev. Biol.
Annual Reviews Inc., 4139 El Camino Way, P.O. Box 10139, Palo Alto, CA 94303-0139, United States, Tel: +1(415)4934400 ISSUES/YR: 1

Annual Review of Clinical Psychology
ISSN: 1548-5943
Annu. Rev. Clin. Psychol.
Annual Reviews Inc., 4139 El Camino Way, P.O. Box 10139, Palo Alto, CA 94303-0139, United States, Tel: +1(415)4934400 ISSUES/YR: 1

- **Annual Review of Genetics**
ISSN: 0066-4197
Annu. Rev. Genet.
Annual Reviews Inc., 4139 El Camino Way, P.O. Box 10139, Palo Alto, CA 94303-0139, United States, Tel: +1(415)4934400 ISSUES/YR: 1

Annual Review of Genomics and Human Genetics
ISSN: 1527-8204
Annu. Rev. Genomics Hum. Genet.
Annual Reviews Inc., 4139 El Camino Way, P.O. Box 10139, Palo Alto, CA 94303-0139, United States, Tel: +1(415)4934400 ISSUES/YR: 1

- **Annual Review of Immunology**
ISSN: 0732-0582
Annu. Rev. Immunol.
Annual Reviews Inc., 4139 El Camino Way, P.O. Box 10139, Palo Alto, CA 94303-0139, United States, Tel: +1(415)4934400 ISSUES/YR: 1

- **Annual Review of Medicine**
ISSN: 0066-4219
Annu. Rev. Med.
Annual Reviews Inc., 4139 El Camino Way, P.O. Box 10139, Palo Alto, CA 94303-0139, United States, Tel: +1(415)4934400 ISSUES/YR: 1

- **Annual Review of Microbiology**
ISSN: 0066-4227
Annu. Rev. Microbiol.
Annual Reviews Inc., 4139 El Camino Way, P.O. Box 10139, Palo Alto, CA 94303-0139, United States, Tel: +1(415)4934400 ISSUES/YR: 1

- **Annual Review of Neuroscience**
ISSN: 0147-006X
Annu. Rev. Neurosci.
Annual Reviews Inc., 4139 El Camino Way, P.O. Box 10139, Palo Alto, CA 94303-0139, United States, Tel: +1(415)4934400 ISSUES/YR: 1

- **Annual Review of Nutrition**
ISSN: 0199-9885
Annu. Rev. Nutr.
Annual Reviews Inc., 4139 El Camino Way, P.O. Box 10139, Palo Alto, CA 94303-0139, United States, Tel: +1(415)4934400 ISSUES/YR: 1

- **Annual Review of Pathology**
ISSN: 0066-4286
Annu. Rev. Pathol.
Annual Reviews Inc., 4139 El Camino Way, P.O. Box 10139, Palo Alto, CA 94303-0139, United States, Tel: +1(415)4934400 ISSUES/YR: 1

- **Annual Review of Pharmacology and Toxicology**
ISSN: 0362-1642
Annu. Rev. Pharmacol. Toxicol.
Annual Reviews Inc., 4139 El Camino Way, P.O. Box 10139, Palo Alto, CA 94303-0139, United States, Tel: +1(415)4934400 ISSUES/YR: 1

- **Annual Review of Physiology**
ISSN: 0066-4278
Annu. Rev. Physiol.
Annual Reviews Inc., 4139 El Camino Way, P.O. Box 10139, Palo Alto, CA 94303-0139, United States, Tel: +1(415)4934400 ISSUES/YR: 1

- **Annual Review of Public Health**
ISSN: 0163-7525
Annu. Rev. Public Health
Annual Reviews Inc., 4139 El Camino Way, P.O. Box 10139, Palo Alto, CA 94303-0139, United States, Tel: +1(415)4934400 ISSUES/YR: 1

- **Anthropology and Medicine**
  ISSN: 1364-8470
  Anthropol. Med.
  Routledge, 4 Park Square, Milton Park, OX14 4RN, Abingdon, Oxfordshire, United Kingdom, Tel: +44(1235)828600, Fax: +44(1235)829000 ISSUES/YR: 3

Anti-Cancer Agents in Medicinal Chemistry
ISSN: 1871-5206
Anti-Cancer Agents Med. Chem.
Bentham Science Publishers B.V., P.O. Box 294, 1400 AG, Bussum, Netherlands, Tel: +31-356257290, Fax: +31-847511310 ISSUES/YR: 6

- **Anti-Cancer Drugs**
  ISSN: 0959-4973
  Anti-Cancer Drugs
  Lippincott Williams and Wilkins, 241 Borough High Street, SE1 1GB, London, United Kingdom, Tel: +44(207)9407500, Fax: +44(207)9407575 ISSUES/YR: 10

Anti-Infective Agents in Medicinal Chemistry
ISSN: 1871-5214
Anti-Infect. Agents Med. Chem.
Bentham Science Publishers B.V., P.O. Box 294, 1400 AG, Bussum, Netherlands, Tel: +31-356257290, Fax: +31-847511310 ISSUES/YR: 4

Anti-Inflammatory and Anti-Allergy Agents in Medicinal Chemistry
ISSN: 1871-5230
Anti-Inflammatory Anti-Allergy Agents Med. Chem.
Bentham Science Publishers B.V., P.O. Box 294, 1400 AG, Bussum, Netherlands, Tel: +31-356257290, Fax: +31-847511310 ISSUES/YR: 4

Antibiotiques
ISSN: 1294-5501
Antibiotiques
Masson, 21 Rue Camille Desmoulins, 92789, Issy les Moulineaux Cedex 9, France, Tel: +33(1)73281634, Fax: +33(1)73281649 ISSUES/YR: 4

Current

- **Anticancer Research**
  ISSN: 0250-7005
  Anticancer Res.
  International Institute of Anticancer Research, 1st km Kapandritiou - Kalamou Rd., P.O. Box 22, GR-19014, Kapandriti, Attiki, Greece, Tel: +30(22950)53389, Fax: +30(22950)53389 ISSUES/YR: 6
- **Antimicrobial Agents and Chemotherapy**
  ISSN: 0066-4804
  Antimicrob. Agents Chemother.
  American Society for Microbiology, 1752 N Street N.W., Washington, DC 20036-2904, United States, Tel: +1(202)942-9244, Fax: +1(202)942-9355 ISSUES/YR: 12
- **Antioxidants and Redox Signaling**
  ISSN: 1523-0864
  Antioxid. Redox Signal.
  Mary Ann Liebert Inc., 140 Huguenot Street, New Rochelle, NY 10801-5215, United States, Tel: +1(914)7402100, Fax: +1(914)7402101 ISSUES/YR: 12
- **Antiviral Chemistry and Chemotherapy**
  ISSN: 0956-3202
  Antiviral Chem. Chemother.
  International Medical Press Ltd, 2-4 Idol Lane, EC3R 5DD, London, United Kingdom, Tel: +44(207)3980700, Fax: +44(207)3980701 ISSUES/YR: 6
- **Antiviral Research**
  ISSN: 0166-3542
  Antiviral Res.
  Elsevier, P.O. Box 211, 1000 AE, Amsterdam, Netherlands, Tel: +31(20)4853757, Fax: +31(20)4853432 ISSUES/YR: 12
- **Antiviral Therapy**
  ISSN: 1359-6535
  Antiviral Ther.
  International Medical Press Ltd, 2-4 Idol Lane, EC3R 5DD, London, United Kingdom, Tel: +44(207)3980700, Fax: +44(207)3980701 ISSUES/YR: 8
- **Antonie van Leeuwenhoek, International Journal of General and Molecular Microbiology**
  ISSN: 0003-6072
  Antonie Van Leeuwenhoek Int. J. Gen. Mol. Microbiol.
  Kluwer Academic Publishers, Van Godewijckstraat 30, 3311 GZ, Dordrecht, Netherlands, Tel: +31(78)6576000, Fax: +31(78)6576254 ISSUES/YR: 8

Anuario de Psicologia
ISSN: 0066-5126
An. Psicol.
Ediciones Paidos Iberica SA, Mariano Cubi 92 entl., 08021, Barcelona, Spain, Tel: +34(93)2419250, Fax: +34(93)2022954 ISSUES/YR: 1

ANZ Journal of Surgery
ISSN: 1445-1433
ANZ J. Surg.
Blackwell Publishing, 550 Swanston Street, VIC 3053, Carlton South, Australia, Tel: +61(3)8359 1011, Fax: +61(3)8359 1120 ISSUES/YR: 12

ANZ Nuclear Medicine
ISSN: 1324-1435
ANZ Nucl. Med.
Australian and New Zealand Soc. of Nuclear Medicine Inc., P.O. Box 7108, VIC 3156, Upper Ferntree Gully, Australia, Tel: +61(3)93479633, Fax: +61(3)93474547 ISSUES/YR: 4

Aphasiology
ISSN: 0268-7038
Aphasiology
Psychology Press Ltd, 4 Park Square, Milton Park, OX14 4RN, Abingdon, Oxfordshire, United Kingdom, Tel: +44(1235)828600, Fax: +44(1235)829000 ISSUES/YR: 12

- **APLAR Journal of Rheumatology**
  ISSN: 0219-0494
  APLAR J. Rheumatol.
  Blackwell Publishing, 550 Swanston Street, VIC 3053, Carlton South, Australia, Tel: +61(3)8359 1011, Fax: +61(3)8359 1120 ISSUES/YR: 3
- **APMIS**
  ISSN: 0903-4641
  APMIS
  Blackwell Munksgaard, 1 Rosenorns Alle, P.O. Box 227, DK-1502, Copenhagen V, Denmark, Tel: +45(77)333333, Fax: +45(77)333377 ISSUES/YR: 12
- **APMIS, Supplement**
  ISSN: 0903-465X
  APMIS Suppl.
  Blackwell Munksgaard, 1 Rosenorns Alle, P.O. Box 227, DK-1502, Copenhagen V, Denmark, Tel: +45(77)333333, Fax: +45(77)333377 ISSUES/YR: 1
- **Apoptosis**
  ISSN: 1360-8185
  Apoptosis
  Kluwer Academic Publishers, Van Godewijckstraat 30, 3311 GZ, Dordrecht, Netherlands, Tel: +31(78)6576000, Fax: +31(78)6576254 ISSUES/YR: 6

Appetite
ISSN: 0195-6663
Appetite
Academic Press, 24-28 Oval Road, NW1 7DX, London, United Kingdom, Tel: +44(207)2674466 ISSUES/YR: 6

Applied and Environmental Microbiology
ISSN: 0099-2240
Appl. Environ. Microbiol.
American Society for Microbiology, 1752 N Street N.W., Washington, DC 20036-2904, United States, Tel: +1(202)942-9244, Fax: +1(202)942-9355 ISSUES/YR: 12

Applied Biochemistry and Biotechnology
ISSN: 0273-2289
Appl. Biochem. Biotechnol.
Humana Press Inc., 999 Riverview Drive, Suite 208, Totowa, NJ 07512-1165, United States, Tel: +1(973)2561699, Fax: +1(973)2568341 ISSUES/YR: 24

Applied Bioinformatics
ISSN: 1175-5636
Appl. Bioinform.
Adis International Ltd, 41 Centorian Drive, Private Bag 65901, Mairangi Bay, 1311, Auckland 10, New Zealand, Tel: +64(9)4770700, Fax: +64(9)4770766 ISSUES/YR: 4

Applied Biotechnology, Food Science and Policy
ISSN: 1175-9534
Appl. Biotechnol. Food. Sci. Policy
Open Mind Journals Ltd, 86C Bush Road, PO Box 300-729, AUCKLAND 1311, Albany, New Zealand, Tel: +64(9)4146417, Fax: +64(9)4146491 ISSUES/YR: 4

• **Applied Cardiopulmonary Pathophysiology**
ISSN: 0920-5268
Appl. Cardiopulm. Pathophysiol.
Pabst Science Publishers, Eichengrund 28, D-49525, Lengerich, Germany, Tel: +49(5484)308, Fax: +49(5484)550 ISSUES/YR: 4

• **Applied Cognitive Psychology**
ISSN: 0888-4080
Appl. Cogn. Psychol.
John Wiley and Sons Ltd, Southern Gate, PO19 8SQ, Chichester, West Sussex, United Kingdom, Tel: +44(1243)779777, Fax: +44(1243)775878 ISSUES/YR: 9

Applied Ergonomics
ISSN: 0003-6870
Appl. Ergon.
Elsevier Ltd, Langford Lane, Kidlington, OX5 1GB, Oxford, United Kingdom, Tel: +44(1865)843000 ISSUES/YR: 6

Applied Genomics and Proteomics
ISSN: 1175-5644
Appl. Genomics Proteomics
Open Mind Journals Ltd, 86C Bush Road, PO Box 300-729, AUCKLAND 1311, Albany, New Zealand, Tel: +64(9)4146417, Fax: +64(9)4146491 ISSUES/YR: 4

Applied Health Economics and Health Policy
ISSN: 1175-5652
Appl. Health Econ. Health Policy
Adis International Ltd, 41 Centorian Drive, Private Bag 65901, Mairangi Bay, 1311, Auckland 10, New Zealand, Tel: +64(9)4770700, Fax: +64(9)4770766 ISSUES/YR: 4

• **Applied Immunohistochemistry and Molecular Morphology**
ISSN: 1541-2016
Appl. Immunohistochem. Mol. Morphol.
Lippincott Williams and Wilkins, 530 Walnut Street, Philadelphia, PA 19106-3621, United States, Tel: +1(215)5218342, Fax: +1(215)5218326 ISSUES/YR: 4

Applied Microbiology and Biotechnology
ISSN: 0175-7598
Appl. Microbiol. Biotechnol.
Springer Verlag, Tiergartenstrasse 17, D-69121, Heidelberg, Germany, Tel: +49(6221)4870, Fax: +49(6221)487366 ISSUES/YR: 18

Applied Neuropsychology
ISSN: 0908-4282
Appl. Neuropsychol.
Lawrence Erlbaum Associates Inc., 10 Industrial Avenue, Mahwah, NJ 07430-2262, United States, Tel: +1(201)2582200, Fax: +1(201)2360072 ISSUES/YR: 4

Applied Psychophysiology Biofeedback
ISSN: 1090-0586
Appl. Psychophysiol. Biofeedback
Springer Science and Business Media Deutschland GmbH, Heidelberger Platz 3, 14197, Berlin, Germany, Tel: +1-30 827875282, Fax: +1-30 827875707 ISSUES/YR: 4

• **Applied Radiation and Isotopes**
ISSN: 0969-8043
Appl. Radiat. Isot.
Elsevier Ltd, Langford Lane, Kidlington, OX5 1GB, Oxford, United Kingdom, Tel: +44(1865)843000, Fax: +44(1865)843010 ISSUES/YR: 12

Applied Radiology
ISSN: 0160-9963
Appl. Radiol.
Anderson Publishing Ltd, 180 Glenside Avenue, Scotch Plains, NJ 07076, United States, Tel: +1(908)301-1995, Fax: +1(908)301-1997 ISSUES/YR: 12

• **Aquatic Toxicology**
ISSN: 0166-445X
Aquatic Toxicol.
Elsevier, P.O. Box 211, 1000 AE, Amsterdam, Netherlands, Tel: +31(20)4853757, Fax: +31(20)4853432 ISSUES/YR: 20

Arbeitsmedizin Sozialmedizin Umweltmedizin
ISSN: 0944-6052
Arbeitsmed. Sozialmed. Umweltmed.
Gentner Verlag, Forststrasse 131, D-70193, Stuttgart, Germany, Tel: +49(711)63672826, Fax: +49(711)63672711 ISSUES/YR: 12

Arbeitsmedizin Sozialmedizin Umweltmedizin, Supplement
ISSN: 0944-6419
Arbeitsmed. Sozialmed. Umweltmed. Suppl.
Gentner Verlag, Forststrasse 131, D-70193, Stuttgart, Germany, Tel: +49(711)63672826, Fax: +49(711)63672711 ISSUES/YR: 1

• **Archiv der Pharmazie**
ISSN: 0365-6233
Arch. Pharm.
Wiley-VCH Verlag, P.O. Box 101161, D-69451, Weinheim, Germany, Tel: +49(6201)606147, Fax: +49(6201)606117 ISSUES/YR: 12

Archiv fur Kriminologie
ISSN: 0003-9225
Arch. Kriminol.
Verlag Schmidt-Romhild, Mengstrasse 16, D-23552, Lubeck, Germany, Tel: +49(451)703101 ISSUES/YR: 12

Archive of Oncology
ISSN: 0354-7310
Arch. Oncol.
Institute of Oncology Sremska Kamenica, Institutski put 4, 21204, Sremska Kamenica, Serbia and Montenegro, Tel: +381(21)615711 ISSUES/YR: 4

Archives de Pediatrie
ISSN: 0929-693X
Arch. Pediatr.
Editions Scientifiques et Medicales Elsevier, 23 rue Linois, 75724, Paris Cedex 15, France, Tel: +33(1)45589826, Fax: +33(1)45589428 ISSUES/YR: 12

Archives des Maladies du Coeur et des Vaisseaux
ISSN: 0003-9683
Arch. Mal. Coeur Vaiss.
J.B. Bailliere Editeur, 2 cite Paradis, 75010, Paris, France, Tel: +33(01)55 336800, Fax: +33(01)55 336815 ISSUES/YR: 12

Archives des Maladies du Coeur et des Vaisseaux - Pratique
ISSN: 1261-694X
Arch. Mal. Coeur Vaiss. Prat.
J.B. Bailliere Editeur, 2 cite Paradis, 75010, Paris, France, Tel: +33(01)55 336800, Fax: +33(01)55 336815 ISSUES/YR: 10

Archives des Maladies Professionnelles et de l'Environnement
ISSN: 1775-8785
Arch. Mal. Prof. Environ.
Masson, 21 Rue Camille Desmoulins, 92789, Issy les Moulineaux Cedex 9, France, Tel: +33(1)73281634, Fax: +33(1)73281649 ISSUES/YR: 8

Archives Italiennes de Biologie
ISSN: 0003-9829
Arch. Ital. Biol.
University of Pisa, Lungarno A. Pacinotti 43, 56100, Pisa, Italy ISSUES/YR: 4

• **Archives of Andrology**
ISSN: 0148-5016
Arch. Androl.
Taylor and Francis Ltd., 4 Park Square, Milton Park, OX14 4RN, Abingdon, Oxfordshire, United Kingdom, Tel: +44(20)70175541, Fax: +44(20)70175198 ISSUES/YR: 6

• **Archives of Biochemistry and Biophysics**
ISSN: 0003-9861
Arch. Biochem. Biophys.
Academic Press Inc., 6277 Sea Harbor Drive, Orlando, FL 32887-4900, United States, Tel: +1(407)3454040, Fax: +1(407)3639661 ISSUES/YR: 24

Archives of Clinical Neuropsychology
ISSN: 0887-6177
Arch. Clin. Neuropsychol.
Elsevier Ltd, Langford Lane, Kidlington, OX5 1GB, Oxford, United Kingdom, Tel: +44(1865)843000, Fax: +44(1865)843010 ISSUES/YR: 8

- **Archives of Dermatological Research**
  ISSN: 0340-3696
  Arch. Dermatol. Res.
  Springer Verlag, Tiergartenstrasse 17, D-69121, Heidelberg, Germany, Tel: +49(6221)4870, Fax: +49(6221)487366 ISSUES/YR: 12
- **Archives of Dermatological Research, Supplement**
  ISSN: 0944-1948
  Arch. Dermatol. Res. Suppl.
  Springer Verlag, Tiergartenstrasse 17, D-69121, Heidelberg, Germany, Tel: +49(6221)4870, Fax: +49(6221)487366 ISSUES/YR: 1
- **Archives of Dermatology**
  ISSN: 0003-987X
  Arch. Dermatol.
  American Medical Association, 515 North State Street, Chicago, IL 60610, United States, Tel: +1(312)4645000 ISSUES/YR: 12
- **Archives of Disease in Childhood**
  ISSN: 0003-9888
  Arch. Dis. Child.
  BMJ Publishing Group, Tavistock Square, WC1H 9JR, London, United Kingdom, Tel: +44(207)3874499, Fax: +44(207)3836668 ISSUES/YR: 12
- **Archives of Disease in Childhood: Education and Practice Edition**
  ISSN: 1743-0585
  Arch. Dis. Child. Educ. Pract. Ed.
  BMJ Publishing Group, Tavistock Square, WC1H 9JR, London, United Kingdom, Tel: +44(207)3874499, Fax: +44(207)3836668 ISSUES/YR: 1
- **Archives of Disease in Childhood: Fetal and Neonatal Edition**
  ISSN: 1359-2998
  Arch. Dis. Child. Fetal Neonatal Ed.
  BMJ Publishing Group, Tavistock Square, WC1H 9JR, London, United Kingdom, Tel: +44(207)3874499, Fax: +44(207)3836668 ISSUES/YR: 6
- **Archives of Environmental Contamination and Toxicology**
  ISSN: 0090-4341
  Arch. Environ. Contam. Toxicol.
  Springer New York LLC, 233 Springer Street, New York, 10013-1578, United States, Tel: +1-212 460 1501, Fax: +1-212 460 1595 ISSUES/YR: 8
- **Archives of Environmental Health**
  ISSN: 0003-9896
  Arch. Environ. Health
  Heldref Publications, 1319 Eighteenth Street NW, Washington, DC 20036-1802, United States, Tel: +1(202)2966267, Fax: +1(202)2936130 ISSUES/YR: 12

Archives of Gastroenterohepatology
ISSN: 0354-2440
Arch. Gastroenterohepatol.
Serbian Medical Association, Dzordza Vasingtona 19, P.O. Box 838, 11000, Belgrade, Serbia and Montenegro, Tel: +381(11)346829 ISSUES/YR: 4

Archives of General Psychiatry
ISSN: 0003-990X
Arch. Gen. Psychiatry
American Medical Association, 515 North State Street, Chicago, IL 60610, United States, Tel: +1(312)4645000 ISSUES/YR: 12

- **Archives of Gerontology and Geriatrics**
  ISSN: 0167-4943
  Arch. Gerontol. Geriatr.
  Elsevier Ireland Ltd, P.O. Box 85, Limerick, Ireland, Tel: +353(61)471944, Fax: +353(61)472144 ISSUES/YR: 6

Archives of Gynecology and Obstetrics
ISSN: 0932-0067
Arch. Gynecol. Obstet.
Springer Verlag, Tiergartenstrasse 17, D-69121, Heidelberg, Germany, Tel: +49(6221)4870, Fax: +49(6221)487366 ISSUES/YR: 4

Archives of Hellenic Medicine
ISSN: 1105-3992
Arch. Hell. Med.
BETA Medical Publishers Ltd, 23 Meandrou Street, GR-11525, Athens, Greece, Tel: +30(210)6714371, Fax: +30(210)6715015 ISSUES/YR: 6

Archives of Histology and Cytology
ISSN: 0914-9465
Arch. Histol. Cytol.
Japan Society of Histological Documentation, Asahi-machi, 951, Niigata, Japan, Tel: +81(25)2236161 ISSUES/YR: 5

- **Archives of Internal Medicine**
  ISSN: 0003-9926
  Arch. Intern. Med.
  American Medical Association, 515 North State Street, Chicago, IL 60610, United States, Tel: +1(312)4645000 ISSUES/YR: 26

Archives of Iranian Medicine
ISSN: 1029-2977
Arch. Iran. Med.
Academy of Medical Sciences of I.R. Iran, P.O. Box 19395.4655, Pasdaran. Av., Tehran, Iran, Tel: +98-212935817, Fax: +98-212932125 ISSUES/YR: 4

Archives of Medical Research
ISSN: 0188-4409
Arch. Med. Res.
Elsevier Inc., 360 Park Avenue South, New York, NY 10010, United States, Tel: +1(212)6333950, Fax: +1(212)6333990 ISSUES/YR: 6

Archives of Medical Science
ISSN: 1734-1922
Arch. Med. Sci.
Termedia Publishing House Ltd., ul Kleberga 8,, 61-615, Poznan, Poland, Tel: +48-61 822 77 81, Fax: +48-61 822 7781 ISSUES/YR: 4

- **Archives of Microbiology**
  ISSN: 0302-8933
  Arch. Microbiol.
  Springer Verlag, Tiergartenstrasse 17, D-69121, Heidelberg, Germany, Tel: +49(6221)4870, Fax: +49(6221)487366 ISSUES/YR: 12
- **Archives of Neurology**
  ISSN: 0003-9942
  Arch. Neurol.
  American Medical Association, 515 North State Street, Chicago, IL 60610, United States, Tel: +1(312)4645000 ISSUES/YR: 12
- **Archives of Ophthalmology**
  ISSN: 0003-9950
  Arch. Ophthalmol.
  American Medical Association, 515 North State Street, Chicago, IL 60610, United States, Tel: +1(312)4645000 ISSUES/YR: 12
- **Archives of Orthopaedic and Trauma Surgery**
  ISSN: 0936-8051
  Arch. Orthop. Trauma Surg.
  Springer Verlag, Tiergartenstrasse 17, D-69121, Heidelberg, Germany, Tel: +49(6221)4870, Fax: +49(6221)487366 ISSUES/YR: 10

Archives of Otolaryngology - Head and Neck Surgery
ISSN: 0886-4470
Arch. Otolaryngol. Head Neck Surg.
American Medical Association, 515 North State Street, Chicago, IL 60610, United States, Tel: +1(312)4645000 ISSUES/YR: 12

Archives of Pathology and Laboratory Medicine
ISSN: 0003-9985
Arch. Pathol. Lab. Med.
College of American Pathologists, 325 Waukegan Road, Northfield, IL 60093-2750, United States, Tel: +1(708)4468800 ISSUES/YR: 12

- **Archives of Pediatrics and Adolescent Medicine**
  ISSN: 1072-4710
  Arch. Pediatr. Adolesc. Med.
  American Medical Association, 515 North State Street, Chicago, IL 60610, United States, Tel: +1(312)4645000 ISSUES/YR: 12

Archives of Physical Medicine and Rehabilitation
ISSN: 0003-9993
Arch. Phys. Med. Rehabil.
W.B. Saunders, Independence Square West, Philadelphia, PA 19106-3399, United States, Tel: +1(215)2387800, Fax: +1(215)2386445 ISSUES/YR: 12

Archives of Physiology and Biochemistry
ISSN: 1381-3455
Arch. Physiol. Biochem.
Taylor and Francis Ltd., 4 Park Square, Milton Park, OX14 4RN, Abingdon, Oxfordshire, United Kingdom, Tel: +44(20)70175541, Fax: +44(20)70175198 ISSUES/YR: 5

Archives of Psychiatry and Psychotherapy
ISSN: 1509-2046
Arch. Psychiatry Psychother.
Polish Psychiatric Association, Lenartowicza 14, 31-138, Cracow, Poland, Tel: +48(12)6331203, Fax: +48(12)6334067 ISSUES/YR: 4

Archives of Public Health
ISSN: 0778-7367
Arch. Public Health
Archives of Public Health, Juliette Wytsmanstraat 14, B-1050, Brussels, Belgium, Tel: +32(2)2104835, Fax: +32(2)2106407 ISSUES/YR: 6

Archives of Sexual Behavior
ISSN: 0004-0002
Arch. Sex. Behav.
Springer Science and Business Media Deutschland GmbH, Heidelberger Platz 3, 14197, Berlin, Germany, Tel: +1-30 827875282, Fax: +1-30 827875707 ISSUES/YR: 6

● **Archives of Surgery**
ISSN: 0004-0010
Arch. Surg.
American Medical Association, 515 North State Street, Chicago, IL 60610, United States, Tel: +1(312)4645000 ISSUES/YR: 12

Archives of the Balkan Medical Union
ISSN: 0041-6940
Arch. Balk. Med. Union
Celsius Publishing House, G. Clemenceau Street 1, 70148, Bucharest, Romania, Tel: +40(1)3137857, Fax: +40(1)3121570 ISSUES/YR: 4

● **Archives of Toxicology**
ISSN: 0340-5761
Arch. Toxicol.
Springer Verlag, Tiergartenstrasse 17, D-69121, Heidelberg, Germany, Tel: +49(6221)4870, Fax: +49(6221)487366 ISSUES/YR: 12

● **Archives of Women's Mental Health**
ISSN: 1434-1816
Arch. Women's Ment. Health
Springer-Verlag Wien, Sachsenplatz 4-6, P.O. Box 89, A-1201, Vienna, Austria, Tel: +43(1)3302415, Fax: +43(1)3302426 ISSUES/YR: 4

Archivio di Ostetricia e Ginecologia
ISSN: 0004-0126
Arch. Ostet. Ginecol.
Giannini Editore, Via Cisterna dell'Olio 6B, 80134, Naples, Italy, Tel: +39(081)5513928 ISSUES/YR: 4

Archivio Italiano di Urologia e Andrologia
ISSN: 1124-3562
Arch. Ital. Urol. Androl.
Edizioni Scripta Manent s.n.c., Via Bassini 41, 20133, Milano, Italy, Tel: +39(02)70608060, Fax: +39(02)70606917 ISSUES/YR: 4

Archivos de Bronconeumologia
ISSN: 0300-2896
Arch. Bronconeumol.
Ediciones Doyma, S.L., Travesera de Gracia 17-21, 08021, Barcelona, Spain, Tel: +34(93)2000711, Fax: +34(93)2091136 ISSUES/YR: 12

Archivos de Cardiologia de Mexico
ISSN: 1405-9940
Arch. Cardiol. Mex.
Instituto Nacional de Cardiologia Ignazio Chavez, Juan Badiano 1, Col. Seccion XVI, Deleg. Tlalpan, 14080, Mexico D.F., Mexico ISSUES/YR: 4

Archivos de la Facultad de Medicina de Zaragoza
ISSN: 0558-6291
Arch. Fac. Med. Zaragoza
Universidad de Zaragoza, C/Domingo Miral s/n, 50009, Zaragoza, Spain, Tel: +34(76)559795 ISSUES/YR: 2

Archivos de Medicina
ISSN: 1698-9465
Arch. Med.
Asociacion Espanola de Medicos Internos Residentes, C Julian Calaveria s-n, 33006, Oviedo, Spain ISSUES/YR: 6

Archivos de Medicina del Deporte
ISSN: 0212-8799
Arch. Med. Deporte
Archivos de Medicina del Deporte, Apartado 1.207, 31080, Pamplona, Spain, Tel: +34(948)265360, Fax: +34(948)171431 ISSUES/YR: 6

Archivos de Neurociencias
ISSN: 0187-4705
Arch. Neurocienc.
Instituto Nacional de Neurologia y Neurocirurgia, Av. Insurgentes Sur 3877, Col. La Fama, Tlalpan, 14269, Mexico DF, Mexico, Tel: +52(55)5-6064532, Fax: +52(55)5-6063295 ISSUES/YR: 4

Archivos de Psiquiatria
ISSN: 1576-0367
Arch. Psiquiatr.
Archivos de Neurobiologia, C/Antonio Palomino 8, 5o izq., 28015, Madrid, Spain, Tel: +34(91)5441266, Fax: +34(91)5441266 ISSUES/YR: 4

Archivum Immunologiae et Therapiae Experimentalis
ISSN: 0004-069X
Arch. Immunol. Ther. Exp.
Birkhauser Verlag AG, Klosterberg 23, P.O. Box 133, CH-4010, Basel, Switzerland, Tel: +41(61)2717400, Fax: +41(61)2050791 ISSUES/YR: 6

Argomenti di Gastroenterologia Clinica
ISSN: 1120-8651
Argomenti Gastroenterol. Clin.
Masson SpA, Via F.lli Bressan 2, 20126, Milan, Italy, Tel: +39(02)270741, Fax: +39(02)27074211 ISSUES/YR: 2

Arhiv za Higijenu Rada i Toksikologiju
ISSN: 0004-1254
Arh. Hig. Rada Toksikol.
Institute for Medical Research and Occupational Health, Ksaverska c. 2, P.O. Box 291, HR-10001, Zagreb, Croatia, Tel: +385(1)434188, Fax: +385(1)274572 ISSUES/YR: 4

Arquivos Brasileiros de Cardiologia
ISSN: 0066-782X
Arq. Bras. Cardiol.
Arquivos Brasileiros de Cardiologia, Rua Beira Rio, 45, 3.o andar, 04548-050, Sao Paolo SP, Brazil, Tel: +55(11)8295337, Fax: +55(11)8295337 ISSUES/YR: 12

Arquivos Brasileiros de Medicina
ISSN: 0365-0723
Arq. Bras. Med.
Editora Cientifica Nacional Ltda, Rua da Gloria 366, 3o andar, P.O. Box 590, 20241-180, Rio de Janeiro, Brazil, Tel: +55(21)2213235, Fax: +55(21)2521691 ISSUES/YR: 6

Arquivos Brasileiros de Oftalmologia
ISSN: 0004-2749
Arq. Bras. Oftalmol.
Conselho Brasileiro De Oftalmologia, Alameda Santos 1343-11.andar-CJ.1110,Cerqueira Cesar, CEP 01419-001, Sao Paulo, Brazil, Tel: +55(11 3266)4000, Fax: +55(11 3171)0953 ISSUES/YR: 6

Arquivos de Gastroenterologia
ISSN: 0004-2803
Arq. Gastroenterol.
IBEPEGE - Inst. Bras. Estudos Pesquisas Gastroent., Rua Dr Seng 320, P.O. Box 6209, 01331-020, Sao Paulo, Brazil, Tel: +55(11)2892768 ISSUES/YR: 4

Arquivos de Geriatria e Gerontologia
ISSN: 1413-7615
Arq. Geriatr. Gerontol.
Editora Cientifica Nacional Ltda, Rua da Gloria 366 - 3o andar, 20241-180 RJ, Rio de Janeiro, Brazil ISSUES/YR: 3

Current

Arquivos de Medicina
ISSN: 0871-3413
Arq. Med.
ArquiMed, Porto Apartado 4514, 4008, Porto, Portugal, Tel: +351-225074377, Fax: +351-225074379 ISSUES/YR: 6

Arquivos de Neuro-Psiquiatria
ISSN: 0004-282X
Arq. Neuro-Psiquiatr.
Arquivos de Neuro-Psiquiatria, Pr. Amadeu Amaral 47/33, 01327-010, Sao Paulo SP, Brazil, Tel: +55(11)2879726, Fax: +55(11)2898879 ISSUES/YR: 4

Ars Pharmaceutica
ISSN: 0004-2927
Ars Pharm.
Editorial Universida de Granada, Campus Universitario de Cartuja, 18071, Granada, Spain, Tel: +34-958 24 39 18 ISSUES/YR: 4

- **Arteriosclerosis, Thrombosis, and Vascular Biology**
ISSN: 1079-5642
Arterioscler. Thromb. Vasc. Biol.
Lippincott Williams and Wilkins, 530 Walnut Street, Philadelphia, PA 19106-3621, United States, Tel: +1(215)5218342, Fax: +1(215)5218326 ISSUES/YR: 12

- **Arthritis and Rheumatism**
ISSN: 0004-3591
Arthritis Rheum.
John Wiley and Sons Inc., P.O.Box 18667, Newark, NJ 07191-8667, United States, Tel: +1(212)8506000, Fax: +1(212)8506012 ISSUES/YR: 12

Arthritis Care and Research
ISSN: 0893-7524
Arthritis Care Res.
John Wiley and Sons Inc., P.O.Box 18667, Newark, NJ 07191-8667, United States, Tel: +1(212)8506000, Fax: +1(212)8506012 ISSUES/YR: 6

Arthritis Research and Therapy
ISSN: 1478-6354
Arthritis Res. Ther.
BioMed Central Ltd., 34 - 42 Cleveland Street, W1T 4LB, London, United Kingdom, Tel: +44(207)3230323, Fax: +44(207)6319926 ISSUES/YR: 4

Arthropod Structure and Development
ISSN: 1467-8039
Arthropod Struct. Dev.
Elsevier Ltd, Langford Lane, Kidlington, OX5 1GB, Oxford, United Kingdom, Tel: +44(1865)843000, Fax: +44(1865)843010 ISSUES/YR: 4

Arthroscopy - Journal of Arthroscopic and Related Surgery
ISSN: 0749-8063
Arthroscopy J. Arthroscopic Relat. Surg.
W.B. Saunders, Independence Square West, Philadelphia, PA 19106-3399, United States, Tel: +1(215)2387800, Fax: +1(215)2386445 ISSUES/YR: 10

Artificial Cells, Blood Substitutes, and Biotechnology
ISSN: not available yet
Artif. Cells Blood Substitutes Biotechnol.
Taylor and Francis Inc., 325 Chestnut St, Suite 800, Philadelphia PA, PA 19106, United States, Tel: +1(800)3541420, Fax: +1(215)6258914 ISSUES/YR: 6

- **Artificial Intelligence in Medicine**
ISSN: 0933-3657
Artif. Intell. Med.
Elsevier, P.O. Box 211, 1000 AE, Amsterdam, Netherlands, Tel: +31(20)4853757, Fax: +31(20)4853432 ISSUES/YR: 9

- **Artificial Organs**
ISSN: 0160-564X
Artif. Organs
Blackwell Publishing Inc., 350 Main Street, Malden, MA 02148, United States, Tel: +1(781)3888200, Fax: +1(781)3888210 ISSUES/YR: 12

Artroscopia
ISSN: 1591-1489
Artroscopia
Mattioli 1885 S.p.A., Via Coduro 1/b, 43036, Fidenza (PR), Italy, Tel: +39(524)84547, Fax: +39(524)84751 ISSUES/YR: 3

- **Arts in Psychotherapy**
ISSN: 0197-4556
Arts Psychother.
Elsevier Ltd, Langford Lane, Kidlington, OX5 1GB, Oxford, United Kingdom, Tel: +44(1865)843000, Fax: +44(1865)843010 ISSUES/YR: 5

Arzneimittel-Forschung/Drug Research
ISSN: 0004-4172
Arzneim.-Forsch. Drug Res.
Editio Cantor Verlag GmbH, P.O. Box 1255, D-88322, Aulendorf, Germany, Tel: +49(7525)9400, Fax: +49(7525)940180 ISSUES/YR: 12

Arztezeitschrift fur Naturheilverfahren und Regulationsmedizin
ISSN: 1614-8339
Arztez. Naturheilverfahren Regul.med.
Medizinisch Literarische Verlagsgesellschaft mbH, P.O. Box 1151/1152, D-29501, Uelzen, Germany, Tel: +49(581)808150, Fax: +49(581)808158 ISSUES/YR: 12

ASAIO Journal
ISSN: 1058-2916
ASAIO J.
Lippincott Williams and Wilkins, 530 Walnut Street, Philadelphia, PA 19106-3621, United States, Tel: +1(215)5218342, Fax: +1(215)5218326 ISSUES/YR: 6

ASGE Clinical Update
ISSN: 1070-7212
ASGE Clin. Update
Mosby Inc., 11830 Westline Industrial Drive, St. Louis, MO 63146, United States, Tel: +1(314)8728370, Fax: +1(314)4321380 ISSUES/YR: 4

- **Asia Pacific Journal of Environmental Law**
ISSN: 1385-2140
Asia Pac. J. Environ. Law
The Australian Centre for Environmental Law (Sydney), 173-175 Philip Street, NSW 2000, Sydney, Australia ISSUES/YR: 4

Asia Pacific Journal of Pharmacology
ISSN: 0217-9687
Asia Pac. J. Pharmacol.
Singapore University Press (Pte) Ltd, AS-01-02,3 Arts Link, 1175569, Singapore, Singapore, Tel: +65-67761148, Fax: +65-67740652 ISSUES/YR: 4

- **Asia-Pacific Journal of Clinical Oncology**
ISSN: 1743-7555
Asia-Pac. J. Clin. Oncol.
Blackwell Publishing Ltd, 9600 Garsington Road, OX4 2XG, Oxford, United Kingdom, Tel: +44(1865)776868, Fax: +44(1865)714591 ISSUES/YR: 4

Asia-Pacific Journal of Public Health
ISSN: 1010-5395
Asia-Pac. J. Public Health
Asia-Pacific Academic Consortium for Public Health, Faculty of Medicine, 50603 KL, University of Malaysia, Malaysia, Tel: +60-379675748, Fax: +60-379675769 ISSUES/YR: 2

Asian Cardiovascular and Thoracic Annals
ISSN: 0218-4923
Asian Cardiovasc. Thorac. Ann.
Asia Publishing Exchange Pte Ltd, 1A Fu Wing Building, 18 Pottinger Street, Hong Kong, Tel: +65-2541 5521, Fax: +65-2815 4811 ISSUES/YR: 4

Asian Journal of Andrology
ISSN: 1008-682X
Asian J. Androl.
Blackwell Publishing Ltd, 9600 Garsington Road, OX4 2XG, Oxford, United Kingdom, Tel: +44(1865)776868, Fax: +44(1865)714591 ISSUES/YR: 4

Asian Journal of Microbiology, Biotechnology and Environmental Sciences
ISSN: 0972-3005
Asian J. Microbiol. Biotechnol. Environ. Sci.
Global Science Publications, 23, Maharshi Dayanand Nagar, Surendera Nagar, 202 001, Aligarh, India ISSUES/YR: 4

Asian Journal of Ophthalmology
ISSN: 1560-2133
Asian J. Ophthalmol.
Scientific Communications International Ltd, Suite C, 10/F Wo On Building, 10 Wo On Lane, Central, Hong Kong, Tel: +852-28 68 9171, Fax: +852-28 68 9269 ISSUES/YR: 4

Asian Journal of Oral and Maxillofacial Surgery
ISSN: 0915-6992
Asian J. Oral. Maxillofac. Surg.
Scientific Communications International Ltd, Suite C, 10/F Wo On Building, 10 Wo On Lane, Central, Hong Kong, Tel: +81-28 68 9171, Fax: +81-28 68 9269 ISSUES/YR: 4

Asian Journal of Surgery
ISSN: 1015-9584
Asian J. Surg.
Elsevier (Singapore) Pte Ltd, 77 Leighton Road, Causeway Bay, Hong Kong, China, Hong Kong, Tel: +852-29651300, Fax: +852-29760778 ISSUES/YR: 4

Asian Oceanian Journal of Radiology
ISSN: 0972-2688
Asian Oceanian J. Radiol.
Asian Oceanian Society of Radiology, 10-B, Kasturba Gandhi Marg, 110 001, New Delhi, India, Tel: +91(11)3329887, Fax: +91(11)3713308 ISSUES/YR: 4

Asian Pacific Journal of Allergy and Immunology
ISSN: 0125-877X
Asian Pac. J. Allergy Immunol.
Allergy and Immunology Society of Thailand, 420/6 Rajvithi Road, 10400, Bangkok, Thailand ISSUES/YR: 4

Assay and Drug Development Technologies
ISSN: 1540-658X
Assay Drug Dev. Technol.
Mary Ann Liebert Inc., 140 Huguenot Street, New Rochelle, NY 10801-5215, United States, Tel: +1(914)7402100, Fax: +1(914)7402101 ISSUES/YR: 6

Assistive Technology
ISSN: 1040-0435
Assistive Technol.
RESNA Press, 1700 N. Moore Street, Suite 1540, Arlington, VA 22209-1903, United States, Tel: +1(703)5246630 ISSUES/YR: 2

Asthma Journal
ISSN: 1363-268X
Asthma J.
Asthma Enterprises Ltd, Providence House, Providence Place, N1 0NT, London, United Kingdom, Tel: +44(207)2262260, Fax: +44(207)7040740 ISSUES/YR: 4

Atemwegs- und Lungenkrankheiten
ISSN: 0341-3055
Atemwegs- Lungenkr.
Dustri-Verlag Dr. Karl Feistle, P.O. Box 1351, D-82032, Deisenhofen/Munchen, Germany, Tel: +49(89)6138610, Fax: +49(89)6135412 ISSUES/YR: 12

Atencion Farmaceutica
ISSN: 1139-7357
Aten. Farm.
Rasgo Editorial S.A., Llanca 16, Dpcho. 4, 08015, Barcelona, Spain, Tel: +34-932 239 982, Fax: +34-902 117 561 ISSUES/YR: 6

Atencion Primaria
ISSN: 0212-6567
Aten. Prim.
Ediciones Doyma, S.L., Travesera de Gracia 17-21, 08021, Barcelona, Spain, Tel: +34(93)2000711, Fax: +34(93)2091136 ISSUES/YR: 19

● **Atherosclerosis**
ISSN: 0021-9150
Atherosclerosis
Elsevier Ireland Ltd, P.O. Box 85, Limerick, Ireland, Tel: +353(61)471944, Fax: +353(61)472144 ISSUES/YR: 12

● **Atherosclerosis Supplements**
ISSN: 1567-5688
Atheroscler. Suppl.
Elsevier Ireland Ltd, P.O. Box 85, Limerick, Ireland, Tel: +353(61)471944, Fax: +353(61)472144 ISSUES/YR: 1

Athletic Therapy Today
ISSN: 1078-7895
Athl. Ther. Today
Human Kinetics Publishers Inc., 1607 N. Market Street, Champaign, IL 61820-2200, United States, Tel: +1(217)3515076, Fax: +1(217)3511549 ISSUES/YR: 6

● **ATLA Alternatives to Laboratory Animals**
ISSN: 0261-1929
ATLA Altern. Lab. Anim.
FRAME, 96-98 North Sherwood Street, NG1 4EE, Nottingham, United Kingdom, Tel: +44(115)9584740, Fax: +44(115)9503570 ISSUES/YR: 6

Atlas of Hand Clinics
ISSN: 1082-3131
Atlas Hand Clin.
W.B. Saunders, Independence Square West, Philadelphia, PA 19106-3399, United States, Tel: +1(215)2387800, Fax: +1(215)2386445 ISSUES/YR: 2

Atlas of the Oral and Maxillofacial Surgery Clinics of North America
ISSN: 1061-3315
Atlas Oral Maxillofac. Surg. Clin. North Am.
W.B. Saunders, Independence Square West, Philadelphia, PA 19106-3399, United States, Tel: +1(215)2387800, Fax: +1(215)2386445 ISSUES/YR: 2

Atlas of the Urologic Clinics of North America
ISSN: 1063-5777
Atlas Urol. Clin. North Am.
W.B. Saunders, Independence Square West, Philadelphia, PA 19106-3399, United States, Tel: +1(215)2387800, Fax: +1(215)2386445 ISSUES/YR: 2

● **Atmospheric Environment**
ISSN: 1352-2310
Atmos. Environ.
Elsevier Ltd, Langford Lane, Kidlington, OX5 1GB, Oxford, United Kingdom, Tel: +44(1865)843000, Fax: +44(1865)843010 ISSUES/YR: 40

● **Attachment and Human Development**
ISSN: 1461-6734
Attachment Human Dev.
Routledge, 4 Park Square, Milton Park, OX14 4RN, Abingdon, Oxfordshire, United Kingdom, Tel: +44(1235)828600, Fax: +44(1235)829000 ISSUES/YR: 4

● **Audiological Medicine**
ISSN: 1651-386X
Audiol. Med.
Taylor and Francis Ltd., 4 Park Square, Milton Park, OX14 4RN, Abingdon, Oxfordshire, United Kingdom, Tel: +44(20)70175541, Fax: +44(20)70175198 ISSUES/YR: 4

● **Audiology and Neurotology**
ISSN: 1420-3030
Audiol. Neurotol.
S. Karger AG, Allschwilerstrasse 10, P.O. Box, CH-4009, Basel, Switzerland, Tel: +41(61)3061111, Fax: +41(61)3061234 ISSUES/YR: 6

Auris Nasus Larynx
ISSN: 0385-8146
Auris Nasus Larynx
Elsevier Ireland Ltd, P.O. Box 85, Limerick, Ireland, Tel: +353(61)471944, Fax: +353(61)472144 ISSUES/YR: 4

Australasian Biotechnology
ISSN: 1036-7128
Australas. Biotechnol.
AusBIOTECH LTD, Level 1, 322 Glenferrie Road, VIC 3144, Malvern, Australia, Tel: +61(3)98281400, Fax: +61(3)92245188 ISSUES/YR: 6

Australasian Journal of Dermatology
ISSN: 0004-8380
Australas. J. Dermatol.
Blackwell Publishing, 550 Swanston Street, VIC 3053, Carlton South, Australia, Tel: +61(3)8359 1011, Fax: +61(3)8359 1120 ISSUES/YR: 4

Current

Australasian Physical and Engineering Sciences in Medicine
ISSN: 0158-9938
Australas. Phys. Eng. Sci. Med.
Australasian Coll. of Phys. Scientists and Engineers, Dept. of Medical Physics, North Terrace, SA 5000, Adelaide, Australia, Tel: +61(8)82225651, Fax: +61(8)82225937 ISSUES/YR: 4

Australasian Psychiatry
ISSN: 1039-8562
Australas. Psychiatry
Blackwell Publishing, 550 Swanston Street, VIC 3053, Carlton South, Australia, Tel: +61(3)8359 1011, Fax: +61(3)8359 1120 ISSUES/YR: 4

● **Australasian Radiology**
ISSN: 0004-8461
Australas. Radiol.
Blackwell Publishing, 550 Swanston Street, VIC 3053, Carlton South, Australia, Tel: +61(3)8359 1011, Fax: +61(3)8359 1120 ISSUES/YR: 6

● **Australia and New Zealand Health Policy**
ISSN: 1743-8462
Aust. New Zealand Health Policy
BioMed Central Ltd., 34 - 42 Cleveland Street, W1T 4LB, London, United Kingdom, Tel: +44(207)3230323, Fax: +44(207)6319926 ISSUES/YR: 1

Australian and New Zealand Journal of Audiology
ISSN: 0157-1532
Aust. New Zealand J. Audiol.
Australian Academic Press Pty. Ltd, 32 Jeays Street, QLD 4006, Bowen Hills, Australia, Tel: +61(7)32571176, Fax: +61(7)32525908 ISSUES/YR: 2

● **Australian and New Zealand Journal of Obstetrics and Gynaecology**
ISSN: 0004-8666
Aust. New Zealand J. Obstet. Gynaecol.
Blackwell Publishing, 550 Swanston Street, VIC 3053, Carlton South, Australia, Tel: +61(3)8359 1011, Fax: +61(3)8359 1120 ISSUES/YR: 6

Australian and New Zealand Journal of Psychiatry
ISSN: 0004-8674
Aust. New Zealand J. Psychiatry
Blackwell Publishing, 550 Swanston Street, VIC 3053, Carlton South, Australia, Tel: +61(3)8359 1011, Fax: +61(3)8359 1120 ISSUES/YR: 12

Australian and New Zealand Journal of Public Health
ISSN: 1326-0200
Aust. New Zealand J. Public Health
Public Health Association of Australia Inc., PO Box 319, ACT 2605, Curtin, Australia, Tel: +61(6)2852373 ISSUES/YR: 6

Australian Journal of Clinical and Experimental Hypnosis
ISSN: 0156-0417
Aust. J. Clin. Exp. Hypn.
Australian Society of Hypnosis Ltd, P.O. Box 5114, 3018, Alphington, Australia, Tel: +61-3 9496 4621, Fax: +61-3 9496 4564 ISSUES/YR: 2

Australian Journal of Clinical Hypnotherapy and Hypnosis
ISSN: 0810-0713
Aust. J. Clin. Hypnother. Hypn.
Australian Academic Press Pty. Ltd, 32 Jeays Street, QLD 4006, Bowen Hills, Australia, Tel: +61(7)32571176, Fax: +61(7)32525908 ISSUES/YR: 2

Australian Journal of Forensic Sciences
ISSN: 0045-0618
Aust. J. Forensic Sci.
McGraw-Hill Co. Aust. Pty. Ltd, 4 Barcoo Street, P.O. Box 239, NSW 2069, East Roseville, Australia, Tel: +61(2)94174288 ISSUES/YR: 2

Australian Journal of Medical Science
ISSN: 1038-1643
Aust. J. Med. Sci.
Australian Institute of Medical Scientists, P.O. Box 1911, QLD 4064, Milton, Australia, Tel: +61(7)33713370, Fax: +61(7)38704857 ISSUES/YR: 4

Australian Journal of Otolaryngology
ISSN: 1037-2105
Aust. J. Otolaryngol.
Australian Society of Otolaryngology Head and Neck Surgery, Suite 403, Level 4, NSW 2061, 68 Alfred Street, Australia, Tel: +61(2)99545856, Fax: +61(2)99576863 ISSUES/YR: 2

Australian Journal of Physiotherapy
ISSN: 0004-9514
Aust. J. Physiother.
Australian Physiotherapy Association, P.O. Box 6465, VIC 8008, St. Kilda Road, Australia, Tel: +61(3)95349400, Fax: +61(3)95349199 ISSUES/YR: 4

Australian Journal of Primary Health
ISSN: 1448-7527
Aust. J. Prim. Health
Australian Institute For Primary Care, La Trobe University, VIC 3083, Bundoora, Australia, Tel: +61(3)9479 3700, Fax: +61(3)9479 5977 ISSUES/YR: 3

● **Australian Journal of Rural Health**
ISSN: 1038-5282
Austr. J. Rural Health
Blackwell Publishing, 550 Swanston Street, VIC 3053, Carlton South, Australia, Tel: +61(3)8359 1011, Fax: +61(3)8359 1120 ISSUES/YR: 6

Australian Occupational Therapy Journal
ISSN: 0045-0766
Aust. Occup. Ther. J.
Blackwell Publishing, 550 Swanston Street, VIC 3053, Carlton South, Australia, Tel: +61(3)8359 1011, Fax: +61(3)8359 1120 ISSUES/YR: 4

Australian Prescriber
ISSN: 0312-8008
Aust. Prescr.
Australian Government Publishing Service, P.O. Box 100, ACT 2606, Woden, Australia, Tel: +61(6)2897038 ISSUES/YR: 6

● **Autism**
ISSN: 1362-3613
Autism
SAGE Publications Ltd, 55 City Road, EC1Y 1SP, London, United Kingdom, Tel: +44(207)3248500, Fax: +44(207)3248600 ISSUES/YR: 4

Autoimmunity
ISSN: 0891-6934
Autoimmunity
Taylor and Francis Ltd., 4 Park Square, Milton Park, OX14 4RN, Abingdon, Oxfordshire, United Kingdom, Tel: +44(20)70175541, Fax: +44(20)70175198 ISSUES/YR: 8

Autoimmunity Reviews
ISSN: 1568-9972
Autoimmun. Rev.
Elsevier, P.O. Box 211, 1000 AE, Amsterdam, Netherlands, Tel: +31(20)4853757, Fax: +31(20)4853432 ISSUES/YR: 8

Autonomic and Autacoid Pharmacology
ISSN: 1474-8665
Auton. Autacoid Pharmacol.
Blackwell Publishing Ltd, 9600 Garsington Road, OX4 2XG, Oxford, United Kingdom, Tel: +44(1865)776868, Fax: +44(1865)714591 ISSUES/YR: 4

● **Autonomic Neuroscience: Basic and Clinical**
ISSN: 1566-0702
Auton. Neurosc. Basic Clin.
Elsevier, P.O. Box 211, 1000 AE, Amsterdam, Netherlands, Tel: +31(20)4853757, Fax: +31(20)4853432 ISSUES/YR: 14

Autophagy
ISSN: 1554-8627
Autophagy
Landes Bioscience, 810 S Church Street, Georgetown, 78626, United States, Tel: +1-512.863.7762, Fax: +1-512.863.0081 ISSUES/YR: 5

Aviation Space and Environmental Medicine
ISSN: 0095-6562
Aviat. Space Environ. Med.
Aerospace Medical Association, 320 S. Henry Street, Alexandria, VA 22314-3579, United States, Tel: +1(703)7392240 ISSUES/YR: 12

AWIC Bulletin
ISSN: 1522-7553
AWIC Bull.
Animal Welfare Information Center, 10301 Baltimore Blvd, 5th Floor, Beltsville, MD 20705-2351, United States, Tel: +1(301)5046212, Fax: +1(301)5047125 ISSUES/YR: 4

Azerbaijan Journal of Metabolism
ISSN: 1683-7460
Azerbaijan J. Metabol.
Azerbaijan Journal of Metabolism, Talibli Street 36, ap. 26, 370012, Baku, Azerbaijan, Tel: +994(12)329243, Fax: +994(12)985525 ISSUES/YR: 2

Azerbaijan Medical Journal
ISSN: 0005-2523
Azerbaijan Med. J.
WHO Office in Azerbaijan, Kicik Daniz kuc. 4, 370014, Baku, Azerbaijan, Fax: +994(12)930711 ISSUES/YR: 4

B-ENT
ISSN: 1781-782X
B-ENT
ASSMB a.s.b.l., Avenue Circulaire 138A, B-1180, Brussels, Belgium, Tel: +32(2)3745158, Fax: +32(2)3749628 ISSUES/YR: 4

BAG - Journal of Basic and Applied Genetics
ISSN: 1666-0390
BAG J. Basic Appl. Genet.
Sociedad Argentina de Genetica, Rodriguez Pena 36, Piso 6o, Dto 'A', Buenos Aires, Argentina ISSUES/YR: 2

Bahrain Medical Bulletin
ISSN: 1012-8298
Bahrain Med. Bull.
Bahrain Medical Bulletin, P.O. Box 32159, Manama, Bahrain, Tel: +973-277209 EXT.548 ISSUES/YR: 4

Balkan Journal of Medical Genetics
ISSN: 1311-0160
Balk. J. Med. Genet.
Macedonian Academy of Sciences and Arts, P.O. Box 428, 91000, Skopje, Macedonia, Tel: +389(91)239061, Fax: +389(91)115434 ISSUES/YR: 4

Bangladesh Journal of Dermatology, Venereology and Leprology
ISSN: 1023-3903
Bangladesh J. Dermatol. Venereol. Leprol.
Bangladesh Dermatological Society, Room 307, Shahbag, Dhaka, Bangladesh ISSUES/YR: 2

Bangladesh Journal of Obstetrics and Gynecology
ISSN: 1018-4287
Bangladesh J. Obstet. Gynecol.
Obstetrical and Gynaecological Society of Bangladesh, House-14/1-D (2nd Floor), Road 4, DHAKA 1205, Dhanmondi R/A, Bangladesh, Tel: +880-2 861 8879, Fax: +880-2 861 8879 ISSUES/YR: 2

Bangladesh Medical Research Council Bulletin
ISSN: 0377-9238
Bangladesh Med. Res. Counc. Bull.
Bangladesh Medical Research Council, Mohakhali, 1212, Dhaka, Bangladesh, Tel: +880(2)8811395, Fax: +880(2)8828820 ISSUES/YR: 3

Bangladesh Renal Journal
ISSN: 1015-0889
Bangladesh Renal J.
Bangladesh Renal Association, 1000, Dacca, Bangladesh, Tel: +880(2)864811, Fax: +880(2)864811 ISSUES/YR: 2

• **Basic and Clinical Pharmacology and Toxicology**
ISSN: 1742-7835
Basic Clin. Pharmacol. Toxicol.
Blackwell Publishing Ltd, 9600 Garsington Road, OX4 2XG, Oxford, United Kingdom, Tel: +44(1865)776868, Fax: +44(1865)714591 ISSUES/YR: 12

• **Basic and Clinical Pharmacology and Toxicology, Supplement**
ISSN: 0901-9928
Basic Clin. Pharmacol. Toxicol. Suppl.
Blackwell Publishing Ltd, 9600 Garsington Road, OX4 2XG, Oxford, United Kingdom, Tel: +44(1865)776868, Fax: +44(1865)714591 ISSUES/YR: 1

Basic Research in Cardiology
ISSN: 0300-8428
Basic Res. Cardiol.
Dr. Dietrich Steinkopff Verlag GmbH and Co. KG, P.O. Box 100462, D-64204, Darmstadt, Germany, Tel: +49(6151)828990, Fax: +49(6151)8289940 ISSUES/YR: 6

Basic Research in Cardiology, Supplement
ISSN: 0175-9418
Basic Res. Cardiol. Suppl.
Dr. Dietrich Steinkopff Verlag GmbH and Co. KG, P.O. Box 100462, D-64204, Darmstadt, Germany, Tel: +49(6151)828990, Fax: +49(6151)8289940 ISSUES/YR: 1

Behavior Genetics
ISSN: 0001-8244
Behav. Genet.
Springer Science and Business Media Deutschland GmbH, Heidelberger Platz 3, 14197, Berlin, Germany, Tel: +1-30 827875282, Fax: +1-30 827875707 ISSUES/YR: 6

Behavior Modification
ISSN: 0145-4455
Behav. Modif.
SAGE Publications Inc., 2455 Teller Road, Thousand Oaks, CA 91320, United States, Tel: +1(805)4990721, Fax: +1(805)4990871 ISSUES/YR: 6

Behavior Therapy
ISSN: 0005-7894
Behav. Ther.
Elsevier Inc., 360 Park Avenue South, New York, NY 10010, United States, Tel: +1(212)6333950, Fax: +1(212)6333990 ISSUES/YR: 4

• **Behavioral and Brain Functions**
ISSN: 1744-9081
Behav. Brain Funct.
BioMed Central Ltd., 34 - 42 Cleveland Street, W1T 4LB, London, United Kingdom, Tel: +44(207)3230323, Fax: +44(207)6319926 ISSUES/YR: 1

Behavioral and Brain Sciences
ISSN: 0140-525X
Behav. Brain Sci.
Cambridge University Press, Shaftesbury Road, CB2 2RU, Cambridge, United Kingdom, Tel: +44(1223)312393, Fax: +44(1223)315052 ISSUES/YR: 6

Behavioral and Cognitive Neuroscience Reviews
ISSN: 1534-5823
Behav. Cogn. Neurosci. Rev.
SAGE Publications Inc., 2455 Teller Road, Thousand Oaks, CA 91320, United States, Tel: +1(805)4990721, Fax: +1(805)4990871 ISSUES/YR: 4

Behavioral Interventions
ISSN: 1072-0847
Behav. Interventions
John Wiley and Sons Ltd, Southern Gate, PO19 8SQ, Chichester, West Sussex, United Kingdom, Tel: +44(1243)779777, Fax: +44(1243)775878 ISSUES/YR: 4

Behavioral Medicine
ISSN: 0896-4289
Behav. Med.
Heldref Publications, 1319 Eighteenth Street NW, Washington, DC 20036-1802, United States, Tel: +1(202)2966267, Fax: +1(202)2936130 ISSUES/YR: 4

Behavioral Neuroscience
ISSN: 0735-7044
Behav. Neurosci.
American Psychological Association Inc., 750 First Street NE, Washington, DC 20002-4242, United States, Tel: +1(202)3365600, Fax: +1(202)3365568 ISSUES/YR: 6

Behavioral Sciences and the Law
ISSN: 0735-3936
Behav. Sci. Law
John Wiley and Sons Ltd, Southern Gate, PO19 8SQ, Chichester, West Sussex, United Kingdom, Tel: +44(1243)779777, Fax: +44(1243)775878 ISSUES/YR: 6

Behaviour Change
ISSN: 0813-4839
Behav. Change
Australian Academic Press Pty. Ltd, 32 Jeays Street, QLD 4006, Bowen Hills, Australia, Tel: +61(7)32571176, Fax: +61(7)32525908 ISSUES/YR: 4

Behaviour Research and Therapy
ISSN: 0005-7967
Behav. Res. Ther.
Elsevier Ltd, Langford Lane, Kidlington, OX5 1GB, Oxford, United Kingdom, Tel: +44(1865)843000, Fax: +44(1865)843010 ISSUES/YR: 12

- **Behavioural and Cognitive Psychotherapy**
ISSN: 1352-4658
Behav. Cognitive Psychother.
Cambridge University Press, Shaftesbury Road, CB2 2RU, Cambridge, United Kingdom, Tel: +44(1223)312393, Fax: +44(1223)315052 ISSUES/YR: 4

- **Behavioural Brain Research**
ISSN: 0166-4328
Behav. Brain Res.
Elsevier, P.O. Box 211, 1000 AE, Amsterdam, Netherlands, Tel: +31(20)4853757, Fax: +31(20)4853432 ISSUES/YR: 20

- **Behavioural Neurology**
ISSN: 0953-4180
Behav. Neurol.
IOS Press, Nieuwe Hemweg 6B, 1013 BG, Amsterdam, Netherlands, Tel: +31(20)6883355, Fax: +31(20)6203419 ISSUES/YR: 4

Behavioural Pharmacology
ISSN: 0955-8810
Behav. Pharmacol.
Lippincott Williams and Wilkins, 241 Borough High Street, SE1 1GB, London, United Kingdom, Tel: +44(207)9407500, Fax: +44(207)9407575 ISSUES/YR: 8

Behavioural Processes
ISSN: 0376-6357
Behav. Processes
Elsevier, P.O. Box 211, 1000 AE, Amsterdam, Netherlands, Tel: +31(20)4853757, Fax: +31(20)4853432 ISSUES/YR: 12

- **Best Practice and Research in Clinical Anaesthesiology**
ISSN: 1521-6896
Best Pract. Res. Clin. Anaesthesiol.
Bailliere Tindall Ltd, 32 Jamestown Road, NW1 7BY, London, United Kingdom, Tel: +44(207)4244200, Fax: +44(207)4854752 ISSUES/YR: 4

Best Practice and Research in Clinical Endocrinology and Metabolism
ISSN: 1521-690X
Best Pract. Res. Clin. Endocrinol. Metab.
Bailliere Tindall Ltd, 32 Jamestown Road, NW1 7BY, London, United Kingdom, Tel: +44(207)4244200, Fax: +44(207)4854752 ISSUES/YR: 4

Best Practice and Research in Clinical Gastroenterology
ISSN: 1521-6918
Best Pract. Res. Clin. Gastroenterol.
Bailliere Tindall Ltd, 32 Jamestown Road, NW1 7BY, London, United Kingdom, Tel: +44(207)4244200, Fax: +44(207)4854752 ISSUES/YR: 6

- **Best Practice and Research in Clinical Haematology**
ISSN: 1521-6926
Best Pract. Res. Clin. Haematol.
Bailliere Tindall Ltd, 32 Jamestown Road, NW1 7BY, London, United Kingdom, Tel: +44(207)4244200, Fax: +44(207)4854752 ISSUES/YR: 4

- **Best Practice and Research in Clinical Obstetrics and Gynaecology**
ISSN: 1521-6934
Best Pract. Res. Clin. Obstet. Gynaecol.
Bailliere Tindall Ltd, 32 Jamestown Road, NW1 7BY, London, United Kingdom, Tel: +44(207)4244200, Fax: +44(207)4854752 ISSUES/YR: 6

- **Best Practice and Research in Clinical Rheumatology**
ISSN: 1521-6942
Best Pract. Res. Clin. Rheumatol.
Bailliere Tindall Ltd, 32 Jamestown Road, NW1 7BY, London, United Kingdom, Tel: +44(207)4244200, Fax: +44(207)4854752 ISSUES/YR: 6

- **Binocular Vision and Strabismus Quarterly**
ISSN: 1088-6281
Binocul. Vis. Strabismus Q.
Binoculus Publishing, P.O. Box 3727, 740 Piney Acres Circle, Dillon, CO 80435-3727, United States, Tel: +1(970)2620753, Fax: +1(970)2620753 ISSUES/YR: 4

- **Bio-Medical Materials and Engineering**
ISSN: 0959-2989
Bio-Med. Mater. Eng.
IOS Press, Nieuwe Hemweg 6B, 1013 BG, Amsterdam, Netherlands, Tel: +31(20)6883355, Fax: +31(20)6203419 ISSUES/YR: 4

Biocatalysis and Biotransformation
ISSN: 1024-2422
Biocatal. Biotransform.
Taylor and Francis Ltd., 4 Park Square, Milton Park, OX14 4RN, Abingdon, Oxfordshire, United Kingdom, Tel: +44(20)70175541, Fax: +44(20)70175198 ISSUES/YR: 6

Biocell
ISSN: 0327-9545
Biocell
Centro Regional de Invest. Cientif. y Tecn., P.O. Box 131, 5500, Mendoza, Argentina, Tel: +54(261)241794 ISSUES/YR: 3

Biochemia Medica
ISSN: 1330-0962
Biochem Med.
Croatian Society of Medical Biochemists, Domagojeva 2, 10000, Zagreb, Croatia, Tel: +385(1)4612606, Fax: +385(1)4612716 ISSUES/YR: 4

- **Biochemical and Biophysical Research Communications**
ISSN: 0006-291X
Biochem. Biophys. Res. Commun.
Academic Press Inc., 6277 Sea Harbor Drive, Orlando, FL 32887-4900, United States, Tel: +1(407)3454040, Fax: +1(407)3639661 ISSUES/YR: 40

- **Biochemical Engineering Journal**
ISSN: 1369-703X
Biochem. Eng. J.
Elsevier, P.O. Box 211, 1000 AE, Amsterdam, Netherlands, Tel: +31(20)4853757, Fax: +31(20)4853432 ISSUES/YR: 6

Biochemical Genetics
ISSN: 0006-2928
Biochem. Genet.
Springer Science and Business Media Deutschland GmbH, Heidelberger Platz 3, 14197, Berlin, Germany, Tel: +1-30 827875282, Fax: +1-30 827875707 ISSUES/YR: 12

- **Biochemical Journal**
ISSN: 0264-6021
Biochem. J.
Portland Press Ltd, 59 Portland Place, W1N 3AJ, London, United Kingdom, Tel: +44(20)75805530, Fax: +44(20)73231136 ISSUES/YR: 24

- **Biochemical Pharmacology**
ISSN: 0006-2952
Biochem. Pharmacol.
Elsevier Inc., 360 Park Avenue South, New York, NY 10010, United States, Tel: +1(212)6333950, Fax: +1(212)6333990 ISSUES/YR: 24

- **Biochemical Society Transactions**
ISSN: 0300-5127
Biochem. Soc. Trans.
Portland Press Ltd, 59 Portland Place, W1N 3AJ, London, United Kingdom, Tel: +44(20)75805530, Fax: +44(20)73231136 ISSUES/YR: 6

- **Biochemistry**
ISSN: 0006-2960
Biochemistry
American Chemical Society, 2540 Olentangy River Road, P.O. Box 3337, Columbus, OH 43210-3337, United States, Tel: +1(614)4473776, Fax: +1(614)4473671 ISSUES/YR: 51

Biochemistry and Cell Biology
ISSN: 0829-8211
Biochem. Cell Biol.
National Research Council of Canada, ONT K1A 0R6, Ottawa, Canada, Tel: +1(613)9939084, Fax: +1(613)9527656 ISSUES/YR: 6

Biochemistry and Molecular Biology Education
ISSN: 1470-8175
Biochem. Mol. Biol. Educ.
American Society for Biochemistry and Molecular Biology Inc., 9650 Rockville Pike, Bethesda, MD 20814, United States, Tel: +1-301 530 7145, Fax: +1-301 571 1824 ISSUES/YR: 6

- **Biochimica et Biophysica Acta - Bioenergetics**
ISSN: 0005-2728
Biochim. Biophys. Acta Bioenerg.
Elsevier, P.O. Box 211, 1000 AE, Amsterdam, Netherlands, Tel: +31(20)4853757, Fax: +31(20)4853432 ISSUES/YR: 1

- **Biochimica et Biophysica Acta - Biomembranes**
ISSN: 0005-2736
Biochim. Biophys. Acta Biomembr.
Elsevier, P.O. Box 211, 1000 AE, Amsterdam, Netherlands, Tel: +31(20)4853757, Fax: +31(20)4853432 ISSUES/YR: 20

- **Biochimica et Biophysica Acta - Gene Structure and Expression**
ISSN: 0167-4781
Biochim. Biophys. Acta Gene Struct. Expr.
Elsevier, P.O. Box 211, 1000 AE, Amsterdam, Netherlands, Tel: +31(20)4853757, Fax: +31(20)4853432 ISSUES/YR: 1

- **Biochimica et Biophysica Acta - General Subjects**
ISSN: 0304-4165
Biochim. Biophys. Acta Gen. Subj.
Elsevier, P.O. Box 211, 1000 AE, Amsterdam, Netherlands, Tel: +31(20)4853757, Fax: +31(20)4853432 ISSUES/YR: 18

- **Biochimica et Biophysica Acta - Molecular and Cell Biology of Lipids**
ISSN: 1388-1981
Biochim. Biophys. Acta Mol. Cell Biol. Lipids
Elsevier, P.O. Box 211, 1000 AE, Amsterdam, Netherlands, Tel: +31(20)4853757, Fax: +31(20)4853432 ISSUES/YR: 18

- **Biochimica et Biophysica Acta - Molecular Basis of Disease**
ISSN: 0925-4439
Biochim. Biophys. Acta Mol. Basis Dis.
Elsevier, P.O. Box 211, 1000 AE, Amsterdam, Netherlands, Tel: +31(20)4853757, Fax: +31(20)4853432 ISSUES/YR: 9

- **Biochimica et Biophysica Acta - Molecular Cell Research**
ISSN: 0167-4889
Biochim. Biophys. Acta Mol. Cell Res.
Elsevier, P.O. Box 211, 1000 AE, Amsterdam, Netherlands, Tel: +31(20)4853757, Fax: +31(20)4853432 ISSUES/YR: 15

- **Biochimica et Biophysica Acta - Proteins and Proteomics**
ISSN: 1570-9639
Biochim. Biophys. Acta Proteins Proteomics
Elsevier, P.O. Box 211, 1000 AE, Amsterdam, Netherlands, Tel: +31(20)4853757, Fax: +31(20)4853432 ISSUES/YR: 16

- **Biochimica et Biophysica Acta - Reviews on Cancer**
ISSN: 0304-419X
Biochim. Biophys. Acta Rev. Cancer
Elsevier, P.O. Box 211, 1000 AE, Amsterdam, Netherlands, Tel: +31(20)4853757, Fax: +31(20)4853432 ISSUES/YR: 4

Biochimie
ISSN: 0300-9084
Biochimie
Elsevier, P.O. Box 211, 1000 AE, Amsterdam, Netherlands, Tel: +31(20)4853757, Fax: +31(20)4853432 ISSUES/YR: 12

Bioconjugate Chemistry
ISSN: 1043-1802
Bioconjugate Chem.
American Chemical Society, 2540 Olentangy River Road, P.O. Box 3337, Columbus, OH 43210-3337, United States, Tel: +1(614)4473776, Fax: +1(614)4473671 ISSUES/YR: 6

Biocontrol Science
ISSN: 1342-4815
Biocontrol Sci.
Society for Antibacterial and Antifungal Agents Japan, 1-13-38 Nishi-honmachi, Nishi-ku, 550, Osaka, Japan, Tel: +81(6)5382166, Fax: +81(6)5382169 ISSUES/YR: 4

- **BioDrugs**
ISSN: 1173-8804
BioDrugs
Adis International Ltd, 41 Centorian Drive, Private Bag 65901, Mairangi Bay, 1311, Auckland 10, New Zealand, Tel: +64(9)4770700, Fax: +64(9)4770766 ISSUES/YR: 6

Bioelectrochemistry
ISSN: 1567-5394
Bioelectrochemistry
Elsevier, P.O. Box 211, 1000 AE, Amsterdam, Netherlands, Tel: +31(20)4853757, Fax: +31(20)4853432 ISSUES/YR: 6

- **Bioelectromagnetics**
ISSN: 0197-8462
Bioelectromagnetics
Wiley-Liss Inc., 111 River Street, Hoboken, NJ 07030-5774, United States, Tel: +1(201)7486645, Fax: +1(201)7486021 ISSUES/YR: 8

BioEssays
ISSN: 0265-9247
BioEssays
John Wiley and Sons Inc., P.O.Box 18667, Newark, NJ 07191-8667, United States, Tel: +1(212)8506000, Fax: +1(212)8506012 ISSUES/YR: 12

Bioethics
ISSN: 0269-9702
Bioethics
Blackwell Publishing Ltd, 9600 Garsington Road, OX4 2XG, Oxford, United Kingdom, Tel: +44(1865)776868, Fax: +44(1865)714591 ISSUES/YR: 6

- **BioFactors**
ISSN: 0951-6433
BioFactors
IOS Press, Nieuwe Hemweg 6B, 1013 BG, Amsterdam, Netherlands, Tel: +31(20)6883355, Fax: +31(20)6203419 ISSUES/YR: 8

- **Biofilms**
ISSN: 1479-0505
Biofilms
Cambridge University Press, Shaftesbury Road, CB2 2RU, Cambridge, United Kingdom, Tel: +44(1223)312393, Fax: +44(1223)315052 ISSUES/YR: 4

Biofutur
ISSN: 0294-3506
Biofutur
Lavoisier, 14 rue de Provigny, F-94236, Cachan Cedex, France, Tel: +33(1)47406700, Fax: +33(1)47406703 ISSUES/YR: 11

Biogenic Amines
ISSN: 0168-8561
Biog. Amines
VSP BV, P.O.Box 9000, 2300 PA, Leiden, Netherlands, Tel: +31-71 535 3500, Fax: +31-71 531 7532 ISSUES/YR: 6

- **Biogerontology**
ISSN: 1389-5729
Biogerontology
Kluwer Academic Publishers, Van Godewijckstraat 30, 3311 GZ, Dordrecht, Netherlands, Tel: +31(78)6576000, Fax: +31(78)6576254 ISSUES/YR: 6

- **Bioinformatics**
ISSN: 1367-4803
Bioinformatics
Oxford University Press, Great Clarendon Street, OX2 6DP, Oxford, United Kingdom, Tel: +44(1865)353907, Fax: +44(1865)353486 ISSUES/YR: 18

Biological and Pharmaceutical Bulletin
ISSN: 0918-6158
Biol. Pharm. Bull.
Pharmaceutical Society of Japan, 2-12-15-201, Shibuya, Shibuya-ku, 150, Tokyo, Japan, Tel: +81(3)34063321, Fax: +81(3)34981835 ISSUES/YR: 12

- **Biological Chemistry**
ISSN: 1431-6730
Biol. Chem.
Walter de Gruyter and Co., Genthiner Strasse 13, D-10785, Berlin, Germany, Tel: +49(30)260-050, Fax: +49(30)260-05251 ISSUES/YR: 12

Biological Procedures Online
ISSN: 1480-9222
Biol. Proc. Online
Biological Procedures Online, 27 Hollyridge Cres., N2N 3L8, Kitchener Ontario, Canada, Tel: +1-519 569 3706, Fax: +1-519 242 0964 ISSUES/YR: 1

- **Biological Psychiatry**
ISSN: 0006-3223
Biol. Psychiatry
Elsevier USA, 6277 Sea Harbor Drive, Orlando, FL 32862 8239, United States, Tel: +1-407 345 4000, Fax: +1-407 363 9661 ISSUES/YR: 24

- **Biological Psychology**
ISSN: 0301-0511
Biol. Psychol.
Elsevier, P.O. Box 211, 1000 AE, Amsterdam, Netherlands, Tel: +31(20)4853757, Fax: +31(20)4853432 ISSUES/YR: 9

Biological Research
ISSN: 0716-9760
Biol. Res.
Society of Biology of Chile, Canada 253, piso 3, Depto. F, P.O. Box 16164, 9, Santiago, Chile, Tel: +56(2)2093503, Fax: +56(2)2258427 ISSUES/YR: 4

Biological Reviews of the Cambridge Philosophical Society
ISSN: 1464-7931
Biol. Rev. Camb. Philos. Soc.
Cambridge University Press, Shaftesbury Road, CB2 2RU, Cambridge, United Kingdom, Tel: +44(1223)312393, Fax: +44(1223)315052 ISSUES/YR: 4

Biological Rhythm Research
ISSN: 0929-1016
Biol. Rhythm Res.
Taylor and Francis Ltd., 4 Park Square, Milton Park, OX14 4RN, Abingdon, Oxfordshire, United Kingdom, Tel: +44(20)70175541, Fax: +44(20)70175198 ISSUES/YR: 6

Biological Trace Element Research
ISSN: 0163-4984
Biol. Trace Elem. Res.
Humana Press Inc., 999 Riverview Drive, Suite 208, Totowa, NJ 07512-1165, United States, Tel: +1(973)2561699, Fax: +1(973)2568341 ISSUES/YR: 18

- **Biologicals**
ISSN: 1045-1056
Biologicals
Academic Press, 24-28 Oval Road, NW1 7DX, London, United Kingdom, Tel: +44(207)2674466 ISSUES/YR: 4

Biologische Medizin
ISSN: 0340-8671
Biol. Med.
Aurelia-Verlag GmbH, Postfach 100045, 76481, Baden-Baden, Germany, Tel: +49(7221)50102, Fax: +49(7221)501420 ISSUES/YR: 4

- **Biology Letters**
ISSN: 1744-9561
Biol. Lett.
Royal Society of London, 6 Carlton House Terrace, SW1Y 5AG, London, United Kingdom, Tel: +44(207)8395561, Fax: +44(207)9761837 ISSUES/YR: 4

Biology of Blood and Marrow Transplantation
ISSN: 1083-8791
Biol. Blood Marrow Transplant.
Elsevier, 360 Park Avenue South, New York, NY 10010, United States, Tel: +1(212)6333950, Fax: +1(212)6333990 ISSUES/YR: 12

- **Biology of Reproduction**
ISSN: 0006-3363
Biol. Reprod.
Society for the Study of Reproduction, 1603 Monroe Street, Madison, WI 53711-2021, United States, Tel: +1(608)2562777, Fax: +1(608)2564610 ISSUES/YR: 12

- **Biology of the Cell**
ISSN: 0248-4900
Biol. Cell
Portland Press Ltd., 59 Portland Place, W1B 1QW, London, United Kingdom, Tel: +44(20)7580-5530, Fax: +44(20)7323-1136 ISSUES/YR: 12

- **Biology of the Neonate**
ISSN: 0006-3126
Biol. Neonate
S. Karger AG, Allschwilerstrasse 10, P.O. Box, CH-4009, Basel, Switzerland, Tel: +41(61)3061111, Fax: +41(61)3061234 ISSUES/YR: 8

- **Biomacromolecules**
ISSN: 1525-7797
Biomacromolecules
American Chemical Society, 2540 Olentangy River Road, P.O. Box 3337, Columbus, OH 43210-3337, United States, Tel: +1(614)4473776, Fax: +1(614)4473671 ISSUES/YR: 6

Biomarkers
ISSN: 1354-750X
Biomarkers
Taylor and Francis Ltd., 4 Park Square, Milton Park, OX14 4RN, Abingdon, Oxfordshire, United Kingdom, Tel: +44(20)70175541, Fax: +44(20)70175198 ISSUES/YR: 6

- **Biomaterials**
ISSN: 0142-9612
Biomaterials
Elsevier Ltd, Langford Lane, Kidlington, OX5 1GB, Oxford, United Kingdom, Tel: +44(1865)843000, Fax: +44(1865)843010 ISSUES/YR: 36

Biomedical Chromatography
ISSN: 0269-3879
Biomed. Chromatogr.
John Wiley and Sons Ltd, Southern Gate, PO19 8SQ, Chichester, West Sussex, United Kingdom, Tel: +44(1243)779777, Fax: +44(1243)775878 ISSUES/YR: 10

Biomedical Digital Libraries
ISSN: 1742-5581
Biomed. Digit. Libr.
BioMed Central Ltd., 34 - 42 Cleveland Street, W1T 4LB, London, United Kingdom, Tel: +44(207)3230323, Fax: +44(207)6319926 ISSUES/YR: 1

Biomedical Engineering
ISSN: 0006-3398
Biomed. Eng.
Springer Science and Business Media Deutschland GmbH, Heidelberger Platz 3, 14197, Berlin, Germany, Tel: +1-30 827875282, Fax: +1-30 827875707 ISSUES/YR: 6

Biomedical Engineering - Applications, Basis and Communications
ISSN: 1016-2356
Biomed. Eng. Appl. Basis Commun.
Institute of Biomedical Engineering, No. 1, Sec. 1 Jen-Ai Road, Taipei R.O.C., Taiwan, Tel: +886(2)23970800 EXT. 1443, Fax: +886(2)23940049 ISSUES/YR: 6

BioMedical Engineering Online
ISSN: 1475-925X
Biomed. Eng. Online
BioMed Central Ltd., 34 - 42 Cleveland Street, W1T 4LB, London, United Kingdom, Tel: +44(207)3230323, Fax: +44(207)6319926 ISSUES/YR: 1

Biomedical Instrumentation and Technology
ISSN: 0899-8205
Biomed. Instrum. Technol.
Association for the Advancement of Medical Instrumentation, 1110 North Glebe Road, Suite 220, Arlington, VA 22201-4795, United States, Tel: +1-800 332 2264, Fax: +1-703 525 1424 ISSUES/YR: 6

Biomedical Materials
ISSN: 1748-6041
Biomed. Mater.
Institute of Physics Publishing, Temple Back, BS1 6BE, Bristol, United Kingdom, Tel: +44(117)9297481, Fax: +44(117)9294318 ISSUES/YR: 4

- **Biomedical Microdevices**
ISSN: 1387-2176
Biomed. Microdevices
Kluwer Academic Publishers, 101 Philip Drive, Assinippi Park, Norwell, MA 02061, United States, Tel: +1(781)8716600, Fax: +1(781)8716528 ISSUES/YR: 4

Current

Biomedical Research
ISSN: 0388-6107
Biomed. Res. (Japan)
Biomedical Research Foundation, Kanda, Chiyoda-ku, P.O. Box 182, 101-91, Tokyo, Japan ISSUES/YR: 6

Biomedical Research
ISSN: 0970-938X
Biomed. Res. (India)
Scientific Publishers of India, Sir Sayed Nagar, 202002, Aligarh, India, Tel: +91(571)502043, Fax: +91(571)400678 ISSUES/YR: 3

Biomedical Reviews
ISSN: 1310-392X
Biomed. Rev.
Bulgarian-American Center, Varna, Bulgaria, Tel: +359(52)454394 ISSUES/YR: 1

Biomedical Sciences Instrumentation
ISSN: 0067-8856
Biomed. Sci. Instrum.
ISA - Instrumentation, Systems, and Automation Society, P.O. Box 12277, Research Triangle Park, NC 27709, United States, Tel: +1(919)549-8411, Fax: +1(919)549-8288 ISSUES/YR: 1

• **Biomedical Signal Processing and Control**
ISSN: 1746-8094
Biomed. Signal Process. Control
Elsevier Ltd, Langford Lane, Kidlington, OX5 1GB, Oxford, United Kingdom, Tel: +44(1865)843000, Fax: +44(1865)843010 ISSUES/YR: 4

Biomedicine
ISSN: 0970-2067
Biomedicine
Indian Association of Biomedical Scientists, 695 011, Trivandrum, India, Tel: +91(471)443152, Fax: +91(471)550728 ISSUES/YR: 3

• **Biomedicine and Pharmacotherapy**
ISSN: 0753-3322
Biomed. Pharmacother.
Editions Scientifiques et Medicales Elsevier, 23 rue Linois, 75724, Paris Cedex 15, France, Tel: +33(1)45589826, Fax: +33(1)45589428 ISSUES/YR: 10

Biomedizinische Technik
ISSN: 0013-5585
Biomed. Tech.
Fachverlag Schiele und Sohn GmbH, Markgrafenstrasse 11, D-10969, Berlin, Germany, Tel: +49(30)2537250 ISSUES/YR: 12

BioMetals
ISSN: 0966-0844
BioMetals
Kluwer Academic Publishers, Van Godewijckstraat 30, 3311 GZ, Dordrecht, Netherlands, Tel: +31(78)6576000, Fax: +31(78)6576254 ISSUES/YR: 6

Biometrics
ISSN: 0006-341X
Biometrics
Blackwell Publishing Inc., 350 Main Street, Malden, MA 02148, United States, Tel: +1(781)3888200, Fax: +1(781)3888210 ISSUES/YR: 4

• **Biomolecular Engineering**
ISSN: 1389-0344
Biomol. Eng.
Elsevier, P.O. Box 211, 1000 AE, Amsterdam, Netherlands, Tel: +31(20)4853757, Fax: +31(20)4853432 ISSUES/YR: 6

Bioorganic and Medicinal Chemistry
ISSN: 0968-0896
Bioorg. Med. Chem.
Elsevier Ltd, Langford Lane, Kidlington, OX5 1GB, Oxford, United Kingdom, Tel: +44(1865)843000, Fax: +44(1865)843010 ISSUES/YR: 24

Bioorganic and Medicinal Chemistry Letters
ISSN: 0960-894X
Bioorg. Med. Chem. Lett.
Elsevier Ltd, Langford Lane, Kidlington, OX5 1GB, Oxford, United Kingdom, Tel: +44(1865)843000, Fax: +44(1865)843010 ISSUES/YR: 24

• **Bioorganic Chemistry**
ISSN: 0045-2068
Bioorg. Chem.
Academic Press Inc., 6277 Sea Harbor Drive, Orlando, FL 32887-4900, United States, Tel: +1(407)3454040, Fax: +1(407)3639661 ISSUES/YR: 6

BioPharm International
ISSN: 1542-166X
BioPharm Int.
Advanstar Communications, 131 West First Street, Duluth, MN 55802-2065, United States, Tel: +1(218)7239200, Fax: +1(218)7239437 ISSUES/YR: 12

Biopharmaceutics and Drug Disposition
ISSN: 0142-2782
Biopharm. Drug Dispos.
John Wiley and Sons Ltd, Southern Gate, PO19 8SQ, Chichester, West Sussex, United Kingdom, Tel: +44(1243)779777, Fax: +44(1243)775878 ISSUES/YR: 9

• **Biophysical Chemistry**
ISSN: 0301-4622
Biophys. Chem.
Elsevier, P.O. Box 211, 1000 AE, Amsterdam, Netherlands, Tel: +31(20)4853757, Fax: +31(20)4853432 ISSUES/YR: 21

Biophysical Journal
ISSN: 0006-3495
Biophys. J.
Biophysical Society, 9650 Rockville Pike, Suite L-0512, Bethesda, MD 20814-3998, United States, Tel: +1(301)5710663, Fax: +1(301)5307133 ISSUES/YR: 12

Biopolymers
ISSN: 0006-3525
Biopolymers
John Wiley and Sons Inc., P.O.Box 18667, Newark, NJ 07191-8667, United States, Tel: +1(212)8506000, Fax: +1(212)8506012 ISSUES/YR: 18

Biopolymers - Peptide Science Section
ISSN: 0006-3525
Biopolym. Pept. Sci. Sect.
John Wiley and Sons Inc., P.O.Box 18667, Newark, NJ 07191-8667, United States, Tel: +1(212)8506000, Fax: +1(212)8506012 ISSUES/YR: 6

• **Bioprocess and Biosystems Engineering**
ISSN: 1615-7591
Bioprocess Biosyst. Eng.
Springer Verlag, Tiergartenstrasse 17, D-69121, Heidelberg, Germany, Tel: +49(6221)4870, Fax: +49(6221)487366 ISSUES/YR: 6

• **Bioresource Technology**
ISSN: 0960-8524
Bioresour. Technol.
Elsevier Ltd, Langford Lane, Kidlington, OX5 1GB, Oxford, United Kingdom, Tel: +44(1865)843000 ISSUES/YR: 18

Biorheology
ISSN: 0006-355X
Biorheology
IOS Press, Nieuwe Hemweg 6B, 1013 BG, Amsterdam, Netherlands, Tel: +31(20)6883355, Fax: +31(20)6203419 ISSUES/YR: 12

Bioscience Reports
ISSN: 0144-8463
Biosci. Rep.
Springer Science and Business Media Deutschland GmbH, Heidelberger Platz 3, 14197, Berlin, Germany, Tel: +1-30 827875282, Fax: +1-30 827875707 ISSUES/YR: 6

Biosciences Biotechnology Research Asia
ISSN: not available yet
Biosci. Biotechnol. Res. Asia
Oriental Scientific Publishing Company, P.O. Box 35, 462 001, G.P.O Bhopal, India ISSUES/YR: 2

Biosensors and Bioelectronics
ISSN: 0956-5663
Biosens. Bioelectron.
Elsevier Ltd, Langford Lane, Kidlington, OX5 1GB, Oxford, United Kingdom, Tel: +44(1865)843000 ISSUES/YR: 12

BioSystems
ISSN: 0303-2647
BioSystems
Elsevier Ireland Ltd, P.O. Box 85, Limerick, Ireland, Tel: +353(61)471944, Fax: +353(61)472144 ISSUES/YR: 15

Biotechnic and Histochemistry
ISSN: 1052-0295
Biotech. Histochem.
Taylor and Francis Ltd., 4 Park Square, Milton Park, OX14 4RN, Abingdon, Oxfordshire, United Kingdom, Tel: +44(20)70175541, Fax: +44(20)70175198 ISSUES/YR: 6

BioTechniques
ISSN: 0736-6205
BioTechniques
Eaton Publishing Company, One Research Drive, Suite 400A, P.O. Box 1070, Westborough, MA 01581-6070, United States, Tel: +1(508)6141414, Fax: +1(508)6162930 ISSUES/YR: 12

Biotechnology and Applied Biochemistry
ISSN: 0885-4513
Biotechnol. Appl. Biochem.
Portland Press Ltd, 59 Portland Place, W1N 3AJ, London, United Kingdom, Tel: +44(20)75805530, Fax: +44(20)73231136 ISSUES/YR: 6

Biotechnology and Bioengineering
ISSN: 0006-3592
Biotechnol. Bioeng.
John Wiley and Sons Inc., P.O.Box 18667, Newark, NJ 07191-8667, United States, Tel: +1(212)8506000, Fax: +1(212)8506012 ISSUES/YR: 28

Biotechnology and Bioprocess Engineering
ISSN: 1226-8372
Biotechnol. Bioprocess Eng.
Korean Society for Biotechnology and Bioengineering, Yeogsam-dong, 135 703, Kangnam-ku Seoul, South Korea ISSUES/YR: 6

Biotechnology and Genetic Engineering Reviews
ISSN: 0264-8725
Biotechnol. Genet. Eng. Rev.
Intercept Ltd, P.O. Box 716, SP10 1YG, Andover, Hampshire, United Kingdom, Tel: +44(1264)334748, Fax: +44(1264)334058 ISSUES/YR: 1

Biotechnology Annual Review
ISSN: 1387-2656
Biotechnol. Annu. Rev.
Elsevier, P.O. Box 211, 1000 AE, Amsterdam, Netherlands, Tel: +31(20)4853757, Fax: +31(20)4853432 ISSUES/YR: 1

Biotecnologia Aplicada
ISSN: 0864-4551
Biotecnol. Apl.
Elfos Scientiae, Ave. 31 e/158 y 190, Cubanacan, Playa, Apdo. 6072, Habana 9, Cuba, Tel: +53(7)218466, Fax: +53(7)218070 ISSUES/YR: 4

Biotherapy
ISSN: 0914-2223
Biotherapy (Japan)
Japanese Society for Cancer Chemotherapy, 2-2-3 Nihonbashi, Chuo-ku, 103, Tokyo, Japan, Tel: +81(3)32810435 ISSUES/YR: 6

- **Bipolar Disorders**
ISSN: 1398-5647
Bipolar Disord.
Blackwell Munksgaard, 1 Rosenorns Alle, P.O. Box 227, DK-1502, Copenhagen V, Denmark, Tel: +45(77)333333, Fax: +45(77)333377 ISSUES/YR: 6

- **Bipolar Disorders, Supplement**
ISSN: 1399-2406
Bipolar Disord. Suppl.
Blackwell Munksgaard, 1 Rosenorns Alle, P.O. Box 227, DK-1502, Copenhagen V, Denmark, Tel: +45(77)333333, Fax: +45(77)333377 ISSUES/YR: 1

- **Birth**
ISSN: 0730-7659
Birth
Blackwell Publishing Inc., 350 Main Street, Malden, MA 02148, United States, Tel: +1(781)3888200, Fax: +1(781)3888210 ISSUES/YR: 4

- **Birth Defects Research Part A - Clinical and Molecular Teratology**
ISSN: 1542-0752
Birth Defects Res. Part A Clin. Mol. Teratol.
Wiley-Liss Inc., 111 River Street, Hoboken, NJ 07030-5774, United States, Tel: +1(201)7486645, Fax: +1(201)7486021 ISSUES/YR: 12

- **Birth Defects Research Part B - Developmental and Reproductive Toxicology**
ISSN: 1542-9733
Birth Defects Res. Part B Dev. Reprod. Toxicol.
Wiley-Liss Inc., 111 River Street, Hoboken, NJ 07030-5774, United States, Tel: +1(201)7486645, Fax: +1(201)7486021 ISSUES/YR: 6

- **Birth Defects Research Part C - Embryo Today: Reviews**
ISSN: 1542-975X
Birth Defects Res. Part C Embryo Today Rev.
Wiley-Liss Inc., 111 River Street, Hoboken, NJ 07030-5774, United States, Tel: +1(201)7486645, Fax: +1(201)7486021 ISSUES/YR: 4

- **BJOG: An International Journal of Obstetrics and Gynaecology**
ISSN: 1470-0328
BJOG Int. J. Obstet. Gynaecol.
Blackwell Publishing Ltd, 9600 Garsington Road, OX4 2XG, Oxford, United Kingdom, Tel: +44(1865)776868, Fax: +44(1865)714591 ISSUES/YR: 12

- **BJU International**
ISSN: 1464-4096
BJU Int.
Blackwell Publishing Ltd, 9600 Garsington Road, OX4 2XG, Oxford, United Kingdom, Tel: +44(1865)776868, Fax: +44(1865)714591 ISSUES/YR: 18

- **Blood**
ISSN: 0006-4971
Blood
American Society of Hematology, 1900 M Street, Suite 2000, Washington, DC 20036, United States, Tel: +1(202)7760544, Fax: +1(202)7760545 ISSUES/YR: 25

- **Blood Cells, Molecules, and Diseases**
ISSN: 1079-9796
Blood Cells Mol. Dis.
Academic Press Inc., 6277 Sea Harbor Drive, Orlando, FL 32887-4900, United States, Tel: +1(407)3454040, Fax: +1(407)3639661 ISSUES/YR: 6

- **Blood Coagulation and Fibrinolysis**
ISSN: 0957-5235
Blood Coagul. Fibrinolysis
Lippincott Williams and Wilkins, 241 Borough High Street, SE1 1GB, London, United Kingdom, Tel: +44(207)9407500, Fax: +44(207)9407575 ISSUES/YR: 8

- **Blood Pressure**
ISSN: 0803-7051
Blood Press.
Taylor and Francis A.S., P.O. Box 2562 Solli, 0202, Oslo, Norway, Tel: +47(22)129880, Fax: +47(22)129890 ISSUES/YR: 6

- **Blood Pressure Monitoring**
ISSN: 1359-5237
Blood Press. Monit.
Lippincott Williams and Wilkins, 241 Borough High Street, SE1 1GB, London, United Kingdom, Tel: +44(207)9407500, Fax: +44(207)9407575 ISSUES/YR: 6

- **Blood Pressure, Supplement**
ISSN: 0803-8023
Blood Press. Suppl.
Taylor and Francis A.S., P.O. Box 2562 Solli, 0202, Oslo, Norway, Tel: +47(22)129880, Fax: +47(22)129890 ISSUES/YR: 1

- **Blood Purification**
ISSN: 0253-5068
Blood Purif.
S. Karger AG, Allschwilerstrasse 10, P.O. Box, CH-4009, Basel, Switzerland, Tel: +41(61)3061111, Fax: +41(61)3061234 ISSUES/YR: 6

- **Blood Reviews**
  ISSN: 0268-960X
  Blood Rev.
  Churchill Livingstone, 1-3 Baxter's Place, Leith Walk, EH1 3AF, Edinburgh, United Kingdom, Tel: +44(131)5562424, Fax: +44(131)5581278 ISSUES/YR: 4

Blutalkohol
ISSN: 0006-5250
Blutalkohol
Steintor-Verlag GmbH, Graphengiesserstrasse 30, D-23556, Lubeck, Germany, Tel: +49(451)8798849, Fax: +49(451)8798837 ISSUES/YR: 6

BMC Anesthesiology
ISSN: 1471-2253
BMC Anesthesiol.
BioMed Central Ltd., 34 - 42 Cleveland Street, W1T 4LB, London, United Kingdom, Tel: +44(207)3230323, Fax: +44(207)6319926 ISSUES/YR: 1

BMC Biochemistry
ISSN: 1471-2091
BMC Biochem.
BioMed Central Ltd., 34 - 42 Cleveland Street, W1T 4LB, London, United Kingdom, Tel: +44(207)3230323, Fax: +44(207)6319926 ISSUES/YR: 1

BMC Bioinformatics
ISSN: 1471-2105
BMC Bioinform.
BioMed Central Ltd., 34 - 42 Cleveland Street, W1T 4LB, London, United Kingdom, Tel: +44(207)3230323, Fax: +44(207)6319926 ISSUES/YR: 1

BMC Biology
ISSN: 1741-7007
BMC Biol.
BioMed Central Ltd., 34 - 42 Cleveland Street, W1T 4LB, London, United Kingdom, Tel: +44(207)3230323, Fax: +44(207)6319926 ISSUES/YR: 1

BMC Biotechnology
ISSN: 1472-6750
BMC Biotechnol.
BioMed Central Ltd., 34 - 42 Cleveland Street, W1T 4LB, London, United Kingdom, Tel: +44(207)3230323, Fax: +44(207)6319926 ISSUES/YR: 1

BMC Blood Disorders
ISSN: 1471-2326
BMC Blood Disord.
BioMed Central Ltd., 34 - 42 Cleveland Street, W1T 4LB, London, United Kingdom, Tel: +44(207)3230323, Fax: +44(207)6319926 ISSUES/YR: 1

BMC Cancer
ISSN: 1471-2407
BMC Cancer
BioMed Central Ltd., 34 - 42 Cleveland Street, W1T 4LB, London, United Kingdom, Tel: +44(207)3230323, Fax: +44(207)6319926 ISSUES/YR: 1

- **BMC Cardiovascular Disorders**
  ISSN: 1471-2261
  BMC Cardiovasc. Disord.
  BioMed Central Ltd., 34 - 42 Cleveland Street, W1T 4LB, London, United Kingdom, Tel: +44(207)3230323, Fax: +44(207)6319926 ISSUES/YR: 1

BMC Cell Biology
ISSN: 1471-2121
BMC Cell Biol.
BioMed Central Ltd., 34 - 42 Cleveland Street, W1T 4LB, London, United Kingdom, Tel: +44(207)3230323, Fax: +44(207)6319926 ISSUES/YR: 1

BMC Chemical Biology
ISSN: 1472-6769
BMC Chem. Biol.
BioMed Central Ltd., 34 - 42 Cleveland Street, W1T 4LB, London, United Kingdom, Tel: +44(207)3230323, Fax: +44(207)6319926 ISSUES/YR: 1

BMC Clinical Pathology
ISSN: 1472-6890
BMC Clin. Pathol.
BioMed Central Ltd., 34 - 42 Cleveland Street, W1T 4LB, London, United Kingdom, Tel: +44(207)3230323, Fax: +44(207)6319926 ISSUES/YR: 1

BMC Clinical Pharmacology
ISSN: 1472-6904
BMC Clin. Parmacol.
BioMed Central Ltd., 34 - 42 Cleveland Street, W1T 4LB, London, United Kingdom, Tel: +44(207)3230323, Fax: +44(207)6319926 ISSUES/YR: 1

BMC Complementary and Alternative Medicine
ISSN: 1472-6882
BMC Complement. Altern. Med.
BioMed Central Ltd., 34 - 42 Cleveland Street, W1T 4LB, London, United Kingdom, Tel: +44(207)3230323, Fax: +44(207)6319926 ISSUES/YR: 1

BMC Dermatology
ISSN: 1471-5945
BMC Dermatol.
BioMed Central Ltd., 34 - 42 Cleveland Street, W1T 4LB, London, United Kingdom, Tel: +44(207)3230323, Fax: +44(207)6319926 ISSUES/YR: 1

BMC Developmental Biology
ISSN: 1471-213X
BMC Dev. Biol.
BioMed Central Ltd., 34 - 42 Cleveland Street, W1T 4LB, London, United Kingdom, Tel: +44(207)3230323, Fax: +44(207)6319926 ISSUES/YR: 1

BMC Ear, Nose and Throat Disorders
ISSN: 1472-6815
BMC Ear Nose Throat Disord.
BioMed Central Ltd., 34 - 42 Cleveland Street, W1T 4LB, London, United Kingdom, Tel: +44(207)3230323, Fax: +44(207)6319926 ISSUES/YR: 1

BMC Emergency Medicine
ISSN: 1471-227X
BMC Emerg. Med.
BioMed Central Ltd., 34 - 42 Cleveland Street, W1T 4LB, London, United Kingdom, Tel: +44(207)3230323, Fax: +44(207)6319926 ISSUES/YR: 1

BMC Endocrine Disorders
ISSN: 1472-6823
BMC Endocr. Disord.
BioMed Central Ltd., 34 - 42 Cleveland Street, W1T 4LB, London, United Kingdom, Tel: +44(207)3230323, Fax: +44(207)6319926 ISSUES/YR: 1

BMC Evolutionary Biology
ISSN: 1471-2148
BMC Evol. Biol.
BioMed Central Ltd., 34 - 42 Cleveland Street, W1T 4LB, London, United Kingdom, Tel: +44(207)3230323, Fax: +44(207)6319926 ISSUES/YR: 1

BMC Family Practice
ISSN: 1471-2296
BMC Fam. Pract.
BioMed Central Ltd., 34 - 42 Cleveland Street, W1T 4LB, London, United Kingdom, Tel: +44(207)3230323, Fax: +44(207)6319926 ISSUES/YR: 1

BMC Gastroenterology
ISSN: 1471-230X
BMC Gastroenterology
BioMed Central Ltd., 34 - 42 Cleveland Street, W1T 4LB, London, United Kingdom, Tel: +44(207)3230323, Fax: +44(207)6319926 ISSUES/YR: 1

BMC Genetics
ISSN: 1471-2156
BMC Genet.
BioMed Central Ltd., 34 - 42 Cleveland Street, W1T 4LB, London, United Kingdom, Tel: +44(207)3230323, Fax: +44(207)6319926 ISSUES/YR: 1

BMC Genomics
ISSN: 1471-2164
BMC Genomics
BioMed Central Ltd., 34 - 42 Cleveland Street, W1T 4LB, London, United Kingdom, Tel: +44(207)3230323, Fax: +44(207)6319926 ISSUES/YR: 1

BMC Geriatrics
ISSN: 1471-2318
BMC Geriatrics
BioMed Central Ltd., 34 - 42 Cleveland Street, W1T 4LB, London, United Kingdom, Tel: +44(207)3230323, Fax: +44(207)6319926 ISSUES/YR: 1

BMC Health Services Research
ISSN: 1472-6963
BMC Health Serv. Res.
BioMed Central Ltd., 34 - 42 Cleveland Street, W1T 4LB, London, United Kingdom, Tel: +44(207)3230323, Fax: +44(207)6319926 ISSUES/YR: 1

BMC Immunology
ISSN: 1471-2172
BMC Immunol.
BioMed Central Ltd., 34 - 42 Cleveland Street, W1T 4LB, London, United Kingdom, Tel: +44(207)3230323, Fax: +44(207)6319926 ISSUES/YR: 1

BMC Infectious Diseases
ISSN: 1471-2334
BMC Infect. Dis.
BioMed Central Ltd., 34 - 42 Cleveland Street, W1T 4LB, London, United Kingdom, Tel: +44(207)3230323, Fax: +44(207)6319926 ISSUES/YR: 1

BMC International Health and Human Rights
ISSN: 1472-698X
BMC Int. Health Hum. Rights
BioMed Central Ltd., 34 - 42 Cleveland Street, W1T 4LB, London, United Kingdom, Tel: +44(207)3230323, Fax: +44(207)6319926 ISSUES/YR: 1

BMC Medical Education
ISSN: 1472-6920
BMC Med. Educ.
BioMed Central Ltd., 34 - 42 Cleveland Street, W1T 4LB, London, United Kingdom, Tel: +44(207)3230323, Fax: +44(207)6319926 ISSUES/YR: 1

BMC Medical Ethics
ISSN: 1472-6939
BMC Med. Ethics
BioMed Central Ltd., 34 - 42 Cleveland Street, W1T 4LB, London, United Kingdom, Tel: +44(207)3230323, Fax: +44(207)6319926 ISSUES/YR: 1

BMC Medical Genetics
ISSN: 1471-2350
BMC Med. Genet.
BioMed Central Ltd., 34 - 42 Cleveland Street, W1T 4LB, London, United Kingdom, Tel: +44(207)3230323, Fax: +44(207)6319926 ISSUES/YR: 1

BMC Medical Imaging
ISSN: 1471-2342
BMC Med. Imaging
BioMed Central Ltd., 34 - 42 Cleveland Street, W1T 4LB, London, United Kingdom, Tel: +44(207)3230323, Fax: +44(207)6319926 ISSUES/YR: 1

BMC Medical Informatics and Decision Making
ISSN: 1472-6947
BMC Med. Informatics Decis. Mak.
BioMed Central Ltd., 34 - 42 Cleveland Street, W1T 4LB, London, United Kingdom, Tel: +44(207)3230323, Fax: +44(207)6319926 ISSUES/YR: 1

BMC Medical Research Methodology
ISSN: 1471-2288
BMC Med. Res. Methodol.
BioMed Central Ltd., 34 - 42 Cleveland Street, W1T 4LB, London, United Kingdom, Tel: +44(207)3230323, Fax: +44(207)6319926 ISSUES/YR: 1

BMC Medicine
ISSN: 1741-7015
BMC Med.
BioMed Central Ltd., 34 - 42 Cleveland Street, W1T 4LB, London, United Kingdom, Tel: +44(207)3230323, Fax: +44(207)6319926 ISSUES/YR: 1

BMC Microbiology
ISSN: 1471-2180
BMC Microbiology
BioMed Central Ltd., 34 - 42 Cleveland Street, W1T 4LB, London, United Kingdom, Tel: +44(207)3230323, Fax: +44(207)6319926 ISSUES/YR: 1

BMC Molecular Biology
ISSN: 1471-2199
BMC Mol. Biol.
BioMed Central Ltd., 34 - 42 Cleveland Street, W1T 4LB, London, United Kingdom, Tel: +44(207)3230323, Fax: +44(207)6319926 ISSUES/YR: 1

BMC Musculoskeletal Disorders
ISSN: 1471-2474
BMC Musuloskeletal Disord.
BioMed Central Ltd., 34 - 42 Cleveland Street, W1T 4LB, London, United Kingdom, Tel: +44(207)3230323, Fax: +44(207)6319926 ISSUES/YR: 1

BMC Nephrology
ISSN: 1471-2369
BMC Nephrology
BioMed Central Ltd., 34 - 42 Cleveland Street, W1T 4LB, London, United Kingdom, Tel: +44(207)3230323, Fax: +44(207)6319926 ISSUES/YR: 1

BMC Neurology
ISSN: 1471-2377
BMC Neurol.
BioMed Central Ltd., 34 - 42 Cleveland Street, W1T 4LB, London, United Kingdom, Tel: +44(207)3230323, Fax: +44(207)6319926 ISSUES/YR: 1

BMC Neuroscience
ISSN: 1471-2202
BMC Neurosci.
BioMed Central Ltd., 34 - 42 Cleveland Street, W1T 4LB, London, United Kingdom, Tel: +44(207)3230323, Fax: +44(207)6319926 ISSUES/YR: 1

BMC Nuclear Medicine
ISSN: 1471-2385
BMC Nucl. Med.
BioMed Central Ltd., 34 - 42 Cleveland Street, W1T 4LB, London, United Kingdom, Tel: +44(207)3230323, Fax: +44(207)6319926 ISSUES/YR: 1

BMC Nursing
ISSN: 1472-6955
BMC Nurs.
BioMed Central Ltd., 34 - 42 Cleveland Street, W1T 4LB, London, United Kingdom, Tel: +44(207)3230323, Fax: +44(207)6319926 ISSUES/YR: 1

BMC Ophthalmology
ISSN: 1471-2415
BMC Ophtalmol.
BioMed Central Ltd., 34 - 42 Cleveland Street, W1T 4LB, London, United Kingdom, Tel: +44(207)3230323, Fax: +44(207)6319926 ISSUES/YR: 1

BMC Oral Health
ISSN: 1472-6831
BMC Oral Health
BioMed Central Ltd., 34 - 42 Cleveland Street, W1T 4LB, London, United Kingdom, Tel: +44(207)3230323, Fax: +44(207)6319926 ISSUES/YR: 1

BMC Palliative Care
ISSN: 1472-684X
BMC Palliative Care
BioMed Central Ltd., 34 - 42 Cleveland Street, W1T 4LB, London, United Kingdom, Tel: +44(207)3230323, Fax: +44(207)6319926 ISSUES/YR: 1

BMC Pediatrics
ISSN: 1471-2431
BMC Pediatr.
BioMed Central Ltd., 34 - 42 Cleveland Street, W1T 4LB, London, United Kingdom, Tel: +44(207)3230323, Fax: +44(207)6319926 ISSUES/YR: 1

BMC Pharmacology
ISSN: 1471-2210
BMC Pharmacol.
BioMed Central Ltd., 34 - 42 Cleveland Street, W1T 4LB, London, United Kingdom, Tel: +44(207)3230323, Fax: +44(207)6319926 ISSUES/YR: 1

BMC Physiology
ISSN: 1472-6793
BMC Physiol.
BioMed Central Ltd., 34 - 42 Cleveland Street, W1T 4LB, London, United Kingdom, Tel: +44(207)3230323, Fax: +44(207)6319926 ISSUES/YR: 1

BMC Pregnancy and Childbirth
ISSN: 1471-2393
BMC Pregnancy Childbirth
BioMed Central Ltd., 34 - 42 Cleveland Street, W1T 4LB, London, United Kingdom, Tel: +44(207)3230323, Fax: +44(207)6319926 ISSUES/YR: 1

BMC Psychiatry
ISSN: 1471-244X
BMC Psychiatry
BioMed Central Ltd., 34 - 42 Cleveland Street, W1T 4LB, London, United Kingdom, Tel: +44(207)3230323, Fax: +44(207)6319926 ISSUES/YR: 1

BMC Public Health
ISSN: 1471-2458
BMC Public Health
BioMed Central Ltd., 34 - 42 Cleveland Street, W1T 4LB, London, United Kingdom, Tel: +44(207)3230323, Fax: +44(207)6319926 ISSUES/YR: 1

BMC Pulmonary Medicine
ISSN: 1471-2466
BMC Pulm. Med.
BioMed Central Ltd., 34 - 42 Cleveland Street, W1T 4LB, London, United Kingdom, Tel: +44(207)3230323, Fax: +44(207)6319926 ISSUES/YR: 1

BMC Structural Biology
ISSN: 1472-6807
BMC Struct. Biol.
BioMed Central Ltd., 34 - 42 Cleveland Street, W1T 4LB, London, United Kingdom, Tel: +44(207)3230323, Fax: +44(207)6319926 ISSUES/YR: 1

BMC Surgery
ISSN: 1471-2482
BMC Surg.
BioMed Central Ltd., 34 - 42 Cleveland Street, W1T 4LB, London, United Kingdom, Tel: +44(207)3230323, Fax: +44(207)6319926 ISSUES/YR: 1

BMC Urology
ISSN: 1471-2490
BMC Urol.
BioMed Central Ltd., 34 - 42 Cleveland Street, W1T 4LB, London, United Kingdom, Tel: +44(207)3230323, Fax: +44(207)6319926 ISSUES/YR: 1

BMC Women's Health
ISSN: 1472-6874
BMC Women's Health
BioMed Central Ltd., 34 - 42 Cleveland Street, W1T 4LB, London, United Kingdom, Tel: +44(207)3230323, Fax: +44(207)6319926 ISSUES/YR: 1

Bolest
ISSN: 1212-0634
Bolest
Tigis s.r.o., Na Habrove 17, 152 00, Prague 5, Czech Republic, Tel: +420-251 813 1924, Fax: +420-251 681 217 ISSUES/YR: 4

Bollettino - Lega Italiana contro l'Epilessia
ISSN: 0394-560X
Boll. Lega Ital. Epilessia
Lega Italiana contro l'Epilessia, Via Di Rudini 8, 20142, Milan, Italy, Tel: +39(11)7718150 (EDITORE), Fax: +39(11)7718183 (EDITORE) ISSUES/YR: 4

Bollettino Chimico Farmaceutico
ISSN: 0006-6648
Boll. Chim. Farm.
Lubner Communication Srl, Via Ausonio 12, 20132, Milan, Italy, Tel: +39(02)89404545, Fax: +39(02)89401168 ISSUES/YR: 6

Bone
ISSN: 8756-3282
Bone
Elsevier Inc., 360 Park Avenue South, New York, NY 10010, United States, Tel: +1(212)6333950, Fax: +1(212)6333990 ISSUES/YR: 12

• **Bone Marrow Transplantation**
ISSN: 0268-3369
Bone Marrow Transplant.
Nature Publishing Group, Houndmills, RG21 6XS, Basingstoke, Hampshire, United Kingdom, Tel: +44(1256)329242, Fax: +44(1256)810526 ISSUES/YR: 24

• **Brachytherapy**
ISSN: 1538-4721
Brachytherapy
Elsevier Inc., 360 Park Avenue South, New York, NY 10010, United States, Tel: +1(212)6333950, Fax: +1(212)6333990 ISSUES/YR: 4

• **Brain**
ISSN: 0006-8950
Brain
Oxford University Press, Great Clarendon Street, OX2 6DP, Oxford, United Kingdom, Tel: +44(1865)353907, Fax: +44(1865)353486 ISSUES/YR: 12

• **Brain and Cognition**
ISSN: 0278-2626
Brain Cogn.
Academic Press Inc., 6277 Sea Harbor Drive, Orlando, FL 32887-4900, United States, Tel: +1(407)3454040, Fax: +1(407)3639661 ISSUES/YR: 9

Brain and Development
ISSN: 0387-7604
Brain Dev.
Elsevier, P.O. Box 211, 1000 AE, Amsterdam, Netherlands, Tel: +31(20)4853757, Fax: +31(20)4853432 ISSUES/YR: 8

Brain and Language
ISSN: 0093-934X
Brain Lang.
Academic Press Inc., 6277 Sea Harbor Drive, Orlando, FL 32887-4900, United States, Tel: +1(407)3454040, Fax: +1(407)3639661 ISSUES/YR: 15

Brain and Nerve
ISSN: 0006-8969
Brain Nerve
Igaku-Shoin Ltd, 5-24-3 Hongo, Bunkyo-ku, 113-91, Tokyo, Japan, Tel: +81(3)38175600 ISSUES/YR: 12

Brain Injury
ISSN: 0269-9052
Brain Inj.
Taylor and Francis Ltd., 4 Park Square, Milton Park, OX14 4RN, Abingdon, Oxfordshire, United Kingdom, Tel: +44(20)70175541, Fax: +44(20)70175198 ISSUES/YR: 14

Brain Pathology
ISSN: 1015-6305
Brain Pathol.
International Society of Neuropathology, 200 Lothrop Street, Pittsburgh, PA 15213, United States, Tel: +1(412)6476617, Fax: +1(412)6475468 ISSUES/YR: 4

• **Brain Research**
ISSN: 0006-8993
Brain Res.
Elsevier, P.O. Box 211, 1000 AE, Amsterdam, Netherlands, Tel: +31(20)4853757, Fax: +31(20)4853432 ISSUES/YR: 60

• **Brain Research Bulletin**
ISSN: 0361-9230
Brain Res. Bull.
Elsevier Inc., 360 Park Avenue South, New York, NY 10010, United States, Tel: +1(212)6333950, Fax: +1(212)6333990 ISSUES/YR: 18

• **Brain Research Reviews**
ISSN: 0165-0173
Brain Res. Rev.
Elsevier, P.O. Box 211, 1000 AE, Amsterdam, Netherlands, Tel: +31(20)4853757, Fax: +31(20)4853432 ISSUES/YR: 9

• **Brain Topography**
ISSN: 0896-0267
Brain Topogr.
Kluwer Academic/Human Sciences Press Inc., 233 Spring Street, New York, NY 10013-1578, United States, Tel: +1(212)6208000, Fax: +1(212)4630742 ISSUES/YR: 4

Brain Tumor Pathology
ISSN: 1433-7398
Brain Tumor Pathol.
Springer-Verlag Tokyo, 3-13, Hongo 3-chome, Bunkyo-ku, 113-8480, Tokyo, Japan, Tel: +81(3)38180861, Fax: +81(3)38180864 ISSUES/YR: 2

• **Brain, Behavior and Evolution**
ISSN: 0006-8977
Brain Behav. Evol.
S. Karger AG, Allschwilerstrasse 10, P.O. Box, CH-4009, Basel, Switzerland, Tel: +41(61)3061111, Fax: +41(61)3061234 ISSUES/YR: 8

● **Brain, Behavior, and Immunity**
ISSN: 0889-1591
Brain Behav. Immun.
Academic Press Inc., 6277 Sea Harbor Drive, Orlando, FL 32887-4900, United States, Tel: +1(407)3454040, Fax: +1(407)3639661 ISSUES/YR: 6

Brazilian Journal of Epilepsy and Clinical Neurophysiology
ISSN: 0104-9275
Braz. J. Epilepsy Clin. Neurophysiol.
Liga Brasileira de Epilepsia, Av. Ipiranga 6690, Sala 322, 90610-000, Porto Alegre/RS, Brazil, Tel: +55(51)3394936, Fax: +55(51)3394936 ISSUES/YR: 4

Brazilian Journal of Infectious Diseases
ISSN: 1413-8670
Braz. J. Infect. Dis.
Contexto, Rua Alfredo Magalhaes 4, 40140-140, Barra, Salvador-Bahia, Brazil, Tel: +55(71)2643602, Fax: +55(71)2643602 ISSUES/YR: 6

Brazilian Journal of Medical and Biological Research
ISSN: 0100-879X
Braz. J. Med. Biol. Res.
Associacao Brasileira de Divulgacao Cientifica, Av. Bandeirantes 3.900, 14049-900 SP, Ribeirao Preto, Brazil, Tel: +55(16)6333825 ISSUES/YR: 12

● **Breast**
ISSN: 0960-9776
Breast
Churchill Livingstone, 1-3 Baxter's Place, Leith Walk, EH1 3AF, Edinburgh, United Kingdom, Tel: +44(131)5562424, Fax: +44(131)5581278 ISSUES/YR: 6

● **Breast Cancer Online**
ISSN: 1470-9031
Breast Cancer Online
Cambridge University Press, Shaftesbury Road, CB2 2RU, Cambridge, United Kingdom, Tel: +44(1223)312393, Fax: +44(1223)315052 ISSUES/YR: 12

Breast Cancer Research
ISSN: 1465-542X
Breast Cancer Res.
BioMed Central Ltd., 34 - 42 Cleveland Street, W1T 4LB, London, United Kingdom, Tel: +44(207)3230323, Fax: +44(207)6319926 ISSUES/YR: 6

● **Breast Cancer Research and Treatment**
ISSN: 0167-6806
Breast Cancer Res. Treat.
Kluwer Academic Publishers, 101 Philip Drive, Assinippi Park, Norwell, MA 02061, United States, Tel: +1(781)8716600, Fax: +1(781)8716528 ISSUES/YR: 18

● **Breast Care**
ISSN: 1661-3791
Breast Care
S. Karger AG, Allschwilerstrasse 10, P.O. Box, CH-4009, Basel, Switzerland, Tel: +41(61)3061111, Fax: +41(61)3061234 ISSUES/YR: 6

● **Breast Disease**
ISSN: 0888-6008
Breast Dis.
IOS Press, Nieuwe Hemweg 6B, 1013 BG, Amsterdam, Netherlands, Tel: +31(20)6883355, Fax: +31(20)6203419 ISSUES/YR: 6

Breast Journal
ISSN: 1075-122X
Breast J.
Blackwell Publishing Inc., 350 Main Street, Malden, MA 02148, United States, Tel: +1(781)3888200, Fax: +1(781)3888210 ISSUES/YR: 6

Bridging Eastern and Western Psychiatry
ISSN: not available yet
Bridging East. West. Psychiatry
Leopolis Agency, 79000 Efrmov str.4/5, Lviv, Ukraine, Tel: +380(322)527151, Fax: +380(322)527151 ISSUES/YR: 2

Briefings in Bioinformatics
ISSN: 1467-5463
Brief. Bioinform.
Oxford University Press, Great Clarendon Street, OX2 6DP, Oxford, United Kingdom, Tel: +44(1865)353907, Fax: +44(1865)353486 ISSUES/YR: 4

Briefings in Functional Genomics and Proteomics
ISSN: 1473-9550
Brief. Funct. Genomics Proteomics
Henry Stewart Publications, 28-30 Little Russel Street, WC1A 2HN, London, United Kingdom, Tel: +44(207)4043040, Fax: +44(207)4042081 ISSUES/YR: 4

● **British Journal of Anaesthesia**
ISSN: 0007-0912
Br. J. Anaesth.
Oxford University Press, Great Clarendon Street, OX2 6DP, Oxford, United Kingdom, Tel: +44(1865)353907, Fax: +44(1865)353486 ISSUES/YR: 12

● **British Journal of Biomedical Science**
ISSN: 0967-4845
Br. J. Biomed. Sci.
Step Publishing Ltd., Tunbridge Wells, TN2 3DR, Kent, United Kingdom, Tel: +44(1892)518 877, Fax: +44(1892)616 177 ISSUES/YR: 4

● **British Journal of Cancer**
ISSN: 0007-0920
Br. J. Cancer
Nature Publishing Group, Houndmills, RG21 6XS, Basingstoke, Hampshire, United Kingdom, Tel: +44(1256)329242, Fax: +44(1256)810526 ISSUES/YR: 24

British Journal of Cardiology
ISSN: 0969-6113
Br. J. Cardiol.
MediNews (Cardiology) Ltd, 3 Duchess Place, B16 8NH, Edgbaston, Birmingham, United Kingdom, Tel: +44(121)4561466, Fax: +44(121)4561467 ISSUES/YR: 6

● **British Journal of Clinical Pharmacology**
ISSN: 0306-5251
Br. J. Clin. Pharmacol.
Blackwell Publishing Ltd, 9600 Garsington Road, OX4 2XG, Oxford, United Kingdom, Tel: +44(1865)776868, Fax: +44(1865)714591 ISSUES/YR: 12

● **British Journal of Clinical Pharmacology, Supplement**
ISSN: 0264-3774
Br. J. Clin. Pharmacol. Suppl.
Blackwell Publishing Ltd, 9600 Garsington Road, OX4 2XG, Oxford, United Kingdom, Tel: +44(1865)776868, Fax: +44(1865)714591 ISSUES/YR: 1

● **British Journal of Clinical Psychology**
ISSN: 0144-6657
Br. J. Clin. Psychol.
British Psychological Society, 48 Princess Road East, LE1 7DR, Leicester, United Kingdom, Tel: +44(116)2549568, Fax: +44(116)2470787 ISSUES/YR: 4

● **British Journal of Dermatology**
ISSN: 0007-0963
Br. J. Dermatol.
Blackwell Publishing Ltd, 9600 Garsington Road, OX4 2XG, Oxford, United Kingdom, Tel: +44(1865)776868, Fax: +44(1865)714591 ISSUES/YR: 12

British Journal of Developmental Disabilities
ISSN: 0969-7950
Br. J. Dev. Disabil.
SEFA (Publications) Ltd, 4 Great William Street, CV37 6RY, Stratford-upon-Avon, United Kingdom ISSUES/YR: 2

British Journal of Diabetes and Vascular Disease
ISSN: 1474-6514
Br. J. Diabetes Vasc. Dis.
MediNews (Diabetes) Ltd, 3 Duchess Place, B16 8NH, Edgbaston, Birmingham, United Kingdom, Tel: +44(121)4544114, Fax: +44(121)4541190 ISSUES/YR: 6

British Journal of General Practice
ISSN: 0960-1643
Br. J. Gen. Pract.
Royal College of General Practitioners, 14 Princes Gate, SW7 1PU, London, United Kingdom, Tel: +44(207)5813232, Fax: +44(20)72253047 ISSUES/YR: 12

● **British Journal of Haematology**
ISSN: 0007-1048
Br. J. Haematol.
Blackwell Publishing Ltd, 9600 Garsington Road, OX4 2XG, Oxford, United Kingdom, Tel: +44(1865)776868, Fax: +44(1865)714591 ISSUES/YR: 23

- **British Journal of Health Psychology**
  ISSN: 1359-107X
  Br. J. Health Psychol.
  British Psychological Society, 48 Princess Road East, LE1 7DR, Leicester, United Kingdom, Tel: +44(116)2549568, Fax: +44(116)2470787 ISSUES/YR: 4

British Journal of Hospital Medicine
ISSN: 0007-1064
Br. J. Hosp. Med.
MA Healthcare Ltd, Dulwich Road, SE24 0PB, London, United Kingdom, Tel: +44(20)7738-5454, Fax: +44(20)7733-2325 ISSUES/YR: 12

British Journal of Intensive Care
ISSN: 0961-7930
Br. J. Intensive Care
Greycoat Publishing, 120 Dawes Road, SW6 7EG, London, United Kingdom, Tel: +44(207)3851515, Fax: +44(207)3857799 ISSUES/YR: 4

- **British Journal of Neurosurgery**
  ISSN: 0268-8697
  Br. J. Neurosurg.
  Taylor and Francis Ltd., 4 Park Square, Milton Park, OX14 4RN, Abingdon, Oxfordshire, United Kingdom, Tel: +44(20)70175541, Fax: +44(20)70175198 ISSUES/YR: 6

British Journal of Nutrition
ISSN: 0007-1145
Br. J. Nutr.
CAB International, Wallingford, OX10 8DE, Oxon, United Kingdom, Tel: +44(1491)832111, Fax: +44(1491)826090 ISSUES/YR: 12

British Journal of Occupational Therapy
ISSN: 0308-0226
Br. J. Occup. Ther.
British Journal of Occupational Therapy, 106-114 Borough High Street, SE1 1LB, London, United Kingdom, Tel: +44(207)4502338, Fax: +44(207)4502350 ISSUES/YR: 12

- **British Journal of Ophthalmology**
  ISSN: 0007-1161
  Br. J. Ophthalmol.
  BMJ Publishing Group, Tavistock Square, WC1H 9JR, London, United Kingdom, Tel: +44(207)3874499, Fax: +44(207)3836668 ISSUES/YR: 12

British Journal of Oral and Maxillofacial Surgery
ISSN: 0266-4356
Br. J. Oral Maxillofac. Surg.
Churchill Livingstone, 1-3 Baxter's Place, Leith Walk, EH1 3AF, Edinburgh, United Kingdom, Tel: +44(131)5562424, Fax: +44(131)5581278 ISSUES/YR: 6

- **British Journal of Pharmacology**
  ISSN: 0007-1188
  Br. J. Pharmacol.
  Nature Publishing Group, Houndmills, RG21 6XS, Basingstoke, Hampshire, United Kingdom, Tel: +44(1256)329242, Fax: +44(1256)810526 ISSUES/YR: 24

British Journal of Psychiatry
ISSN: 0007-1250
Br. J. Psychiatry
Royal College of Psychiatrists, 17 Belgrave Square, SW1X 8PG, London, United Kingdom, Tel: +44(20)72352351, Fax: +44(20)72451231 ISSUES/YR: 12

British Journal of Psychotherapy
ISSN: 0265-9883
Br. J. Psychother.
Artesian Books Ltd, P. O. Box 310, TW11 0LZ, Teddington, United Kingdom, Tel: +44-208 287 1128 ISSUES/YR: 4

British Journal of Radiology
ISSN: 0007-1285
Br. J. Radiol.
British Institute of Radiology, 36 Portland Place, W1N 1AT, London, United Kingdom, Tel: +44(207)5804189 ISSUES/YR: 12

British Journal of Sports Medicine
ISSN: 0306-3674
Br. J. Sports Med.
BMJ Publishing Group, Tavistock Square, WC1H 9JR, London, United Kingdom, Tel: +44(207)3874499, Fax: +44(207)3836668 ISSUES/YR: 12

- **British Journal of Surgery**
  ISSN: 0007-1323
  Br. J. Surg.
  John Wiley and Sons Ltd, Southern Gate, PO19 8SQ, Chichester, West Sussex, United Kingdom, Tel: +44(1243)779777, Fax: +44(1243)775878 ISSUES/YR: 12
- **British Medical Bulletin**
  ISSN: 0007-1420
  Br. Med. Bull.
  Oxford University Press, Great Clarendon Street, OX2 6DP, Oxford, United Kingdom, Tel: +44(1865)353907, Fax: +44(1865)353486 ISSUES/YR: 4
- **British Medical Journal**
  ISSN: 0959-8146
  Br. Med. J.
  BMJ Publishing Group, Tavistock Square, WC1H 9JR, London, United Kingdom, Tel: +44(207)3874499, Fax: +44(207)3836668 ISSUES/YR: 51

Bulgarian Medicine
ISSN: 0861-9883
Bulg. Med.
Academy of Medicine, 1 G. Sofiiski Boul., 1431, Sofia, Bulgaria, Tel: +359(2)523171, Fax: +359(2)522393 ISSUES/YR: 6

Bull RGKMC
ISSN: 0971-8001
Bull RGKMC
R.G. Kar Medical College, P.P. Box 12328, 7000 004, Calcutta, India, Tel: +91(33)5557656, Fax: +91(33)5557669 ISSUES/YR: 4

Bulletin du Cancer
ISSN: 0007-4551
Bull. Cancer
John Libbey Eurotext, 127, avenue de la Republique, 92120, Montrouge, France, Tel: +33(01)46730660, Fax: +33(01)40840999 ISSUES/YR: 12

Bulletin of Environmental Contamination and Toxicology
ISSN: 0007-4861
Bull. Environ. Contam. Toxicol.
Springer New York LLC, 233 Springer Street, New York, 10013-1578, United States, Tel: +1-212 460 1501, Fax: +1-212 460 1595 ISSUES/YR: 12

Bulletin of Experimental Biology and Medicine
ISSN: 0007-4888
Bull. Exp. Biol. Med.
Springer Science and Business Media Deutschland GmbH, Heidelberger Platz 3, 14197, Berlin, Germany, Tel: +1-30 827875282, Fax: +1-30 827875707 ISSUES/YR: 12

Bulletin of Hunan Medical University
ISSN: 1000-5625
Bull. Hunan Med. Univ.
Hunan Medical University, 410078, Changsha, Hunan, China, Tel: +86(010)88018116, Fax: +86(010)88018128 ISSUES/YR: 6

Bulletin of National Institute of Health Sciences
ISSN: 1343-4292
Bull. Natl. Inst. Health Sci.
National Institute of Health Sciences, 18-1 Kamiyoga, 1-chome, Setagaya-ku, 158, Tokyo, Japan ISSUES/YR: 1

Bulletin of the Chemical Society of Japan
ISSN: 0009-2673
Bull. Chem. Soc. Jpn.
Chemical Society of Japan, 1-5 Kanda-Surugadai, Chiyoda-ku, 101-0062, Tokyo, Japan, Tel: +81(3)32926169, Fax: +81(3)32926317 ISSUES/YR: 12

Bulletin of the Menninger Clinic
ISSN: 0025-9284
Bull. Menninger Clin.
Guilford Publications, 72 Spring Street, New York, NY 10012, United States, Tel: +1(212)4319800 ISSUES/YR: 4

Bulletin of the Physical Fitness Research Institute
ISSN: 0389-9071
Bull. Phys. Fitness Res. Inst.
Meiji Life Foundation of Health and Welfare, 150 Tobukicho, Hachioji City, 192, Tokyo, Japan, Tel: +81(426)911163 ISSUES/YR: 1

Bulletin of the World Health Organization
ISSN: 0042-9686
Bull. WHO
World Health Organization, 20 Ave. Appia, CH-1211, Geneva 27, Switzerland, Tel: +41(22)7912111, Fax: +41(22)7914857 ISSUES/YR: 12

Bulletin on the Rheumatic Diseases
ISSN: 0007-5248
Bull. Rheum. Dis.
Arthritis Foundation, 1330 West Peachtree Street, Atlanta, GA 30309, United States, Tel: +1(404)8727100 ISSUES/YR: 12

Bulletin, Postgraduate Institute of Medical Education and Research, Chandigarh
ISSN: 0302-2404
Bull. Postgrad. Inst. Med. Educ. Res. Chandigarh
Postgraduate Institute of Medical Education and Research, 160012, Chandigarh, India, Tel: +91(172)540403 ISSUES/YR: 4

Bulletin: Hospital for Joint Diseases
ISSN: 0018-5647
Bull. Hosp. Jt. Dis.
J. Michael Ryan Publishing Inc., 24 Crescent Drive North, Andover, NJ 07821, United States, Tel: +1(973)7867777, Fax: +1(973)7867776 ISSUES/YR: 4

Burns
ISSN: 0305-4179
Burns
Elsevier Ltd, Langford Lane, Kidlington, OX5 1GB, Oxford, United Kingdom, Tel: +44(1865)843000 ISSUES/YR: 8

• **Ca-A Cancer Journal for Clinicians**
ISSN: 0007-9235
Ca Cancer J. Clin.
Lippincott Williams and Wilkins, 530 Walnut Street, Philadelphia, PA 19106-3621, United States, Tel: +1(215)5218342, Fax: +1(215)5218326 ISSUES/YR: 6

Cahiers de Nutrition et de Dietetique
ISSN: 0007-9960
Cah. Nutr. Diet.
Masson, 21 Rue Camille Desmoulins, 92789, Issy les Moulineaux Cedex 9, France, Tel: +33(1)73281634, Fax: +33(1)73281649 ISSUES/YR: 6

Cahiers Sante
ISSN: 1157-5999
Cah. Sante
John Libbey Eurotext, 127, avenue de la Republique, 92120, Montrouge, France, Tel: +33(01)46730660, Fax: +33(01)40840999 ISSUES/YR: 4

• **Calcified Tissue International**
ISSN: 0171-967X
Calcif. Tissue Int.
Springer New York LLC, 233 Springer Street, New York, 10013-1578, United States, Tel: +1-212 460 1501, Fax: +1-212 460 1595 ISSUES/YR: 12

Canadian Association of Radiologists Journal
ISSN: 0846-5371
Can. Assoc. Radiol. J.
Canadian Medical Association, 1740 boul.Cote-Vertu Blvd, H4L 2A4, St-Laurent, Canada, Tel: +1-514 738 3111, Fax: +1-514 738 5199 ISSUES/YR: 5

Canadian Family Physician
ISSN: 0008-350X
Can. Fam. Phys.
College of Family Physicians of Canada, 2630 Skymark Avenue, ONT L4W 5A4, Mississauga Ont., Canada, Tel: +1(905)6290900, Fax: +1(905)6290893 ISSUES/YR: 12

• **Canadian Journal of Anesthesia**
ISSN: 0832-610X
Can. J. Anesth.
Canadian Anaesthetists' Society, 1 Eglinton Avenue East, Suite 208, ONT M4P 3A1, Toronto, Canada, Tel: +1(416)4800602 ISSUES/YR: 10

Canadian Journal of Cardiology
ISSN: 0828-282X
Can. J. Cardiol.
Pulsus Group Inc., 2902 South Sheridan Way, ONT L6J 7L6, Oakville, Canada, Tel: +1(905)8294770, Fax: +1(905)8294799 ISSUES/YR: 14

Canadian Journal of Chemistry
ISSN: 0008-4042
Can. J. Chem.
National Research Council of Canada, ONT K1A 0R6, Ottawa, Canada, Tel: +1(613)9939084, Fax: +1(613)9527656 ISSUES/YR: 12

Canadian Journal of Clinical Pharmacology
ISSN: 1198-581X
Can. J. Clin. Pharmacol.
Canadian Society of Clinical Pharmacology, St. Joseph's Healthcare Hamilton, L8N 1G6, 105 Main Street E. Level 1. Hamilton, Ontario, Canada, Tel: +1-905 522 1155 EXT: 4276, Fax: +1-905 528 7386 ISSUES/YR: 4

Canadian Journal of Community Mental Health
ISSN: 0713-3936
Can. J. Community Ment. Health
Wilfrid Laurier University Press, 75 University Avenue West, N2L 3C5, Waterloo, Ontario, Canada, Tel: +1(519) 884-0710, Fax: +1(519) 725-1399 ISSUES/YR: 2

Canadian Journal of Diabetes
ISSN: 1499-2671
Can. J. Diabetes
Canadian Diabetes Association, 15 Toronto Street, Suite 800, ONT M5C 2E3, Toronto, Canada, Tel: +1(416)3633373, Fax: +1(416)3637465 ISSUES/YR: 4

Canadian Journal of Emergency Medicine
ISSN: 1481-8035
Can. J. Emerg. Med.
Canadian Medical Association, 1740 boul.Cote-Vertu Blvd, H4L 2A4, St-Laurent, Canada, Tel: +1-514 738 3111, Fax: +1-514 738 5199 ISSUES/YR: 6

• **Canadian Journal of Gastroenterology**
ISSN: 0835-7900
Can. J. Gastroenterol.
Pulsus Group Inc., 2902 South Sheridan Way, ONT L6J 7L6, Oakville, Canada, Tel: +1(905)8294770, Fax: +1(905)8294799 ISSUES/YR: 12

Canadian Journal of Hospital Pharmacy
ISSN: 0008-4123
Can. J. Hosp. Pharm.
Canadian Society of Hospital Pharmacists, 1145 Hunt Club Road, Suite 350, ONT K1V 0Y3, Ottawa, Canada, Tel: +1(613)7369733 ISSUES/YR: 4

Canadian Journal of Human Sexuality
ISSN: 1188-4517
Can. J. Hum. Sex.
SIECCAN, 850 Coxwell Avenue, ONT M4C 5R1, Toronto, Canada, Tel: +1(416)4665304 ISSUES/YR: 4

Canadian Journal of Infectious Diseases and Medical Microbiology
ISSN: 1712-9532
Can. J. Infect. Dis. Med. Microbiol.
Pulsus Group Inc., 2902 South Sheridan Way, ONT L6J 7L6, Oakville, Canada, Tel: +1(905)8294770, Fax: +1(905)8294799 ISSUES/YR: 6

• **Canadian Journal of Microbiology**
ISSN: 0008-4166
Can. J. Microbiol.
National Research Council of Canada, ONT K1A 0R6, Ottawa, Canada, Tel: +1(613)9939084, Fax: +1(613)9527656 ISSUES/YR: 12

• **Canadian Journal of Neurological Sciences**
ISSN: 0317-1671
Can. J. Neurol. Sci.
Canadian Journal of Neurological Sciences, P.O. Box 5456, Station A, AB T2H 1X8, Calgary, Canada, Tel: +1(403)2299575, Fax: +1(403)2291661 ISSUES/YR: 4

Canadian Journal of Occupational Therapy
ISSN: 0008-4174
Can. J. Occup. Ther.
Canadian Association of Occupational Therapists, 1125 Colonel By Dr, Suite 3400, ONT K1S 5R1, Ottowa, Canada, Tel: +1(613)5232268, Fax: +1(613)5232552 ISSUES/YR: 5

Canadian Journal of Ophthalmology
ISSN: 0008-4182
Can. J. Ophthalmol.
Canadian Ophthalmological Society, 1525 Carling Avenue, Suite 610, ONT K1Z 8R9, Ottawa, Canada, Tel: +1(613)7296779, Fax: +1(613)7297209 ISSUES/YR: 6

• **Canadian Journal of Physiology and Pharmacology**
ISSN: 0008-4212
Can. J. Physiol. Pharmacol.
National Research Council of Canada, ONT K1A 0R6, Ottawa, Canada, Tel: +1(613)9939084, Fax: +1(613)9527656 ISSUES/YR: 12

Canadian Journal of Psychiatry
ISSN: 0706-7437
Can. J. Psychiatry
Canadian Psychiatric Association, 141 Laurier Avenue West, Suite 702, K1P 5J3, Ottawa, ON, Canada, Tel: +1(613)2342815, Fax: +1(613)234-9857 ISSUES/YR: 10

Canadian Journal of Public Health
ISSN: 0008-4263
Can. J. Public Health
Canadian Public Health Association, 1565 Carling Avenue, Suite 400, ONT K1Z 8R1, Ottawa, Canada, Tel: +1(613)7253769 ISSUES/YR: 6

Canadian Journal of Respiratory Therapy
ISSN: 1205-9838
Can. J. Resp. Ther.
Canadian Society of Respiratory Therapists, 1785 Alta Vista Dr., Suite 102, ON K1G 3Y6, Ottawa, Canada, Tel: +1(613)7313164, Fax: +1(613)5214314 ISSUES/YR: 4

Canadian Journal of Surgery
ISSN: 0008-428X
Can. J. Surg.
Canadian Medical Association, 1740 boul.Cote-Vertu Blvd, H4L 2A4, St-Laurent, Canada, Tel: +1-514 738 3111, Fax: +1-514 738 5199 ISSUES/YR: 6

Canadian Journal of Veterinary Research
ISSN: 0830-9000
Can. J. Vet. Res.
Canadian Veterinary Medical Association, 339 Booth Street, ONT K1R 7K1, Ottawa, Canada, Tel: +1(613)2361162, Fax: +1(613)2369681 ISSUES/YR: 4

Canadian Journal on Aging
ISSN: 0714-9808
Can. J. Aging
Canadian Association on Gerontology, 5201 Dufferin Street, ON M3H 5T8, North York, Canada, Tel: +1-416 667 7810, Fax: +1-416 667 7810 ISSUES/YR: 4

Canadian Medical Association Journal
ISSN: 0820-3946
Can. Med. Assoc. J.
Canadian Medical Association, 1740 boul.Cote-Vertu Blvd, H4L 2A4, St-Laurent, Canada, Tel: +1-514 738 3111, Fax: +1-514 738 5199 ISSUES/YR: 24

Canadian Pharmacists Journal
ISSN: 1715-1635
Can. Pharm. J.
Canadian Pharmacists Association, 1785 Alta Vista Drive, ON K1G 3Y6, Ottawa, Canada, Tel: +1(613)5237877 EXT 224, Fax: +1(613)7397765 ISSUES/YR: 6

• **Canadian Respiratory Journal**
ISSN: 1198-2241
Can. Respir. J.
Pulsus Group Inc., 2902 South Sheridan Way, ONT L6J 7L6, Oakville, Canada, Tel: +1(905)8294770, Fax: +1(905)8294799 ISSUES/YR: 8

Canadian Veterinary Journal
ISSN: 0008-5286
Can. Vet. J.
Canadian Veterinary Medical Association, 339 Booth Street, ONT K1R 7K1, Ottawa, Canada, Tel: +1(613)2361162, Fax: +1(613)2369681 ISSUES/YR: 12

• **Cancer**
ISSN: 0008-543X
Cancer
John Wiley and Sons Inc., P.O.Box 18667, Newark, NJ 07191-8667, United States, Tel: +1(212)8506000, Fax: +1(212)8506012 ISSUES/YR: 24

• **Cancer and Metastasis Reviews**
ISSN: 0167-7659
Cancer Metastasis Rev.
Kluwer Academic Publishers, Van Godewijckstraat 30, 3311 GZ, Dordrecht, Netherlands, Tel: +31(78)6576000, Fax: +31(78)6576254 ISSUES/YR: 4

Cancer Biology and Therapy
ISSN: 1538-4047
Cancer Biol. Ther.
Landes Bioscience, 810 S Church Street, Georgetown, 78626, United States, Tel: +1-512.863.7762, Fax: +1-512.863.0081 ISSUES/YR: 6

• **Cancer Biotherapy and Radiopharmaceuticals**
ISSN: 1084-9785
Cancer Biother. Radiopharm.
Mary Ann Liebert Inc., 140 Huguenot Street, New Rochelle, NY 10801-5215, United States, Tel: +1(914)7402100, Fax: +1(914)7402101 ISSUES/YR: 6

• **Cancer Causes and Control**
ISSN: 0957-5243
Cancer Causes Control
Kluwer Academic Publishers, Van Godewijckstraat 30, 3311 GZ, Dordrecht, Netherlands, Tel: +31(78)6576000, Fax: +31(78)6576254 ISSUES/YR: 10

• **Cancer Cell**
ISSN: 1535-6108
Cancer Cell
Cell Press, 1100 Massachusetts Avenue, Cambridge, MA 02138-9957, United States, Tel: +1(617)6617057, Fax: +1(617)6617061 ISSUES/YR: 12

Cancer Cell International
ISSN: 1475-2867
Cancer Cell Int.
BioMed Central Ltd., 34 - 42 Cleveland Street, W1T 4LB, London, United Kingdom, Tel: +44(207)3230323, Fax: +44(207)6319926 ISSUES/YR: 1

• **Cancer Chemotherapy and Pharmacology**
ISSN: 0344-5704
Cancer Chemother. Pharmacol.
Springer Verlag, Tiergartenstrasse 17, D-69121, Heidelberg, Germany, Tel: +49(6221)4870, Fax: +49(6221)487366 ISSUES/YR: 12

• **Cancer Chemotherapy and Pharmacology, Supplement**
ISSN: 0943-9404
Cancer Chemother. Pharmacol. Suppl.
Springer Verlag, Tiergartenstrasse 17, D-69121, Heidelberg, Germany, Tel: +49(6221)4870, Fax: +49(6221)487366 ISSUES/YR: 1

Cancer Control
ISSN: 1073-2748
Cancer Control
H. Lee Moffitt Cancer Center and Research Institute, 12902 Magnolia Drive, Tampa, FL 33612-9497, United States, Tel: +1(813)6321349 ISSUES/YR: 6

• **Cancer Detection and Prevention**
ISSN: 0361-090X
Cancer Detect. Prev.
Elsevier Ltd, Langford Lane, Kidlington, OX5 1GB, Oxford, United Kingdom, Tel: +44(1865)843000, Fax: +44(1865)843010 ISSUES/YR: 6

• **Cancer Epidemiology Biomarkers and Prevention**
ISSN: 1055-9965
Cancer Epidemiol. Biomarkers Prev.
American Association for Cancer Research Inc., 150 South Independence Mall West, Philadelphia, PA 19106-3483, United States, Tel: +1(215)4409300, Fax: +1(215)4409354 ISSUES/YR: 12

Cancer Forum
ISSN: 0311-306X
Cancer Forum
The Cancer Council Australia, Camperdown NSW, NSW 2001, GPO Box 4708, Sydney, Australia, Tel: +61-2 9036 3100, Fax: +61-2 9036 3101 ISSUES/YR: 3

- **Cancer Gene Therapy**
  ISSN: 0929-1903
  Cancer Gene Ther.
  Nature Publishing Group, 345 Park Avenue South, New York, NY 10010-1707, United States, Tel: +1(212)7269200, Fax: +1(212)6960052 ISSUES/YR: 12
- **Cancer Genetics and Cytogenetics**
  ISSN: 0165-4608
  Cancer Genet. Cytogenet.
  Elsevier Inc., 360 Park Avenue South, New York, NY 10010, United States, Tel: +1(212)6333950, Fax: +1(212)6333990 ISSUES/YR: 16

Cancer Genomics and Proteomics
ISSN: 1109-6535
Cancer Genomics Proteomics
International Institute of Anticancer Research, 1st km Kapandritiou - Kalamou Rd., P.O. Box 22, GR-19014, Kapandriti, Attiki, Greece, Tel: +30(22950)53389, Fax: +30(22950)53389 ISSUES/YR: 6

Cancer Immunity
ISSN: 1424-9634
Cancer Immun.
Academy of Cancer Immunology, Chemin des Boveresses 155, CH-1066, Epalinges, Switzerland ISSUES/YR: 1

- **Cancer Immunology, Immunotherapy**
  ISSN: 0340-7004
  Cancer Immunol. Immunother.
  Springer Verlag, Tiergartenstrasse 17, D-69121, Heidelberg, Germany, Tel: +49(6221)4870, Fax: +49(6221)487366 ISSUES/YR: 12

Cancer Informatics
ISSN: 1176-9351
Cancer Informatics
Libertas Academicas Ltd., PO Box 302-624, 1330, North Harbor, Auckland, New Zealand, Tel: +64(21)740006 ISSUES/YR: 1

- **Cancer Investigation**
  ISSN: 0735-7907
  Cancer Invest.
  Taylor and Francis Inc., 325 Chestnut St, Suite 800, Philadelphia PA, PA 19106, United States, Tel: +1(800)3541420, Fax: +1(215)6258914 ISSUES/YR: 6
- **Cancer Journal**
  ISSN: 1528-9117
  Cancer J.
  Jones and Bartlett Publishers, 40 Tall Pine Drive, Sudbury, MA 01776, United States, Tel: +1(978)4435000, Fax: +1(978)4438000 ISSUES/YR: 6
- **Cancer Letters**
  ISSN: 0304-3835
  Cancer Lett.
  Elsevier Ireland Ltd, P.O. Box 85, Limerick, Ireland, Tel: +353(61)471944, Fax: +353(61)472144 ISSUES/YR: 26

Cancer Molecular Biology
ISSN: 1110-5313
Cancer Mol. Biol.
Ain Shams Faculty of Medicine, Abbassia, Cairo, Egypt, Tel: +20(202)2858940, Fax: +20(202)2859928 ISSUES/YR: 4

- **Cancer Nursing**
  ISSN: 0162-220X
  Cancer Nurs.
  Lippincott Williams and Wilkins, 530 Walnut Street, Philadelphia, PA 19106-3621, United States, Tel: +1(215)5218342, Fax: +1(215)5218326 ISSUES/YR: 6
- **Cancer Research**
  ISSN: 0008-5472
  Cancer Res.
  American Association for Cancer Research Inc., 150 South Independence Mall West, Philadelphia, PA 19106-3483, United States, Tel: +1(215)4409300, Fax: +1(215)4409354 ISSUES/YR: 24
- **Cancer Science**
  ISSN: 1347-9032
  Cancer Sci.
  Blackwell Publishing Ltd, 9600 Garsington Road, OX4 2XG, Oxford, United Kingdom, Tel: +44(1865)776868, Fax: +44(1865)714591 ISSUES/YR: 12

Cancer Treatment Reviews
ISSN: 0305-7372
Cancer Treat. Rev.
W.B. Saunders Ltd, 32 Jamestown Road, NW1 7BY, London, United Kingdom, Tel: +44(20)74244200, Fax: +44(20)74854752 ISSUES/YR: 8

- **Cancer/Radiotherapie**
  ISSN: 1278-3218
  Cancer Radiother.
  Editions Scientifiques et Medicales Elsevier, 23 rue Linois, 75724, Paris Cedex 15, France, Tel: +33(1)45589826, Fax: +33(1)45589428 ISSUES/YR: 6
- **Carbohydrate Research**
  ISSN: 0008-6215
  Carbohydr. Res.
  Elsevier Ltd, Langford Lane, Kidlington, OX5 1GB, Oxford, United Kingdom, Tel: +44(1865)843000, Fax: +44(1865)843010 ISSUES/YR: 18
- **Carcinogenesis**
  ISSN: 0143-3334
  Carcinogenesis
  Oxford University Press, Great Clarendon Street, OX2 6DP, Oxford, United Kingdom, Tel: +44(1865)353907, Fax: +44(1865)353486 ISSUES/YR: 12
- **Cardiology**
  ISSN: 0008-6312
  Cardiology
  S. Karger AG, Allschwilerstrasse 10, P.O. Box, CH-4009, Basel, Switzerland, Tel: +41(61)3061111, Fax: +41(61)3061234 ISSUES/YR: 8
- **Cardiology Clinics**
  ISSN: 0733-8651
  Cardiol. Clin.
  W.B. Saunders, Independence Square West, Philadelphia, PA 19106-3399, United States, Tel: +1(215)2387800, Fax: +1(215)2386445 ISSUES/YR: 4

Cardiology in Review
ISSN: 1061-5377
Cardiol. Rev.
Lippincott Williams and Wilkins, 351 West Camden Street, Baltimore, MD 21201-2436, United States, Tel: +1(410)5284000 ISSUES/YR: 6

Cardiology in the Young
ISSN: 1047-9511
Cardiol. Young
Cambridge University Press, Shaftesbury Road, CB2 2RU, Cambridge, United Kingdom, Tel: +44(1223)312393, Fax: +44(1223)315052 ISSUES/YR: 6

Cardiology International
ISSN: 1468-8581
Cardiol. Int.
Greycoat Publishing, 120 Dawes Road, SW6 7EG, London, United Kingdom, Tel: +44(207)3851515, Fax: +44(207)3857799 ISSUES/YR: 4

Cardiology Review
ISSN: 1092-6607
Cardiol. Rev.
MRA Publications Inc., 241 Forsgate Dr., Jamesburg, NJ 08831, United States, Tel: +1(732)6561140 ISSUES/YR: 12

Cardiovascular and Hematological Agents in Medicinal Chemistry
ISSN: 1871-5257
cardiovasc. Hematol. Agents Med. Chem.
Bentham Science Publishers B.V., P.O. Box 294, 1400 AG, Bussum, Netherlands, Tel: +31-356257290, Fax: +31-847511310 ISSUES/YR: 4

Cardiovascular and Hematological Disorders - Drug Targets
ISSN: 1871-529X
Cardiovasc. Hematol. Disord. Drug Targets
Bentham Science Publishers B.V., P.O. Box 294, 1400 AG, Bussum, Netherlands, Tel: +31-356257290, Fax: +31-847511310 ISSUES/YR: 4

- **CardioVascular and Interventional Radiology**
  ISSN: 0174-1551
  Cardiovasc. Intervent. Radiol.
  Springer New York LLC, 233 Springer Street, New York, 10013-1578, United States, Tel: +1-212 460 1501, Fax: +1-212 460 1595 ISSUES/YR: 6

Current

Cardiovascular Diabetology
ISSN: 1475-2840
Cardiovasc. Diabetol.
BioMed Central Ltd., 34 - 42 Cleveland Street, W1T 4LB, London, United Kingdom, Tel: +44(207)3230323, Fax: +44(207)6319926 ISSUES/YR: 1

- **Cardiovascular Drug Reviews**
ISSN: 0897-5957
Cardiovasc. Drug Rev.
Blackwell Publishing Ltd, 9600 Garsington Road, OX4 2XG, Oxford, United Kingdom, Tel: +44(1865)776868, Fax: +44(1865)714591 ISSUES/YR: 4

- **Cardiovascular Drugs and Therapy**
ISSN: 0920-3206
Cardiovasc. Drugs Ther.
Kluwer Academic Publishers, 101 Philip Drive, Assinippi Park, Norwell, MA 02061, United States, Tel: +1(781)8716600, Fax: +1(781)8716528 ISSUES/YR: 6

Cardiovascular Engineering
ISSN: 1567-8822
Cardiovasc. Eng.
Kluwer Academic Publishers, Van Godewijckstraat 30, 3311 GZ, Dordrecht, Netherlands, Tel: +31(78)6576000, Fax: +31(78)6576254 ISSUES/YR: 4

Cardiovascular Journal of South Africa
ISSN: 1015-9657
Cardiovasc. J. South Afr.
South African Medical Association, Private Bag X1, 7430, Pinelands, South Africa, Tel: +27(21)5313081, Fax: +27(21)5314126 ISSUES/YR: 5

- **Cardiovascular Pathology**
ISSN: 1054-8807
Cardiovasc. Pathol.
Elsevier Inc., 360 Park Avenue South, New York, NY 10010, United States, Tel: +1(212)6333950, Fax: +1(212)6333990 ISSUES/YR: 6

- **Cardiovascular Radiation Medicine**
ISSN: 1522-1865
Cardiovasc. Radiat. Med.
Elsevier Inc., 360 Park Avenue South, New York, NY 10010, United States, Tel: +1(212)6333950, Fax: +1(212)6333990 ISSUES/YR: 4

- **Cardiovascular Research**
ISSN: 0008-6363
Cardiovasc. Res.
Elsevier, P.O. Box 211, 1000 AE, Amsterdam, Netherlands, Tel: +31(20)4853757, Fax: +31(20)4853432 ISSUES/YR: 12

- **Cardiovascular Revascularization Medicine**
ISSN: 1553-8389
Cardiovasc. Revasc. Med.
Elsevier Inc., 360 Park Avenue South, New York, NY 10010, United States, Tel: +1(212)6333950, Fax: +1(212)6333990 ISSUES/YR: 4

- **Cardiovascular Toxicology**
ISSN: 1530-7905
Cardiovasc. Toxicol.
Humana Press Inc., 999 Riverview Drive, Suite 208, Totowa, NJ 07512-1165, United States, Tel: +1(973)2561699, Fax: +1(973)2568341 ISSUES/YR: 4

Cardiovascular Ultrasound
ISSN: 1476-7120
Cardiovasc. Ultrasound
BioMed Central Ltd., 34 - 42 Cleveland Street, W1T 4LB, London, United Kingdom, Tel: +44(207)3230323, Fax: +44(207)6319926 ISSUES/YR: 1

Care Management Journals
ISSN: 1521-0987
Care Man. J.
Springer Publishing Company, 11 West 42nd Street, 15th Floor, New York, NY 10036, United States, Tel: +1(212)4314370, Fax: +1(212)9417842 ISSUES/YR: 4

Care of the Critically Ill
ISSN: 0266-0970
Care Crit. Ill
Nature Publishing Group, Houndmills, RG21 6XS, Basingstoke, Hampshire, United Kingdom, Tel: +44(1256)329242, Fax: +44(1256)810526 ISSUES/YR: 6

Case Reports and Clinical Practice Review
ISSN: 1507-6164
Case Rep. Clin. Pract. Rev.
Medical Science International Ltd., ul. Ustrzycka 11, 02-141 Warsaw, Poland, Tel: +48(22)8681265, Fax: +48(22)7813232 ISSUES/YR: 1

Casopis Lekaru Ceskych
ISSN: 0008-7335
Cas. Lek. Cesk.
Czech Medical Association J.E. Purkyne, Sokolska 31, P.O. Box 88, 120 26, Praha 2, Czech Republic, Tel: +420(2)24266201-4, Fax: +420(2)24266212 ISSUES/YR: 24

Catheterization and Cardiovascular Interventions
ISSN: 1522-1946
Catheter. Cardiovasc. Interventions
Wiley-Liss Inc., 111 River Street, Hoboken, NJ 07030-5774, United States, Tel: +1(201)7486645, Fax: +1(201)7486021 ISSUES/YR: 12

- **Cell**
ISSN: 0092-8674
Cell
Cell Press, 1100 Massachusetts Avenue, Cambridge, MA 02138-9957, United States, Tel: +1(617)6617057, Fax: +1(617)6617061 ISSUES/YR: 26

Cell and Chromosome
ISSN: 1475-9268
Cell Chromosome
BioMed Central Ltd., 34 - 42 Cleveland Street, W1T 4LB, London, United Kingdom, Tel: +44(207)3230323, Fax: +44(207)6319926 ISSUES/YR: 1

- **Cell and Tissue Banking**
ISSN: 1389-9333
Cell Tissue Bank.
Kluwer Academic Publishers, Van Godewijckstraat 30, 3311 GZ, Dordrecht, Netherlands, Tel: +31(78)6576000, Fax: +31(78)6576254 ISSUES/YR: 4

- **Cell and Tissue Research**
ISSN: 0302-766X
Cell Tissue Res.
Springer Verlag, Tiergartenstrasse 17, D-69121, Heidelberg, Germany, Tel: +49(6221)4870, Fax: +49(6221)487366 ISSUES/YR: 12

- **Cell Biochemistry and Function**
ISSN: 0263-6484
Cell Biochem. Funct.
John Wiley and Sons Ltd, Southern Gate, PO19 8SQ, Chichester, West Sussex, United Kingdom, Tel: +44(1243)779777, Fax: +44(1243)775878 ISSUES/YR: 4

- **Cell Biology and Toxicology**
ISSN: 0742-2091
Cell Biol. Toxicol.
Kluwer Academic Publishers, Van Godewijckstraat 30, 3311 GZ, Dordrecht, Netherlands, Tel: +31(78)6576000, Fax: +31(78)6576254 ISSUES/YR: 6

Cell Biology Education
ISSN: 1536-7509
Cell Biol. Educ.
American Society for Cell Biology, 8120 Woodmont Avenue, Suite 750, Bethesda, MD 20814-2755, United States, Tel: +1(301)3479300, Fax: +1(301)3479310 ISSUES/YR: 4

Cell Biology International
ISSN: 1065-6995
Cell Biol. Int.
Academic Press, 24-28 Oval Road, NW1 7DX, London, United Kingdom, Tel: +44(207)2674466 ISSUES/YR: 12

- **Cell Calcium**
ISSN: 0143-4160
Cell Calcium
Churchill Livingstone, 1-3 Baxter's Place, Leith Walk, EH1 3AF, Edinburgh, United Kingdom, Tel: +44(131)5562424, Fax: +44(131)5581278 ISSUES/YR: 12

- **Cell Communication and Adhesion**
  ISSN: 1541-9061
  Cell Commun. Adhes.
  Taylor and Francis Ltd., 4 Park Square, Milton Park, OX14 4RN, Abingdon, Oxfordshire, United Kingdom, Tel: +44(20)70175541, Fax: +44(20)70175198 ISSUES/YR: 6
- **Cell Communication and Signaling**
  ISSN: 1478-811X
  Cell Commun. Signal.
  BioMed Central Ltd., 34 - 42 Cleveland Street, W1T 4LB, London, United Kingdom, Tel: +44(207)3230323, Fax: +44(207)6319926 ISSUES/YR: 1

Cell Cycle
ISSN: 1538-4101
Cell Cycle
Landes Bioscience, 810 S Church Street, Georgetown, 78626, United States, Tel: +1-512.863.7762, Fax: +1-512.863.0081 ISSUES/YR: 12

- **Cell Death and Differentiation**
  ISSN: 1350-9047
  Cell Death Differ.
  Nature Publishing Group, Houndmills, RG21 6XS, Basingstoke, Hampshire, United Kingdom, Tel: +44(1256)329242, Fax: +44(1256)810526 ISSUES/YR: 12
- **Cell Metabolism**
  ISSN: 1550-4131
  Cell Metab.
  Cell Press, 1100 Massachusetts Avenue, Cambridge, MA 02138-9957, United States, Tel: +1(617)6617057, Fax: +1(617)6617061 ISSUES/YR: 6
- **Cell Motility and the Cytoskeleton**
  ISSN: 0886-1544
  Cell Motil. Cytoskeleton
  Wiley-Liss Inc., 111 River Street, Hoboken, NJ 07030-5774, United States, Tel: +1(201)7486645, Fax: +1(201)7486021 ISSUES/YR: 12
- **Cell Preservation Technology**
  ISSN: 1538-344X
  Cell Preserv. Technol.
  Mary Ann Liebert Inc., 140 Huguenot Street, New Rochelle, NY 10801-5215, United States, Tel: +1(914)7402100, Fax: +1(914)7402101 ISSUES/YR: 4

Cell Proliferation
ISSN: 0960-7722
Cell Prolif.
Blackwell Publishing Ltd, 9600 Garsington Road, OX4 2XG, Oxford, United Kingdom, Tel: +44(1865)776868, Fax: +44(1865)714591 ISSUES/YR: 6

Cell Proliferation, Supplement
ISSN: 1477-8408
Cell Prolif. Suppl.
Blackwell Publishing Ltd, 9600 Garsington Road, OX4 2XG, Oxford, United Kingdom, Tel: +44(1865)776868, Fax: +44(1865)714591 ISSUES/YR: 1

- **Cell Stress and Chaperones**
  ISSN: 1355-8145
  Cell Stress Chaperones
  Cell Stress and Chaperones, 75 N. Eagleville Road, Storrs, CT 06269-3044, United States, Tel: +1(860)4864257, Fax: +1(860)4865709 ISSUES/YR: 4
- **Cell Structure and Function**
  ISSN: 0386-7196
  Cell Struct. Funct.
  Japan Society for Cell Biology, Shimotachiuri, Ogawa Higashi, Kamikyo-ku, 602, Tokyo, Japan, Fax: +81(75)4413159 ISSUES/YR: 6
- **Cell Transplantation**
  ISSN: 0963-6897
  Cell Transplant.
  Cognizant Communication Corporation, 3 Hartsdale Road, Elmsford, NY 10523-3701, United States, Tel: +1(914)5927720, Fax: +1(914)5928981 ISSUES/YR: 8
- **Cells and Materials**
  ISSN: 1051-6794
  Cells Mater.
  Scanning Microscopy International, 1034 Alabama Drive, Elk Grove Village, IL 60007-2920, United States, Tel: +1(708)5296677 ISSUES/YR: 4
- **Cells Tissues Organs**
  ISSN: 1422-6405
  Cells Tissues Organs
  S. Karger AG, Allschwilerstrasse 10, P.O. Box, CH-4009, Basel, Switzerland, Tel: +41(61)3061111, Fax: +41(61)3061234 ISSUES/YR: 12

Cellular and Molecular Biology
ISSN: 0145-5680
Cell. Mol. Biol.
Cellular and Molecular Biology Association, Batiment Mecanique, 86022, Poitiers Cedez, France, Tel: +33-549 36 63 97, Fax: +33-549 45 36 41 ISSUES/YR: 8

Cellular and Molecular Biology Letters
ISSN: 1425-8153
Cell. Mol. Biol. Lett.
University of Wroclaw, Przybyszewskiego 63/77, 51-148, Wroclaw, Poland, Tel: +48(71)252930, Fax: +48(71)252930 ISSUES/YR: 4

Cellular and Molecular Life Sciences
ISSN: 1420-682X
Cell. Mol. Life Sci.
Birkhauser Verlag AG, Klosterberg 23, P.O. Box 133, CH-4010, Basel, Switzerland, Tel: +41(61)2717400, Fax: +41(61)2050791 ISSUES/YR: 24

- **Cellular and Molecular Neurobiology**
  ISSN: 0272-4340
  Cell. Mol. Neurobiol.
  Springer Science and Business Media Deutschland GmbH, Heidelberger Platz 3, 14197, Berlin, Germany, Tel: +1-30 827875282, Fax: +1-30 827875707 ISSUES/YR: 6
- **Cellular Immunology**
  ISSN: 0008-8749
  Cell. Immunol.
  Academic Press Inc., 6277 Sea Harbor Drive, Orlando, FL 32887-4900, United States, Tel: +1(407)3454040, Fax: +1(407)3639661 ISSUES/YR: 16
- **Cellular Microbiology**
  ISSN: 1462-5814
  Cell. Microbiol.
  Blackwell Publishing Ltd, 9600 Garsington Road, OX4 2XG, Oxford, United Kingdom, Tel: +44(1865)776868, Fax: +44(1865)714591 ISSUES/YR: 12
- **Cellular Oncology**
  ISSN: 1570-5870
  Cell. Oncol.
  IOS Press, Nieuwe Hemweg 6B, 1013 BG, Amsterdam, Netherlands, Tel: +31(20)6883355, Fax: +31(20)6203419 ISSUES/YR: 6
- **Cellular Physiology and Biochemistry**
  ISSN: 1015-8987
  Cell. Physiol. Biochem.
  S. Karger AG, Allschwilerstrasse 10, P.O. Box, CH-4009, Basel, Switzerland, Tel: +41(61)3061111, Fax: +41(61)3061234 ISSUES/YR: 6
- **Cellular Signalling**
  ISSN: 0898-6568
  Cell. Signal.
  Elsevier Inc., 360 Park Avenue South, New York, NY 10010, United States, Tel: +1(212)6333950, Fax: +1(212)6333990 ISSUES/YR: 12

Central European Journal of Public Health
ISSN: 1210-7778
Cent. Eur. J. Public Health
Tigis s.r.o., Na Habrove 17, 152 00, Prague 5, Czech Republic, Tel: +420-251 813 1924, Fax: +420-251 681 217 ISSUES/YR: 4

Central Nervous System Agents in Medicinal Chemistry
ISSN: 1871-5249
Cent. Nerv. Syst. Agents Med. Chem.
Bentham Science Publishers B.V., P.O. Box 294, 1400 AG, Bussum, Netherlands, Tel: +31-356257290, Fax: +31-847511310 ISSUES/YR: 4

Central-European Journal of Immunology
ISSN: 1426-3912
Cent.-Eur. J. Immunol.
Termedia Publishing House Ltd., ul Kleberga 8,, 61-615, Poznan, Poland, Tel: +48-61 822 77 81, Fax: +48-61 822 7781 ISSUES/YR: 4

Current

Cephalalgia
ISSN: 0333-1024
Cephalalgia
Blackwell Publishing Ltd, 9600 Garsington Road, OX4 2XG, Oxford, United Kingdom, Tel: +44(1865)776868, Fax: +44(1865)714591 ISSUES/YR: 12

Cephalalgia, Supplement
ISSN: 0800-1952
Cephalalgia Suppl.
Blackwell Publishing Ltd, 9600 Garsington Road, OX4 2XG, Oxford, United Kingdom, Tel: +44(1865)776868, Fax: +44(1865)714591 ISSUES/YR: 1

- **Cerebellum**
ISSN: 1473-4222
Cerebellum
Taylor and Francis A.S., P.O. Box 2562 Solli, 0202, Oslo, Norway, Tel: +44(22)129880, Fax: +44(22)129890 ISSUES/YR: 4

- **Cerebral Cortex**
ISSN: 1047-3211
Cereb. Cortex
Oxford University Press, 2001 Evans Road, Cary, NC 27513, United States, Tel: +1(919)6770977, Fax: +1(919)6771714 ISSUES/YR: 12

- **Cerebrospinal Fluid Research**
ISSN: 1743-8454
Cerebrospinal Fluid Res.
BioMed Central Ltd., 34 - 42 Cleveland Street, W1T 4LB, London, United Kingdom, Tel: +44(207)3230323, Fax: +44(207)6319926 ISSUES/YR: 1

- **Cerebrovascular Diseases**
ISSN: 1015-9770
Cerebrovasc. Dis.
S. Karger AG, Allschwilerstrasse 10, P.O. Box, CH-4009, Basel, Switzerland, Tel: +41(61)3061111, Fax: +41(61)3061234 ISSUES/YR: 8

Ceska a Slovenska Farmacie
ISSN: 1210-7816
Ceska Slov. Farm.
Czech Medical Association J.E. Purkyne, Sokolska 31, P.O. Box 88, 120 26, Praha 2, Czech Republic, Tel: +420(2)24266201-4, Fax: +420(2)24266212 ISSUES/YR: 6

Ceska a Slovenska Gastroenterologie a Hepatologie
ISSN: 1213-323X
Ceska Slov. Gastroenterol. Hepatol.
Czech Medical Association J.E. Purkyne, Sokolska 31, P.O. Box 88, 120 26, Praha 2, Czech Republic, Tel: +420(2)24266201-4, Fax: +420(2)24266212 ISSUES/YR: 6

Ceska a Slovenska Neurologie a Neurochirurgie
ISSN: 1210-7859
Ceska Slov. Neurol. Neurochir.
Czech Medical Association J.E. Purkyne, Sokolska 31, P.O. Box 88, 120 26, Praha 2, Czech Republic, Tel: +420(2)24266201-4, Fax: +420(2)24266212 ISSUES/YR: 6

Ceska a Slovenska Oftalmologie
ISSN: 1211-9059
Ceska Slov. Oftalmol.
Czech Medical Association J.E. Purkyne, Sokolska 31, P.O. Box 88, 120 26, Praha 2, Czech Republic, Tel: +420(2)24266201-4, Fax: +420(2)24266212 ISSUES/YR: 6

Ceska a Slovenska Psychiatrie
ISSN: 1212-0383
Ceska Slov. Psychiatr.
Czech Medical Association J.E. Purkyne, Sokolska 31, P.O. Box 88, 120 26, Praha 2, Czech Republic, Tel: +420(2)24266201-4, Fax: +420(2)24266212 ISSUES/YR: 8

Ceska Gynekologie
ISSN: 1210-7832
Ceska Gynekol.
Czech Medical Association J.E. Purkyne, Sokolska 31, P.O. Box 88, 120 26, Praha 2, Czech Republic, Tel: +420(2)24266201-4, Fax: +420(2)24266212 ISSUES/YR: 6

Ceska Radiologie
ISSN: 1210-7883
Ceska Radiol.
Czech Medical Association J.E. Purkyne, Sokolska 31, P.O. Box 88, 120 26, Praha 2, Czech Republic, Tel: +420(2)24266201-4, Fax: +420(2)24266212 ISSUES/YR: 6

Ceska Revmatologie
ISSN: 1210-7905
Ceska Revmatol.
Czech Medical Association J.E. Purkyne, Sokolska 31, P.O. Box 88, 120 26, Praha 2, Czech Republic, Tel: +420(2)24266201-4, Fax: +420(2)24266212 ISSUES/YR: 4

Ceske Pracovni Lekarstvi
ISSN: 1212-6721
Ceske Prac. Lek.
Tigis s.r.o., Na Habrove 17, 152 00, Prague 5, Czech Republic, Tel: +420-251 813 1924, Fax: +420-251 681 217 ISSUES/YR: 4

Cesko-Slovenska Dermatologie
ISSN: 0009-0514
Cesko-Slov. Dermatol.
Czech Medical Association J.E. Purkyne, Sokolska 31, P.O. Box 88, 120 26, Praha 2, Czech Republic, Tel: +420(2)24266201-4, Fax: +420(2)24266212 ISSUES/YR: 6

Cesko-Slovenska Pediatrie
ISSN: 0069-2328
Cesko-Slov. Pediatr.
Czech Medical Association J.E. Purkyne, Sokolska 31, P.O. Box 88, 120 26, Praha 2, Czech Republic, Tel: +420(2)24266201-4, Fax: +420(2)24266212 ISSUES/YR: 12

Chang Gung Medical Journal
ISSN: 0255-8270
Chang Gung Med. J.
Chang Gung Medical Journal, 199 Tung Hwa North Road, 10591, Taipei, Taiwan, Tel: +886(2)7135211 ISSUES/YR: 12

- **ChemBioChem**
ISSN: 1439-4227
ChemBioChem
Wiley-VCH Verlag, P.O. Box 101161, D-69451, Weinheim, Germany, Tel: +49(6201)606147, Fax: +49(6201)606117 ISSUES/YR: 12

Chemical and Pharmaceutical Bulletin
ISSN: 0009-2363
Chem. Pharm. Bull.
Pharmaceutical Society of Japan, 2-12-15-201, Shibuya, Shibuya-ku, 150, Tokyo, Japan, Tel: +81(3)34063321, Fax: +81(3)34981835 ISSUES/YR: 12

- **Chemical Biology and Drug Design**
ISSN: 1747-0277
Chem. Biol. Drug Des.
Blackwell Publishing Ltd, 9600 Garsington Road, OX4 2XG, Oxford, United Kingdom, Tel: +44(1865)776868, Fax: +44(1865)714591 ISSUES/YR: 12

Chemical Communications
ISSN: 1359-7345
Chem. Commun.
Royal Society of Chemistry, Milton Road, CB4 OWF, Cambridge, United Kingdom, Tel: +44(1223)420066, Fax: +44(1223)420247 ISSUES/YR: 48

- **Chemical Immunology and Allergy**
ISSN: 1660-2242
Chem. Immunol. Allergy
S. Karger AG, Allschwilerstrasse 10, P.O. Box, CH-4009, Basel, Switzerland, Tel: +41(61)3061111, Fax: +41(61)3061234 ISSUES/YR: 1

Chemical Research in Toxicology
ISSN: 0893-228X
Chem. Res. Toxicol.
American Chemical Society, 2540 Olentangy River Road, P.O. Box 3337, Columbus, OH 43210-3337, United States, Tel: +1(614)4473776, Fax: +1(614)4473671 ISSUES/YR: 12

- **Chemical Senses**
ISSN: 0379-864X
Chem. Senses
Oxford University Press, Great Clarendon Street, OX2 6DP, Oxford, United Kingdom, Tel: +44(1865)353907, Fax: +44(1865)353486 ISSUES/YR: 9

Current

Chemico-Biological Interactions
ISSN: 0009-2797
Chem.-Biol. Interact.
Elsevier Ireland Ltd, P.O. Box 85, Limerick, Ireland, Tel: +353(61)471944, Fax: +353(61)472144 ISSUES/YR: 18

Chemistry and Biodiversity
ISSN: 1612-1872
Chem. Biodiversity
Wiley-VCH Verlag, P.O. Box 101161, D-69451, Weinheim, Germany, Tel: +49(6201)606147, Fax: +49(6201)606117 ISSUES/YR: 12

● **Chemistry and Physics of Lipids**
ISSN: 0009-3084
Chem. Phys. Lipids
Elsevier Ireland Ltd, P.O. Box 85, Limerick, Ireland, Tel: +353(61)471944, Fax: +353(61)472144 ISSUES/YR: 12

● **Chemometrics and Intelligent Laboratory Systems**
ISSN: 0169-7439
Chemometr. Intelligent Lab. Syst.
Elsevier, P.O. Box 211, 1000 AE, Amsterdam, Netherlands, Tel: +31(20)4853757, Fax: +31(20)4853432 ISSUES/YR: 8

Chemosphere
ISSN: 0045-6535
Chemosphere
Elsevier Ltd, Langford Lane, Kidlington, OX5 1GB, Oxford, United Kingdom, Tel: +44(1865)843000, Fax: +44(1865)843010 ISSUES/YR: 44

Chemotaxis and Migration
ISSN: 1608-2265
Chemotaxis Migr.
VICER Publishing, P.O. Box 14, A-1097, Vienna, Austria, Tel: +43(676)9568085, Fax: +43(676)9568086 ISSUES/YR: 4

Current

Chemotherapie Journal
ISSN: 0940-6735
Chemother. J.
Wissenschaftliche Verlagsgesellschaft mbH, Birkenwaldstrasse 44, D-70191, Stuttgart, Germany, Tel: +49(711)25820 ISSUES/YR: 6

Chemotherapie Journal, Supplement
ISSN: 0944-6486
Chemother. J. Suppl.
Wissenschaftliche Verlagsgesellschaft mbH, Birkenwaldstrasse 44, D-70191, Stuttgart, Germany, Tel: +49(711)25820 ISSUES/YR: 1

● **Chemotherapy**
ISSN: 0009-3157
Chemotherapy
S. Karger AG, Allschwilerstrasse 10, P.O. Box, CH-4009, Basel, Switzerland, Tel: +41(61)3061111, Fax: +41(61)3061234 ISSUES/YR: 6

Chemtracts
ISSN: 1431-9268
Chemtracts
Data Trace Publishing Company, 110 West Road, Ste. 227, Towson, MD 21204-2316, United States, Tel: +1(410)4944994, Fax: +1(410)4940515 ISSUES/YR: 14

● **Chest**
ISSN: 0012-3692
Chest
American College of Chest Physicians, 3300 Dundee Road, Northbrook, IL 60062-2348, United States, Tel: +1(847)4981400 ISSUES/YR: 12

Child Abuse and Neglect
ISSN: 0145-2134
Child Abuse Negl.
Elsevier Ltd, Langford Lane, Kidlington, OX5 1GB, Oxford, United Kingdom, Tel: +44(1865)843000, Fax: +44(1865)843010 ISSUES/YR: 12

● **Child and Adolescent Mental Health**
ISSN: 1475-357X
Child Adolesc. Ment. Health
Blackwell Publishing Ltd, 9600 Garsington Road, OX4 2XG, Oxford, United Kingdom, Tel: +44(1865)776868, Fax: +44(1865)714591 ISSUES/YR: 4

● **Child and Adolescent Psychiatric Clinics of North America**
ISSN: 1056-4993
Child Adolesc. Psychiatr. Clin. North Am.
W.B. Saunders, Independence Square West, Philadelphia, PA 19106-3399, United States, Tel: +1(215)2387800, Fax: +1(215)2386445 ISSUES/YR: 4

Child Neuropsychology
ISSN: 0929-7049
Child Neuropsychol.
Psychology Press Ltd, 4 Park Square, Milton Park, OX14 4RN, Abingdon, Oxfordshire, United Kingdom, Tel: +44(1235)828600, Fax: +44(1235)829000 ISSUES/YR: 6

Child Psychiatry and Human Development
ISSN: 0009-398X
Child Psychiatry Hum. Dev.
Kluwer Academic/Human Sciences Press Inc., 233 Spring Street, New York, NY 10013-1578, United States, Tel: +1(212)6208000, Fax: +1(212)4630742 ISSUES/YR: 4

● **Child Psychology and Psychiatry Review**
ISSN: 1360-6417
Child Psychol. Psychiatry Rev.
Cambridge University Press, Shaftesbury Road, CB2 2RU, Cambridge, United Kingdom, Tel: +44(1223)312393, Fax: +44(1223)315052 ISSUES/YR: 4

● **Child's Nervous System**
ISSN: 0256-7040
Child's Nerv. Syst.
Springer Verlag, Tiergartenstrasse 17, D-69121, Heidelberg, Germany, Tel: +49(6221)4870, Fax: +49(6221)487366 ISSUES/YR: 12

● **Child: Care, Health and Development**
ISSN: 0305-1862
Child Care Health Dev.
Blackwell Publishing Ltd, 9600 Garsington Road, OX4 2XG, Oxford, United Kingdom, Tel: +44(1865)776868, Fax: +44(1865)714591 ISSUES/YR: 6

Children's Health Care
ISSN: 0273-9615
Child. Health Care
Lawrence Erlbaum Associates Inc., 10 Industrial Avenue, Mahwah, NJ 07430-2262, United States, Tel: +1(201)2582200, Fax: +1(201)2360072 ISSUES/YR: 4

Children's Pulmonology
ISSN: 0354-4265
Child. Pulmonol.
Children's Hospital for Pulmonary Diseases and TBC, H. Milana Tepica 1, 11000, Belgrade, Serbia and Montenegro, Tel: +381(11)660384, Fax: +381(11)660532 ISSUES/YR: 2

Chinese Critical Care Medicine
ISSN: 1003-0603
Chin. Crit. Care Med.
Chinese Critical Care Medicine, Journal Office, 122 Munan Street, 300050, Tianjin, Heping District, China, Tel: +86(022)23306917, Fax: +86(022)23306917 ISSUES/YR: 12

Chinese Journal of Andrology
ISSN: 1008-0848
Chin. J. Androl.
Shanghai Di-Er Yike Daxue, Shanghai, China ISSUES/YR: 6

Chinese Journal of Antibiotics
ISSN: 1001-8689
Chin. J. Antibiot.
Editorial Office of Chinese Journal of Antibiotics, 9 Shanbanqiao Road, Chengdu, Sichuan, China ISSUES/YR: 12

Chinese Journal of Biomedical Engineering
ISSN: 0258-8021
Chin. J. Biomed. Eng.
Chinese Academy of Medical Sciences, 9 Dong Dan San Tiao, 100730, Beijing, China, Tel: +86(1)5133074 ISSUES/YR: 6

Chinese Journal of Cancer Research
ISSN: 1000-9604
Chin. J. Cancer Res.
Beijing Institute for Cancer Research, Da-Hong-Luo-Chang Street, Western District, 100034, Beijing, China, Fax: +86(10)66175832 ISSUES/YR: 4

Chinese Journal of Cerebrovascular Diseases
ISSN: 1672-5921
Chin. J. Cerebrovasc. Dis.
Society of China University Journals in Natural Sciences, Beijing Normal University, 100083, Beijing, China, Fax: +86-10 62337605 ISSUES/YR: 12

Chinese Journal of Clinical Nutrition
ISSN: 1008-5882
Chin. J. Clin. Nutr.
Editorial Office of Chinese Journal of, 9 Dongdan Santiao, 100730, Beijing, China, Tel: +86-10 65237951, Fax: +86-10 65133074 ISSUES/YR: 6

Chinese Journal of Clinical Oncology
ISSN: 1000-8179
Chin. J. Clin. Oncol.
China Anti-Cancer Association, Huan-Hu-Xi Road, Ti-Yuan-Bei, 300060, Tianjin, China, Tel: +86(22)319929 EXT.205 ISSUES/YR: 24

Chinese Journal of Clinical Rehabilitation
ISSN: 1671-5926
Chin. J. Clin. Rehab.
Chinese Journal of Clinical Rehabilitation, c/o 16, Tenth Latitude Road, Heping District, 110003, Shenyang, China, Tel: +86(24)22714373, Fax: +86(24)22830178 ISSUES/YR: 32

Chinese Journal of Contemporary Neurology and Neurosurgery
ISSN: 1672-6731
Chin. J. Contemp. Neurol. Neurosurg.
Tianjin Huanhu Hospital, 122 Qixiangtai Road, 300060, Tianjin, China, Tel: +86-2223552328 ISSUES/YR: 6

Chinese Journal of Contemporary Pediatrics
ISSN: 1008-8830
Chin. J. Contemp. Pediatr.
Xiangya Hospital of CSU, 141 Xiangya Road, 410008, Changsha Hunan, China, Tel: +86(731)4327402, Fax: +86(731)4327402 ISSUES/YR: 6

• **Chinese Journal of Digestive Diseases**
ISSN: 1443-9611
Chin. J. Dig. Dis.
Blackwell Publishing, 550 Swanston Street, VIC 3053, Carlton South, Australia, Tel: +61(3)8359 1011, Fax: +61(3)8359 1120 ISSUES/YR: 4

Chinese Journal of Emergency Medicine
ISSN: 1671-0282
Chin. J. Emerg. Med.
Chinese Medical Association, 20 Ave. Appia, CH-1211, Geneva 27, Switzerland, Tel: +86(22)7912111, Fax: +86(22)7914857 ISSUES/YR: 12

Chinese Journal of Endemiology
ISSN: 1000-4955
Chin. J. Endemiol.
Chinese Medical Association, 42 Dongsi Xidajie, P.O. Box 2258, 100710, Beijing, China, Tel: +86(10)65244339, Fax: +86(10)65134886 ISSUES/YR: 6

Chinese Journal of Evidence-Based Medicine
ISSN: 1672-2531
Chin. J. Evid.-Based Med.
West China University of Medical Science, 37 Wainan Guoxue Xiang, 610041, Chengdu, Sichuan, China ISSUES/YR: 12

Chinese Journal of Forensic Medicine
ISSN: 1001-5728
Chin. J. Forensic Med.
Chinese Forensic Medicine Association, P.O. Box 3817, 100038, Beijing PRC, China, Tel: +86(10)63495531 ISSUES/YR: 6

Chinese Journal of Gastroenterology
ISSN: 1008-7125
Chin. J. Gastroenterol.
Shanghai Institute of Digestive Diseases, 145 Shan-Dong Zhong Road, 200001, Shanghai, China ISSUES/YR: 6

Chinese Journal of Infection and Chemotherapy
ISSN: 1009-7708
Chin. J. Infect. Chemother.
Editorial Department of Chinese Journal of Infection, 12 Wulumuqi Zhong Road, 200040, Shanghai, China, Tel: +86-021 62489999 EXT. 6457, Fax: +86-021 62484347 ISSUES/YR: 6

• **Chinese Journal of Integrative Medicine**
ISSN: 1672-0415
Chin. J. Integr. Med.
Chinese Journal of Integrated Traditional and Western Medicine Press, 1 Caochang, Xiyuan, 100091, Beijing, China, Tel: +86(10)6287-7592, Fax: +86(10)6287-7592 ISSUES/YR: 4

Chinese Journal of Interventional Imaging and Therapy
ISSN: 1672-8475
Chin. J. Intervent. Imaging Ther.
Editorial Office - Chinese Journal of Interventional, 1-301, Hongjialiyuan, Luozhuangnanli, 100088, Haidian District, Beijing, China ISSUES/YR: 6

Chinese Journal of Lung Cancer
ISSN: 1009-3419
Chin. J. Lung Cancer
Chinese Journal of Lung Cancer, 37 Guoxue Xiang, Chengdu, 610041, Sichuan, China, Tel: +86(28)5422587, Fax: +86(28)5422743 ISSUES/YR: 6

Chinese Journal of Medical Genetics
ISSN: 1003-9406
Chin. J. Med. Genet.
West China University of Medical Sciences, Renmin Nanlu, 610044, Chengdu, Sichuan, China ISSUES/YR: 6

Chinese Journal of Medical Imaging Technology
ISSN: 1003-3289
Chin. J. Med. Imaging Technol.
Editorial Board of Chinese Journal of Medical Imaging Technology, PO Box 2712, 100080, Beijing, China, Tel: +86(1)6253-6197, Fax: +86(1)6254-2307 ISSUES/YR: 12

Chinese Journal of Microbiology and Immunology
ISSN: 0254-5101
Chin. J. Microbiol. Immunol.
Society of Microbiology and Immunology, Chaoyangqu, 100024, Beijing, China ISSUES/YR: 12

Chinese Journal of Natural Medicines
ISSN: 1672-3651
Chin. J. Nat. Med.
Chinese Journal of Natural Medicines Editorial Board, 24 Tong-jia Xiang, 210009, Nanjing, China, Tel: +86(25)3271565, Fax: +86(25)3271279 ISSUES/YR: 6

Chinese Journal of Neurology
ISSN: 1006-7876
Chin. J. Neurol.
Chinese Medical Association, 42 Dongsi Xidajie, P.O. Box 2258, 100710, Beijing, China, Tel: +86(10)65244339, Fax: +86(10)65134886 ISSUES/YR: 12

Chinese Journal of Neuroscience
ISSN: 1008-0872
Chin. J. Neurosci.
Science Press, 18,Shuangqing Street,Haidian, 100085, Beijing, China, Tel: +86(10)62941099, Fax: +86(10)62923563 ISSUES/YR: 4

Chinese Journal of Ophthalmology
ISSN: 0412-4081
Chin. J. Ophthalmol.
Chinese Medical Association, 20 Ave. Appia, CH-1211, Geneva 27, Switzerland, Tel: +86(22)7912111, Fax: +86(22)7914857 ISSUES/YR: 12

Chinese Journal of Pathology
ISSN: 0529-5807
Chin. J. Pathol.
Chinese Medical Association, 42 Dongsi Xidajie, P.O. Box 2258, 100710, Beijing, China, Tel: +86(10)65244339, Fax: +86(10)65134886 ISSUES/YR: 12

Chinese Journal of Pharmacology and Toxicology
ISSN: 1000-3002
Chin. J. Pharmacol. Toxicol.
Chinese Journal of Pharmacology and Toxicology, 27 Taiping Road, 100850, Beijing, China ISSUES/YR: 6

Chinese Journal of Physiology
ISSN: 0304-4920
Chin. J. Physiol.
Chinese Physiological Society, 11529, Taipei, Taiwan, Tel: +886(2)7899125, Fax: +886(2)7822835 ISSUES/YR: 4

Chinese Journal of Radiology
ISSN: 1018-8940
Chin. J. Radiol.
Radiological Society of the R.O.C., 201, Sect. 2, Shih-Pai Road, 11217, Taipei, Taiwan, Fax: +886(2)8733643 ISSUES/YR: 6

Chinese Journal of Radiology
ISSN: 1005-1201
Chin. J. Radiol. (Zhonghua Fangshexue Zazhi)
Chinese Medical Association, 42 Dongsi Xidajie, P.O. Box 2258, 100710, Beijing, China, Tel: +86(10)65244339, Fax: +86(10)65134886 ISSUES/YR: 12

Chinese Journal of Rehabilitation Medicine
ISSN: 1001-1242
Chin. J. Rehabil. Med.
China-Japan Friendship Hospital, Yinghuayuan Street East, Hepingjie, 100029, Beijing PRC, China, Tel: +86(010)64218095, Fax: +86(010)64218095 ISSUES/YR: 4

Chinese Journal of Traumatology - English Edition
ISSN: 1008-1275
Chin. J. Traumatol. Eng. Ed.
Chinese Journal of Traumatology (English Edition), 10 Changjiang Zhilu, Daping, 400042, Chongqing, China, Tel: +86(23)68711977, Fax: +86(23)68711977 ISSUES/YR: 6

Chinese Medical Journal
ISSN: 0366-6999
Chin. Med. J.
Chinese Medical Association, 42 Dongsi Xidajie, P.O. Box 2258, 100710, Beijing, China, Tel: +86(10)65244339, Fax: +86(10)65134886 ISSUES/YR: 12

Chinese Medical Sciences Journal
ISSN: 1001-9294
Chin. Med. Sci. J.
Chinese Academy of Medical Sciences, 9 Dong Dan San Tiao, 100730, Beijing, China, Tel: +86(1)5133074 ISSUES/YR: 4

Chinese Ophthalmic Research
ISSN: 1003-0808
Chin. Ophthalmic Res.
Henan Institute of Ophthalmology, 7 Weiwu Road, 450003, Zhengzhou, China ISSUES/YR: 6

Chinese Pharmaceutical Journal
ISSN: 1001-2494
Chin. Pharm. J. (China)
Chinese Medical Association, 42 Dongsi Xidajie, P.O. Box 2258, 100710, Beijing, China, Tel: +86(10)65244339, Fax: +86(10)65134886 ISSUES/YR: 24

Chinese Pharmaceutical Journal
ISSN: 1016-1015
Chin. Pharm. J. (Taiwan)
Pharmaceutical Society of Republic of China, No. 1 Jen-Ai Road, Sec. 1, Taipei, Taiwan, Tel: +886(2)3416347 ISSUES/YR: 6

Chinese Pharmacological Bulletin
ISSN: 1001-1978
Chin. Pharmacol. Bull.
Institute of Clinical Pharmacology, 21 Mei-Shan Road, 230032, Hefei, Anhui, China ISSUES/YR: 12

Chinese Traditional and Herbal Drugs
ISSN: 0253-2670
Chin. Trad. Herbal Drugs
Editorial Office of Chinese Traditional and, No. 308 Anshanxidao, Nankai District, 300193, Tianjin, China, Tel: +86-22 27474913, Fax: +86-22 23006821 ISSUES/YR: 12

Chinesische Medizin
ISSN: 0930-2786
Chinesische Med.
Urban und Vogel GmbH, Neumarkter Strasse 43, D-81673, Munchen, Germany, Tel: +49(89)43721300, Fax: +49(89)43721399 ISSUES/YR: 4

• **Chirality**
ISSN: 0899-0042
Chirality
Wiley-Liss Inc., 111 River Street, Hoboken, NJ 07030-5774, United States, Tel: +1(201)7486645, Fax: +1(201)7486021 ISSUES/YR: 9

• **Chiropractic and Osteopathy**
ISSN: 1746-1340
Chiropr. Osteopathy
BioMed Central Ltd., 34 - 42 Cleveland Street, W1T 4LB, London, United Kingdom, Tel: +44(207)3230323, Fax: +44(207)6319926 ISSUES/YR: 1

Chirurgia
ISSN: 0394-9508
Chirurgia
Edizioni Minerva Medica S.p.A., Corso Bramante 83-85, 10126, Turin, Italy, Tel: +39(011)678282, Fax: +39(011)674502 ISSUES/YR: 6

Chirurgia del Piede
ISSN: 0392-0771
Chir. Piede
Edizioni Minerva Medica S.p.A., Corso Bramante 83-85, 10126, Turin, Italy, Tel: +39(011)678282, Fax: +39(011)674502 ISSUES/YR: 3

Chirurgia Polska
ISSN: 1507-5524
Chir. Pol.
Via Medica, Ul. Swietokrzyska 73, 80-180, Gdansk, Poland, Tel: +48(58)3209494, Fax: +48(58)3209460 ISSUES/YR: 4

Chirurgie de la Main
ISSN: 1297-3203
Chir. Main
Editions Scientifiques et Medicales Elsevier, 23 rue Linois, 75724, Paris Cedex 15, France, Tel: +33(1)45589826, Fax: +33(1)45589428 ISSUES/YR: 6

Chirurgische Gastroenterologie Interdisziplinar
ISSN: 0177-9990
Chir. Gastroenterol. Interdiszip.
S. Karger AG, Allschwilerstrasse 10, P.O. Box, CH-4009, Basel, Switzerland, Tel: +41(61)3061111, Fax: +41(61)3061234 ISSUES/YR: 4

Chirurgische Praxis
ISSN: 0009-4846
Chir. Prax.
Hans Marseille Verlag GmbH, P.O. Box 22 13 41, 80503, Munchen, Germany, Tel: +49(89)227988, Fax: +49(89)2904643 ISSUES/YR: 4

• **Chromatographia**
ISSN: 0009-5893
Chromatographia
Friedr. Vieweg und Sohn Verlags GmbH, P.O. Box 1546, D-65005, Wiesbaden, Germany, Tel: +49(611)534380, Fax: +49(611)7878439 ISSUES/YR: 24

Chromosoma
ISSN: 0009-5915
Chromosoma
Springer Verlag, Tiergartenstrasse 17, D-69121, Heidelberg, Germany, Tel: +49(6221)4870, Fax: +49(6221)487366 ISSUES/YR: 8

• **Chromosome Research**
ISSN: 0967-3849
Chromosome Res.
Kluwer Academic Publishers, Van Godewijckstraat 30, 3311 GZ, Dordrecht, Netherlands, Tel: +31(78)6576000, Fax: +31(78)6576254 ISSUES/YR: 8

Chronic Diseases in Canada
ISSN: 0228-8699
Chronic Dis. Can.
Laboratory Centre for Disease Control, Tunney's Pasture, Address Locator: 0602C3, ONT K1A 0L2, Ottawa, Canada, Tel: +1(613)9411291, Fax: +1(613)9527009 ISSUES/YR: 4

Chronic Illness
ISSN: 1742-3953
Chronic Illn.
Maney Publishing, Hudson Road, LS9 7DL, Leeds, United Kingdom, Tel: +44(113)2497481, Fax: +44(113)2486983 ISSUES/YR: 4

- **Chronic Respiratory Disease**
  ISSN: 1479-9723
  Chronic Respir. Dis.
  Hodder Arnold Journals, 338 Euston Road, NW1 3BH, London, United Kingdom, Tel: +44-+44 (0)20 7873 6337, Fax: +44-+44 (0)20 7873 6376 ISSUES/YR: 4

Chronobiology International
ISSN: 0742-0528
Chronobiol. Int.
Taylor and Francis Inc., 325 Chestnut St, Suite 800, Philadelphia PA, PA 19106, United States, Tel: +1(800)3541420, Fax: +1(215)6258914 ISSUES/YR: 6

Ciencia Ginecologika
ISSN: 1137-2273
Cienc. Ginecol.
Alpe Editores S.A., Pedro Rico 27, oficinas 11-12, 28029, Madrid, Spain, Tel: +34(91)7338811, Fax: +34(91)3159652 ISSUES/YR: 6

Ciencia y Tecnologia Pharmaceutica
ISSN: 1575-3409
Cienc. Tecnol. Pharm.
Alpe Editores S.A., Pedro Rico 27, oficinas 11-12, 28029, Madrid, Spain, Tel: +34(91)7338811, Fax: +34(91)3159652 ISSUES/YR: 4

Circular Farmaceutica
ISSN: 0009-7314
Circ. Farm.
Col-legi de Farmaceutics de la Provincia de Barcelona, Girona no's 64 and 66, 08009, Barcelona, Spain, Tel: +34(93)2440710, Fax: +34(93)2313671 ISSUES/YR: 4

- **Circulation**
  ISSN: 0009-7322
  Circulation
  Lippincott Williams and Wilkins, 530 Walnut Street, Philadelphia, PA 19106-3621, United States, Tel: +1(215)5218342, Fax: +1(215)5218326 ISSUES/YR: 50

Circulation Journal
ISSN: 1346-9843
Circ. J.
Japanese Circulation Society, 14 Yoshida Kawaharacho, Sakyo-ku, 606, Kyoto, Japan, Tel: +81(75)7518643 ISSUES/YR: 12

- **Circulation Research**
  ISSN: 0009-7330
  Circ. Res.
  Lippincott Williams and Wilkins, 530 Walnut Street, Philadelphia, PA 19106-3621, United States, Tel: +1(215)5218342, Fax: +1(215)5218326 ISSUES/YR: 24

Cirugia Plastica Ibero-Latinoamericana
ISSN: 0376-7892
Cir. Plast. Ibero-Latinoam.
Sociedad Espanola de Cirugia Plastica Reparadora y Estetica (SECPRE), Villanueva 11 - 3e planta, 28001, Madrid, Spain, Tel: +34(91)5765995, Fax: +34(91)4315153 ISSUES/YR: 4

CLB Chemie in Labor und Biotechnik
ISSN: 0943-6677
CLB Chem. Labor Biotech.
Zeitschrift CLB, Rolf Kickuth, 69251, Bammentaler Strabe 6-8, Gaiberg, Germany, Tel: +49(69)2600699, Fax: +49(69)2600609 ISSUES/YR: 12

- **Cleft Palate-Craniofacial Journal**
  ISSN: 1055-6656
  Cleft Palate-Craniofac. J.
  American Cleft Palate Craniofacial Association, 1504 East Franklin St. Suite 102, Chapel Hill, NC 27514, United States, Tel: +1-919 933 9044, Fax: +1-919 933 9604 ISSUES/YR: 6

Cleveland Clinic Journal of Medicine
ISSN: 0891-1150
Clevel. Clin. J. Med.
Cleveland Clinic Educational Foundation, 9500 Euclid Avenue, Cleveland, OH 44195-5058, United States, Tel: +1(216)4442661 ISSUES/YR: 12

- **Climacteric**
  ISSN: 1369-7137
  Climacteric
  Taylor and Francis Ltd., 4 Park Square, Milton Park, OX14 4RN, Abingdon, Oxfordshire, United Kingdom, Tel: +44(20)70175541, Fax: +44(20)70175198 ISSUES/YR: 6
- **Clinica Chimica Acta**
  ISSN: 0009-8981
  Clin. Chim. Acta
  Elsevier, P.O. Box 211, 1000 AE, Amsterdam, Netherlands, Tel: +31(20)4853757, Fax: +31(20)4853432 ISSUES/YR: 24

Clinica e Investigacion en Ginecologia y Obstetricia
ISSN: 0210-573X
Clin. Invest. Ginecol. Obstet.
Ediciones Doyma, S.L., Travesera de Gracia 17-21, 08021, Barcelona, Spain, Tel: +34(93)2000711, Fax: +34(93)2091136 ISSUES/YR: 10

Clinica Terapeutica
ISSN: 0009-9074
Clin. Ter.
Societa Editrice Universo, Via G. B. Morgagni 1, 00161, Roma, Italy, Tel: +39(06)4402053/54 ISSUES/YR: 6

Clinical Advances in Hematology and Oncology
ISSN: 1543-0790
Clin. Adv. Hematol. Oncol.
Millennium Medical Publishing, Inc., 611 Broadway, Suite 828, New York, NY 10012, United States, Tel: +1(212) 995-2211, Fax: +1(212) 995-5572 ISSUES/YR: 12

- **Clinical Anatomy**
  ISSN: 0897-3806
  Clin. Anat.
  Wiley-Liss Inc., 111 River Street, Hoboken, NJ 07030-5774, United States, Tel: +1(201)7486645, Fax: +1(201)7486021 ISSUES/YR: 8
- **Clinical and Applied Immunology Reviews**
  ISSN: 1529-1049
  Clin. Appl. Immunol. Rev.
  Elsevier Inc., 360 Park Avenue South, New York, NY 10010, United States, Tel: +1(212)6333950, Fax: +1(212)6333990 ISSUES/YR: 6
- **Clinical and Applied Thrombosis/Hemostasis**
  ISSN: 1076-0296
  Clin. Appl. Thromb. Hemost.
  Westminster Publications Inc., 708 Glen Cove Avenue, Glen Head, NY 11545, United States, Tel: +1(516)7590025, Fax: +1(516)7595524 ISSUES/YR: 4
- **Clinical and Developmental Immunology**
  ISSN: 1740-2522
  Clin. Dev. Immunol.
  Taylor and Francis Ltd., 4 Park Square, Milton Park, OX14 4RN, Abingdon, Oxfordshire, United Kingdom, Tel: +44(20)70175541, Fax: +44(20)70175198 ISSUES/YR: 4
- **Clinical and Diagnostic Laboratory Immunology**
  ISSN: 1071-412X
  Clin. Diagn. Lab. Immunol.
  American Society for Microbiology, 1752 N Street N.W., Washington, DC 20036-2904, United States, Tel: +1(202)942-9244, Fax: +1(202)942-9355 ISSUES/YR: 6
- **Clinical and Experimental Allergy**
  ISSN: 0954-7894
  Clin. Exp. Allergy
  Blackwell Publishing Ltd, 9600 Garsington Road, OX4 2XG, Oxford, United Kingdom, Tel: +44(1865)776868, Fax: +44(1865)714591 ISSUES/YR: 12
- **Clinical and Experimental Allergy Reviews**
  ISSN: 1472-9725
  Clin. Exp. Allergy Suppl. Rev.
  Blackwell Publishing Ltd, 9600 Garsington Road, OX4 2XG, Oxford, United Kingdom, Tel: +44(1865)776868, Fax: +44(1865)714591 ISSUES/YR: 1
- **Clinical and Experimental Dermatology**
  ISSN: 0307-6938
  Clin. Exp. Dermatol.
  Blackwell Publishing Ltd, 9600 Garsington Road, OX4 2XG, Oxford, United Kingdom, Tel: +44(1865)776868, Fax: +44(1865)714591 ISSUES/YR: 6

- **Clinical and Experimental Dermatology, Supplement**
  ISSN: 0966-3487
  Clin. Exp. Dermatol. Suppl.
  Blackwell Publishing Ltd, 9600 Garsington Road, OX4 2XG, Oxford, United Kingdom, Tel: +44(1865)776868, Fax: +44(1865)714591 ISSUES/YR: 1

Clinical and Experimental Hypertension
ISSN: 1064-1963
Clin. Exp. Hypertens.
Taylor and Francis Inc., 325 Chestnut St, Suite 800, Philadelphia PA, PA 19106, United States, Tel: +1(800)3541420, Fax: +1(215)6258914 ISSUES/YR: 8

- **Clinical and Experimental Immunology**
  ISSN: 0009-9104
  Clin. Exp. Immunol.
  Blackwell Publishing Ltd, 9600 Garsington Road, OX4 2XG, Oxford, United Kingdom, Tel: +44(1865)776868, Fax: +44(1865)714591 ISSUES/YR: 12

- **Clinical and Experimental Medicine**
  ISSN: 1591-8890
  Clin. Exp. Med.
  Springer Verlag, Tiergartenstrasse 17, D-69121, Heidelberg, Germany, Tel: +49(6221)4870, Fax: +49(6221)487366 ISSUES/YR: 4

Clinical and Experimental Metastasis
ISSN: 0262-0898
Clin. Exp. Metastasis
Kluwer Academic Publishers, Van Godewijckstraat 30, 3311 GZ, Dordrecht, Netherlands, Tel: +31(78)6576000, Fax: +31(78)6576254 ISSUES/YR: 8

Clinical and Experimental Nephrology
ISSN: 1342-1751
Clin. Exp. Nephrol.
Springer-Verlag Tokyo, 3-13, Hongo 3-chome, Bunkyo-ku, 113-8480, Tokyo, Japan, Tel: +81(3)38180861, Fax: +81(3)38180864 ISSUES/YR: 4

Clinical and Experimental Obstetrics and Gynecology
ISSN: 0390-6663
Clin. Exp. Obstet. Gynecol.
S.O.G. CANADA Inc., 4900 Cote St. Luc, Apt. 212, QUE H3W 2H3, Montreal, Canada, Tel: +1(514)4893242, Fax: +1(514)4854513 ISSUES/YR: 4

Clinical and Experimental Ophthalmology
ISSN: 1442-6404
Clin. Exp. Ophthalmol.
Blackwell Publishing, 550 Swanston Street, VIC 3053, Carlton South, Australia, Tel: +61(3)8359 1011, Fax: +61(3)8359 1120 ISSUES/YR: 6

Clinical and Experimental Pharmacology and Physiology
ISSN: 0305-1870
Clin. Exp. Pharmacol. Physiol.
Blackwell Publishing, 550 Swanston Street, VIC 3053, Carlton South, Australia, Tel: +61(3)8359 1011, Fax: +61(3)8359 1120 ISSUES/YR: 12

- **Clinical and Experimental Rheumatology**
  ISSN: 0392-856X
  Clin. Exp. Rheumatol.
  Clinical and Experimental Rheumatology S.A.S., Via Santa Maria 31, 56126, Pisa, Italy, Tel: +39(050)40124 ISSUES/YR: 6

- **Clinical and Investigative Medicine**
  ISSN: 0147-958X
  Clin. Invest. Med.
  The Canadian Society for Clinical Investigation, 774 Echo Drive Ottawa, K1S 5N8, Ontario, Canada, Tel: +1-613 730 6240, Fax: +1-613 730 1116 ISSUES/YR: 6

- **Clinical and Laboratory Haematology**
  ISSN: 0141-9854
  Clin. Lab. Haematol.
  Blackwell Publishing Ltd, 9600 Garsington Road, OX4 2XG, Oxford, United Kingdom, Tel: +44(1865)776868, Fax: +44(1865)714591 ISSUES/YR: 6

- **Clinical and Laboratory Haematology, Supplement**
  ISSN: 0960-3964
  Clin. Lab. Haematol. Suppl.
  Blackwell Publishing Ltd, 9600 Garsington Road, OX4 2XG, Oxford, United Kingdom, Tel: +44(1865)776868, Fax: +44(1865)714591 ISSUES/YR: 1

Clinical and Molecular Allergy
ISSN: 1476-7961
Clin. Mol. Allergy
BioMed Central Ltd., 34 - 42 Cleveland Street, W1T 4LB, London, United Kingdom, Tel: +44(207)3230323, Fax: +44(207)6319926 ISSUES/YR: 1

Clinical and Refractive Optometry
ISSN: 1705-4850
Clin. Refractive Optom.
Mediconcept Inc., 3333 Cote Vertu Boulevard, Suite 300, QUE H4R 2N1, St. Laurent, Canada, Tel: +1(514)3314561, Fax: +1(514)3361129 ISSUES/YR: 10

Clinical and Surgical Ophthalmology
ISSN: 1705-4842
Clin. Surg. Ophthalmol.
Mediconcept Inc., 3333 Cote Vertu Blvd. Suite 300, H4R 2N1, Montreal, Quebec, Canada, Tel: +1-514 3314561, Fax: +1-514 3361129 ISSUES/YR: 12

Clinical and Transfusion Haematology
ISSN: 0861-7880
Clin. Transf. Haematol.
Medical Information Center, 1 Sv. G. Sofijski str., 1431, Sofia, Bulgaria, Tel: +359-522342, Fax: +359-522393 ISSUES/YR: 4

Clinical and Translational Oncology
ISSN: 1699-048X
Clin. Transl. Oncol.
Ediciones Doyma, S.L., Travesera de Gracia 17-21, 08021, Barcelona, Spain, Tel: +34(93)2000711, Fax: +34(93)2091136 ISSUES/YR: 6

- **Clinical and Vaccine Immunology**
  ISSN: 1556-6811
  Clin. Vaccine Immunol.
  American Society for Microbiology, 1752 N Street N.W., Washington, DC 20036-2904, United States, Tel: +1(202)942-9244, Fax: +1(202)942-9355 ISSUES/YR: 12

Clinical Application of Immunology
ISSN: 1312-0832
Clin. Appl. Immunol.
Publishing House of University School of Medicine, 1, St. Kliment Ohridski Street, 5800, Pleven, Bulgaria, Tel: +359(64)30294, Fax: +359(64)801603 ISSUES/YR: 4

Clinical Autonomic Research
ISSN: 0959-9851
Clin. Auton. Res.
Dr. Dietrich Steinkopff Verlag GmbH and Co. KG, P.O. Box 100462, D-64204, Darmstadt, Germany, Tel: +49(6151)828990, Fax: +49(6151)8289940 ISSUES/YR: 6

- **Clinical Biochemistry**
  ISSN: 0009-9120
  Clin. Biochem.
  Elsevier Inc., 360 Park Avenue South, New York, NY 10010, United States, Tel: +1(212)6333950, Fax: +1(212)6333990 ISSUES/YR: 12

- **Clinical Biomechanics**
  ISSN: 0268-0033
  Clin. Biomech.
  Elsevier Ltd, Langford Lane, Kidlington, OX5 1GB, Oxford, United Kingdom, Tel: +44(1865)843000 ISSUES/YR: 10

Clinical Breast Cancer
ISSN: 1526-8209
Clin. Breast Cancer
Cancer Information Group, LP, 3535 Worth Street, Dallas, TX 75246, United States, Tel: +1(214)8207126, Fax: +1(214)8208844 ISSUES/YR: 4

- **Clinical Cancer Research**
  ISSN: 1078-0432
  Clin. Cancer Res.
  American Association for Cancer Research Inc., 150 South Independence Mall West, Philadelphia, PA 19106-3483, United States, Tel: +1(215)4409300, Fax: +1(215)4409354 ISSUES/YR: 24

Clinical Cardiology
ISSN: 0160-9289
Clin. Cardiol.
Foundation for Advances in Medicine and Science Inc., P.O. Box 832, Mahwah, NJ 07430-0832, United States, Tel: +1(201)8181010, Fax: +1(201)8180086 ISSUES/YR: 12

Clinical Case Studies
ISSN: 1534-6501
Clin. Case Stud.
SAGE Publications Inc., 2455 Teller Road, Thousand Oaks, CA 91320, United States, Tel: +1(805)4990721, Fax: +1(805)4990871 ISSUES/YR: 4

Clinical Cases in Mineral and Bone Metabolism
ISSN: 1724-8914
Clin. Cases Miner. Bone Metab.
CIC Edizioni Internazionali s.r.l., Corso Trieste 42, 00198, Roma, Italy, Tel: +39(06)8412673, Fax: +39(06)8412688 ISSUES/YR: 3

Clinical Chemistry
ISSN: 0009-9147
Clin. Chem.
American Association for Clinical Chemistry Inc., 2101 L Street NW, Suite 202, Washington, DC 20037-1526, United States, Tel: +1(202)8570717 ISSUES/YR: 12

• **Clinical Chemistry and Laboratory Medicine**
ISSN: 1434-6621
Clin. Chem. Lab. Med.
Walter de Gruyter and Co., Genthiner Strasse 13, D-10785, Berlin, Germany, Tel: +49(30)260-050, Fax: +49(30)260-05251 ISSUES/YR: 12

Clinical Child Psychology and Psychiatry
ISSN: 1359-1045
Clin. Child Psychol. Psychiatry
SAGE Publications Ltd, 55 City Road, EC1Y 1SP, London, United Kingdom, Tel: +1(207)3248500, Fax: +1(207)3248600 ISSUES/YR: 4

Clinical Colorectal Cancer
ISSN: 1533-0028
Clin. Colorectal Cancer
Cancer Information Group, LP, 3535 Worth Street, Dallas, TX 75246, United States, Tel: +1(214)8207126, Fax: +1(214)8208844 ISSUES/YR: 4

Clinical Cornerstone
ISSN: 1098-3597
Clin. Cornerstone
Excerpta Medica Inc., 105 Raider Blvd, Suite 101, Hillsborough, NJ 08844, United States, Tel: +1(908)8748550, Fax: +1(908)8745633 ISSUES/YR: 4

Clinical Dermatology: Retinoids and other Treatments
ISSN: 1750-306X
Clin. Dermatol. Retinoids Treat.
Mediscript Ltd, 1 Mountview Court,310 Friern Barnet Lande, N20 0LD, London, United Kingdom, Tel: +44-208 446 8898, Fax: +44-208 446 9194 ISSUES/YR: 4

Clinical Diabetes
ISSN: 0891-8929
Clin. Diabetes
American Diabetes Association Inc., 1701 North Beauregard St., Alexandria, VA 22311, United States, Tel: +1(703)5491500, Fax: +1-703 5496995 ISSUES/YR: 4

• **Clinical Drug Investigation**
ISSN: 1173-2563
Clin. Drug Invest.
Adis International Ltd, 41 Centorian Drive, Private Bag 65901, Mairangi Bay, 1311, Auckland 10, New Zealand, Tel: +64(9)4770700, Fax: +64(9)4770766 ISSUES/YR: 12

• **Clinical Dysmorphology**
ISSN: 0962-8827
Clin. Dysmorphol.
Lippincott Williams and Wilkins, 241 Borough High Street, SE1 1GB, London, United Kingdom, Tel: +44(207)9407500, Fax: +44(207)9407575 ISSUES/YR: 4

• **Clinical EEG and Neuroscience**
ISSN: 0009-9155
Clin. EEG Neurosci.
EEG and Clinical Neuroscience Society (ECNS), 805 W. Liberty Drive, Wheaton, IL 60187, United States, Tel: +1(888)3662203, Fax: +1(630)6536233 ISSUES/YR: 4

• **Clinical Endocrinology**
ISSN: 0300-0664
Clin. Endocrinol.
Blackwell Publishing Ltd, 9600 Garsington Road, OX4 2XG, Oxford, United Kingdom, Tel: +44(1865)776868, Fax: +44(1865)714591 ISSUES/YR: 12

Clinical Gastroenterology and Hepatology
ISSN: 1542-3565
Clin. Gastroenterol. Hepatol.
W.B. Saunders, Independence Square West, Philadelphia, PA 19106-3399, United States, Tel: +1(215)2387800, Fax: +1(215)2386445 ISSUES/YR: 12

• **Clinical Genetics**
ISSN: 0009-9163
Clin. Genet.
Blackwell Publishing Ltd, 9600 Garsington Road, OX4 2XG, Oxford, United Kingdom, Tel: +44(1865)776868, Fax: +44(1865)714591 ISSUES/YR: 12

Clinical Genitourinary Cancer
ISSN: 1558-7673
Clin. Genitourin. Cancer
Cancer Information Group, LP, 3535 Worth Street, Dallas, TX 75246, United States, Tel: +1(214)8207126, Fax: +1(214)8208844 ISSUES/YR: 4

Clinical Geriatrics
ISSN: 1095-1598
Clin. Geriatr.
HMP Communications LLP, 4365 U.S. Highway 1 Suite 250, Princeton, NJ 08540, United States, Tel: +1(609)452 9200, Fax: +1(609)452 8241 ISSUES/YR: 12

Clinical Gerontologist
ISSN: 0731-7115
Clin. Gerontol.
Haworth Press Inc., 10 Alice Street, Binghamton, NY 13904-1580, United States, Tel: +1(607)7225857, Fax: +1(607)7710012 ISSUES/YR: 12

• **Clinical Governance**
ISSN: 1477-7274
Clin. Gov.
Emerald, 60/62 Toller Lane, BD8 9BY, Bradford, West Yorkshire, United Kingdom, Tel: +44(1274)777700, Fax: +44(1274)785200 ISSUES/YR: 4

Clinical Hemorheology and Microcirculation
ISSN: 1386-0291
Clin. Hemorheol. Microcirc.
IOS Press, Nieuwe Hemweg 6B, 1013 BG, Amsterdam, Netherlands, Tel: +31(20)6883355, Fax: +31(20)6203419 ISSUES/YR: 8

• **Clinical Imaging**
ISSN: 0899-7071
Clin. Imaging
Elsevier Inc., 360 Park Avenue South, New York, NY 10010, United States, Tel: +1(212)6333950, Fax: +1(212)6333990 ISSUES/YR: 6

• **Clinical Immunology**
ISSN: 1521-6616
Clin. Immunol.
Academic Press Inc., 6277 Sea Harbor Drive, Orlando, FL 32887-4900, United States, Tel: +1(407)3454040, Fax: +1(407)3639661 ISSUES/YR: 12

• **Clinical Infectious Diseases**
ISSN: 1058-4838
Clin. Infect. Dis.
University of Chicago Press, 1427 E. 60th Street, Chicago, IL 60637-2954, United States, Tel: +1(773)7027600, Fax: +1(773)7020694 ISSUES/YR: 24

Clinical Intensive Care
ISSN: 0956-3075
Clin. Intensive Care
Taylor and Francis Ltd., 4 Park Square, Milton Park, OX14 4RN, Abingdon, Oxfordshire, United Kingdom, Tel: +44(20)70175541, Fax: +44(20)70175198 ISSUES/YR: 4

• **Clinical Journal of Pain**
ISSN: 0749-8047
Clin. J. Pain
Lippincott Williams and Wilkins, 530 Walnut Street, Philadelphia, PA 19106-3621, United States, Tel: +1(215)5218342, Fax: +1(215)5218326 ISSUES/YR: 6

• **Clinical Journal of Sport Medicine**
ISSN: 1050-642X
Clin. J. Sport Med.
Lippincott Williams and Wilkins, 530 Walnut Street, Philadelphia, PA 19106-3621, United States, Tel: +1(215)5218342, Fax: +1(215)5218326 ISSUES/YR: 6

Clinical Kinesiology
ISSN: 0896-9620
Clin. Kinesiology
Turnkey Publishing Inc., 11450 Bristle Oak Trail, Austin, TX 78750, United States, Tel: +1-512 335 2288, Fax: +1-512 335 3083 ISSUES/YR: 4

Clinical Laboratory
ISSN: 1433-6510
Clin. Lab.
Verlag Klinisches Labor GmbH, Im Breitspiel 15, D-69126, Heidelberg, Germany, Tel: +49(6221)3432115 ISSUES/YR: 12

• **Clinical Linguistics and Phonetics**
ISSN: 0269-9206
Clin. Linguist. Phon.
Taylor and Francis Ltd., 4 Park Square, Milton Park, OX14 4RN, Abingdon, Oxfordshire, United Kingdom, Tel: +44(20)70175541, Fax: +44(20)70175198 ISSUES/YR: 10

Clinical Lung Cancer
ISSN: 1525-7304
Clin. Lung Cancer
Cancer Information Group, LP, 3535 Worth Street, Dallas, TX 75246, United States, Tel: +1(214)8207126, Fax: +1(214)8208844 ISSUES/YR: 6

Clinical Lymphoma and Myeloma
ISSN: 1557-9190
Clin. Lymphoma Myeloma
Cancer Information Group, LP, 3535 Worth Street, Dallas, TX 75246, United States, Tel: +1(214)8207126, Fax: +1(214)8208844 ISSUES/YR: 4

Clinical Medicine and Research
ISSN: 1539-4182
Clin. Med. Res.
Marshfield Clinic, 1000 North Oak Avenue, Marshfield, WI 54449, United States, Tel: +1-7153875511 ISSUES/YR: 4

Clinical Medicine, Journal of the Royal College of Physicians of London
ISSN: 1470-2118
Clin. Med. J. R. Coll. Phys. Lond.
Royal College of Physicians, 11 St Andrews Place, Regents Park, NW1 4LE, London, United Kingdom, Tel: +44(207)9351174, Fax: +44(207)4875218 ISSUES/YR: 6

• **Clinical Microbiology and Infection**
ISSN: 1198-743X
Clin. Microbiol. Infect.
Blackwell Publishing Ltd, 9600 Garsington Road, OX4 2XG, Oxford, United Kingdom, Tel: +44(1865)776868, Fax: +44(1865)714591 ISSUES/YR: 12

• **Clinical Microbiology and Infection, Supplement**
ISSN: 1470-9465
Clin. Microbiol. Infect. Suppl.
Blackwell Publishing Ltd, 9600 Garsington Road, OX4 2XG, Oxford, United Kingdom, Tel: +44(1865)776868, Fax: +44(1865)714591 ISSUES/YR: 1

Clinical Microbiology Newsletter
ISSN: 0196-4399
Clin. Microbiol. Newsl.
Elsevier Inc., 360 Park Avenue South, New York, NY 10010, United States, Tel: +1(212)6333950, Fax: +1(212)6333990 ISSUES/YR: 24

Clinical Microbiology Reviews
ISSN: 0893-8512
Clin. Microbiol. Rev.
American Society for Microbiology, 1752 N Street N.W., Washington, DC 20036-2904, United States, Tel: +1(202)942-9244, Fax: +1(202)942-9355 ISSUES/YR: 4

Clinical MRI
ISSN: 0961-9275
Clin. MRI
Lanmark Press, P.O.Box 317, Beaconsfield, HP9 2LR, Buckinghamshire, United Kingdom, Tel: +44-1494 675535, Fax: +44-1494 670588 ISSUES/YR: 4

Clinical Nephrology
ISSN: 0301-0430
Clin. Nephrol.
Dustri-Verlag Dr. Karl Feistle, P.O. Box 1351, D-82032, Deisenhofen/Munchen, Germany, Tel: +49(89)6138610, Fax: +49(89)6135412 ISSUES/YR: 12

Clinical Neurology
ISSN: 0009-918X
Clin. Neurol.
Societas Neurologica Japonica, 31-21 Yushima 2-chome, Bunkyo-ku, 113, Tokyo, Japan, Tel: +81(3)38151080 ISSUES/YR: 12

Clinical Neurology and Neurosurgery
ISSN: 0303-8467
Clin. Neurol. Neurosurg.
Elsevier, P.O. Box 211, 1000 AE, Amsterdam, Netherlands, Tel: +31(20)4853757, Fax: +31(20)4853432 ISSUES/YR: 4

• **Clinical Neuropathology**
ISSN: 0722-5091
Clin. Neuropathol.
Dustri-Verlag Dr. Karl Feistle, P.O. Box 1351, D-82032, Deisenhofen/Munchen, Germany, Tel: +49(89)6138610, Fax: +49(89)6135412 ISSUES/YR: 6

• **Clinical Neuropharmacology**
ISSN: 0362-5664
Clin. Neuropharmacol.
Lippincott Williams and Wilkins, 530 Walnut Street, Philadelphia, PA 19106-3621, United States, Tel: +1(215)5218342, Fax: +1(215)5218326 ISSUES/YR: 6

• **Clinical Neurophysiology**
ISSN: 1388-2457
Clin. Neurophysiol.
Elsevier Ireland Ltd, P.O. Box 85, Limerick, Ireland, Tel: +353(61)471944, Fax: +353(61)472144 ISSUES/YR: 12

Clinical Neuropsychiatry
ISSN: 1724-4935
Clin. Neuropsychiatry
Giovanni Fioriti Editore, Via Archimede 179, 00197, Rome, Italy, Tel: +39(06)8072063, Fax: +39(06)80664609 ISSUES/YR: 6

Clinical Neuropsychologist
ISSN: 1385-4046
Clin. Neuropsychol.
Psychology Press Ltd, 4 Park Square, Milton Park, OX14 4RN, Abingdon, Oxfordshire, United Kingdom, Tel: +44(1235)828600, Fax: +44(1235)829000 ISSUES/YR: 4

• **Clinical Neuroscience Research**
ISSN: 1566-2772
Clin. Neurosc. Res.
Elsevier Ltd, Langford Lane, Kidlington, OX5 1GB, Oxford, United Kingdom, Tel: +44(1865)843000, Fax: +44(1865)843010 ISSUES/YR: 6

Clinical Nuclear Medicine
ISSN: 0363-9762
Clin. Nucl. Med.
Lippincott Williams and Wilkins, 530 Walnut Street, Philadelphia, PA 19106-3621, United States, Tel: +1(215)5218342, Fax: +1(215)5218326 ISSUES/YR: 12

Clinical Nutrition
ISSN: 0261-5614
Clin. Nutr.
Churchill Livingstone, 1-3 Baxter's Place, Leith Walk, EH1 3AF, Edinburgh, United Kingdom, Tel: +44(131)5562424, Fax: +44(131)5581278 ISSUES/YR: 6

Clinical Nutrition, Supplement
ISSN: 1744-1161
Clin. Nutr. Suppl.
Churchill Livingstone, 1-3 Baxter's Place, Leith Walk, EH1 3AF, Edinburgh, United Kingdom, Tel: +44(131)5562424, Fax: +44(131)5581278 ISSUES/YR: 1

Clinical Obstetrics and Gynecology
ISSN: 0009-9201
Clin. Obstet. Gynecol.
Lippincott Williams and Wilkins, 530 Walnut Street, Philadelphia, PA 19106-3621, United States, Tel: +1(215)5218342, Fax: +1(215)5218326 ISSUES/YR: 4

• **Clinical Oncology**
ISSN: 0936-6555
Clin. Oncol.
Elsevier, P.O. Box 211, 1000 AE, Amsterdam, Netherlands, Tel: +44(20)4853757, Fax: +44(20)4853432 ISSUES/YR: 8

- **Clinical Orthopaedics and Related Research**
ISSN: 0009-921X
Clin. Orthop. Relat. Res.
Lippincott Williams and Wilkins, 530 Walnut Street, Philadelphia, PA 19106-3621, United States, Tel: +1(215)5218342, Fax: +1(215)5218326 ISSUES/YR: 12
- **Clinical Otolaryngology**
ISSN: 0307-7772
Clin. Otolaryngol.
Blackwell Publishing Ltd, 9600 Garsington Road, OX4 2XG, Oxford, United Kingdom, Tel: +44(1865)776868, Fax: +44(1865)714591 ISSUES/YR: 6

Clinical Pediatric Emergency Medicine
ISSN: 1522-8401
Clin. Pediatr. Emerg. Med.
W.B. Saunders Ltd, 32 Jamestown Road, NW1 7BY, London, United Kingdom, Tel: +44(20)74244200, Fax: +44(20)74854752 ISSUES/YR: 4

Clinical Pediatric Endocrinology
ISSN: 0918-5739
Clin. Pediatr. Endocrinol.
Jeff Corporation Co. Ltd, 2-3-12 Shiba, Minato-ku, 105, Tokyo, Japan ISSUES/YR: 2

Clinical Pediatrics
ISSN: 0009-9228
Clin. Pediatr.
Westminster Publications Inc., 708 Glen Cove Avenue, Glen Head, NY 11545, United States, Tel: +1(516)7590025, Fax: +1(516)7595524 ISSUES/YR: 9

- **Clinical Pharmacokinetics**
ISSN: 0312-5963
Clin. Pharmacokinet.
Adis International Ltd, 41 Centorian Drive, Private Bag 65901, Mairangi Bay, 1311, Auckland 10, New Zealand, Tel: +64(9)4770700, Fax: +64(9)4770766 ISSUES/YR: 15
- **Clinical Pharmacology and Therapeutics**
ISSN: 0009-9236
Clin. Pharmacol. Ther.
Mosby Inc., 11830 Westline Industrial Drive, St. Louis, MO 63146, United States, Tel: +1(314)8728370, Fax: +1(314)4321380 ISSUES/YR: 12
- **Clinical Physiology and Functional Imaging**
ISSN: 1475-0961
Clin. Physiol. Funct. Imaging
Blackwell Publishing Ltd, 9600 Garsington Road, OX4 2XG, Oxford, United Kingdom, Tel: +44(1865)776868, Fax: +44(1865)714591 ISSUES/YR: 6

Clinical Practice and Epidemiology in Mental Health
ISSN: 1745-0179
Clin. Pract. Epidemiol. Ment. Health
BioMed Central Ltd., 34 - 42 Cleveland Street, W1T 4LB, London, United Kingdom, Tel: +44(207)3230323, Fax: +44(207)6319926 ISSUES/YR: 1

- **Clinical Proteomics**
ISSN: 1542-6416
Clin. Proteomics
Humana Press Inc., 999 Riverview Drive, Suite 208, Totowa, NJ 07512-1165, United States, Tel: +1(973)2561699, Fax: +1(973)2568341 ISSUES/YR: 4
- **Clinical Psychology and Psychotherapy**
ISSN: 1063-3995
Clin. Psychol. Psychother.
John Wiley and Sons Ltd, Southern Gate, PO19 8SQ, Chichester, West Sussex, United Kingdom, Tel: +44(1243)779777, Fax: +44(1243)775878 ISSUES/YR: 6

Clinical Psychology Forum
ISSN: 1747-5732
Clin. Psychol. Forum
British Psychological Society, 48 Princess Road East, LE1 7DR, Leicester, United Kingdom, Tel: +44(116)2549568, Fax: +44(116)2470787 ISSUES/YR: 12

Clinical Psychology Review
ISSN: 0272-7358
Clin. Psychol. Rev.
Elsevier Inc., 360 Park Avenue South, New York, NY 10010, United States, Tel: +1(212)6333950, Fax: +1(212)6333990 ISSUES/YR: 8

Clinical Psychology: Science and Practice
ISSN: 0969-5893
Clin. Psychol. Sci. Pract.
Blackwell Publishing Inc., 350 Main Street, Malden, MA 02148, United States, Tel: +1(781)3888200, Fax: +1(781)3888210 ISSUES/YR: 4

Clinical Psychopharmacology and Neuroscience
ISSN: 1738-1088
Clin. Psychopharmacol. Neurosci.
Korean College of Neuropsychopharmacology, #320 Life Officetel, 150-010, 61-3 Youido-dong, Yeongdeungpo-gu, Seoul, South Korea, Tel: +82-2-784-2742, Fax: +82-2-784-5542 ISSUES/YR: 4

Clinical Pulmonary Medicine
ISSN: 1068-0640
Clin. Pulm. Med.
Lippincott Williams and Wilkins, 351 West Camden Street, Baltimore, MD 21201-2436, United States, Tel: +1(410)5284000 ISSUES/YR: 6

- **Clinical Radiology**
ISSN: 0009-9260
Clin. Radiol.
W.B. Saunders Ltd, 32 Jamestown Road, NW1 7BY, London, United Kingdom, Tel: +44(20)74244200, Fax: +44(20)74854752 ISSUES/YR: 12
- **Clinical Radiology Extra**
ISSN: 1477-6804
Clin. Radiol. Extra
W.B. Saunders Ltd, 32 Jamestown Road, NW1 7BY, London, United Kingdom, Tel: +44(20)74244200, Fax: +44(20)74854752 ISSUES/YR: 12

Clinical Rehabilitation
ISSN: 0269-2155
Clin. Rehabil.
Arnold, 338 Euston Road, NW1 3BH, London, United Kingdom, Tel: +44(20)78736000, Fax: +44(20)78736376 ISSUES/YR: 8

Clinical Research and Regulatory Affairs
ISSN: 1060-1333
Clin. Res. Regul. Aff.
Taylor and Francis Inc., 325 Chestnut St, Suite 800, Philadelphia PA, PA 19106, United States, Tel: +1(800)3541420, Fax: +1(215)6258914 ISSUES/YR: 4

Clinical Research in Cardiology
ISSN: 1861-0684
Clin. Res. Cardiol.
Dr. Dietrich Steinkopff Verlag GmbH and Co. KG, P.O. Box 100462, D-64204, Darmstadt, Germany, Tel: +49(6151)828990, Fax: +49(6151)8289940 ISSUES/YR: 12

Clinical Reviews in Allergy and Immunology
ISSN: 1080-0549
Clin. Rev. Allergy Immunol.
Humana Press Inc., 999 Riverview Drive, Suite 208, Totowa, NJ 07512-1165, United States, Tel: +1(973)2561699, Fax: +1(973)2568341 ISSUES/YR: 6

Clinical Reviews in Bone and Mineral Metabolism
ISSN: 1534-8644
Clin. Rev. Bone Miner. Metab.
Humana Press Inc., 999 Riverview Drive, Suite 208, Totowa, NJ 07512-1165, United States, Tel: +1(973)2561699, Fax: +1(973)2568341 ISSUES/YR: 4

- **Clinical Rheumatology**
ISSN: 0770-3198
Clin. Rheumatol.
Springer-Verlag London Ltd, Catteshall Road, Farncombe, GU7 3DJ, Godalming, Surrey, United Kingdom, Tel: +44(1483)414142, Fax: +44(1483)415144 ISSUES/YR: 6

Clinical Risk
ISSN: 1356-2622
Clin. Risk
Royal Society of Medicine Press Ltd, P.O. Box 9002, W1A 0ZA, London, United Kingdom, Tel: +44(207)2902927, Fax: +44(207)2902929 ISSUES/YR: 6

- **Clinical Science**
  ISSN: 0143-5221
  Clin. Sci.
  Portland Press Ltd, 59 Portland Place, W1N 3AJ, London, United Kingdom, Tel: +44(20)75805530, Fax: +44(20)73231136 ISSUES/YR: 12
  Clinical Teacher
  ISSN: 1743-4971
  Clin. Teach.
  Blackwell Publishing Ltd, 9600 Garsington Road, OX4 2XG, Oxford, United Kingdom, Tel: +44(1865)776868, Fax: +44(1865)714591 ISSUES/YR: 2
  Clinical Therapeutics
  ISSN: 0149-2918
  Clin. Ther.
  Excerpta Medica Inc., 105 Raider Blvd, Suite 101, Hillsborough, NJ 08844, United States, Tel: +1(908)8748550, Fax: +1(908)8745633 ISSUES/YR: 12
  Clinical Toxicology
  ISSN: 1556-3650
  Clin. Toxicol.
  Taylor and Francis Inc., 325 Chestnut St, Suite 800, Philadelphia PA, PA 19106, United States, Tel: +1(800)3541420, Fax: +1(215)6258914 ISSUES/YR: 7
- **Clinical Transplantation**
  ISSN: 0902-0063
  Clin. Transplant.
  Blackwell Publishing Ltd, 9600 Garsington Road, OX4 2XG, Oxford, United Kingdom, Tel: +44(1865)776868, Fax: +44(1865)714591 ISSUES/YR: 6
- **Clinical Transplantation, Supplement**
  ISSN: 1399-6738
  Clin. Transplant. Suppl.
  Blackwell Publishing Ltd, 9600 Garsington Road, OX4 2XG, Oxford, United Kingdom, Tel: +44(1865)776868, Fax: +44(1865)714591 ISSUES/YR: 1
- **Clinical Trials**
  ISSN: 1740-7745
  Clin. Trials
  Hodder Arnold Journals, 338 Euston Road, NW1 3BH, London, United Kingdom, Tel: +44-+44 (0)20 7873 6337, Fax: +44-+44 (0)20 7873 6376 ISSUES/YR: 6
- **Clinician in Management**
  ISSN: 0965-5751
  Clin. Manage.
  Radcliffe Publishing Ltd, 18 Marcham Road, OX14 1AA, Abingdon, Oxon, United Kingdom, Tel: +44(1235)528820, Fax: +44(1235)528830 ISSUES/YR: 4
- **Clinics in Chest Medicine**
  ISSN: 0272-5231
  Clin. Chest Med.
  W.B. Saunders, Independence Square West, Philadelphia, PA 19106-3399, United States, Tel: +1(215)2387800, Fax: +1(215)2386445 ISSUES/YR: 4
  Clinics in Colon and Rectal Surgery
  ISSN: 1531-0043
  Clin. Colon Rectal Surg.
  Thieme Medical Publishers, Inc., 333 7th Avenue, New York, NY 10001-5004, United States, Tel: +1(212)7600888, Fax: +1(212)7799020 ISSUES/YR: 4
- **Clinics in Dermatology**
  ISSN: 0738-081X
  Clin. Dermatol.
  Elsevier Inc., 360 Park Avenue South, New York, NY 10010, United States, Tel: +1(212)6333950, Fax: +1(212)6333990 ISSUES/YR: 6
  Clinics in Family Practice
  ISSN: 1522-5720
  Clin. Fam. Pract.
  W.B. Saunders, Independence Square West, Philadelphia, PA 19106-3399, United States, Tel: +1(215)2387800, Fax: +1(215)2386445 ISSUES/YR: 4
  Clinics in Geriatric Medicine
  ISSN: 0749-0690
  Clin. Geriatr. Med.
  W.B. Saunders, Independence Square West, Philadelphia, PA 19106-3399, United States, Tel: +1(215)2387800, Fax: +1(215)2386445 ISSUES/YR: 4
- **Clinics in Laboratory Medicine**
  ISSN: 0272-2712
  Clin. Lab. Med.
  W.B. Saunders, Independence Square West, Philadelphia, PA 19106-3399, United States, Tel: +1(215)2387800, Fax: +1(215)2386445 ISSUES/YR: 4
  Clinics in Liver Disease
  ISSN: 1089-3261
  Clin. Liver Dis.
  W.B. Saunders, Independence Square West, Philadelphia, PA 19106-3399, United States, Tel: +1(215)2387800, Fax: +1(215)2386445 ISSUES/YR: 4
  Clinics in Occupational and Environmental Medicine
  ISSN: 1526-0046
  Clin. Occup. Environ. Med.
  W.B. Saunders, Independence Square West, Philadelphia, PA 19106-3399, United States, Tel: +1(215)2387800, Fax: +1(215)2386445 ISSUES/YR: 4
- **Clinics in Perinatology**
  ISSN: 0095-5108
  Clin. Perinatol.
  W.B. Saunders, Independence Square West, Philadelphia, PA 19106-3399, United States, Tel: +1(215)2387800, Fax: +1(215)2386445 ISSUES/YR: 4
  Clinics in Plastic Surgery
  ISSN: 0094-1298
  Clin. Plast. Surg.
  W.B. Saunders, Independence Square West, Philadelphia, PA 19106-3399, United States, Tel: +1(215)2387800, Fax: +1(215)2386445 ISSUES/YR: 4
- **Clinics in Podiatric Medicine and Surgery**
  ISSN: 0891-8422
  Clin. Podiatr. Med. Surg.
  W.B. Saunders, Independence Square West, Philadelphia, PA 19106-3399, United States, Tel: +1(215)2387800, Fax: +1(215)2386445 ISSUES/YR: 4
  Clinics in Sports Medicine
  ISSN: 0278-5919
  Clin. Sports Med.
  W.B. Saunders, Independence Square West, Philadelphia, PA 19106-3399, United States, Tel: +1(215)2387800, Fax: +1(215)2386445 ISSUES/YR: 4
- **Cloning and Stem Cells**
  ISSN: 1536-2302
  Cloning Stem Cells
  Mary Ann Liebert Inc., 140 Huguenot Street, New Rochelle, NY 10801-5215, United States, Tel: +1(914)7402100, Fax: +1(914)7402101 ISSUES/YR: 4
  CME Bulletin Endocrinology and Diabetes
  ISSN: 1367-8973
  CME Bull. Endocrinol. Diabetes
  Rila Publications Ltd, 73 Newman Street, W1A 4PG, London, United Kingdom, Tel: +44(207)6311299, Fax: +44(207)5807166 ISSUES/YR: 3
  CME Bulletin Otorhinolaryngology Head and Neck Surgery
  ISSN: 1364-8829
  CME Bull. Otorhinolaryngol. Head Neck Surg.
  Rila Publications Ltd, 73 Newman Street, W1A 4PG, London, United Kingdom, Tel: +44(207)6311299, Fax: +44(207)5807166 ISSUES/YR: 3
  CME Cancer Medicine
  ISSN: 1475-8075
  CME Cancer Med.
  Rila Publications Ltd, 73 Newman Street, W1A 4PG, London, United Kingdom, Tel: +44(207)6311299, Fax: +44(207)5807166 ISSUES/YR: 3
  CME Journal Gastroenterology, Hepatology and Nutrition
  ISSN: 1367-9015
  CME J. Gastroenterol. Hepatol. Nutr.
  Rila Publications Ltd, 73 Newman Street, W1A 4PG, London, United Kingdom, Tel: +44(207)6311299, Fax: +44(207)5807166 ISSUES/YR: 3
  CME Journal Geriatric Medicine
  ISSN: 1475-1453
  CME J. Geriatr. Med.
  Rila Publications Ltd, 73 Newman Street, W1A 4PG, London, United Kingdom, Tel: +44(207)6311299, Fax: +44(207)5807166 ISSUES/YR: 3

CME Journal of Gynecologic Oncology
ISSN: 1219-9087
CME J. Gynecol. Oncol.
Primed-X Press, P.O. Box 46, 1301, Budapest, Hungary, Tel: +36(1)2752172, Fax: +36(1)2752172 ISSUES/YR: 3

CME Journal Ophthalmology
ISSN: 1367-0131
CME J. Ophthalmol.
Rila Publications Ltd, 73 Newman Street, W1A 4PG, London, United Kingdom, Tel: +44(207)6311299, Fax: +44(207)5807166 ISSUES/YR: 3

CME Journal Radiology
ISSN: 1466-8386
CME J. Radiol.
Rila Publications Ltd, 73 Newman Street, W1A 4PG, London, United Kingdom, Tel: +44(207)6311299, Fax: +44(207)5807166 ISSUES/YR: 3

CME Orthopaedics
ISSN: 1367-8957
CME Orthop.
Rila Publications Ltd, 73 Newman Street, W1A 4PG, London, United Kingdom, Tel: +44(207)6311299, Fax: +44(207)5807166 ISSUES/YR: 3

CNS and Neurological Disorders - Drug Targets
ISSN: 1871-5273
CNS Neurol. Disord. Drug Targets
Bentham Science Publishers B.V., P.O. Box 294, 1400 AG, Bussum, Netherlands, Tel: +31-356257290, Fax: +31-847511310 ISSUES/YR: 6

CNS Drug Reviews
ISSN: 1080-563X
CNS Drug Rev.
Blackwell Publishing Ltd, 9600 Garsington Road, OX4 2XG, Oxford, United Kingdom, Tel: +44(1865)776868, Fax: +44(1865)714591 ISSUES/YR: 4

• **CNS Drugs**
ISSN: 1172-7047
CNS Drugs
Adis International Ltd, 41 Centorian Drive, Private Bag 65901, Mairangi Bay, 1311, Auckland 10, New Zealand, Tel: +64(9)4770700, Fax: +64(9)4770766 ISSUES/YR: 12

• **CNS Spectrums**
ISSN: 1092-8529
CNS Spectr.
MBL Communications, 333 Hudson St. 7th Floor, NY 10013, United States, Tel: +1-212 328 0800 EXT: 222, Fax: +1-212 328 0600 ISSUES/YR: 12

Cochlear Implants International
ISSN: 1467-0100
Cochlear Implants Int.
John Wiley and Sons Ltd, Southern Gate, PO19 8SQ, Chichester, West Sussex, United Kingdom, Tel: +44(1243)779777, Fax: +44(1243)775878 ISSUES/YR: 2

Cochrane Database of Systematic Reviews
ISSN: 1469-493X
Cochrane Database Syst. Rev.
John Wiley and Sons Ltd, Southern Gate, PO19 8SQ, Chichester, West Sussex, United Kingdom, Tel: +44(1243)779777, Fax: +44(1243)775878 ISSUES/YR: 4

Cocuk Cerrahisi Dergisi
ISSN: 1305-5194
Cocuk Cerrahisi Derg.
Logos Medical Publishing, Yildiz Posta Cad. No.36, 80280, D.66-67 Gayrettepe, Istanbul, Turkey, Tel: +90(212)2880541, Fax: +90(212)2116185 ISSUES/YR: 3

Cocuk Sagligi ve Hastaliklari Dergisi
ISSN: 0010-0161
Cocuk Sagligi Hast. Derg.
Cocuk Sagligi ve Hastaliklan Dergisi, Samanpazari, P.O. Box 66, 06240, Ankara, Turkey, Tel: +90(312)3244291 ISSUES/YR: 4

• **Cognition**
ISSN: 0010-0277
Cognition
Elsevier, P.O. Box 211, 1000 AE, Amsterdam, Netherlands, Tel: +31(20)4853757, Fax: +31(20)4853432 ISSUES/YR: 12

• **Cognition and Emotion**
ISSN: 0269-9931
Cogn. Emot.
Psychology Press Ltd, 4 Park Square, Milton Park, OX14 4RN, Abingdon, Oxfordshire, United Kingdom, Tel: +44(1235)828600, Fax: +44(1235)829000 ISSUES/YR: 8

• **Cognitive and Behavioral Neurology**
ISSN: 1543-3633
Cogn. Behav. Neurol.
Lippincott Williams and Wilkins, 530 Walnut Street, Philadelphia, PA 19106-3621, United States, Tel: +1(215)5218342, Fax: +1(215)5218326 ISSUES/YR: 4

• **Cognitive and Behavioral Practice**
ISSN: 1077-7229
Cogn. Behav. Pract.
Elsevier Inc., 360 Park Avenue South, New York, NY 10010, United States, Tel: +1(212)6333950, Fax: +1(212)6333990 ISSUES/YR: 4

Cognitive Behaviour Therapy
ISSN: 1650-6073
Cogn. Behav. Ther.
Taylor and Francis A.S., P.O. Box 2562 Solli, 0202, Oslo, Norway, Tel: +47(22)129880, Fax: +47(22)129890 ISSUES/YR: 4

• **Cognitive Neuropsychiatry**
ISSN: 1354-6805
Cogn. Neuropsych.
Psychology Press Ltd, 4 Park Square, Milton Park, OX14 4RN, Abingdon, Oxfordshire, United Kingdom, Tel: +44(1235)828600, Fax: +44(1235)829000 ISSUES/YR: 6

• **Cognitive Neuropsychology**
ISSN: 0264-3294
Cogn. Neuropsychol.
Psychology Press Ltd, 4 Park Square, Milton Park, OX14 4RN, Abingdon, Oxfordshire, United Kingdom, Tel: +44(1235)828600, Fax: +44(1235)829000 ISSUES/YR: 8

• **Cognitive Processing**
ISSN: 1612-4782
Cogn. Process.
Springer Verlag, Tiergartenstrasse 17, D-69121, Heidelberg, Germany, Tel: +49(6221)4870, Fax: +49(6221)487366 ISSUES/YR: 4

• **Cognitive Systems Research**
ISSN: not available yet
Cogn. Sys. Res.
Elsevier, P.O. Box 211, 1000 AE, Amsterdam, Netherlands, Tel: +31(20)4853757, Fax: +31(20)4853432 ISSUES/YR: 4

Cognitive Therapy and Research
ISSN: 0147-5916
Cogn. Ther. Res.
Springer Science and Business Media Deutschland GmbH, Heidelberger Platz 3, 14197, Berlin, Germany, Tel: +1-30 827875282, Fax: +1-30 827875707 ISSUES/YR: 6

Cognitive, Affective and Behavioral Neuroscience
ISSN: 1530-7026
Cogn. Affective Behav. Neurosci.
Psychonomic Society Inc., 1710 Fortview Road, Austin, TX 78704, United States, Tel: +1(512)4622442 ISSUES/YR: 4

• **Cold Spring Harbor Symposia on Quantitative Biology**
ISSN: 0091-7451
Cold Spring Harbor Symp. Quant. Biol.
Cold Spring Harbor Laboratory Press, 500 Sunnyside Boulevard, Woodbury New York, NY 11797, United States, Tel: +1(516)3491930, Fax: +1(516)3491946 ISSUES/YR: 1

• **Colloid and Polymer Science**
ISSN: 0303-402X
Colloid Polym. Sci.
Springer Verlag, Tiergartenstrasse 17, D-69121, Heidelberg, Germany, Tel: +49(6221)4870, Fax: +49(6221)487366 ISSUES/YR: 12

• **Colloids and Surfaces A: Physicochemical and Engineering Aspects**
ISSN: 0927-7757
Colloids Surf. A Physicochem. Eng. Asp.
Elsevier, P.O. Box 211, 1000 AE, Amsterdam, Netherlands, Tel: +31(20)4853757, Fax: +31(20)4853432 ISSUES/YR: 36

- **Colloids and Surfaces B: Biointerfaces**
  ISSN: 0927-7765
  Colloids Surf. B Biointerfaces
  Elsevier, P.O. Box 211, 1000 AE, Amsterdam, Netherlands, Tel: +31(20)4853757, Fax: +31(20)4853432 ISSUES/YR: 12

Colombia Medica
ISSN: 1657-9534
Colomb. Med.
Corporacion Editora Medica del Valle, Calle 4B, 36-00, Cali, Valle, Colombia, Tel: +57-2 558-1939 ISSUES/YR: 3

Coloproctology
ISSN: 0174-2442
Coloproctology
Urban und Vogel GmbH, Neumarkter Strasse 43, D-81673, Munchen, Germany, Tel: +49(89)43721300, Fax: +49(89)43721399 ISSUES/YR: 6

- **Colorectal Disease**
  ISSN: 1462-8910
  Colorectal Dis.
  Blackwell Publishing Ltd, 9600 Garsington Road, OX4 2XG, Oxford, United Kingdom, Tel: +44(1865)776868, Fax: +44(1865)714591 ISSUES/YR: 6
- **Combinatorial Chemistry and High Throughput Screening**
  ISSN: 1386-2073
  Comb. Chem. High Throughput Screen.
  Bentham Science Publishers B.V., P.O. Box 294, 1400 AG, Bussum, Netherlands, Tel: +31-356257290, Fax: +31-847511310 ISSUES/YR: 8
- **Combustion and Flame**
  ISSN: 0010-2180
  Combust. Flame
  Elsevier Inc., 360 Park Avenue South, New York, NY 10010, United States, Tel: +1(212)6333950, Fax: +1(212)6333990 ISSUES/YR: 12

Communication and Medicine
ISSN: 1612-1783
Commun. Med.
Walter de Gruyter and Co., Genthiner Strasse 13, D-10785, Berlin, Germany, Tel: +49(30)260-050, Fax: +49(30)260-05251 ISSUES/YR: 2

Community Eye Health Journal
ISSN: 0953-6833
Community Eye Health J.
International Centre for Eye Health, Keppel St., WC1E 7HT, London, United Kingdom, Tel: +44-207 612 7972, Fax: +44(207)2503207 ISSUES/YR: 4

- **Community Genetics**
  ISSN: 1422-2795
  Community Genet.
  S. Karger AG, Allschwilerstrasse 10, P.O. Box, CH-4009, Basel, Switzerland, Tel: +41(61)3061111, Fax: +41(61)3061234 ISSUES/YR: 4

Community Mental Health Journal
ISSN: 0010-3853
Community Ment. Health J.
Kluwer Academic Publishers, Van Godewijckstraat 30, 3311 GZ, Dordrecht, Netherlands, Tel: +31(78)6576000, Fax: +31(78)6576254 ISSUES/YR: 6

Community Oncology
ISSN: 1548-5315
Community Oncol.
Biolink Communications, 46 Green Street, 2nd Floor, Huntington, NY 11743, United States, Tel: +1(631)4248900, Fax: +1(631)4248905 ISSUES/YR: 6

- **Comparative and Functional Genomics**
  ISSN: 1531-6912
  Comp. Funct. Genomics
  Hindawi Publishing Corporation, 410 Park Avenue, 15th Floor, 287 pmb, New York, NY 10022, United States, Tel: +1(20)2-2706352, Fax: +1(20)2-2706351 ISSUES/YR: 8

Comparative Biochemistry and Physiology - A Molecular and Integrative Physiology
ISSN: 1095-6433
Comp. Biochem. Physiol. A Mol. Integr. Physiol.
Elsevier Inc., 360 Park Avenue South, New York, NY 10010, United States, Tel: +1(212)6333950, Fax: +1(212)6333990 ISSUES/YR: 12

- **Comparative Biochemistry and Physiology - B Biochemistry and Molecular Biology**
  ISSN: 1096-4959
  Comp. Biochem. Physiol. B Biochem. Mol. Biol.
  Elsevier Inc., 360 Park Avenue South, New York, NY 10010, United States, Tel: +1(212)6333950, Fax: +1(212)6333990 ISSUES/YR: 12
- **Comparative Biochemistry and Physiology - C Toxicology and Pharmacology**
  ISSN: 1532-0456
  Comp. Biochem. Physiol. C Toxicol. Pharmacol.
  Elsevier Inc., 360 Park Avenue South, New York, NY 10010, United States, Tel: +1(212)6333950, Fax: +1(212)6333990 ISSUES/YR: 12
- **Comparative Biochemistry and Physiology - Part D: Genomics and Proteomics**
  ISSN: 1744-117X
  Comp. Biochem. Physiol. Part D Genomics Proteomics
  Elsevier Inc., 360 Park Avenue South, New York, NY 10010, United States, Tel: +1(212)6333950, Fax: +1(212)6333990 ISSUES/YR: 4

Comparative Clinical Pathology
ISSN: 1618-5641
Comp. Clin. Pathol.
Springer-Verlag London Ltd, Catteshall Road, Farncombe, GU7 3DJ, Godalming, Surrey, United Kingdom, Tel: +44(1483)414142, Fax: +44(1483)415144 ISSUES/YR: 4

Comparative Hepatology
ISSN: 1476-5926
Comp. Hepatol.
BioMed Central Ltd., 34 - 42 Cleveland Street, W1T 4LB, London, United Kingdom, Tel: +44(207)3230323, Fax: +44(207)6319926 ISSUES/YR: 1

Comparative Immunology, Microbiology and Infectious Diseases
ISSN: 0147-9571
Comp. Immunol. Microbiol. Infect. Dis.
Elsevier Ltd, Langford Lane, Kidlington, OX5 1GB, Oxford, United Kingdom, Tel: +44(1865)843000, Fax: +44(1865)843010 ISSUES/YR: 6

Comparative Medicine
ISSN: 1532-0820
Comp. Med.
American Association for Laboratory Animal Science, 9190 Crestwyn Hills Drive, Memphis, TN 38125, United States, Tel: +1(901)7548620, Fax: +1(901)7595849 ISSUES/YR: 6

Complementary Therapies in Medicine
ISSN: 0965-2299
Complement. Ther. Med.
Churchill Livingstone, 1-3 Baxter's Place, Leith Walk, EH1 3AF, Edinburgh, United Kingdom, Tel: +44(131)5562424, Fax: +44(131)5581278 ISSUES/YR: 4

- **Complexus**
  ISSN: 1424-8492
  Complexus
  S. Karger AG, Allschwilerstrasse 10, P.O. Box, CH-4009, Basel, Switzerland, Tel: +41(61)3061111, Fax: +41(61)3061234 ISSUES/YR: 4

Comprehensive Ophthalmology Update
ISSN: 1527-7313
Compr. Ophthalmol. Update
Comprehensive Ophthalmology Update, 7 Kent Street, Brookline MA 02445, United States, Tel: +1-617 5662138, Fax: +1-617 5667673 ISSUES/YR: 6

Comprehensive Psychiatry
ISSN: 0010-440X
Compr. Psychiatry
W.B. Saunders, Independence Square West, Philadelphia, PA 19106-3399, United States, Tel: +1(215)2387800, Fax: +1(215)2386445 ISSUES/YR: 6

Comprehensive Therapy
ISSN: 0098-8243
Compr. Ther.
Humana Press Inc., 999 Riverview Drive, Suite 208, Totowa, NJ 07512-1165, United States, Tel: +1(973)2561699, Fax: +1(973)2568341 ISSUES/YR: 4

Comptes Rendus - Biologies
ISSN: 1631-0691
C. R. Biol.
Editions Scientifiques et Medicales Elsevier, 23 rue Linois, 75724, Paris Cedex 15, France, Tel: +33(1)45589826, Fax: +33(1)45589428 ISSUES/YR: 12

• **Computer Aided Surgery**
ISSN: 1092-9088
Comput. Aided Surg.
Taylor and Francis Ltd., 4 Park Square, Milton Park, OX14 4RN, Abingdon, Oxfordshire, United Kingdom, Tel: +44(20)70175541, Fax: +44(20)70175198 ISSUES/YR: 6

Computer Methods and Programs in Biomedicine
ISSN: 0169-2607
Comput. Methods Programs Biomed.
Elsevier Ireland Ltd, P.O. Box 85, Limerick, Ireland, Tel: +353(61)471944, Fax: +353(61)472144 ISSUES/YR: 9

• **Computerized Medical Imaging and Graphics**
ISSN: 0895-6111
Comput. Med. Imaging Graph.
Elsevier Ltd, Langford Lane, Kidlington, OX5 1GB, Oxford, United Kingdom, Tel: +44(1865)843000, Fax: +44(1865)843010 ISSUES/YR: 8

• **Computers in Biology and Medicine**
ISSN: 0010-4825
Comput. Biol. Med.
Elsevier Ltd, Langford Lane, Kidlington, OX5 1GB, Oxford, United Kingdom, Tel: +44(1865)843000, Fax: +44(1865)843010 ISSUES/YR: 10

Confinia Cephalalgica
ISSN: 1122-0279
Confin. Cephalalgica
Fondazione CIRNA, via L. Porta, 5-27100 Pavia, Italy, Tel: +39-0382 303 044 ISSUES/YR: 3

Confrontations Psychiatriques
ISSN: 0153-9329
Confront. Psychiatr.
SPECIA Rhone-Poulenc Rorer, 15 rue de la Vanne, 92545, Montrouge Cdx, France, Tel: +33(01)55715571 ISSUES/YR: 2

• **Congenital Anomalies**
ISSN: 0914-3505
Congen. Anomal.
Blackwell Publishing, 550 Swanston Street, VIC 3053, Carlton South, Australia, Tel: +61(3)8359 1011, Fax: +61(3)8359 1120 ISSUES/YR: 4

• **Congenital Heart Disease**
ISSN: 1747-079X
Congenit. Heart Dis.
Blackwell Publishing Ltd, 9600 Garsington Road, OX4 2XG, Oxford, United Kingdom, Tel: +44(1865)776868, Fax: +44(1865)714591 ISSUES/YR: 6

Connecticut Medicine
ISSN: 0010-6178
Conn. Med.
Connecticut State Medical Society, 160 St. Ronan Street, New Haven, CT 06511, United States, Tel: +1(203)8650587 ISSUES/YR: 10

Connective Tissue Research
ISSN: 0300-8207
Connect. Tissue Res.
Taylor and Francis Ltd., 4 Park Square, Milton Park, OX14 4RN, Abingdon, Oxfordshire, United Kingdom, Tel: +44(20)70175541, Fax: +44(20)70175198 ISSUES/YR: 5

Consciousness and Cognition
ISSN: 1053-8100
Conscious. Cogn.
Academic Press Inc., 6277 Sea Harbor Drive, Orlando, FL 32887-4900, United States, Tel: +1(407)3454040, Fax: +1(407)3639661 ISSUES/YR: 4

• **Consultant**
ISSN: 0010-7069
Consultant
Cliggott Publishing Co., 330 Boston Post Road, Box 4027, Darien, CT 06820-4027, United States, Tel: +1(203)6626400 ISSUES/YR: 14

Consultant Pharmacist
ISSN: 0888-5109
Consult. Pharm.
American Society of Consultant Pharmacists, 1321 Duke Street, Alexandria, VA 22314-3563, United States, Tel: +1(703)739-1300, Fax: +1(703)7391500 ISSUES/YR: 12

Contact Dermatitis
ISSN: 0105-1873
Contact Dermatitis
Blackwell Publishing Ltd, 9600 Garsington Road, OX4 2XG, Oxford, United Kingdom, Tel: +44(1865)776868, Fax: +44(1865)714591 ISSUES/YR: 12

• **Contact Lens and Anterior Eye**
ISSN: 1367-0484
Contact Lens Anterior Eye
Elsevier, P.O. Box 211, 1000 AE, Amsterdam, Netherlands, Tel: +31(20)4853757, Fax: +31(20)4853432 ISSUES/YR: 4

Contagion: Reports, Cases, and Commentaries in HIV and Infectious Disease Research
ISSN: 1553-538X
Contagion Rep. Cases Comment. HIV Infect. Dis. Res.
TyWin Communications, 318 Cleveland Avenue, Unit 1 Highland Park, New Jersey, 08904, United States, Tel: +1(732)2490800, Fax: +1(732)2490071 ISSUES/YR: 12

Contemporary Clinical Trials
ISSN: 1551-7144
Contemp. Clin. Trials
Elsevier Inc., 360 Park Avenue South, New York, NY 10010, United States, Tel: +1(212)6333950, Fax: +1(212)6333990 ISSUES/YR: 6

Contemporary Hypnosis
ISSN: 0960-5290
Contemp. Hypn.
John Wiley and Sons Ltd, Southern Gate, PO19 8SQ, Chichester, West Sussex, United Kingdom, Tel: +44(1243)779777, Fax: +44(1243)775878 ISSUES/YR: 4

Contemporary Topics in Laboratory Animal Science
ISSN: 1060-0558
Contemp. Top. Lab. Anim. Sci.
American Association for Laboratory Animal Science, 9190 Crestwyn Hills Drive, Memphis, TN 38125, United States, Tel: +1(901)7548620, Fax: +1(901)7595849 ISSUES/YR: 6

Continuing Medical Education - Cardiology
ISSN: 1463-0486
Contin. Med. Edu. Cardiol.
Rila Publications Ltd, 73 Newman Street, W1A 4PG, London, United Kingdom, Tel: +44(207)6311299, Fax: +44(207)5807166 ISSUES/YR: 3

CONTINUUM Lifelong Learning in Neurology
ISSN: 1080-2371
CONTINUUM Lifelong Learn. Neurol.
Lippincott Williams and Wilkins, 530 Walnut Street, Philadelphia, PA 19106-3621, United States, Tel: +1(215)5218342, Fax: +1(215)5218326 ISSUES/YR: 6

Contraception
ISSN: 0010-7824
Contraception
Elsevier USA, 6277 Sea Harbor Drive, Orlando, FL 32862 8239, United States, Tel: +1-407 345 4000, Fax: +1-407 363 9661 ISSUES/YR: 12

• **COPD: Journal of Chronic Obstructive Pulmonary Disease**
ISSN: 1541-2555
COPD J. Chron. Obstruct. Pulm. Dis.
Taylor and Francis Inc., 325 Chestnut St, Suite 800, Philadelphia PA, PA 19106, United States, Tel: +1(800)3541420, Fax: +1(215)6258914 ISSUES/YR: 4

Cor et Vasa
ISSN: 0010-8650
Cor Vasa
MedProGO s.r.o., Anglicka 19, CZ-120 00, Prague, Czech Republic, Tel: +420-222512649, Fax: +420-222247568 ISSUES/YR: 6

Core Evidence
ISSN: 1555-1741
Core Evid.
Core Medical Publishing, 445 Hamilton Avenue, Suite 1102, White Plains, NY 10601, United States, Tel: +1(1 914)576-5734, Fax: +1(1 914)632-1733 ISSUES/YR: 4

- **Cornea**
  ISSN: 0277-3740
  Cornea
  Lippincott Williams and Wilkins, 530 Walnut Street, Philadelphia, PA 19106-3621, United States, Tel: +1(215)5218342, Fax: +1(215)5218326 ISSUES/YR: 8
- **Coronary Artery Disease**
  ISSN: 0954-6928
  Coron. Artery Dis.
  Lippincott Williams and Wilkins, 241 Borough High Street, SE1 1GB, London, United Kingdom, Tel: +44(207)9407500, Fax: +44(207)9407575 ISSUES/YR: 8

Cortex
ISSN: 0010-9452
Cortex
Masson SpA, Via F.lli Bressan 2, 20126, Milan, Italy, Tel: +39(02)270741, Fax: +39(02)27074211 ISSUES/YR: 5

Cosmetic Dermatology
ISSN: 1041-3766
Cosmet. Dermatol.
Quadrant Healthcom Inc., 26 Main Street, Chatham, NJ 07928-2402, United States, Tel: +1(973)7018900, Fax: +1(973)7018895 ISSUES/YR: 12

COSSMA
ISSN: 1439-7676
COSSMA
HMS Eurocom B.V., P.O.Box 82, 7710 AB, Niewleusen, Netherlands, Tel: +31-529 480 538, Fax: +31-529 484 893 ISSUES/YR: 12

Cost Effectiveness and Resource Allocation
ISSN: 1478-7547
Cost Eff. Resour. Allocat.
BioMed Central Ltd., 34 - 42 Cleveland Street, W1T 4LB, London, United Kingdom, Tel: +44(207)3230323, Fax: +44(207)6319926 ISSUES/YR: 1

Counselling Psychology Quarterly
ISSN: 0951-5070
Couns. Psychol. Q.
Routledge, 4 Park Square, Milton Park, OX14 4RN, Abingdon, Oxfordshire, United Kingdom, Tel: +44(1235)828600, Fax: +44(1235)829000 ISSUES/YR: 4

CPD Anaesthesia
ISSN: 1466-2922
CPD Anaesth.
Rila Publications Ltd, 73 Newman Street, W1A 4PG, London, United Kingdom, Tel: +44(207)6311299, Fax: +44(207)5807166 ISSUES/YR: 3

CPD Bulletin Cellular Pathology
ISSN: 1464-2506
CPD Bull. Cell. Pathol.
Rila Publications Ltd, 73 Newman Street, W1A 4PG, London, United Kingdom, Tel: +44(207)6311299, Fax: +44(207)5807166 ISSUES/YR: 3

CPD Bulletin Clinical Biochemistry
ISSN: 1367-8930
CPD Bull. Clin. Biochem.
Rila Publications Ltd, 73 Newman Street, W1A 4PG, London, United Kingdom, Tel: +44(207)6311299, Fax: +44(207)5807166 ISSUES/YR: 3

CPD Bulletin Immunology and Allergy
ISSN: 1367-8949
CPD Bull. Immunol. Allergy
Rila Publications Ltd, 73 Newman Street, W1A 4PG, London, United Kingdom, Tel: +44(207)6311299, Fax: +44(207)5807166 ISSUES/YR: 3

CPD Bulletin Old Age Psychiatry
ISSN: 1464-2514
CPD Bull. Old Age Psychiatry
Rila Publications Ltd, 73 Newman Street, W1A 4PG, London, United Kingdom, Tel: +44(207)6311299, Fax: +44(207)5807166 ISSUES/YR: 3

CPD Infection
ISSN: 1468-1668
CPD Infect.
Rila Publications Ltd, 73 Newman Street, W1A 4PG, London, United Kingdom, Tel: +44(207)6311299, Fax: +44(207)5807166 ISSUES/YR: 3

Criminal Behaviour and Mental Health
ISSN: 0957-9664
Crim. Behav. Ment. Health
John Wiley and Sons Ltd, Southern Gate, PO19 8SQ, Chichester, West Sussex, United Kingdom, Tel: +44(1243)779777, Fax: +44(1243)775878 ISSUES/YR: 4

- **Critical Care**
  ISSN: 1364-8535
  Crit. Care
  BioMed Central Ltd., 34 - 42 Cleveland Street, W1T 4LB, London, United Kingdom, Tel: +44(207)3230323, Fax: +44(207)6319926 ISSUES/YR: 6

Critical Care and Shock
ISSN: 1410-7767
Crit. Care Shock
Indonesian Society of Critical Care Medicine, Jl.S. Parman Kav. 87, Slipi, 11420, Jakarta, Indonesia, Tel: +62(21)568408590 EXT.1242, Fax: +62(21)56961530 ISSUES/YR: 4

- **Critical Care Clinics**
  ISSN: 0749-0704
  Crit. Care Clin.
  W.B. Saunders, Independence Square West, Philadelphia, PA 19106-3399, United States, Tel: +1(215)2387800, Fax: +1(215)2386445 ISSUES/YR: 4
- **Critical Care Medicine**
  ISSN: 0090-3493
  Crit. Care Med.
  Lippincott Williams and Wilkins, 351 West Camden Street, Baltimore, MD 21201-2436, United States, Tel: +1(410)5284000 ISSUES/YR: 12

Critical Care Nursing Quarterly
ISSN: 0887-9303
Crit. Care Nurs. Q.
Lippincott Williams and Wilkins, 351 West Camden Street, Baltimore, MD 21201-2436, United States, Tel: +1(410)5284000 ISSUES/YR: 4

- **Critical Pathways in Cardiology**
  ISSN: 1535-282X
  Crit. Pathways Cardiol.
  Lippincott Williams and Wilkins, 530 Walnut Street, Philadelphia, PA 19106-3621, United States, Tel: +1(215)5218342, Fax: +1(215)5218326 ISSUES/YR: 4

Critical Public Health
ISSN: 0958-1596
Crit. Pub. Health
Routledge, 4 Park Square, Milton Park, OX14 4RN, Abingdon, Oxfordshire, United Kingdom, Tel: +44(1235)828600, Fax: +44(1235)829000 ISSUES/YR: 4

- **Critical Reviews in Biochemistry and Molecular Biology**
  ISSN: 1040-9238
  Crit. Rev. Biochem. Mol. Biol.
  Taylor and Francis Inc., 325 Chestnut St, Suite 800, Philadelphia PA, PA 19106, United States, Tel: +1(800)3541420, Fax: +1(215)6258914 ISSUES/YR: 6
- **Critical Reviews in Biomedical Engineering**
  ISSN: 0278-940X
  Crit. Rev. Biomed. Eng.
  Begell House Inc., 145 Madison Avenue 6th Floor, New York, NY 10016, United States, Tel: +1(212)7251999, Fax: +1(212)2138368 ISSUES/YR: 6
- **Critical Reviews in Biotechnology**
  ISSN: 0738-8551
  Crit. Rev. Biotechnol.
  Taylor and Francis Inc., 325 Chestnut St, Suite 800, Philadelphia PA, PA 19106, United States, Tel: +1(800)3541420, Fax: +1(215)6258914 ISSUES/YR: 4
- **Critical Reviews in Clinical Laboratory Sciences**
  ISSN: 1040-8363
  Crit. Rev. Clin. Lab. Sci.
  Taylor and Francis Inc., 325 Chestnut St, Suite 800, Philadelphia PA, PA 19106, United States, Tel: +1(800)3541420, Fax: +1(215)6258914 ISSUES/YR: 6

- **Critical Reviews in Computed Tomography**
  ISSN: 1040-8371
  Crit. Rev. Comput. Tomography
  Taylor and Francis Inc., 325 Chestnut St, Suite 800, Philadelphia PA, PA 19106, United States, Tel: +1(800)3541420, Fax: +1(215)6258914 ISSUES/YR: 6

  Critical Reviews in Environmental Science and Technology
  ISSN: 1064-3389
  Crit. Rev. Environ. Sci. Technol.
  Taylor and Francis Inc., 325 Chestnut St, Suite 800, Philadelphia PA, PA 19106, United States, Tel: +1(800)3541420, Fax: +1(215)6258914 ISSUES/YR: 6

- **Critical Reviews in Eukaryotic Gene Expression**
  ISSN: 1045-4403
  Crit. Rev. Eukaryotic Gene Expr.
  Begell House Inc., 145 Madison Avenue 6th Floor, New York, NY 10016, United States, Tel: +1(212)7251999, Fax: +1(212)2138368 ISSUES/YR: 4

- **Critical Reviews in Immunology**
  ISSN: 1040-8401
  Crit. Rev. Immunol.
  Begell House Inc., 145 Madison Avenue 6th Floor, New York, NY 10016, United States, Tel: +1(212)7251999, Fax: +1(212)2138368 ISSUES/YR: 6

- **Critical Reviews in Microbiology**
  ISSN: 1040-841X
  Crit. Rev. Microbiol.
  Taylor and Francis Inc., 325 Chestnut St, Suite 800, Philadelphia PA, PA 19106, United States, Tel: +1(800)3541420, Fax: +1(215)6258914 ISSUES/YR: 4

  Critical Reviews in Neurobiology
  ISSN: 0892-0915
  Crit. Rev. Neurobiol.
  Begell House Inc., 145 Madison Avenue 6th Floor, New York, NY 10016, United States, Tel: +1(212)7251999, Fax: +1(212)2138368 ISSUES/YR: 4

  Critical Reviews in Oncology/Hematology
  ISSN: 1040-8428
  Crit. Rev. Oncol. Hematol.
  Elsevier Ireland Ltd, P.O. Box 85, Limerick, Ireland, Tel: +353(61)471944, Fax: +353(61)472144 ISSUES/YR: 12

- **Critical Reviews in Oral Biology and Medicine**
  ISSN: 1045-4411
  Crit. Rev. Oral Biol. Med.
  Intern. and American Associations for Dental Research, 1619 Duke Street, Alexandria, VA 22314-3406, United States, Tel: +1(703)5480066, Fax: +1(703)5481883 ISSUES/YR: 6

- **Critical Reviews in Physical and Rehabilitation Medicine**
  ISSN: 0896-2960
  Crit. Rev. Phys. Rehabil. Med.
  Begell House Inc., 145 Madison Avenue 6th Floor, New York, NY 10016, United States, Tel: +1(212)7251999, Fax: +1(212)2138368 ISSUES/YR: 4

  Critical Reviews in Therapeutic Drug Carrier Systems
  ISSN: 0743-4863
  Crit. Rev. Ther. Drug Carrier Syst.
  Begell House Inc., 145 Madison Avenue 6th Floor, New York, NY 10016, United States, Tel: +1(212)7251999, Fax: +1(212)2138368 ISSUES/YR: 6

  Critical Reviews in Toxicology
  ISSN: 1040-8444
  Crit. Rev. Toxicol.
  Taylor and Francis Inc., 325 Chestnut St, Suite 800, Philadelphia PA, PA 19106, United States, Tel: +1(800)3541420, Fax: +1(215)6258914 ISSUES/YR: 6

  Croatian Journal of Gastroenterology and Hepatology
  ISSN: 0353-9296
  Croat. J. Gastroenterol. Hepatol.
  Medicinska Naklada Zagreb, Vlaska 69, 10000, Zagreb, Croatia ISSUES/YR: 4

  Croatian Medical Journal
  ISSN: 0353-9504
  Croat. Med. J.
  Pabst Science Publishers, Eichengrund 28, D-49525, Lengerich, Germany, Tel: +385(5484)308, Fax: +385(5484)550 ISSUES/YR: 6

- **Cryobiology**
  ISSN: 0011-2240
  Cryobiology
  Academic Press Inc., 6277 Sea Harbor Drive, Orlando, FL 32887-4900, United States, Tel: +1(407)3454040, Fax: +1(407)3639661 ISSUES/YR: 6

  Cuadernos de Medicina Reproductiva
  ISSN: 1135-0970
  Cuad. Med. Reprod.
  Adalia Farma, C/Peru, 6-Edificio Twin Golf 2, 28290, Madrid, Spain, Tel: +34-91 630 59 80, Fax: +34-91 636 84 22 ISSUES/YR: 2

  Cultural Diversity and Ethnic Minority Psychology
  ISSN: 1099-9809
  Cult. Diversity Ethnic Minor. Psychol.
  American Psychological Association Inc., 750 First Street NE, Washington, DC 20002-4242, United States, Tel: +1(202)3365600, Fax: +1(202)3365568 ISSUES/YR: 4

- **Culture, Health and Sexuality**
  ISSN: 1369-1058
  Cult. Health Sexual.
  Routledge, 4 Park Square, Milton Park, OX14 4RN, Abingdon, Oxfordshire, United Kingdom, Tel: +44(1235)828600, Fax: +44(1235)829000 ISSUES/YR: 6

  Current Allergy and Asthma Reports
  ISSN: 1529-7322
  Curr. Allergy Asthma Rep.
  Current Science Ltd, 34-42 Cleveland Street, W1P 6LB, London, United Kingdom, Tel: +44(207)3230323, Fax: +44(207)5801938 ISSUES/YR: 6

- **Current Alzheimer Research**
  ISSN: 1567-2050
  Curr. Alzheimer Res.
  Bentham Science Publishers B.V., P.O. Box 294, 1400 AG, Bussum, Netherlands, Tel: +31-356257290, Fax: +31-847511310 ISSUES/YR: 5

- **Current Anaesthesia and Critical Care**
  ISSN: 0953-7112
  Curr. Anaesth. Crit. Care
  Churchill Livingstone, 1-3 Baxter's Place, Leith Walk, EH1 3AF, Edinburgh, United Kingdom, Tel: +44(131)5562424, Fax: +44(131)5581278 ISSUES/YR: 6

  Current Atherosclerosis Reports
  ISSN: 1523-3804
  Curr. Atheroscler. Rep.
  Current Science Ltd, 34-42 Cleveland Street, W1P 6LB, London, United Kingdom, Tel: +44(207)3230323, Fax: +44(207)5801938 ISSUES/YR: 6

  Current Cancer Drug Targets
  ISSN: 1568-0096
  Curr. Cancer Drug Targets
  Bentham Science Publishers B.V., P.O. Box 294, 1400 AG, Bussum, Netherlands, Tel: +31-356257290, Fax: +31-847511310 ISSUES/YR: 8

- **Current Cancer Therapy Reviews**
  ISSN: 1573-3947
  Curr. Cancer Ther. Rev.
  Bentham Science Publishers B.V., P.O. Box 294, 1400 AG, Bussum, Netherlands, Tel: +31-356257290, Fax: +31-847511310 ISSUES/YR: 3

  Current Cardiology Reports
  ISSN: 1523-3782
  Curr. Cardiol. Rep.
  Current Science Ltd, 34-42 Cleveland Street, W1P 6LB, London, United Kingdom, Tel: +44(207)3230323, Fax: +44(207)5801938 ISSUES/YR: 6

- **Current Cardiology Reviews**
  ISSN: 1573-3947
  Curr. Cardiol. Rev.
  Bentham Science Publishers B.V., P.O. Box 294, 1400 AG, Bussum, Netherlands, Tel: +31-356257290, Fax: +31-847511310 ISSUES/YR: 3

  Current Computer-Aided Drug Design
  ISSN: 1573-4099
  Curr. Comput.-Aided Drug Des.
  Bentham Science Publishers B.V., P.O. Box 294, 1400 AG, Bussum, Netherlands, Tel: +31-356257290, Fax: +31-847511310 ISSUES/YR: 4

Current Controlled Trials in Cardiovascular Medicine
ISSN: 1468-6694
Curr. Control. Trials Cardiovasc. Med.
BioMed Central Ltd., 34 - 42 Cleveland Street, W1T 4LB, London, United Kingdom, Tel: +44(207)3230323, Fax: +44(207)6319926 ISSUES/YR: 6

Current Diabetes Reports
ISSN: 1534-4827
Curr. Diabetes Rep.
Current Science Ltd, 34-42 Cleveland Street, W1P 6LB, London, United Kingdom, Tel: +44(207)3230323, Fax: +44(207)5801938 ISSUES/YR: 6

● **Current Diabetes Reviews**
ISSN: 1573-3998
Curr. Diabetes Rev.
Bentham Science Publishers B.V., P.O. Box 294, 1400 AG, Bussum, Netherlands, Tel: +31-356257290, Fax: +31-847511310 ISSUES/YR: 3

Current Diagnostic Pathology
ISSN: 0968-6053
Curr. Diagn. Pathol.
Churchill Livingstone, 1-3 Baxter's Place, Leith Walk, EH1 3AF, Edinburgh, United Kingdom, Tel: +44(131)5562424, Fax: +44(131)5581278 ISSUES/YR: 6

● **Current Drug Delivery**
ISSN: 1567-2018
Curr. Drug Deliv.
Bentham Science Publishers B.V., P.O. Box 294, 1400 AG, Bussum, Netherlands, Tel: +31-356257290, Fax: +31-847511310 ISSUES/YR: 4

● **Current Drug Discovery Technologies**
ISSN: 1570-1638
Curr. Drug Discov. Technol.
Bentham Science Publishers B.V., P.O. Box 294, 1400 AG, Bussum, Netherlands, Tel: +31-356257290, Fax: +31-847511310 ISSUES/YR: 4

Current Drug Metabolism
ISSN: 1389-2002
Curr. Drug Metab.
Bentham Science Publishers B.V., P.O. Box 294, 1400 AG, Bussum, Netherlands, Tel: +31-356257290, Fax: +31-847511310 ISSUES/YR: 6

Current Drug Targets
ISSN: 1389-4501
Curr. Drug Targets
Bentham Science Publishers B.V., P.O. Box 294, 1400 AG, Bussum, Netherlands, Tel: +31-356257290, Fax: +31-847511310 ISSUES/YR: 8

Current Enzyme Inhibition
ISSN: 1573-4080
Curr. Enzyme Inhib.
Bentham Science Publishers B.V., P.O. Box 294, 1400 AG, Bussum, Netherlands, Tel: +31-356257290, Fax: +31-847511310 ISSUES/YR: 3

● **Current Eye Research**
ISSN: 0271-3683
Curr. Eye Res.
Taylor and Francis Ltd., 4 Park Square, Milton Park, OX14 4RN, Abingdon, Oxfordshire, United Kingdom, Tel: +44(20)70175541, Fax: +44(20)70175198 ISSUES/YR: 12

Current Gastroenterology Reports
ISSN: 1522-8037
Curr. Gastroenterol. Rep.
Current Science Ltd, 34-42 Cleveland Street, W1P 6LB, London, United Kingdom, Tel: +44(207)3230323, Fax: +44(207)5801938 ISSUES/YR: 6

Current Gene Therapy
ISSN: 1566-5232
Curr. Gene Ther.
Bentham Science Publishers B.V., P.O. Box 294, 1400 AG, Bussum, Netherlands, Tel: +31-356257290, Fax: +31-847511310 ISSUES/YR: 6

● **Current Genetics**
ISSN: 0172-8083
Curr. Genet.
Springer Verlag, Tiergartenstrasse 17, D-69121, Heidelberg, Germany, Tel: +49(6221)4870, Fax: +49(6221)487366 ISSUES/YR: 12

Current Genomics
ISSN: 1389-2029
Curr. Genomics
Bentham Science Publishers B.V., P.O. Box 294, 1400 AG, Bussum, Netherlands, Tel: +31-356257290, Fax: +31-847511310 ISSUES/YR: 8

Current Heart Failure Reports
ISSN: 1546-9530
Curr. Heart Fail. Rep.
Current Science, Inc., 400 Market Street, Suite 700, Philadelphia, PA 19106, United States, Tel: +1(215)5742266, Fax: +1(215)5742225 ISSUES/YR: 4

Current Hematology Reports
ISSN: 1540-3408
Curr. Hematol. Rep.
Current Science, Inc., 400 Market Street, Suite 700, Philadelphia, PA 19106, United States, Tel: +1(215)5742266, Fax: +1(215)5742225 ISSUES/YR: 6

Current Hepatitis Reports
ISSN: 1540-3416
Curr. Hepatitis Rep.
Current Science, Inc., 400 Market Street, Suite 700, Philadelphia, PA 19106, United States, Tel: +1(215)5742266, Fax: +1(215)5742225 ISSUES/YR: 4

Current HIV Research
ISSN: 1570-162X
Curr. HIV Res.
Bentham Science Publishers B.V., P.O. Box 294, 1400 AG, Bussum, Netherlands, Tel: +31-356257290, Fax: +31-847511310 ISSUES/YR: 4

Current HIV/AIDS Reports
ISSN: 1548-3568
Curr. HIV/AIDS Rep.
Current Science, Inc., 400 Market Street, Suite 700, Philadelphia, PA 19106, United States, Tel: +1(215)5742266, Fax: +1(215)5742225 ISSUES/YR: 6

Current Hypertension Reports
ISSN: 1522-6417
Curr. Hypertens. Rep.
Current Science Ltd, 34-42 Cleveland Street, W1P 6LB, London, United Kingdom, Tel: +44(207)3230323, Fax: +44(207)5801938 ISSUES/YR: 6

● **Current Hypertension Reviews**
ISSN: 1573-4021
Curr. Hypertens. Rev.
Bentham Science Publishers B.V., P.O. Box 294, 1400 AG, Bussum, Netherlands, Tel: +31-356257290, Fax: +31-847511310 ISSUES/YR: 3

● **Current Immunology Reviews**
ISSN: 1573-3955
Curr. Immunol. Rev.
Bentham Science Publishers B.V., P.O. Box 294, 1400 AG, Bussum, Netherlands, Tel: +31-356257290, Fax: +31-847511310 ISSUES/YR: 3

Current Infectious Disease Reports
ISSN: 1523-3847
Curr. Infect. Dis. Rep.
Current Science Ltd, 34-42 Cleveland Street, W1P 6LB, London, United Kingdom, Tel: +44(207)3230323, Fax: +44(207)5801938 ISSUES/YR: 6

Current Issues in Intestinal Microbiology
ISSN: 1466-531X
Curr. Issues Intest. Microbiol.
Horizon Scientific Press, P.O. Box 1, NR18 0EH, Wymondham, Norfolk, United Kingdom, Tel: +44(1953)601106, Fax: +44(1953)603068 ISSUES/YR: 2

Current Issues in Molecular Biology
ISSN: 1467-3037
Curr. Issues Mol. Biol.
Caister Academic Press, 32 Hewitts Lane, NR18 0JA, Wymondham, Norfolk, United Kingdom, Tel: +44(1953)601106, Fax: +44(1953)603068 ISSUES/YR: 2

● **Current Medical Imaging Reviews**
ISSN: 1573-4056
Curr. Med. Imaging Rev.
Bentham Science Publishers B.V., P.O. Box 294, 1400 AG, Bussum, Netherlands, Tel: +31-356257290, Fax: +31-847511310 ISSUES/YR: 3

Current Medical Research and Opinion
ISSN: 0300-7995
Curr. Med. Res. Opin.
Librapharm, Venture West, New Greenham Park, RG19 6HX, Newbury, Berks., United Kingdom, Tel: +44(1635)522651, Fax: +44(1635)36294 ISSUES/YR: 12

Current Medical Research and Opinion, Supplement
ISSN: 0141-9951
Curr. Med. Res. Opin. Suppl.
Librapharm, Venture West, New Greenham Park, RG19 6HX, Newbury, Berks., United Kingdom, Tel: +44(1635)522651, Fax: +44(1635)36294 ISSUES/YR: 1

Current Medicinal Chemistry
ISSN: 0929-8673
Curr. Med. Chem.
Bentham Science Publishers B.V., P.O. Box 294, 1400 AG, Bussum, Netherlands, Tel: +31-356257290, Fax: +31-847511310 ISSUES/YR: 26

• **Current Microbiology**
ISSN: 0343-8651
Curr. Microbiol.
Springer New York LLC, 233 Springer Street, New York, 10013-1578, United States, Tel: +1-212 460 1501, Fax: +1-212 460 1595 ISSUES/YR: 12

Current Molecular Medicine
ISSN: 1566-5240
Curr. Mol. Med.
Bentham Science Publishers B.V., P.O. Box 294, 1400 AG, Bussum, Netherlands, Tel: +31-356257290, Fax: +31-847511310 ISSUES/YR: 8

• **Current Nanoscience**
ISSN: 1573-4137
Curr. Nanosci.
Bentham Science Publishers B.V., P.O. Box 294, 1400 AG, Bussum, Netherlands, Tel: +31-356257290, Fax: +31-847511310 ISSUES/YR: 3

Current Neurology and Neuroscience Reports
ISSN: 1528-4042
Curr. Neurol. Neurosci. Rep.
Current Science Ltd, 34-42 Cleveland Street, W1P 6LB, London, United Kingdom, Tel: +44(207)3230323, Fax: +44(207)5801938 ISSUES/YR: 6

Current Neuropharmacology
ISSN: 1570-159X
Curr. Neuropharmacol.
Bentham Science Publishers B.V., P.O. Box 294, 1400 AG, Bussum, Netherlands, Tel: +31-356257290, Fax: +31-847511310 ISSUES/YR: 4

• **Current Neurovascular Research**
ISSN: 1567-2026
Curr. Neurovasc. Res.
Bentham Science Publishers B.V., P.O. Box 294, 1400 AG, Bussum, Netherlands, Tel: +31-356257290, Fax: +31-847511310 ISSUES/YR: 5

Current News in Medicine
ISSN: 1608-2222
Curr. News Med.
VICER Publishing, P.O. Box 14, A-1097, Vienna, Austria, Tel: +43(676)9568085, Fax: +43(676)9568086 ISSUES/YR: 4

• **Current Nutrition and Food Science**
ISSN: 1573-4013
Curr. Nutr. Food Sci.
Bentham Science Publishers B.V., P.O. Box 294, 1400 AG, Bussum, Netherlands, Tel: +31-356257290, Fax: +31-847511310 ISSUES/YR: 3

Current Obstetrics and Gynaecology
ISSN: 0957-5847
Curr. Obstet. Gynaecol.
Churchill Livingstone, 1-3 Baxter's Place, Leith Walk, EH1 3AF, Edinburgh, United Kingdom, Tel: +44(131)5562424, Fax: +44(131)5581278 ISSUES/YR: 6

Current Oncology
ISSN: 1198-0052
Curr. Oncol.
Multimed Inc., 66 Martin Street, ONT L9T 2R2, Milton, Canada, Tel: +1(905)8752456, Fax: +1(905)8752864 ISSUES/YR: 4

Current Oncology Reports
ISSN: 1523-3790
Curr. Oncol. Rep.
Current Science Ltd, 34-42 Cleveland Street, W1P 6LB, London, United Kingdom, Tel: +44(207)3230323, Fax: +44(207)5801938 ISSUES/YR: 6

• **Current Opinion in Allergy and Clinical Immunology**
ISSN: 1528-4050
Curr. Opin. Allergy Clin. Immunol.
Lippincott Williams and Wilkins, 530 Walnut Street, Philadelphia, PA 19106-3621, United States, Tel: +1(215)5218342, Fax: +1(215)5218326 ISSUES/YR: 6

• **Current Opinion in Anaesthesiology**
ISSN: 0952-7907
Curr. Opin. Anaesthesiol.
Lippincott Williams and Wilkins, 241 Borough High Street, SE1 1GB, London, United Kingdom, Tel: +44(207)9407500, Fax: +44(207)9407575 ISSUES/YR: 6

• **Current Opinion in Biotechnology**
ISSN: 0958-1669
Curr. Opin. Biotechnol.
Elsevier Ltd, Langford Lane, Kidlington, OX5 1GB, Oxford, United Kingdom, Tel: +44(1865)843000 ISSUES/YR: 6

• **Current Opinion in Cardiology**
ISSN: 0268-4705
Curr. Opin. Cardiol.
Lippincott Williams and Wilkins, 530 Walnut Street, Philadelphia, PA 19106-3621, United States, Tel: +1(215)5218342, Fax: +1(215)5218326 ISSUES/YR: 6

• **Current Opinion in Cell Biology**
ISSN: 0955-0674
Curr. Opin. Cell Biol.
Elsevier Ltd, Langford Lane, Kidlington, OX5 1GB, Oxford, United Kingdom, Tel: +44(1865)843000 ISSUES/YR: 6

Current Opinion in Chemical Biology
ISSN: 1367-5931
Curr. Opin. Chem. Biol.
Elsevier Ltd, Langford Lane, Kidlington, OX5 1GB, Oxford, United Kingdom, Tel: +44(1865)843000 ISSUES/YR: 6

Current Opinion in Clinical Experimental Research
ISSN: 1563-311X
Curr. Opin. Clin. Exp. Res.
VICER Publishing, P.O. Box 14, A-1097, Vienna, Austria, Tel: +43(676)9568085, Fax: +43(676)9568086 ISSUES/YR: 4

Current Opinion in Clinical Nutrition and Metabolic Care
ISSN: 1363-1950
Curr. Opin. Clin. Nutr. Metab. Care
Lippincott Williams and Wilkins, 530 Walnut Street, Philadelphia, PA 19106-3621, United States, Tel: +1(215)5218342, Fax: +1(215)5218326 ISSUES/YR: 6

Current Opinion in Colloid and Interface Science
ISSN: 1359-0294
Curr. Opin. Colloid Interface Sci.
Elsevier Ltd, Langford Lane, Kidlington, OX5 1GB, Oxford, United Kingdom, Tel: +44(1865)843000, Fax: +44(1865)843010 ISSUES/YR: 6

Current Opinion in Critical Care
ISSN: 1070-5295
Curr. Opin. Crit. Care
Lippincott Williams and Wilkins, 530 Walnut Street, Philadelphia, PA 19106-3621, United States, Tel: +1(215)5218342, Fax: +1(215)5218326 ISSUES/YR: 6

Current Opinion in Drug Discovery and Development
ISSN: 1367-6733
Curr. Opin. Drug Discov. Dev.
Current Drugs Ltd, 34-42 Cleveland Street, W1P 6LB, London, United Kingdom, Tel: +44(207)5808393, Fax: +44(207)5805646 ISSUES/YR: 6

Current Opinion in Endocrinology and Diabetes
ISSN: 1068-3097
Curr. Opin. Endocrinol. Diabetes
Lippincott Williams and Wilkins, 530 Walnut Street, Philadelphia, PA 19106-3621, United States, Tel: +1(215)5218342, Fax: +1(215)5218326 ISSUES/YR: 6

Current Opinion in European Medicine
ISSN: 1607-6524
Curr. Opin. Eur. Med.
VICER Publishing, P.O. Box 14, A-1097, Vienna, Austria, Tel: +43(676)9568085, Fax: +43(676)9568086 ISSUES/YR: 3

Current Opinion in Gastroenterology
ISSN: 0267-1379
Curr. Opin. Gastroenterol.
Lippincott Williams and Wilkins, 530 Walnut Street, Philadelphia, PA 19106-3621, United States, Tel: +1(215)5218342, Fax: +1(215)5218326 ISSUES/YR: 6

• **Current Opinion in Genetics and Development**
ISSN: 0959-437X
Curr. Opin. Genet. Dev.
Elsevier Ltd, Langford Lane, Kidlington, OX5 1GB, Oxford, United Kingdom, Tel: +44(1865)843000 ISSUES/YR: 6

• **Current Opinion in Hematology**
ISSN: 1065-6251
Curr. Opin. Hematol.
Lippincott Williams and Wilkins, 530 Walnut Street, Philadelphia, PA 19106-3621, United States, Tel: +1(215)5218342, Fax: +1(215)5218326 ISSUES/YR: 6

Current Opinion in Immunology
ISSN: 0952-7915
Curr. Opin. Immunol.
Elsevier Ltd, Langford Lane, Kidlington, OX5 1GB, Oxford, United Kingdom, Tel: +44(1865)843000 ISSUES/YR: 6

Current Opinion in Infectious Diseases
ISSN: 0951-7375
Curr. Opin. Infect. Dis.
Lippincott Williams and Wilkins, 241 Borough High Street, SE1 1GB, London, United Kingdom, Tel: +44(207)9407500, Fax: +44(207)9407575 ISSUES/YR: 6

Current Opinion in Investigational Drugs
ISSN: 1472-4472
Curr. Opin. Invest. Drugs
Thomson Scientific Ltd., 14 Great Queen Street, WC2B 5DF, London, United Kingdom, Tel: +44-20 7344 2800, Fax: +44-20 7344 2900 ISSUES/YR: 12

• **Current Opinion in Lipidology**
ISSN: 0957-9672
Curr. Opin. Lipidology
Lippincott Williams and Wilkins, 241 Borough High Street, SE1 1GB, London, United Kingdom, Tel: +44(207)9407500, Fax: +44(207)9407575 ISSUES/YR: 6

Current Opinion in Microbiology
ISSN: 1369-5274
Curr. Opin. Microbiol.
Elsevier Ltd, Langford Lane, Kidlington, OX5 1GB, Oxford, United Kingdom, Tel: +44(1865)843000 ISSUES/YR: 6

Current Opinion in Molecular Therapeutics
ISSN: 1464-8431
Curr. Opin. Mol. Ther.
Current Drugs Ltd, 34-42 Cleveland Street, W1P 6LB, London, United Kingdom, Tel: +44(207)5808393, Fax: +44(207)5805646 ISSUES/YR: 6

• **Current Opinion in Nephrology and Hypertension**
ISSN: 1062-4821
Curr. Opin. Nephrol. Hypertens.
Lippincott Williams and Wilkins, 241 Borough High Street, SE1 1GB, London, United Kingdom, Tel: +44(207)9407500, Fax: +44(207)9407575 ISSUES/YR: 6

• **Current Opinion in Neurobiology**
ISSN: 0959-4388
Curr. Opin. Neurobiol.
Elsevier Ltd, Langford Lane, Kidlington, OX5 1GB, Oxford, United Kingdom, Tel: +44(1865)843000 ISSUES/YR: 6

Current Opinion in Neurology
ISSN: 1350-7540
Curr. Opin. Neurol.
Lippincott Williams and Wilkins, 241 Borough High Street, SE1 1GB, London, United Kingdom, Tel: +44(207)9407500, Fax: +44(207)9407575 ISSUES/YR: 6

• **Current Opinion in Obstetrics and Gynecology**
ISSN: 1040-872X
Curr. Opin. Obstet. Gynecol.
Lippincott Williams and Wilkins, 241 Borough High Street, SE1 1GB, London, United Kingdom, Tel: +44(207)9407500, Fax: +44(207)9407575 ISSUES/YR: 6

• **Current Opinion in Oncology**
ISSN: 1040-8746
Curr. Opin. Oncol.
Lippincott Williams and Wilkins, 530 Walnut Street, Philadelphia, PA 19106-3621, United States, Tel: +1(215)5218342, Fax: +1(215)5218326 ISSUES/YR: 6

• **Current Opinion in Ophthalmology**
ISSN: 1040-8738
Curr. Opin. Ophthalmol.
Lippincott Williams and Wilkins, 530 Walnut Street, Philadelphia, PA 19106-3621, United States, Tel: +1(215)5218342, Fax: +1(215)5218326 ISSUES/YR: 6

Current Opinion in Organ Transplantation
ISSN: 1087-2418
Curr. Opin. Organ Transplant.
Lippincott Williams and Wilkins, 530 Walnut Street, Philadelphia, PA 19106-3621, United States, Tel: +1(215)5218342, Fax: +1(215)5218326 ISSUES/YR: 4

• **Current Opinion in Orthopaedics**
ISSN: 1041-9918
Curr. Opin. Orthop.
Lippincott Williams and Wilkins, 530 Walnut Street, Philadelphia, PA 19106-3621, United States, Tel: +1(215)5218342, Fax: +1(215)5218326 ISSUES/YR: 6

• **Current Opinion in Otolaryngology and Head and Neck Surgery**
ISSN: 1068-9508
Curr. Opin. Otolaryngol. Head Neck Surg.
Lippincott Williams and Wilkins, 530 Walnut Street, Philadelphia, PA 19106-3621, United States, Tel: +1(215)5218342, Fax: +1(215)5218326 ISSUES/YR: 6

• **Current Opinion in Pediatrics**
ISSN: 1040-8703
Curr. Opin. Pediatr.
Lippincott Williams and Wilkins, 530 Walnut Street, Philadelphia, PA 19106-3621, United States, Tel: +1(215)5218342, Fax: +1(215)5218326 ISSUES/YR: 6

• **Current Opinion in Pharmacology**
ISSN: 1471-4892
Curr. Opin. Pharmacol.
Elsevier Ltd, Langford Lane, Kidlington, OX5 1GB, Oxford, United Kingdom, Tel: +44(1865)843000, Fax: +44(1865)843010 ISSUES/YR: 6

Current Opinion in Psychiatry
ISSN: 0951-7367
Curr. Opin. Psychiatry
Lippincott Williams and Wilkins, 241 Borough High Street, SE1 1GB, London, United Kingdom, Tel: +44(207)9407500, Fax: +44(207)9407575 ISSUES/YR: 6

Current Opinion in Pulmonary Medicine
ISSN: 1070-5287
Curr. Opin. Pulm. Med.
Lippincott Williams and Wilkins, 530 Walnut Street, Philadelphia, PA 19106-3621, United States, Tel: +1(215)5218342, Fax: +1(215)5218326 ISSUES/YR: 6

• **Current Opinion in Rheumatology**
ISSN: 1040-8711
Curr. Opin. Rheumatol.
Lippincott Williams and Wilkins, 530 Walnut Street, Philadelphia, PA 19106-3621, United States, Tel: +1(215)5218342, Fax: +1(215)5218326 ISSUES/YR: 6

• **Current Opinion in Structural Biology**
ISSN: 0959-440X
Curr. Opin. Struct. Biol.
Elsevier Ltd, Langford Lane, Kidlington, OX5 1GB, Oxford, United Kingdom, Tel: +44(1865)843000 ISSUES/YR: 6

• **Current Opinion in Urology**
ISSN: 0963-0643
Curr. Opin. Urol.
Lippincott Williams and Wilkins, 241 Borough High Street, SE1 1GB, London, United Kingdom, Tel: +44(207)9407500, Fax: +44(207)9407575 ISSUES/YR: 6

Current Organic Chemistry
ISSN: 1385-2728
Curr. Org. Chem.
Bentham Science Publishers B.V., P.O. Box 294, 1400 AG, Bussum, Netherlands, Tel: +31-356257290, Fax: +31-847511310 ISSUES/YR: 18

• **Current Orthopaedics**
ISSN: 0268-0890
Curr. Orthop.
Churchill Livingstone, 1-3 Baxter's Place, Leith Walk, EH1 3AF, Edinburgh, United Kingdom, Tel: +44(131)5562424, Fax: +44(131)5581278 ISSUES/YR: 6

Current Osteoporosis Reports
ISSN: 1544-1873
Curr. Osteoporosis Rep.
Current Science, Inc., 400 Market Street, Suite 700, Philadelphia, PA 19106, United States, Tel: +1(215)5742266, Fax: +1(215)5742225 ISSUES/YR: 4

Current Paediatrics
ISSN: 0957-5839
Curr. Paediatr.
Churchill Livingstone, 1-3 Baxter's Place, Leith Walk, EH1 3AF, Edinburgh, United Kingdom, Tel: +44(131)5562424, Fax: +44(131)5581278 ISSUES/YR: 7

Current Pain and Headache Reports
ISSN: 1531-3433
Curr. Pain Headache Rep.
Current Science Ltd, 34-42 Cleveland Street, W1P 6LB, London, United Kingdom, Tel: +44(207)3230323, Fax: +44(207)5801938 ISSUES/YR: 6

Current Pediatric Research
ISSN: 0971-9032
Curr. Paediatr. Res.
Scientific Publishers of India, Sir Sayed Nagar, 202002, Aligarh, India, Tel: +91(571)502043, Fax: +91(571)400678 ISSUES/YR: 2

• **Current Pediatric Reviews**
ISSN: 1573-3963
Curr. Pediatr. Rev.
Bentham Science Publishers B.V., P.O. Box 294, 1400 AG, Bussum, Netherlands, Tel: +31-356257290, Fax: +31-847511310 ISSUES/YR: 3

• **Current Pharmaceutical Analysis**
ISSN: 1573-4129
Curr. Pharm. Anal.
Bentham Science Publishers B.V., P.O. Box 294, 1400 AG, Bussum, Netherlands, Tel: +31-356257290, Fax: +31-847511310 ISSUES/YR: 3

Current Pharmaceutical Biotechnology
ISSN: 1389-2010
Curr. Pharm. Biotechnol.
Bentham Science Publishers B.V., P.O. Box 294, 1400 AG, Bussum, Netherlands, Tel: +31-356257290, Fax: +31-847511310 ISSUES/YR: 6

• **Current Pharmaceutical Design**
ISSN: 1381-6128
Curr. Pharm. Des.
Bentham Science Publishers B.V., P.O. Box 294, 1400 AG, Bussum, Netherlands, Tel: +31-356257290, Fax: +31-847511310 ISSUES/YR: 32

Current Pharmacogenomics
ISSN: 1570-1603
Curr. Pharmacogenomics
Bentham Science Publishers B.V., P.O. Box 294, 1400 AG, Bussum, Netherlands, Tel: +31-356257290, Fax: +31-847511310 ISSUES/YR: 4

Current Problems in Cancer
ISSN: 0147-0272
Curr. Probl. Cancer
Mosby Inc., 11830 Westline Industrial Drive, St. Louis, MO 63146, United States, Tel: +1(314)8728370, Fax: +1(314)4321380 ISSUES/YR: 6

Current Problems in Cardiology
ISSN: 0146-2806
Curr. Probl. Cardiol.
Mosby Inc., 11830 Westline Industrial Drive, St. Louis, MO 63146, United States, Tel: +1(314)8728370, Fax: +1(314)4321380 ISSUES/YR: 12

Current Problems in Diagnostic Radiology
ISSN: 0363-0188
Curr. Probl. Diagn. Radiol.
Mosby Inc., 11830 Westline Industrial Drive, St. Louis, MO 63146, United States, Tel: +1(314)8728370, Fax: +1(314)4321380 ISSUES/YR: 6

Current Problems in Pediatric and Adolescent Health Care
ISSN: 1538-5442
Curr. Probl. Pediatr. Adolesc. Health Care
Mosby Inc., 11830 Westline Industrial Drive, St. Louis, MO 63146, United States, Tel: +1(314)8728370, Fax: +1(314)4321380 ISSUES/YR: 10

Current Problems in Surgery
ISSN: 0011-3840
Curr. Probl. Surg.
Mosby Inc., 11830 Westline Industrial Drive, St. Louis, MO 63146, United States, Tel: +1(314)8728370, Fax: +1(314)4321380 ISSUES/YR: 12

Current Prostate Reports
ISSN: 1544-225X
Curr. Prostate Rep.
Current Science, Inc., 400 Market Street, Suite 700, Philadelphia, PA 19106, United States, Tel: +1(215)5742266, Fax: +1(215)5742225 ISSUES/YR: 4

Current Protein and Peptide Science
ISSN: 1389-2037
Curr. Protein Pept. Sci.
Bentham Science Publishers B.V., P.O. Box 294, 1400 AG, Bussum, Netherlands, Tel: +31-356257290, Fax: +31-847511310 ISSUES/YR: 6

Current Psychiatry Reports
ISSN: 1523-3812
Curr. Psychiatry Rep.
Current Science Ltd, 34-42 Cleveland Street, W1P 6LB, London, United Kingdom, Tel: +44(207)3230323, Fax: +44(207)5801938 ISSUES/YR: 6

• **Current Psychiatry Reviews**
ISSN: 1573-4005
Curr. Psychiatry Rev.
Bentham Science Publishers B.V., P.O. Box 294, 1400 AG, Bussum, Netherlands, Tel: +31-356257290, Fax: +31-847511310 ISSUES/YR: 3

Current Psychosis and Therapeutics Reports
ISSN: 1545-8083
Curr. Psychosis Ther. Rep.
Current Science, Inc., 400 Market Street, Suite 700, Philadelphia, PA 19106, United States, Tel: +1(215)5742266, Fax: +1(215)5742225 ISSUES/YR: 4

• **Current Respiratory Medicine Reviews**
ISSN: 1573-398X
Curr. Respir. Med. Rev.
Bentham Science Publishers B.V., P.O. Box 294, 1400 AG, Bussum, Netherlands, Tel: +31-356257290, Fax: +31-847511310 ISSUES/YR: 3

Current Rheumatology Reports
ISSN: 1523-3774
Curr. Rheumatol. Rep.
Current Science, Inc., 400 Market Street, Suite 700, Philadelphia, PA 19106, United States, Tel: +1(215)5742266, Fax: +1(215)5742225 ISSUES/YR: 4

• **Current Rheumatology Reviews**
ISSN: 1573-3971
Curr. Rheumatol. Rev.
Bentham Science Publishers B.V., P.O. Box 294, 1400 AG, Bussum, Netherlands, Tel: +31-356257290, Fax: +31-847511310 ISSUES/YR: 3

Current Sexual Health Reports
ISSN: 1548-3584
Curr. Sex. Health Rep.
Current Science, Inc., 400 Market Street, Suite 700, Philadelphia, PA 19106, United States, Tel: +1(215)5742266, Fax: +1(215)5742225 ISSUES/YR: 4

Current Sports Medicine Reports
ISSN: 1537-890X
Curr. Sports Med. Rep.
Current Science, Inc., 400 Market Street, Suite 700, Philadelphia, PA 19106, United States, Tel: +1(215)5742266, Fax: +1(215)5742225 ISSUES/YR: 6

- **Current Studies in Hematology and Blood Transfusion**
  ISSN: 0258-0330
  Curr. Stud. Hematol. Blood Transfus.
  S. Karger AG, Allschwilerstrasse 10, P.O. Box, CH-4009, Basel, Switzerland, Tel: +41(61)3061111, Fax: +41(61)3061234 ISSUES/YR: 1
- **Current Surgery**
  ISSN: 0149-7944
  Curr. Surg.
  Elsevier Inc., 360 Park Avenue South, New York, NY 10010, United States, Tel: +1(212)6333950, Fax: +1(212)6333990 ISSUES/YR: 6
- **Current Therapeutic Research - Clinical and Experimental**
  ISSN: 0011-393X
  Curr. Ther. Res. Clin. Exp.
  Excerpta Medica Inc., 105 Raider Blvd, Suite 101, Hillsborough, NJ 08844, United States, Tel: +1(908)8748550, Fax: +1(908)8745633 ISSUES/YR: 6

Current Topics in Medicinal Chemistry
ISSN: 1568-0266
Curr. Top. Med. Chem.
Bentham Science Publishers B.V., P.O. Box 294, 1400 AG, Bussum, Netherlands, Tel: +31-356257290, Fax: +31-847511310 ISSUES/YR: 16

- **Current Topics in Microbiology and Immunology**
  ISSN: 0070-217X
  Curr. Top. Microbiol. Immunol.
  Springer Verlag, Tiergartenstrasse 17, D-69121, Heidelberg, Germany, Tel: +49(6221)4870, Fax: +49(6221)487366 ISSUES/YR: 1

Current Topics in Neurochemistry
ISSN: 0972-477X
Curr. Top. Neurochem.
Research Trends, T.C. 17/250(3), Chadiyara Road, 695012, Poojapura, Trivandrum, India, Tel: +91-471 2344424, Fax: +91-471 2344423 ISSUES/YR: 1

Current Topics in Nutraceutical Research
ISSN: 1540-7535
Curr. Top. Nutraceutical Res.
New Century Health Publishers, PO BOX 50702, New Orleans, United States, Fax: +1-5042427690 ISSUES/YR: 4

Current Topics in Peptide and Protein Research
ISSN: 0972-4524
Curr. Top. Pept. Protein Res.
Research Trends, T.C. 17/250(3), Chadiyara Road, 695012, Poojapura, Trivandrum, India, Tel: +91-471 2344424, Fax: +91-471 2344423 ISSUES/YR: 1

Current Topics in Pharmacology
ISSN: 0972-4559
Curr. Top. Pharmacol.
Research Trends, T.C. 17/250(3), Chadiyara Road, 695012, Poojapura, Trivandrum, India, Tel: +91-471 2344424, Fax: +91-471 2344423 ISSUES/YR: 1

Current Topics in Toxicology
ISSN: 0972-8228
Curr. Top. Toxicol.
Research Trends, T.C. 17/250(3), Chadiyara Road, 695012, Poojapura, Trivandrum, India, Tel: +91-471 2344424, Fax: +91-471 2344423 ISSUES/YR: 1

Current Treatment Options in Cardiovascular Medicine
ISSN: 1092-8464
Curr. Treat. Options Cardiovasc. Med.
Current Science Ltd, 34-42 Cleveland Street, W1P 6LB, London, United Kingdom, Tel: +44(207)3230323, Fax: +44(207)5801938 ISSUES/YR: 6

Current Treatment Options in Gastroenterology
ISSN: 1092-8472
Curr. Treat. Options Gastroenterol.
Current Science Ltd, 34-42 Cleveland Street, W1P 6LB, London, United Kingdom, Tel: +44(207)3230323, Fax: +44(207)5801938 ISSUES/YR: 6

Current Treatment Options in Neurology
ISSN: 1092-8480
Curr. Treat. Options Neurol.
Current Science Ltd, 34-42 Cleveland Street, W1P 6LB, London, United Kingdom, Tel: +44(207)3230323, Fax: +44(207)5801938 ISSUES/YR: 6

Current Treatment Options in Oncology
ISSN: 1527-2729
Curr. Treat. Options Oncol.
Current Science Ltd, 34-42 Cleveland Street, W1P 6LB, London, United Kingdom, Tel: +44(207)3230323, Fax: +44(207)5801938 ISSUES/YR: 6

Current Trends in Immunology
ISSN: 0972-4567
Curr. Trends Immunol.
Research Trends, T.C. 17/250(3), Chadiyara Road, 695012, Poojapura, Trivandrum, India, Tel: +91-471 2344424, Fax: +91-471 2344423 ISSUES/YR: 1

Current Urology Reports
ISSN: 1527-2737
Curr. Urol. Rep.
Current Science, Inc., 400 Market Street, Suite 700, Philadelphia, PA 19106, United States, Tel: +1(215)5742266, Fax: +1(215)5742225 ISSUES/YR: 6

Current Vascular Pharmacology
ISSN: 1570-1611
Curr. Vasc. Pharmacol.
Bentham Science Publishers B.V., P.O. Box 294, 1400 AG, Bussum, Netherlands, Tel: +31-356257290, Fax: +31-847511310 ISSUES/YR: 4

- **Current Women's Health Reviews**
  ISSN: 1573-4048
  Curr. Women's Health Rev.
  Bentham Science Publishers B.V., P.O. Box 294, 1400 AG, Bussum, Netherlands, Tel: +31-356257290, Fax: +31-847511310 ISSUES/YR: 3

Cutaneous and Ocular Toxicology
ISSN: 1556-9527
Cutaneous Ocul. Toxicol.
Taylor and Francis Inc., 325 Chestnut St, Suite 800, Philadelphia PA, PA 19106, United States, Tel: +1(800)3541420, Fax: +1(215)6258914 ISSUES/YR: 4

Cyberpsychology and Behavior
ISSN: 1094-9313
Cyberpsychol. Behav.
Mary Ann Liebert Inc., 140 Huguenot Street, New Rochelle, NY 10801-5215, United States, Tel: +1(914)7402100, Fax: +1(914)7402101 ISSUES/YR: 6

- **Cytogenetic and Genome Research**
  ISSN: 1424-8581
  Cytogenet. Genome Res.
  S. Karger AG, Allschwilerstrasse 10, P.O. Box, CH-4009, Basel, Switzerland, Tel: +41(61)3061111, Fax: +41(61)3061234 ISSUES/YR: 16

CytoJournal
ISSN: 1742-6413
CytoJournal
BioMed Central Ltd., 34 - 42 Cleveland Street, W1T 4LB, London, United Kingdom, Tel: +44(207)3230323, Fax: +44(207)6319926 ISSUES/YR: 1

- **Cytokine**
  ISSN: 1043-4666
  Cytokine
  Academic Press, 24-28 Oval Road, NW1 7DX, London, United Kingdom, Tel: +44(207)2674466 ISSUES/YR: 24
- **Cytokine and Growth Factor Reviews**
  ISSN: 1359-6101
  Cytokine Growth Factor Rev.
  Elsevier Ltd, Langford Lane, Kidlington, OX5 1GB, Oxford, United Kingdom, Tel: +44(1865)843000, Fax: +44(1865)843010 ISSUES/YR: 6
- **Cytometry Part A**
  ISSN: 1552-4922
  Cytometry Part A
  Wiley-Liss Inc., 111 River Street, Hoboken, NJ 07030-5774, United States, Tel: +1(201)7486645, Fax: +1(201)7486021 ISSUES/YR: 12
- **Cytometry Part B - Clinical Cytometry**
  ISSN: 1552-4949
  Cytometry Part B Clin. Cytometry
  Wiley-Liss Inc., 111 River Street, Hoboken, NJ 07030-5774, United States, Tel: +1(201)7486645, Fax: +1(201)7486021 ISSUES/YR: 6

• **Cytopathology**
ISSN: 0956-5507
Cytopathology
Blackwell Publishing Ltd, 9600 Garsington Road, OX4 2XG, Oxford, United Kingdom, Tel: +44(1865)776868, Fax: +44(1865)714591 ISSUES/YR: 6

Cytopathology, Supplement
ISSN: 1350-4037
Cytopathol. Suppl.
Blackwell Publishing Ltd, 9600 Garsington Road, OX4 2XG, Oxford, United Kingdom, Tel: +44(1865)776868, Fax: +44(1865)714591 ISSUES/YR: 1

Cytotechnology
ISSN: 0920-9069
Cytotechnology
Kluwer Academic Publishers, Van Godewijckstraat 30, 3311 GZ, Dordrecht, Netherlands, Tel: +31(78)6576000, Fax: +31(78)6576254 ISSUES/YR: 9

• **Cytotherapy**
ISSN: 1465-3249
Cytotherapy
Taylor and Francis Ltd., 4 Park Square, Milton Park, OX14 4RN, Abingdon, Oxfordshire, United Kingdom, Tel: +44(20)70175541, Fax: +44(20)70175198 ISSUES/YR: 6

Danish Medical Bulletin
ISSN: 1603-9629
Dan. Med. Bull.
Danish Medical Association, Esplanaden 8 C, 3., DK-1263, Copenhagen, Denmark, Tel: +45(3)5448500/3-5448375, Fax: +45(3)5448370 ISSUES/YR: 4

Daru
ISSN: 1560-8115
Daru
Tehran University of Medical Sciences, P.O. Box 14155-6451, Tehran, Iran, Fax: +98(21)6461178 ISSUES/YR: 4

Death Studies
ISSN: 0748-1187
Death Stud.
Taylor and Francis Inc., 325 Chestnut St, Suite 800, Philadelphia PA, PA 19106, United States, Tel: +1(800)3541420, Fax: +1(215)6258914 ISSUES/YR: 10

• **Dementia and Geriatric Cognitive Disorders**
ISSN: 1420-8008
Dementia Geriatr. Cogn. Disord.
S. Karger AG, Allschwilerstrasse 10, P.O. Box, CH-4009, Basel, Switzerland, Tel: +41(61)3061111, Fax: +41(61)3061234 ISSUES/YR: 8

Dengue Bulletin
ISSN: 1020-895X
Dengue Bull.
World Health Organization, Mahatma Gandhi Road, 110002, New Delhi, India, Tel: +91(11)3370804, Fax: +91(11)3379507 ISSUES/YR: 1

Dentomaxillofacial Radiology
ISSN: 0250-832X
Dentomaxillofac. Radiol.
British Institute of Radiology, 36 Portland Place, W1N 1AT, London, United Kingdom, Tel: +44(207)5804189 ISSUES/YR: 6

• **Depression and Anxiety**
ISSN: 1091-4269
Depression Anxiety
Wiley-Liss Inc., 111 River Street, Hoboken, NJ 07030-5774, United States, Tel: +1(201)7486645, Fax: +1(201)7486021 ISSUES/YR: 8

Deri Hastaliklari ve Frengi Arsivi
ISSN: 1019-214X
Deri Hast. Frengi Ars.
Istanbul Assoc. of Dermatology and Venerology, Dermatoloji Anabilim Dali, 34303, Cerrahpasa-Istanbul, Turkey, Tel: +90(212)5245487 ISSUES/YR: 4

• **Dermatitis**
ISSN: 1710-3568
Dermatitis
BC Decker Inc., P.O. Box 620, LCD 1, L8N 3K7, Hamilton, Ontario, Canada, Tel: +1(905)522-7017, Fax: +1(905)522-7839 ISSUES/YR: 4

Dermatologia Clinica
ISSN: 0392-1395
Dermatol. Clin.
CIC Edizioni Internazionali s.r.l., Corso Trieste 42, 00198, Roma, Italy, Tel: +39(06)8412673, Fax: +39(06)8412688 ISSUES/YR: 4

Dermatologia Cosmetica, Medica y Quirurgica
ISSN: 1665-4390
Dermatol. Cosmet. Med. Quir.
Medipiel, Av. Rio Churubusco 316-15, 04100 D.F., Col. Del Carmen - Coyoacan, Mexico, Tel: +52-52 55 5659 9416, Fax: +52-52 55 5659 4824 ISSUES/YR: 4

Dermatologia Kliniczna
ISSN: 1730-7201
Dematol. Kliniczna
Cornetis, ul. Dlugosza 2-6, 51-162, Wroclaw, Poland, Tel: +48-071 326 3101, Fax: +48-071 325 2803 ISSUES/YR: 4

Dermatologia Revista Mexicana
ISSN: 0185-4038
Dermatol. Rev. Mex.
Comunicaciones Cientificas Mexicanas S.A. de C.V., Calz. Gen. Anaya 330, Col. de Carmen Coyoacan, CP 04100, Mexico D.F., Mexico, Tel: +52(55)5-659-3863, Fax: +52(55)5-554-6054 ISSUES/YR: 6

• **Dermatologic Clinics**
ISSN: 0733-8635
Dermatol. Clin.
W.B. Saunders, Independence Square West, Philadelphia, PA 19106-3399, United States, Tel: +1(215)2387800, Fax: +1(215)2386445 ISSUES/YR: 4

• **Dermatologic Surgery**
ISSN: 1076-0512
Dermatol. Surg.
Blackwell Publishing Ltd, 9600 Garsington Road, OX4 2XG, Oxford, United Kingdom, Tel: +44(1865)776868, Fax: +44(1865)714591 ISSUES/YR: 12

Dermatologic Therapy
ISSN: 1396-0296
Dermatol. Ther.
Blackwell Publishing Inc., 350 Main Street, Malden, MA 02148, United States, Tel: +1(781)3888200, Fax: +1(781)3888210 ISSUES/YR: 6

Dermatologica Sinica
ISSN: 1027-8117
Dermatol. Sin.
The Chinese Dermatological Society, Chung Shan South Road, 7, Taipei, Taiwan, Tel: +886(2)23928763, Fax: +886(2)23951186 ISSUES/YR: 4

Dermatologie in Beruf und Umwelt
ISSN: 1438-776X
Dermatol. Beruf Umwelt
Dustri-Verlag Dr. Karl Feistle, P.O. Box 1351, D-82032, Deisenhofen/Munchen, Germany, Tel: +49(89)6138610, Fax: +49(89)6135412 ISSUES/YR: 4

• **Dermatology**
ISSN: 1018-8665
Dermatology
S. Karger AG, Allschwilerstrasse 10, P.O. Box, CH-4009, Basel, Switzerland, Tel: +41(61)3061111, Fax: +41(61)3061234 ISSUES/YR: 8

Dermatology Online Journal
ISSN: 1087-2108
Dermatol. Online J.
Dermatology Online Journal, United States ISSUES/YR: 1

Deutsche Apotheker Zeitung
ISSN: 0011-9857
Dtsch. Apoth. Ztg.
Deutscher Apotheker Verlag, P.O. Box 10 10 61, D-70 009, Stuttgart, Germany, Tel: +49(711)25820 ISSUES/YR: 52

Deutsche Medizinische Wochenschrift
ISSN: 0012-0472
Dtsch. Med. Wochenschr.
Georg Thieme Verlag, Rudigerstrasse 14, D-70469, Stuttgart, Germany, Tel: +49(711)8931-321, Fax: +49(711)8931-422 ISSUES/YR: 52

Current

Deutsche Zeitschrift fur Akupunktur
ISSN: 0415-6412
Dtsch. Z. Akupunkt.
Urban und Fischer Verlag Jena, P.O. Box 100537, 07705, Jena, Germany, Tel: +49(3641)6263, Fax: +49(3641)626500 ISSUES/YR: 4

Deutsche Zeitschrift fur Onkologie
ISSN: 1617-5891
Dtsch. Z. Onkol.
Georg Thieme Verlag, Rudigerstrasse 14, D-70469, Stuttgart, Germany, Tel: +49(711)8931-321, Fax: +49(711)8931-422 ISSUES/YR: 4

Deutsches Arzteblatt
ISSN: 0012-1207
Dtsch. Arztebl.
Deutscher Arzte-Verlag GmbH, Dieselstrasse 2 (Postfach 0254), 50859, Cologne, Germany, Tel: +49(22)2370110, Fax: +49(22)347011460 ISSUES/YR: 52

• **Developing World Bioethics**
ISSN: 1471-8731
Dev. World Bioethics
Blackwell Publishing Ltd, 9600 Garsington Road, OX4 2XG, Oxford, United Kingdom, Tel: +44(1865)776868, Fax: +44(1865)714591 ISSUES/YR: 2

• **Development**
ISSN: 0950-1991
Development (Cambridge)
Company of Biologists Ltd, 140 Cowley Road, CB4 4DL, Cambridge, United Kingdom, Tel: +44(1223)426164, Fax: +44(1223)423353 ISSUES/YR: 24

• **Development and Psychopathology**
ISSN: 0954-5794
Dev. Psychopathol.
Cambridge University Press, Shaftesbury Road, CB2 2RU, Cambridge, United Kingdom, Tel: +44(1223)312393, Fax: +44(1223)315052 ISSUES/YR: 4

• **Development Genes and Evolution**
ISSN: 0949-944X
Dev. Genes Evol.
Springer Verlag, Tiergartenstrasse 17, D-69121, Heidelberg, Germany, Tel: +49(6221)4870, Fax: +49(6221)487366 ISSUES/YR: 12

Development Growth and Differentiation
ISSN: 0012-1592
Dev. Growth Differ.
Blackwell Publishing, 550 Swanston Street, VIC 3053, Carlton South, Australia, Tel: +61(3)8359 1011, Fax: +61(3)8359 1120 ISSUES/YR: 9

• **Developmental and Comparative Immunology**
ISSN: 0145-305X
Dev. Comp. Immunol.
Elsevier Ltd, Langford Lane, Kidlington, OX5 1GB, Oxford, United Kingdom, Tel: +44(1865)843000, Fax: +44(1865)843010 ISSUES/YR: 10

• **Developmental Biology**
ISSN: 0012-1606
Dev. Biol.
Academic Press Inc., 6277 Sea Harbor Drive, Orlando, FL 32887-4900, United States, Tel: +1(407)3454040, Fax: +1(407)3639661 ISSUES/YR: 24

• **Developmental Cell**
ISSN: 1534-5807
Dev. Cell
Cell Press, 1100 Massachusetts Avenue, Cambridge, MA 02138-9957, United States, Tel: +1(617)6617057, Fax: +1(617)6617061 ISSUES/YR: 12

• **Developmental Dynamics**
ISSN: 1058-8388
Dev. Dyn.
Wiley-Liss Inc., 111 River Street, Hoboken, NJ 07030-5774, United States, Tel: +1(201)7486645, Fax: +1(201)7486021 ISSUES/YR: 12

• **Developmental Medicine and Child Neurology**
ISSN: 0012-1622
Dev. Med. Child Neurol.
Cambridge University Press, Shaftesbury Road, CB2 2RU, Cambridge, United Kingdom, Tel: +44(1223)312393, Fax: +44(1223)315052 ISSUES/YR: 12

• **Developmental Neuropsychology**
ISSN: 8756-5641
Dev. Neuropsychol.
Lawrence Erlbaum Associates Inc., 10 Industrial Avenue, Mahwah, NJ 07430-2262, United States, Tel: +1(201)2582200, Fax: +1(201)2360072 ISSUES/YR: 6

• **Developmental Neuroscience**
ISSN: 0378-5866
Dev. Neurosci.
S. Karger AG, Allschwilerstrasse 10, P.O. Box, CH-4009, Basel, Switzerland, Tel: +41(61)3061111, Fax: +41(61)3061234 ISSUES/YR: 6

Developmental Psychobiology
ISSN: 0012-1630
Dev. Psychobiol.
John Wiley and Sons Inc., P.O.Box 18667, Newark, NJ 07191-8667, United States, Tel: +1(212)8506000, Fax: +1(212)8506012 ISSUES/YR: 8

• **Diabetes**
ISSN: 0012-1797
Diabetes
American Diabetes Association Inc., 1701 North Beauregard St., Alexandria, VA 22311, United States, Tel: +1(703)5491500, Fax: +1-703 5496995 ISSUES/YR: 12

Diabetes and Metabolism
ISSN: 1262-3636
Diabetes Metab.
Masson, 21 Rue Camille Desmoulins, 92789, Issy les Moulineaux Cedex 9, France, Tel: +33(1)73281634, Fax: +33(1)73281649 ISSUES/YR: 6

Diabetes and Vascular Disease Research
ISSN: 1479-1641
Diabets Vasc. Dis. Res.
Sherbonre Gibbs Limited, 3 Duchess Place, B16 8NH, Edgbaston Birmingham, United Kingdom, Tel: +44(121)4544114, Fax: +44(121)4541190 ISSUES/YR: 4

Diabetes Care
ISSN: 0149-5992
Diabetes Care
American Diabetes Association Inc., 1701 North Beauregard St., Alexandria, VA 22311, United States, Tel: +1(703)5491500, Fax: +1-703 5496995 ISSUES/YR: 12

Diabetes Research and Clinical Practice
ISSN: 0168-8227
Diabetes Res. Clin. Pract.
Elsevier Ireland Ltd, P.O. Box 85, Limerick, Ireland, Tel: +353(61)471944, Fax: +353(61)472144 ISSUES/YR: 12

Diabetes Spectrum
ISSN: 1040-9165
Diabetes Spectr.
American Diabetes Association Inc., 1701 North Beauregard St., Alexandria, VA 22311, United States, Tel: +1(703)5491500, Fax: +1-703 5496995 ISSUES/YR: 4

• **Diabetes Technology and Therapeutics**
ISSN: 1520-9156
Diabetes Technol. Ther.
Mary Ann Liebert Inc., 140 Huguenot Street, New Rochelle, NY 10801-5215, United States, Tel: +1(914)7402100, Fax: +1(914)7402101 ISSUES/YR: 6

Diabetes und Stoffwechsel
ISSN: 0942-0037
Diabetes Stoffwechsel
Verlag Kirchheim and Co. GmbH, Postfach 2524, D-55015, Mainz, Germany, Tel: +49(6131)960700, Fax: +49(6131)9607070 ISSUES/YR: 6

• **Diabetes, Nutrition and Metabolism - Clinical and Experimental**
ISSN: 0394-3402
Diabetes Nutr. Metab. Clin. Exp.
Editrice Kurtis s.r.l., Via Luigi Zoja 30, 20153, Milan, Italy, Tel: +39(02)48202740, Fax: +39(02)48201219 ISSUES/YR: 6

Diabetes, Obesity and Metabolism
ISSN: 1462-8902
Diabestes Obes. Metab.
Blackwell Publishing Ltd, 9600 Garsington Road, OX4 2XG, Oxford, United Kingdom, Tel: +44(1865)776868, Fax: +44(1865)714591 ISSUES/YR: 6

Current

Diabetes, Obesity and Metabolism, Supplement
ISSN: 1465-7740
Diabetes Obes. Metab. Suppl.
Blackwell Publishing Ltd, 9600 Garsington Road, OX4 2XG, Oxford, United Kingdom, Tel: +44(1865)776868, Fax: +44(1865)714591 ISSUES/YR: 1

- **Diabetes/Metabolism Research and Reviews**
ISSN: 1520-7552
Diabetes Metab. Res. Rev.
John Wiley and Sons Ltd, Southern Gate, PO19 8SQ, Chichester, West Sussex, United Kingdom, Tel: +44(1243)779777, Fax: +44(1243)775878 ISSUES/YR: 6

Diabetic Medicine
ISSN: 0742-3071
Diabetic Med.
Blackwell Publishing Ltd, 9600 Garsington Road, OX4 2XG, Oxford, United Kingdom, Tel: +44(1865)776868, Fax: +44(1865)714591 ISSUES/YR: 12

- **Diabetologia**
ISSN: 0012-186X
Diabetologia
Springer Verlag, Tiergartenstrasse 17, D-69121, Heidelberg, Germany, Tel: +49(6221)4870, Fax: +49(6221)487366 ISSUES/YR: 12

Diabetologia Croatica
ISSN: 0351-0042
Diabetol. Croat.
Vuk Vrhovac Institute, Dugi Dol 4a, P.O. Box 958, HR-10000, Zagreb, Croatia, Tel: +385(1)2332222, Fax: +385(1)2331515 ISSUES/YR: 4

Diabetologia Doswiadczalna i Kliniczna
ISSN: 1643-3165
Diabetol. Dosw. Klin.
Via Medica, Ul. Swietokrzyska 73, 80-180, Gdansk, Poland, Tel: +48(58)3209494, Fax: +48(58)3209460 ISSUES/YR: 6

Diabetologia Polska
ISSN: 1233-4723
Diabetol. Pol.
PTBPC, ul. Kopcinskiego 22, 90-153, Lodz, Poland ISSUES/YR: 4

Diabetologie Metabolismus Endokrinologie Vyziva
ISSN: 1211-9326
Diabetol. Metabol. Endokrinol. Vyz.
Tigis s.r.o., Na Habrove 17, 152 00, Prague 5, Czech Republic, Tel: +420-251 813 1924, Fax: +420-251 681 217 ISSUES/YR: 4

Diagnostic and Interventional Radiology
ISSN: 1305-3825
Diagn. Intervention. Radiol.
Turkish Society of Radiology, Cankaya Evleri, F/2, 06540, Ankara, Turkey, Tel: +90(312)4423653, Fax: +90(312)4423654 ISSUES/YR: 4

- **Diagnostic Cytopathology**
ISSN: 8755-1039
Diagn. Cytopathol.
Wiley-Liss Inc., 111 River Street, Hoboken, NJ 07030-5774, United States, Tel: +1(201)7486645, Fax: +1(201)7486021 ISSUES/YR: 12

- **Diagnostic Microbiology and Infectious Disease**
ISSN: 0732-8893
Diagn. Microbiol. Infect. Dis.
Elsevier Inc., 360 Park Avenue South, New York, NY 10010, United States, Tel: +1(212)6333950, Fax: +1(212)6333990 ISSUES/YR: 12

- **Diagnostic Molecular Pathology**
ISSN: 1052-9551
Diagn. Mol. Pathol.
Lippincott Williams and Wilkins, 530 Walnut Street, Philadelphia, PA 19106-3621, United States, Tel: +1(215)5218342, Fax: +1(215)5218326 ISSUES/YR: 4

Dialogues in Cardiovascular Medicine
ISSN: 1272-9949
Dialogues Cardiovasc. Med.
Servier International, 192 Avenue Charles de Gaulle, 92587, Neuilly sur Seine Cedex, France, Tel: +33(1)55726000, Fax: +33(1)55726888 ISSUES/YR: 4

Dialogues in Clinical Neuroscience
ISSN: 1294-8322
Dialogues Clin. Neurosci.
Servier International, 192 Avenue Charles de Gaulle, 92587, Neuilly sur Seine Cedex, France, Tel: +33(1)55726000, Fax: +33(1)55726888 ISSUES/YR: 4

- **Dialysis and Transplantation**
ISSN: 0090-2934
Dial. Transplant.
John Wiley and Sons Inc., P.O.Box 18667, Newark, NJ 07191-8667, United States, Tel: +1(212)8506000, Fax: +1(212)8506012 ISSUES/YR: 12

- **Differentiation**
ISSN: 0301-4681
Differentiation
Blackwell Publishing Ltd, 9600 Garsington Road, OX4 2XG, Oxford, United Kingdom, Tel: +44(1865)776868, Fax: +44(1865)714591 ISSUES/YR: 10

Difficult Airways
ISSN: 1609-2961
Difficult Airways
VICER Publishing, P.O. Box 14, A-1097, Vienna, Austria, Tel: +43(676)9568085, Fax: +43(676)9568086 ISSUES/YR: 4

- **Digestion**
ISSN: 0012-2823
Digestion
S. Karger AG, Allschwilerstrasse 10, P.O. Box, CH-4009, Basel, Switzerland, Tel: +41(61)3061111, Fax: +41(61)3061234 ISSUES/YR: 8

Digestive and Liver Disease
ISSN: 1590-8658
Dig. Liver Dis.
W.B. Saunders Ltd, 32 Jamestown Road, NW1 7BY, London, United Kingdom, Tel: +1(20)74244200, Fax: +1(20)74854752 ISSUES/YR: 12

- **Digestive Diseases**
ISSN: 0257-2753
Dig. Dis.
S. Karger AG, Allschwilerstrasse 10, P.O. Box, CH-4009, Basel, Switzerland, Tel: +41(61)3061111, Fax: +41(61)3061234 ISSUES/YR: 4

- **Digestive Diseases and Sciences**
ISSN: 0163-2116
Dig. Dis. Sci.
Springer Science and Business Media Deutschland GmbH, Heidelberger Platz 3, 14197, Berlin, Germany, Tel: +1-30 827875282, Fax: +1-30 827875707 ISSUES/YR: 12

Digestive Endoscopy
ISSN: 0915-5635
Dig. Endosc.
Blackwell Publishing Ltd, 9600 Garsington Road, OX4 2XG, Oxford, United Kingdom, Tel: +44(1865)776868, Fax: +44(1865)714591 ISSUES/YR: 4

Digestive Surgery
ISSN: 0253-4886
Dig. Surg.
S. Karger AG, Allschwilerstrasse 10, P.O. Box, CH-4009, Basel, Switzerland, Tel: +41(61)3061111, Fax: +41(61)3061234 ISSUES/YR: 6

- **Disability and Rehabilitation**
ISSN: 0379-0797
Disabil. Rehabil.
Taylor and Francis Ltd., 4 Park Square, Milton Park, OX14 4RN, Abingdon, Oxfordshire, United Kingdom, Tel: +44(20)70175541, Fax: +44(20)70175198 ISSUES/YR: 24

Disease Management
ISSN: 1093-507X
Dis. Manage.
Mary Ann Liebert Inc., 140 Huguenot Street, New Rochelle, NY 10801-5215, United States, Tel: +1(914)7402100, Fax: +1(914)7402101 ISSUES/YR: 4

Disease Management and Health Outcomes
ISSN: 1173-8790
Dis. Manage. Health Outcomes
Adis International Ltd, 41 Centorian Drive, Private Bag 65901, Mairangi Bay, 1311, Auckland 10, New Zealand, Tel: +64(9)4770700, Fax: +64(9)4770766 ISSUES/YR: 6

Current

- **Disease Markers**
  ISSN: 0278-0240
  Dis. Markers
  IOS Press, Nieuwe Hemweg 6B, 1013 BG, Amsterdam, Netherlands, Tel: +31(20)6883355, Fax: +31(20)6203419 ISSUES/YR: 6

Disease-a-Month
ISSN: 0011-5029
Dis. Mon.
Mosby Inc., 11830 Westline Industrial Drive, St. Louis, MO 63146, United States, Tel: +1(314)8728370, Fax: +1(314)4321380 ISSUES/YR: 12

Diseases of the Colon and Rectum
ISSN: 0012-3706
Dis. Colon Rectum
Springer New York LLC, 233 Springer Street, New York, 10013-1578, United States, Tel: +1-212 460 1501, Fax: +1-212 460 1595 ISSUES/YR: 12

- **Diseases of the Esophagus**
  ISSN: 1120-8694
  Dis. Esophagus
  Blackwell Publishing, 550 Swanston Street, VIC 3053, Carlton South, Australia, Tel: +61(3)8359 1011, Fax: +61(3)8359 1120 ISSUES/YR: 4

Diving and Hyperbaric Medicine
ISSN: 1833-3516
Diving Hyperbaric Med.
South Pacific Underwater Medicine Society, 630 St. Kildare Road, VIC 3004, Melbourne, Australia, Tel: +61(3)98194898, Fax: +61(3)98195298 ISSUES/YR: 4

- **DNA and Cell Biology**
  ISSN: 1044-5498
  DNA Cell Biol.
  Mary Ann Liebert Inc., 140 Huguenot Street, New Rochelle, NY 10801-5215, United States, Tel: +1(914)7402100, Fax: +1(914)7402101 ISSUES/YR: 12

- **DNA Repair**
  ISSN: 1568-7864
  DNA Repair
  Elsevier, P.O. Box 211, 1000 AE, Amsterdam, Netherlands, Tel: +31(20)4853757, Fax: +31(20)4853432 ISSUES/YR: 12

DNA Research
ISSN: 1340-2838
DNA Res.
Oxford University Press, Great Clarendon Street, OX2 6DP, Oxford, United Kingdom, Tel: +44(1865)353907, Fax: +44(1865)353486 ISSUES/YR: 6

- **DNA Sequence - Journal of DNA Sequencing and Mapping**
  ISSN: 1042-5179
  DNA Sequence J. DNA Sequencing Mapp.
  Taylor and Francis Ltd., 4 Park Square, Milton Park, OX14 4RN, Abingdon, Oxfordshire, United Kingdom, Tel: +44(20)70175541, Fax: +44(20)70175198 ISSUES/YR: 6

Documenta Ophthalmologica
ISSN: 0012-4486
Doc. Ophthalmol.
Kluwer Academic Publishers, Van Godewijckstraat 30, 3311 GZ, Dordrecht, Netherlands, Tel: +31(78)6576000, Fax: +31(78)6576254 ISSUES/YR: 6

Dokkyo Journal of Medical Sciences
ISSN: 0385-5023
Dokkyo J. Med. Sci.
Dokkyo Journal of Medical Sciences, 880 Kitakobayashi, Mibu, Shimotsugu-gun, 321-02, Tochigi, Japan, Tel: +81(282)872110 ISSUES/YR: 3

DOLOR
ISSN: 0214-0659
DOLOR
Publicaciones Permanyer, Mallorca 310, E-08037, Barcelona, Spain, Tel: +34(93)2075920, Fax: +34(93)4576642 ISSUES/YR: 4

Douleur et Analgesie
ISSN: 1011-288X
Douleur Analg.
Editions Medecine et Hygiene, 7, Avenue de la Roseraie, P.O. Box 456, CH-1211, Geneva 4, Switzerland, Tel: +41(22)7029311, Fax: +41(22)7029355 ISSUES/YR: 4

Douleurs
ISSN: 1624-5687
Douleurs
Masson, 21 Rue Camille Desmoulins, 92789, Issy les Moulineaux Cedex 9, France, Tel: +33(1)73281634, Fax: +33(1)73281649 ISSUES/YR: 6

- **Drug and Alcohol Dependence**
  ISSN: 0376-8716
  Drug Alcohol Depend.
  Elsevier Ireland Ltd, P.O. Box 85, Limerick, Ireland, Tel: +353(61)471944, Fax: +353(61)472144 ISSUES/YR: 12

Drug and Chemical Toxicology
ISSN: 0148-0545
Drug Chem. Toxicol.
Taylor and Francis Inc., 325 Chestnut St, Suite 800, Philadelphia PA, PA 19106, United States, Tel: +1(800)3541420, Fax: +1(215)6258914 ISSUES/YR: 4

Drug and Therapeutics Bulletin
ISSN: 0012-6543
Drug Ther. Bull.
Consumers' Association, 2 Marylebone Road, NW1 4DF, London, United Kingdom, Tel: +44(207)4865544 ISSUES/YR: 12

Drug Benefit Trends
ISSN: 1080-5826
Drug Benefit Trends
CMP Healthcare Media LLC, 330 Boston Post Road, Darien, CT 06820-4027, United States, Tel: +1(203)6626400, Fax: +1(212)6311600 ISSUES/YR: 12

- **Drug Delivery**
  ISSN: 1071-7544
  Drug Deliv.
  Taylor and Francis Ltd., 4 Park Square, Milton Park, OX14 4RN, Abingdon, Oxfordshire, United Kingdom, Tel: +44(20)70175541, Fax: +44(20)70175198 ISSUES/YR: 6

- **Drug Design Reviews Online**
  ISSN: 1567-2697
  Drug Des. Rev. Online
  Bentham Science Publishers B.V., P.O. Box 294, 1400 AG, Bussum, Netherlands, Tel: +31-356257290, Fax: +31-847511310 ISSUES/YR: 8

Drug Development and Industrial Pharmacy
ISSN: 0363-9045
Drug Dev. Ind. Pharm.
Taylor and Francis Inc., 325 Chestnut St, Suite 800, Philadelphia PA, PA 19106, United States, Tel: +1(800)3541420, Fax: +1(215)6258914 ISSUES/YR: 10

Drug Development Research
ISSN: 0272-4391
Drug Dev. Res.
Wiley-Liss Inc., 111 River Street, Hoboken, NJ 07030-5774, United States, Tel: +1(201)7486645, Fax: +1(201)7486021 ISSUES/YR: 12

Drug Discovery Today
ISSN: 1359-6446
Drug Discov. Today
Elsevier Ltd, Langford Lane, Kidlington, OX5 1GB, Oxford, United Kingdom, Tel: +44(1865)843300 ISSUES/YR: 24

Drug Discovery Today: BIOSILICO
ISSN: 1741-8364
Drug Discov. Today Biosilico
Elsevier Ltd, Langford Lane, Kidlington, OX5 1GB, Oxford, United Kingdom, Tel: +44(1865)843300 ISSUES/YR: 6

Drug Discovery Today: Disease Mechanisms
ISSN: 1740-6765
Drug Discov.Today Dis. Mech.
Elsevier Ltd, Langford Lane, Kidlington, OX5 1GB, Oxford, United Kingdom, Tel: +44(1865)843300 ISSUES/YR: 4

Drug Discovery Today: Disease Models
ISSN: 1740-6757
Drug Discov. Today Dis. Models
Elsevier Ltd, Langford Lane, Kidlington, OX5 1GB, Oxford, United Kingdom, Tel: +44(1865)843300 ISSUES/YR: 4

Drug Discovery Today: TARGETS
ISSN: 1741-8372
Drug Discov. Today Targets
Elsevier Ltd, Langford Lane, Kidlington, OX5 1GB, Oxford, United Kingdom, Tel: +44(1865)843300 ISSUES/YR: 6

Current

Drug Discovery Today: Technologies
ISSN: 1740-6749
Drug Discov. Today Techn.
Elsevier Ltd, Langford Lane, Kidlington, OX5 1GB, Oxford, United Kingdom, Tel: +44(1865)843300 ISSUES/YR: 6

Drug Discovery Today: Therapeutic Strategies
ISSN: 1740-6773
Drug Discov. Today Ther. Strateg.
Elsevier Ltd, Langford Lane, Kidlington, OX5 1GB, Oxford, United Kingdom, Tel: +44(1865)843300 ISSUES/YR: 4

Drug Discovery World
ISSN: 1469-4344
Drug Discov. World
RJ Communications and Media Ltd., 39 Vineyard Path, SW14 8ET, Mortlake London, United Kingdom, Tel: +44(208)487-5656, Fax: +44(208)487-5666 ISSUES/YR: 4

• **Drug Information Journal**
ISSN: 0092-8615
Drug Inf. J.
Drug Information Association, 800 Enterprise Road, Suite 200, Horsham, PA 19044-3595, United States, Tel: +1(215)442 6100, Fax: +1(215)442 6199 ISSUES/YR: 4

• **Drug Metabolism and Disposition**
ISSN: 0090-9556
Drug Metab. Dispos.
American Society for Pharmacology and Experimental Therapy, 9650 Rockville Pike, Bethesda, MD 20814, United States, Tel: +1(301)5307024, Fax: +1(301)5715728 ISSUES/YR: 12

Drug Metabolism and Drug Interactions
ISSN: 0792-5077
Drug Metab. Drug Interact.
Freund Publishing House Ltd, 61 Nachmani Street, P.O. Box 35010, 61350, Tel Aviv, Israel, Tel: +972(3)5628540, Fax: +972(3)5628538 ISSUES/YR: 4

Drug Metabolism Reviews
ISSN: 0360-2532
Drug Metab. Rev.
Taylor and Francis Inc., 325 Chestnut St, Suite 800, Philadelphia PA, PA 19106, United States, Tel: +1(800)3541420, Fax: +1(215)6258914 ISSUES/YR: 4

Drug News and Perspectives
ISSN: 0214-0934
Drug News Perspect.
Prous Science, P.O. Box 540, 08080, Barcelona, Spain, Tel: +34(93)4592220, Fax: +34(93)4581535 ISSUES/YR: 10

Drug R and D Backgrounders
ISSN: not available yet
Drug R D Backgr.
Prous Science, P.O. Box 540, 08080, Barcelona, Spain, Tel: +34(93)4592220, Fax: +34(93)4581535 ISSUES/YR: 6

• **Drug Resistance Updates**
ISSN: 1368-7646
Drug Resist. Updates
Churchill Livingstone, 1-3 Baxter's Place, Leith Walk, EH1 3AF, Edinburgh, United Kingdom, Tel: +44(131)5562424, Fax: +44(131)5581278 ISSUES/YR: 6

• **Drug Safety**
ISSN: 0114-5916
Drug Saf.
Adis International Ltd, 41 Centorian Drive, Private Bag 65901, Mairangi Bay, 1311, Auckland 10, New Zealand, Tel: +64(9)4770700, Fax: +64(9)4770766 ISSUES/YR: 15

Drug Topics
ISSN: 0012-6616
Drug Topics
Medical Economics Publishing, 5 Paragon Drive, Montvale, NJ 07645-1752, United States, Tel: +1(201)3587200 ISSUES/YR: 23

Drugs
ISSN: 0012-6667
Drugs
Adis International Ltd, 41 Centorian Drive, Private Bag 65901, Mairangi Bay, 1311, Auckland 10, New Zealand, Tel: +64(9)4770700, Fax: +64(9)4770766 ISSUES/YR: 24

• **Drugs and Aging**
ISSN: 1170-229X
Drugs Aging
Adis International Ltd, 41 Centorian Drive, Private Bag 65901, Mairangi Bay, 1311, Auckland 10, New Zealand, Tel: +64(9)4770700, Fax: +64(9)4770766 ISSUES/YR: 15

Drugs and Therapy Perspectives
ISSN: 1172-0360
Drugs Ther. Perspect.
Adis International Ltd, 41 Centorian Drive, Private Bag 65901, Mairangi Bay, 1311, Auckland 10, New Zealand, Tel: +64(9)4770700, Fax: +64(9)4770766 ISSUES/YR: 12

Drugs in Context
ISSN: 1745-1981
Drugs Context
CSF Medical Communications Ltd, 1 Bankside, Lodge Road, OX29 8LJ, Oxfordshire, United Kingdom, Tel: +44-44 1993 885370, Fax: +44-44 1993 881868 ISSUES/YR: 1

Drugs in Context (US)
ISSN: 1745-8412
Drugs Context US
CSF Medical Communications Ltd, 1 Bankside, Lodge Road, OX29 8LJ, Oxfordshire, United Kingdom, Tel: +44-44 1993 885370, Fax: +44-44 1993 881868 ISSUES/YR: 1

• **Drugs in R and D**
ISSN: 1174-5886
Drugs R D
Adis International Ltd, 41 Centorian Drive, Private Bag 65901, Mairangi Bay, 1311, Auckland 10, New Zealand, Tel: +64(9)4770700, Fax: +64(9)4770766 ISSUES/YR: 6

Drugs of the Future
ISSN: 0377-8282
Drugs Future
Prous Science, P.O. Box 540, 08080, Barcelona, Spain, Tel: +34(93)4592220, Fax: +34(93)4581535 ISSUES/YR: 12

Drugs of Today
ISSN: 0025-7656
Drugs Today
Prous Science, P.O. Box 540, 08080, Barcelona, Spain, Tel: +34(93)4592220, Fax: +34(93)4581535 ISSUES/YR: 12

Drugs under Experimental and Clinical Research
ISSN: 0378-6501
Drugs Exp. Clin. Res.
Bioscience Ediprint Inc., 16 Rue Alexandre-Gavard, 1227, Carouge-Geneva, Switzerland, Tel: +41(22)3003383, Fax: +41(22)3002489 ISSUES/YR: 6

Dynamic Medicine
ISSN: 1476-5918
Dyn. Med.
BioMed Central Ltd., 34 - 42 Cleveland Street, W1T 4LB, London, United Kingdom, Tel: +44(207)3230323, Fax: +44(207)6319926 ISSUES/YR: 1

Dyslexia
ISSN: 1076-9242
Dyslexia
John Wiley and Sons Ltd, Southern Gate, PO19 8SQ, Chichester, West Sussex, United Kingdom, Tel: +44(1243)779777, Fax: +44(1243)775878 ISSUES/YR: 4

• **Dysphagia**
ISSN: 0179-051X
Dysphagia
Springer New York LLC, 233 Springer Street, New York, 10013-1578, United States, Tel: +1-212 460 1501, Fax: +1-212 460 1595 ISSUES/YR: 4

Ear and Hearing
ISSN: 0196-0202
Ear Hear.
Lippincott Williams and Wilkins, 351 West Camden Street, Baltimore, MD 21201-2436, United States, Tel: +1(410)5284000 ISSUES/YR: 6

Ear, Nose and Throat Journal
ISSN: 0145-5613
Ear Nose Throat J.
Medquest Communications LLC, 3800 Lakeside Avenue, Suite 201, Cleveland, OH 44114, United States, Tel: +1(216)3919100, Fax: +1(216)3919200 ISSUES/YR: 12

Early Human Development
ISSN: 0378-3782
Early Hum. Dev.
Elsevier Ireland Ltd, P.O. Box 85, Limerick, Ireland, Tel: +353(61)471944, Fax: +353(61)472144 ISSUES/YR: 12

East and Central African Journal of Pharmaceutical Sciences
ISSN: 1026-552X
East Centr. Afr. J. Pharm. Sci.
University of Nairobi, Faculty of Pharmacy, P.O. Box 19676-00202, Nairobi, Kenya ISSUES/YR: 3

Eastern Journal of Medicine
ISSN: 1301-0883
East. J. Med.
Yuzuncu Yil Universitesi Tip Fakultesi, Maras Cad. 65200, Van, Turkey, Tel: +90(432)2164706, Fax: +90(432)2167519 ISSUES/YR: 2

Eastern Mediterranean Health Journal
ISSN: 1020-3397
East. Mediterr. Health J.
World Health Organization, P.O. Box 1517, 21563, Alexandria, Egypt, Tel: +20(3)4830090, Fax: +20(3)4838916 ISSUES/YR: 6

Eating and Weight Disorders
ISSN: 1124-4909
Eating Weight Disord.
Editrice Kurtis s.r.l., Via Luigi Zoja 30, 20153, Milan, Italy, Tel: +39(02)48202740, Fax: +39(02)48201219 ISSUES/YR: 3

• **Eating Behaviors**
ISSN: 1471-0153
Eating Behav.
Elsevier Ltd, Langford Lane, Kidlington, OX5 1GB, Oxford, United Kingdom, Tel: +44(1865)843000, Fax: +44(1865)843010 ISSUES/YR: 4

Eating Disorders
ISSN: 1064-0266
Eating Disord.
Routledge, 4 Park Square, Milton Park, OX14 4RN, Abingdon, Oxfordshire, United Kingdom, Tel: +44(1235)828600, Fax: +44(1235)829000 ISSUES/YR: 5

EAU Update Series
ISSN: 1570-9124
EAU Update Ser.
Elsevier, P.O. Box 211, 1000 AE, Amsterdam, Netherlands, Tel: +31(20)4853757, Fax: +31(20)4853432 ISSUES/YR: 4

EBR - European Biopharmaceutical Review
ISSN: 1364-369X
EBR Eur. Biopharm. Rev.
Samedan Ltd, 16 Hampden Gurney Street, W1H 5AL, London, United Kingdom, Tel: +44(20)77245444, Fax: +44(20)77242632 ISSUES/YR: 4

• **Echocardiography**
ISSN: 0742-2822
Echocardiography
Blackwell Publishing Inc., 350 Main Street, Malden, MA 02148, United States, Tel: +1(781)3888200, Fax: +1(781)3888210 ISSUES/YR: 8

• **EcoHealth**
ISSN: 1612-9202
EcoHealth
Springer New York LLC, 233 Springer Street, New York, 10013-1578, United States, Tel: +1-212 460 1501, Fax: +1-212 460 1595 ISSUES/YR: 4

Economics and Human Biology
ISSN: 1570-677X
Econ. Hum. Biol.
Elsevier, P.O. Box 211, 1000 AE, Amsterdam, Netherlands, Tel: +31(20)4853757, Fax: +31(20)4853432 ISSUES/YR: 3

• **Ecotoxicology**
ISSN: 0963-9292
Ecotoxicology
Kluwer Academic Publishers, Van Godewijckstraat 30, 3311 GZ, Dordrecht, Netherlands, Tel: +31(78)6576000, Fax: +31(78)6576254 ISSUES/YR: 8

Ecotoxicology and Environmental Safety
ISSN: 0147-6513
Ecotoxicol. Environ. Saf.
Academic Press, 24-28 Oval Road, NW1 7DX, London, United Kingdom, Tel: +44(207)2674466 ISSUES/YR: 9

EDTNA-ERCA Journal
ISSN: 1019-083X
EDTNA ERCA J.
EDTNA-ERCA, Pilatusstrasse 35, P.O. Box 3052, CH-6002, Luzern, Switzerland, Tel: +41-4407555, Fax: +41-4403962 ISSUES/YR: 4

• **Education for Health: Change in Learning and Practice**
ISSN: 1357-6283
Educ. Health
Taylor and Francis Ltd., 4 Park Square, Milton Park, OX14 4RN, Abingdon, Oxfordshire, United Kingdom, Tel: +44(20)70175541, Fax: +44(20)70175198 ISSUES/YR: 3

Education for Primary Care
ISSN: 1473-9879
Educ. Prim. Care
Radcliffe Publishing Ltd, 18 Marcham Road, OX14 1AA, Abingdon, Oxon, United Kingdom, Tel: +44(1235)528820, Fax: +44(1235)528830 ISSUES/YR: 4

Egyptian Journal of Anaesthesia
ISSN: 1110-1849
Egypt. J. Anaesth.
Central Society of Egyptian Anaesthesiologists, P.O. Box 167, Panorama October 11811, Nasr City, Cairo, Egypt ISSUES/YR: 4

Electrolyte and Blood Pressure
ISSN: 1738-5997
Electrolyte Blood Press.
Korean Society of Electrolyte and Blood Pressure Research, 249-1 Gyomoon-dong, 471-071, Guri, Gyeonggi-do, South Korea, Tel: +82-31 5602861, Fax: +82-31 5660801 ISSUES/YR: 1

Electromyography and Clinical Neurophysiology
ISSN: 0301-150X
Electromyogr. Clin. Neurophysiol.
Nauwelaerts Publishing Company, Rue de l'Eglise St-Sulpice 19, B-1320, Beauvechain, Belgium, Tel: +32(10)866737 ISSUES/YR: 8

Electronic Journal of Biotechnology
ISSN: 0717-3458
Electron. J. Biotechnol.
Electronic Journal of Biotechnology, Casilla 4059, Valparaiso, Chile, Tel: +56(32)273000, Fax: +56(32)212746 ISSUES/YR: 3

Electronic Journal of Communicative Psychoanalysis
ISSN: 1492-1103
Electron. J. Commun. Psychoanal.
Electronic Journal of Communicative Psychoanalysis, 2302 W 33rd Ave, BC V6M 1C3, Vancouver, Canada, Tel: +1(604)3251878 ISSUES/YR: 4

Electronic Journal of Human Sexuality
ISSN: 1545-5556
Electron. J. Human Sex.
Institute for Advanced Study of Human Sexuality, San Francisco, United States ISSUES/YR: 1

Electrophoresis
ISSN: 0173-0835
Electrophoresis
Wiley-VCH Verlag, P.O. Box 101161, D-69451, Weinheim, Germany, Tel: +49(6201)606147, Fax: +49(6201)606117 ISSUES/YR: 24

• **EMA - Emergency Medicine Australasia**
ISSN: 1742-6731
EMA Emerg. Med. Australas.
Blackwell Publishing, 550 Swanston Street, VIC 3053, Carlton South, Australia, Tel: +61(3)8359 1011, Fax: +61(3)8359 1120 ISSUES/YR: 6

• **EMBO Journal**
ISSN: 0261-4189
EMBO J.
Nature Publishing Group, Houndmills, RG21 6XS, Basingstoke, Hampshire, United Kingdom, Tel: +44(1256)329242, Fax: +44(1256)810526 ISSUES/YR: 24

Current

- **EMBO Reports**
ISSN: 1469-221X
EMBO Rep.
Nature Publishing Group, Houndmills, RG21 6XS, Basingstoke, Hampshire, United Kingdom, Tel: +44(1256)329242, Fax: +44(1256)810526 ISSUES/YR: 12
- **Emergency Medicine Clinics of North America**
ISSN: 0733-8627
Emerg. Med. Clin. North Am.
W.B. Saunders, Independence Square West, Philadelphia, PA 19106-3399, United States, Tel: +1(215)2387800, Fax: +1(215)2386445 ISSUES/YR: 4
- **Emergency Medicine Journal**
ISSN: 1472-0205
Emerg. Med. J.
BMJ Publishing Group, Tavistock Square, WC1H 9JR, London, United Kingdom, Tel: +44(207)3874499, Fax: +44(207)3836668 ISSUES/YR: 12
- **Emergency Radiology**
ISSN: 1070-3004
Emerg. Radiol.
Springer New York LLC, 233 Springer Street, New York, 10013-1578, United States, Tel: +1-212 460 1501, Fax: +1-212 460 1595 ISSUES/YR: 6

Emerging Infectious Diseases
ISSN: 1080-6040
Emerg. Infect. Dis.
Centers for Disease Control and Prevention (CDC), 1600 Clifton Road, Mailstop C-12, Atlanta, GA 30333, United States, Tel: +1(404)6393967, Fax: +1(404)6393039 ISSUES/YR: 12

Emerging Themes in Epidemiology
ISSN: 1742-7622
Emerg. Themes Epidemiol.
BioMed Central Ltd., 34 - 42 Cleveland Street, W1T 4LB, London, United Kingdom, Tel: +44(207)3230323, Fax: +44(207)6319926 ISSUES/YR: 1

Emirates Medical Journal
ISSN: 0250-6882
Emirates Med. J.
Emirates Medical Association, P.O. Box 6600, Dubai, United Arab Emirates, Tel: +971(4)377377, Fax: +971(4)344082 ISSUES/YR: 3

Emotion
ISSN: 1528-3542
Emotion
American Psychological Association Inc., 750 First Street NE, Washington, DC 20002-4242, United States, Tel: +1(202)3365600, Fax: +1(202)3365568 ISSUES/YR: 4

- **Emotional and Behavioural Difficulties**
ISSN: 1363-2752
Emot. Behav. Difficulties
Routledge, 325 Chestnut Street, Philadelphia, PA 19106, United States, Tel: +1(215)62508900 ISSUES/YR: 4

Encephale
ISSN: 0013-7006
Encephale
Masson, 21 Rue Camille Desmoulins, 92789, Issy les Moulineaux Cedex 9, France, Tel: +33(1)73281634, Fax: +33(1)73281649 ISSUES/YR: 6

- **Endocrine**
ISSN: 0969-711X
Endocrine
Humana Press Inc., 999 Riverview Drive, Suite 208, Totowa, NJ 07512-1165, United States, Tel: +1(973)2561699, Fax: +1(973)2568341 ISSUES/YR: 9

Endocrine Journal
ISSN: 0918-8959
Endocr. J.
Japan Endocrine Society, Yoshida Kawaramachi 14, Sayko-ku, 606-8305, Kyoto-shi, Kyoto, Japan, Tel: +81(75)7522955, Fax: +81(75)7522963 ISSUES/YR: 6

- **Endocrine Pathology**
ISSN: 1046-3976
Endocr. Pathol.
Humana Press Inc., 999 Riverview Drive, Suite 208, Totowa, NJ 07512-1165, United States, Tel: +1(973)2561699, Fax: +1(973)2568341 ISSUES/YR: 4

Endocrine Practice
ISSN: 1530-891X
Endocr. Pract.
Endocrine Practice, 1000 Riverside Avenue, Suite 205, Jacksonville, FL 32204, United States, Tel: +1(904)3537878, Fax: +1(904)3538185 ISSUES/YR: 6

Endocrine Regulations
ISSN: 1210-0668
Endocr. Regul.
Institute of Experimental Endocrinology, Vlarska 3, 83306, Bratislava, Slovakia ISSUES/YR: 4

- **Endocrine Research**
ISSN: 0743-5800
Endocr. Res.
Taylor and Francis Inc., 325 Chestnut St, Suite 800, Philadelphia PA, PA 19106, United States, Tel: +1(800)3541420, Fax: +1(215)6258914 ISSUES/YR: 4
- **Endocrine Reviews**
ISSN: 0163-769X
Endocr. Rev.
Endocrine Society, 8401 Connecticut Ave. Suite 900, Chevy Chase, MD 20815, United States, Tel: +1-888-363-6274 ISSUES/YR: 6

Endocrine, Metabolic and Immune Disorders - Drug Targets
ISSN: 1871-5303
Endocr. Metab. Immune Disord. Drug Targets
Bentham Science Publishers B.V., P.O. Box 294, 1400 AG, Bussum, Netherlands, Tel: +31-356257290, Fax: +31-847511310 ISSUES/YR: 4

Endocrine-Related Cancer
ISSN: 1351-0088
Endocr.-Relat. Cancer
Society for Endocrinology, 22 Apex Court, Woodlands, Bradley Stoke, BS32 4JT, Bristol, United Kingdom, Tel: +44(1454)642220, Fax: +44(1454)642222 ISSUES/YR: 4

Endocrinologia
ISSN: 1310-8131
Endocrinologia (Bulgaria)
Medical Information Center, 1 Sv. G. Sofijski str., 1431, Sofia, Bulgaria, Tel: +359-522342, Fax: +359-522393 ISSUES/YR: 4

Endocrinologia y Nutricion
ISSN: 1575-0922
Endocrinol. Nutr.
Ediciones Doyma, S.L., Travesera de Gracia 17-21, 08021, Barcelona, Spain, Tel: +34(93)2000711, Fax: +34(93)2091136 ISSUES/YR: 10

Endocrinologist
ISSN: 1051-2144
Endocrinologist
Lippincott Williams and Wilkins, 351 West Camden Street, Baltimore, MD 21201-2436, United States, Tel: +1(410)5284000 ISSUES/YR: 6

- **Endocrinology**
ISSN: 0013-7227
Endocrinology
Endocrine Society, 8401 Connecticut Ave. Suite 900, Chevy Chase, MD 20815, United States, Tel: +1-888-363-6274 ISSUES/YR: 12
- **Endocrinology and Metabolism Clinics of North America**
ISSN: 0889-8529
Endocrinol. Metab. Clin. North Am.
W.B. Saunders, Independence Square West, Philadelphia, PA 19106-3399, United States, Tel: +1(215)2387800, Fax: +1(215)2386445 ISSUES/YR: 4

Endokrynologia, Diabetologia i Choroby Przemiany Materii Wieku Rozwojowego
ISSN: 1234-625X
Endokrynol. Diabetol. Choroby Przemiany Materii Wieku Rozwojowego
Cornetis, ul. Dlugosza 2-6, 51-162, Wroclaw, Poland, Tel: +48-071 326 3101, Fax: +48-071 325 2803 ISSUES/YR: 4

Endoscopic Forum for Digestive Disease
ISSN: 0912-0505
Endosc. Forum Dig. Dis.
Cancer and Chemotherapy Publ. Inc., 2-2-3 Nihonbashi, Chuo-ku, 103, Tokyo, Japan, Tel: +81(3)32780052 ISSUES/YR: 2

Current

- **Endoscopy**
  ISSN: 0013-726X
  Endoscopy
  Georg Thieme Verlag, Rudigerstrasse 14, D-70469, Stuttgart, Germany, Tel: +49(711)8931-321, Fax: +49(711)8931-422 ISSUES/YR: 12

Endoskopie
ISSN: 1211-1074
Endoskopie
Czech Medical Association J.E. Purkyne, Sokolska 31, P.O. Box 88, 120 26, Praha 2, Czech Republic, Tel: +420(2)24266201-4, Fax: +420(2)24266212 ISSUES/YR: 4

- **Endothelium: Journal of Endothelial Cell Research**
  ISSN: 1062-3329
  Endothelium J. Endothelial Cell Res.
  Taylor and Francis Ltd., 4 Park Square, Milton Park, OX14 4RN, Abingdon, Oxfordshire, United Kingdom, Tel: +44(20)70175541, Fax: +44(20)70175198 ISSUES/YR: 6

Enfermedades Emergentes
ISSN: 1575-4723
Enferm. Emergentes
Nexus Medica, c/Caracas 33, 08030, Barcelona, Spain, Tel: +34(93)4593492, Fax: +34(93)4590283 ISSUES/YR: 4

Enfermedades Infecciosas y Microbiologia
ISSN: 1405-0994
Enferm. Infecc. Microbiol.
Comunicaciones Cientificas Mexicanas S.A. de C.V., Calz. Gen. Anaya 330, Col. de Carmen Coyoacan, CP 04100, Mexico D.F., Mexico, Tel: +52(55)5-659-3863, Fax: +52(55)5-554-6054 ISSUES/YR: 4

Enfermedades Infecciosas y Microbiologia Clinica
ISSN: 0213-005X
Enferm. Infecc. Microbiol. Clin.
Ediciones Doyma, S.L., Travesera de Gracia 17-21, 08021, Barcelona, Spain, Tel: +34(93)2000711, Fax: +34(93)2091136 ISSUES/YR: 10

Engineering in Life Sciences
ISSN: 1618-0240
Eng. Life Sci.
Wiley-VCH Verlag, P.O. Box 101161, D-69451, Weinheim, Germany, Tel: +49(6201)606147, Fax: +49(6201)606117 ISSUES/YR: 6

- **Enhancer - Biotherapy of Cancer**
  ISSN: 1572-0004
  Enhancer. Biother. Cancer
  Excerpta Medica, Radarweg 29, 1043 NX, Amsterdam, Netherlands ISSUES/YR: 3

Environment
ISSN: 0013-9157
Environment
Heldref Publications, 1319 Eighteenth Street NW, Washington, DC 20036-1802, United States, Tel: +1(202)2966267, Fax: +1(202)2936130 ISSUES/YR: 10

- **Environment International**
  ISSN: 0160-4120
  Environ. Int.
  Elsevier Ltd, Langford Lane, Kidlington, OX5 1GB, Oxford, United Kingdom, Tel: +44(1865)843000, Fax: +44(1865)843010 ISSUES/YR: 8

Environmental and Molecular Mutagenesis
ISSN: 0893-6692
Environ. Mol. Mutagen.
Wiley-Liss Inc., 111 River Street, Hoboken, NJ 07030-5774, United States, Tel: +1(201)7486645, Fax: +1(201)7486021 ISSUES/YR: 10

Environmental Conservation
ISSN: 0376-8929
Environ. Conserv.
Cambridge University Press, Shaftesbury Road, CB2 2RU, Cambridge, United Kingdom, Tel: +44(1223)312393, Fax: +44(1223)315052 ISSUES/YR: 4

Environmental Engineering Science
ISSN: 1092-8758
Environ. Eng. Sci.
Mary Ann Liebert Inc., 140 Huguenot Street, New Rochelle, NY 10801-5215, United States, Tel: +1(914)7402100, Fax: +1(914)7402101 ISSUES/YR: 6

Environmental Health and Preventive Medicine
ISSN: 1342-078X
Environ. Health Prev. Med.
Japanese Society for Hygiene, N15W7, 060, Sapporo, Japan, Tel: +81(11)7065553, Fax: +81(11)7171140 ISSUES/YR: 4

Environmental Health Criteria
ISSN: 0250-863X
Environ. Health Criteria
World Health Organization, 20 Ave. Appia, CH-1211, Geneva 27, Switzerland, Tel: +41(22)7912111, Fax: +41(22)7914857 ISSUES/YR: 1

- **Environmental Health Perspectives**
  ISSN: 0091-6765
  Environ. Health Perspect.
  Public Health Services, US Dept of Health and Human Services, P.O. Box 12233, Research Triangle Park, NC 27709, United States, Tel: +1(919)5413406 ISSUES/YR: 16

Environmental Health: A Global Access Science Source
ISSN: 1476-069X
Environ. Health Global Access Sci. Sour.
BioMed Central Ltd., 34 - 42 Cleveland Street, W1T 4LB, London, United Kingdom, Tel: +44(207)3230323, Fax: +44(207)6319926 ISSUES/YR: 1

Environmental Management
ISSN: 0364-152X
Environ. Manage.
Springer New York LLC, 233 Springer Street, New York, 10013-1578, United States, Tel: +1-212 460 1501, Fax: +1-212 460 1595 ISSUES/YR: 12

Environmental Monitoring and Assessment
ISSN: 0167-6369
Environ. Monit. Assess.
Kluwer Academic Publishers, Van Godewijckstraat 30, 3311 GZ, Dordrecht, Netherlands, Tel: +31(78)6576000, Fax: +31(78)6576254 ISSUES/YR: 24

Environmental Pollution
ISSN: 0269-7491
Environ. Pollut.
Elsevier Ltd, Langford Lane, Kidlington, OX5 1GB, Oxford, United Kingdom, Tel: +44(1865)843000 ISSUES/YR: 12

- **Environmental Research**
  ISSN: 0013-9351
  Environ. Res.
  Academic Press Inc., 6277 Sea Harbor Drive, Orlando, FL 32887-4900, United States, Tel: +1(407)3454040, Fax: +1(407)3639661 ISSUES/YR: 9

- **Environmental Science and Policy**
  ISSN: 1462-9011
  Environ. Sci. Policy
  Elsevier Ltd, Langford Lane, Kidlington, OX5 1GB, Oxford, United Kingdom, Tel: +44(1865)843000, Fax: +44(1865)843010 ISSUES/YR: 6

Environmental Science and Technology
ISSN: 0013-936X
Environ. Sci. Technol.
American Chemical Society, 2540 Olentangy River Road, P.O. Box 3337, Columbus, OH 43210-3337, United States, Tel: +1(614)4473776, Fax: +1(614)4473671 ISSUES/YR: 24

Environmental Technology
ISSN: 0959-3330
Environ. Technol.
Selper Ltd, Unit 66, 2 Old Brompton Road, SW7 2DQ, London, United Kingdom, Tel: +44(207)5946013, Fax: +44(207)5966016 ISSUES/YR: 12

- **Environmental Toxicology**
  ISSN: 1520-4081
  Environ. Toxicol.
  John Wiley and Sons Inc., P.O.Box 18667, Newark, NJ 07191-8667, United States, Tel: +1(212)8506000, Fax: +1(212)8506012 ISSUES/YR: 6

- **Environmental Toxicology and Chemistry**
  ISSN: 0730-7268
  Environ. Toxicol. Chem.
  SETAC Press, 1010 North 12th Avenue, Pensacola, FL 32501-3367, United States, Tel: +1(850)4691500, Fax: +1(850)4699778 ISSUES/YR: 12

Current

- **Environmental Toxicology and Pharmacology**
ISSN: 1382-6689
Environ. Toxicol. Pharmacol.
Elsevier, P.O. Box 211, 1000 AE, Amsterdam, Netherlands, Tel: +31(20)4853757, Fax: +31(20)4853432 ISSUES/YR: 9

Environnement, Risques et Sante
ISSN: 1635-0421
Envir. Risques Sante
John Libbey Eurotext, 127, avenue de la Republique, 92120, Montrouge, France, Tel: +33(01)46730660, Fax: +33(01)40840999 ISSUES/YR: 6

Enzyme and Microbial Technology
ISSN: 0141-0229
Enzyme Microb. Technol.
Elsevier Inc., 360 Park Avenue South, New York, NY 10010, United States, Tel: +1(212)6333950, Fax: +1(212)6333990 ISSUES/YR: 14

EOS Rivista di Immunologia ed Immunofarmacologia
ISSN: 0392-6699
EOS Riv. Immunol. Immunofarmacol.
Sigma-Tau s.p.a., Viale Shakespeare 47, 00144, Roma, Italy ISSUES/YR: 4

Epidemiologia e Psichiatria Sociale
ISSN: 1121-189X
Epidemiol. Psychiatr. Soc.
Il Pensiero Scientifico Editore s.r.l., Via Bradano 3/C, 00199, Roma, Italy, Tel: +39(06)86207158, Fax: +39(06)86207160 ISSUES/YR: 4

Epidemiologic Perspectives and Innovations
ISSN: 1742-5573
Epidemiol. Perspect. Innov.
BioMed Central Ltd., 34 - 42 Cleveland Street, W1T 4LB, London, United Kingdom, Tel: +44(207)3230323, Fax: +44(207)6319926 ISSUES/YR: 1

- **Epidemiologic Reviews**
ISSN: 0193-936X
Epidemiol. Rev.
Oxford University Press, 2001 Evans Road, Cary, NC 27513, United States, Tel: +44(919)6770977, Fax: +44(919)6771714 ISSUES/YR: 1

Epidemiologie, Mikrobiologie, Imunologie
ISSN: 1210-7913
Epidemiol. Mikrobiol. Imunol.
Czech Medical Association J.E. Purkyne, Sokolska 31, P.O. Box 88, 120 26, Praha 2, Czech Republic, Tel: +420(2)24266201-4, Fax: +420(2)24266212 ISSUES/YR: 4

- **Epidemiology**
ISSN: 1044-3983
Epidemiology
Lippincott Williams and Wilkins, 351 West Camden Street, Baltimore, MD 21201-2436, United States, Tel: +1(410)5284000 ISSUES/YR: 6

Epidemiology and Infection
ISSN: 0950-2688
Epidemiol. Infect.
Cambridge University Press, Shaftesbury Road, CB2 2RU, Cambridge, United Kingdom, Tel: +44(1223)312393, Fax: +44(1223)315052 ISSUES/YR: 6

- **Epilepsia**
ISSN: 0013-9580
Epilepsia
Blackwell Publishing Inc., 350 Main Street, Malden, MA 02148, United States, Tel: +1(781)3888200, Fax: +1(781)3888210 ISSUES/YR: 12

Epilepsies
ISSN: 1149-6576
Epilepsies
John Libbey Eurotext, 127, avenue de la Republique, 92120, Montrouge, France, Tel: +33(01)46730660, Fax: +33(01)40840999 ISSUES/YR: 4

Epilepsy and Behavior
ISSN: 1525-5050
Epilepsy Behav.
Academic Press Inc., 6277 Sea Harbor Drive, Orlando, FL 32887-4900, United States, Tel: +1(407)3454040, Fax: +1(407)3639661 ISSUES/YR: 4

- **Epilepsy Research**
ISSN: 0920-1211
Epilepsy Res.
Elsevier, P.O. Box 211, 1000 AE, Amsterdam, Netherlands, Tel: +31(20)4853757, Fax: +31(20)4853432 ISSUES/YR: 12

- **Epileptic Disorders**
ISSN: 1294-9361
Epileptic Disord.
John Libbey Eurotext, 127, avenue de la Republique, 92120, Montrouge, France, Tel: +33(01)46730660, Fax: +33(01)40840999 ISSUES/YR: 4

Epitheorese Klinikes Farmakologias kai Farmakokinetikes
ISSN: 1011-6575
Epitheorese Klin. Farmakol. Farmakokinet.
PHARMAKON-Press, 20 Daskalaki Str., GR-115 26, Athens, Greece, Tel: +30(210)7778101 ISSUES/YR: 3

Equilibrium Research
ISSN: 0385-5716
Equilib. Res.
Japan Society for Equilibrium Research, 39 Kawabata Higash Maruta-cho, Sakyo-ku, 606, Kyoto, Japan, Tel: +81(75)7510068 ISSUES/YR: 6

Equilibrium Research, Supplement
ISSN: 0916-0337
Equilib. Res. Suppl.
Japan Society for Equilibrium Research, 39 Kawabata Higash Maruta-cho, Sakyo-ku, 606, Kyoto, Japan, Tel: +81(75)7510068 ISSUES/YR: 1

Erciyes Tip Dergisi
ISSN: 1300-199X
Erciyes Tip Derg.
Erciyes University, 38039, Kayseri, Turkey, Tel: +90(352)4374911, Fax: +90(352)4375285 ISSUES/YR: 4

Ergonomics
ISSN: 0014-0139
Ergonomics
Taylor and Francis Ltd., 4 Park Square, Milton Park, OX14 4RN, Abingdon, Oxfordshire, United Kingdom, Tel: +44(20)70175541, Fax: +44(20)70175198 ISSUES/YR: 15

Ergotherapie und Rehabilitation
ISSN: 0942-8623
Ergother. Rehabil.
Schulz-Kirchner Verlag GmbH, Mollweg 2, D-65510, Idstein, Germany, Tel: +49(6126)93200, Fax: +49(6126)932050 ISSUES/YR: 12

Ernahrungs Umschau
ISSN: 0014-021X
Ernahr. Umsch.
Umschau Zeitschriftenverlag Breidenstein GmbH, Bruningstr. 580, 65929, Frankfurt, Germany, Tel: +49(69)2600-692, Fax: +49(69)2600-609 ISSUES/YR: 12

- **ESHRE Monographs**
ISSN: 1477-741X
ESHRE Monogr.
Oxford University Press, Great Clarendon Street, OX2 6DP, Oxford, United Kingdom, Tel: +44(1865)353907, Fax: +44(1865)353486 ISSUES/YR: 1

- **Esophagus**
ISSN: 1612-9059
Esophagus
Springer-Verlag Tokyo, 3-13, Hongo 3-chome, Bunkyo-ku, 113-8480, Tokyo, Japan, Tel: +81(3)38180861, Fax: +81(3)38180864 ISSUES/YR: 4

- **Ethical Human Psychology and Psychiatry**
ISSN: 1523-150X
Eth. Human Psychol. Psychiatry
Springer Publishing Company, 11 West 42nd Street, 15th Floor, New York, NY 10036, United States, Tel: +1(212)4314370, Fax: +1(212)9417842 ISSUES/YR: 3

Ethique et Sante
ISSN: 1765-4629
Ethique Sante
Masson, 21 Rue Camille Desmoulins, 92789, Issy les Moulineaux Cedex 9, France, Tel: +33(1)73281634, Fax: +33(1)73281649 ISSUES/YR: 4

Current

Ethnicity and Disease
ISSN: 1049-510X
Ethn. Dis.
Ethnicity and Disease, 2045 Manchester St NE, Atlanta, GA 30324, United States, Tel: +1-404 875 6263, Fax: +1-404 875 6334 ISSUES/YR: 4

Ethnicity and Health
ISSN: 1355-7858
Ethn. Health
Routledge, 4 Park Square, Milton Park, OX14 4RN, Abingdon, Oxfordshire, United Kingdom, Tel: +44(1235)828600, Fax: +44(1235)829000 ISSUES/YR: 4

Europa Medicophysica
ISSN: 0014-2573
Eur. Medicophys.
Edizioni Minerva Medica S.p.A., Corso Bramante 83-85, 10126, Turin, Italy, Tel: +39(011)678282, Fax: +39(011)674502 ISSUES/YR: 4

• **Europace**
ISSN: 1099-5129
Europace
Oxford University Press, Great Clarendon Street, OX2 6DP, Oxford, United Kingdom, Tel: +44(1865)353907, Fax: +44(1865)353486 ISSUES/YR: 6

• **European Addiction Research**
ISSN: 1022-6877
Eur. Addict. Res.
S. Karger AG, Allschwilerstrasse 10, P.O. Box, CH-4009, Basel, Switzerland, Tel: +41(61)3061111, Fax: +41(61)3061234 ISSUES/YR: 4

European Annals of Allergy and Clinical Immunology
ISSN: 0397-9148
Eur. Ann. Allergy Clinical Immunol.
Meditions Carline, 1-3 rue du Depart, 75014, Paris, France, Tel: +33(1)40640075, Fax: +33(1)43222699 ISSUES/YR: 10

• **European Archives of Oto-Rhino-Laryngology**
ISSN: 0937-4477
Eur. Arch. Oto-Rhino-Laryngol.
Springer Verlag, Tiergartenstrasse 17, D-69121, Heidelberg, Germany, Tel: +49(6221)4870, Fax: +49(6221)487366 ISSUES/YR: 10

• **European Archives of Psychiatry and Clinical Neuroscience**
ISSN: 0940-1334
Eur. Arch. Psychiatry Clin. Neurosci.
Dr. Dietrich Steinkopff Verlag GmbH and Co. KG, P.O. Box 100462, D-64204, Darmstadt, Germany, Tel: +49(6151)828990, Fax: +49(6151)8289940 ISSUES/YR: 6

European Biophysics Journal
ISSN: 0175-7571
Eur. Biophys. J.
Springer Verlag, Tiergartenstrasse 17, D-69121, Heidelberg, Germany, Tel: +49(6221)4870, Fax: +49(6221)487366 ISSUES/YR: 8

European Cells and Materials
ISSN: 1473-2262
Eur. Cells and Mater.
Swiss Society for Biomaterials, Clavadelstrasse, CH-7270, Davos, Switzerland, Tel: +44(0)814142397 ISSUES/YR: 2

European Child and Adolescent Psychiatry
ISSN: 1018-8827
Eur. Child Adolesc. Psychiatry
Dr. Dietrich Steinkopff Verlag GmbH and Co. KG, P.O. Box 100462, D-64204, Darmstadt, Germany, Tel: +49(6151)828990, Fax: +49(6151)8289940 ISSUES/YR: 6

European Child and Adolescent Psychiatry, Supplement
ISSN: 1433-5719
Eur. Child Adolesc. Psychiatry Suppl.
Dr. Dietrich Steinkopff Verlag GmbH and Co. KG, P.O. Box 100462, D-64204, Darmstadt, Germany, Tel: +49(6151)828990, Fax: +49(6151)8289940 ISSUES/YR: 1

European Cytokine Network
ISSN: 1148-5493
Eur. Cytokine Netw.
John Libbey Eurotext, 127, avenue de la Republique, 92120, Montrouge, France, Tel: +33(01)46730660, Fax: +33(01)40840999 ISSUES/YR: 4

European Eating Disorders Review
ISSN: 1072-4133
Eur. Eating Disord. Rev.
John Wiley and Sons Ltd, Southern Gate, PO19 8SQ, Chichester, West Sussex, United Kingdom, Tel: +44(1243)779777, Fax: +44(1243)775878 ISSUES/YR: 6

• **European Heart Journal**
ISSN: 0195-668X
Eur. Heart J.
Oxford University Press, Great Clarendon Street, OX2 6DP, Oxford, United Kingdom, Tel: +44(1865)353907, Fax: +44(1865)353486 ISSUES/YR: 24

• **European Heart Journal, Supplement**
ISSN: 1520-765X
Eur. Heart J. Suppl.
Oxford University Press, Great Clarendon Street, OX2 6DP, Oxford, United Kingdom, Tel: +44(1865)353907, Fax: +44(1865)353486 ISSUES/YR: 1

European Journal of Acupuncture
ISSN: 1609-4557
Eur. J. Acupunct.
VICER Publishing, P.O. Box 14, A-1097, Vienna, Austria, Tel: +43(676)9568085, Fax: +43(676)9568086 ISSUES/YR: 4

European Journal of Allied Health
ISSN: 1609-0152
Eur. J. Allied Health
VICER Publishing, P.O. Box 14, A-1097, Vienna, Austria, Tel: +43(676)9568085, Fax: +43(676)9568086 ISSUES/YR: 4

European Journal of Anaesthesiology
ISSN: 0265-0215
Eur. J. Anaesthesiol.
Cambridge University Press, Shaftesbury Road, CB2 2RU, Cambridge, United Kingdom, Tel: +44(1223)312393, Fax: +44(1223)315052 ISSUES/YR: 12

European Journal of Anaesthesiology, Supplement
ISSN: 0952-1941
Eur. J. Anaesthesiol. Suppl.
Cambridge University Press, Shaftesbury Road, CB2 2RU, Cambridge, United Kingdom, Tel: +44(1223)312393, Fax: +44(1223)315052 ISSUES/YR: 1

European Journal of Anatomy
ISSN: 1136-4890
Eur. J. Anat.
Sociedad Anatomica Espanola, Avda. Campo Charro s/n, 37007, Salamanca, Spain, Fax: +34(23)294559 ISSUES/YR: 3

• **European Journal of Applied Physiology**
ISSN: 1439-6319
Eur. J. Appl. Physiol.
Springer Verlag, Tiergartenstrasse 17, D-69121, Heidelberg, Germany, Tel: +49(6221)4870, Fax: +49(6221)487366 ISSUES/YR: 12

• **European Journal of Cancer**
ISSN: 0959-8049
Eur. J. Cancer
Elsevier Ltd, Langford Lane, Kidlington, OX5 1GB, Oxford, United Kingdom, Tel: +44(1865)843000, Fax: +44(1865)843010 ISSUES/YR: 18

• **European Journal of Cancer Prevention**
ISSN: 0959-8278
Eur.J. Cancer Prev.
Lippincott Williams and Wilkins, 241 Borough High Street, SE1 1GB, London, United Kingdom, Tel: +44(207)9407500, Fax: +44(207)9407575 ISSUES/YR: 6

• **European Journal of Cancer, Supplement**
ISSN: 1359-6349
Eur. J. Cancer Suppl.
Elsevier Ltd, Langford Lane, Kidlington, OX5 1GB, Oxford, United Kingdom, Tel: +44(1865)843000, Fax: +44(1865)843010 ISSUES/YR: 1

• **European Journal of Cardio-thoracic Surgery**
ISSN: 1010-7940
Eur. J. Cardio-thorac. Surg.
Elsevier, P.O. Box 211, 1000 AE, Amsterdam, Netherlands, Tel: +31(20)4853757, Fax: +31(20)4853432 ISSUES/YR: 12

- **European Journal of Cardiovascular Nursing**
  ISSN: 1474-5151
  Eur. J. Cardiovasc. Nurs.
  Elsevier, P.O. Box 211, 1000 AE, Amsterdam, Netherlands, Tel: +31(20)4853757, Fax: +31(20)4853432 ISSUES/YR: 4
- **European Journal of Cardiovascular Prevention and Rehabilitation**
  ISSN: 1741-8267
  Eur. J. Cardiovasc. Prev. Rehabil.
  Lippincott Williams and Wilkins, 241 Borough High Street, SE1 1GB, London, United Kingdom, Tel: +44(207)9407500, Fax: +44(207)9407575 ISSUES/YR: 6
- **European Journal of Cell Biology**
  ISSN: 0171-9335
  Eur. J. Cell Biol.
  Urban und Fischer Verlag GmbH und Co. KG, P.O. Box 201930, 80019, Munchen, Germany, Tel: +49(89)53830, Fax: +49(89)5383221 ISSUES/YR: 12
- **European Journal of Clinical Investigation**
  ISSN: 0014-2972
  Eur. J. Clin. Invest.
  Blackwell Publishing Ltd, 9600 Garsington Road, OX4 2XG, Oxford, United Kingdom, Tel: +44(1865)776868, Fax: +44(1865)714591 ISSUES/YR: 12
- **European Journal of Clinical Microbiology and Infectious Diseases**
  ISSN: 0934-9723
  Eur. J. Clin. Microbiol. Infect. Dis.
  Springer Verlag, Tiergartenstrasse 17, D-69121, Heidelberg, Germany, Tel: +49(6221)4870, Fax: +49(6221)487366 ISSUES/YR: 12

European Journal of Clinical Nutrition
ISSN: 0954-3007
Eur. J. Clin. Nutr.
Nature Publishing Group, Houndmills, RG21 6XS, Basingstoke, Hampshire, United Kingdom, Tel: +44(1256)329242, Fax: +44(1256)810526 ISSUES/YR: 12

- **European Journal of Clinical Pharmacology**
  ISSN: 0031-6970
  Eur. J. Clin. Pharmacol.
  Springer Verlag, Tiergartenstrasse 17, D-69121, Heidelberg, Germany, Tel: +49(6221)4870, Fax: +49(6221)487366 ISSUES/YR: 12
- **European Journal of Cognitive Psychology**
  ISSN: 0954-1446
  Eur. J. Cogn. Psychol.
  Psychology Press Ltd, 4 Park Square, Milton Park, OX14 4RN, Abingdon, Oxfordshire, United Kingdom, Tel: +44(1235)828600, Fax: +44(1235)829000 ISSUES/YR: 6
- **European Journal of Contraception and Reproductive Health Care**
  ISSN: 1362-5187
  Eur. J. Contracept. Reprod. Health Care
  Taylor and Francis Ltd., 4 Park Square, Milton Park, OX14 4RN, Abingdon, Oxfordshire, United Kingdom, Tel: +44(20)70175541, Fax: +44(20)70175198 ISSUES/YR: 4

European Journal of Dermatology
ISSN: 1167-1122
Eur. J. Dermatol.
John Libbey Eurotext, 127, avenue de la Republique, 92120, Montrouge, France, Tel: +33(01)46730660, Fax: +33(01)40840999 ISSUES/YR: 6

European Journal of Drug Metabolism and Pharmacokinetics
ISSN: 0378-7966
Eur. J. Drug Metab. Pharmacokinet.
Editions Medecine et Hygiene, 7, Avenue de la Roseraie, P.O. Box 456, CH-1211, Geneva 4, Switzerland, Tel: +41(22)7029311, Fax: +41(22)7029355 ISSUES/YR: 4

- **European Journal of Echocardiography**
  ISSN: 1525-2167
  Eur. J. Echocardiogr.
  W.B. Saunders Ltd, 32 Jamestown Road, NW1 7BY, London, United Kingdom, Tel: +44(20)74244200, Fax: +44(20)74854752 ISSUES/YR: 6
- **European Journal of Emergency Medicine**
  ISSN: 0969-9546
  Eur. J. Emerg. Med.
  Lippincott Williams and Wilkins, 530 Walnut Street, Philadelphia, PA 19106-3621, United States, Tel: +1(215)5218342, Fax: +1(215)5218326 ISSUES/YR: 6
- **European Journal of Endocrinology**
  ISSN: 0804-4643
  Eur. J. Endocrinol.
  BioScientifica Ltd., 22 Apex Court, Woodlands, Bradley Stoke, BS32 4JT, Bristol, United Kingdom, Tel: +44(1454)642240, Fax: +44(1454)642222 ISSUES/YR: 12

European Journal of Endocrinology, Supplement
ISSN: 0804-4635
Eur. J. Endocrinol. Suppl.
BioScientifica Ltd., 22 Apex Court, Woodlands, Bradley Stoke, BS32 4JT, Bristol, United Kingdom, Tel: +44(1454)642240, Fax: +44(1454)642222 ISSUES/YR: 1

European Journal of Epidemiology
ISSN: 0393-2990
Eur. J. Epidemiol.
Kluwer Academic Publishers, Van Godewijckstraat 30, 3311 GZ, Dordrecht, Netherlands, Tel: +31(78)6576000, Fax: +31(78)6576254 ISSUES/YR: 12

- **European Journal of Gastroenterology and Hepatology**
  ISSN: 0954-691X
  Eur. J. Gastroenterol. Hepatol.
  Lippincott Williams and Wilkins, 241 Borough High Street, SE1 1GB, London, United Kingdom, Tel: +44(207)9407500, Fax: +44(207)9407575 ISSUES/YR: 12

European Journal of General Medicine
ISSN: 1304-3889
Eur. J. Gen. Med.
TIP ARASTIRMALARI DERNEGI, PO Box 33, 65200, VAN, Turkey, Tel: +90-0 90 532 3426134, Fax: +90-0 90 432 2121867 ISSUES/YR: 4

European Journal of General Practice
ISSN: 1381-4788
Eur. J. Gen. Pract.
Mediselect BV, P.O. Box 63, 3830 AB, Leusden, Netherlands, Tel: +31(33)4229900, Fax: +31(33)4229922 ISSUES/YR: 4

European Journal of Gynaecological Oncology
ISSN: 0392-2936
Eur. J. Gynaecol. Oncol.
S.O.G. CANADA Inc., 4900 Cote St. Luc, Apt. 212, QUE H3W 2H3, Montreal, Canada, Tel: +1(514)4893242, Fax: +1(514)4854513 ISSUES/YR: 6

- **European Journal of Haematology**
  ISSN: 0902-4441
  Eur. J. Haematol.
  Blackwell Publishing Ltd, 9600 Garsington Road, OX4 2XG, Oxford, United Kingdom, Tel: +44(1865)776868, Fax: +44(1865)714591 ISSUES/YR: 12
- **European Journal of Haematology, Supplement**
  ISSN: 0902-4506
  Eur. J. Haematol. Suppl.
  Blackwell Munksgaard, 1 Rosenorns Alle, P.O. Box 227, DK-1502, Copenhagen V, Denmark, Tel: +45(77)333333, Fax: +45(77)333377 ISSUES/YR: 1
- **European Journal of Health Economics**
  ISSN: 1618-7598
  Eur. J. Health Econ.
  Springer Verlag, Tiergartenstrasse 17, D-69121, Heidelberg, Germany, Tel: +49(6221)4870, Fax: +49(6221)487366 ISSUES/YR: 4

European Journal of Health Law
ISSN: 0929-0273
Eur. J. Health Law
Brill Academic Publishers, P.O. Box 9000, 2300 PA, Leiden, Netherlands, Tel: +31(71)5353500, Fax: +31(71)5317532 ISSUES/YR: 4

- **European Journal of Heart Failure**
  ISSN: 1388-9842
  Eur. J. Heart Fail.
  Elsevier, P.O. Box 211, 1000 AE, Amsterdam, Netherlands, Tel: +31(20)4853757, Fax: +31(20)4853432 ISSUES/YR: 8

- **European Journal of Heart Failure, Supplement**
ISSN: 1567-4215
Eur. J. Heart Fail. Suppl.
Elsevier, P.O. Box 211, 1000 AE, Amsterdam, Netherlands, Tel: +31(20)4853757, Fax: +31(20)4853432 ISSUES/YR: 1

European Journal of Herbal Medicine
ISSN: 1352-4755
Eur. J. Herbal Med.
National Institute of Medical Herbalists Ltd, Elm House 54 Mary Arches Street, EX4 3BA, Exeter Devon, United Kingdom, Tel: +44(1392)426022, Fax: +44(1392)498963 ISSUES/YR: 3

European Journal of Histochemistry
ISSN: 1121-760X
Eur. J. Histochem.
Luigi Ponzio e figlio Editori, Viale Monte Grappa 2/A, 27100, Pavia, Italy, Tel: +39(0382)576600, Fax: +39(0382)574200 ISSUES/YR: 4

- **European Journal of Human Genetics**
ISSN: 1018-4813
Eur. J. Hum. Genet.
Nature Publishing Group, Houndmills, RG21 6XS, Basingstoke, Hampshire, United Kingdom, Tel: +44(1256)329242, Fax: +44(1256)810526 ISSUES/YR: 12

- **European Journal of Immunology**
ISSN: 0014-2980
Eur. J. Immunol.
Wiley-VCH Verlag, P.O. Box 101161, D-69451, Weinheim, Germany, Tel: +49(6201)606147, Fax: +49(6201)606117 ISSUES/YR: 12

European Journal of Inflammation
ISSN: 1721-727X
Eur. J. Inflamm.
Biolife, 39 Bis, 64029 Silvi Marina, Italy ISSUES/YR: 3

European Journal of Internal Medicine
ISSN: 0953-6205
Eur. J. Intern. Med.
Elsevier, P.O. Box 211, 1000 AE, Amsterdam, Netherlands, Tel: +31(20)4853757, Fax: +31(20)4853432 ISSUES/YR: 8

European Journal of Lymphology and Related Problems
ISSN: 0778-5569
Eur. J. Lymphol. Relat. Probl.
Service de Kinesitherapie, Avenue Paul Heger 28, P.O. Box 168, 1050, Brussels, Belgium, Tel: +32(2)6502470 ISSUES/YR: 4

European Journal of Medical Genetics
ISSN: 1769-7212
Eur. J. Med. Genet.
Editions Scientifiques et Medicales Elsevier, 23 rue Linois, 75724, Paris Cedex 15, France, Tel: +33(1)45589826, Fax: +33(1)45589428 ISSUES/YR: 4

European Journal of Medical Research
ISSN: 0949-2321
Eur. J. Med. Res.
Holzapfel Publishers Munich, Rauchstr. 8, D-81679, Munich, Germany, Tel: +49(89) 13 99 87 32, Fax: +49(89) 13 99 87 31 ISSUES/YR: 12

European Journal of Medicinal Chemistry
ISSN: 0223-5234
Eur. J. Med. Chem.
Editions Scientifiques et Medicales Elsevier, 23 rue Linois, 75724, Paris Cedex 15, France, Tel: +33(1)45589826, Fax: +33(1)45589428 ISSUES/YR: 12

- **European Journal of Morphology**
ISSN: 0924-3860
Eur. J. Morphol.
Taylor and Francis Ltd., 4 Park Square, Milton Park, OX14 4RN, Abingdon, Oxfordshire, United Kingdom, Tel: +44(20)70175541, Fax: +44(20)70175198 ISSUES/YR: 5

- **European Journal of Neurology**
ISSN: 1351-5101
Eur. J. Neurol.
Blackwell Publishing Ltd, 9600 Garsington Road, OX4 2XG, Oxford, United Kingdom, Tel: +44(1865)776868, Fax: +44(1865)714591 ISSUES/YR: 12

- **European Journal of Neurology, Supplement**
ISSN: 1471-0552
Eur. J. Neurol. Suppl.
Blackwell Publishing Ltd, 9600 Garsington Road, OX4 2XG, Oxford, United Kingdom, Tel: +44(1865)776868, Fax: +44(1865)714591 ISSUES/YR: 1

- **European Journal of Neuroscience**
ISSN: 0953-816X
Eur. J. Neurosci.
Blackwell Publishing Ltd, 9600 Garsington Road, OX4 2XG, Oxford, United Kingdom, Tel: +44(1865)776868, Fax: +44(1865)714591 ISSUES/YR: 24

European Journal of Nuclear Medicine and Molecular Imaging
ISSN: 1619-7070
Eur. J. Nucl. Med. Mol. Imaging
Springer Verlag, Tiergartenstrasse 17, D-69121, Heidelberg, Germany, Tel: +49(6221)4870, Fax: +49(6221)487366 ISSUES/YR: 12

European Journal of Nutrition
ISSN: 1436-6207
Eur. J. Nutr.
Dr. Dietrich Steinkopff Verlag GmbH and Co. KG, P.O. Box 100462, D-64204, Darmstadt, Germany, Tel: +49(6151)828990, Fax: +49(6151)8289940 ISSUES/YR: 6

- **European Journal of Obstetrics Gynecology and Reproductive Biology**
ISSN: 0301-2115
Eur. J. Obstet. Gynecol. Reprod. Biol.
Elsevier Ireland Ltd, P.O. Box 85, Limerick, Ireland, Tel: +353(61)471944, Fax: +353(61)472144 ISSUES/YR: 12

European Journal of Oncology
ISSN: 1128-6598
Eur. J. Oncol.
Mattioli 1885 S.p.A., Via Coduro 1/b, 43036, Fidenza (PR), Italy, Tel: +39(524)84547, Fax: +39(524)84751 ISSUES/YR: 4

- **European Journal of Ophthalmology**
ISSN: 1120-6721
Eur. J. Ophthalmol.
Wichtig Editore s.r.l., Via Friuli 72/74, 20135, Milan, Italy, Tel: +39(02)55195443, Fax: +39(02)55195971 ISSUES/YR: 6

European Journal of Oriental Medicine
ISSN: 1351-6647
Eu. J. Oriental Med.
British Acupuncture Council, 63 Jeddo Road, W12 9HQ, London, United Kingdom, Tel: +44(181)7491300, Fax: +44(181)7350477 ISSUES/YR: 2

- **European Journal of Orthopaedic Surgery and Traumatology**
ISSN: 0948-4817
Eur. J. Orthop. Surg. Traumatol.
Springer-Verlag France, 1 rue Paul Cezanne, 75008, Paris, France, Tel: +33(1)53933628, Fax: +33(1)53933683 ISSUES/YR: 4

- **European Journal of Paediatric Neurology**
ISSN: 1090-3798
Eur. J. Paediatr. Neurol.
W.B. Saunders Ltd, 32 Jamestown Road, NW1 7BY, London, United Kingdom, Tel: +44(20)74244200, Fax: +44(20)74854752 ISSUES/YR: 6

- **European Journal of Pain**
ISSN: 1090-3801
Eur. J. Pain
W.B. Saunders Ltd, 32 Jamestown Road, NW1 7BY, London, United Kingdom, Tel: +44(20)74244200, Fax: +44(20)74854752 ISSUES/YR: 6

European Journal of Pediatric Dermatology
ISSN: 1122-7672
Eur. J. Pediatr. Dermatol.
European Society for Pediatric Dermatology, Via Bitritto 131, I-70124, Bari, Italy, Tel: +39(080)5054467, Fax: +39(080)5054467 ISSUES/YR: 4

European Journal of Pediatric Surgery
ISSN: 0939-7248
Eur. J. Pediatr. Surg.
Hippokrates Verlag GmbH, Rudigerstrasse 14, D-70469, Stuttgart, Germany, Tel: +49(711)89310, Fax: +49(711)8931453 ISSUES/YR: 6

European Journal of Pediatric Surgery, Supplement
ISSN: 0939-6764
Eur. J. Pediatr. Surg. Suppl.
Hippokrates Verlag GmbH, Rudigerstrasse 14, D-70469, Stuttgart, Germany, Tel: +49(711)89310, Fax: +49(711)8931453 ISSUES/YR: 1

● **European Journal of Pediatrics**
ISSN: 0340-6199
Eur. J. Pediatr.
Springer Verlag, Tiergartenstrasse 17, D-69121, Heidelberg, Germany, Tel: +49(6221)4870, Fax: +49(6221)487366 ISSUES/YR: 12

● **European Journal of Pediatrics, Supplement**
ISSN: 0943-9676
Eur. J. Pediatr. Suppl.
Springer Verlag, Tiergartenstrasse 17, D-69121, Heidelberg, Germany, Tel: +49(6221)4870, Fax: +49(6221)487366 ISSUES/YR: 1

● **European Journal of Pharmaceutical Sciences**
ISSN: 0928-0987
Eur. J. Pharm. Sci.
Elsevier, P.O. Box 211, 1000 AE, Amsterdam, Netherlands, Tel: +31(20)4853757, Fax: +31(20)4853432 ISSUES/YR: 10

European Journal of Pharmaceutics and Biopharmaceutics
ISSN: 0939-6411
Eur. J. Pharm. Biopharm.
Elsevier, P.O. Box 211, 1000 AE, Amsterdam, Netherlands, Tel: +31(20)4853757, Fax: +31(20)4853432 ISSUES/YR: 6

● **European Journal of Pharmacology**
ISSN: 0014-2999
Eur. J. Pharmacol.
Elsevier, P.O. Box 211, 1000 AE, Amsterdam, Netherlands, Tel: +31(20)4853757, Fax: +31(20)4853432 ISSUES/YR: 69

● **European Journal of Plastic Surgery**
ISSN: 0930-343X
Eur. J. Plast. Surg.
Springer Verlag, Tiergartenstrasse 17, D-69121, Heidelberg, Germany, Tel: +49(6221)4870, Fax: +49(6221)487366 ISSUES/YR: 8

European Journal of Psychiatry
ISSN: 0213-6163
Eur. J. Psychiatry
University of Zaragoza, Avda. San Juan Bosco 15, P.O. Box 6029, 50009, Zaragoza, Spain, Tel: +34(76)559795 ISSUES/YR: 4

● **European Journal of Public Health**
ISSN: 1101-1262
Eur. J. Public Health
Oxford University Press, Great Clarendon Street, OX2 6DP, Oxford, United Kingdom, Tel: +44(1865)353907, Fax: +44(1865)353486 ISSUES/YR: 4

● **European Journal of Radiology**
ISSN: 0720-048X
Eur. J. Radiol.
Elsevier Ireland Ltd, P.O. Box 85, Limerick, Ireland, Tel: +353(61)471944, Fax: +353(61)472144 ISSUES/YR: 12

● **European Journal of Special Needs Education**
ISSN: 0885-6257
Europ. J. Spec. Needs Educ.
Routledge, 4 Park Square, Milton Park, OX14 4RN, Abingdon, Oxfordshire, United Kingdom, Tel: +44(1235)828600, Fax: +44(1235)829000 ISSUES/YR: 4

● **European Journal of Surgical Oncology**
ISSN: 0748-7983
Eur. J. Surg. Oncol.
W.B. Saunders Ltd, 32 Jamestown Road, NW1 7BY, London, United Kingdom, Tel: +44(20)74244200, Fax: +44(20)74854752 ISSUES/YR: 10

European Journal of Trauma
ISSN: 1439-0590
Eur. J. Trauma
Urban und Vogel GmbH, Neumarkter Strasse 43, D-81673, Munchen, Germany, Tel: +49(89)43721300, Fax: +49(89)43721399 ISSUES/YR: 6

● **European Journal of Vascular and Endovascular Surgery**
ISSN: 1078-5884
Eur. J. Vasc. Endovasc. Surg.
W.B. Saunders Ltd, 32 Jamestown Road, NW1 7BY, London, United Kingdom, Tel: +44(20)74244200, Fax: +44(20)74854752 ISSUES/YR: 12

● **European Neurology**
ISSN: 0014-3022
Eur. Neurol.
S. Karger AG, Allschwilerstrasse 10, P.O. Box, CH-4009, Basel, Switzerland, Tel: +41(61)3061111, Fax: +41(61)3061234 ISSUES/YR: 8

● **European Neuropsychopharmacology**
ISSN: 0924-977X
Eur. Neuropsychopharmacol.
Elsevier, P.O. Box 211, 1000 AE, Amsterdam, Netherlands, Tel: +31(20)4853757, Fax: +31(20)4853432 ISSUES/YR: 6

European Pharmaceutical Contractor
ISSN: 1364-369X
Eur. Pharm. Contract.
Samedan Ltd, 16 Hampden Gurney Street, W1H 5AL, London, United Kingdom, Tel: +44(20)77245444, Fax: +44(20)77242632 ISSUES/YR: 4

● **European Psychiatry**
ISSN: 0924-9338
Eur. Psychiatry
Editions Scientifiques et Medicales Elsevier, 23 rue Linois, 75724, Paris Cedex 15, France, Tel: +33(1)45589826, Fax: +33(1)45589428 ISSUES/YR: 8

● **European Radiology**
ISSN: 0938-7994
Eur. Radiol.
Springer Verlag, Tiergartenstrasse 17, D-69121, Heidelberg, Germany, Tel: +49(6221)4870, Fax: +49(6221)487366 ISSUES/YR: 12

● **European Radiology, Supplement**
ISSN: 1613-3749
Eur. Radiol. Suppl.
Springer Verlag, Tiergartenstrasse 17, D-69121, Heidelberg, Germany, Tel: +49(6221)4870, Fax: +49(6221)487366 ISSUES/YR: 1

● **European Respiratory Journal**
ISSN: 0903-1936
Eur. Respir. J.
European Respiratory Society, 4 Ave Sainte-Luce, CH-1003, Lausanne, Switzerland, Tel: +41-21 213 0101, Fax: +41-21 213 0100 ISSUES/YR: 12

● **European Respiratory Journal, Supplement**
ISSN: 0904-1850
Eur. Respir. J. Suppl.
European Respiratory Society, 4 Ave Sainte-Luce, CH-1003, Lausanne, Switzerland, Tel: +45-21 213 0101, Fax: +45-21 213 0100 ISSUES/YR: 1

European Respiratory Review
ISSN: 0905-9180
Eur. Respir. Rev.
European Respiratory Society, 4 Ave Sainte-Luce, CH-1003, Lausanne, Switzerland, Tel: +41-21 213 0101, Fax: +41-21 213 0100 ISSUES/YR: 4

European Review for Medical and Pharmacological Sciences
ISSN: 1128-3602
Eur. Rev. Med. Pharmacol. Sci.
Verduci Editore, Via Gregorio VII 186, 00165, Roma, Italy, Tel: +39(06)39375224, Fax: +39(06)6385672 ISSUES/YR: 6

● **European Spine Journal**
ISSN: 0940-6719
Eur. Spine J.
Springer Verlag, Tiergartenstrasse 17, D-69121, Heidelberg, Germany, Tel: +49(6221)4870, Fax: +49(6221)487366 ISSUES/YR: 8

European Surgery - Acta Chirurgica Austriaca
ISSN: 1682-1769
Eur. Surg. Acta Chir. Austriaca
Springer-Verlag Wien, Sachsenplatz 4-6, P.O. Box 89, A-1201, Vienna, Austria, Tel: +43(1)3302415, Fax: +43(1)3302426 ISSUES/YR: 6

- **European Surgical Research**
ISSN: 0014-312X
Eur. Surg. Res.
S. Karger AG, Allschwilerstrasse 10, P.O. Box, CH-4009, Basel, Switzerland, Tel: +41(61)3061111, Fax: +41(61)3061234 ISSUES/YR: 6

- **European Urology**
ISSN: 0302-2838
Eur. Urol.
Elsevier, P.O. Box 211, 1000 AE, Amsterdam, Netherlands, Tel: +31(20)4853757, Fax: +31(20)4853432 ISSUES/YR: 12

- **European Urology, Supplements**
ISSN: 1569-9056
Eur. Urol. Suppl.
Elsevier, P.O. Box 211, 1000 AE, Amsterdam, Netherlands, Tel: +31(20)4853757, Fax: +31(20)4853432 ISSUES/YR: 1

Eurorehab
ISSN: 1210-0366
Eurorehab
VOPD Prudy, Oracska 20, 83106 Bratislava, Slovakia ISSUES/YR: 4

Evaluation and the Health Professions
ISSN: 0163-2787
Eval. Health Prof.
SAGE Publications Inc., 2455 Teller Road, Thousand Oaks, CA 91320, United States, Tel: +1(805)4990721, Fax: +1(805)4990871 ISSUES/YR: 4

- **Evidence-based Cardiovascular Medicine**
ISSN: 1361-2611
Evid.-based Cardiovasc. Med.
Churchill Livingstone, 1-3 Baxter's Place, Leith Walk, EH1 3AF, Edinburgh, United Kingdom, Tel: +44(131)5562424, Fax: +44(131)5581278 ISSUES/YR: 4

- **Evidence-based Complementary and Alternative Medicine**
ISSN: 1741-427X
Evid.-Based Complement. Altern. Med.
Oxford University Press, Great Clarendon Street, OX2 6DP, Oxford, United Kingdom, Tel: +44(1865)353907, Fax: +44(1865)353486 ISSUES/YR: 3

Evidence-Based Gastroenterology
ISSN: 1527-8557
Evid.-Based Gastroenterol.
Lippincott Williams and Wilkins, 530 Walnut Street, Philadelphia, PA 19106-3621, United States, Tel: +1(215)5218342, Fax: +1(215)5218326 ISSUES/YR: 4

- **Evidence-Based Healthcare and Public Health**
ISSN: 1744-2249
Evid.-Based Healthc. Public Health
Churchill Livingstone, 1-3 Baxter's Place, Leith Walk, EH1 3AF, Edinburgh, United Kingdom, Tel: +44(131)5562424, Fax: +44(131)5581278 ISSUES/YR: 6

Evidence-Based Integrative Medicine
ISSN: 1176-2330
Evid.-Based Integr. Med.
Adis International Ltd, 41 Centorian Drive, Private Bag 65901, Mairangi Bay, 1311, Auckland 10, New Zealand, Tel: +64(9)4770700, Fax: +64(9)4770766 ISSUES/YR: 4

Evidence-Based Medicine
ISSN: 1356-5524
Evid.-Based Med.
BMJ Publishing Group, Tavistock Square, WC1H 9JR, London, United Kingdom, Tel: +44(207)3874499, Fax: +44(207)3836668 ISSUES/YR: 6

- **Evidence-based Obstetrics and Gynecology**
ISSN: 1361-259X
Evid.-based Obstet. Gynecol.
Churchill Livingstone, 1-3 Baxter's Place, Leith Walk, EH1 3AF, Edinburgh, United Kingdom, Tel: +44(131)5562424, Fax: +44(131)5581278 ISSUES/YR: 4

- **Evidence-Based Ophthalmology**
ISSN: 1555-9203
Evid.-Based Ophthalmol.
Lippincott Williams and Wilkins, 530 Walnut Street, Philadelphia, PA 19106-3621, United States, Tel: +1(215)5218342, Fax: +1(215)5218326 ISSUES/YR: 4

Evolution Psychiatrique
ISSN: 0014-3855
Evol. Psychiatr.
Editions Scientifiques et Medicales Elsevier, 23 rue Linois, 75724, Paris Cedex 15, France, Tel: +33(1)45589826, Fax: +33(1)45589428 ISSUES/YR: 4

EVRS Educational Electronic Journal
ISSN: 1765-3169
EVRS Educ. Electron. J.
(EVRS) European VitreoRetinal Society, 8 Rue Camille Flammarion, 44000, Nantes, France, Tel: +33-33 (0)1 46 41 27 09, Fax: +33-33 (0)1 46 24 49 38 ISSUES/YR: 4

- **Exercise and Sport Sciences Reviews**
ISSN: 0091-6331
Exercise Sport Sci. Rev.
Lippincott Williams and Wilkins, 530 Walnut Street, Philadelphia, PA 19106-3621, United States, Tel: +1(215)5218342, Fax: +1(215)5218326 ISSUES/YR: 4

- **Exogenous Dermatology**
ISSN: 1424-4616
Exogenous Dermatol.
S. Karger AG, Allschwilerstrasse 10, P.O. Box, CH-4009, Basel, Switzerland, Tel: +41(61)3061111, Fax: +41(61)3061234 ISSUES/YR: 6

Experimental Aging Research
ISSN: 0361-073X
Exp. Aging Res.
Taylor and Francis Inc., 325 Chestnut St, Suite 800, Philadelphia PA, PA 19106, United States, Tel: +1(800)3541420, Fax: +1(215)6258914 ISSUES/YR: 4

Experimental and Clinical Cardiology
ISSN: 1205-6626
Exp. Clin. Cardiol.
Pulsus Group Inc., 2902 South Sheridan Way, ONT L6J 7L6, Oakville, Canada, Tel: +1(905)8294770, Fax: +1(905)8294799 ISSUES/YR: 3

- **Experimental and Clinical Endocrinology and Diabetes**
ISSN: 0947-7349
Exp. Clin. Endocrinol. Diabetes
MVS Medizinverlage Stuttgart, Steiermarkerstr. 3-5, D-70469, Stuttgart, Germany, Tel: +49(711)8931-0, Fax: +49(711)8931706 ISSUES/YR: 10

Experimental and Clinical Psychopharmacology
ISSN: 1064-1297
Exp. Clin. Psychopharmacol.
American Psychological Association Inc., 750 First Street NE, Washington, DC 20002-4242, United States, Tel: +1(202)3365600, Fax: +1(202)3365568 ISSUES/YR: 4

Experimental and Clinical Transplantation
ISSN: 1304-0855
Exp. Clin. Transplant.
Baskent University, Taskent Caddesi No:77, Kat 4, 06490, Bahcelievler, Ankara, Turkey, Tel: +90-90 312 222 23 25, Fax: +90-90 312 215 08 35 ISSUES/YR: 2

Experimental and Molecular Medicine
ISSN: 1226-3613
Exp. Mol. Med.
Korean Society of Med. Biochemistry and Mol. Biology, 812 KOFST, 635-4 Yoksam-dong Kangnam-ku, 135-703, Seoul, South Korea, Tel: +82(2)5651621, Fax: +82(2)5651622 ISSUES/YR: 6

Experimental and Molecular Pathology
ISSN: 0014-4800
Exp. Mol. Pathol.
Academic Press Inc., 6277 Sea Harbor Drive, Orlando, FL 32887-4900, United States, Tel: +1(407)3454040, Fax: +1(407)3639661 ISSUES/YR: 6

Current

Experimental and Toxicologic Pathology
ISSN: 0940-2993
Exp. Toxicol. Pathol.
Urban und Fischer Verlag Jena, P.O. Box 100537, 07705, Jena, Germany, Tel: +49(3641)6263, Fax: +49(3641)626500 ISSUES/YR: 6

Experimental Animals
ISSN: 1341-1357
Exp. Anim.
International Press Editing Centre Incorporation, 1-2-3 Sugamo, Toshima-ku, 170-0002, Tokyo, Japan, Tel: +81(3)59784067, Fax: +81(3)59784068 ISSUES/YR: 4

Experimental Biology and Medicine
ISSN: 1535-3702
Exp. Biol. Med.
Society for Experimental Biology and Medicine, 195 W. Spring Valley Avenue, Maywood, NJ 07607-1727, United States, Tel: +1(201)2919080, Fax: +1(201)2912988 ISSUES/YR: 11

• **Experimental Brain Research**
ISSN: 0014-4819
Exp. Brain Res.
Springer Verlag, Tiergartenstrasse 17, D-69121, Heidelberg, Germany, Tel: +49(6221)4870, Fax: +49(6221)487366 ISSUES/YR: 24

• **Experimental Cell Research**
ISSN: 0014-4827
Exp. Cell Res.
Academic Press Inc., 6277 Sea Harbor Drive, Orlando, FL 32887-4900, United States, Tel: +1(407)3454040, Fax: +1(407)3639661 ISSUES/YR: 20

Experimental Dermatology
ISSN: 0906-6705
Exp. Dermatol.
Blackwell Publishing Ltd, 9600 Garsington Road, OX4 2XG, Oxford, United Kingdom, Tel: +44(1865)776868, Fax: +44(1865)714591 ISSUES/YR: 12

Experimental Dermatology, Supplement
ISSN: 1479-1250
Exp. Dermatol. Suppl.
Blackwell Publishing Ltd, 9600 Garsington Road, OX4 2XG, Oxford, United Kingdom, Tel: +44(1865)776868, Fax: +44(1865)714591 ISSUES/YR: 1

• **Experimental Diabesity Research**
ISSN: 1543-8600
Exp. Diabesity Res.
Hindawi Publishing Corporation, 410 Park Avenue, 15th Floor, 287 pmb, New York, NY 10022, United States, Tel: +1(20)2-2706352, Fax: +1(20)2-2706351 ISSUES/YR: 4

• **Experimental Eye Research**
ISSN: 0014-4835
Exp. Eye Res.
Academic Press, 24-28 Oval Road, NW1 7DX, London, United Kingdom, Tel: +44(207)2674466 ISSUES/YR: 12

• **Experimental Gerontology**
ISSN: 0531-5565
Exp. Gerontol.
Elsevier Inc., 360 Park Avenue South, New York, NY 10010, United States, Tel: +1(212)6333950, Fax: +1(212)6333990 ISSUES/YR: 12

• **Experimental Hematology**
ISSN: 0301-472X
Exp. Hematol.
Elsevier Inc., 360 Park Avenue South, New York, NY 10010, United States, Tel: +1(212)6333950, Fax: +1(212)6333990 ISSUES/YR: 12

• **Experimental Lung Research**
ISSN: 0190-2148
Exp. Lung Res.
Taylor and Francis Inc., 325 Chestnut St, Suite 800, Philadelphia PA, PA 19106, United States, Tel: +1(800)3541420, Fax: +1(215)6258914 ISSUES/YR: 8

• **Experimental Neurology**
ISSN: 0014-4886
Exp. Neurol.
Academic Press Inc., 6277 Sea Harbor Drive, Orlando, FL 32887-4900, United States, Tel: +1(407)3454040, Fax: +1(407)3639661 ISSUES/YR: 12

Experimental Oncology
ISSN: 1812-9269
Exp. Oncol.
Academy of Sciences of the Ukraine, 45, Vasilkovskaya Street, 252022, Kiev, Ukraine, Tel: +380(44)2666510, Fax: +380(44)2671656 ISSUES/YR: 4

• **Experimental Parasitology**
ISSN: 0014-4894
Exp. Parasitol.
Academic Press Inc., 6277 Sea Harbor Drive, Orlando, FL 32887-4900, United States, Tel: +1(407)3454040, Fax: +1(407)3639661 ISSUES/YR: 12

Experimental Physiology
ISSN: 0958-0670
Exp. Physiol.
Blackwell Publishing Ltd, 9600 Garsington Road, OX4 2XG, Oxford, United Kingdom, Tel: +44(1865)776868, Fax: +44(1865)714591 ISSUES/YR: 6

Expert Opinion on Biological Therapy
ISSN: 1471-2598
Expert Opin. Biol. Ther.
Ashley Publications Ltd, 2 Albert Place, Finchley Central, N3 1QB, London, United Kingdom, Tel: +44(208)3433883, Fax: +44(208)3432313 ISSUES/YR: 12

Expert Opinion on Drug Delivery
ISSN: 1742-5247
Expert Opin. Drug Deliv.
Ashley Publications Ltd, 2 Albert Place, Finchley Central, N3 1QB, London, United Kingdom, Tel: +44(208)3433883, Fax: +44(208)3432313 ISSUES/YR: 6

Expert Opinion on Drug Metabolism and Toxicology
ISSN: 1742-5255
Expert Opin. Drug Metab. Toxicol.
Ashley Publications Ltd, 2 Albert Place, Finchley Central, N3 1QB, London, United Kingdom, Tel: +44(208)3433883, Fax: +44(208)3432313 ISSUES/YR: 4

Expert Opinion on Drug Safety
ISSN: 1474-0338
Expert Opin. Drug Saf.
Ashley Publications Ltd, 2 Albert Place, Finchley Central, N3 1QB, London, United Kingdom, Tel: +44(208)3433883, Fax: +44(208)3432313 ISSUES/YR: 6

Expert Opinion on Emerging Drugs
ISSN: 1472-8214
Expert Opin. Emerg. Drugs
Ashley Publications Ltd, 2 Albert Place, Finchley Central, N3 1QB, London, United Kingdom, Tel: +44(208)3433883, Fax: +44(208)3432313 ISSUES/YR: 4

Expert Opinion on Investigational Drugs
ISSN: 1354-3784
Expert Opin. Invest. Drugs
Ashley Publications Ltd, 2 Albert Place, Finchley Central, N3 1QB, London, United Kingdom, Tel: +44(208)3433883, Fax: +44(208)3432313 ISSUES/YR: 12

Expert Opinion on Pharmacotherapy
ISSN: 1465-6566
Expert Opin. Pharmacother.
Ashley Publications Ltd, 2 Albert Place, Finchley Central, N3 1QB, London, United Kingdom, Tel: +44(208)3433883, Fax: +44(208)3432313 ISSUES/YR: 16

Expert Opinion on Therapeutic Patents
ISSN: 1354-3776
Expert Opin. Ther. Pat.
Ashley Publications Ltd, 2 Albert Place, Finchley Central, N3 1QB, London, United Kingdom, Tel: +44(208)3433883, Fax: +44(208)3432313 ISSUES/YR: 12

Expert Opinion on Therapeutic Targets
ISSN: 1472-8222
Expert Opin. Ther. Targets
Ashley Publications Ltd, 2 Albert Place, Finchley Central, N3 1QB, London, United Kingdom, Tel: +44(208)3433883, Fax: +44(208)3432313 ISSUES/YR: 6

Expert Review of Anti-Infective Therapy
ISSN: 1478-7210
Expert Rev. Anti-Inf.
Future Drugs Ltd, 2 Albert Place, Finchley Central, N3 1QB, London, United Kingdom, Tel: +44(20)83492033, Fax: +44(20)83432313 ISSUES/YR: 6

Expert Review of Anticancer Therapy
ISSN: 1473-7140
Expert Rev. Anticancer Ther.
Future Drugs Ltd, 2 Albert Place, Finchley Central, N3 1QB, London, United Kingdom, Tel: +44(20)83492033, Fax: +44(20)83432313 ISSUES/YR: 12

Expert Review of Cardiovascular Therapy
ISSN: 1477-9072
Exp. Rev. Cardiovasc. Ther.
Future Drugs Ltd, 2 Albert Place, Finchley Central, N3 1QB, London, United Kingdom, Tel: +44(20)83492033, Fax: +44(20)83432313 ISSUES/YR: 6

Expert Review of Clinical Immunology
ISSN: 1744-666X
Expert Rev. Clin. Immunol.
Future Drugs Ltd, 2 Albert Place, Finchley Central, N3 1QB, London, United Kingdom, Tel: +44(20)83492033, Fax: +44(20)83432313 ISSUES/YR: 4

Expert Review of Medical Devices
ISSN: 1743-4440
Expert Rev. Med. Devices
Future Drugs Ltd, 2 Albert Place, Finchley Central, N3 1QB, London, United Kingdom, Tel: +44(20)83492033, Fax: +44(20)83432313 ISSUES/YR: 6

Expert Review of Molecular Diagnostics
ISSN: 1473-7159
Expert Rev. Mol. Diagn.
Future Drugs Ltd, 2 Albert Place, Finchley Central, N3 1QB, London, United Kingdom, Tel: +44(20)83492033, Fax: +44(20)83432313 ISSUES/YR: 6

Expert Review of Neurotherapeutics
ISSN: 1473-7175
Expert Rev. Neurother.
Future Drugs Ltd, 2 Albert Place, Finchley Central, N3 1QB, London, United Kingdom, Tel: +44(20)83492033, Fax: +44(20)83432313 ISSUES/YR: 6

Expert Review of Pharmacoeconomics and Outcomes Research
ISSN: 1473-7167
Expert Rev. Pharmacoecon. Outcomes Res.
Future Drugs Ltd, 2 Albert Place, Finchley Central, N3 1QB, London, United Kingdom, Tel: +44(20)83492033, Fax: +44(20)83432313 ISSUES/YR: 6

Expert Review of Proteomics
ISSN: 1478-9450
Expert Rev. Proteomics
Future Drugs Ltd, 2 Albert Place, Finchley Central, N3 1QB, London, United Kingdom, Tel: +44(20)83492033, Fax: +44(20)83432313 ISSUES/YR: 6

● **Expert Review of Vaccines**
ISSN: 1476-0584
Expert Rev. Vaccines
Future Drugs Ltd, 2 Albert Place, Finchley Central, N3 1QB, London, United Kingdom, Tel: +44(20)83492033, Fax: +44(20)83432313 ISSUES/YR: 6

● **Expert Reviews in Molecular Medicine**
ISSN: 1462-3994
Expert Rev. Mol. Med.
Cambridge University Press, Shaftesbury Road, CB2 2RU, Cambridge, United Kingdom, Tel: +44(1223)312393, Fax: +44(1223)315052 ISSUES/YR: 1

● **Explore: The Journal of Science and Healing**
ISSN: 1550-8307
Explor. J. Sci. Heal.
Elsevier Inc., 170 S Independence Mall W 300 E, Philadelphia, PA 19106-3399, United States ISSUES/YR: 6

Eye
ISSN: 0950-222X
Eye
Nature Publishing Group, Houndmills, RG21 6XS, Basingstoke, Hampshire, United Kingdom, Tel: +44(1256)329242, Fax: +44(1256)810526 ISSUES/YR: 12

● **Eye and Contact Lens**
ISSN: 1542-2321
Eye Contact Lens
Lippincott Williams and Wilkins, 530 Walnut Street, Philadelphia, PA 19106-3621, United States, Tel: +1(215)5218342, Fax: +1(215)5218326 ISSUES/YR: 4

Fabad Journal of Pharmaceutical Sciences
ISSN: 1300-4182
Fabad J. Pharm. Sci.
Ankara University, Tandogan, 06100, Ankara, Turkey, Tel: +90(312)2126805, Fax: +90(312)2131081 ISSUES/YR: 4

Facial Plastic Surgery
ISSN: 0736-6825
Facial Plast. Surg.
Thieme Medical Publishers, Inc., 333 7th Avenue, New York, NY 10001-5004, United States, Tel: +1(212)7600888, Fax: +1(212)7799020 ISSUES/YR: 4

Facial Plastic Surgery Clinics of North America
ISSN: 1064-7406
Facial Plast. Surg. Clin. North Am.
W.B. Saunders, Independence Square West, Philadelphia, PA 19106-3399, United States, Tel: +1(215)2387800, Fax: +1(215)2386445 ISSUES/YR: 4

● **Familial Cancer**
ISSN: 1389-9600
Fam. Cancer
Kluwer Academic Publishers, Van Godewijckstraat 30, 3311 GZ, Dordrecht, Netherlands, Tel: +31(78)6576000, Fax: +31(78)6576254 ISSUES/YR: 4

Familiendynamik
ISSN: 0342-2747
Familiendynamik
J.C. Cotta'sche Buchhandlung Nachvolger GmbH, P.O. Box 106016, 70049, Stuttgart, Germany, Tel: +49(711)66720, Fax: +49(711)66722030 ISSUES/YR: 4

Families, Systems and Health
ISSN: 1091-7527
Fam. Syst. Health
American Psychological Association Inc., 750 First Street NE, Washington, DC 20002-4242, United States, Tel: +1(202)3365600, Fax: +1(202)3365568 ISSUES/YR: 4

Family Medicine
ISSN: 0742-3225
Fam. Med.
Society of Teachers of Family Medicine, Ste 540, Leawood, KS 66211, United States, Tel: +1(913)906 6000, Fax: +1(913)906 6096 ISSUES/YR: 10

Family Medicine and Primary Care Review
ISSN: 1734-3402
Fam. Med. Prim. Care Rev.
Wydawnictwo Continuo, ul.lelewela 4, 53505, Wroclaw woj.dolnoslaskie, Poland, Tel: +48-717912030, Fax: +48-713439018 ISSUES/YR: 4

Family Practice
ISSN: 0263-2136
Fam. Pract.
Oxford University Press, Great Clarendon Street, OX2 6DP, Oxford, United Kingdom, Tel: +44(1865)353907, Fax: +44(1865)353486 ISSUES/YR: 6

Farmaceuticky Obzor
ISSN: 0014-8172
Farm. Obz.
Postgraduate Medical and Pharmaceutical School, Limbova 12, 833 03, Bratislava, Slovakia, Tel: +421(7)374560 ISSUES/YR: 12

Farmaceutico Hospitales
ISSN: 0214-4697
Farm. Hosp.
Ediciones Mayo S.A., Aribau 185-187 2e planta, 08021, Barcelona, Spain, Tel: +34(93)2090255, Fax: +34(93)2020643 ISSUES/YR: 10

Farmaceutisch Tijdschrift voor Belgie
ISSN: 0771-2367
Farm. Tijdschr. Belg.
APB Algemene Pharmaceutische Bond, Archimedesstraat 11, B-1000, Brussels, Belgium, Tel: +32(2)2854200, Fax: +32(2)2854285 ISSUES/YR: 4

Farmaceutski Glasnik
ISSN: 0014-8202
Farm. Glas.
Hrvatsko Farmaceutsko Drustvo, Masarykova 2/II, HR-10000, Zagreb, Croatia, Tel: +385(41)427944 ISSUES/YR: 12

Farmacevtski Vestnik
ISSN: 0014-8229
Farm. Vestn.
Slovenian Pharmaceutical Society, Dunajska 184 A, 1000, Ljubljana, Slovenia, Tel: +386-01 569 2601, Fax: +386-01 569 2602 ISSUES/YR: 4

Farmacia Hospitalaria
ISSN: 1130-6343
Farm. Hosp.
ARAN Ediciones S.A., Castello 128 - 1o, 28006, Madrid, Spain, Tel: +34(91)7820035, Fax: +34(91)5615787 ISSUES/YR: 6

• **FASEB Journal**
ISSN: 0892-6638
FASEB J.
FASEB, 9650 Rockville Pike, Bethesda, MD 20814-3998, United States, Tel: +1(301)5307100 ISSUES/YR: 14

• **FEBS Journal**
ISSN: 1742-464X
FEBS J.
Blackwell Publishing Ltd, 9600 Garsington Road, OX4 2XG, Oxford, United Kingdom, Tel: +44(1865)776868, Fax: +44(1865)714591 ISSUES/YR: 24

• **FEBS Letters**
ISSN: 0014-5793
FEBS Lett.
Elsevier, P.O. Box 211, 1000 AE, Amsterdam, Netherlands, Tel: +31(20)4853757, Fax: +31(20)4853432 ISSUES/YR: 30

• **FEMS Immunology and Medical Microbiology**
ISSN: 0928-8244
FEMS Immunol. Med. Microbiol.
Blackwell Publishing Ltd, 9600 Garsington Road, OX4 2XG, Oxford, United Kingdom, Tel: +44(1865)776868, Fax: +44(1865)714591 ISSUES/YR: 12

• **FEMS Microbiology Ecology**
ISSN: 0168-6496
FEMS Microbiol. Ecol.
Blackwell Publishing Ltd, 9600 Garsington Road, OX4 2XG, Oxford, United Kingdom, Tel: +44(1865)776868, Fax: +44(1865)714591 ISSUES/YR: 12

• **FEMS Microbiology Letters**
ISSN: 0378-1097
FEMS Microbiol. Lett.
Blackwell Publishing Ltd, 9600 Garsington Road, OX4 2XG, Oxford, United Kingdom, Tel: +44(1865)776868, Fax: +44(1865)714591 ISSUES/YR: 24

FEMS Microbiology Reviews
ISSN: 0168-6445
FEMS Microbiol. Rev.
Blackwell Publishing Ltd, 9600 Garsington Road, OX4 2XG, Oxford, United Kingdom, Tel: +44(1865)776868, Fax: +44(1865)714591 ISSUES/YR: 5

FEMS Yeast Research
ISSN: 1567-1356
FEMS Yeast Res.
Blackwell Publishing Ltd, 9600 Garsington Road, OX4 2XG, Oxford, United Kingdom, Tel: +44(1865)776868, Fax: +44(1865)714591 ISSUES/YR: 12

• **Fertility and Sterility**
ISSN: 0015-0282
Fertil. Steril.
Elsevier Inc., 360 Park Avenue South, New York, NY 10010, United States, Tel: +1(212)6333950, Fax: +1(212)6333990 ISSUES/YR: 12

Fetal and Maternal Medicine Review
ISSN: 0965-5395
Fetal Matern. Med. Rev.
Cambridge University Press, Shaftesbury Road, CB2 2RU, Cambridge, United Kingdom, Tel: +44(1223)312393, Fax: +44(1223)315052 ISSUES/YR: 4

• **Fetal and Pediatric Pathology**
ISSN: 1551-3815
Fetal Pediatr. Pathol.
Taylor and Francis Ltd., 4 Park Square, Milton Park, OX14 4RN, Abingdon, Oxfordshire, United Kingdom, Tel: +44(20)70175541, Fax: +44(20)70175198 ISSUES/YR: 6

• **Fetal Diagnosis and Therapy**
ISSN: 1015-3837
Fetal Diagn. Ther.
S. Karger AG, Allschwilerstrasse 10, P.O. Box, CH-4009, Basel, Switzerland, Tel: +41(61)3061111, Fax: +41(61)3061234 ISSUES/YR: 6

Feuillets de Radiologie
ISSN: 0181-9801
Feuill. Radiol.
Masson, 21 Rue Camille Desmoulins, 92789, Issy les Moulineaux Cedex 9, France, Tel: +33(1)73281634, Fax: +33(1)73281649 ISSUES/YR: 6

Filaria Journal
ISSN: 1475-2883
Filaria J.
BioMed Central Ltd., 34 - 42 Cleveland Street, W1T 4LB, London, United Kingdom, Tel: +44(207)3230323, Fax: +44(207)6319926 ISSUES/YR: 1

Fish Physiology and Biochemistry
ISSN: 0920-1742
Fish Physiol. Biochem.
Kluwer Academic Publishers, Van Godewijckstraat 30, 3311 GZ, Dordrecht, Netherlands, Tel: +31(78)6576000, Fax: +31(78)6576254 ISSUES/YR: 8

• **Fitoterapia**
ISSN: 0367-326X
Fitoterapia
Elsevier, P.O. Box 211, 1000 AE, Amsterdam, Netherlands, Tel: +31(20)4853757, Fax: +31(20)4853432 ISSUES/YR: 8

Fizjoterapia
ISSN: 1230-8323
Fizjoterapia
Academy of Physical Education, Rzezbiarska 4, 51-629, Wroclaw, Poland, Tel: +48-71 347 36 36 ISSUES/YR: 4

Fizjoterapia Polska
ISSN: 1642-0136
Fizjoter. Pol.
Agencja Wydawnicza Medsportpress, ul. Marymoncka 34 (skr. 23), 01-813, Warsaw, Poland, Tel: +48(22)8346772, Fax: +48(22)8346772 ISSUES/YR: 4

Fizyoterapi Rehabilitasyon
ISSN: 1300-8757
Fiz. Rehab.
Turkish Physical Therapy Association, Samanpazari, 06100, Anakara, Turkey, Tel: +90(312)3051577, Fax: +90(312)3243847 ISSUES/YR: 3

Flavour and Fragrance Journal
ISSN: 0882-5734
Flavour Fragrance J.
John Wiley and Sons Ltd, Southern Gate, PO19 8SQ, Chichester, West Sussex, United Kingdom, Tel: +44(1243)779777, Fax: +44(1243)775878 ISSUES/YR: 6

Fluoride
ISSN: 0015-4725
Fluoride
International Society for Fluoride Research, 17 Pioneer Crescent, Dunedin 9001, New Zealand, Fax: +64(3)4767689 ISSUES/YR: 4

Focus on Alternative and Complementary Therapies
ISSN: 1465-3753
Focus Altern. Complement. Ther.
Pharmaceutical Press, 1 Lambeth High Street, SE1 7JN, London, United Kingdom, Tel: +44(207)7359141, Fax: +44(207)7355085 ISSUES/YR: 4

Focus on Parkinson's Disease
ISSN: 0924-2015
Focus Parkinson's Dis.
Excerpta Medica, Radarweg 29, 1043 NX, Amsterdam, Netherlands ISSUES/YR: 1

Folia Biologica
ISSN: 0015-5497
Folia Biol. (Poland)
Institute of Systematics and Evolution of Animals, Slawkowska 17, 31-016, Krakow, Poland, Tel: +48(12)6227066 EXT.238 ISSUES/YR: 4

Folia Biologica
ISSN: 0015-5500
Folia Biol. (Czech Republic)
Academy of Sciences of the Czech Republic, Flemingovo nam. 2, 166 37, Prague 6, Czech Republic ISSUES/YR: 6

Folia Cardiologica
ISSN: 1507-4145
Folia Cardiol.
Via Medica, Ul. Swietokrzyska 73, 80-180, Gdansk, Poland, Tel: +48(58)3209494, Fax: +48(58)3209460 ISSUES/YR: 6

Folia Histochemica et Cytobiologica
ISSN: 0239-8508
Folia Histochem. Cytobiol.
Polish Histochemical and Cytochemical Society, P.O. Box 843, 30-960, Krakow, Poland, Tel: +48(12)6227027, Fax: +48(12)6227027 ISSUES/YR: 4

Folia Morphologica
ISSN: 0015-5659
Folia Morphol.
Via Medica, Ul. Swietokrzyska 73, 80-180, Gdansk, Poland, Tel: +48(58)3209494, Fax: +48(58)3209460 ISSUES/YR: 4

Folia Neuropathologica
ISSN: 1641-4640
Folia Neuropathol.
Via Medica, Ul. Swietokrzyska 73, 80-180, Gdansk, Poland, Tel: +48(58)3209494, Fax: +48(58)3209460 ISSUES/YR: 4

Folia Ophthalmologica Japonica
ISSN: 0015-5667
Folia Ophthalmol. Jpn.
Association of Folia Ophthalmologica Japonica, 3-6 Mihogaoka, 567, Ibaraki, Japan, Tel: +81(726)237878, Fax: +81(726)236060 ISSUES/YR: 12

Folia Pharmacologica Japonica
ISSN: 0015-5691
Folia Pharmacol. Jpn.
Japanese Pharmacological Society, Gokomachi-Ebisugawa, Nakagyo-ku, 604-0982, Kyoto, Japan, Tel: +81(75)2524641, Fax: +81(75)2524618 ISSUES/YR: 12

Current

• **Folia Phoniatrica et Logopaedica**
ISSN: 1021-7762
Folia Phoniatr. Logop.
S. Karger AG, Allschwilerstrasse 10, P.O. Box, CH-4009, Basel, Switzerland, Tel: +41(61)3061111, Fax: +41(61)3061234 ISSUES/YR: 6

• **Food Additives and Contaminants**
ISSN: 0265-203X
Food Addit. Contam.
Taylor and Francis Ltd., 4 Park Square, Milton Park, OX14 4RN, Abingdon, Oxfordshire, United Kingdom, Tel: +44(20)70175541, Fax: +44(20)70175198 ISSUES/YR: 12

Food and Chemical Toxicology
ISSN: 0278-6915
Food Chem. Toxicol.
Elsevier Ltd, Langford Lane, Kidlington, OX5 1GB, Oxford, United Kingdom, Tel: +44(1865)843000, Fax: +44(1865)843010 ISSUES/YR: 12

Food and Drug Law Journal
ISSN: 1064-590X
Food Drug Law J.
Food and Drug Law Institute, 1000 Vermont Avenue N.W., Suite 1200, Washington, DC 20005-4903, United States, Tel: +1(202)3711420 ISSUES/YR: 3

Food Chemistry
ISSN: 0308-8146
Food Chem.
Elsevier Ltd, Langford Lane, Kidlington, OX5 1GB, Oxford, United Kingdom, Tel: +44(1865)843000 ISSUES/YR: 16

• **Foodborne Pathogens and Disease**
ISSN: 1535-3141
Foodborne Pathog. Dis.
Mary Ann Liebert Inc., 140 Huguenot Street, New Rochelle, NY 10801-5215, United States, Tel: +1(914)7402100, Fax: +1(914)7402101 ISSUES/YR: 4

• **Foot**
ISSN: 0958-2592
Foot
Churchill Livingstone, 1-3 Baxter's Place, Leith Walk, EH1 3AF, Edinburgh, United Kingdom, Tel: +44(131)5562424, Fax: +44(131)5581278 ISSUES/YR: 4

• **Foot and Ankle International**
ISSN: 1071-1007
Foot Ankle Int.
AOFAS - American Orthopaedic Foot and Ankle Society, 2517 Eastlake Ave. E, Ste 200, Seattle, WA 98102-3278, United States, Tel: +1(206)2231120, Fax: +1(206)2231178 ISSUES/YR: 12

• **Foot and Ankle Surgery**
ISSN: 1268-7731
Foot Ankle Surg.
Elsevier Ltd, Langford Lane, Kidlington, OX5 1GB, Oxford, United Kingdom, Tel: +44(1865)843000 ISSUES/YR: 4

• **Forensic Science International**
ISSN: 0379-0738
Forensic Sci. Int.
Elsevier Ireland Ltd, P.O. Box 85, Limerick, Ireland, Tel: +353(61)471944, Fax: +353(61)472144 ISSUES/YR: 24

Formulary
ISSN: 1082-801X
Formulary
Advanstar Communications, 131 West First Street, Duluth, MN 55802-2065, United States, Tel: +1(218)7239200, Fax: +1(218)7239437 ISSUES/YR: 12

Forschende Komplementarmedizin
ISSN: 1661-4119
Forsch. Komplementarmed.
S. Karger AG, Allschwilerstrasse 10, P.O. Box, CH-4009, Basel, Switzerland, Tel: +41(61)3061111, Fax: +41(61)3061234 ISSUES/YR: 6

Fortschritte der Neurologie Psychiatrie
ISSN: 0720-4299
Fortschr. Neurol. Psychiatr.
Georg Thieme Verlag, Rudigerstrasse 14, D-70469, Stuttgart, Germany, Tel: +49(711)8931-321, Fax: +49(711)8931-422 ISSUES/YR: 12

FORUM - Trends in Experimental and Clinical Medicine
ISSN: 1121-8142
FORUM Trends Exp. Clin. Med.
Scuola Internazionale di Oncologia e Medicina Sperimentale, Via Martin Piaggio 17/6, 16122, Genova, Italy, Tel: +39(010)5458611 ISSUES/YR: 4

Forum der Psychoanalyse
ISSN: 0178-7667
Forum Psychoanal.
Springer Verlag, Tiergartenstrasse 17, D-69121, Heidelberg, Germany, Tel: +49(6221)4870, Fax: +49(6221)487366 ISSUES/YR: 4

Forum for Nordic Dermato-Venerology
ISSN: 1402-2915
Forum Nordic Derm.-Venerol.
Forum for Nordic Dermato-Venerology, Tradgardsgatan 14, SE-751 85, Uppsala, Sweden, Tel: +46(18)6115095, Fax: +46(18)557332 ISSUES/YR: 4

Forum Logopadie
ISSN: 0932-0547
Forum Logop.
Schulz-Kirchner Verlag GmbH, Mollweg 2, D-65510, Idstein, Germany, Tel: +49(6126)93200, Fax: +49(6126)932050 ISSUES/YR: 6

• **Free Radical Biology and Medicine**
ISSN: 0891-5849
Free Radic. Biol. Med.
Elsevier Inc., 360 Park Avenue South, New York, NY 10010, United States, Tel: +1(212)6333950, Fax: +1(212)6333990 ISSUES/YR: 24

Free Radical Research
ISSN: 1071-5762
Free Radic. Res.
Taylor and Francis Ltd., 4 Park Square, Milton Park, OX14 4RN, Abingdon, Oxfordshire, United Kingdom, Tel: +44(20)70175541, Fax: +44(20)70175198 ISSUES/YR: 12

Frontiers in Drug Design and Discovery
ISSN: 1567-2042
Front. Drug Des. and Discov.
Bentham Science Publishers B.V., P.O. Box 294, 1400 AG, Bussum, Netherlands, Tel: +31-356257290, Fax: +31-847511310 ISSUES/YR: 1

Frontiers in Medicinal Chemistry
ISSN: 1574-0889
Front. Med. Chem.
Bentham Science Publishers B.V., P.O. Box 294, 1400 AG, Bussum, Netherlands, Tel: +31-356257290, Fax: +31-847511310 ISSUES/YR: 1

- **Frontiers in Neuroendocrinology**
ISSN: 0091-3022
Front. Neuroendocrinol.
Academic Press Inc., 6277 Sea Harbor Drive, Orlando, FL 32887-4900, United States, Tel: +1(407)3454040, Fax: +1(407)3639661 ISSUES/YR: 4

Fudan University Journal of Medical Sciences
ISSN: 0257-8131
Fudan Univ. J. Med. Sci.
Fudan University, P.O. Box 285, 138 Yi Xue Yuan Road, 200032, Shanghai, China, Tel: +86(21)4311900, Fax: +86(21)64042261 ISSUES/YR: 6

- **Functional and Integrative Genomics**
ISSN: 1438-793X
Funct. Integr. Genomics
Springer Verlag, Tiergartenstrasse 17, D-69121, Heidelberg, Germany, Tel: +49(6221)4870, Fax: +49(6221)487366 ISSUES/YR: 4

Functional Neurology
ISSN: 0393-5264
Funct. Neurol.
CIC Edizioni Internazionali s.r.l., Corso Trieste 42, 00198, Roma, Italy, Tel: +39(06)8412673, Fax: +39(06)8412688 ISSUES/YR: 4

- **Fundamental and Clinical Pharmacology**
ISSN: 0767-3981
Fundam. Clin. Pharmacol.
Blackwell Publishing Ltd, 9600 Garsington Road, OX4 2XG, Oxford, United Kingdom, Tel: +44(1865)776868, Fax: +44(1865)714591 ISSUES/YR: 6

- **Fungal Genetics and Biology**
ISSN: 1087-1845
Fungal Genet. Biol.
Academic Press Inc., 6277 Sea Harbor Drive, Orlando, FL 32887-4900, United States, Tel: +1(407)3454040, Fax: +1(407)3639661 ISSUES/YR: 12

Fuss und Sprunggelenk
ISSN: 1619-9987
Fuss Sprunggelenk
Dr. Dietrich Steinkopff Verlag GmbH and Co. KG, P.O. Box 100462, D-64204, Darmstadt, Germany, Tel: +49(6151)828990, Fax: +49(6151)8289940 ISSUES/YR: 4

- **Future Cardiology**
ISSN: 1479-6678
Future Cardiol.
Future Medicine Ltd., 2nd Albert Place, Finchley Central, N3 1QB, London, United Kingdom, Tel: +44-208 343 3883, Fax: +44-208 343 2313 ISSUES/YR: 6

Future of Children
ISSN: 1054-8289
Future Child.
Center for the Future of Children, 300 Second Street, Suite 102, Los Altos, CA 94022, United States, Tel: +1(415)9177111 ISSUES/YR: 1

- **Future Oncology**
ISSN: 1479-6694
Future Oncol.
Future Medicine Ltd., 2nd Albert Place, Finchley Central, N3 1QB, London, United Kingdom, Tel: +44-208 343 3883, Fax: +44-208 343 2313 ISSUES/YR: 6

Gaceta Medica de Mexico
ISSN: 0016-3813
Gac. Med. Mex.
Academia Nacional de Medicina, Ave. Cuauhtemoc 330, 06725, Mexico DF, Mexico, Tel: +52(55)5578-2044, Fax: +52(55)5578-4271 ISSUES/YR: 6

- **Gait and Posture**
ISSN: 0966-6362
Gait Posture
Elsevier Ireland Ltd, P.O. Box 85, Limerick, Ireland, Tel: +31(61)471944, Fax: +31(61)472144 ISSUES/YR: 6

- **Gann Monographs on Cancer Research**
ISSN: 0072-0151
Gann Monogr. Cancer Res.
S. Karger AG, Allschwilerstrasse 10, P.O. Box, CH-4009, Basel, Switzerland, Tel: +41(61)3061111, Fax: +41(61)3061234 ISSUES/YR: 1

- **Gastric Cancer**
ISSN: 1436-3291
Gastric Cancer
Springer-Verlag Tokyo, 3-13, Hongo 3-chome, Bunkyo-ku, 113-8480, Tokyo, Japan, Tel: +81(3)38180861, Fax: +81(3)38180864 ISSUES/YR: 4

Gastroenterologia Polska
ISSN: 1232-9886
Gastroenterol. Pol.
Cornetis, ul. Dlugosza 2-6, 51-162, Wroclaw, Poland, Tel: +48-071 326 3101, Fax: +48-071 325 2803 ISSUES/YR: 6

Gastroenterologia y Hepatologia
ISSN: 0210-5705
Gastroenterol. Hepatol.
Ediciones Doyma, S.L., Travesera de Gracia 17-21, 08021, Barcelona, Spain, Tel: +34(93)2000711, Fax: +34(93)2091136 ISSUES/YR: 10

Gastroenterological Endoscopy
ISSN: 0387-1207
Gastroenterol. Endosc.
Japan Gastroenterological Endoscopy Society, 3-22 Kanda Ogawa-machi, Chiyoda-ku, 101, Tokyo, Japan, Tel: +81(3)32914111, Fax: +81(3)32915568 ISSUES/YR: 12

Gastroenterologie Clinique et Biologique
ISSN: 0399-8320
Gastroenterol. Clin. Biol.
Masson, 21 Rue Camille Desmoulins, 92789, Issy les Moulineaux Cedex 9, France, Tel: +33(1)73281634, Fax: +33(1)73281649 ISSUES/YR: 12

- **Gastroenterology**
ISSN: 0016-5085
Gastroenterology
W.B. Saunders, Independence Square West, Philadelphia, PA 19106-3399, United States, Tel: +1(215)2387800, Fax: +1(215)2386445 ISSUES/YR: 12

Gastroenterology and Hepatology
ISSN: 1554-7914
Gastroenterol. Hepatol.
Gastro-Hep Communications, Inc., 611 Broadway, Suite 828, New York, NY 10012, United States ISSUES/YR: 12

Gastroenterology Clinics of North America
ISSN: 0889-8553
Gastroenterol. Clin. North Am.
W.B. Saunders, Independence Square West, Philadelphia, PA 19106-3399, United States, Tel: +1(215)2387800, Fax: +1(215)2386445 ISSUES/YR: 4

- **Gastroenterology International**
ISSN: 0950-5911
Gastroenterol. Int.
International University Press, Via Dora 1, 00198, Roma, Italy, Tel: +39(06)85300906, Fax: +39(06)85300924 ISSUES/YR: 4

- **Gastrointestinal Endoscopy**
ISSN: 0016-5107
Gastrointest. Endosc.
Mosby Inc., 11830 Westline Industrial Drive, St. Louis, MO 63146, United States, Tel: +1(314)8728370, Fax: +1(314)4321380 ISSUES/YR: 6

- **Gastrointestinal Endoscopy Clinics of North America**
ISSN: 1052-5157
Gastrointest. Endosc. Clin. North Am.
W.B. Saunders, Independence Square West, Philadelphia, PA 19106-3399, United States, Tel: +1(215)2387800, Fax: +1(215)2386445 ISSUES/YR: 4

Gazi Medical Journal
ISSN: 1300-056X
Gazi Med. J.
Gazi Universitesi, Besevler, 06500, Ankara, Turkey ISSUES/YR: 4

Gazi Universitesi Eczacilik Fakultesi Dergisi
ISSN: 1015-9592
Gazi Univ. Eczacilik Fak. Derg.
Gazi Universitesi, 06330, Ankara, Turkey ISSUES/YR: 2

Gazzetta Medica Italiana Archivio per le Scienze Mediche
ISSN: 0393-3660
Gazz. Med. Ital. Arch. Sci. Med.
Edizioni Minerva Medica S.p.A., Corso Bramante 83-85, 10126, Turin, Italy, Tel: +39(011)678282, Fax: +39(011)674502 ISSUES/YR: 6

Geburtshilfe und Frauenheilkunde
ISSN: 0016-5751
Geburtshilfe Frauenheilkd.
Georg Thieme Verlag, Rudigerstrasse 14, D-70469, Stuttgart, Germany, Tel: +49(711)8931-321, Fax: +49(711)8931-422 ISSUES/YR: 12

GED - Gastrenterologia Endoscopia Digestiva
ISSN: 0101-7772
GED Gastrentologia Endosc. Dig.
Redprint Editora Ltda, Rua Domingos de Morais 2.777 - 13.o, CEP 04035-001, Sao Paulo, Brazil, Tel: +55(11)55724813, Fax: +55(11)55711719 ISSUES/YR: 6

Gefahrstoffe Reinhaltung der Luft
ISSN: 0949-8036
Gefahrstoffe Reinhalt. Luft
Springer-VDI Verlag GmbH und Co. KG, Postfach 101022, 40001, Dusseldorf, Germany, Tel: +49(211)61030, Fax: +49-211 6103 414 ISSUES/YR: 12

Gefasschirurgie
ISSN: 0948-7034
Gefasschirurgie
Springer Verlag, Tiergartenstrasse 17, D-69121, Heidelberg, Germany, Tel: +49(6221)4870, Fax: +49(6221)487366 ISSUES/YR: 4

Gender and Psychoanalysis
ISSN: 1091-6318
Gender Psychoanal.
International Universities Press Inc., P.O. Box 389, Guilford, CT 0643-0389, United States, Tel: +1(203)2454000, Fax: +1(203)2450775 ISSUES/YR: 4

● **Gender Medicine**
ISSN: 1550-8579
Gender Med.
Excerpta Medica Inc., 105 Raider Blvd, Suite 101, Hillsborough, NJ 08844, United States, Tel: +1(908)8748550, Fax: +1(908)8745633 ISSUES/YR: 4

● **Gene**
ISSN: 0378-1119
Gene
Elsevier, P.O. Box 211, 1000 AE, Amsterdam, Netherlands, Tel: +31(20)4853757, Fax: +31(20)4853432 ISSUES/YR: 36

Gene Expression
ISSN: 1052-2166
Gene Expr.
Cognizant Communication Corporation, 3 Hartsdale Road, Elmsford, NY 10523-3701, United States, Tel: +1(914)5927720, Fax: +1(914)5928981 ISSUES/YR: 6

● **Gene Expression Patterns**
ISSN: 1567-133X
Gene Expr. Patterns
Elsevier, P.O. Box 211, 1000 AE, Amsterdam, Netherlands, Tel: +31(20)4853757, Fax: +31(20)4853432 ISSUES/YR: 6

● **Gene Therapy**
ISSN: 0969-7128
Gene Ther.
Nature Publishing Group, Houndmills, RG21 6XS, Basingstoke, Hampshire, United Kingdom, Tel: +44(1256)329242, Fax: +44(1256)810526 ISSUES/YR: 24

Gene Therapy and Regulation
ISSN: 1388-9532
Gene Ther. Regul.
VSP BV, P.O.Box 9000, 2300 PA, Leiden, Netherlands, Tel: +31-71 535 3500, Fax: +31-71 531 7532 ISSUES/YR: 4

Geneeskunde en Sport
ISSN: 0016-6448
Geneeskd. Sport
Stichting Geneeskunde en Sport, Postbus 52, 3720 AB, Bilthoven, Netherlands, Tel: +31(030)2252290, Fax: +31(030)2252498 ISSUES/YR: 6

Geneesmiddelenbulletin
ISSN: 0304-4629
Geneesmiddelenbulletin
Stichting Geneesmiddelenbulletin, Lomanlaan 85, 3526 XC, Utrecht, Netherlands, Tel: +31(30)2252290, Fax: +31(30)2252498 ISSUES/YR: 12

● **General and Comparative Endocrinology**
ISSN: 0016-6480
Gen. Comp. Endocrinol.
Academic Press Inc., 6277 Sea Harbor Drive, Orlando, FL 32887-4900, United States, Tel: +1(407)3454040, Fax: +1(407)3639661 ISSUES/YR: 12

General Hospital Psychiatry
ISSN: 0163-8343
Gen. Hosp. Psychiatry
Elsevier Inc., 360 Park Avenue South, New York, NY 10010, United States, Tel: +1(212)6333950, Fax: +1(212)6333990 ISSUES/YR: 6

General Medicine
ISSN: 1311-1817
Gen. Med.
Medical Information Center, 1 Sv. G. Sofijski str., 1431, Sofia, Bulgaria, Tel: +359-522342, Fax: +359-522393 ISSUES/YR: 4

General Physiology and Biophysics
ISSN: 0231-5882
Gen. Physiol. Biophys.
R and D print, Hrobakova 22, 851 02, Bratislava, Slovakia, Tel: +421-2 5932 7705 ISSUES/YR: 4

● **Genes and Development**
ISSN: 0890-9369
Genes Dev.
Cold Spring Harbor Laboratory Press, 500 Sunnyside Boulevard, Woodbury New York, NY 11797, United States, Tel: +1(516)3491930, Fax: +1(516)3491946 ISSUES/YR: 24

Genes and Genetic Systems
ISSN: 1341-7568
Genes Genet. Syst.
Genetics Society of Japan, Yata, Mishima, 411, Shizuoka, Japan, Tel: +81(559)816736, Fax: +81(559)816736 ISSUES/YR: 6

● **Genes and Immunity**
ISSN: 1466-4879
Genes Immun.
Nature Publishing Group, Houndmills, RG21 6XS, Basingstoke, Hampshire, United Kingdom, Tel: +44(1256)329242, Fax: +44(1256)810526 ISSUES/YR: 8

● **Genes Chromosomes and Cancer**
ISSN: 1045-2257
Genes Chromosomes Cancer
Wiley-Liss Inc., 111 River Street, Hoboken, NJ 07030-5774, United States, Tel: +1(201)7486645, Fax: +1(201)7486021 ISSUES/YR: 12

● **Genes to Cells**
ISSN: 1356-9597
Genes Cells
Blackwell Publishing Ltd, 9600 Garsington Road, OX4 2XG, Oxford, United Kingdom, Tel: +44(1865)776868, Fax: +44(1865)714591 ISSUES/YR: 12

● **Genes, Brain and Behavior**
ISSN: 1601-1848
Genes Brain Behav.
Blackwell Publishing Ltd, 9600 Garsington Road, OX4 2XG, Oxford, United Kingdom, Tel: +44(1865)776868, Fax: +44(1865)714591 ISSUES/YR: 6

Current

- **Genesis**
ISSN: 1526-954X
Genesis
Wiley-Liss Inc., 111 River Street, Hoboken, NJ 07030-5774, United States, Tel: +1(201)7486645, Fax: +1(201)7486021 ISSUES/YR: 12

Genetic Counseling
ISSN: 1015-8146
Genet. Couns.
Editions Medecine et Hygiene, 7, Avenue de la Roseraie, P.O. Box 456, CH-1211, Geneva 4, Switzerland, Tel: +41(22)7029311, Fax: +41(22)7029355 ISSUES/YR: 4

Genetic Epidemiology
ISSN: 0741-0395
Genet. Epidemiol.
Wiley-Liss Inc., 111 River Street, Hoboken, NJ 07030-5774, United States, Tel: +1(201)7486645, Fax: +1(201)7486021 ISSUES/YR: 8

Genetic Testing
ISSN: 1090-6576
Genet. Test.
Mary Ann Liebert Inc., 140 Huguenot Street, New Rochelle, NY 10801-5215, United States, Tel: +1(914)7402100, Fax: +1(914)7402101 ISSUES/YR: 4

Genetic Vaccines and Therapy
ISSN: 1479-0556
Genet. Vaccines Ther.
BioMed Central Ltd., 34 - 42 Cleveland Street, W1T 4LB, London, United Kingdom, Tel: +44(207)3230323, Fax: +44(207)6319926 ISSUES/YR: 1

Genetical Research
ISSN: 0016-6723
Genet. Res.
Cambridge University Press, Shaftesbury Road, CB2 2RU, Cambridge, United Kingdom, Tel: +44(1223)312393, Fax: +44(1223)315052 ISSUES/YR: 6

- **Genetics**
ISSN: 0016-6731
Genetics
Genetics Society of America, 9650 Rockville Pike, Bethesda, MD 20814-3998, United States, Tel: +1(301)5711825 ISSUES/YR: 12

Genetics and Molecular Biology
ISSN: 1415-4757
Genet. Mol. Biol.
Brazilian Journal of Genetics, Ribeirao Preto, 14025-670, Sao Paolo, Brazil, Tel: +55(16) 621-8540 ISSUES/YR: 4

Genetics and Molecular Research
ISSN: 1676-5680
Genet. Mol. Res.
Fundacao de Pesquisas Cientificas de Ribeirao Preto, Rua Hudson 655, 14024-000, Jardim Canada, Ribeirao Preto, Brazil ISSUES/YR: 4

Genetics in Medicine
ISSN: 1098-3600
Gen. Med.
Lippincott Williams and Wilkins, 351 West Camden Street, Baltimore, MD 21201-2436, United States, Tel: +1(410)5284000 ISSUES/YR: 9

Genome Biology
ISSN: 1474-760X
Genome Biol.
BioMed Central Ltd., 34 - 42 Cleveland Street, W1T 4LB, London, United Kingdom, Tel: +44(207)3230323, Fax: +44(207)6319926 ISSUES/YR: 12

Genome Letters
ISSN: 1537-3053
Genome Let.
American Scientific Publishers, 25650 North Lewis Way, Stevenson Ranch, CA 9138-1439, United States, Tel: +1(661)2540807, Fax: +1(661)2541207 ISSUES/YR: 4

- **Genome Research**
ISSN: 1088-9051
Genome Res.
Cold Spring Harbor Laboratory Press, 500 Sunnyside Boulevard, Woodbury New York, NY 11797, United States, Tel: +1(516)3491930, Fax: +1(516)3491946 ISSUES/YR: 12

- **Genomics**
ISSN: 0888-7543
Genomics
Academic Press Inc., 6277 Sea Harbor Drive, Orlando, FL 32887-4900, United States, Tel: +1(407)3454040, Fax: +1(407)3639661 ISSUES/YR: 12

Genomics, Proteomics and Bioinformatics
ISSN: 1672-0229
Genomics. Proteomics Bioinform.
Beijing Genomics Institute, Building 6, 101300, Beijing Airport Industrial Zone B, Beijing, China, Tel: +86-10-80494199, Fax: +86-10-80498676 ISSUES/YR: 4

Geriatric and Medical Intelligence
ISSN: 1121-8460
Geriatr. Med. Intell.
Wichtig Editore s.r.l., Via Friuli 72/74, 20135, Milan, Italy, Tel: +39(02)55195443, Fax: +39(02)55195971 ISSUES/YR: 2

Geriatrics
ISSN: 0016-867X
Geriatrics
Advanstar Communications, 131 West First Street, Duluth, MN 55802-2065, United States, Tel: +1(218)7239200, Fax: +1(218)7239437 ISSUES/YR: 12

Geriatrics and Aging
ISSN: 1488-8408
Geriatr. Aging
Geriatrics and Aging, 20 Eglinton Avenue West, Suite 1109, Box 2083, ON M4R 1K8, Toronto, Canada, Tel: +1(416)4809478, Fax: +1(416)4802740 ISSUES/YR: 10

Geriatrics Today: Journal of the Canadian Geriatrics Society
ISSN: 1496-3892
Geriatr. Today J. Can. Geriatr. Soc.
Kenilworth Media Inc., 15 Wertheim Court, Suite 710, L4B 3H7, Richmond Hill, Ontario, Canada, Tel: +1-905 771 7333, Fax: +1-905 771 7336 ISSUES/YR: 4

Geriatrika
ISSN: 0212-9744
Geriatrika
Alpe Editores S.A., Pedro Rico 27, oficinas 11-12, 28029, Madrid, Spain, Tel: +34(91)7338811, Fax: +34(91)3159652 ISSUES/YR: 6

German Journal of Psychiatry
ISSN: 1433-1055
Germ. J. Psychiatry
University of Goettingen, von-Siebold-Str. 5, D-37075, Goettingen, Germany, Tel: +49(551)396607, Fax: +49(551)392004 ISSUES/YR: 1

Gerokomos
ISSN: 1134-928X
Gerokomos
Spanish Publishers Associate, Edificio Vertice - Antonio Lopez 249- 1 piso, 28041, Madrid, Spain, Tel: +34-91500 2075, Fax: +34-91500 2075 ISSUES/YR: 4

Gerontologist
ISSN: 0016-9013
Gerontologist
Gerontological Society of America, 1030 15th Street, NW Suite 250, Washington, DC 20005-1503, United States, Tel: +1(202)4083375, Fax: +1(202)8421150 ISSUES/YR: 6

- **Gerontology**
ISSN: 0304-324X
Gerontology
S. Karger AG, Allschwilerstrasse 10, P.O. Box, CH-4009, Basel, Switzerland, Tel: +41(61)3061111, Fax: +41(61)3061234 ISSUES/YR: 6

Gesundheitsokonomie und Qualitatsmanagement
ISSN: 1432-2625
Gesundh.okon. Qual.manage.
Georg Thieme Verlag, Rudigerstrasse 14, D-70469, Stuttgart, Germany, Tel: +49(711)8931-321, Fax: +49(711)8931-422 ISSUES/YR: 6

Gesundheitswesen
ISSN: 0941-3790
Gesundheitswesen
Georg Thieme Verlag, Rudigerstrasse 14, D-70469, Stuttgart, Germany, Tel: +49(711)8931-321, Fax: +49(711)8931-422 ISSUES/YR: 12

Gesundheitswesen, Supplement
ISSN: 0949-7013
Gesundheitswes. Suppl.
Georg Thieme Verlag, Rudigerstrasse 14, D-70469, Stuttgart, Germany, Tel: +49(711)8931-321, Fax: +49(711)8931-422 ISSUES/YR: 1

GIMT - Giornale Italiano delle Malattie del Torace
ISSN: 1127-0810
GIMT G. Ital. Mal. Torace
Mattioli 1885 S.p.A., Via Coduro 1/b, 43036, Fidenza (PR), Italy, Tel: +39(524)84547, Fax: +39(524)84751 ISSUES/YR: 6

Ginecologia y Obstetricia Clinica
ISSN: 1695-3827
Ginecol. Obstet. Clin.
Nexus Medica, c/Caracas 33, 08030, Barcelona, Spain, Tel: +34(93)4593492, Fax: +34(93)4590283 ISSUES/YR: 4

Ginecologia y Obstetricia de Mexico
ISSN: 0300-9041
Ginecol. Obstet. Mex.
Asociacion Mexicana de Ginecologia y Obstetricia, Av. Baja California 311, Hipdromo, -, Mexico D.F., Mexico, Tel: +52(55)55153668 ISSUES/YR: 12

Ginekologia Onkologiczna
ISSN: 1731-5379
Ginekol. Onkol.
Medical Communications, Ul.Ojcowska 11, 02-918, Warszawa, Poland, Tel: +48-226519783, Fax: +48-8425363 ISSUES/YR: 4

Ginekologia Praktyczna
ISSN: 1231-6407
Ginekol. Praktyczna
Termedia Publishing House Ltd., ul Kleberga 8,, 61-615, Poznan, Poland, Tel: +48-61 822 77 81, Fax: +48-61 822 7781 ISSUES/YR: 6

Giornale di Gerontologia
ISSN: 0017-0305
G. Gerontol.
Pacini Editore s.r.l., Via A. Gherardesca 1, 56121, Ospedaletto (Pisa), Italy, Tel: +39(050)313011, Fax: +39(050)3130300 ISSUES/YR: 6

Giornale Italiano di Cardiologia
ISSN: not available yet
G. Ital. Cardiol.
CEPI s.r.l., Via A. Ristori 38, 00197, Rome, Italy, Tel: +39-+39.06809681, Fax: +39-+39.068072458 ISSUES/YR: 1

Giornale Italiano di Dermatologia e Venereologia
ISSN: 0026-4741
G. Ital. Dermatol. Venereol.
Edizioni Minerva Medica S.p.A., Corso Bramante 83-85, 10126, Turin, Italy, Tel: +39(011)678282, Fax: +39(011)674502 ISSUES/YR: 6

Giornale Italiano di Diabetologia e Metabolismo
ISSN: 1593-6104
G. Ital. Diabetol. Metab.
UTET Periodici Scientifici srl, via Montefeltro 6/A, 20156, Milano, Italy, Tel: +39(02)6241171, Fax: +39(02)62411748 ISSUES/YR: 4

Giornale Italiano di Endoscopia Digestiva
ISSN: 0394-0225
G. Ital. Endosc. Dig.
Area Qualita Srl, Via comelico 3, 20135, Milano, Italy ISSUES/YR: 4

Giornale Italiano di Farmacia Clinica
ISSN: 1120-3749
G. Ital. Farm. Clin.
Il Pensiero Scientifico Editore s.r.l., Via Bradano 3/C, 00199, Roma, Italy, Tel: +39(06)86207158, Fax: +39(06)86207160 ISSUES/YR: 4

Giornale Italiano di Medicina del Lavoro ed Ergonomia
ISSN: 0391-9889
G. Ital. Med. Lav. Ergon.
Tipografia PI-ME Editrice Srl, Via Vigentina 136, 27100, Pavia, Italy, Tel: +39(382)572169, Fax: +39(382)572102 ISSUES/YR: 4

Giornale Italiano di Medicina Tropicale
ISSN: 0394-3445
G. Ital. Med. Trop.
Societa Italiana di Medicina Tropicale, Viale Regina Elena 299, 00161, Rome, Italy, Tel: +39(06)49903380, Fax: +39(06)49387065 ISSUES/YR: 4

Giornale Italiano di Ostetricia e Ginecologia
ISSN: 0391-9013
G. Ital. Ostet. Ginecol.
CIC Edizioni Internazionali s.r.l., Corso Trieste 42, 00198, Roma, Italy, Tel: +39(06)8412673, Fax: +39(06)8412688 ISSUES/YR: 12

• **GLIA**
ISSN: 0894-1491
GLIA
John Wiley and Sons Inc., P.O.Box 18667, Newark, NJ 07191-8667, United States, Tel: +1(212)8506000, Fax: +1(212)8506012 ISSUES/YR: 12

• **Globalization and Health**
ISSN: 1744-8603
Globalization Health
BioMed Central Ltd., 34 - 42 Cleveland Street, W1T 4LB, London, United Kingdom, Tel: +44(207)3230323, Fax: +44(207)6319926 ISSUES/YR: 1

• **Glycobiology**
ISSN: 0959-6658
Glycobiology
Oxford University Press, Great Clarendon Street, OX2 6DP, Oxford, United Kingdom, Tel: +44(1865)353907, Fax: +44(1865)353486 ISSUES/YR: 12

• **Glycoconjugate Journal**
ISSN: 0282-0080
Glycoconjugate J.
Kluwer Academic Publishers, Van Godewijckstraat 30, 3311 GZ, Dordrecht, Netherlands, Tel: +31(78)6576000, Fax: +31(78)6576254 ISSUES/YR: 9

Gogus-Kalp-Damar Anestezi ve Yogun Bakim Dernegi Dergisi
ISSN: not available yet
Gogus-Kalp-Damar Anestezi Yogun Bakim Dernegi Derg.
Turkish Anaesthesiology and Intensive Care Society, Inonu Cad. 7/4 Zambak apt., Gumussuyu Taksim. Istanbul, Turkey ISSUES/YR: 4

Good Clinical Practice Journal
ISSN: 1350-0961
Good Clin. Pract. J.
PJB Publications, Sheepen Place, C03 3LP, Colchester, United Kingdom, Tel: +44-20 70175540 ISSUES/YR: 12

GOZE
ISSN: 1577-3051
GOZE
Gogo Osasunaren Zientzietarako Erakundea, Ercilla 4, 48009, Bilbao, Spain, Tel: +34(94)4230049 ISSUES/YR: 3

Goztepe Tip Dergisi
ISSN: 1300-526X
Goztepe Tip Dergisi
Logos Medical Publishing, Yildiz Posta Cad. No.36, 80280, D.66-67 Gayrettepe, Istanbul, Turkey, Tel: +90(212)2880541, Fax: +90(212)2116185 ISSUES/YR: 12

• **Graefe's Archive for Clinical and Experimental Ophthalmology**
ISSN: 0721-832X
Graefe's Arch. Clin. Exp. Ophthalmol.
Springer Verlag, Tiergartenstrasse 17, D-69121, Heidelberg, Germany, Tel: +49(6221)4870, Fax: +49(6221)487366 ISSUES/YR: 12

Group Analysis
ISSN: 0533-3164
Group Anal.
SAGE Publications Ltd, 55 City Road, EC1Y 1SP, London, United Kingdom, Tel: +44(207)3248500, Fax: +44(207)3248600 ISSUES/YR: 4

Group Dynamics
ISSN: 1089-2699
Group Dyn.
Educational Publishing Foundation, 750 First Street NE, Washington, DC 20002-4242, United States, Tel: +1(202)3365540, Fax: +1(202)3365549 ISSUES/YR: 4

● **Growth Factors**
ISSN: 0897-7194
Growth Factors
Taylor and Francis Ltd., 4 Park Square, Milton Park, OX14 4RN, Abingdon, Oxfordshire, United Kingdom, Tel: +44(20)70175541, Fax: +44(20)70175198 ISSUES/YR: 4

● **Growth Hormone and IGF Research**
ISSN: 1096-6374
Growth Horm. IGF Res.
Churchill Livingstone, 1-3 Baxter's Place, Leith Walk, EH1 3AF, Edinburgh, United Kingdom, Tel: +44(131)5562424, Fax: +44(131)5581278 ISSUES/YR: 6

● **Growth, Development and Aging**
ISSN: 0017-4793
Growth Dev. Aging
Growth Publishing Co. Inc., P.O. Box 205, Hulls Cove, ME 04644-0205, United States, Tel: +1(207)2883533, Fax: +1(207)2885079 ISSUES/YR: 2

Gulhane Medical Journal
ISSN: 1302-0471
Gulhane Med. J.
Gulhane Askeri Tip Akademisi, Etlik, 06018, Ankara, Turkey, Tel: +90(312)3212063 ISSUES/YR: 4

● **Gut**
ISSN: 0017-5749
Gut
BMJ Publishing Group, Tavistock Square, WC1H 9JR, London, United Kingdom, Tel: +44(207)3874499, Fax: +44(207)3836668 ISSUES/YR: 12

Gynaecologia et Perinatologia
ISSN: 1330-0091
Gynaecol. Perinatol.
Naprijed, Petrova 13, 10000, Zagreb, Croatia, Tel: +385(1)460-4698, Fax: +385(1)463-3512 ISSUES/YR: 4

Gynaecologia et Perinatologia, Supplement
ISSN: 1331-0151
Gynaecol. Perinatol. Suppl.
Naprijed, Petrova 13, 10000, Zagreb, Croatia, Tel: +385(1)460-4698, Fax: +385(1)463-3512 ISSUES/YR: 1

Gynakologe
ISSN: 0017-5994
Gynakologe
Springer Verlag, Tiergartenstrasse 17, D-69121, Heidelberg, Germany, Tel: +49(6221)4870, Fax: +49(6221)487366 ISSUES/YR: 12

Gynakologie fur Hausarzte
ISSN: 1420-6811
Gynakol. Hausarzte
Rosenfluh Publikationen, Schaffhauserstrasse 13, 8212, Neuhausen, Switzerland, Tel: +41(52)6727822, Fax: +41(52)6727823 ISSUES/YR: 5

Gynakologisch-geburtshilfliche Rundschau
ISSN: 1018-8843
Gynakol.-geburtshilfliche Rundsch.
S. Karger AG, Allschwilerstrasse 10, P.O. Box, CH-4009, Basel, Switzerland, Tel: +41(61)3061111, Fax: +41(61)3061234 ISSUES/YR: 4

Gynakologische Endokrinologie
ISSN: 1610-2894
Gynakol. Endokrinol.
Springer Verlag, Tiergartenstrasse 17, D-69121, Heidelberg, Germany, Tel: +49(6221)4870, Fax: +49(6221)487366 ISSUES/YR: 4

Gynakologische Praxis
ISSN: 0341-8677
Gynakol. Prax.
Hans Marseille Verlag GmbH, P.O. Box 22 13 41, 80503, Munchen, Germany, Tel: +49(89)227988, Fax: +49(89)2904643 ISSUES/YR: 4

● **Gynecologic and Obstetric Investigation**
ISSN: 0378-7346
Gynecol. Obstet. Invest.
S. Karger AG, Allschwilerstrasse 10, P.O. Box, CH-4009, Basel, Switzerland, Tel: +41(61)3061111, Fax: +41(61)3061234 ISSUES/YR: 8

● **Gynecologic Oncology**
ISSN: 0090-8258
Gynecol. Oncol.
Academic Press Inc., 6277 Sea Harbor Drive, Orlando, FL 32887-4900, United States, Tel: +1(407)3454040, Fax: +1(407)3639661 ISSUES/YR: 12

● **Gynecological Endocrinology**
ISSN: 0951-3590
Gynecol. Endocrinol.
Taylor and Francis Ltd., 4 Park Square, Milton Park, OX14 4RN, Abingdon, Oxfordshire, United Kingdom, Tel: +44(20)70175541, Fax: +44(20)70175198 ISSUES/YR: 6

Gynecological Surgery
ISSN: 1613-2076
Gynecol. Surg.
Springer Verlag, Tiergartenstrasse 17, D-69121, Heidelberg, Germany, Tel: +49(6221)4870, Fax: +49(6221)487366 ISSUES/YR: 4

Gynecologie Obstetrique Fertilite
ISSN: 1297-9589
Gynecol. Obstet. Fertil.
Editions Scientifiques et Medicales Elsevier, 23 rue Linois, 75724, Paris Cedex 15, France, Tel: +33(1)45589826, Fax: +33(1)45589428 ISSUES/YR: 12

HAEMA
ISSN: 1108-2682
HAEMA
HAEMA Journal, Laiko General Hospital, GR-115 27, Athens, Greece, Fax: +30(210)7788830 ISSUES/YR: 4

Haematologica
ISSN: 0390-6078
Haematologica
Haematologica Journal Office, Strada Nuova 134, 27100, Pavia, Italy, Tel: +39(03)82531182, Fax: +39(03)8227721 ISSUES/YR: 12

Haematologica, Supplement
ISSN: 1592-8721
Haematologica. Suppl.
Haematologica Journal Office, Strada Nuova 134, 27100, Pavia, Italy, Tel: +39(03)82531182, Fax: +39(03)8227721 ISSUES/YR: 1

● **Haemophilia**
ISSN: 1351-8216
Haemophilia
Blackwell Publishing Ltd, 9600 Garsington Road, OX4 2XG, Oxford, United Kingdom, Tel: +44(1865)776868, Fax: +44(1865)714591 ISSUES/YR: 6

● **Haemophilia, Supplement**
ISSN: 1355-0691
Haemophilia, Suppl.
Blackwell Publishing Ltd, 9600 Garsington Road, OX4 2XG, Oxford, United Kingdom, Tel: +44(1865)776868, Fax: +44(1865)714591 ISSUES/YR: 1

Hallym International Journal of Aging
ISSN: 1535-6523
Hallym Int. J. Aging
Baywood Publishing Co. Inc., 26 Austin Avenue, P.O. Box 337, Amityville, NY 11701, United States, Tel: +1(516)6911270, Fax: +1(516)6911770 ISSUES/YR: 2

Hamostaseologie
ISSN: 0720-9355
Hamostaseologie
F.K. Schattauer Verlagsgesellschaft mbH, P.O. Box 10 45 45, D-70040, Stuttgart, Germany, Tel: +49(711)229870, Fax: +49(711)2298750 ISSUES/YR: 4

Hand Clinics
ISSN: 0749-0712
Hand Clin.
W.B. Saunders, Independence Square West, Philadelphia, PA 19106-3399, United States, Tel: +1(215)2387800, Fax: +1(215)2386445 ISSUES/YR: 4

Handchirurgie Mikrochirurgie Plastische Chirurgie
ISSN: 0722-1819
Handchir. Mikrochir. Plast. Chir.
Georg Thieme Verlag, Rudigerstrasse 14, D-70469, Stuttgart, Germany, Tel: +49(711)8931-321, Fax: +49(711)8931-422 ISSUES/YR: 6

Harefuah
ISSN: 0017-7768
Harefuah
Israel Medical Association, PO Box 3604, 52136, Ramat Gan, Israel, Tel: +972-3 610 0475, Fax: +972-3 575 1616 ISSUES/YR: 12

Harm Reduction Journal
ISSN: 1477-7517
Harm Reduct. J.
BioMed Central Ltd., 34 - 42 Cleveland Street, W1T 4LB, London, United Kingdom, Tel: +44(207)3230323, Fax: +44(207)6319926 ISSUES/YR: 1

Hart Bulletin
ISSN: 0301-8202
Hart Bull.
Mediselect BV, P.O. Box 63, 3830 AB, Leusden, Netherlands, Tel: +31(33)4229900, Fax: +31(33)4229922 ISSUES/YR: 6

• **Harvard Review of Psychiatry**
ISSN: 1067-3229
Harv. Rev. Psychiatry
Routledge, 4 Park Square, Milton Park, OX14 4RN, Abingdon, Oxfordshire, United Kingdom, Tel: +44(1235)828600, Fax: +44(1235)829000 ISSUES/YR: 6

Haut
ISSN: 0938-2216
Haut
Viavital Verlag GmbH, Riehlerstrasse 36, 50668, Koln - Cologne, Germany, Tel: +49(221)7764200, Fax: +49(221)7764201 ISSUES/YR: 7

Hautarzt
ISSN: 0017-8470
Hautarzt
Springer Verlag, Tiergartenstrasse 17, D-69121, Heidelberg, Germany, Tel: +49(6221)4870, Fax: +49(6221)487366 ISSUES/YR: 12

• **Head and Neck**
ISSN: 1043-3074
Head Neck
John Wiley and Sons Inc., P.O.Box 18667, Newark, NJ 07191-8667, United States, Tel: +1(212)8506000, Fax: +1(212)8506012 ISSUES/YR: 12

• **Headache**
ISSN: 0017-8748
Headache
Blackwell Publishing Inc., 350 Main Street, Malden, MA 02148, United States, Tel: +1(781)3888200, Fax: +1(781)3888210 ISSUES/YR: 10

• **Headache and Pain: Diagnostic Challenges, Current Therapy**
ISSN: 1545-8261
Headache Pain Diagn. Chall. Curr. Ther.
Cliggott Publishing Co., 330 Boston Post Road, Box 4027, Darien, CT 06820-4027, United States, Tel: +1(203)6626400 ISSUES/YR: 4

• **Headache Currents**
ISSN: 1743-5005
Headache Curr.
Blackwell Publishing Ltd, 9600 Garsington Road, OX4 2XG, Oxford, United Kingdom, Tel: +44(1865)776868, Fax: +44(1865)714591 ISSUES/YR: 6

• **Health and Place**
ISSN: 1353-8292
Health Place
Elsevier Ltd, Langford Lane, Kidlington, OX5 1GB, Oxford, United Kingdom, Tel: +44(1865)843000, Fax: +44(1865)843010 ISSUES/YR: 4

Health and Population: Perspectives and Issues
ISSN: 0253-6803
Health Popul. Perspect. Issues
National Institute of Health and Family Welfare, New Mehrauli Road, Munirka, 110 067, New Dehli, India ISSUES/YR: 4

Health and Quality of Life Outcomes
ISSN: 1477-7525
Health Qual. Life Outcomes
BioMed Central Ltd., 34 - 42 Cleveland Street, W1T 4LB, London, United Kingdom, Tel: +44(207)3230323, Fax: +44(207)6319926 ISSUES/YR: 1

Health Care Analysis
ISSN: 1065-3058
Health Care Anal.
Kluwer Academic Publishers, Van Godewijckstraat 30, 3311 GZ, Dordrecht, Netherlands, Tel: +31(78)6576000, Fax: +31(78)6576254 ISSUES/YR: 4

Health Care Financing Review
ISSN: 0195-8631
Health Care Financ. Rev.
Health Care Financing Administration, 7500 Security Boulevard C3-24-07, Baltimore, MD 21207, United States, Tel: +1-410 786 6572, Fax: +1-410 786 6511 ISSUES/YR: 4

Health Care Management Review
ISSN: 0361-6274
Health Care Manage. Rev.
Lippincott Williams and Wilkins, 241 Borough High Street, SE1 1GB, London, United Kingdom, Tel: +1(207)9407500, Fax: +1(207)9407575 ISSUES/YR: 4

• **Health Economics**
ISSN: 1057-9230
Health Econ.
John Wiley and Sons Ltd, Southern Gate, PO19 8SQ, Chichester, West Sussex, United Kingdom, Tel: +44(1243)779777, Fax: +44(1243)775878 ISSUES/YR: 12

• **Health Education Journal**
ISSN: 0017-8969
Health Educ. J.
SAGE Publications Ltd, 55 City Road, EC1Y 1SP, London, United Kingdom, Tel: +44(207)3248500, Fax: +44(207)3248600 ISSUES/YR: 4

• **Health Education Research**
ISSN: 0268-1153
Health Educ. Res.
Oxford University Press, Great Clarendon Street, OX2 6DP, Oxford, United Kingdom, Tel: +44(1865)353907, Fax: +44(1865)353486 ISSUES/YR: 6

• **Health Expectations**
ISSN: 1369-6513
Health Expect.
Blackwell Publishing Ltd, 9600 Garsington Road, OX4 2XG, Oxford, United Kingdom, Tel: +44(1865)776868, Fax: +44(1865)714591 ISSUES/YR: 4

Health Informatics Journal
ISSN: 1460-4582
Health Informatics J.
SAGE Publications Ltd, 55 City Road, EC1Y 1SP, London, United Kingdom, Tel: +44(207)3248500, Fax: +44(207)3248600 ISSUES/YR: 4

• **Health Physics**
ISSN: 0017-9078
Health Phys.
Lippincott Williams and Wilkins, 351 West Camden Street, Baltimore, MD 21201-2436, United States, Tel: +1(410)5284000 ISSUES/YR: 12

Health Policy
ISSN: 0168-8510
Health Policy
Elsevier Ireland Ltd, P.O. Box 85, Limerick, Ireland, Tel: +353(61)471944, Fax: +353(61)472144 ISSUES/YR: 12

Health Policy and Planning
ISSN: 0268-1080
Health Policy Plann.
Oxford University Press, Great Clarendon Street, OX2 6DP, Oxford, United Kingdom, Tel: +44(1865)353907, Fax: +44(1865)353486 ISSUES/YR: 4

- **Health Promotion International**
ISSN: 0957-4824
Health Promot. Int.
Oxford University Press, Great Clarendon Street, OX2 6DP, Oxford, United Kingdom, Tel: +44(1865)353907, Fax: +44(1865)353486 ISSUES/YR: 4

Health Psychology
ISSN: 0278-6133
Health Psychol.
American Psychological Association Inc., 750 First Street NE, Washington, DC 20002-4242, United States, Tel: +1(202)3365600, Fax: +1(202)3365568 ISSUES/YR: 6

Health Research Policy and Systems
ISSN: 1478-4505
Health Res. Policy Syst.
BioMed Central Ltd., 34 - 42 Cleveland Street, W1T 4LB, London, United Kingdom, Tel: +44(207)3230323, Fax: +44(207)6319926 ISSUES/YR: 1

- **Health Services and Outcomes Research Methodology**
ISSN: 1387-3741
Health Serv. Outcomes Res. Methodol.
Kluwer Academic Publishers, Van Godewijckstraat 30, 3311 GZ, Dordrecht, Netherlands, Tel: +31(78)6576000, Fax: +31(78)6576254 ISSUES/YR: 4

- **Health Services Management Research**
ISSN: 0951-4848
Health Serv. Manage. Res.
Royal Society of Medicine Press Ltd, P.O. Box 9002, W1A 0ZA, London, United Kingdom, Tel: +44(207)2902927, Fax: +44(207)2902929 ISSUES/YR: 4

Health Services Research
ISSN: 0017-9124
Health Serv. Res.
Blackwell Publishing Inc., 350 Main Street, Malden, MA 02148, United States, Tel: +1(781)3888200, Fax: +1(781)3888210 ISSUES/YR: 6

Health Technology Assessment
ISSN: 1366-5278
Health Technol. Assess.
National Co-ordinating Centre for HTA, Bouldrewood, Mail Point 728, Highfield, Southhampton, United Kingdom, Tel: +44(1703)595586 ISSUES/YR: 20

Health, Risk and Society
ISSN: 1369-8575
Health Risk Soc.
Routledge, 4 Park Square, Milton Park, OX14 4RN, Abingdon, Oxfordshire, United Kingdom, Tel: +44(1235)828600, Fax: +44(1235)829000 ISSUES/YR: 4

Healthcare Review Online
ISSN: 1173-7956
Healthc. Rev. Online
Enigma Publishing Ltd., 100 Bush Rd, Albany, Auckland, New Zealand, Tel: +64(9)4864005, Fax: +64(9)4860383 ISSUES/YR: 4

Hearing Journal
ISSN: 0745-7472
Hear. J.
Lippincott Williams and Wilkins, 530 Walnut Street, Philadelphia, PA 19106-3621, United States, Tel: +1(215)5218342, Fax: +1(215)5218326 ISSUES/YR: 12

- **Hearing Research**
ISSN: 0378-5955
Hear. Res.
Elsevier, P.O. Box 211, 1000 AE, Amsterdam, Netherlands, Tel: +31(20)4853757, Fax: +31(20)4853432 ISSUES/YR: 24

- **Heart**
ISSN: 1355-6037
Heart
BMJ Publishing Group, Tavistock Square, WC1H 9JR, London, United Kingdom, Tel: +44(207)3874499, Fax: +44(207)3836668 ISSUES/YR: 12

- **Heart and Lung: Journal of Acute and Critical Care**
ISSN: 0147-9563
Heart Lung J. Acute Crit. Care
Mosby Inc., 11830 Westline Industrial Drive, St. Louis, MO 63146, United States, Tel: +1(314)8728370, Fax: +1(314)4321380 ISSUES/YR: 6

- **Heart and Metabolism**
ISSN: 1566-0338
Heart Metab.
ADIS International Ltd, Chowley Oak Lane, Tattenhall, CH3 9GA, Chester, United Kingdom, Tel: +44(1829)771155, Fax: +44(1829)770330 ISSUES/YR: 4

- **Heart and Vessels**
ISSN: 0910-8327
Heart Vessels
Springer-Verlag Tokyo, 3-13, Hongo 3-chome, Bunkyo-ku, 113-8480, Tokyo, Japan, Tel: +81(3)38180861, Fax: +81(3)38180864 ISSUES/YR: 4

Heart Failure Clinics
ISSN: 1551-7136
Heart Fail. Clin.
W.B. Saunders Ltd, 32 Jamestown Road, NW1 7BY, London, United Kingdom, Tel: +1(20)74244200, Fax: +1(20)74854752 ISSUES/YR: 4

Heart Failure Reviews
ISSN: 1382-4147
Heart Fail. Rev.
Kluwer Academic Publishers, Van Godewijckstraat 30, 3311 GZ, Dordrecht, Netherlands, Tel: +31(78)6576000, Fax: +31(78)6576254 ISSUES/YR: 4

- **Heart Lung and Circulation**
ISSN: 1443-9506
Heart Lung Circul.
Australasian Society of Cardiac Thoracic Surgeons, Commercial Road, 3004, Melbourne, Australia, Tel: +61(3)92763691, Fax: +61(3)92762317 ISSUES/YR: 3

- **Heart Rhythm**
ISSN: 1547-5271
Heart Rhythm
Elsevier, P.O. Box 211, 1000 AE, Amsterdam, Netherlands, Tel: +31(20)4853757, Fax: +31(20)4853432 ISSUES/YR: 12

- **Heart Surgery Forum**
ISSN: 1098-3511
Heart Surg. Forum
Forum Multimedia Publishing LLC, 375 Greenbrier Drive, Suite 100, Charlottesville, VA 22901, United States, Tel: +1(434)8172000, Fax: +1(434)8172020 ISSUES/YR: 4

HEC Forum
ISSN: 0956-2737
HEC Forum
Kluwer Academic Publishers, Van Godewijckstraat 30, 3311 GZ, Dordrecht, Netherlands, Tel: +31(78)6576000, Fax: +31(78)6576254 ISSUES/YR: 4

- **Helicobacter**
ISSN: 1083-4389
Helicobacter
Blackwell Publishing Ltd, 9600 Garsington Road, OX4 2XG, Oxford, United Kingdom, Tel: +44(1865)776868, Fax: +44(1865)714591 ISSUES/YR: 6

- **Helicobacter, Supplement**
ISSN: 1478-4041
Helicobacter Suppl.
Blackwell Publishing Ltd, 9600 Garsington Road, OX4 2XG, Oxford, United Kingdom, Tel: +44(1865)776868, Fax: +44(1865)714591 ISSUES/YR: 1

Hellenic Journal of Cardiology
ISSN: 1011-7970
Hell. J. Cardiol.
Hellenic Cardiological Society, Potamianou 8, GR-11528, Athens, Greece, Tel: +30(210)7221633, Fax: +30(210)7226139 ISSUES/YR: 6

Hellenic Journal of Gastroenterology
ISSN: 1012-0424
Hell. J. Gastroenterol.
BETA Medical Publishers Ltd, 23 Meandrou Street, GR-11525, Athens, Greece, Tel: +30(210)6714371, Fax: +30(210)6715015 ISSUES/YR: 4

Hellenic Journal of Nuclear Medicine
ISSN: 1108-1430
Hell. J. Nucl. Med.
P.Ziti and Co, Ermou 51, Thessaloniki 546 23, Greece, Tel: +30(23920)72222, Fax: +30(23920)72229 ISSUES/YR: 3

- **Helvetica Chimica Acta**
  ISSN: 0018-019X
  Helv. Chim. Acta
  Verlag Helvetica Chimica Acta AG, Hofwiesenstrasse 26. Postfach, CH-8042, Zurich, Switzerland, Tel: +41(1)3602434, Fax: +41(1)3602435 ISSUES/YR: 12
- **Hematological Oncology**
  ISSN: 0278-0232
  Hematol. Oncol.
  John Wiley and Sons Ltd, Southern Gate, PO19 8SQ, Chichester, West Sussex, United Kingdom, Tel: +44(1243)779777, Fax: +44(1243)775878 ISSUES/YR: 4

  Hematologie
  ISSN: 1264-7527
  Hematologie
  John Libbey Eurotext, 127, avenue de la Republique, 92120, Montrouge, France, Tel: +33(01)46730660, Fax: +33(01)40840999 ISSUES/YR: 6
- **Hematology**
  ISSN: 1024-5332
  Hematology
  Taylor and Francis Ltd., 4 Park Square, Milton Park, OX14 4RN, Abingdon, Oxfordshire, United Kingdom, Tel: +44(20)70175541, Fax: +44(20)70175198 ISSUES/YR: 6

  Hematology Journal
  ISSN: 1466-4860
  Hematol. J.
  European Hematology Association, Westblaak 71, 3012 KE, Rotterdam, Netherlands, Tel: +44(10)4361760, Fax: +44(10)4361817 ISSUES/YR: 6
- **Hematology/Oncology Clinics of North America**
  ISSN: 0889-8588
  Hematol. Oncol. Clin. North Am.
  W.B. Saunders, Independence Square West, Philadelphia, PA 19106-3399, United States, Tel: +1(215)2387800, Fax: +1(215)2386445 ISSUES/YR: 6

  Hemodialysis International
  ISSN: 1492-7535
  Hemodial. Int.
  Blackwell Publishing Inc., 350 Main Street, Malden, MA 02148, United States, Tel: +1(781)3888200, Fax: +1(781)3888210 ISSUES/YR: 4

  Hemoglobin
  ISSN: 0363-0269
  Hemoglobin
  Taylor and Francis Inc., 325 Chestnut St, Suite 800, Philadelphia PA, PA 19106, United States, Tel: +1(800)3541420, Fax: +1(215)6258914 ISSUES/YR: 4

  Hepato-Gastro
  ISSN: 1253-7020
  Hepato-Gastro
  John Libbey Eurotext, 127, avenue de la Republique, 92120, Montrouge, France, Tel: +33(01)46730660, Fax: +33(01)40840999 ISSUES/YR: 6
- **Hepato-Gastroenterology**
  ISSN: 0172-6390
  Hepato-Gastroenterology
  H.G.E. Update Medical Publishing Ltd., P.O. Box 17257, GR-10024, Athens, Greece, Tel: +30(210)7213949, Fax: +30(210)7245988 ISSUES/YR: 6

  Hepatobiliary and Pancreatic Diseases International
  ISSN: 1499-3872
  Hepatobiliary Pancreatic Dis. Int.
  Firts Affiliated Hospital, 79 Qingchun Road, Hangzhou, 310003, Zhejiang, China, Tel: +86(571)87236865, Fax: +86(571)87236600 ISSUES/YR: 4

  Hepatologia Polska
  ISSN: 1232-9878
  Hepatol. Pol.
  Volumed s.c., ul. Olsztynska 3, 51-423, Wroclaw, Poland, Tel: +48(71)3253561, Fax: +48(71)3253561 ISSUES/YR: 4
- **Hepatology**
  ISSN: 0270-9139
  Hepatology
  John Wiley and Sons Ltd, Southern Gate, PO19 8SQ, Chichester, West Sussex, United Kingdom, Tel: +1(1243)779777, Fax: +1(1243)775878 ISSUES/YR: 12
- **Hepatology Research**
  ISSN: 1386-6346
  Hepatol. Res.
  Elsevier Ireland Ltd, P.O. Box 85, Limerick, Ireland, Tel: +353(61)471944, Fax: +353(61)472144 ISSUES/YR: 12

  Herbals and Complementary Medicine
  ISSN: 1609-4573
  Herbals Complement. Med.
  VICER Publishing, P.O. Box 14, A-1097, Vienna, Austria, Tel: +43(676)9568085, Fax: +43(676)9568086 ISSUES/YR: 4

  Hereditary Cancer in Clinical Practice
  ISSN: 1731-2302
  Hered. Cancer Clin. Pract.
  Termedia Publishing House Ltd., ul Kleberga 8,, 61-615, Poznan, Poland, Tel: +48-61 822 77 81, Fax: +48-61 822 7781 ISSUES/YR: 4

  Hereditas
  ISSN: 0018-0661
  Hereditas
  Blackwell Publishing Ltd, 9600 Garsington Road, OX4 2XG, Oxford, United Kingdom, Tel: +44(1865)776868, Fax: +44(1865)714591 ISSUES/YR: 6
- **Hernia**
  ISSN: 1265-4906
  Hernia
  Springer-Verlag France, 1 rue Paul Cezanne, 75008, Paris, France, Tel: +33(1)53933628, Fax: +33(1)53933683 ISSUES/YR: 4
- **Heroin Addiction and Related Clinical Problems**
  ISSN: 1592-1638
  Heroin Addict. Relat. Clin. Probl.
  Pacini Editore s.r.l., Via A. Gherardesca 1, 56121, Ospedaletto (Pisa), Italy, Tel: +39(050)313011, Fax: +39(050)3130300 ISSUES/YR: 2
- **Herpes**
  ISSN: 0969-7667
  Herpes
  Cambridge Medical Publications, High Street, BN11 1DJ, Worthing, West Sussex, United Kingdom, Tel: +44(1903)205884, Fax: +44(1903)234862 ISSUES/YR: 3

  Herz
  ISSN: 0340-9937
  Herz
  Urban und Vogel GmbH, Neumarkter Strasse 43, D-81673, Munchen, Germany, Tel: +49(89)43721300, Fax: +49(89)43721399 ISSUES/YR: 8

  Herz, Supplement
  ISSN: 0946-1299
  Herz Suppl.
  Urban und Vogel GmbH, Neumarkter Strasse 43, D-81673, Munchen, Germany, Tel: +49(89)43721300, Fax: +49(89)43721399 ISSUES/YR: 1

  Herzschrittmachertherapie und Elektrophysiologie
  ISSN: 0938-7412
  Herzschrittmacherther. Elektrophysiol.
  Dr. Dietrich Steinkopff Verlag GmbH and Co. KG, P.O. Box 100462, D-64204, Darmstadt, Germany, Tel: +49(6151)828990, Fax: +49(6151)8289940 ISSUES/YR: 4
- **High Altitude Medicine and Biology**
  ISSN: 1527-0297
  High Alt. Med. Biol.
  Mary Ann Liebert Inc., 140 Huguenot Street, New Rochelle, NY 10801-5215, United States, Tel: +1(914)7402100, Fax: +1(914)7402101 ISSUES/YR: 4
- **High Blood Pressure and Cardiovascular Prevention**
  ISSN: 1120-9879
  High Blood Press. Cardiovasc. Prev.
  Adis International Ltd, 41 Centorian Drive, Private Bag 65901, Mairangi Bay, 1311, Auckland 10, New Zealand, Tel: +64(9)4770700, Fax: +64(9)4770766 ISSUES/YR: 4

  HIP International
  ISSN: 1120-7000
  HIP Int.
  Wichtig Editore s.r.l., Via Friuli 72/74, 20135, Milan, Italy, Tel: +39(02)55195443, Fax: +39(02)55195971 ISSUES/YR: 4

- **Hippocampus**
ISSN: 1050-9631
Hippocampus
Wiley-Liss Inc., 111 River Street, Hoboken, NJ 07030-5774, United States, Tel: +1(201)7486645, Fax: +1(201)7486021 ISSUES/YR: 6

Hippokratia
ISSN: 1108-4189
Hippokratia
Hippokratia General Hospital of Thessaloniki, 49, Konstantinoupoleos str., 546 42, Thessaloniki, Greece, Tel: +30-310 892839, Fax: +30-310 818254 ISSUES/YR: 4

Hirosaki Medical Journal
ISSN: 0439-1721
Hirosaki Med. J.
Hirosaki University School of Medicine, 5 Zaifu-cho, 036, Hirosaki, Japan, Tel: +81(172)735111 ISSUES/YR: 4

Hiroshima Journal of Medical Sciences
ISSN: 0018-2052
Hiroshima J. Med. Sci.
Hiroshima University Medical Press, Kasumi 1-2-3, Minami-ku, 734, Hiroshima, Japan, Tel: +81(82)2575122, Fax: +81(82)2575124 ISSUES/YR: 4

- **Hispanic Healthcare International**
ISSN: 1540-4153
Hisp. Healthc. Int.
Springer Publishing Company, 11 West 42nd Street, 15th Floor, New York, NY 10036, United States, Tel: +1(212)4314370, Fax: +1(212)9417842 ISSUES/YR: 3

- **Histochemistry and Cell Biology**
ISSN: 0948-6143
Histochem. Cell Biol.
Springer Verlag, Tiergartenstrasse 17, D-69121, Heidelberg, Germany, Tel: +49(6221)4870, Fax: +49(6221)487366 ISSUES/YR: 12

Histology and Histopathology
ISSN: 0213-3911
Histol. Histopathol.
Histology and Histopathology, Plaza Fuensanta 2 - 7.o C, 30008, Murcia, Spain, Tel: +34-243694 ISSUES/YR: 4

- **Histopathology**
ISSN: 0309-0167
Histopathology
Blackwell Publishing Ltd, 9600 Garsington Road, OX4 2XG, Oxford, United Kingdom, Tel: +44(1865)776868, Fax: +44(1865)714591 ISSUES/YR: 12

HIV and AIDS Review
ISSN: 1730-1270
HIV AIDS Rev.
Medical Science International Ltd., ul. Ustrzycka 11, 02-141 Warsaw, Poland, Tel: +48(22)8681265, Fax: +48(22)7813232 ISSUES/YR: 4

- **HIV Clinical Trials**
ISSN: 1528-4336
HIV Clin. Trials
Thomas Land Publishers Inc., 255 Jefferson Road, St. Louis, MO 63119, United States, Tel: +1(314)9637445, Fax: +1(314)9639345 ISSUES/YR: 6

- **HIV Medicine**
ISSN: 1464-2662
HIV Med.
Blackwell Publishing Ltd, 9600 Garsington Road, OX4 2XG, Oxford, United Kingdom, Tel: +44(1865)776868, Fax: +44(1865)714591 ISSUES/YR: 4

HNO
ISSN: 0017-6192
HNO
Springer Verlag, Tiergartenstrasse 17, D-69121, Heidelberg, Germany, Tel: +49(6221)4870, Fax: +49(6221)487366 ISSUES/YR: 12

Hokkaido Journal of Orthopaedics and Traumatology
ISSN: 1343-3873
Hokkaido J. Orthop. Traumatol.
Hokkaido Orthopedic and Traumatic Surgery Society, North 14-jo, West 5-chome, Kita-ku, Sapporo-shi, Hokkaido, Japan, Tel: +81(11)7161161 ISSUES/YR: 2

Hokuriku Journal of Anesthesiology
ISSN: 0367-5947
Hokuriku J. Anesthesiol.
Hokuriku Branch of Japanese Society of Anesthesiology, 13-1 Takaramachi, Kanazawa-shi, 920, Ishikawa, Japan, Tel: +81(762)628151 ISSUES/YR: 1

Homeopathy
ISSN: 1475-4916
Homeopathy
Churchill Livingstone, 1-3 Baxter's Place, Leith Walk, EH1 3AF, Edinburgh, United Kingdom, Tel: +44(131)5562424, Fax: +44(131)5581278 ISSUES/YR: 4

Homeostasis in Health and Disease
ISSN: 0960-7560
Homeost. Health Dis.
CIANS, Srobarova 48, 100 42, Prague 10, Czech Republic, Tel: +420(2)743149 ISSUES/YR: 6

Hong Kong Dermatology and Venereology Bulletin
ISSN: 1028-9623
Hong Kong Dermatol. Venereol. Bul.
Medcom Limited, Olympia Plaza, 255 King's Road, North Point, Hong Kong, Tel: +852-25783833, Fax: +852-25783929 ISSUES/YR: 4

Hong Kong Journal of Dermatology and Venereology
ISSN: 1814-7453
Hong Kong J. Dermatol. Venereol.
Medcom Limited, Olympia Plaza, 255 King's Road, North Point, Hong Kong, Tel: +852-25783833, Fax: +852-25783929 ISSUES/YR: 4

Hong Kong Journal of Emergency Medicine
ISSN: 1024-9079
Hong Kong J. Emerg. Med.
Medcom Limited, Olympia Plaza, 255 King's Road, North Point, Hong Kong, Tel: +61-25783833, Fax: +61-25783929 ISSUES/YR: 4

- **Hong Kong Journal of Nephrology**
ISSN: 1561-5413
Hong Kong J. Nephrol.
Elsevier (Singapore) Pte Ltd, 77 Leighton Road, Causeway Bay, Hong Kong, China, Hong Kong, Tel: +852-29651300, Fax: +852-29760778 ISSUES/YR: 2

- **Hong Kong Journal of Occupational Therapy**
ISSN: 1569-1861
Hong Kong J. Occup. Ther.
Elsevier (Singapore) Pte Ltd, 77 Leighton Road, Causeway Bay, Hong Kong, China, Hong Kong, Tel: +852-29651300, Fax: +852-29760778 ISSUES/YR: 1

Hong Kong Journal of Paediatrics
ISSN: 1013-9923
Hong Kong J. Paediatr.
Medcom Limited, Olympia Plaza, 255 King's Road, North Point, Hong Kong, Tel: +852-25783833, Fax: +852-25783929 ISSUES/YR: 4

Hong Kong Journal of Psychiatry
ISSN: 1026-2121
Hong Kong J. Psychiatry
Scientific Communications International Ltd, Suite C, 10/F Wo On Building, 10 Wo On Lane, Central, Hong Kong, Tel: +852-28 68 9171, Fax: +852-28 68 9269 ISSUES/YR: 4

Hong Kong Medical Journal
ISSN: 1024-2708
Hong Kong Med. J.
Scientific Communications International Ltd, Suite C, 10/F Wo On Building, 10 Wo On Lane, Central, Hong Kong, Tel: +852-28 68 9171, Fax: +852-28 68 9269 ISSUES/YR: 6

Hong Kong Physiotherapy Journal
ISSN: 1013-7025
Hong Kong Physiother. J.
Elsevier (Singapore) Pte Ltd, 77 Leighton Road, Causeway Bay, Hong Kong, China, Hong Kong, Tel: +852-29651300, Fax: +852-29760778 ISSUES/YR: 1

Hong Kong Practitioner
ISSN: 1027-3948
Hong Kong Pract.
Hong Kong College of Family Physicians, Aberdeen, Hong Kong, China, Hong Kong, Tel: +852-25286618, Fax: +852-28660616 ISSUES/YR: 12

- **Hormone and Metabolic Research**
ISSN: 0018-5043
Horm. Metab. Res.
Georg Thieme Verlag, Rudigerstrasse 14, D-70469, Stuttgart, Germany, Tel: +49(711)8931-321, Fax: +49(711)8931-422 ISSUES/YR: 12
- **Hormone and Metabolic Research, Supplement**
ISSN: 0170-5903
Horm. Metab. Res. Suppl.
Georg Thieme Verlag, Rudigerstrasse 14, D-70469, Stuttgart, Germany, Tel: +49(711)8931-321, Fax: +49(711)8931-422 ISSUES/YR: 1
- **Hormone Research**
ISSN: 0301-0163
Horm. Res.
S. Karger AG, Allschwilerstrasse 10, P.O. Box, CH-4009, Basel, Switzerland, Tel: +41(61)3061111, Fax: +41(61)3061234 ISSUES/YR: 12
- **Hormones and Behavior**
ISSN: 0018-506X
Horm. Behav.
Academic Press Inc., 6277 Sea Harbor Drive, Orlando, FL 32887-4900, United States, Tel: +1(407)3454040, Fax: +1(407)3639661 ISSUES/YR: 10

Hospital Pharmacist
ISSN: 1352-7967
Hosp. Pharm.
Pharmaceutical Press, 1 Lambeth High Street, SE1 7JN, London, United Kingdom, Tel: +44(207)7359141, Fax: +44(207)7355085 ISSUES/YR: 11

Hospital Pharmacy
ISSN: 0018-5787
Hosp. Pharm.
Facts and Comparisons, 111 W. Port Plaza, Ste. 300, St. Louis, MO 6314603098, United States, Fax: +1(314)8785563 ISSUES/YR: 12

Hospitals and Health Networks
ISSN: 1068-8838
Hosp. Health Netw.
Health Forum Inc., 29th Floor, Chicago, IL 60606, United States, Tel: +1(800)821 2039, Fax: +1(312)422 4506 ISSUES/YR: 12

HPB
ISSN: 1365-182X
HPB
Taylor and Francis Ltd., 4 Park Square, Milton Park, OX14 4RN, Abingdon, Oxfordshire, United Kingdom, Tel: +44(20)70175541, Fax: +44(20)70175198 ISSUES/YR: 6

HPB, Supplement
ISSN: 1651-5323
HPB Suppl.
Taylor and Francis Ltd., 4 Park Square, Milton Park, OX14 4RN, Abingdon, Oxfordshire, United Kingdom, Tel: +44(20)70175541, Fax: +44(20)70175198 ISSUES/YR: 1

Hrvatska Revija Za Rehabilitacijska Istrazivanja
ISSN: 1331-3010
Hrvat. Rev. Rehabil. Istraz.
University of Zagreb, Kuslanova 59a, 10000, Zagreb, Croatia, Tel: +385(1)2338022 ISSUES/YR: 2

Huisarts en Wetenschap
ISSN: 0018-7070
Huisarts Wet.
Bohn Stafleu Van Loghum bv, P.O. Box 246, 3990 GA, Houten, Netherlands, Tel: +31(030)6383830, Fax: +31(030)6383999 ISSUES/YR: 13

- **Human and Experimental Toxicology**
ISSN: 0960-3271
Hum. Exp. Toxicol.
Arnold, 338 Euston Road, NW1 3BH, London, United Kingdom, Tel: +44(20)78736000, Fax: +44(20)78736376 ISSUES/YR: 12
- **Human Antibodies**
ISSN: 1093-2607
Hum. Antibodies
IOS Press, Nieuwe Hemweg 6B, 1013 BG, Amsterdam, Netherlands, Tel: +31(20)6883355, Fax: +31(20)6203419 ISSUES/YR: 4

Human Biology
ISSN: 0018-7143
Hum. Biol.
Wayne State University Press, 4809 Woodward Avenue, Detroit, MI 48201-1309, United States, Tel: +1(313)5776119 ISSUES/YR: 6

- **Human Brain Mapping**
ISSN: 1065-9471
Hum. Brain Mapp.
Wiley-Liss Inc., 111 River Street, Hoboken, NJ 07030-5774, United States, Tel: +1(201)7486645, Fax: +1(201)7486021 ISSUES/YR: 12
- **Human Development**
ISSN: 0018-716X
Hum. Dev.
S. Karger AG, Allschwilerstrasse 10, P.O. Box, CH-4009, Basel, Switzerland, Tel: +41(61)3061111, Fax: +41(61)3061234 ISSUES/YR: 6

Human Factors
ISSN: 0018-7208
Hum. Factors
Human Factors and Ergonomics Society, P.O. Box 1369, Santa Monica, CA 90406, United States, Tel: +1(310)3941811 ISSUES/YR: 4

Human Fertility
ISSN: 1464-7273
Hum. Fertil.
Taylor and Francis Ltd., 4 Park Square, Milton Park, OX14 4RN, Abingdon, Oxfordshire, United Kingdom, Tel: +44(20)70175541, Fax: +44(20)70175198 ISSUES/YR: 4

Human Gene Therapy
ISSN: 1043-0342
Hum. Gene Ther.
Mary Ann Liebert Inc., 140 Huguenot Street, New Rochelle, NY 10801-5215, United States, Tel: +1(914)7402100, Fax: +1(914)7402101 ISSUES/YR: 12

- **Human Genetics**
ISSN: 0340-6717
Hum. Genet.
Springer Verlag, Tiergartenstrasse 17, D-69121, Heidelberg, Germany, Tel: +49(6221)4870, Fax: +49(6221)487366 ISSUES/YR: 12

Human Genomics
ISSN: 1479-7364
Hum. Genomics
Henry Stewart Publications, 28-30 Little Russel Street, WC1A 2HN, London, United Kingdom, Tel: +44(207)4043040, Fax: +44(207)4042081 ISSUES/YR: 6

Human Heredity
ISSN: 0001-5652
Hum. Hered.
S. Karger AG, Allschwilerstrasse 10, P.O. Box, CH-4009, Basel, Switzerland, Tel: +41(61)3061111, Fax: +41(61)3061234 ISSUES/YR: 8

- **Human Immunology**
ISSN: 0198-8859
Hum. Immunol.
Elsevier Inc., 360 Park Avenue South, New York, NY 10010, United States, Tel: +1(212)6333950, Fax: +1(212)6333990 ISSUES/YR: 12
- **Human Molecular Genetics**
ISSN: 0964-6906
Hum. Mol. Genet.
Oxford University Press, Great Clarendon Street, OX2 6DP, Oxford, United Kingdom, Tel: +44(1865)353907, Fax: +44(1865)353486 ISSUES/YR: 26

Human Movement
ISSN: 1508-5287
Hum. Mov.
Wydawnictwo AWF, ul. Banacha 11, 51-617, Wroclaw, Poland, Tel: +48(071)3483071 ISSUES/YR: 4

Human Movement Science
ISSN: 0167-9457
Hum. Mov. Sci.
Elsevier, P.O. Box 211, 1000 AE, Amsterdam, Netherlands, Tel: +31(20)4853757, Fax: +31(20)4853432 ISSUES/YR: 6

- **Human Mutation**
  ISSN: 1059-7794
  Hum. Mutat.
  Wiley-Liss Inc., 111 River Street, Hoboken, NJ 07030-5774, United States, Tel: +1(201)7486645, Fax: +1(201)7486021 ISSUES/YR: 12

Human Pathology
ISSN: 0046-8177
Hum. Pathol.
W.B. Saunders, Independence Square West, Philadelphia, PA 19106-3399, United States, Tel: +1(215)2387800, Fax: +1(215)2386445 ISSUES/YR: 12

- **Human Psychopharmacology**
  ISSN: 0885-6222
  Hum. Psychopharmacol.
  John Wiley and Sons Ltd, Southern Gate, PO19 8SQ, Chichester, West Sussex, United Kingdom, Tel: +44(1243)779777, Fax: +44(1243)775878 ISSUES/YR: 8

Human Reproduction
ISSN: 0268-1161
Hum. Reprod.
Oxford University Press, Great Clarendon Street, OX2 6DP, Oxford, United Kingdom, Tel: +44(1865)353907, Fax: +44(1865)353486 ISSUES/YR: 12

- **Human Reproduction Update**
  ISSN: 1355-4786
  Hum. Reprod. Update
  Oxford University Press, Great Clarendon Street, OX2 6DP, Oxford, United Kingdom, Tel: +44(1865)353907, Fax: +44(1865)353486 ISSUES/YR: 6

Human Resources for Health
ISSN: 1478-4491
Hum. Resour. Health
BioMed Central Ltd., 34 - 42 Cleveland Street, W1T 4LB, London, United Kingdom, Tel: +44(207)3230323, Fax: +44(207)6319926 ISSUES/YR: 1

Human Vaccines
ISSN: 1554-8600
Hum. Vaccines
Landes Bioscience, 810 S Church Street, Georgetown, 78626, United States, Tel: +1-512.863.7762, Fax: +1-512.863.0081 ISSUES/YR: 6

- **Hybridoma**
  ISSN: 1554-0014
  Hybridoma
  Mary Ann Liebert Inc., 140 Huguenot Street, New Rochelle, NY 10801-5215, United States, Tel: +1(914)7402100, Fax: +1(914)7402101 ISSUES/YR: 6

Hygiena
ISSN: 1210-7840
Hygiena
Czech Medical Association J.E. Purkyne, Sokolska 31, P.O. Box 88, 120 26, Praha 2, Czech Republic, Tel: +420(2)24266201-4, Fax: +420(2)24266212 ISSUES/YR: 4

Hygiene + Medizin
ISSN: 0172-3790
Hyg. Med.
mhp-Verlag GmbH, Markplatz 13, D-65183, Wiesbaden, Germany, Tel: +49(611)505930, Fax: +49(611)5059311 ISSUES/YR: 12

- **Hypertension**
  ISSN: 0194-911X
  Hypertension
  Lippincott Williams and Wilkins, 530 Walnut Street, Philadelphia, PA 19106-3621, United States, Tel: +1(215)5218342, Fax: +1(215)5218326 ISSUES/YR: 12

Hypertension in Pregnancy
ISSN: 1064-1955
Hypertens. Pregnancy
Taylor and Francis Inc., 325 Chestnut St, Suite 800, Philadelphia PA, PA 19106, United States, Tel: +1(800)3541420, Fax: +1(215)6258914 ISSUES/YR: 3

Hypertension Research
ISSN: 0916-9636
Hypertens. Res.
Japanese Society of Hypertension, 4-2 Shinsenri-higashi-machi, 1-chome, 560-0082, Toyonaka, Japan, Tel: +81(6)68732301, Fax: +81(6)68732300 ISSUES/YR: 12

- **IARC Monographs on the Evaluation of Carcinogenic Risks to Humans**
  ISSN: 1017-1606
  IARC Monogr. Eval. Carcinog. Risks Hum.
  International Agency for Research on Cancer, 150 Cours Albert-Thomas, 69372, Lyon, France, Tel: +33(4)72738485, Fax: +33(4)72738302 ISSUES/YR: 1

Iatreia
ISSN: 0121-0793
Iatreia
Universidad de Antioquia, Apartado Aereo 1226, Medellin, Colombia, Tel: +57(4)5106919, Fax: +57(94)2630253 ISSUES/YR: 4

IDrugs
ISSN: 1369-7056
IDrugs
Thomson Scientific Ltd., 14 Great Queen Street, WC2B 5DF, London, United Kingdom, Tel: +44-20 7344 2800, Fax: +44-20 7344 2900 ISSUES/YR: 12

IEE Proceedings: Systems Biology
ISSN: 1741-2471
IEE Proc. Syst. Biol.
Institution of Electrical Engineers, P.O. Box 8, SG1 1HQ, Stevenage, Hertfordshire, United Kingdom, Tel: +44(1438)313311 ISSUES/YR: 4

IEEE Engineering in Medicine and Biology Magazine
ISSN: 0739-5175
IEEE Eng. Med. Biol. Mag.
Institute of Electrical and Electronics Engineers Inc., 3 Park Avenue, 17th Floor, New York, NY 10016-5997, United States, Tel: +1(212)7057900, Fax: +1(212)7524929 ISSUES/YR: 6

IEEE Transactions on Biomedical Engineering
ISSN: 0018-9294
IEEE Trans. Biomed. Eng.
Institute of Electrical and Electronics Engineers Computer Society, 445 Hoes Lane - P.O.Box 1331, Piscataway, NJ 08855-1331, United States, Tel: +1(732)981 0060, Fax: +1(732)981 9967 ISSUES/YR: 12

- **IEEE Transactions on Neural Systems and Rehabilitation Engineering**
  ISSN: 1534-4320
  IEEE Trans. Neural Syst. Rehabil. Eng.
  Institute of Electrical and Electronics Engineers Inc., 3 Park Avenue, 17th Floor, New York, NY 10016-5997, United States, Tel: +1(212)7057900, Fax: +1(212)7524929 ISSUES/YR: 4

IJCI - International Journal of Clinical Investigation
ISSN: 1590-7120
IJCI Int. J. Clin. Invest.
Editrice DOCTOR'S, Via S. Ignazio di Loyola 280, 80131, Napels, Italy, Tel: +39(81)5875849, Fax: +39(81)5877027 ISSUES/YR: 1

ILAR Journal
ISSN: 1084-2020
ILAR J.
Institute for Laboratory Animal Research, 2101 Constitution Avenue NW, NAS 347, Washington, DC 20418, United States, Tel: +1(202)3342590, Fax: +1(202)3341687 ISSUES/YR: 4

IM - Internal Medicine
ISSN: 1056-9286
IM Intern. Med.
Medical Economics Publishing, 5 Paragon Drive, Montvale, NJ 07645-1752, United States, Tel: +1(201)3587200 ISSUES/YR: 12

Imagerie de la Femme
ISSN: 1163-1961
Imag. Femme
Masson, 21 Rue Camille Desmoulins, 92789, Issy les Moulineaux Cedex 9, France, Tel: +33(1)73281634, Fax: +33(1)73281649 ISSUES/YR: 4

- **Imaging**
  ISSN: 0965-6812
  Imaging
  British Institute of Radiology, 36 Portland Place, W1N 1AT, London, United Kingdom, Tel: +44(207)5804189 ISSUES/YR: 4

Current

- **Imaging Decisions MRI**
  ISSN: 1433-3317
  Imaging Decis. MRI
  Blackwell Publishing Ltd, 9600 Garsington Road, OX4 2XG, Oxford, United Kingdom, Tel: +44(1865)776868, Fax: +44(1865)714591 ISSUES/YR: 4

Imago
ISSN: 1124-3031
Imago
Istituto Cortivo Editore, Via Cortivo 23, 35133, Padova, Italy, Tel: +39(049)702311, Fax: +39(049)702311 ISSUES/YR: 4

- **Immunity**
  ISSN: 1074-7613
  Immunity
  Cell Press, 1100 Massachusetts Avenue, Cambridge, MA 02138-9957, United States, Tel: +1(617)6617057, Fax: +1(617)6617061 ISSUES/YR: 12
- **Immunity and Ageing**
  ISSN: 1742-4933
  Immun. Ageing
  BioMed Central Ltd., 34 - 42 Cleveland Street, W1T 4LB, London, United Kingdom, Tel: +44(207)3230323, Fax: +44(207)6319926 ISSUES/YR: 1
- **Immuno-Analyse et Biologie Specialisee**
  ISSN: 0923-2532
  Immuno-Anal. Biol. Spec.
  Editions Scientifiques et Medicales Elsevier, 23 rue Linois, 75724, Paris Cedex 15, France, Tel: +33(1)45589826, Fax: +33(1)45589428 ISSUES/YR: 6
- **Immunobiology**
  ISSN: 0171-2985
  Immunobiology
  Urban und Fischer Verlag GmbH und Co. KG, P.O. Box 201930, 80019, Munchen, Germany, Tel: +49(89)53830, Fax: +49(89)5383221 ISSUES/YR: 10
- **Immunogenetics**
  ISSN: 0093-7711
  Immunogenetics
  Springer Verlag, Tiergartenstrasse 17, D-69121, Heidelberg, Germany, Tel: +49(6221)4870, Fax: +49(6221)487366 ISSUES/YR: 12
- **Immunohematology**
  ISSN: 0894-203X
  Immunohematology
  American Red Cross, 700 Spring Garden Street, Philadelphia, PA 19123-3594, United States, Tel: +1(215)4514902 ISSUES/YR: 4
- **Immunologic Research**
  ISSN: 0257-277X
  Immunol. Res.
  Humana Press Inc., 999 Riverview Drive, Suite 208, Totowa, NJ 07512-1165, United States, Tel: +1(973)2561699, Fax: +1(973)2568341 ISSUES/YR: 6
- **Immunological Investigations**
  ISSN: 0882-0139
  Immunol. Invest.
  Taylor and Francis Inc., 325 Chestnut St, Suite 800, Philadelphia PA, PA 19106, United States, Tel: +1(800)3541420, Fax: +1(215)6258914 ISSUES/YR: 4
- **Immunological Reviews**
  ISSN: 0105-2896
  Immunol. Rev.
  Blackwell Publishing Ltd, 9600 Garsington Road, OX4 2XG, Oxford, United Kingdom, Tel: +44(1865)776868, Fax: +44(1865)714591 ISSUES/YR: 6

Immunologiya
ISSN: 0206-4952
Immunologiya
Ministry of Public Health and Medical Industry of Russia, 24-2 Kashirskoye Shosse, 115478, Moscow, Russian Federation, Tel: +7(95)1164925 ISSUES/YR: 6

- **Immunology**
  ISSN: 0019-2805
  Immunology
  Blackwell Publishing Ltd, 9600 Garsington Road, OX4 2XG, Oxford, United Kingdom, Tel: +44(1865)776868, Fax: +44(1865)714591 ISSUES/YR: 12
- **Immunology and Allergy Clinics of North America**
  ISSN: 0889-8561
  Immunol. Allergy Clin. North Am.
  W.B. Saunders, Independence Square West, Philadelphia, PA 19106-3399, United States, Tel: +1(215)2387800, Fax: +1(215)2386445 ISSUES/YR: 4

Immunology and Cell Biology
ISSN: 0818-9641
Immunol. Cell Biol.
Blackwell Publishing, 550 Swanston Street, VIC 3053, Carlton South, Australia, Tel: +61(3)8359 1011, Fax: +61(3)8359 1120 ISSUES/YR: 6

- **Immunology Letters**
  ISSN: 0165-2478
  Immunol. Lett.
  Elsevier, P.O. Box 211, 1000 AE, Amsterdam, Netherlands, Tel: +31(20)4853757, Fax: +31(20)4853432 ISSUES/YR: 12

Immunology, Endocrine and Metabolic Agents in Medicinal Chemistry
ISSN: 1871-5214
Immunol. Endocr. Metab. Agents Med. Chem.
Bentham Science Publishers B.V., P.O. Box 294, 1400 AG, Bussum, Netherlands, Tel: +31-356257290, Fax: +31-847511310 ISSUES/YR: 4

- **Immunology, Supplement**
  ISSN: 0953-4954
  Immunol. Suppl.
  Blackwell Publishing Ltd, 9600 Garsington Road, OX4 2XG, Oxford, United Kingdom, Tel: +44(1865)776868, Fax: +44(1865)714591 ISSUES/YR: 1
- **Immunopharmacology and Immunotoxicology**
  ISSN: 0892-3973
  Immunopharmacol. Immunotoxicol.
  Taylor and Francis Inc., 325 Chestnut St, Suite 800, Philadelphia PA, PA 19106, United States, Tel: +1(800)3541420, Fax: +1(215)6258914 ISSUES/YR: 4

Impegno Ospedaliero, Sezione Scientifica
ISSN: 0393-0394
Impegno Osp. Sez. Sci.
A.N.M.I.R.S., Via Constantino Morin 45, 00195, Rome, Italy, Tel: +39(06)37514089 ISSUES/YR: 6

In Practice
ISSN: 0263-841X
In Practice
British Veterinary Association, 7 Mansfield Street, W1G 9NQ, London, United Kingdom, Tel: +44(20)76366541, Fax: +44(20)76370620 ISSUES/YR: 10

In Silico Biology
ISSN: 1386-6338
In Silico Biol.
IOS Press, Nieuwe Hemweg 6B, 1013 BG, Amsterdam, Netherlands, Tel: +31(20)6883355, Fax: +31(20)6203419 ISSUES/YR: 4

In Vitro Cellular and Developmental Biology - Animal
ISSN: 1071-2690
In Vitro Cell. Dev. Biol. Anim.
Society for In Vitro Biology, 13000-F York Road 304, Charlotte, NC 28278, United States, Tel: +1(301)3245054, Fax: +1(301)3245057 ISSUES/YR: 10

In Vivo
ISSN: 0258-851X
In Vivo
International Institute of Anticancer Research, 1st km Kapandritiou - Kalamou Rd., P.O. Box 22, GR-19014, Kapandriti, Attiki, Greece, Tel: +30(22950)53389, Fax: +30(22950)53389 ISSUES/YR: 6

Indian Drugs
ISSN: 0019-462X
Indian Drugs
Indian Drug Manufacturers' Association, Dr. Annie Besant Road, Worli, 400 018, Mumbai, India, Tel: +91(22)4926308, Fax: +91(22)4950723 ISSUES/YR: 12

Indian Heart Journal
ISSN: 0019-4832
Indian Heart J.
Cardiological Society of India, 110 002, New Delhi, India ISSUES/YR: 6

Indian Journal of Biotechnology
ISSN: 0972-5849
Indian J. Biotechnol.
National Institute of Science Communication and Information Resources (NISCAIR), Dr. K. S. Krishnan Marg (Near Pusa Gate), 110-012, New Delhi, India, Tel: +91(11)5786301-07, Fax: +91(11)5787062 ISSUES/YR: 4

Indian Journal of Clinical Biochemistry
ISSN: 0970-1915
Indian J. Clin. Biochem.
All India Institute of Medical Sciences, Ansari Nagar, 110 029, New Delhi, India, Tel: +91(11)6863002, Fax: +91(11)6862663 ISSUES/YR: 2

Indian Journal of Critical Care Medicine
ISSN: 0972-5229
Indian J. Crit. Care Med.
Medknow Publication, 12 Manisha Plaza, MN Rd, 400 070 MUMBAI, Kurla, India, Tel: +91(22)25032970, Fax: +91(22)25032398 ISSUES/YR: 4

Indian Journal of Dermatology, Venereology and Leprology
ISSN: 0378-6323
Indian J. Dermatol. Venereol. Leprol.
Medknow Publication, 12 Manisha Plaza, MN Rd, 400 070 MUMBAI, Kurla, India, Tel: +91(22)25032970, Fax: +91(22)25032398 ISSUES/YR: 6

Indian Journal of Gastroenterology
ISSN: 0254-8860
Indian J. Gastroenterol.
Indian Society of Gastroenterology, 534 Sandhurst Bridge, 400 007, Mumbai, India, Tel: +91(22)3613344 ISSUES/YR: 6

Indian Journal of Hematology and Blood Transfusion
ISSN: 0971-4502
Indian J. Hematol. Blood Transfus.
Indian Society of Hematology and Blood Transfusion, Parel, 400 012, Mumbai, India, Tel: +91(22)4161413 EXT. 4213, Fax: +91(22)4146937 ISSUES/YR: 4

Indian Journal of Human Genetics
ISSN: 0971-6866
Indian J. Hum. Genet.
Medknow Publication, 12 Manisha Plaza, MN Rd, 400 070 MUMBAI, Kurla, India, Tel: +91(22)25032970, Fax: +91(22)25032398 ISSUES/YR: 2

Indian Journal of Leprosy
ISSN: 0254-9395
Indian J. Lepr.
Hind Kusht Nivaran Sangh (Indian Leprosy Association), 1 Red Cross Road, 110 001, New Delhi, India ISSUES/YR: 4

Indian Journal of Medical Microbiology
ISSN: 0255-0857
Indian J. Med. Microbiol.
Medknow Publication, 12 Manisha Plaza, MN Rd, 400 070 MUMBAI, Kurla, India, Tel: +91(22)25032970, Fax: +91(22)25032398 ISSUES/YR: 4

Indian Journal of Medical Research
ISSN: 0971-5916
Indian J. Med. Res.
Indian Council of Medical Research, Ansari Nagar, P.O. Box 4911, 110029, New Delhi, India, Tel: +91(11)663001 ISSUES/YR: 12

Indian Journal of Medical Research, Supplement
ISSN: 0367-9012
Indian J. Med. Res. Suppl.
Indian Council of Medical Research, Ansari Nagar, P.O. Box 4911, 110029, New Delhi, India, Tel: +91(11)663001 ISSUES/YR: 1

Indian Journal of Medical Sciences
ISSN: 0019-5359
Indian J. Med. Sci.
Medknow Publication, 12 Manisha Plaza, MN Rd, 400 070 MUMBAI, Kurla, India, Tel: +91(22)25032970, Fax: +91(22)25032398 ISSUES/YR: 12

Indian Journal of Occupational and Environmental Medicine
ISSN: 0019-5278
Indian J. Occup. Environ. Med.
Medknow Publication, 12 Manisha Plaza, MN Rd, 400 070 MUMBAI, Kurla, India, Tel: +91(22)25032970, Fax: +91(22)25032398 ISSUES/YR: 2

Indian Journal of Ophthalmology
ISSN: 0301-4738
Indian J. Ophthalmol.
Medknow Publication, 12 Manisha Plaza, MN Rd, 400 070 MUMBAI, Kurla, India, Tel: +91(22)25032970, Fax: +91(22)25032398 ISSUES/YR: 4

Indian Journal of Otolaryngology and Head and Neck Surgery
ISSN: 0019-5421
Indian J. Otolaryngol. Head Neck Surg.
Medknow Publication, 12 Manisha Plaza, MN Rd, 400 070 MUMBAI, Kurla, India, Tel: +91(22)25032970, Fax: +91(22)25032398 ISSUES/YR: 4

Indian Journal of Otology
ISSN: 0971-7749
Indian J. Otol.
Indian Journal of Otology, 281 Civil Lines North, Railway Road, 251001, Muzaffarnagar, India, Tel: +91(131)402555, Fax: +91(131)46266 ISSUES/YR: 4

Indian Journal of Pathology and Microbiology
ISSN: 0377-4929
Indian J. Pathol. Microbiol.
Indian Association of Pathologists and Microbiologists, Sector 32-A, 160 030, Chandigarh, India, Tel: +91(172)665253-57 EXT 1050, Fax: +91(172)609360 ISSUES/YR: 4

Indian Journal of Pediatrics
ISSN: 0019-5456
Indian J. Pediatr.
The Indian Journal of Pediatrics, 125 II Floor, Gautam Nagar, 110 049, New Delhi,, India, Tel: +91(11)6568098, Fax: +91(11)6857587 ISSUES/YR: 12

Indian Journal of Pharmacology
ISSN: 0253-7613
Indian J. Pharmacol.
Medknow Publication, 12 Manisha Plaza, MN Rd, 400 070 MUMBAI, Kurla, India, Tel: +91(22)25032970, Fax: +91(22)25032398 ISSUES/YR: 6

Indian Journal of Physiology and Pharmacology
ISSN: 0019-5499
Indian J. Physiol. Pharmacol.
Association of Physiologists and Pharmacologists of India, 110 029, New Delhi, Ansari Nagar, India, Tel: +91(11)661123 ISSUES/YR: 4

Indian Journal of Practical Pediatrics
ISSN: not available yet
Indian J. Pract. Pediatr.
Indian Journal of Practical Pediatrics, 33 Halls Road, Egmore, 600 008, Chennai, India, Tel: +91(44)8235032 ISSUES/YR: 4

Indian Journal of Radiology and Imaging
ISSN: 0971-3026
Indian J. Radiol. Imaging
Indian Radiological and Imaging Association, Jaslok Hospital and Research Centre, 400 026, 15 Dr. G. Deshmukh Marg, Mumbai, India, Tel: +91(22)4933333 ISSUES/YR: 4

Indian Pacing and Electrophysiology Journal
ISSN: 0972-6292
Indian Pacing Electrophysiol. J.
Indian Pacing and Electrophysiology Group, Pulikkottil Thondayad, 673017, Calicut Kerala, India, Tel: +91(495)2356888 ISSUES/YR: 4

Indian Pediatrics
ISSN: 0019-6061
Indian Pediatr.
Indian Academy of Pediatrics, 110 002, New Delhi, India ISSUES/YR: 12

- **Indoor and Built Environment**
  ISSN: 1420-326X
  Indoor Built Environ.
  SAGE Publications Ltd, 55 City Road, EC1Y 1SP, London, United Kingdom, Tel: +44(207)3248500, Fax: +44(207)3248600 ISSUES/YR: 6

Industrial Health
ISSN: 0019-8366
Ind. Health
National Institute of Industrial Health, 21-1 Nagao 6-chome, Tama-ku, Kawasaki-shi, 214, Kanagawa, Japan, Tel: +81(44)8656111, Fax: +81(44)8656116 ISSUES/YR: 4

Infancy
ISSN: 1525-0008
Infancy
Lawrence Erlbaum Associates Inc., 10 Industrial Avenue, Mahwah, NJ 07430-2262, United States, Tel: +1(201)2582200, Fax: +1(201)2360072 ISSUES/YR: 6

- **Infant Behavior and Development**
  ISSN: 0163-6383
  Infant Behav. Dev.
  Elsevier Ltd, Langford Lane, Kidlington, OX5 1GB, Oxford, United Kingdom, Tel: +44(1865)843000, Fax: +44(1865)843010 ISSUES/YR: 4

- **Infant Mental Health Journal**
  ISSN: 0163-9641
  Infant Ment. Health J.
  John Wiley and Sons Inc., P.O.Box 18667, Newark, NJ 07191-8667, United States, Tel: +1(212)8506000, Fax: +1(212)8506012 ISSUES/YR: 6

Infanto
ISSN: 1413-0270
Infanto
Lemos Editorial e Graficos Ltda, Rua Rui Barbosa 70, Bela Vista, 01326-010, Sao Paulo SP, Brazil, Tel: +55(11)33711855, Fax: +55(11)33711855 ISSUES/YR: 3

Infection
ISSN: 0300-8126
Infection
Urban und Vogel GmbH, Neumarkter Strasse 43, D-81673, Munchen, Germany, Tel: +49(89)43721300, Fax: +49(89)43721399 ISSUES/YR: 6

- **Infection and Immunity**
  ISSN: 0019-9567
  Infect. Immun.
  American Society for Microbiology, 1752 N Street N.W., Washington, DC 20036-2904, United States, Tel: +1(202)942-9244, Fax: +1(202)942-9355 ISSUES/YR: 12

Infection Control and Hospital Epidemiology
ISSN: 0899-823X
Infect. Control Hosp. Epidemiol.
Slack Incorporated, 6900 Grove Road, Thorofare, NJ 08086-9447, United States, Tel: +1(856)8481000, Fax: +1(856)8535991 ISSUES/YR: 12

- **Infection, Genetics and Evolution**
  ISSN: 1567-1348
  Infec. Genet. Evol.
  Elsevier, P.O. Box 211, 1000 AE, Amsterdam, Netherlands, Tel: +31(20)4853757, Fax: +31(20)4853432 ISSUES/YR: 4

Infection, Supplement
ISSN: 0173-2129
Infect. Suppl.
Urban und Vogel GmbH, Neumarkter Strasse 43, D-81673, Munchen, Germany, Tel: +49(89)43721300, Fax: +49(89)43721399 ISSUES/YR: 1

Infections in Medicine
ISSN: 0749-6524
Infect. Med.
CMP Healthcare Media LLC, 330 Boston Post Road, Darien, CT 06820-4027, United States, Tel: +1(203)6626400, Fax: +1(212)6311600 ISSUES/YR: 12

Infectious Disease Clinics of North America
ISSN: 0891-5520
Infect. Dis. Clin. North Am.
W.B. Saunders, Independence Square West, Philadelphia, PA 19106-3399, United States, Tel: +1(215)2387800, Fax: +1(215)2386445 ISSUES/YR: 4

Infectious Diseases in Clinical Practice
ISSN: 1056-9103
Infect. Dis. Clin. Pract.
Lippincott Williams and Wilkins, 351 West Camden Street, Baltimore, MD 21201-2436, United States, Tel: +1(410)5284000 ISSUES/YR: 6

- **Infectious Diseases in Obstetrics and Gynecology**
  ISSN: 1064-7449
  Infect. Dis. Obstet. Gynecol.
  Hindawi Publishing Corporation, 410 Park Avenue, 15th Floor, 287 pmb, New York, NY 10022, United States, Tel: +1(20)2-2706352, Fax: +1(20)2-2706351 ISSUES/YR: 4

Infectious Disorders - Drug Targets
ISSN: 1871-5265
Infect. Disord. Drug Targets
Bentham Science Publishers B.V., P.O. Box 294, 1400 AG, Bussum, Netherlands, Tel: +31-356257290, Fax: +31-847511310 ISSUES/YR: 4

Infektoloski Glasnik
ISSN: 1331-2820
Infektol. Glas.
University Hospital of Infectious Diseases, Mirogojska 8, 10000, Zagreb, Croatia, Tel: +385(1)4603222, Fax: +385(1)4678235 ISSUES/YR: 4

Infezioni in Medicina
ISSN: 1124-9390
Infez. Med.
EDIMES Edizioni Medico Scientifiche, Via Riviera 39, 27100, Pavia, Italy, Tel: +39(0382)526253, Fax: +39(0384)423120 ISSUES/YR: 4

Inflammation
ISSN: 0360-3997
Inflammation
Springer Science and Business Media Deutschland GmbH, Heidelberger Platz 3, 14197, Berlin, Germany, Tel: +1-30 827875282, Fax: +1-30 827875707 ISSUES/YR: 6

Inflammation and Allergy - Drug Targets
ISSN: 1871-5281
Inflamm. Allergy Drug Targets
Bentham Science Publishers B.V., P.O. Box 294, 1400 AG, Bussum, Netherlands, Tel: +31-356257290, Fax: +31-847511310 ISSUES/YR: 4

Inflammation Research
ISSN: 1023-3830
Inflamm. Res.
Birkhauser Verlag AG, Klosterberg 23, P.O. Box 133, CH-4010, Basel, Switzerland, Tel: +41(61)2717400, Fax: +41(61)2050791 ISSUES/YR: 12

- **Inflammatory Bowel Diseases**
  ISSN: 1078-0998
  Inflammatory Bowel Dis.
  Lippincott Williams and Wilkins, 530 Walnut Street, Philadelphia, PA 19106-3621, United States, Tel: +1(215)5218342, Fax: +1(215)5218326 ISSUES/YR: 6

- **Inflammopharmacology**
  ISSN: 0925-4692
  Inflammopharmacology
  VSP BV, P.O.Box 9000, 2300 PA, Leiden, Netherlands, Tel: +31-71 535 3500, Fax: +31-71 531 7532 ISSUES/YR: 4

Informacao Psiquiatrica
ISSN: 0101-4331
Inf. Psiquiatr.
Editora Cientifica Nacional Ltda, Rua da Gloria 366, 3o andar, P.O. Box 590, 20241-180, Rio de Janeiro, Brazil, Tel: +55(21)2213235, Fax: +55(21)2521691 ISSUES/YR: 4

Informatics in Primary Care
ISSN: 1476-0320
Informatics Prim. Care
Radcliffe Publishing Ltd, 18 Marcham Road, OX14 1AA, Abingdon, Oxon, United Kingdom, Tel: +44(1235)528820, Fax: +44(1235)528830 ISSUES/YR: 4

Information Psychiatrique
ISSN: 0020-0204
Inf. Psychiatr.
John Libbey Eurotext, 127, avenue de la Republique, 92120, Montrouge, France, Tel: +33(01)46730660, Fax: +33(01)40840999 ISSUES/YR: 10

Current

Informe Medico
ISSN: 1316-9688
Inf. Med.
Centro Empresarial Los Ruices, Av. Principal de Los Ruices, Piso 5, Ofic. 514, 1050 A, Caracas, Venezuela, Tel: +58(212)2325293, Fax: +58(212)2397129 ISSUES/YR: 6

• **Inhalation Toxicology**
ISSN: 0895-8378
Inhal. Toxicol.
Taylor and Francis Inc., 325 Chestnut St, Suite 800, Philadelphia PA, PA 19106, United States, Tel: +44(800)3541420, Fax: +44(215)6258914 ISSUES/YR: 14

• **Injury**
ISSN: 0020-1383
Injury
Elsevier Ltd, Langford Lane, Kidlington, OX5 1GB, Oxford, United Kingdom, Tel: +44(1865)843000 ISSUES/YR: 12

Inmunologia
ISSN: 0213-9626
Inmunologia
Ediciones Ergon SA, C/Arboleda 1, 28220, Majadahonda (Madrid), Spain, Tel: +34(91)6362937, Fax: +34(91)6362930 ISSUES/YR: 4

Inquiry
ISSN: 0046-9580
Inquiry (USA)
Blue Cross and Blue Shield Association, P.O. Box 527, Glenview, IL 60025-9917, United States, Tel: +1(847)7249280 ISSUES/YR: 4

• **Insulin**
ISSN: 1557-0843
Insulin
Elsevier Inc., 360 Park Avenue South, New York, NY 10010, United States, Tel: +1(212)6333950, Fax: +1(212)6333990 ISSUES/YR: 4

• **Integrative Cancer Therapies**
ISSN: 1534-7354
Integr. Cancer Ther.
SAGE Publications Inc., 2455 Teller Road, Thousand Oaks, CA 91320, United States, Tel: +1(805)4990721, Fax: +1(805)4990871 ISSUES/YR: 4

Integrative Medicine
ISSN: 1546-993X
Integr. Med.
InnoVision Communications, 169 Saxony Road, Suite 103, Encinitas, CA 92024, United States, Tel: +1(760)633 3910, Fax: +1(760)633 3918 ISSUES/YR: 6

Intensiv- und Notfallbehandlung
ISSN: 0947-5362
Intensiv- Notfallbehandl.
Dustri-Verlag Dr. Karl Feistle, P.O. Box 1351, D-82032, Deisenhofen/Munchen, Germany, Tel: +49(89)6138610, Fax: +49(89)6135412 ISSUES/YR: 4

Intensive Care Medicine
ISSN: 0342-4642
Intensive Care Med.
Springer Verlag, Tiergartenstrasse 17, D-69121, Heidelberg, Germany, Tel: +49(6221)4870, Fax: +49(6221)487366 ISSUES/YR: 12

Intensive Care Medicine, Supplement
ISSN: 0935-1701
Intensive Care Med. Suppl.
Springer Verlag, Tiergartenstrasse 17, D-69121, Heidelberg, Germany, Tel: +49(6221)4870, Fax: +49(6221)487366 ISSUES/YR: 1

Intensivmedizin und Notfallmedizin
ISSN: 0175-3851
Intensivmed. Notf.med.
Dr. Dietrich Steinkopff Verlag GmbH and Co. KG, P.O. Box 100462, D-64204, Darmstadt, Germany, Tel: +49(6151)828990, Fax: +49(6151)8289940 ISSUES/YR: 8

• **Interactive Cardiovascular and Thoracic Surgery**
ISSN: 1569-9293
Interact. Cardiovasc. Thorac. Surg.
European Association for Cardio-Thoracic Surgery, 3 Park Street, SL4 1LU, Windsor, Berkshire, United Kingdom, Tel: +44-1753 832 166, Fax: +44-1753 620 407 ISSUES/YR: 4

Internal Medicine
ISSN: 0918-2918
Intern. Med.
Japanese Society of Internal Medicine, 34-3, 3-chome, Hongo, Bunkyo-ku, 113, Tokyo, Japan, Tel: +81(3)38135991 ISSUES/YR: 12

Internal Medicine Clinical and Laboratory
ISSN: 1590-9271
Intern. Med. Clin. Lab.
Pacini Editore s.r.l., Via A. Gherardesca 1, 56121, Ospedaletto (Pisa), Italy, Tel: +39(050)313011, Fax: +39(050)3130300 ISSUES/YR: 3

• **Internal Medicine Journal**
ISSN: 1444-0903
Intern. Med. J.
Blackwell Publishing, 550 Swanston Street, VIC 3053, Carlton South, Australia, Tel: +61(3)8359 1011, Fax: +61(3)8359 1120 ISSUES/YR: 12

• **International Anesthesiology Clinics**
ISSN: 0020-5907
Int. Anesthesiol. Clin.
Lippincott Williams and Wilkins, 530 Walnut Street, Philadelphia, PA 19106-3621, United States, Tel: +1(215)5218342, Fax: +1(215)5218326 ISSUES/YR: 4

International Angiology
ISSN: 0392-9590
Int. Angiol.
Edizioni Minerva Medica S.p.A., Corso Bramante 83-85, 10126, Turin, Italy, Tel: +39(011)678282, Fax: +39(011)674502 ISSUES/YR: 4

• **International Archives of Allergy and Immunology**
ISSN: 1018-2438
Int. Arch. Allergy Immunol.
S. Karger AG, Allschwilerstrasse 10, P.O. Box, CH-4009, Basel, Switzerland, Tel: +41(61)3061111, Fax: +41(61)3061234 ISSUES/YR: 12

International Archives of Occupational and Environmental Health
ISSN: 0340-0131
Int. Arch. Occup. Environ. Health
Springer Verlag, Tiergartenstrasse 17, D-69121, Heidelberg, Germany, Tel: +49(6221)4870, Fax: +49(6221)487366 ISSUES/YR: 8

International Braz J Urol
ISSN: 1677-5538
Int. Braz. J. Urol.
Brazilian Society of Urology, Rua Visconde Silva 52, sl. 805, 22271-090, Rio de Janeiro, RJ, Brazil, Tel: +55(21)5396787, Fax: +55(21)5354947 ISSUES/YR: 6

• **International Clinical Psychopharmacology**
ISSN: 0268-1315
Int. Clin. Psychopharmacol.
Lippincott Williams and Wilkins, 241 Borough High Street, SE1 1GB, London, United Kingdom, Tel: +44(207)9407500, Fax: +44(207)9407575 ISSUES/YR: 6

• **International Forum of Psychoanalysis**
ISSN: 0803-706X
Int. Forum Psychoanal.
Routledge, 4 Park Square, Milton Park, OX14 4RN, Abingdon, Oxfordshire, United Kingdom, Tel: +44(1235)828600, Fax: +44(1235)829000 ISSUES/YR: 4

• **International Heart Journal**
ISSN: 1349-2365
Int. Heart J.
International Heart Journal Association, 7-3-1 Hongo, Bunkyo-ku, 113-8655, Tokyo, Japan, Tel: +81(3)58006797, Fax: +81(3)38152087 ISSUES/YR: 6

• **International Immunology**
ISSN: 0953-8178
Int. Immunol.
Oxford University Press, Great Clarendon Street, OX2 6DP, Oxford, United Kingdom, Tel: +44(1865)353907, Fax: +44(1865)353486 ISSUES/YR: 12

- **International Immunopharmacology**
  ISSN: 1567-5769
  Int. Immunopharmacol.
  Elsevier, P.O. Box 211, 1000 AE, Amsterdam, Netherlands, Tel: +31(20)4853757, Fax: +31(20)4853432 ISSUES/YR: 14

International Journal for Equity in Health
ISSN: 1475-9276
Int. J. Equity Health
BioMed Central Ltd., 34 - 42 Cleveland Street, W1T 4LB, London, United Kingdom, Tel: +44(207)3230323, Fax: +44(207)6319926 ISSUES/YR: 1

International Journal for Parasitology
ISSN: 0020-7519
Int. J. Parasitol.
Elsevier Ltd, Langford Lane, Kidlington, OX5 1GB, Oxford, United Kingdom, Tel: +44(1865)843000, Fax: +44(1865)843010 ISSUES/YR: 14

- **International Journal for Quality in Health Care**
  ISSN: 1353-4505
  Int. J. Qual. Health Care
  Oxford University Press, Great Clarendon Street, OX2 6DP, Oxford, United Kingdom, Tel: +44(1865)353907, Fax: +44(1865)353486 ISSUES/YR: 6

International Journal for Vitamin and Nutrition Research
ISSN: 0300-9831
Int. J. Vitam. Nutr. Res.
Verlag Hans Huber AG, Langgassstrasse 76, CH-3000, Bern, Switzerland, Tel: +41(31)3004500, Fax: +41(31)3004590 ISSUES/YR: 6

International Journal of Adolescent Medicine and Health
ISSN: 0334-0139
Int. J. Adolesc. Med. Health
Freund Publishing House Ltd, 61 Nachmani Street, P.O. Box 35010, 61350, Tel Aviv, Israel, Tel: +972(3)5628540, Fax: +972(3)5628538 ISSUES/YR: 4

International Journal of Aging and Human Development
ISSN: 0091-4150
Int. J. Aging Hum. Dev.
Baywood Publishing Co. Inc., 26 Austin Avenue, P.O. Box 337, Amityville, NY 11701, United States, Tel: +1(516)6911270, Fax: +1(516)6911770 ISSUES/YR: 8

- **International Journal of Andrology**
  ISSN: 0105-6263
  Int. J. Androl.
  Blackwell Publishing Ltd, 9600 Garsington Road, OX4 2XG, Oxford, United Kingdom, Tel: +44(1865)776868, Fax: +44(1865)714591 ISSUES/YR: 6

- **International Journal of Andrology, Supplement**
  ISSN: 1350-4908
  Int. J. Androl. Suppl.
  Blackwell Publishing Ltd, 9600 Garsington Road, OX4 2XG, Oxford, United Kingdom, Tel: +44(1865)776868, Fax: +44(1865)714591 ISSUES/YR: 1

- **International Journal of Angiology**
  ISSN: 1061-1711
  Int. J. Angiol.
  Springer New York LLC, 233 Springer Street, New York, 10013-1578, United States, Tel: +1-212 460 1501, Fax: +1-212 460 1595 ISSUES/YR: 4

- **International Journal of Antimicrobial Agents**
  ISSN: 0924-8579
  Int. J. Antimicrob. Agents
  Elsevier, P.O. Box 211, 1000 AE, Amsterdam, Netherlands, Tel: +31(20)4853757, Fax: +31(20)4853432 ISSUES/YR: 12

International Journal of Aromatherapy
ISSN: 0962-4562
Int. J. Aromather.
Churchill Livingstone, 1-3 Baxter's Place, Leith Walk, EH1 3AF, Edinburgh, United Kingdom, Tel: +44(131)5562424, Fax: +44(131)5581278 ISSUES/YR: 4

International Journal of Artificial Organs
ISSN: 0391-3988
Int. J. Artif. Organs
Wichtig Editore s.r.l., Via Friuli 72/74, 20135, Milan, Italy, Tel: +39(02)55195443, Fax: +39(02)55195971 ISSUES/YR: 12

- **International Journal of Audiology**
  ISSN: 1499-2027
  Int. J. Audiol.
  Taylor and Francis Ltd., 4 Park Square, Milton Park, OX14 4RN, Abingdon, Oxfordshire, United Kingdom, Tel: +44(20)70175541, Fax: +44(20)70175198 ISSUES/YR: 12

International Journal of Behavioral Development
ISSN: 0165-0254
Int. J. Behav. Dev.
Psychology Press Ltd, 4 Park Square, Milton Park, OX14 4RN, Abingdon, Oxfordshire, United Kingdom, Tel: +44(1235)828600, Fax: +44(1235)829000 ISSUES/YR: 6

International Journal of Behavioral Medicine
ISSN: 1070-5503
Int. J. Behav. Med.
Lawrence Erlbaum Associates Inc., 10 Industrial Avenue, Mahwah, NJ 07430-2262, United States, Tel: +1(201)2582200, Fax: +1(201)2360072 ISSUES/YR: 4

International Journal of Behavioral Nutrition and Physical Activity
ISSN: 1479-5868
Int. J. Behav. Nutr. Phys. Act.
BioMed Central Ltd., 34 - 42 Cleveland Street, W1T 4LB, London, United Kingdom, Tel: +44(207)3230323, Fax: +44(207)6319926 ISSUES/YR: 1

International Journal of Biochemistry and Cell Biology
ISSN: 1357-2725
Int. J. Biochem. Cell Biol.
Elsevier Ltd, Langford Lane, Kidlington, OX5 1GB, Oxford, United Kingdom, Tel: +44(1865)843000, Fax: +44(1865)843010 ISSUES/YR: 12

International Journal of Bioinformatics Research and Applications
ISSN: 1744-5458
Int. J. Bioinformatics Res. Appl.
Inderscience Enterprises Ltd., P.O. Box 896, CH-1215, Geneve 15, Switzerland, Fax: +41(22)7910885 ISSUES/YR: 4

International Journal of Biological Macromolecules
ISSN: 0141-8130
Int. J. Biol. Macromol.
Elsevier, P.O. Box 211, 1000 AE, Amsterdam, Netherlands, Tel: +31(20)4853757, Fax: +31(20)4853432 ISSUES/YR: 10

International Journal of Biological Markers
ISSN: 0393-6155
Int. J. Biol. Markers
Wichtig Editore s.r.l., Via Friuli 72/74, 20135, Milan, Italy, Tel: +39(02)55195443, Fax: +39(02)55195971 ISSUES/YR: 4

International Journal of Biological Sciences
ISSN: 1449-2288
Int. J. Biol. Sci.
Ivyspring International Publisher, P.O. Box 4546, NSW 2263, Lake Haven, Australia, Tel: +61(2)43905688, Fax: +61(2)43905660 ISSUES/YR: 4

International Journal of Biostatistics
ISSN: 1557-4679
Int. J. Biostat.
Berkeley Electronic Press, 2809 Telegraph Avenue, Suite 202, Berkeley, CA 94705, United States, Tel: +1(510)6651200, Fax: +1(510)6651201 ISSUES/YR: 1

International Journal of Biotechnology
ISSN: 0963-6048
Int. J. Biotechnol.
Inderscience Enterprises Ltd., P.O. Box 896, CH-1215, Geneve 15, Switzerland, Fax: +41(22)7910885 ISSUES/YR: 4

- **International Journal of Cancer**
  ISSN: 0020-7136
  Int. J. Cancer
  Wiley-Liss Inc., 111 River Street, Hoboken, NJ 07030-5774, United States, Tel: +1(201)7486645, Fax: +1(201)7486021 ISSUES/YR: 30

Current

International Journal of Cancer Research
ISSN: 1811-9727
Int. J. Cancer Res.
Asian Network for Scientific Information, 308-Lasani Town, Sargodha Road, Faisalabad, Pakistan, Tel: +92(41)2001145-46, Fax: +92(21)5206036 ISSUES/YR: 2

● **International Journal of Cardiology**
ISSN: 0167-5273
Int. J. Cardiol.
Elsevier Ireland Ltd, P.O. Box 85, Limerick, Ireland, Tel: +353(61)471944, Fax: +353(61)472144 ISSUES/YR: 15

International Journal of Cardiovascular Imaging
ISSN: 1569-5794
Int. J. Card. Imaging
Kluwer Academic Publishers, Van Godewijckstraat 30, 3311 GZ, Dordrecht, Netherlands, Tel: +31(78)6576000, Fax: +31(78)6576254 ISSUES/YR: 6

International Journal of Clinical and Experimental Hypnosis
ISSN: 0020-7144
Int. J. Clin. Exp. Hypn.
Routledge, 4 Park Square, Milton Park, OX14 4RN, Abingdon, Oxfordshire, United Kingdom, Tel: +44(1235)828600, Fax: +44(1235)829000 ISSUES/YR: 4

● **International Journal of Clinical Oncology**
ISSN: 1341-9625
Int. J. Clin. Oncol.
Springer-Verlag Tokyo, 3-13, Hongo 3-chome, Bunkyo-ku, 113-8480, Tokyo, Japan, Tel: +81(3)38180861, Fax: +81(3)38180864 ISSUES/YR: 6

International Journal of Clinical Pharmacology and Therapeutics
ISSN: 0946-1965
Int. J. Clin. Pharmacol. Ther.
Dustri-Verlag Dr. Karl Feistle, P.O. Box 1351, D-82032, Deisenhofen/Munchen, Germany, Tel: +49(89)6138610, Fax: +49(89)6135412 ISSUES/YR: 12

International Journal of Clinical Pharmacology Research
ISSN: 0251-1649
Int. J. Clin. Pharmacol. Res.
Bioscience Ediprint Inc., 16 Rue Alexandre-Gavard, 1227, Carouge-Geneva, Switzerland, Tel: +41(22)3003383, Fax: +41(22)3002489 ISSUES/YR: 4

● **International Journal of Clinical Practice**
ISSN: 1368-5031
Int. J. Clin. Pract.
Blackwell Publishing Ltd, 9600 Garsington Road, OX4 2XG, Oxford, United Kingdom, Tel: +44(1865)776868, Fax: +44(1865)714591 ISSUES/YR: 12

● **International Journal of Clinical Practice, Supplement**
ISSN: 1368-504X
Int. J. Clin. Pract. Suppl.
Blackwell Publishing Ltd, 9600 Garsington Road, OX4 2XG, Oxford, United Kingdom, Tel: +44(1865)776868, Fax: +44(1865)714591 ISSUES/YR: 1

● **International Journal of Colorectal Disease**
ISSN: 0179-1958
Int. J. Colorectal Dis.
Springer Verlag, Tiergartenstrasse 17, D-69121, Heidelberg, Germany, Tel: +49(6221)4870, Fax: +49(6221)487366 ISSUES/YR: 6

International Journal of Cosmetic Science
ISSN: 0142-5463
Int. J. Cosmet. Sci.
Blackwell Publishing Ltd, 9600 Garsington Road, OX4 2XG, Oxford, United Kingdom, Tel: +44(1865)776868, Fax: +44(1865)714591 ISSUES/YR: 6

International Journal of Cosmetic Surgery and Aesthetic Dermatology
ISSN: 1530-8200
Int. J. Cosmet. Surg. Aesthet. Dermatol.
Mary Ann Liebert Inc., 140 Huguenot Street, New Rochelle, NY 10801-5215, United States, Tel: +1(914)7402100, Fax: +1(914)7402101 ISSUES/YR: 4

● **International Journal of Dairy Technology**
ISSN: 1364-727X
Int. J. Dairy Technol.
Blackwell Publishing Ltd, 9600 Garsington Road, OX4 2XG, Oxford, United Kingdom, Tel: +44(1865)776868, Fax: +44(1865)714591 ISSUES/YR: 4

International Journal of Dermatology
ISSN: 0011-9059
Int. J. Dermatol.
Blackwell Publishing Ltd, 9600 Garsington Road, OX4 2XG, Oxford, United Kingdom, Tel: +44(1865)776868, Fax: +44(1865)714591 ISSUES/YR: 12

● **International Journal of Developmental Biology**
ISSN: 0214-6282
Int. J. Dev. Biol.
University of the Basque Country Press, P.O. Box 1397, E-48080, Bilbao, Vizcaya, Spain ISSUES/YR: 10

● **International Journal of Developmental Neuroscience**
ISSN: 0736-5748
Int. J. Dev. Neurosci.
Elsevier Ltd, Langford Lane, Kidlington, OX5 1GB, Oxford, United Kingdom, Tel: +44(1865)843000, Fax: +44(1865)843010 ISSUES/YR: 8

International Journal of Diabetes and Metabolism
ISSN: 1606-7754
Int. J. Diabetes and Metabol.
United Arab Emirates University, P.O. Box 17666, Al Ain, United Arab Emirates, Tel: +971(3)672000 ISSUES/YR: 3

● **International Journal of Disaster Medicine**
ISSN: 1503-1438
Int. J. Disaster Med.
Taylor and Francis A.S., P.O. Box 2562 Solli, 0202, Oslo, Norway, Tel: +47(22)129880, Fax: +47(22)129890 ISSUES/YR: 4

● **International Journal of Drug Policy**
ISSN: 0955-3959
Int. J. Drug Policy
Elsevier, P.O. Box 211, 1000 AE, Amsterdam, Netherlands, Tel: +31(20)4853757, Fax: +31(20)4853432 ISSUES/YR: 6

● **International Journal of Eating Disorders**
ISSN: 0276-3478
Int. J. Eating Disord.
John Wiley and Sons Inc., P.O.Box 18667, Newark, NJ 07191-8667, United States, Tel: +1(212)8506000, Fax: +1(212)8506012 ISSUES/YR: 8

International Journal of Electronic Healthcare
ISSN: 1741-8453
Int. J. Electron. Healthc.
Inderscience Enterprises Ltd., P.O. Box 896, CH-1215, Geneve 15, Switzerland, Fax: +41(22)7910885 ISSUES/YR: 4

International Journal of Emergency Management
ISSN: 1471-4825
Int. J. Emerg. Manage.
Inderscience Enterprises Ltd., P.O. Box 896, CH-1215, Geneve 15, Switzerland, Fax: +41(22)7910885 ISSUES/YR: 4

International Journal of Environment and Pollution
ISSN: 0957-4352
Int. J. Environ. Pollut.
Inderscience Enterprises Ltd., P.O. Box 896, CH-1215, Geneve 15, Switzerland, Fax: +41(22)7910885 ISSUES/YR: 12

International Journal of Environmental Health Research
ISSN: 0960-3123
Int. J. Environ. Health Res.
Taylor and Francis Ltd., 4 Park Square, Milton Park, OX14 4RN, Abingdon, Oxfordshire, United Kingdom, Tel: +44(20)70175541, Fax: +44(20)70175198 ISSUES/YR: 6

International Journal of Environmental Research and Public Health
ISSN: 1660-4601
Int. J. Environ. Res. Public Health
Molecular Diversity Preservation International, Matthaeusstrasse 11, CH-4057, Basel, Switzerland, Tel: +41-79 322 3379, Fax: +41-61 302 8918 ISSUES/YR: 4

Current

- **International Journal of Epidemiology**
  ISSN: 0300-5771
  Int. J. Epidemiol.
  Oxford University Press, Great Clarendon Street, OX2 6DP, Oxford, United Kingdom, Tel: +44(1865)353907, Fax: +44(1865)353486 ISSUES/YR: 6
- **International Journal of Experimental Pathology**
  ISSN: 0959-9673
  Int. J. Exp. Pathol.
  Blackwell Publishing Ltd, 9600 Garsington Road, OX4 2XG, Oxford, United Kingdom, Tel: +44(1865)776868, Fax: +44(1865)714591 ISSUES/YR: 6
- **International Journal of Fertility and Women's Medicine**
  ISSN: 1069-3130
  Int. J. Fertil. Women's Med.
  Medical Science Publishing International Inc., 405 Main Street, Port Washington, NY 11050, United States, Tel: +1(516)9447340 ISSUES/YR: 6

International Journal of Food Microbiology
ISSN: 0168-1605
Int. J. Food Microbiol.
Elsevier, P.O. Box 211, 1000 AE, Amsterdam, Netherlands, Tel: +31(20)4853757, Fax: +31(20)4853432 ISSUES/YR: 21

International Journal of Food Sciences and Nutrition
ISSN: 0963-7486
Int. J. Food Sci. Nutr.
Taylor and Francis Ltd., 4 Park Square, Milton Park, OX14 4RN, Abingdon, Oxfordshire, United Kingdom, Tel: +44(20)70175541, Fax: +44(20)70175198 ISSUES/YR: 8

- **International Journal of Gastrointestinal Cancer**
  ISSN: 0169-4197
  Int. J. Gastrointest. Cancer
  Humana Press Inc., 999 Riverview Drive, Suite 208, Totowa, NJ 07512-1165, United States, Tel: +1(973)2561699, Fax: +1(973)2568341 ISSUES/YR: 6

International Journal of Geriatric Psychiatry
ISSN: 0885-6230
Int. J. Geriatr. Psychiatry
John Wiley and Sons Ltd, Southern Gate, PO19 8SQ, Chichester, West Sussex, United Kingdom, Tel: +44(1243)779777, Fax: +44(1243)775878 ISSUES/YR: 12

International Journal of Group Psychotherapy
ISSN: 0020-7284
Int. J. Group Psychother.
Guilford Publications, 72 Spring Street, New York, NY 10012, United States, Tel: +1(212)4319800 ISSUES/YR: 4

- **International Journal of Gynecological Cancer**
  ISSN: 1048-891X
  Int. J. Gynecol. Cancer
  Blackwell Publishing Inc., 350 Main Street, Malden, MA 02148, United States, Tel: +1(781)3888200, Fax: +1(781)3888210 ISSUES/YR: 6
- **International Journal of Gynecological Pathology**
  ISSN: 0277-1691
  Int. J. Gynecol. Pathol.
  Lippincott Williams and Wilkins, 530 Walnut Street, Philadelphia, PA 19106-3621, United States, Tel: +1(215)5218342, Fax: +1(215)5218326 ISSUES/YR: 4
- **International Journal of Gynecology and Obstetrics**
  ISSN: 0020-7292
  Int. J. Gynecol. Obstet.
  Elsevier Ireland Ltd, P.O. Box 85, Limerick, Ireland, Tel: +353(61)471944, Fax: +353(61)472144 ISSUES/YR: 12

International Journal of Health Geographics
ISSN: 1476-072X
Int. J. Health Geogr.
BioMed Central Ltd., 34 - 42 Cleveland Street, W1T 4LB, London, United Kingdom, Tel: +44(207)3230323, Fax: +44(207)6319926 ISSUES/YR: 1

International Journal of Health Planning and Management
ISSN: 0749-6753
Int. J. Health Plann. Manage.
John Wiley and Sons Ltd, Southern Gate, PO19 8SQ, Chichester, West Sussex, United Kingdom, Tel: +44(1243)779777, Fax: +44(1243)775878 ISSUES/YR: 4

International Journal of Health Promotion and Education
ISSN: 1463-5240
Int. J. Health Promot. Edu.
University Dental Hospital, Higher Cambridge Street, M15 6FH, Manchester, United Kingdom, Tel: +44(61)2756610, Fax: +44(61)2756776 ISSUES/YR: 4

International Journal of Health Services
ISSN: 0020-7314
Int. J. Health Serv.
Baywood Publishing Co. Inc., 26 Austin Avenue, P.O. Box 337, Amityville, NY 11701, United States, Tel: +1(516)6911270, Fax: +1(516)6911770 ISSUES/YR: 4

International Journal of Healthcare Technology and Management
ISSN: 1368-2156
Int. J. Healthc. Technol. Manage.
Inderscience Enterprises Ltd., P.O. Box 896, CH-1215, Geneve 15, Switzerland, Fax: +44(22)7910885 ISSUES/YR: 6

International Journal of Hematology
ISSN: 0925-5710
Int. J. Hematol.
Carden Jennings Publishing Co. Ltd, 375 Greenbrier Drive #100, Charlottesville, VA 22901-1618, United States, Tel: +1(434)8172000, Fax: +1(434)8172020 ISSUES/YR: 10

International Journal of Hygiene and Environmental Health
ISSN: 1438-4639
Int. J. Hyg. Environ. Health
Urban und Fischer Verlag Jena, P.O. Box 100537, 07705, Jena, Germany, Tel: +49(3641)6263, Fax: +49(3641)626500 ISSUES/YR: 6

International Journal of Hyperthermia
ISSN: 0265-6736
Int. J. Hyperthermia
Taylor and Francis Ltd., 4 Park Square, Milton Park, OX14 4RN, Abingdon, Oxfordshire, United Kingdom, Tel: +44(20)70175541, Fax: +44(20)70175198 ISSUES/YR: 8

- **International Journal of Immunogenetics**
  ISSN: 1744-3121
  Int. J. Immunogenet.
  Blackwell Publishing Ltd, 9600 Garsington Road, OX4 2XG, Oxford, United Kingdom, Tel: +44(1865)776868, Fax: +44(1865)714591 ISSUES/YR: 6
- **International Journal of Immunopathology and Pharmacology**
  ISSN: 0394-6320
  Int. J. Immunopathol. Pharmacol.
  Biomedical Research Press s.a.s., Via Lucania 10, 65016, Montesilvano (PE), Italy, Tel: +39(0871)355293 ISSUES/YR: 3
- **International Journal of Impotence Research**
  ISSN: 0955-9930
  Int. J. Impotence Res.
  Nature Publishing Group, Houndmills, RG21 6XS, Basingstoke, Hampshire, United Kingdom, Tel: +44(1256)329242, Fax: +44(1256)810526 ISSUES/YR: 6
- **International Journal of Industrial Ergonomics**
  ISSN: 0169-8141
  Int. J. Ind. Ergon.
  Elsevier, P.O. Box 211, 1000 AE, Amsterdam, Netherlands, Tel: +31(20)4853757, Fax: +31(20)4853432 ISSUES/YR: 12

International Journal of Infectious Diseases
ISSN: 1201-9712
Int. J. Infect. Dis.
Elsevier, P.O. Box 211, 1000 AE, Amsterdam, Netherlands, Tel: +31(20)4853757, Fax: +31(20)4853432 ISSUES/YR: 6

International Journal of Intensive Care
ISSN: 1350-2794
Int. J. Intensive Care
Greycoat Publishing, 120 Dawes Road, SW6 7EG, London, United Kingdom, Tel: +44(207)3851515, Fax: +44(207)3857799 ISSUES/YR: 4

International Journal of Language and Communication Disorders
ISSN: 1368-2822
Int. J. Lang. Commun. Disord.
Taylor and Francis Ltd., 4 Park Square, Milton Park, OX14 4RN, Abingdon, Oxfordshire, United Kingdom, Tel: +44(20)70175541, Fax: +44(20)70175198 ISSUES/YR: 6

International Journal of Law and Psychiatry
ISSN: 0160-2527
Int. J. Law Psychiatry
Elsevier Ltd, Langford Lane, Kidlington, OX5 1GB, Oxford, United Kingdom, Tel: +44(1865)843000, Fax: +44(1865)843010 ISSUES/YR: 6

• **International Journal of Legal Medicine**
ISSN: 0937-9827
Int. J. Leg. Med.
Springer Verlag, Tiergartenstrasse 17, D-69121, Heidelberg, Germany, Tel: +49(6221)4870, Fax: +49(6221)487366 ISSUES/YR: 6

International Journal of Leprosy and Other Mycobacterial Diseases
ISSN: 0148-916X
Int. J. Lepr. Other Mycobact. Dis.
Allen Press Inc., P.O. Box 7075, Lawrence, KS 66044-8897, United States, Tel: +1(785)8431234, Fax: +1(785)8431244 ISSUES/YR: 4

• **International Journal of Life Cycle Assessment**
ISSN: 0948-3349
Int. J. Life Cycle Assess.
Ecomed Publishers, Rudolf-Diesel-Strasse 3, D-86899, Landsberg, Germany, Tel: +49(8191)125802, Fax: +49(8191)125492 ISSUES/YR: 6

International Journal of Low Radiation
ISSN: 1477-6545
Int. J. Low Radiat.
Inderscience Enterprises Ltd., P.O. Box 896, CH-1215, Geneve 15, Switzerland, Fax: +41(22)7910885 ISSUES/YR: 4

• **International Journal of Lower Extremity Wounds**
ISSN: 1534-7346
Int. J. Lower Extremity Wounds
SAGE Publications Inc., 2455 Teller Road, Thousand Oaks, CA 91320, United States, Tel: +1(805)4990721, Fax: +1(805)4990871 ISSUES/YR: 4

• **International Journal of Medical Informatics**
ISSN: 1386-5056
Int. J. Med. Informatics
Elsevier Ireland Ltd, P.O. Box 85, Limerick, Ireland, Tel: +353(61)471944, Fax: +353(61)472144 ISSUES/YR: 12

International Journal of Medical Microbiology
ISSN: 1438-4221
Int. J. Med. Microbiol.
Urban und Fischer Verlag GmbH und Co. KG, P.O. Box 201930, 80019, Munchen, Germany, Tel: +49(89)53830, Fax: +49(89)5383221 ISSUES/YR: 8

International Journal of Medical Robotics and Computer Assisted Surgery
ISSN: 1478-5951
Int. J. Med. Rob. Comput. Assisted Surg.
John Wiley and Sons Ltd, Southern Gate, PO19 8SQ, Chichester, West Sussex, United Kingdom, Tel: +44(1243)779777, Fax: +44(1243)775878 ISSUES/YR: 4

International Journal of Medical Sciences
ISSN: 1449-1907
Int. J. Med. Sci.
Ivyspring International Publisher, P.O. Box 4546, NSW 2263, Lake Haven, Australia, Tel: +61(2)43905688, Fax: +61(2)43905660 ISSUES/YR: 3

International Journal of Medical Toxicology and Legal Medicine
ISSN: 0972-0448
Int. J. Med. Toxicol. Legal Med.
All India Institute of Medical Sciences, Ansari Nagar, 110 029, New Delhi, India, Tel: +91(11)6863002, Fax: +91(11)6862663 ISSUES/YR: 2

International Journal of Medicinal Mushrooms
ISSN: 1521-9437
Int. J. Med. Mushrooms
Begell House Inc., 145 Madison Avenue 6th Floor, New York, NY 10016, United States, Tel: +1(212)7251999, Fax: +1(212)2138368 ISSUES/YR: 4

International Journal of Medicine
ISSN: 1468-3814
Int. J. Med.
Kingscos Medical Publishers, P.O. Box 18265, EC2A 3TT, London, United Kingdom, Tel: +44(20)77497245, Fax: +44(20)77398683 ISSUES/YR: 4

International Journal of Medicine, Biology and the Environment
ISSN: 1128-935X
Int. J. Med. Biol. Environ.
Institute Nationale Ecology Cancerology (INEC), Piazza Botta 10, 27100, Pavia, Italy, Tel: +39(0382)506333 ISSUES/YR: 2

• **International Journal of Men's Health**
ISSN: 1532-6306
Int. J. Men's Health
Men's Studies Press, One Campus Road, Staten Island, NY 10301, United States, Tel: +1(423)3692375, Fax: +1(423)3691125 ISSUES/YR: 3

International Journal of Methods in Psychiatric Research
ISSN: 1049-8931
Int. J. Methods Psychiatr. Res.
John Wiley and Sons Ltd, Southern Gate, PO19 8SQ, Chichester, West Sussex, United Kingdom, Tel: +44(1243)779777, Fax: +44(1243)775878 ISSUES/YR: 4

International Journal of Molecular Sciences
ISSN: 1422-0067
Int. J. Mol. Sci.
Molecular Diversity Preservation International, Matthaeusstrasse 11, CH-4057, Basel, Switzerland, Tel: +41-79 322 3379, Fax: +41-61 302 8918 ISSUES/YR: 12

International Journal of Neuroprotection and Neurogeneration
ISSN: 1745-1183
Int. J. Neuroprot. Neurogener.
RPA Publishing Inc. Nottingham, UK, NG3 5DE, Nottingham, United Kingdom, Tel: +44(115)9692016, Fax: +44(115)9692017 ISSUES/YR: 3

• **International Journal of Neuropsychopharmacology**
ISSN: 1461-1457
Int. J. Neuropsychopharmacol.
Cambridge University Press, Shaftesbury Road, CB2 2RU, Cambridge, United Kingdom, Tel: +44(1223)312393, Fax: +44(1223)315052 ISSUES/YR: 4

International Journal of Neuroscience
ISSN: 0020-7454
Int. J. Neurosci.
Taylor and Francis Ltd., 4 Park Square, Milton Park, OX14 4RN, Abingdon, Oxfordshire, United Kingdom, Tel: +44(20)70175541, Fax: +44(20)70175198 ISSUES/YR: 12

International Journal of Nursing Education Scholarship
ISSN: 1548-923X
Int. J. Nurs. Educ. Scholarsh.
Berkeley Electronic Press, 2809 Telegraph Avenue, Suite 202, Berkeley, CA 94705, United States, Tel: +1(510)6651200, Fax: +1(510)6651201 ISSUES/YR: 1

• **International Journal of Obesity**
ISSN: 0307-0565
Int. J. Obes.
Nature Publishing Group, Houndmills, RG21 6XS, Basingstoke, Hampshire, United Kingdom, Tel: +44(1256)329242, Fax: +44(1256)810526 ISSUES/YR: 12

International Journal of Obstetric Anesthesia
ISSN: 0959-289X
Int. J. Obstet. Anesth.
Churchill Livingstone, 1-3 Baxter's Place, Leith Walk, EH1 3AF, Edinburgh, United Kingdom, Tel: +44(131)5562424, Fax: +44(131)5581278 ISSUES/YR: 4

- **International Journal of Occupational and Environmental Health**
  ISSN: 1077-3525
  int. J. Occup. Environ. Health
  Abel Publications Services Inc., 1611 Aquinas Court, Burlington, NC 27215, United States, Tel: +1(336)5851850, Fax: +1(336)5851851 ISSUES/YR: 4

International Journal of Occupational Medicine and Environmental Health
ISSN: 1232-1087
Int. J. Occup. Med. Environ. Health
Nofer Institute of Occupational Medicine, P.O. Box 199, 90-950, Lodz, Poland, Tel: +48(42)631-4502, Fax: +48(42)631-4719 ISSUES/YR: 4

International Journal of Offender Therapy and Comparative Criminology
ISSN: 0306-624X
Int. J. Offender Ther. Comp. Criminol.
SAGE Publications Inc., 2455 Teller Road, Thousand Oaks, CA 91320, United States, Tel: +1(805)4990721, Fax: +1(805)4990871 ISSUES/YR: 6

International Journal of Ophthalmology
ISSN: 1672-5123
Int. J. Ophthalmol.
International Journal of Ophthalmology (c/o Editorial Office), No. 269 Youyi East Road, 710054, Xi'an, China, Tel: +86-86 29 83085628, Fax: +86-86 29 82245172 ISSUES/YR: 6

International Journal of Oral and Maxillofacial Surgery
ISSN: 0901-5027
Int. J. Oral Maxillofac. Surg.
Churchill Livingstone, 1-3 Baxter's Place, Leith Walk, EH1 3AF, Edinburgh, United Kingdom, Tel: +44(131)5562424, Fax: +44(131)5581278 ISSUES/YR: 8

International Journal of Osteopathic Medicine
ISSN: 1746-0689
Int. J. Ostheopath. Med.
Elsevier Ltd, Langford Lane, Kidlington, OX5 1GB, Oxford, United Kingdom, Tel: +44(1865)843000 ISSUES/YR: 2

- **International Journal of Pediatric Otorhinolaryngology**
  ISSN: 0165-5876
  Int. J. Pediatr. Otorhinolaryngol.
  Elsevier Ireland Ltd, P.O. Box 85, Limerick, Ireland, Tel: +353(61)471944, Fax: +353(61)472144 ISSUES/YR: 12

International Journal of Peptide Research and Therapeutics
ISSN: 1573-3149
Int. J. Pept. Res. Ther.
Springer New York LLC, 233 Springer Street, New York, 10013-1578, United States, Tel: +1-212 460 1501, Fax: +1-212 460 1595 ISSUES/YR: 4

International Journal of Pharmaceutical Compounding
ISSN: 1092-4221
Int. J. Pharm. Compd.
International Journal of Pharmaceutical Compounding, 122 N. Bryant, Edmond, OK 73034, United States, Tel: +1(800)757-4572, Fax: +1(405)330-5622 ISSUES/YR: 6

- **International Journal of Pharmaceutical Medicine**
  ISSN: 1364-9027
  Int. J. Pharm. Med.
  ADIS International Ltd, Chowley Oak Lane, Tattenhall, CH3 9GA, Chester, United Kingdom, Tel: +64(1829)771155, Fax: +64(1829)770330 ISSUES/YR: 6
- **International Journal of Pharmaceutics**
  ISSN: 0378-5173
  Int. J. Pharm.
  Elsevier, P.O. Box 211, 1000 AE, Amsterdam, Netherlands, Tel: +31(20)4853757, Fax: +31(20)4853432 ISSUES/YR: 32

International Journal of Pharmacology
ISSN: 1811-7775
Int. J. Pharmacol.
Asian Network for Scientific Information, 308-Lasani Town, Sargodha Road, Faisalabad, Pakistan, Tel: +92(41)2001145-46, Fax: +92(21)5206036 ISSUES/YR: 4

- **International Journal of Pharmacy Practice**
  ISSN: 0961-7671
  Int. J. Pharm. Pract.
  Pharmaceutical Press, 1 Lambeth High Street, SE1 7JN, London, United Kingdom, Tel: +44(207)7359141, Fax: +44(207)7355085 ISSUES/YR: 4

International Journal of Prisoner Health
ISSN: 1744-9200
Int. J. Prison. Health
Taylor and Francis Ltd., 4 Park Square, Milton Park, OX14 4RN, Abingdon, Oxfordshire, United Kingdom, Tel: +44(20)70175541, Fax: +44(20)70175198 ISSUES/YR: 4

- **International Journal of Psychiatry in Clinical Practice**
  ISSN: 1365-1501
  Int. J. Psychiatry Clin. Pract.
  Taylor and Francis Ltd., 4 Park Square, Milton Park, OX14 4RN, Abingdon, Oxfordshire, United Kingdom, Tel: +44(20)70175541, Fax: +44(20)70175198 ISSUES/YR: 4

International Journal of Psychiatry in Medicine
ISSN: 0091-2174
Int. J. Psychiatry Med.
Baywood Publishing Co. Inc., 26 Austin Avenue, P.O. Box 337, Amityville, NY 11701, United States, Tel: +1(516)6911270, Fax: +1(516)6911770 ISSUES/YR: 4

International Journal of Psychoanalysis
ISSN: 0020-7578
Int. J. Psychoanal.
Institute of Psycho-Analysis, 112A Shirland Road, W9 2EQ, London, United Kingdom, Tel: +44(207) 563 5000, Fax: +44(207) 563 5001 ISSUES/YR: 6

International Journal of Psychophysiology
ISSN: 0167-8760
Int. J. Psychophysiol.
Elsevier, P.O. Box 211, 1000 AE, Amsterdam, Netherlands, Tel: +31(20)4853757, Fax: +31(20)4853432 ISSUES/YR: 12

- **International Journal of Psychotherapy**
  ISSN: 1356-9082
  Int. J. Psychother.
  European Association for Psychotherapy, Rosenbursenstrasse 8/3/7, A-1010, Wien, Austria, Tel: +43-1 512 5297, Fax: +43-1 512 2604 ISSUES/YR: 3
- **International Journal of Radiation Biology**
  ISSN: 0955-3002
  Int. J. Radiat. Biol.
  Taylor and Francis Ltd., 4 Park Square, Milton Park, OX14 4RN, Abingdon, Oxfordshire, United Kingdom, Tel: +44(20)70175541, Fax: +44(20)70175198 ISSUES/YR: 12
- **International Journal of Radiation Oncology Biology Physics**
  ISSN: 0360-3016
  Int. J. Radiat. Oncol. Biol. Phys.
  Elsevier Inc., 360 Park Avenue South, New York, NY 10010, United States, Tel: +1(212)6333950, Fax: +1(212)6333990 ISSUES/YR: 15
- **International Journal of Rehabilitation Research**
  ISSN: 0342-5282
  Int. J. Rehabil. Res.
  Lippincott Williams and Wilkins, 241 Borough High Street, SE1 1GB, London, United Kingdom, Tel: +44(207)9407500, Fax: +44(207)9407575 ISSUES/YR: 4

International Journal of Respiratory Care
ISSN: 1747-1273
Int. J. Respir. Care
Greycoat Publishing, 120 Dawes Road, SW6 7EG, London, United Kingdom, Tel: +44(207)3851515, Fax: +44(207)3857799 ISSUES/YR: 4

International Journal of Risk and Safety in Medicine
ISSN: 0924-6479
Int. J. Risk Saf. Med.
IOS Press, Nieuwe Hemweg 6B, 1013 BG, Amsterdam, Netherlands, Tel: +31(20)6883355, Fax: +31(20)6203419 ISSUES/YR: 4

Current

International Journal of Social Psychiatry
ISSN: 0020-7640
Int. J. Soc. Psychiatry
SAGE Publications Ltd, 55 City Road, EC1Y 1SP, London, United Kingdom, Tel: +44(207)3248500, Fax: +44(207)3248600 ISSUES/YR: 4

International Journal of Sport Nutrition and Exercise Metabolism
ISSN: 1526-484X
Int. J. Sport Nutr. Exer. Metabol.
Human Kinetics Publishers Inc., 1607 N. Market Street, Champaign, IL 61820-2200, United States, Tel: +1(217)3515076, Fax: +1(217)3511549 ISSUES/YR: 6

International Journal of Sports Medicine
ISSN: 0172-4622
Int. J. Sports Med.
Georg Thieme Verlag, Rudigerstrasse 14, D-70469, Stuttgart, Germany, Tel: +49(711)8931-321, Fax: +49(711)8931-422 ISSUES/YR: 10

International Journal of Sports Medicine, Supplement
ISSN: 0943-917X
Int. J. Sports Med. Suppl.
Georg Thieme Verlag, Rudigerstrasse 14, D-70469, Stuttgart, Germany, Tel: +49(711)8931-321, Fax: +49(711)8931-422 ISSUES/YR: 1

• **International Journal of STD and AIDS**
ISSN: 0956-4624
Int. J. STD AIDS
Royal Society of Medicine Press Ltd, P.O. Box 9002, W1A 0ZA, London, United Kingdom, Tel: +44(207)2902927, Fax: +44(207)2902929 ISSUES/YR: 12

• **International Journal of Stroke**
ISSN: 1747-4930
Int. J. Stroke
Blackwell Publishing Ltd, 9600 Garsington Road, OX4 2XG, Oxford, United Kingdom, Tel: +44(1865)776868, Fax: +44(1865)714591 ISSUES/YR: 4

• **International Journal of Surgery**
ISSN: 1743-9191
Int. J. Surg.
Elsevier Ltd, Langford Lane, Kidlington, OX5 1GB, Oxford, United Kingdom, Tel: +44(1865)843000, Fax: +44(1865)843010 ISSUES/YR: 4

• **International Journal of Surgical Pathology**
ISSN: 1066-8969
Int. J. Surg. Pathol.
Westminster Publications Inc., 708 Glen Cove Avenue, Glen Head, NY 11545, United States, Tel: +1(516)7590025, Fax: +1(516)7595524 ISSUES/YR: 4

• **International Journal of Systematic and Evolutionary Microbiology**
ISSN: 1466-5026
Int. J. Syst. Evol. Microbiol.
Society for General Microbiology, Basingstoke Road, Spencers Wood, RG7 1AE, Reading, Berkshire, United Kingdom, Tel: +44(118)9885577, Fax: +44(118)9885656 ISSUES/YR: 6

International Journal of Technology Assessment in Health Care
ISSN: 0266-4623
Int. J. Technol. Assess. Health Care
Cambridge University Press, 40 West 20th Street, New York, NY 10011-4211, United States, Tel: +1(212)9243900, Fax: +1(212)6913239 ISSUES/YR: 4

International Journal of Tissue Reactions
ISSN: 0250-0868
Int. J. Tissue React.
Bioscience Ediprint Inc., 16 Rue Alexandre-Gavard, 1227, Carouge-Geneva, Switzerland, Tel: +41(22)3003383, Fax: +41(22)3002489 ISSUES/YR: 4

International Journal of Toxicology
ISSN: 1091-5818
Int. J. Toxicol.
Taylor and Francis Ltd., 4 Park Square, Milton Park, OX14 4RN, Abingdon, Oxfordshire, United Kingdom, Tel: +44(20)70175541, Fax: +44(20)70175198 ISSUES/YR: 6

International Journal of Transgenderism
ISSN: 1553-2739
Int. J. Transgenderism
Haworth Press Inc., 10 Alice Street, Binghamton, NY 13904-1580, United States, Tel: +1(607)7225857, Fax: +1(607)7710012 ISSUES/YR: 4

• **International Journal of Tuberculosis and Lung Disease**
ISSN: 1027-3719
Int. J. Tuberc. Lung Dis.
International Union against Tubercul. and Lung Dis., 68 boulevard Saint-Michel, 75006, Paris, France, Tel: +33(1)44320360, Fax: +33(1)43299087 ISSUES/YR: 12

• **International Journal of Urology**
ISSN: 0919-8172
Int. J. Urol.
Blackwell Publishing, 550 Swanston Street, VIC 3053, Carlton South, Australia, Tel: +61(3)8359 1011, Fax: +61(3)8359 1120 ISSUES/YR: 12

International Journal of Virology
ISSN: 1816-4900
Int. J. Virol.
Asian Network for Scientific Information, 308-Lasani Town, Sargodha Road, Faisalabad, Pakistan, Tel: +92(41)2001145-46, Fax: +92(21)5206036 ISSUES/YR: 2

International Medical Journal
ISSN: 1341-2051
Int. Med. J.
Japan International Cultural Exchange Foundation, 2-11-32 Mishuku, Setagaya-ku, 154, Tokyo, Japan, Tel: +81(3)34249090 ISSUES/YR: 4

International MS Journal
ISSN: 1352-8963
Int. MS J.
Cambridge Medical Publications, High Street, BN11 1DJ, Worthing, West Sussex, United Kingdom, Tel: +44(1903)205884, Fax: +44(1903)234862 ISSUES/YR: 3

International Ophthalmology
ISSN: 0165-5701
Int. Ophthalmol.
Kluwer Academic Publishers, Van Godewijckstraat 30, 3311 GZ, Dordrecht, Netherlands, Tel: +31(78)6576000, Fax: +31(78)6576254 ISSUES/YR: 6

• **International Ophthalmology Clinics**
ISSN: 0020-8167
Int. Ophthalmol. Clin.
Lippincott Williams and Wilkins, 530 Walnut Street, Philadelphia, PA 19106-3621, United States, Tel: +1(215)5218342, Fax: +1(215)5218326 ISSUES/YR: 4

• **International Orthopaedics**
ISSN: 0341-2695
Int. Orthop.
Springer Verlag, Tiergartenstrasse 17, D-69121, Heidelberg, Germany, Tel: +49(6221)4870, Fax: +49(6221)487366 ISSUES/YR: 6

International Pediatrics
ISSN: 0885-6265
Int. Pediatr.
Miami Children's Hospital Medical Journal Inc., 3100 Southwest 62nd Avenue, Miami, FL 33155, United States, Tel: +1(305)6636823 ISSUES/YR: 4

International Psychogeriatrics
ISSN: 1041-6102
Int. Psychogeriatr.
Cambridge University Press, Shaftesbury Road, CB2 2RU, Cambridge, United Kingdom, Tel: +1(1223)312393, Fax: +1(1223)315052 ISSUES/YR: 4

International Review of Allergology and Clinical Immunology
ISSN: 1232-9142
Int. Rev. Allergol. Clin. Immunol.
Medpress, ul. Slonki 67, 02-814, Warsaw, Poland, Tel: +48(22)6444820, Fax: +48(22)6444820 ISSUES/YR: 4

• **International Review of Cytology**
ISSN: 0074-7696
Int. Rev. Cytol.
Academic Press Inc., 525 B Street, Suite 1900, San Diego, CA 92101-4495, United States ISSUES/YR: 1

- **International Review of Psychiatry**
  ISSN: 0954-0261
  Int. Rev. Psychiatry
  Routledge, 4 Park Square, Milton Park, OX14 4RN, Abingdon, Oxfordshire, United Kingdom, Tel: +44(1235)828600, Fax: +44(1235)829000 ISSUES/YR: 4
- **International Reviews of Immunology**
  ISSN: 0883-0185
  Int. Rev. Immunol.
  Taylor and Francis Ltd., 4 Park Square, Milton Park, OX14 4RN, Abingdon, Oxfordshire, United Kingdom, Tel: +44(20)70175541, Fax: +44(20)70175198 ISSUES/YR: 6

International Seminars in Surgical Oncology
ISSN: 1477-7800
Int. Semin. Surg. Oncol.
BioMed Central Ltd., 34 - 42 Cleveland Street, W1T 4LB, London, United Kingdom, Tel: +44(207)3230323, Fax: +44(207)6319926 ISSUES/YR: 1

- **International Tinnitus Journal**
  ISSN: 0946-5448
  Int. Tinnitus J.
  Tinnitus Center, Box 1239, 450 Clarkson Avenue, Brooklyn, NY 11203, United States, Tel: +1(718)7738888, Fax: +1(718)4653669 ISSUES/YR: 2
- **International Urogynecology Journal and Pelvic Floor Dysfunction**
  ISSN: 0937-3462
  Int. Urogynecol. J. Pelvic Floor Dysfunct.
  Springer-Verlag London Ltd, Catteshall Road, Farncombe, GU7 3DJ, Godalming, Surrey, United Kingdom, Tel: +44(1483)414142, Fax: +44(1483)415144 ISSUES/YR: 6

International Urology and Nephrology
ISSN: 0301-1623
Int. Urol. Nephrol.
Kluwer Academic Publishers, Van Godewijckstraat 30, 3311 GZ, Dordrecht, Netherlands, Tel: +31(78)6576000, Fax: +31(78)6576254 ISSUES/YR: 4

- **International Wound Journal**
  ISSN: 1742-4801
  Int. Wound J.
  Blackwell Publishing Ltd, 9600 Garsington Road, OX4 2XG, Oxford, United Kingdom, Tel: +44(1865)776868, Fax: +44(1865)714591 ISSUES/YR: 4

Internist
ISSN: 0020-9554
Internist
Springer Verlag, Tiergartenstrasse 17, D-69121, Heidelberg, Germany, Tel: +49(6221)4870, Fax: +49(6221)487366 ISSUES/YR: 12

Internista
ISSN: 1121-9017
Internista
Mattioli 1885 S.p.A., Via Coduro 1/b, 43036, Fidenza (PR), Italy, Tel: +39(524)84547, Fax: +39(524)84751 ISSUES/YR: 4

Internistische Praxis
ISSN: 0020-9570
Internist. Prax.
Hans Marseille Verlag GmbH, P.O. Box 22 13 41, 80503, Munchen, Germany, Tel: +49(89)227988, Fax: +49(89)2904643 ISSUES/YR: 4

Interventional Neuroradiology
ISSN: 1123-9344
Intervent. Neuroradiol.
Edizioni del Centauro, Via del Pratello, 8, 40122, Bologna, Italy, Tel: +39(0)51227634, Fax: +39(0)51220099 ISSUES/YR: 4

- **Intervirology**
  ISSN: 0300-5526
  Intervirology
  S. Karger AG, Allschwilerstrasse 10, P.O. Box, CH-4009, Basel, Switzerland, Tel: +41(61)3061111, Fax: +41(61)3061234 ISSUES/YR: 6
- **Invertebrate Neuroscience**
  ISSN: 1354-2516
  Invertebr. Neurosci.
  Springer Verlag, Tiergartenstrasse 17, D-69121, Heidelberg, Germany, Tel: +49(6221)4870, Fax: +49(6221)487366 ISSUES/YR: 4

Investigacion Cardiovascular
ISSN: 1139-2096
Invest. Cardiovasc.
Fundacion Mapfre Medicina, Ctra. Pozuelo a Majadahonda, s/n, 28220, Majadahonda (Madrid), Spain, Tel: +34(91)6265704, Fax: +34(91)6265825 ISSUES/YR: 2

Investigacion Clinica
ISSN: 0535-5133
Invest. Clin.
Instituto de Investigaciones Clinicas, P.O. Box 1151, 4001-A, Maracaibo, Zulia, Venezuela, Tel: +58(61)916053, Fax: +58(61)412145 ISSUES/YR: 4

Investigacion Medica Internacional
ISSN: 0185-2108
Invest. Med. Int.
Mundo Medico S.A. de C.V., Ejercito Nacional 381, Col. Granada, 11520, Mexico D.F., Mexico, Tel: +52(55)5-5451798, Fax: +52(55)5-2036418 ISSUES/YR: 2

- **Investigational New Drugs**
  ISSN: 0167-6997
  Invest. New Drugs
  Kluwer Academic Publishers, 101 Philip Drive, Assinippi Park, Norwell, MA 02061, United States, Tel: +1(781)8716600, Fax: +1(781)8716528 ISSUES/YR: 4
- **Investigative Ophthalmology and Visual Science**
  ISSN: 0146-0404
  Invest. Ophthalmol. Vis. Sci.
  Association for Research in Vision and Opthalmology Inc., 9650 Rockville Pike - Suite 1610, Bethesda, MD 20814-3998, United States, Tel: +1(301)5711844 ISSUES/YR: 12
- **Investigative Radiology**
  ISSN: 0020-9996
  Invest. Radiol.
  Lippincott Williams and Wilkins, 530 Walnut Street, Philadelphia, PA 19106-3621, United States, Tel: +1(215)5218342, Fax: +1(215)5218326 ISSUES/YR: 12

IPCS Concise International Chemical Assessment Documents
ISSN: 1020-6167
IPCS Concise Int. Chem. Assess. Doc.
World Health Organization, 20 Ave. Appia, CH-1211, Geneva 27, Switzerland, Tel: +41(22)7912111, Fax: +41(22)7914857 ISSUES/YR: 1

Iranian Biomedical Journal
ISSN: 1028-852X
Iran. Biomed. J.
Pasteur Institute of Iran, Pasteur Ave., 13164, Tehran, Iran, Tel: +98(21)6492596, Fax: +98(21)6492596 ISSUES/YR: 4

Iranian Journal of Allergy, Asthma and Immunology
ISSN: 1735-1502
Iran. J. Allergy Asthma Immunol.
Iranian Journal of Allergy, Asthma and Immunology, P.O. Box 13145-955, 14194, Teheran, Iran, Tel: +98(21)66938545, Fax: +98(21)66938545 ISSUES/YR: 4

Iranian Journal of Diabetes and Lipid Disorders
ISSN: 1726-7544
Iran. J. Diabetes Lipid Disord.
Tehran Universtiy of Medical Science, North Kargar Avenue, 14114, Tehran, Iran, Tel: +98-+9821802 69 02 03, Fax: +98-+98 821 802 93 99 ISSUES/YR: 2

Iranian Journal of Immunology
ISSN: 1735-365X
Iran. J. Immunol.
Iranian Journal of Allergy, Asthma and Immunology, P.O. Box 13145-955, 14194, Teheran, Iran, Tel: +98(21)66938545, Fax: +98(21)66938545 ISSUES/YR: 4

Iranian Journal of Medical Sciences
ISSN: 0253-0716
Iran. J. Med. Sci.
Shiraz University of Medical Sciences, Nemazee Hospital, 71934, Shiraz, Iran, Tel: +98-7116260134, Fax: +98-7116263052 ISSUES/YR: 4

Iranian Journal of Nuclear Medicine
ISSN: 1681-2824
Iran. J. Nucl. Med.
Tehran University of Medical Sciences, P.O. Box 14155-6451, Tehran, Iran, Fax: +98(21)6461178 ISSUES/YR: 2

Iranian Journal of Pharmacology and Therapeutics
ISSN: 1735-2657
Iran. J. Pharmacol. Ther.
Academy of Medical Sciences of I.R. Iran, P.O. Box 19395.4655, Pasdaran. Av., Tehran, Iran, Tel: +98-212935817, Fax: +98-212932125 ISSUES/YR: 2

Iranian Journal of Public Health
ISSN: 0304-4556
Iran. J. Public Health
Iranian Journal of Public Health, P.O. Box 6446, 14155, Tehran, Iran, Tel: +98-8951402, Fax: +98-8951397 ISSUES/YR: 4

Iranian Journal of Radiation Research
ISSN: 1728-4554
Iran. J. Radiat. Res.
Novim Medical Radiation Institute, P.O.Box: 14665/599, Tehran, Iran, Tel: +98(21)8086783, Fax: +98(21)8086782 ISSUES/YR: 4

Irish Journal of Medical Science
ISSN: 0021-1265
Ir. J. Med. Sci.
Royal Academy of Medicine in Ireland, 6 Kildare Street, Dublin 2, Ireland, Tel: +353(1)6767650, Fax: +353(1)6611684 ISSUES/YR: 4

Irish Journal of Psychological Medicine
ISSN: 0790-9667
Ir. J. Psychol. Med.
MedMedia Ltd, 25 Adelaide Street, Dun Laoghaire, Co Dublin, Ireland, Tel: +353(1)2803967, Fax: +353(1)2807076 ISSUES/YR: 4

Irish Medical Journal
ISSN: 0332-3102
Ir. Med. J.
Irish Medical Association, 10 Fitzwilliam Place, Dublin 2, Ireland, Tel: +353(1)6767273, Fax: +353(1)6622818 ISSUES/YR: 10

IRYO - Japanese Journal of National Medical Services
ISSN: 0021-1699
IRYO Jpn. J. Natl. Med. Serv.
Japanese Society of National Medical Services, 1-2-2 Kasumigaseki, Chiyoda-ku, 100-45, Tokyo, Japan, Tel: +81(3)35031711 EXT.2632 ISSUES/YR: 12

Isokinetics and Exercise Science
ISSN: 0959-3020
Isokinetics Exerc. Sci.
IOS Press, Nieuwe Hemweg 6B, 1013 BG, Amsterdam, Netherlands, Tel: +31(20)6883355, Fax: +31(20)6203419 ISSUES/YR: 4

Israel Medical Association Journal
ISSN: 1565-1088
Isr. Med. Assoc. J.
Israel Medical Association, 2 Twin Towers,11th Floor,35 Jabotinsky Street,PO Box 3566, 52135, Ramat Gan, Israel, Tel: +972-3 610 0475, Fax: +972-3 575 1616 ISSUES/YR: 12

Italian Heart Journal
ISSN: 1129-471X
Ital. Heart J.
CEPI s.r.l., Via A. Ristori 38, 00197, Rome, Italy, Tel: +39-+39.06809681, Fax: +39-+39.068072458 ISSUES/YR: 12

Italian Heart Journal Supplement
ISSN: 1129-4728
Ital. Heart J. Suppl.
CEPI s.r.l., Via A. Ristori 38, 00197, Rome, Italy, Tel: +39-+39.06809681, Fax: +39-+39.068072458 ISSUES/YR: 1

Italian Heart Journal, Supplement
ISSN: 1590-3796
Ital. Heart J. Suppl.
CEPI s.r.l., Via A. Ristori 38, 00197, Rome, Italy, Tel: +39-+39.06809681, Fax: +39-+39.068072458 ISSUES/YR: 1

Italian Journal of Allergy and Clinical Immunology
ISSN: 1120-6373
Ital. J. Allergy Clin. Immunol.
Italian Society of Allergy and Clinical Immunology, GRANELLI-Via F. Sforza, 35, 20122, Milano, Italy, Tel: +39(02)55192920, Fax: +39(02)55194671 ISSUES/YR: 4

Italian Journal of Gynaecology and Obstetrics
ISSN: 1121-8339
Ital. J. Gynaecol. Obstet.
S.I.G.O. Publishing Co, Via dei Soldati 25, 00186, Rome, Italy ISSUES/YR: 4

Italian Journal of Pediatrics
ISSN: 1720-8424
Ital. J. Pediat.
Pacini Editore s.r.l., Via A. Gherardesca 1, 56121, Ospedaletto (Pisa), Italy, Tel: +39(050)313011, Fax: +39(050)3130300 ISSUES/YR: 6

Italian Journal of Psychiatry and Behavioural Sciences
ISSN: 1122-2247
Ital. J. Psychiatry Behav. Sci.
IDELSON, Via A. de Gaspari 55, 80133, Napels, Italy, Tel: +39(081)5524733, Fax: +39(081)5518295 ISSUES/YR: 3

Italian Journal of Psychopathology
ISSN: 1592-1107
Ital. J. Psychopatol.
Pacini Editore s.r.l., Via A. Gherardesca 1, 56121, Ospedaletto (Pisa), Italy, Tel: +39(050)313011, Fax: +39(050)3130300 ISSUES/YR: 4

Italian Journal of Vascular and Endovascular Surgery
ISSN: 1824-4777
Ital. J. Vasc. Endovasc. Surg.
Edizioni Minerva Medica S.p.A., Corso Bramante 83-85, 10126, Turin, Italy, Tel: +39(011)678282, Fax: +39(011)674502 ISSUES/YR: 4

ITBM-RBM
ISSN: 1297-9562
ITBM-RBM
Editions Scientifiques et Medicales Elsevier, 23 rue Linois, 75724, Paris Cedex 15, France, Tel: +33(1)45589826, Fax: +33(1)45589428 ISSUES/YR: 6

ITBM-RBM News
ISSN: 1297-9570
ITBM-RBM News
Editions Scientifiques et Medicales Elsevier, 23 rue Linois, 75724, Paris Cedex 15, France, Tel: +33(1)45589826, Fax: +33(1)45589428 ISSUES/YR: 6

IUBMB Life
ISSN: 1521-6543
IUBMB Life
Taylor and Francis Ltd., 4 Park Square, Milton Park, OX14 4RN, Abingdon, Oxfordshire, United Kingdom, Tel: +44(20)70175541, Fax: +44(20)70175198 ISSUES/YR: 12

JACR Journal of the American College of Radiology
ISSN: 1546-1440
J. Am. Coll. Radiol.
Elsevier, P.O. Box 211, 1000 AE, Amsterdam, Netherlands, Tel: +31(20)4853757, Fax: +31(20)4853432 ISSUES/YR: 12

JALA - Journal of the Association for Laboratory Automation
ISSN: 1535-5535
JALA J. Assoc. Lab. Autom.
Elsevier, P.O. Box 211, 1000 AE, Amsterdam, Netherlands, Tel: +1(20)4853757, Fax: +1(20)4853432 ISSUES/YR: 6

Japan Journal of Logopedics and Phoniatrics
ISSN: 0030-2813
Jpn. J. Logop. Phoniatr.
Japan Society of Logopedics and Phoniatrics, 2-3-10 Kohraku, Bunkyo-ku, 112, Tokyo, Japan, Fax: +81(3)56845954 ISSUES/YR: 4

Japan Medical Association Journal
ISSN: 1346-8650
Japan Med. Assoc. J.
Japan Medical Association, 2-28-16 Hongkomagome, Bunkyo-Ku, 113-8621, tokyo, Japan, Tel: +81(3)3946-2121, Fax: +81(3)3946-6295 ISSUES/YR: 12

Current

- **Japanese Heart Journal**
ISSN: 0021-4868
Jpn. Heart J.
International Heart Journal Association, 7-3-1 Hongo, Bunkyo-ku, 113-8655, Tokyo, Japan, Tel: +81(3)58006797, Fax: +81(3)38152087 ISSUES/YR: 6

Japanese Journal of Aerospace and Environmental Medicine
ISSN: 0387-0723
Jpn. J. Aerosp. Environ. Med.
Japan Society of Aerospace and Environmental Medicine, 3-25-28 Nishi-shimbashi, Minato-ku, 105, Tokyo, Japan, Tel: +81(3)3433-1111, Fax: +81(3)3431-2666 ISSUES/YR: 4

Japanese Journal of Allergology
ISSN: 0021-4884
Jpn. J. Allergol.
Japanese Society of Allergology, 1-35-26 Hongo, Bunkyo-ku, 113, Tokyo, Japan, Tel: +81(3)38160280, Fax: +81(3)38160219 ISSUES/YR: 12

Japanese Journal of Anesthesiology
ISSN: 0021-4892
Jpn. J. Anesthesiol.
Kokuseido Publishing Co. Ltd, 23-5-202 Hongo, 3-chome, Bunkyo-ku, 113, Tokyo, Japan, Tel: +81(3)38110995 ISSUES/YR: 12

Japanese Journal of Antibiotics
ISSN: 0368-2781
Jpn. J. Antibiot.
Japan Antibiotics Research Association, 2-20-8 Kamiosaki, Shinagawa-ku, 141, Tokyo, Japan, Tel: +81(3)34910181 ISSUES/YR: 6

Japanese Journal of Chemotherapy
ISSN: 1340-7007
Jpn. J. Chemother.
Japan Society of Chemotherapy, 20-8 Kami-osaki 2-chome, 141-0021, Shinagawa-ku, Tokyo, Japan, Tel: +81(3)34937129, Fax: +81(3)54340843 ISSUES/YR: 12

Japanese Journal of Chest Diseases
ISSN: 0385-3667
Jpn. J. Chest Dis.
Kokuseido Publishing Co. Ltd, 23-5-202 Hongo, 3-chome, Bunkyo-ku, 113, Tokyo, Japan, Tel: +81(3)38110995 ISSUES/YR: 12

Japanese Journal of Clinical Chemistry
ISSN: 0370-5633
Jpn. J. Clin. Chem.
Japanese Journal of Clinical Chemistry, 1-4-2 Shinsenrihigashi-machi, 565, Toyonaka, Japan, Tel: +81(6)8732301, Fax: +81(6)8732300 ISSUES/YR: 4

Japanese Journal of Clinical Oncology
ISSN: 0368-2811
Jpn. J. Clin. Oncol.
Oxford University Press, Great Clarendon Street, OX2 6DP, Oxford, United Kingdom, Tel: +44(1865)353907, Fax: +44(1865)353486 ISSUES/YR: 12

Japanese Journal of Clinical Ophthalmology
ISSN: 0370-5579
Jpn. J. Clin. Ophthalmol.
Igaku-Shoin Ltd, 5-24-3 Hongo, Bunkyo-ku, 113-91, Tokyo, Japan, Tel: +81(3)38175600 ISSUES/YR: 13

Japanese Journal of Clinical Pharmacology and Therapeutics
ISSN: 0388-1601
Jpn. J. Clin. Pharmacol. Ther.
Japanese Society of Clinical Pharmacology and Therapeutics, 2-3-10 Kanda, Surugadai, Chuo-ku, 101, Tokyo, Japan, Tel: +81(3)32947311 ISSUES/YR: 6

Japanese Journal of Clinical Radiology
ISSN: 0009-9252
Jpn. J. Clin. Radiol.
Kanehara Shuppan Co. Ltd, P.O. Box 1, Hongo Post Office, 113-91, Tokyo, Japan, Tel: +81(3)38117163 ISSUES/YR: 13

Japanese Journal of Clinical Urology
ISSN: 0385-2393
Jpn. J. Clin. Urol.
Igaku-Shoin Ltd, 5-24-3 Hongo, Bunkyo-ku, 113-91, Tokyo, Japan, Tel: +81(3)38175600 ISSUES/YR: 13

Japanese Journal of Fertility and Sterility
ISSN: 0029-0629
Jpn. J. Fertil. Steril.
Nihon Funin Gakukai, 5-4 Kojimachi, 102, Tokyo, Japan, Tel: +81(3)32887266 ISSUES/YR: 4

Japanese Journal of Forensic Toxicology
ISSN: 0915-9606
Jpn. J. Forensic Toxicol.
Japanese Association of Forensic Toxicology, 1-5-8 Hatanodai, Shinagawa-ku, 142, Tokyo, Japan, Tel: +81(3)37848205 ISSUES/YR: 3

Japanese Journal of Gastroenterological Surgery
ISSN: 0386-9768
Jpn. J. Gastroenterol. Surg.
Japanese Society of Gastroenterological Surgery, 2-2-4 Kudan Minami, Chiyoda-ku, 102, Tokyo, Japan ISSUES/YR: 12

Japanese Journal of Gastroenterology
ISSN: 0446-6586
Jpn. J. Gastroent.
Japanese Society of Gastroenterology, 8-9-13 Ginza, Chuo-ku, 104, Tokyo, Japan, Tel: +81(3)35734297 ISSUES/YR: 12

Japanese Journal of Infectious Diseases
ISSN: 1344-6304
Jpn. J. Infect. Dis.
National Institute of Health, Toyama 1-23-1, Shinjuku-ku, 162-8640, Tokyo, Japan, Tel: +81(3)52851111, Fax: +81(3)52851192 ISSUES/YR: 6

Japanese Journal of Interventional Cardiology
ISSN: 0914-8922
Jpn. J. Intervent. Cardiol.
Japanese Society of Interventional Cardiology, 1-1, Kifune-machi, Kokurakita-ku, Kitakyushu-shi, 802, Tokyo, Japan, Tel: +81(3)38167796 ISSUES/YR: 6

Japanese Journal of Legal Medicine
ISSN: 0047-1887
Jpn. J. Leg. Med.
Medico-Legal Society of Japan, 3-1 Hongo 7-chome, Bunkyo-ku, 113, Tokyo, Japan, Tel: +81(3)58005416 ISSUES/YR: 2

Japanese Journal of Leprosy
ISSN: 1342-3681
Jpn. J. Lepr.
Japanese Leprosy Association, 2-1 Aobacho 4-chome, Higashimurayama-shi, 189-0002, Tokyo, Japan, Tel: +81(42)391-8211, Fax: +81(42)394-9092 ISSUES/YR: 3

Japanese Journal of Lung Cancer
ISSN: 0386-9628
Jpn. J. Lung Cancer
Japan Lung Cancer Society, 4-9-8-401 Chuo Chuo-ku Chibasi, 260-0013, Chiba, Japan, Tel: +81(43)2223110, Fax: +81(43)2220655 ISSUES/YR: 7

Japanese Journal of Medical Electronics and Biological Engineering
ISSN: 0021-3292
Jpn. J. Med. Electron. Biol. Eng.
Japan Soc. of Med. Electronics and Biol. Engineering, 16-9, Honkomagome 5-Chome, Bunkyo-ku, 113, Tokyo, Japan, Tel: +81(3)58145811 ISSUES/YR: 4

Japanese Journal of Medical Mycology
ISSN: 0916-4804
Jpn. J. Med. Mycol.
Japanese Society for Medical Mycology, 1st Genesis Bldg 5F, 2-31-14, Yushima,, 113-0034, Bunkyo-ku, Tokyo, Japan, Tel: +81(3)58051901, Fax: +81(3)58051092 ISSUES/YR: 4

Japanese Journal of Nephrology
ISSN: 0385-2385
Jpn. J. Nephrol.
Japanese Society of Nephrology, 2-8-8 Kyobashi Chuo-ku, 104-0031, Tokyo, Japan, Tel: +81(3)5159 1051, Fax: +81(3)3535 8253 ISSUES/YR: 8

Japanese Journal of Neuropsychopharmacology
ISSN: 1340-2544
Jpn. J. Neuropsychopharmacol.
Japanese Society of Neuropsychopharmacology, 1433 Nogawa, Miyamae-ku, 216, Kawasaki, Japan, Tel: +81(44)7544422 ISSUES/YR: 6

Current

Japanese Journal of Neurosurgery
ISSN: 0917-950X
Jpn. J. Neurosurg.
Japanese Congress of Neurological Surgeons, 2-1-1 Hongo, Bunkyo-ku, 113-0033, Tokyo, Japan, Tel: +81(3)38133111 ISSUES/YR: 12

• **Japanese Journal of Ophthalmology**
ISSN: 0021-5155
Jpn. J. Ophthalmol.
Springer-Verlag Tokyo, 3-13, Hongo 3-chome, Bunkyo-ku, 113-8480, Tokyo, Japan, Tel: +81(3)38180861, Fax: +81(3)38180864 ISSUES/YR: 6

Japanese Journal of Physiology
ISSN: 0021-521X
Jpn. J. Physiol.
The Physiological Society of Japan, 3-30-10 Bunkyo-Ku, 113-0033, Tokyo, Japan, Tel: +81(3)58145811, Fax: +81(3)58145822 ISSUES/YR: 6

Japanese Journal of Plastic and Reconstructive Surgery
ISSN: 0021-5228
Jpn. J. Plast. Reconstr. Surg.
Kokuseido Publishing Co. Ltd, 23-5-202 Hongo, 3-chome, Bunkyo-ku, 113, Tokyo, Japan, Tel: +81(3)38110995 ISSUES/YR: 12

Japanese Journal of Thoracic and Cardiovascular Surgery
ISSN: 1344-4964
Jpn. J. Thorac. Cardiovasc. Surg.
Japanese Association for Thoracic Surgery, 2-3-10 Koraku, Bunkyo-ku, 112, Tokyo, Japan, Tel: +81(3)38124253, Fax: +81(3)38164560 ISSUES/YR: 12

Japanese Journal of Urology
ISSN: 0021-5287
Jpn. J. Urol.
Japanese Urological Association, 14-10 Hongo 3-chome, Bunkyo-ku, 113, Tokyo, Japan, Tel: +81(3)38147921 ISSUES/YR: 7

Japanese Journal of Veterinary Research
ISSN: 0047-1917
Jpn. J. Vet. Res.
Hokkaido University, North 18-jo, West 060, 9-chome, Kita-ku, 060, Sapporo City, Hokkaido, Japan, Tel: +81(11)7162111 ISSUES/YR: 4

Japanese Pharmacology and Therapeutics
ISSN: 0386-3603
Jpn. Pharmacol. Ther.
Life Science Publishing Co. Ltd, 11-7 Nihonbashi, Kobunacho, Chuo-ku, 103, Tokyo, Japan, Tel: +81(3)36647750, Fax: +81(3)36647734 ISSUES/YR: 12

• **JARO - Journal of the Association for Research in Otolaryngology**
ISSN: 1525-3961
JARO J. Assoc. Res. Otolaryngol.
Springer New York LLC, 233 Springer Street, New York, 10013-1578, United States, Tel: +1-212 460 1501, Fax: +1-212 460 1595 ISSUES/YR: 4

JAVA - Journal of the Association for Vascular Access
ISSN: 1552-8855
JAVA J. Assoc. Vasc. Access
Association for Vascular Access, 11441 South State Street, Suite A, #113, Draper, UT 84020, United States, Tel: +1(801)5761824, Fax: +1(801)5539137 ISSUES/YR: 4

JDDG - Journal of the German Society of Dermatology
ISSN: 1610-0379
JDDG J. German Soc. Dermatol.
Blackwell Publishing Ltd, 9600 Garsington Road, OX4 2XG, Oxford, United Kingdom, Tel: +44(1865)776868, Fax: +44(1865)714591 ISSUES/YR: 12

• **JEUR**
ISSN: 0993-9857
JEUR
Masson, 21 Rue Camille Desmoulins, 92789, Issy les Moulineaux Cedex 9, France, Tel: +33(1)73281634, Fax: +33(1)73281649 ISSUES/YR: 4

Jinekoloji ve Obstetri Bulteni
ISSN: 1300-0438
Jinekoloji Obstet. Bul.
Hekimler Yayin Birligi, Bilkent Plaza, AE Blok, 21-24, 06533, Bilkent, Ankara, Turkey, Tel: +90(312)2665566, Fax: +90(312)2665577 ISSUES/YR: 4

Jinekoloji ve Obstetrik Dergisi
ISSN: 1016-5126
Jinekol. Obstet. Derg.
Turkish Anaesthesiology and Intensive Care Society, Inonu Cad. 7/4 Zambak apt., Gumussuyu Taksim. Istanbul, Turkey ISSUES/YR: 4

JK Practitioner
ISSN: 0971-8834
JK Pract.
JK Practitioner, P.O. Box 884, 190001, Srinagar, Kashmir, India, Tel: +91(194)474174, Fax: +91(194)455592 ISSUES/YR: 4

JK Science
ISSN: 0972-1177
JK Science
JK Science, Shiv Bhawan, Hari Market, P.O. Box 158, 180 001, Jammu, India, Tel: +91(191)544721, Fax: +91(191)462167 ISSUES/YR: 4

JMS - Journal of Medical Society
ISSN: 0972-4958
JMS J. Med. Soc.
Regional Institute of Medical Sciences, 795 004, Imphal, Manipur, India, Tel: +91(385)310493, Fax: +91(385)310625 ISSUES/YR: 3

JNCCN Journal of the National Comprehensive Cancer Network
ISSN: 1540-1405
JNCCN J. Nat. Compr. Cancer Netw.
Jones and Bartlett Publishers, 40 Tall Pine Drive, Sudbury, MA 01776, United States, Tel: +1(978)4435000, Fax: +1(978)4438000 ISSUES/YR: 4

Joint Bone Spine
ISSN: 1297-319X
Jt. Bone Spine
Editions Scientifiques et Medicales Elsevier, 23 rue Linois, 75724, Paris Cedex 15, France, Tel: +33(1)45589826, Fax: +33(1)45589428 ISSUES/YR: 6

Jordan Journal of Applied Sciences - Natural Sciences
ISSN: 1605-2587
Jordan J. Appl. Sci. Nat. Sci.
Applied Science University (Private), Shafa Badran, 11931, Amman, Jordan, Tel: +962(6)5237181, Fax: +962(6)5232899 ISSUES/YR: 2

Jordan Medical Journal
ISSN: 0446-9283
Jordan Med. J.
Jordan Medical Association, P.O. Box 915, Amman, Jordan, Tel: +962(6)66164 ISSUES/YR: 2

Jornal Brasileiro de Psiquiatria
ISSN: 0047-2085
J. Bras. Psiquiatr.
Editora Cientifica Nacional Ltda, Rua da Gloria 366, 3o andar, P.O. Box 590, 20241-180, Rio de Janeiro, Brazil, Tel: +55(21)2213235, Fax: +55(21)2521691 ISSUES/YR: 4

Jornal Brasileiro de Reproducao Assistida
ISSN: 1517-5693
J. Bras. Reproducao Assistida
Sociedade Brasileira de Reproducao Assistida, Avenida das Americas 4666, Salas 312/313, CEP 22649-900, Barra da Tijuca-RJ, Brazil, Tel: +55(21)2430 9060, Fax: +55(21)2430 9070 ISSUES/YR: 5

Jornal de Pediatria
ISSN: 0021-7557
J. Pediatr.
Sociedade Brasileira de Pediatria, Rua Santa Clara 292, Copacabana, Rio de Janeiro, Brazil, Tel: +55(21)2351999 ISSUES/YR: 6

Jornal Vascular Brasileiro
ISSN: 1677-5449
J. Vasc. Bras.
Sociedade Brasileira de Angiologia e Cirurgia Vascular, Avenida Venezuela 131 grupo 509/510, Cep 20081-311 Rio de Janeiro, Brazil, Tel: +55-2122831228, Fax: +55-2122634884 ISSUES/YR: 1

- **Journal Belge de Radiologie**
ISSN: 0021-7646
J. Belge Radiol.
JBR-BTR, Leopoldlaan 17 A, 9900, Eeklo, Belgium, Tel: +32(9)3784757 ISSUES/YR: 6

Journal d'Economie Medicale
ISSN: 0294-0736
J. Econ. Med.
Editions Alexandre Lacassagne, 68, Rue Montgolfier, 69006, Lyon, France, Tel: +33(04)72440606, Fax: +33(04)72449908 ISSUES/YR: 8

Journal de Gynecologie Obstetrique et Biologie de la Reproduction
ISSN: 0368-2315
J. Gynecol. Obstet. Biol. Reprod.
Masson, 21 Rue Camille Desmoulins, 92789, Issy les Moulineaux Cedex 9, France, Tel: +33(1)73281634, Fax: +33(1)73281649 ISSUES/YR: 8

Journal de Medecine Legale Droit Medical
ISSN: 0249-6208
J. Med. Leg. Droit Med.
Editions Alexandre Lacassagne, 68, Rue Montgolfier, 69006, Lyon, France, Tel: +33(04)72440606, Fax: +33(04)72449908 ISSUES/YR: 8

Journal de Mycologie Medicale
ISSN: 1156-5233
J. Mycol. Med.
Elsevier, P.O. Box 211, 1000 AE, Amsterdam, Netherlands, Tel: +31(20)4853757, Fax: +31(20)4853432 ISSUES/YR: 4

Journal de Pediatrie et de Puericulture
ISSN: 0987-7983
J. Pediatr. Pueric.
Editions Scientifiques et Medicales Elsevier, 23 rue Linois, 75724, Paris Cedex 15, France, Tel: +33(1)45589826, Fax: +33(1)45589428 ISSUES/YR: 8

Journal de Pharmacie Clinique
ISSN: 0291-1981
J. Pharm. Clin.
John Libbey Eurotext, 127, avenue de la Republique, 92120, Montrouge, France, Tel: +33(01)46730660, Fax: +33(01)40840999 ISSUES/YR: 4

Journal de Pharmacie de Belgique
ISSN: 0047-2166
J. Pharm. Belg.
APB Algemene Pharmaceutische Bond, Archimedesstraat 11, B-1000, Brussels, Belgium, Tel: +32(2)2854200, Fax: +32(2)2854285 ISSUES/YR: 4

Journal de Radiologie
ISSN: 0221-0363
J. Radiol.
Masson, 21 Rue Camille Desmoulins, 92789, Issy les Moulineaux Cedex 9, France, Tel: +33(1)73281634, Fax: +33(1)73281649 ISSUES/YR: 12

Journal de Traumatologie du Sport
ISSN: 0762-915X
J. Traumatol. Sport
Masson, 21 Rue Camille Desmoulins, 92789, Issy les Moulineaux Cedex 9, France, Tel: +33(1)73281634, Fax: +33(1)73281649 ISSUES/YR: 4

Journal des Maladies Vasculaires
ISSN: 0398-0499
J. Mal. Vasc.
Masson, 21 Rue Camille Desmoulins, 92789, Issy les Moulineaux Cedex 9, France, Tel: +33(1)73281634, Fax: +33(1)73281649 ISSUES/YR: 5

- **Journal for Vascular Ultrasound**
ISSN: 1544-3167
J. Vasc. Ultrasound
Society of Vascular Ultrasound, 4601 President's Drive, Lanham, MD 20706-4365, United States, Tel: +1(301)4597550 ISSUES/YR: 4

Journal Francais d'Ophtalmologie
ISSN: 0181-5512
J. Fr. Ophtalmol.
Masson, 21 Rue Camille Desmoulins, 92789, Issy les Moulineaux Cedex 9, France, Tel: +33(1)73281634, Fax: +33(1)73281649 ISSUES/YR: 10

Journal fur Anasthesie und Intensivbehandlung
ISSN: 0941-4223
J. Anasth. Intensivbehandl.
Pabst Science Publishers, Eichengrund 28, D-49525, Lengerich, Germany, Tel: +49(5484)308, Fax: +49(5484)550 ISSUES/YR: 4

Journal fur Ernahrungsmedizin
ISSN: 1563-2873
J. Ernahrungsmed.
Krause und Pachernegg GmbH, Mozartgasse 10, A-3003, Gablitz, Austria, Tel: +43(2231)612580, Fax: +43(2231)6125810 ISSUES/YR: 4

Journal fur Fertilitat und Reproduktion
ISSN: 1019-066X
J. Fertil. Reprod.
Krause und Pachernegg GmbH, Mozartgasse 10, A-3003, Gablitz, Austria, Tel: +43(2231)612580, Fax: +43(2231)6125810 ISSUES/YR: 4

Journal fur Gastroenterologische und Hepatologische Erkrankungen
ISSN: 1728-6263
J. Gastroenterol. Hepatol. Erkr.
Krause und Pachernegg GmbH, Mozartgasse 10, A-3003, Gablitz, Austria, Tel: +43(2231)612580, Fax: +43(2231)6125810 ISSUES/YR: 4

Journal fur Hypertonie
ISSN: 1028-2327
J. Hypertonie
Krause und Pachernegg GmbH, Mozartgasse 10, A-3003, Gablitz, Austria, Tel: +43(2231)612580, Fax: +43(2231)6125810 ISSUES/YR: 4

Journal fur Kardiologie
ISSN: 1024-0098
J. Kardiol.
Krause und Pachernegg GmbH, Mozartgasse 10, A-3003, Gablitz, Austria, Tel: +43(2231)612580, Fax: +43(2231)6125810 ISSUES/YR: 10

Journal fur Menopause
ISSN: 1023-0904
J. Menopause
Krause und Pachernegg GmbH, Mozartgasse 10, A-3003, Gablitz, Austria, Tel: +43(2231)612580, Fax: +43(2231)6125810 ISSUES/YR: 4

Journal fur Mineralstoffwechsel
ISSN: 1023-7763
J. Mineralstoffwechsel
Krause und Pachernegg GmbH, Mozartgasse 10, A-3003, Gablitz, Austria, Tel: +43(2231)612580, Fax: +43(2231)6125810 ISSUES/YR: 4

Journal fur Neurologie, Neurochirurgie und Psychiatrie
ISSN: 1608-1587
J. Neurol. Neurochir. Psychiatrie
Krause und Pachernegg GmbH, Mozartgasse 10, A-3003, Gablitz, Austria, Tel: +43(2231)612580, Fax: +43(2231)6125810 ISSUES/YR: 4

Journal fur Reproduktionsmedizin und Endokrinologie
ISSN: 1810-2107
J. Reproduktionsmed. Endokrinol.
Krause und Pachernegg GmbH, Mozartgasse 10, A-3003, Gablitz, Austria, Tel: +43(2231)612580, Fax: +43(2231)6125810 ISSUES/YR: 4

Journal fur Urologie und Urogynakologie
ISSN: 1023-6090
J. Urol. Urogynakol.
Krause und Pachernegg GmbH, Mozartgasse 10, A-3003, Gablitz, Austria, Tel: +43(2231)612580, Fax: +43(2231)6125810 ISSUES/YR: 4

Journal International de Bioethique
ISSN: 1145-0762
J. Int. Bioethique
Editions Alexandre Lacassagne, 68, Rue Montgolfier, 69006, Lyon, France, Tel: +33(04)72440606, Fax: +33(04)72449908 ISSUES/YR: 4

Journal International Medical Sciences Academy
ISSN: 0971-071X
J. Int. Med. Sci. Acad.
International Medical Sciences Academy, Ansari Nagar, Ring Road, 110029, New Delhi, India, Tel: +91(11)654660 ISSUES/YR: 4

Journal Medical Libanais
ISSN: 0023-9852
J. Med. Liban.
Lebanese Order of Physicians, P.O. Box 11-640, Beirut, Lebanon, Tel: +961(1)453051, Fax: +961(1)451480 ISSUES/YR: 4

• **Journal of AAPOS**
ISSN: 1091-8531
J. AAPOS
Mosby Inc., 11830 Westline Industrial Drive, St. Louis, MO 63146, United States, Tel: +1(314)8728370, Fax: +1(314)4321380 ISSUES/YR: 6

Journal of Abnormal Child Psychology
ISSN: 0091-0627
J. Abnorm. Child Psychol.
Springer Science and Business Media Deutschland GmbH, Heidelberger Platz 3, 14197, Berlin, Germany, Tel: +1-30 827875282, Fax: +1-30 827875707 ISSUES/YR: 6

Journal of Abnormal Psychology
ISSN: 0021-843X
J. Abnorm. Psychol.
American Psychological Association Inc., 750 First Street NE, Washington, DC 20002-4242, United States, Tel: +1(202)3365600, Fax: +1(202)3365568 ISSUES/YR: 4

• **Journal of Acquired Immune Deficiency Syndromes**
ISSN: 1525-4135
J. Acquired Immune Defic. Syndr.
Lippincott Williams and Wilkins, 530 Walnut Street, Philadelphia, PA 19106-3621, United States, Tel: +1(215)5218342, Fax: +1(215)5218326 ISSUES/YR: 15

Journal of Addictive Diseases
ISSN: 1055-0887
J. Addict. Dis.
Haworth Press Inc., 10 Alice Street, Binghamton, NY 13904-1580, United States, Tel: +1(607)7225857 ISSUES/YR: 4

Journal of Adolescence
ISSN: 0140-1971
J. Adolesc.
Academic Press, 24-28 Oval Road, NW1 7DX, London, United Kingdom, Tel: +44(207)2674466 ISSUES/YR: 6

• **Journal of Adolescent Health**
ISSN: 1054-139X
J. Adolesc. Health
Elsevier USA, 6277 Sea Harbor Drive, Orlando, FL 32862 8239, United States, Tel: +1-407 345 4000, Fax: +1-407 363 9661 ISSUES/YR: 12

Journal of Aerosol Medicine: Deposition, Clearance, and Effects in the Lung
ISSN: 0894-2684
J. Aerosol Med. Deposition Clear. Eff. Lung
Mary Ann Liebert Inc., 140 Huguenot Street, New Rochelle, NY 10801-5215, United States, Tel: +1(914)7402100, Fax: +1(914)7402101 ISSUES/YR: 4

• **Journal of Aerosol Science**
ISSN: 0021-8502
J. Aerosol Sci.
Elsevier Ltd, Langford Lane, Kidlington, OX5 1GB, Oxford, United Kingdom, Tel: +44(1865)843000, Fax: +44(1865)843010 ISSUES/YR: 12

• **Journal of Affective Disorders**
ISSN: 0165-0327
J. Affective Disord.
Elsevier, P.O. Box 211, 1000 AE, Amsterdam, Netherlands, Tel: +31(20)4853757, Fax: +31(20)4853432 ISSUES/YR: 15

Journal of Aggression, Maltreatment and Trauma
ISSN: 1092-6771
J. Aggression Maltreat. Trauma
Haworth Press Inc., 10 Alice Street, Binghamton, NY 13904-1580, United States, Tel: +1(607)7225857, Fax: +1(607)7221424 ISSUES/YR: 8

Journal of Aging and Health
ISSN: 0898-2643
J. Aging Health
SAGE Publications Inc., 2455 Teller Road, Thousand Oaks, CA 91320, United States, Tel: +1(805)4990721, Fax: +1(805)4990871 ISSUES/YR: 5

Journal of Aging and Pharmacotherapy
ISSN: 1540-5303
J. Aging Pharmacother.
Haworth Press Inc., 10 Alice Street, Binghamton, NY 13904-1580, United States, Tel: +1(607)7225857, Fax: +1(607)7221424 ISSUES/YR: 4

Journal of Aging and Physical Activity
ISSN: 1063-8652
J. Aging Phys. Act.
Human Kinetics Publishers Inc., 1607 N. Market Street, Champaign, IL 61820-2200, United States, Tel: +1(217)3515076, Fax: +1(217)3511549 ISSUES/YR: 4

• **Journal of Allergy and Clinical Immunology**
ISSN: 0091-6749
J. Allergy Clin. Immunol.
Mosby Inc., 11830 Westline Industrial Drive, St. Louis, MO 63146, United States, Tel: +1(314)8728370, Fax: +1(314)4321380 ISSUES/YR: 12

• **Journal of Alternative and Complementary Medicine**
ISSN: 1075-5535
J. Altern. Complement. Med.
Mary Ann Liebert Inc., 140 Huguenot Street, New Rochelle, NY 10801-5215, United States, Tel: +1(914)7402100, Fax: +1(914)7402101 ISSUES/YR: 6

• **Journal of Alzheimer's Disease**
ISSN: 1387-2877
J. Alzheimer's Dis.
IOS Press, Nieuwe Hemweg 6B, 1013 BG, Amsterdam, Netherlands, Tel: +31(20)6883355, Fax: +31(20)6203419 ISSUES/YR: 8

Journal of Ambulatory Care Management
ISSN: 0148-9917
J. Ambul. Care Manage.
Lippincott Williams and Wilkins, 530 Walnut Street, Philadelphia, PA 19106-3621, United States, Tel: +1(215)5218342, Fax: +1(215)5218326 ISSUES/YR: 4

Journal of Anaesthesiology Clinical Pharmacology
ISSN: 0970-9185
J. Anaesthesiol. Clin. Pharmacol.
Journal of Anaesthesiology Clinical Pharmacology, Dayanand Medical College and Hospital, 141001, Ludhiana, India, Tel: +91(161)2304242, Fax: +91(161)2302620 ISSUES/YR: 4

Journal of Analytical Psychology
ISSN: 0021-8774
J. Anal. Psychol.
Blackwell Publishing Ltd, 9600 Garsington Road, OX4 2XG, Oxford, United Kingdom, Tel: +44(1865)776868, Fax: +44(1865)714591 ISSUES/YR: 5

Journal of Analytical Toxicology
ISSN: 0146-4760
J. Anal. Toxicol.
Preston Publications, 6600 W. Touhy Avenue, Niles, IL 60714-4588, United States, Tel: +1(847)6472900, Fax: +1(847)6471155 ISSUES/YR: 8

• **Journal of Anatomy**
ISSN: 0021-8782
J. Anat.
Blackwell Publishing Ltd, 9600 Garsington Road, OX4 2XG, Oxford, United Kingdom, Tel: +44(1865)776868, Fax: +44(1865)714591 ISSUES/YR: 12

- **Journal of Andrology**
  ISSN: 0196-3635
  J. Androl.
  American Society of Andrology Inc., 74 New Montgomery, Suite 230, San Francisco, CA 94105, United States ISSUES/YR: 6

Journal of Anesthesia
ISSN: 0913-8668
J. Anesth.
Springer-Verlag Tokyo, 3-13, Hongo 3-chome, Bunkyo-ku, 113-8480, Tokyo, Japan, Tel: +81(3)38180861, Fax: +81(3)38180864 ISSUES/YR: 4

- **Journal of Antibiotics**
  ISSN: 0021-8820
  J. Antibiot.
  Japan Antibiotics Research Association, 2-20-8 Kamiosaki, Shinagawa-ku, 141, Tokyo, Japan, Tel: +81(3)34910181 ISSUES/YR: 12

Journal of Antimicrobial Chemotherapy
ISSN: 0305-7453
J. Antimicrob. Chemother.
Oxford University Press, Great Clarendon Street, OX2 6DP, Oxford, United Kingdom, Tel: +44(1865)353907, Fax: +44(1865)353486 ISSUES/YR: 12

Journal of Anxiety Disorders
ISSN: 0887-6185
J. Anxiety Disord.
Elsevier Ltd, Langford Lane, Kidlington, OX5 1GB, Oxford, United Kingdom, Tel: +44(1865)843000, Fax: +44(1865)843010 ISSUES/YR: 6

Journal of Applied Animal Welfare Science
ISSN: 1088-8705
J. Appl. Anim. Welf. Sci.
Lawrence Erlbaum Associates Inc., 10 Industrial Avenue, Mahwah, NJ 07430-2262, United States, Tel: +1(201)2582200, Fax: +1(201)2360072 ISSUES/YR: 4

Journal of Applied Biomaterials and Biomechanics
ISSN: 1722-6899
J. Appl. Biomater. Biomech.
Wichtig Editore s.r.l., Via Friuli 72/74, 20135, Milan, Italy, Tel: +39(02)55195443, Fax: +39(02)55195971 ISSUES/YR: 3

Journal of Applied Biomechanics
ISSN: 1065-8483
J. Appl. Biomech.
Human Kinetics Publishers Inc., 1607 N. Market Street, Champaign, IL 61820-2200, United States, Tel: +1(217)3515076, Fax: +1(217)3511549 ISSUES/YR: 4

Journal of Applied Biomedicine
ISSN: 1214-021X
J. App. Biomed.
University of South Bohemia, Jirovcova 24, 370 04, Ceske Budejovic, Czech Republic, Tel: +420-420 38 5102921, Fax: +420-420 038 5102919 ISSUES/YR: 4

Journal of Applied Cosmetology
ISSN: 0392-8543
J. Appl. Cosmetol.
International Ediemme, Via Innocenzo XI 41, 00165, Roma, Italy, Tel: +39(06)9286261, Fax: +39(06)6380839 ISSUES/YR: 4

Journal of Applied Gerontology
ISSN: 0733-4648
J. Appl. Gerontol.
SAGE Publications Inc., 2455 Teller Road, Thousand Oaks, CA 91320, United States, Tel: +1(805)4990721, Fax: +1(805)4990871 ISSUES/YR: 4

Journal of Applied Microbiology
ISSN: 1364-5072
J. Appl. Microbiol.
Blackwell Publishing Ltd, 9600 Garsington Road, OX4 2XG, Oxford, United Kingdom, Tel: +44(1865)776868, Fax: +44(1865)714591 ISSUES/YR: 12

- **Journal of Applied Physiology**
  ISSN: 8750-7587
  J. Appl. Physiol.
  American Physiological Society, 9650 Rockville Pike, Bethesda, MD 20814-3991, United States, Tel: +1(301)6347180, Fax: +1(301)6347241 ISSUES/YR: 12

Journal of Applied Research
ISSN: 1537-064X
J. Appl. Res.
Therapeutic Solutions LLC, P.O. Box 463, Newtown, PA 18940, United States, Tel: +1(267)5660270 ISSUES/YR: 4

- **Journal of Applied Toxicology**
  ISSN: 0260-437X
  J. Appl. Toxicol.
  John Wiley and Sons Ltd, Southern Gate, PO19 8SQ, Chichester, West Sussex, United Kingdom, Tel: +44(1243)779777, Fax: +44(1243)775878 ISSUES/YR: 6

Journal of Arthroplasty
ISSN: 0883-5403
J. Arthroplasty
Churchill Livingstone Inc., 650 Avenue of the Americas, New York, NY 10011, United States, Tel: +1(212)2065000, Fax: +1(212)7277808 ISSUES/YR: 8

- **Journal of Artificial Organs**
  ISSN: 1434-7229
  J. Artif. Organs
  Springer-Verlag Tokyo, 3-13, Hongo 3-chome, Bunkyo-ku, 113-8480, Tokyo, Japan, Tel: +81(3)38180861, Fax: +81(3)38180864 ISSUES/YR: 4

Journal of Asian Natural Products Research
ISSN: 1028-6020
J. Asian Nat. Prod. Res.
Taylor and Francis Ltd., 4 Park Square, Milton Park, OX14 4RN, Abingdon, Oxfordshire, United Kingdom, Tel: +44(20)70175541, Fax: +44(20)70175198 ISSUES/YR: 8

- **Journal of Assisted Reproduction and Genetics**
  ISSN: 1058-0468
  J. Assisted Reprod. Genet.
  Springer Science and Business Media Deutschland GmbH, Heidelberger Platz 3, 14197, Berlin, Germany, Tel: +1-30 827875282, Fax: +1-30 827875707 ISSUES/YR: 12

Journal of Asthma
ISSN: 0277-0903
J. Asthma
Taylor and Francis Inc., 325 Chestnut St, Suite 800, Philadelphia PA, PA 19106, United States, Tel: +1(800)3541420, Fax: +1(215)6258914 ISSUES/YR: 8

Journal of Athletic Training
ISSN: 1062-6050
J. Athl. Train.
National Athletic Trainers' Association Inc., 2952 Stemmons Freeway, Dallas, TX 75247, United States ISSUES/YR: 4

Journal of Atmospheric Chemistry
ISSN: 0167-7764
J. Atmos. Chem.
Kluwer Academic Publishers, Van Godewijckstraat 30, 3311 GZ, Dordrecht, Netherlands, Tel: +31(78)6576000, Fax: +31(78)6576254 ISSUES/YR: 9

- **Journal of Autism and Developmental Disorders**
  ISSN: 0162-3257
  J. Autism Dev. Disord.
  Springer Science and Business Media Deutschland GmbH, Heidelberger Platz 3, 14197, Berlin, Germany, Tel: +1-30 827875282, Fax: +1-30 827875707 ISSUES/YR: 6

Journal of Autoimmune Diseases
ISSN: 1740-2557
J.Autoimmune Dis.
BioMed Central Ltd., 34 - 42 Cleveland Street, W1T 4LB, London, United Kingdom, Tel: +44(207)3230323, Fax: +44(207)6319926 ISSUES/YR: 1

- **Journal of Autoimmunity**
  ISSN: 0896-8411
  J. Autoimmun.
  Academic Press, 24-28 Oval Road, NW1 7DX, London, United Kingdom, Tel: +44(207)2674466 ISSUES/YR: 8

- **Journal of Automated Methods and Management in Chemistry**
  ISSN: 1463-9246
  J. Autom. Methods Manage. Chem.
  Hindawi Publishing Corporation, 410 Park Avenue, 15th Floor, 287 pmb, New York, NY 10022, United States, Tel: +1(20)2-2706352, Fax: +1(20)2-2706351 ISSUES/YR: 4

Current

Journal of B.U.ON.
ISSN: 1107-0625
J. B.U.ON.
Journal of B.U.ON., 51 Botassi Street, 185 37, Piraeus, Greece, Tel: +30(1)4285000, Fax: +30(1)4538953 ISSUES/YR: 4

• **Journal of Back and Musculoskeletal Rehabilitation**
ISSN: 1053-8127
J. Back Musculoskelet. Rehabil.
IOS Press, Nieuwe Hemweg 6B, 1013 BG, Amsterdam, Netherlands, Tel: +31(20)6883355, Fax: +31(20)6203419 ISSUES/YR: 4

• **Journal of Bacteriology**
ISSN: 0021-9193
J. Bacteriol.
American Society for Microbiology, 1752 N Street N.W., Washington, DC 20036-2904, United States, Tel: +1(202)942-9244, Fax: +1(202)942-9355 ISSUES/YR: 24

Journal of Bacteriology and Virology
ISSN: 1598-2467
J. Bacteriol. Virol.
Chonnam National University Medical School, 5-1 Hak-dong Dong-ku, 501-190, Kwangju, South Korea, Tel: +82(62)2204132, Fax: +82(62)2287294 ISSUES/YR: 4

Journal of Basic and Clinical Physiology and Pharmacology
ISSN: 0334-1534
J. Basic Clin. Physiol. Pharmacol.
Freund Publishing House Ltd, 61 Nachmani Street, P.O. Box 35010, 61350, Tel Aviv, Israel, Tel: +972(3)5628540, Fax: +972(3)5628538 ISSUES/YR: 4

Journal of Behavior Therapy and Experimental Psychiatry
ISSN: 0005-7916
J. Behav. Ther. Exp. Psychiatry
Elsevier Ltd, Langford Lane, Kidlington, OX5 1GB, Oxford, United Kingdom, Tel: +44(1865)843000, Fax: +44(1865)843010 ISSUES/YR: 4

Journal of Behavioral Health Services and Research
ISSN: 1094-3412
J. Behav. Health Serv. Res.
Springer New York LLC, 233 Springer Street, New York, 10013-1578, United States, Tel: +1-212 460 1501, Fax: +1-212 460 1595 ISSUES/YR: 4

Journal of Behavioral Medicine
ISSN: 0160-7715
J. Behav. Med.
Springer Science and Business Media Deutschland GmbH, Heidelberger Platz 3, 14197, Berlin, Germany, Tel: +1-30 827875282, Fax: +1-30 827875707 ISSUES/YR: 6

Journal of Bioactive and Compatible Polymers
ISSN: 0883-9115
J. Bioact. Compat. Polym.
SAGE Publications Ltd, 55 City Road, EC1Y 1SP, London, United Kingdom, Tel: +44(207)3248500, Fax: +44(207)3248600 ISSUES/YR: 6

• **Journal of Biochemical and Biophysical Methods**
ISSN: 0165-022X
J. Biochem. Biophys. Methods
Elsevier, P.O. Box 211, 1000 AE, Amsterdam, Netherlands, Tel: +31(20)4853757, Fax: +31(20)4853432 ISSUES/YR: 12

Journal of Biochemical and Molecular Toxicology
ISSN: 1095-6670
J. Biochem. Mol. Toxicol.
John Wiley and Sons Inc., P.O.Box 18667, Newark, NJ 07191-8667, United States, Tel: +1(212)8506000, Fax: +1(212)8506012 ISSUES/YR: 6

Journal of Biochemistry
ISSN: 0021-924X
J. Biochem.
Japanese Biochemical Society, 25-16 Hongo 5-chome, Bunkyo-ku, 113-0033, Tokyo, Japan, Tel: +81(3)38151913, Fax: +81(3)38151934 ISSUES/YR: 12

Journal of Bioenergetics and Biomembranes
ISSN: 0145-479X
J. Bioenerg. Biomembr.
Springer Science and Business Media Deutschland GmbH, Heidelberger Platz 3, 14197, Berlin, Germany, Tel: +1-30 827875282, Fax: +1-30 827875707 ISSUES/YR: 6

Journal of Bioinformatics and Computational Biology
ISSN: 0219-7200
J. Bioinformatics Comput. Biol.
World Scientific Publishing Co. Pte Ltd, 5 Toh Tuck Link, 596224, Singapore, Singapore, Tel: +65(6466)5775, Fax: +65(6467)7667 ISSUES/YR: 4

Journal of Biolaw and Business
ISSN: 1095-5127
J. Biolaw Bus.
Aspen Law and Business, 1185 Avenue of the Americas, New York, NY 10036, United States, Tel: +1(212)5970200, Fax: +1(212)5970336 ISSUES/YR: 4

• **Journal of Biological Chemistry**
ISSN: 0021-9258
J. Biol. Chem.
American Society for Biochemistry and Molecular Biology Inc., 9650 Rockville Pike, Bethesda, MD 20814, United States, Tel: +1-301 530 7145, Fax: +1-301 571 1824 ISSUES/YR: 52

• **Journal of Biological Inorganic Chemistry**
ISSN: 0949-8257
J. Biol. Inorg. Chem.
Springer Verlag, Tiergartenstrasse 17, D-69121, Heidelberg, Germany, Tel: +49(6221)4870, Fax: +49(6221)487366 ISSUES/YR: 8

• **Journal of Biological Physics**
ISSN: 0092-0606
J. Biol. Phys.
Kluwer Academic Publishers, Van Godewijckstraat 30, 3311 GZ, Dordrecht, Netherlands, Tel: +31(78)6576000, Fax: +31(78)6576254 ISSUES/YR: 4

Journal of Biological Regulators and Homeostatic Agents
ISSN: 0393-974X
J. Biol. Regul. Homeostatic Agents
Wichtig Editore s.r.l., Via Friuli 72/74, 20135, Milan, Italy, Tel: +39(02)55195443, Fax: +39(02)55195971 ISSUES/YR: 4

• **Journal of Biological Rhythms**
ISSN: 0748-7304
J. Biol. Rhythms
SAGE Publications Inc., 2455 Teller Road, Thousand Oaks, CA 91320, United States, Tel: +1(805)4990721, Fax: +1(805)4990871 ISSUES/YR: 6

Journal of Biological Sciences
ISSN: 1727-3048
J. Biol. Sci.
Asian Network for Scientific Information, 308-Lasani Town, Sargodha Road, Faisalabad, Pakistan, Tel: +92(41)2001145-46, Fax: +92(21)5206036 ISSUES/YR: 4

• **Journal of Biology**
ISSN: 1475-4924
J. Biol.
BioMed Central Ltd., 34 - 42 Cleveland Street, W1T 4LB, London, United Kingdom, Tel: +44(207)3230323, Fax: +44(207)6319926 ISSUES/YR: 1

• **Journal of Biomaterials Science, Polymer Edition**
ISSN: 0920-5063
J. Biomater. Sci. Polym. Ed.
VSP BV, P.O.Box 9000, 2300 PA, Leiden, Netherlands, Tel: +31-71 535 3500, Fax: +31-71 531 7532 ISSUES/YR: 12

Journal of Biomechanical Engineering
ISSN: 0148-0731
J. Biomech. Eng.
American Society of Mechanical Engineers, Three Park Avenue, New York, NY 10016-5990, United States, Tel: +1(212)5917722, Fax: +1(212)5917674 ISSUES/YR: 6

• **Journal of Biomechanics**
ISSN: 0021-9290
J. Biomech.
Elsevier Ltd, Langford Lane, Kidlington, OX5 1GB, Oxford, United Kingdom, Tel: +44(1865)843000, Fax: +44(1865)843010 ISSUES/YR: 12

• **Journal of Biomedical Informatics**
ISSN: 1532-0464
J. Biomed. Informatics
Academic Press Inc., 6277 Sea Harbor Drive, Orlando, FL 32887-4900, United States, Tel: +1(407)3454040, Fax: +1(407)3639661 ISSUES/YR: 6

Journal of Biomedical Materials Research - Part A
ISSN: 0021-9304
J. Biomed. Mater. Res. Part A
John Wiley and Sons Inc., P.O.Box 18667, Newark, NJ 07191-8667, United States, Tel: +1(212)8506000, Fax: +1(212)8506012 ISSUES/YR: 16

Journal of Biomedical Materials Research - Part B Applied Biomaterials
ISSN: 0021-9304
J. Biomed. Mater. Res. Part B Appl. Biomater.
John Wiley and Sons Inc., P.O.Box 18667, Newark, NJ 07191-8667, United States, Tel: +1(212)8506000, Fax: +1(212)8506012 ISSUES/YR: 8

- **Journal of Biomedical Science**
ISSN: 1021-7770
J. Biomed. Sci.
Springer Verlag, Tiergartenstrasse 17, D-69121, Heidelberg, Germany, Tel: +41(6221)4870, Fax: +41(6221)487366 ISSUES/YR: 6

Journal of Biomedicine and Biotechnology
ISSN: 1110-7243
J. Biomed. Biotechnol.
Hindawi Publishing Corporation, 410 Park Avenue, 15th Floor, 287 pmb, New York, NY 10022, United States, Tel: +1(20)2-2706352, Fax: +1(20)2-2706351 ISSUES/YR: 4

- **Journal of Biomolecular NMR**
ISSN: 0925-2738
J. Biomol. NMR
Kluwer Academic Publishers, Van Godewijckstraat 30, 3311 GZ, Dordrecht, Netherlands, Tel: +31(78)6576000, Fax: +31(78)6576254 ISSUES/YR: 12

- **Journal of Biomolecular Screening**
ISSN: 1087-0571
J. Biomol. Screen.
SAGE Publications Inc., 2455 Teller Road, Thousand Oaks, CA 91320, United States, Tel: +1(805)4990721, Fax: +1(805)4990871 ISSUES/YR: 8

- **Journal of Biomolecular Structure and Dynamics**
ISSN: 0739-1102
J. Biomol. Struct. Dyn.
Adenine Press, 2066 Central Avenue, Schenectady, NY 12304, United States, Tel: +1(518)4560784, Fax: +1(518)4524955 ISSUES/YR: 6

- **Journal of Biopharmaceutical Statistics**
ISSN: 1054-3406
J. Biopharm. Stat.
Taylor and Francis Inc., 325 Chestnut St, Suite 800, Philadelphia PA, PA 19106, United States, Tel: +1(800)3541420, Fax: +1(215)6258914 ISSUES/YR: 6

Journal of Bioscience and Bioengineering
ISSN: 1389-1723
J. Biosci. Bioeng.
Elsevier, P.O. Box 211, 1000 AE, Amsterdam, Netherlands, Tel: +31(20)4853757, Fax: +31(20)4853432 ISSUES/YR: 12

Journal of Biosciences
ISSN: 0250-5991
J. Biosci.
Indian Academy of Sciences, C.V. Raman Avenue, P.O. Box 8005, 560 080, Bangalore, India, Tel: +91(80)3342546, Fax: +91(80)3346094 ISSUES/YR: 5

Journal of Biosocial Science
ISSN: 0021-9320
J. Biosoc. Sci.
Cambridge University Press, Shaftesbury Road, CB2 2RU, Cambridge, United Kingdom, Tel: +44(1223)312393, Fax: +44(1223)315052 ISSUES/YR: 6

- **Journal of Biotechnology**
ISSN: 0168-1656
J. Biotechnol.
Elsevier, P.O. Box 211, 1000 AE, Amsterdam, Netherlands, Tel: +31(20)4853757, Fax: +31(20)4853432 ISSUES/YR: 24

Journal of Bodywork and Movement Therapies
ISSN: 1360-8592
J. Bodywork Mov. Ther.
Churchill Livingstone, 1-3 Baxter's Place, Leith Walk, EH1 3AF, Edinburgh, United Kingdom, Tel: +44(131)5562424, Fax: +44(131)5581278 ISSUES/YR: 4

- **Journal of Bone and Joint Surgery - Series A**
ISSN: 0021-9355
J. Bone Jt. Surg. Ser. A
Journal of Bone and Joint Surgery Inc., 20 Pickering Street, Needham, MA 02492, United States, Tel: +1(781)4499780, Fax: +1(781)4499742 ISSUES/YR: 12

- **Journal of Bone and Joint Surgery - Series B**
ISSN: 0301-620X
J. Bone Jt. Surg. Ser. B
British Editorial Society of Bone and Joint Surgery, 22 Buckingham Street, WC2N 6ET, London, United Kingdom, Tel: +44(207)7820010, Fax: +44(207)7820995 ISSUES/YR: 8

- **Journal of Bone and Mineral Metabolism**
ISSN: 0914-8779
J. Bone Miner. Metab.
Springer-Verlag Tokyo, 3-13, Hongo 3-chome, Bunkyo-ku, 113-8480, Tokyo, Japan, Tel: +81(3)38180861, Fax: +81(3)38180864 ISSUES/YR: 6

Journal of Bone and Mineral Research
ISSN: 0884-0431
J. Bone Miner. Res.
American Society for Bone and Mineral Research, 2025 M Street NW, Suite 800, Washington, DC 20036-3309, United States, Tel: +1(202)3671161, Fax: +1(202)3672161 ISSUES/YR: 12

Journal of Brain Science
ISSN: 1341-5301
J. Brain Sci.
PJD Publications Ltd, P.O. Box 966, Westbury, NY 11590-0966, United States, Tel: +1(516)6260650, Fax: +1(516)6265546 ISSUES/YR: 4

- **Journal of Brief Therapy**
ISSN: 1533-2098
J. Brief Ther.
Springer Publishing Company, 11 West 42nd Street, 15th Floor, New York, NY 10036, United States, Tel: +1(212)4314370, Fax: +1(212)9417842 ISSUES/YR: 2

- **Journal of Bronchology**
ISSN: 1070-8030
J. Bronchol.
Lippincott Williams and Wilkins, 530 Walnut Street, Philadelphia, PA 19106-3621, United States, Tel: +1(215)5218342, Fax: +1(215)5218326 ISSUES/YR: 4

Journal of Burn Care and Rehabilitation
ISSN: 0273-8481
J. Burn Care Rehabil.
Lippincott Williams and Wilkins, 530 Walnut Street, Philadelphia, PA 19106-3621, United States, Tel: +1(215)5218342, Fax: +1(215)5218326 ISSUES/YR: 6

- **Journal of Cancer Education**
ISSN: 0885-8195
J. Cancer Educ.
Lawrence Erlbaum Associates Inc., 10 Industrial Avenue, Mahwah, NJ 07430-2262, United States, Tel: +1(201)2582200, Fax: +1(201)2360072 ISSUES/YR: 4

Journal of Cancer Integrative Medicine
ISSN: 1544-6301
J. Cancer Integr. Med.
Prime National Publishing Corp., 470 Boston Post Road, Weston, MA 02193-1529, United States, Tel: +1(781)8992702, Fax: +1(781)8994900 ISSUES/YR: 4

- **Journal of Cancer Research and Clinical Oncology**
ISSN: 0171-5216
J. Cancer Res. Clin. Oncol.
Springer Verlag, Tiergartenstrasse 17, D-69121, Heidelberg, Germany, Tel: +49(6221)4870, Fax: +49(6221)487366 ISSUES/YR: 12

- **Journal of Cancer Research and Clinical Oncology, Supplement**
ISSN: 0943-9382
J. Cancer Res. Clin. Oncol. Suppl.
Springer Verlag, Tiergartenstrasse 17, D-69121, Heidelberg, Germany, Tel: +49(6221)4870, Fax: +49(6221)487366 ISSUES/YR: 1

Journal of Carcinogenesis
ISSN: 1477-3163
J. Carcinogenesis
BioMed Central Ltd., 34 - 42 Cleveland Street, W1T 4LB, London, United Kingdom, Tel: +44(207)3230323, Fax: +44(207)6319926 ISSUES/YR: 1

- **Journal of Cardiac Failure**
ISSN: 1071-9164
J. Card. Fail.
Churchill Livingstone Inc., 650 Avenue of the Americas, New York, NY 10011, United States, Tel: +1(212)2065000, Fax: +1(212)7277808 ISSUES/YR: 9

Journal of Cardiac Surgery
ISSN: 0886-0440
J. Card. Surg.
Blackwell Publishing Inc., 350 Main Street, Malden, MA 02148, United States, Tel: +1(781)3888200, Fax: +1(781)3888210 ISSUES/YR: 6

Journal of Cardiology
ISSN: 0914-5087
J. Cardiol.
Japanese College of Cardiology (Nippon-Sinzobyo-Gakkai), Hongo 4-9-22, Bunkyo-ku, 113, Tokyo, Japan, Tel: +81(3)58020112, Fax: +81(3)58020118 ISSUES/YR: 12

Journal of Cardiopulmonary Rehabilitation
ISSN: 0883-9212
J. Cardiopulm. Rehabil.
Lippincott Williams and Wilkins, 530 Walnut Street, Philadelphia, PA 19106-3621, United States, Tel: +1(215)5218342, Fax: +1(215)5218326 ISSUES/YR: 6

- **Journal of Cardiothoracic and Vascular Anesthesia**
ISSN: 1053-0770
J. Cardiothorac. Vasc. Anesth.
W.B. Saunders, Independence Square West, Philadelphia, PA 19106-3399, United States, Tel: +1(215)2387800, Fax: +1(215)2386445 ISSUES/YR: 6

Journal of Cardiothoracic-Renal Research
ISSN: 1574-0668
J. Cardiothorac.-Renal Res.
Elsevier, P.O. Box 211, 1000 AE, Amsterdam, Netherlands, Tel: +31(20)4853757, Fax: +31(20)4853432 ISSUES/YR: 4

- **Journal of Cardiovascular Electrophysiology**
ISSN: 1045-3873
J. Cardiovasc. Electrophysiol.
Blackwell Publishing Inc., 350 Main Street, Malden, MA 02148, United States, Tel: +1(781)3888200, Fax: +1(781)3888210 ISSUES/YR: 12

- **Journal of Cardiovascular Magnetic Resonance**
ISSN: 1097-6647
J. Cardiovasc. Magn. Reson.
Taylor and Francis Inc., 325 Chestnut St, Suite 800, Philadelphia PA, PA 19106, United States, Tel: +1(800)3541420, Fax: +1(215)6258914 ISSUES/YR: 6

Journal of Cardiovascular Medicine
ISSN: 1558-2027
J. Cardiovasc. Med.
Lippincott Williams and Wilkins, 345 Hudson St., 16th Fl., New York, NY 10014-4502, United States, Tel: +1-212 886 1200, Fax: +1-212 886 1215 ISSUES/YR: 12

- **Journal of Cardiovascular Pharmacology**
ISSN: 0160-2446
J. Cardiovasc. Pharmacol.
Lippincott Williams and Wilkins, 530 Walnut Street, Philadelphia, PA 19106-3621, United States, Tel: +1(215)5218342, Fax: +1(215)5218326 ISSUES/YR: 12

- **Journal of Cardiovascular Pharmacology and Therapeutics**
ISSN: 1074-2484
J. Cardiovasc. Pharmacol. Ther.
Westminster Publications Inc., 708 Glen Cove Avenue, Glen Head, NY 11545, United States, Tel: +1(516)7590025, Fax: +1(516)7595524 ISSUES/YR: 4

Journal of Cardiovascular Surgery
ISSN: 0021-9509
J. Cardiovasc. Surg.
Edizioni Minerva Medica S.p.A., Corso Bramante 83-85, 10126, Turin, Italy, Tel: +39(011)678282, Fax: +39(011)674502 ISSUES/YR: 6

- **Journal of Cataract and Refractive Surgery**
ISSN: 0886-3350
J. Cataract Refractive Surg.
Elsevier Inc., 360 Park Avenue South, New York, NY 10010, United States, Tel: +1(212)6333950, Fax: +1(212)6333990 ISSUES/YR: 12

- **Journal of Cell Biology**
ISSN: 0021-9525
J. Cell Biol.
Rockefeller University Press, 1114 First Avenue, 4th Floor, New York, NY 10021-8325, United States, Tel: +1(212)3278572, Fax: +1(212)3277944 ISSUES/YR: 26

- **Journal of Cell Science**
ISSN: 0021-9533
J. Cell Sci.
Company of Biologists Ltd, 140 Cowley Road, CB4 4DL, Cambridge, United Kingdom, Tel: +44(1223)426164, Fax: +44(1223)423353 ISSUES/YR: 24

- **Journal of Cellular Biochemistry**
ISSN: 0730-2312
J. Cell. Biochem.
Wiley-Liss Inc., 111 River Street, Hoboken, NJ 07030-5774, United States, Tel: +1(201)7486645, Fax: +1(201)7486021 ISSUES/YR: 18

Journal of Cellular Pathology
ISSN: 1359-7388
J. Cell. Pathol.
Greenwich Medical Media Ltd, 137 Euston Road, NW1 2AA, London, United Kingdom, Tel: +44(207)3885444, Fax: +44(207)3835445 ISSUES/YR: 4

- **Journal of Cellular Physiology**
ISSN: 0021-9541
J. Cell. Physiol.
Wiley-Liss Inc., 111 River Street, Hoboken, NJ 07030-5774, United States, Tel: +1(201)7486645, Fax: +1(201)7486021 ISSUES/YR: 12

Journal of Central South University (Medical Sciences)
ISSN: 1672-7347
J. Cent. South Univ. Med. Sci.
Journal of Central South University (Medical Sciences), P.O. Box 75, 410078, 110 Xiangya Road, Changsha Hunan, China ISSUES/YR: 6

- **Journal of Cerebral Blood Flow and Metabolism**
ISSN: 0271-678X
J. Cereb. Blood Flow Metab.
Nature Publishing Group, Houndmills, RG21 6XS, Basingstoke, Hampshire, United Kingdom, Tel: +1(1256)329242, Fax: +1(1256)810526 ISSUES/YR: 12

- **Journal of Chemical Neuroanatomy**
ISSN: 0891-0618
J. Chem. Neuroanat.
Elsevier, P.O. Box 211, 1000 AE, Amsterdam, Netherlands, Tel: +31(20)4853757, Fax: +31(20)4853432 ISSUES/YR: 8

Journal of Chemical Technology and Biotechnology
ISSN: 0268-2575
J. Chem. Technol. Biotechnol.
John Wiley and Sons Ltd, Southern Gate, PO19 8SQ, Chichester, West Sussex, United Kingdom, Tel: +44(1243)779777, Fax: +44(1243)775878 ISSUES/YR: 12

Journal of Chemotherapy
ISSN: 1120-009X
J. Chemother.
E.I.F.T. srl, Via XX Settembre 102, 50129, Florence, Italy, Tel: +39(055)486147 ISSUES/YR: 6

Journal of Child and Adolescent Mental Health
ISSN: 1728-0583
J. Child Adolesc. Ment. Health
NISC Pty Ltd, 19 Worcester Street, 6140, Grahamstown, South Africa, Tel: +27(46)622 9698, Fax: +27(46)622 9550 ISSUES/YR: 2

Journal of Child and Adolescent Psychopharmacology
ISSN: 1044-5463
J. Child Adolesc. Psychopharmacol.
Mary Ann Liebert Inc., 140 Huguenot Street, New Rochelle, NY 10801-5215, United States, Tel: +1(914)7402100, Fax: +1(914)7402101 ISSUES/YR: 6

- **Journal of Child Neurology**
ISSN: 0883-0738
J. Child Neurol.
BC Decker Inc., P.O. Box 620, LCD 1, L8N 3K7, Hamilton, Ontario, Canada, Tel: +1(905)522-7017, Fax: +1(905)522-7839 ISSUES/YR: 12

Journal of Child Psychology and Psychiatry and Allied Disciplines
ISSN: 0021-9630
J. Child Psychol. Psychiatry Allied Discip.
Blackwell Publishing Ltd, 9600 Garsington Road, OX4 2XG, Oxford, United Kingdom, Tel: +44(1865)776868, Fax: +44(1865)714591 ISSUES/YR: 11

Journal of Child Psychotherapy
ISSN: 0075-417X
J. Child Psychother.
Routledge, 4 Park Square, Milton Park, OX14 4RN, Abingdon, Oxfordshire, United Kingdom, Tel: +44(1235)828600, Fax: +44(1235)829000 ISSUES/YR: 3

Journal of Child Sexual Abuse
ISSN: 1053-8712
J. Child Sex. Abuse
Haworth Press Inc., 10 Alice Street, Binghamton, NY 13904-1580, United States, Tel: +1(607)7225857, Fax: +1(607)7710012 ISSUES/YR: 4

Journal of China Pharmaceutical University
ISSN: 1000-5048
J. China Pharm. Univ.
China Pharmaceutical University, 24 Tongjia Xiang, 210009, Nanjing, Jiangsu, China ISSUES/YR: 6

Journal of Chinese Integrative Medicine
ISSN: 1672-1977
J. Chin. Integr. Med.
Editorial Board of Journal of Chinese Integrative Medicine, 174, Changhai Road, 200433, Shanghai, China ISSUES/YR: 6

Journal of Chinese Medicine
ISSN: 0143-8042
J. Chin. Med.
Journal of Chinese Medicine, 22 Cromwell Road, BN3 3EB, Hove, Sussex, United Kingdom, Tel: +44(1273)777760, Fax: +44(1273)748588 ISSUES/YR: 3

- **Journal of Chromatography A**
ISSN: 0021-9673
J. Chromatogr. A
Elsevier, P.O. Box 211, 1000 AE, Amsterdam, Netherlands, Tel: +31(20)4853757, Fax: +31(20)4853432 ISSUES/YR: 80

- **Journal of Chromatography B: Analytical Technologies in the Biomedical and Life Sciences**
ISSN: 1570-0232
J. Chromatogr. B Anal. Technol. Biomed. Life Sci.
Elsevier, P.O. Box 211, 1000 AE, Amsterdam, Netherlands, Tel: +31(20)4853757, Fax: +31(20)4853432 ISSUES/YR: 30

Journal of Chronic Fatigue Syndrome
ISSN: 1057-3321
J. Chronic Fatigue Syndr.
Haworth Press Inc., 10 Alice Street, Binghamton, NY 13904-1580, United States, Tel: +1(607)7225857 ISSUES/YR: 4

Journal of Circadian Rhythms
ISSN: 1740-3391
J. Circadian Rhythms
BioMed Central Ltd., 34 - 42 Cleveland Street, W1T 4LB, London, United Kingdom, Tel: +44(207)3230323, Fax: +44(207)6319926 ISSUES/YR: 1

- **Journal of Clinical and Basic Cardiology**
ISSN: 1561-2775
J. Clin. Basic Cardiol.
Krause und Pachernegg GmbH, Mozartgasse 10, A-3003, Gablitz, Austria, Tel: +43(2231)612580, Fax: +43(2231)6125810 ISSUES/YR: 4

- **Journal of Clinical and Experimental Neuropsychology**
ISSN: 1380-3395
J. Clin. Exp. Neuropsychol.
Psychology Press Ltd, 4 Park Square, Milton Park, OX14 4RN, Abingdon, Oxfordshire, United Kingdom, Tel: +44(1235)828600, Fax: +44(1235)829000 ISSUES/YR: 8

- **Journal of Clinical Anesthesia**
ISSN: 0952-8180
J. Clin. Anesth.
Elsevier Inc., 360 Park Avenue South, New York, NY 10010, United States, Tel: +1(212)6333950, Fax: +1(212)6333990 ISSUES/YR: 8

Journal of Clinical Apheresis
ISSN: 0733-2459
J. Clin. Apheresis
Wiley-Liss Inc., 111 River Street, Hoboken, NJ 07030-5774, United States, Tel: +1(201)7486645, Fax: +1(201)7486021 ISSUES/YR: 4

Journal of Clinical Biochemistry and Nutrition
ISSN: 0912-0009
J. Clin. Biochem. Nutr.
Institute of Applied Biochemistry, 16-9 Hongokomagome, 5-chome, Bunkyo-ku, 113-8622, Tokyo, Japan, Tel: +81(3)58145811, Fax: +81(3)58145822 ISSUES/YR: 6

- **Journal of Clinical Densitometry**
ISSN: 1094-6950
J. Clin. Densitometry
Elsevier Inc., 360 Park Avenue South, New York, NY 10010, United States, Tel: +1(212)6333950, Fax: +1(212)6333990 ISSUES/YR: 4

Journal of Clinical Dermatology
ISSN: 1000-4963
J. Clin. Dermatol.
Editorial Board of Journal of Clinical, 300, Guangzhou Road, 210029, Nanjing, China, Tel: +86-25 83729951, Fax: +86-25 83713795 ISSUES/YR: 12

- **Journal of Clinical Endocrinology and Metabolism**
ISSN: 0021-972X
J. Clin. Endocrinol. Metab.
Endocrine Society, 8401 Connecticut Ave. Suite 900, Chevy Chase, MD 20815, United States, Tel: +1-888-363-6274 ISSUES/YR: 12

Journal of Clinical Engineering
ISSN: 0363-8855
J. Clin. Eng.
Lippincott Williams and Wilkins, 530 Walnut Street, Philadelphia, PA 19106-3621, United States, Tel: +1(215)5218342, Fax: +1(215)5218326 ISSUES/YR: 6

- **Journal of Clinical Epidemiology**
ISSN: 0895-4356
J. Clin. Epidemiol.
Elsevier USA, 6277 Sea Harbor Drive, Orlando, FL 32862 8239, United States, Tel: +1-407 345 4000, Fax: +1-407 363 9661 ISSUES/YR: 12

Journal of Clinical Forensic Medicine
ISSN: 1353-1131
J. Clin. Forensic Med.
Churchill Livingstone, 1-3 Baxter's Place, Leith Walk, EH1 3AF, Edinburgh, United Kingdom, Tel: +44(131)5562424, Fax: +44(131)5581278 ISSUES/YR: 6

- **Journal of Clinical Gastroenterology**
ISSN: 0192-0790
J. Clin. Gastroenterol.
Lippincott Williams and Wilkins, 530 Walnut Street, Philadelphia, PA 19106-3621, United States, Tel: +1(215)5218342, Fax: +1(215)5218326 ISSUES/YR: 10

- **Journal of Clinical Immunology**
ISSN: 0271-9142
J. Clin. Immunol.
Springer Science and Business Media Deutschland GmbH, Heidelberger Platz 3, 14197, Berlin, Germany, Tel: +1-30 827875282, Fax: +1-30 827875707 ISSUES/YR: 6

- **Journal of Clinical Investigation**
ISSN: 0021-9738
J. Clin. Invest.
The American Society for Clinical Investigation, P.O. Box 7226, Ann Arbor, MI 48107, United States, Tel: +1(734)222-6050, Fax: +1(734)222-6058 ISSUES/YR: 12

Current

Journal of Clinical Laboratory Analysis
ISSN: 0887-8013
J. Clin. Lab. Anal.
Wiley-Liss Inc., 111 River Street, Hoboken, NJ 07030-5774, United States, Tel: +1(201)7486645, Fax: +1(201)7486021 ISSUES/YR: 6

Journal of Clinical Ligand Assay
ISSN: 1081-1672
J. Clin. Ligand Assay
Clinical Ligand Assay Society Inc., 3139 S. Wayne Rd, Wayne, MI 48184, United States, Tel: +1(313)7226290 ISSUES/YR: 4

• **Journal of Clinical Microbiology**
ISSN: 0095-1137
J. Clin. Microbiol.
American Society for Microbiology, 1752 N Street N.W., Washington, DC 20036-2904, United States, Tel: +1(202)942-9244, Fax: +1(202)942-9355 ISSUES/YR: 12

• **Journal of Clinical Monitoring and Computing**
ISSN: 1387-1307
J. Clin. Monit. Comput.
Kluwer Academic Publishers, Van Godewijckstraat 30, 3311 GZ, Dordrecht, Netherlands, Tel: +31(78)6576000, Fax: +31(78)6576254 ISSUES/YR: 6

• **Journal of Clinical Neuromuscular Disease**
ISSN: 1522-0443
J. Clin. Neuromuscular Dis.
Lippincott Williams and Wilkins, 530 Walnut Street, Philadelphia, PA 19106-3621, United States, Tel: +1(215)5218342, Fax: +1(215)5218326 ISSUES/YR: 4

Journal of Clinical Neurophysiology
ISSN: 0736-0258
J. Clin. Neurophysiol.
Lippincott Williams and Wilkins, 530 Walnut Street, Philadelphia, PA 19106-3621, United States, Tel: +1(215)5218342, Fax: +1(215)5218326 ISSUES/YR: 6

• **Journal of Clinical Neuroscience**
ISSN: 0967-5868
J. Clin. Neurosci.
Churchill Livingstone, 1-3 Baxter's Place, Leith Walk, EH1 3AF, Edinburgh, United Kingdom, Tel: +44(131)5562424, Fax: +44(131)5581278 ISSUES/YR: 8

• **Journal of Clinical Oncology**
ISSN: 0732-183X
J. Clin. Oncol.
American Society of Clinical Oncology, 330 John Carlyle Street, Suite 300, Alexandria, VA 22314, United States, Tel: +1(703)5191430, Fax: +1(703)5188155 ISSUES/YR: 24

Journal of Clinical Outcomes Management
ISSN: 1079-6533
J. Clin. Outcomes Manage.
Turner White Communications Inc., 125 Strafford Avenue, Suite 220, Wayne, PA 19087-3391, United States, Tel: +1(610)9754541, Fax: +1(610)9754564 ISSUES/YR: 12

• **Journal of Clinical Pathology**
ISSN: 0021-9746
J. Clin. Pathol.
BMJ Publishing Group, Tavistock Square, WC1H 9JR, London, United Kingdom, Tel: +44(207)3874499, Fax: +44(207)3836668 ISSUES/YR: 12

Journal of Clinical Pharmacology
ISSN: 0091-2700
J. Clin. Pharmacol.
SAGE Publications Inc., 2455 Teller Road, Thousand Oaks, CA 91320, United States, Tel: +1(805)4990721, Fax: +1(805)4990871 ISSUES/YR: 12

Journal of Clinical Pharmacy and Therapeutics
ISSN: 0269-4727
J. Clin. Pharm. Ther.
Blackwell Publishing Ltd, 9600 Garsington Road, OX4 2XG, Oxford, United Kingdom, Tel: +44(1865)776868, Fax: +44(1865)714591 ISSUES/YR: 6

• **Journal of Clinical Psychiatry**
ISSN: 0160-6689
J. Clin. Psychiatry
Physicians Postgraduate Press Inc., P.O. Box 752870, Memphis, TN 38119-2870, United States, Tel: +1(901)7513800 ISSUES/YR: 12

Journal of Clinical Psychoanalysis
ISSN: 1076-044X
J. Clin. Psychoanal.
International Universities Press Inc., P.O. Box 389, Guilford, CT 0643-0389, United States, Tel: +1(203)2454000, Fax: +1(203)2450775 ISSUES/YR: 4

Journal of Clinical Psychology
ISSN: 0021-9762
J. Clin. Psychol.
John Wiley and Sons Inc., P.O.Box 18667, Newark, NJ 07191-8667, United States, Tel: +1(212)8506000, Fax: +1(212)8506012 ISSUES/YR: 12

• **Journal of Clinical Psychology in Medical Settings**
ISSN: 1068-9583
J. Clin. Psychol. Med. Settings
Springer Science and Business Media Deutschland GmbH, Heidelberger Platz 3, 14197, Berlin, Germany, Tel: +1-30 827875282, Fax: +1-30 827875707 ISSUES/YR: 4

• **Journal of Clinical Psychopharmacology**
ISSN: 0271-0749
J. Clin. Psychopharmacol.
Lippincott Williams and Wilkins, 351 West Camden Street, Baltimore, MD 21201-2436, United States, Tel: +1(410)5284000 ISSUES/YR: 6

Journal of Clinical Research
ISSN: 1369-5207
J. Clin. Res.
PJB Publications, Sheepen Place, C03 3LP, Colchester, United Kingdom, Tel: +44-20 70175540 ISSUES/YR: 1

• **Journal of Clinical Rheumatology**
ISSN: 1076-1608
J. Clin. Rheumatol.
Lippincott Williams and Wilkins, 351 West Camden Street, Baltimore, MD 21201-2436, United States, Tel: +1(410)5284000 ISSUES/YR: 6

Journal of Clinical Sleep Medicine
ISSN: 1550-9389
J. Clin. Sleep Med.
American Academy of Sleep Medicine, 1 Westbrook Corporate Center, Suite 920, Westchester, IL 60154, United States, Tel: +1(708)4920930, Fax: +1(708)4920943 ISSUES/YR: 4

Journal of Clinical Ultrasound
ISSN: 0091-2751
J. Clin. Ultrasound
John Wiley and Sons Inc., P.O.Box 18667, Newark, NJ 07191-8667, United States, Tel: +1(212)8506000, Fax: +1(212)8506012 ISSUES/YR: 9

• **Journal of Clinical Virology**
ISSN: 1386-6532
J. Clin. Virol.
Elsevier, P.O. Box 211, 1000 AE, Amsterdam, Netherlands, Tel: +31(20)4853757, Fax: +31(20)4853432 ISSUES/YR: 12

Journal of Cognition and Development
ISSN: 1524-8372
J. Cogn. Dev.
Lawrence Erlbaum Associates Inc., 10 Industrial Avenue, Mahwah, NJ 07430-2262, United States, Tel: +1(201)2582200, Fax: +1(201)2360072 ISSUES/YR: 4

• **Journal of Cognitive Neuroscience**
ISSN: 0898-929X
J. Cogn. Neurosci.
MIT Press, 238 Main Street, Suite 500, Cambridge, MA 02142-1046, United States, Tel: +1(617)2535646, Fax: +1(617)5771545 ISSUES/YR: 10

Journal of Cognitive Psychotherapy
ISSN: 0889-8391
J. Cogn. Psychother.
Springer Publishing Company, 11 West 42nd Street, 15th Floor, New York, NY 10036, United States, Tel: +1(212)4314370, Fax: +1(212)9417842 ISSUES/YR: 4

• **Journal of Colloid and Interface Science**
ISSN: 0021-9797
J. Colloid Interface Sci.
Academic Press Inc., 6277 Sea Harbor Drive, Orlando, FL 32887-4900, United States, Tel: +1(407)3454040, Fax: +1(407)3639661 ISSUES/YR: 24

Journal of Commercial Biotechnology
ISSN: 1462-8732
J. Commer. Biotechnol.
Palgrave Macmillan Ltd., Houndmills, RG21 6XS, Basingstoke, Hants., United Kingdom, Tel: +44(1256)329242, Fax: +44(1256)810526 ISSUES/YR: 4

Journal of Communication Disorders
ISSN: 0021-9924
J. Commun. Disord.
Elsevier Inc., 360 Park Avenue South, New York, NY 10010, United States, Tel: +1(212)6333950, Fax: +1(212)6333990 ISSUES/YR: 6

Journal of Community Health
ISSN: 0094-5145
J. Community Health
Kluwer Academic Publishers, Van Godewijckstraat 30, 3311 GZ, Dordrecht, Netherlands, Tel: +31(78)6576000, Fax: +31(78)6576254 ISSUES/YR: 6

Journal of Community Health Nursing
ISSN: 0737-0016
J. Community Health Nurs.
Lawrence Erlbaum Associates Inc., 10 Industrial Avenue, Mahwah, NJ 07430-2262, United States, Tel: +1(201)2582200, Fax: +1(201)2360072 ISSUES/YR: 4

Journal of Comparative Family Studies
ISSN: 0047-2328
J. Comp. Fam. Stud.
University of Calgary, 2500 University Drive NW, ALB TN2 1N4, Calgary, Canada, Tel: +1(403)2207317 ISSUES/YR: 4

• **Journal of Comparative Neurology**
ISSN: 0021-9967
J. Comp. Neurol.
Wiley-Liss Inc., 111 River Street, Hoboken, NJ 07030-5774, United States, Tel: +1(201)7486645, Fax: +1(201)7486021 ISSUES/YR: 36

Journal of Comparative Pathology
ISSN: 0021-9975
J. Comp. Pathol.
W.B. Saunders Ltd, 32 Jamestown Road, NW1 7BY, London, United Kingdom, Tel: +44(20)74244200, Fax: +44(20)74854752 ISSUES/YR: 8

Journal of Complementary and Integrative Medicine
ISSN: 1553-3840
J. Complement. Integr. Med.
Berkeley Electronic Press, 2809 Telegraph Avenue, Suite 202, Berkeley, CA 94705, United States, Tel: +1(510)6651200, Fax: +1(510)6651201 ISSUES/YR: 3

• **Journal of Computational Biology**
ISSN: 1066-5277
J. Comput. Biol.
Mary Ann Liebert Inc., 140 Huguenot Street, New Rochelle, NY 10801-5215, United States, Tel: +1(914)7402100, Fax: +1(914)7402101 ISSUES/YR: 6

• **Journal of Computational Neuroscience**
ISSN: 0929-5313
J. Comput. Neurosci.
Kluwer Academic Publishers, Van Godewijckstraat 30, 3311 GZ, Dordrecht, Netherlands, Tel: +31(78)6576000, Fax: +31(78)6576254 ISSUES/YR: 6

• **Journal of Computer Assisted Tomography**
ISSN: 0363-8715
J. Comput. Assisted Tomogr.
Lippincott Williams and Wilkins, 530 Walnut Street, Philadelphia, PA 19106-3621, United States, Tel: +1(215)5218342, Fax: +1(215)5218326 ISSUES/YR: 6

• **Journal of Computer-Aided Molecular Design**
ISSN: 0920-654X
J. Comp.-Aided Mol. Des.
Kluwer Academic Publishers, Van Godewijckstraat 30, 3311 GZ, Dordrecht, Netherlands, Tel: +31(78)6576000, Fax: +31(78)6576254 ISSUES/YR: 12

Journal of Consulting and Clinical Psychology
ISSN: 0022-006X
J. Consult. Clin. Psychol.
American Psychological Association Inc., 750 First Street NE, Washington, DC 20002-4242, United States, Tel: +1(202)3365600, Fax: +1(202)3365568 ISSUES/YR: 6

• **Journal of Contaminant Hydrology**
ISSN: 0169-7722
J. Contam. Hydrol.
Elsevier, P.O. Box 211, 1000 AE, Amsterdam, Netherlands, Tel: +31(20)4853757, Fax: +31(20)4853432 ISSUES/YR: 20

Journal of Contemporary Dental Practice
ISSN: 1526-3711
J. Contemp. Dental Pract.
Procter and Gamble, 1 Procter and Gamble Plaza, Cincinnati, OH 45201, United States, Tel: +1(513)9831100 ISSUES/YR: 4

Journal of Contemporary Psychotherapy
ISSN: 0022-0116
J. Contemp. Psychother.
Kluwer Academic/Human Sciences Press Inc., 233 Spring Street, New York, NY 10013-1578, United States, Tel: +1(212)6208000, Fax: +1(212)4630742 ISSUES/YR: 4

• **Journal of Controlled Release**
ISSN: 0168-3659
J. Control. Release
Elsevier, P.O. Box 211, 1000 AE, Amsterdam, Netherlands, Tel: +31(20)4853757, Fax: +31(20)4853432 ISSUES/YR: 21

• **Journal of Cosmetic and Laser Therapy**
ISSN: 1476-4172
J. Cosmet. Laser Ther.
Taylor and Francis Ltd., 4 Park Square, Milton Park, OX14 4RN, Abingdon, Oxfordshire, United Kingdom, Tel: +44(20)70175541, Fax: +44(20)70175198 ISSUES/YR: 4

• **Journal of Cosmetic Dermatology**
ISSN: 1473-2130
J. Cosmet. Dermatol.
Blackwell Publishing Ltd, 9600 Garsington Road, OX4 2XG, Oxford, United Kingdom, Tel: +44(1865)776868, Fax: +44(1865)714591 ISSUES/YR: 4

Journal of Counseling Psychology
ISSN: 0022-0167
J. Couns. Psychol.
American Psychological Association Inc., 750 First Street NE, Washington, DC 20002-4242, United States, Tel: +1(202)3365600, Fax: +1(202)3365568 ISSUES/YR: 4

Journal of Cranio-Maxillofacial Surgery
ISSN: 1010-5182
J. Cranio-Maxillofac. Surg.
Churchill Livingstone, 1-3 Baxter's Place, Leith Walk, EH1 3AF, Edinburgh, United Kingdom, Tel: +44(131)5562424, Fax: +44(131)5581278 ISSUES/YR: 6

• **Journal of Craniofacial Surgery**
ISSN: 1049-2275
J. Craniofac. Surg.
Lippincott Williams and Wilkins, 530 Walnut Street, Philadelphia, PA 19106-3621, United States, Tel: +1(215)5218342, Fax: +1(215)5218326 ISSUES/YR: 6

Journal of Critical Care
ISSN: 0883-9441
J. Crit. Care
W.B. Saunders, Independence Square West, Philadelphia, PA 19106-3399, United States, Tel: +1(215)2387800, Fax: +1(215)2386445 ISSUES/YR: 4

Journal of Critical Illness
ISSN: 1040-0257
J. Crit. Illn.
Cliggott Publishing Co., 330 Boston Post Road, Box 4027, Darien, CT 06820-4027, United States, Tel: +1(203)6626400 ISSUES/YR: 12

• **Journal of Cutaneous Medicine and Surgery**
ISSN: 1203-4754
J. Cutaneous Med. Surg.
Springer New York LLC, 233 Springer Street, New York, 10013-1578, United States, Tel: +1-212 460 1501, Fax: +1-212 460 1595 ISSUES/YR: 6

Journal of Cutaneous Pathology
ISSN: 0303-6987
J. Cutaneous Pathol.
Blackwell Publishing Ltd, 9600 Garsington Road, OX4 2XG, Oxford, United Kingdom, Tel: +44(1865)776868, Fax: +44(1865)714591 ISSUES/YR: 10

Journal of Cystic Fibrosis
ISSN: 1569-1993
J. Cyst. Fibrosis
Elsevier, P.O. Box 211, 1000 AE, Amsterdam, Netherlands, Tel: +31(20)4853757, Fax: +31(20)4853432 ISSUES/YR: 4

Journal of Dementia Care
ISSN: 1351-8372
J. Dementia Care
Hawker Publications Ltd, Culvert Road, SW11 5DH, London, United Kingdom, Tel: +44(207)7202108, Fax: +44(207)4983023 ISSUES/YR: 6

● **Journal of Dermatological Science**
ISSN: 0923-1811
J. Dermatol. Sci.
Elsevier Ireland Ltd, P.O. Box 85, Limerick, Ireland, Tel: +353(61)471944, Fax: +353(61)472144 ISSUES/YR: 9

● **Journal of Dermatological Treatment**
ISSN: 0954-6634
J. Dermatol. Treat.
Taylor and Francis Ltd., 4 Park Square, Milton Park, OX14 4RN, Abingdon, Oxfordshire, United Kingdom, Tel: +44(20)70175541, Fax: +44(20)70175198 ISSUES/YR: 6

Journal of Dermatology
ISSN: 0385-2407
J. Dermatol.
Blackwell Publishing Ltd, 9600 Garsington Road, OX4 2XG, Oxford, United Kingdom, Tel: +44(1865)776868, Fax: +44(1865)714591 ISSUES/YR: 12

● **Journal of Developmental and Behavioral Pediatrics**
ISSN: 0196-206X
J. Dev. Behav. Pediatr.
Lippincott Williams and Wilkins, 530 Walnut Street, Philadelphia, PA 19106-3621, United States, Tel: +1(215)5218342, Fax: +1(215)5218326 ISSUES/YR: 6

● **Journal of Diabetes and its Complications**
ISSN: 1056-8727
J. Diabetes Complications
Elsevier Inc., 360 Park Avenue South, New York, NY 10010, United States, Tel: +1(212)6333950, Fax: +1(212)6333990 ISSUES/YR: 6

● **Journal of Diagnostic Medical Sonography**
ISSN: 8756-4793
J. Diagn. Med. Sonogr.
SAGE Publications Inc., 2455 Teller Road, Thousand Oaks, CA 91320, United States, Tel: +1(805)4990721, Fax: +1(805)4990871 ISSUES/YR: 6

Journal of Diagnostic Radiography and Imaging
ISSN: 1460-4728
J. Diagn. Radiogr. Imaging
Cambridge University Press, Shaftesbury Road, CB2 2RU, Cambridge, United Kingdom, Tel: +44(1223)312393, Fax: +44(1223)315052 ISSUES/YR: 4

● **Journal of Digital Imaging**
ISSN: 0897-1889
J. Digit. Imaging
Springer New York LLC, 233 Springer Street, New York, 10013-1578, United States, Tel: +1-212 460 1501, Fax: +1-212 460 1595 ISSUES/YR: 4

Journal of Drug Delivery Science and Technology
ISSN: 1157-1489
J. Drug Deliv. Sci. Technol.
Editions de Sante, 49 rue Galilee, 75116, Paris, France, Tel: +33(1)40701615, Fax: +33(1)40701614 ISSUES/YR: 6

Journal of Drug Education
ISSN: 0047-2379
J. Drug Educ.
Baywood Publishing Co. Inc., 26 Austin Avenue, P.O. Box 337, Amityville, NY 11701, United States, Tel: +1(516)6911270, Fax: +1(516)6911770 ISSUES/YR: 4

Journal of Drug Issues
ISSN: 0022-0426
J. Drug Issues
Journal of Drug Issues Inc., Hecht House, 634 W. Call Street, Tallahassee, FL 32306-1127, United States, Tel: +1(850)6647368, Fax: +1(850)6449614 ISSUES/YR: 4

● **Journal of Drug Targeting**
ISSN: 1061-186X
J. Drug Targeting
Taylor and Francis Ltd., 4 Park Square, Milton Park, OX14 4RN, Abingdon, Oxfordshire, United Kingdom, Tel: +44(20)70175541, Fax: +44(20)70175198 ISSUES/YR: 10

Journal of Dual Diagnosis
ISSN: 1550-4263
J. Dual Diagn.
Haworth Press Inc., 10 Alice Street, Binghamton, NY 13904-1580, United States, Tel: +1(607)7225857 ISSUES/YR: 4

Journal of Ecophysiology and Occupational Health
ISSN: 0972-4397
J. Ecophysiol. Occup. Health
Journal of Ecophysiology and Occupational Health, P.O. Box 80, M.G. Marg, 226001, Lucknow, India, Tel: +91(522)220207, Fax: +91(522)228227 ISSUES/YR: 4

Journal of ECT
ISSN: 1095-0680
J. ECT
Lippincott Williams and Wilkins, 530 Walnut Street, Philadelphia, PA 19106-3621, United States, Tel: +1(215)5218342, Fax: +1(215)5218326 ISSUES/YR: 4

● **Journal of Electrocardiology**
ISSN: 0022-0736
J. Electrocardiol.
Churchill Livingstone Inc., 650 Avenue of the Americas, New York, NY 10011, United States, Tel: +1(212)2065000, Fax: +1(212)7277808 ISSUES/YR: 4

● **Journal of Electromyography and Kinesiology**
ISSN: 1050-6411
J. Electromyogr. Kinesiology
Elsevier Ltd, Langford Lane, Kidlington, OX5 1GB, Oxford, United Kingdom, Tel: +44(1865)843000 ISSUES/YR: 6

● **Journal of Emergency Medicine**
ISSN: 0736-4679
J. Emerg. Med.
Elsevier USA, 6277 Sea Harbor Drive, Orlando, FL 32862 8239, United States, Tel: +1-407 345 4000, Fax: +1-407 363 9661 ISSUES/YR: 8

Journal of Emergency Primary Health Care
ISSN: 1447-4999
J. Emerg. Prim. Health Care
Centre for Ambulance and Paramedic Studies, McMahons Road, 3199, Frankston, Victoria, Australia, Tel: +61(39904)4198, Fax: +61(39904)4168 ISSUES/YR: 4

Journal of Emotional Abuse
ISSN: 1092-6798
J. Emot. Abuse
Haworth Press Inc., 10 Alice Street, Binghamton, NY 13904-1580, United States, Tel: +1(607)7225857, Fax: +1(607)7221424 ISSUES/YR: 4

Journal of Endocrine Genetics
ISSN: 1565-012X
J. Endocr. Genet.
Freund Publishing House Ltd, 61 Nachmani Street, P.O. Box 35010, 61350, Tel Aviv, Israel, Tel: +972(3)5628540, Fax: +972(3)5628538 ISSUES/YR: 4

Journal of Endocrinological Investigation
ISSN: 0391-4097
J. Endocrinol. Invest.
Editrice Kurtis s.r.l., Via Luigi Zoja 30, 20153, Milan, Italy, Tel: +39(02)48202740, Fax: +39(02)48201219 ISSUES/YR: 11

● **Journal of Endocrinology**
ISSN: 0022-0795
J. Endocrinol.
Society for Endocrinology, 22 Apex Court, Woodlands, Bradley Stoke, BS32 4JT, Bristol, United Kingdom, Tel: +44(1454)642220, Fax: +44(1454)642222 ISSUES/YR: 12

Journal of Endocrinology, Metabolism and Diabetes of South Africa
ISSN: 1608-9677
J. Endocrinol. Metab. Diabetes S. Afr.
South African Medical Association, Private Bag X1, 7430, Pinelands, South Africa, Tel: +27(21)5313081, Fax: +27(21)5314126 ISSUES/YR: 3

Journal of Endotoxin Research
ISSN: 0968-0519
J. Endotoxin Res.
Maney Publishing, Hudson Road, LS9 7DL, Leeds, United Kingdom, Tel: +44(113)2497481, Fax: +44(113)2486983 ISSUES/YR: 6

- **Journal of Endourology**
ISSN: 0892-7790
J. Endourol.
Mary Ann Liebert Inc., 140 Huguenot Street, New Rochelle, NY 10801-5215, United States, Tel: +1(914)7402100, Fax: +1(914)7402101 ISSUES/YR: 10

- **Journal of Endovascular Therapy**
ISSN: 1526-6028
J. Endovasc. Ther.
International Society of Endovascular Specialists, PMB 605, 1928 East Highland Street, #F104, Phoenix, 85016, United States, Tel: +1(602)2405833, Fax: +1(602)2666018 ISSUES/YR: 6

Journal of Environmental Biology
ISSN: 0254-8704
J. Environ. Biol.
JEB Foundation, 771 Civil Lines (South), 251 001, Muzaffarnagar, India, Tel: +91(1312)405306 ISSUES/YR: 4

Journal of Environmental Health
ISSN: 0022-0892
J. Environ. Health
National Environmental Health Association, 720 S. Colorado Blvd, Suite 970 S. Tower, Denver, CO 80222, United States, Tel: +1(303)7569090, Fax: +1(303)6819490 ISSUES/YR: 10

Journal of Environmental Management
ISSN: 0301-4797
J. Environ. Manage.
Academic Press, 24-28 Oval Road, NW1 7DX, London, United Kingdom, Tel: +44(207)2674466 ISSUES/YR: 12

- **Journal of Environmental Monitoring**
ISSN: 1464-0325
J. Environ. Monit.
Royal Society of Chemistry, Milton Road, CB4 OWF, Cambridge, United Kingdom, Tel: +44(1223)420066, Fax: +44(1223)420247 ISSUES/YR: 12

- **Journal of Environmental Pathology, Toxicology and Oncology**
ISSN: 0731-8898
J. Environ. Pathol. Toxicol. Oncol.
Begell House Inc., 145 Madison Avenue 6th Floor, New York, NY 10016, United States, Tel: +1(212)7251999, Fax: +1(212)2138368 ISSUES/YR: 4

Journal of Environmental Quality
ISSN: 0047-2425
J. Environ. Qual.
ASA/CSSA/SSSA, 677 South Segoe Road, Madison, WI 53711, United States ISSUES/YR: 6

Journal of Environmental Radioactivity
ISSN: 0265-931X
J. Environ. Radioact.
Elsevier Ltd, Langford Lane, Kidlington, OX5 1GB, Oxford, United Kingdom, Tel: +44(1865)843000 ISSUES/YR: 12

Journal of Environmental Science and Health - Part A Toxic/Hazardous Substances and Environmental Engineering
ISSN: 1093-4529
J. Environ. Sci. Health Part A Toxic Hazard. Subst. Environ. Eng.
Taylor and Francis Inc., 325 Chestnut St, Suite 800, Philadelphia PA, PA 19106, United States, Tel: +1(800)3541420, Fax: +1(215)6258914 ISSUES/YR: 12

Journal of Environmental Science and Health - Part B Pesticides, Food Contaminants, and Agricultural Wastes
ISSN: 0360-1234
J. Environ. Sci. Health Part B Pestic. Food Contamin. Agric. Wastes
Taylor and Francis Inc., 325 Chestnut St, Suite 800, Philadelphia PA, PA 19106, United States, Tel: +1(800)3541420, Fax: +1(215)6258914 ISSUES/YR: 8

Journal of Environmental Science and Health - Part C Environmental Carcinogenesis and Ecotoxicology Reviews
ISSN: 1059-0501
J. Environ. Sci. Health Part C Environ. Carcinog. Ecotoxicol. Rev.
Taylor and Francis Inc., 325 Chestnut St, Suite 800, Philadelphia PA, PA 19106, United States, Tel: +1(800)3541420, Fax: +1(215)6258914 ISSUES/YR: 2

- **Journal of Enzyme Inhibition and Medicinal Chemistry**
ISSN: 1475-6366
J. Enzyme Inhib. Med. Chem.
Taylor and Francis Ltd., 4 Park Square, Milton Park, OX14 4RN, Abingdon, Oxfordshire, United Kingdom, Tel: +44(20)70175541, Fax: +44(20)70175198 ISSUES/YR: 6

Journal of Epidemiology and Community Health
ISSN: 0143-005X
J. Epidemiol. Community Health
BMJ Publishing Group, Tavistock Square, WC1H 9JR, London, United Kingdom, Tel: +44(207)3874499, Fax: +44(207)3836668 ISSUES/YR: 12

- **Journal of Ethnicity in Substance Abuse**
ISSN: 1533-2640
J. Ethn. Subst. Abuse
Haworth Press Inc., 10 Alice Street, Binghamton, NY 13904-1580, United States, Tel: +1(607)7225857, Fax: +1(607)7710012 ISSUES/YR: 4

Journal of Ethnopharmacology
ISSN: 0378-8741
J. Ethnopharmacol.
Elsevier Ireland Ltd, P.O. Box 85, Limerick, Ireland, Tel: +353(61)471944, Fax: +353(61)472144 ISSUES/YR: 12

- **Journal of Evaluation in Clinical Practice**
ISSN: 1356-1294
J. Eval. Clin. Pract.
Blackwell Publishing Ltd, 9600 Garsington Road, OX4 2XG, Oxford, United Kingdom, Tel: +44(1865)776868, Fax: +44(1865)714591 ISSUES/YR: 4

Journal of Exercise Physiology Online
ISSN: 1097-9751
J. Exerc. Physiol. Online
American Society of Exercise Physiologists, 1200 Kenwood Ave, Duluth, MN 55811, United States, Tel: +1(218)7236297, Fax: +1(218)7236472 ISSUES/YR: 4

Journal of Experimental and Clinical Assisted Reproduction
ISSN: 1743-1050
J. Exp. Clin. Assisted Reprod.
BioMed Central Ltd., 34 - 42 Cleveland Street, W1T 4LB, London, United Kingdom, Tel: +44(207)3230323, Fax: +44(207)6319926 ISSUES/YR: 1

- **Journal of Experimental and Clinical Cancer Research**
ISSN: 0392-9078
J. Exp. Clin. Cancer Res.
APSIT, Via delle Messi d'Oro 156, 00158, Roma, Italy, Tel: +39(06)49852536 ISSUES/YR: 4

Journal of Experimental Animal Science
ISSN: 0939-8600
J. Exp. Anim. Sci.
Urban und Fischer Verlag Jena, P.O. Box 100537, 07705, Jena, Germany, Tel: +49(3641)6263, Fax: +49(3641)626500 ISSUES/YR: 4

- **Journal of Experimental Medicine**
ISSN: 0022-1007
J. Exp. Med.
Rockefeller University Press, 1114 First Avenue, 4th Floor, New York, NY 10021-8325, United States, Tel: +1(212)3278572, Fax: +1(212)3277944 ISSUES/YR: 24

- **Journal of Experimental Therapeutics and Oncology**
ISSN: 1359-4117
J. Exp. Ther. Oncol.
Old City Publishing, 628 North 2nd Street, Philadelphia, PA 19123, United States, Tel: +44-215 925 4390, Fax: +44-215 925 4371 ISSUES/YR: 4

Current

Journal of Exposure Science and Environmental Epidemiology
ISSN: 1559-0631
J. Expos. Sci. Environ. Epidemiol.
Nature Publishing Group, 345 Park Avenue South, New York, NY 10010-1707, United States, Tel: +1(212)7269200, Fax: +1(212)6960052 ISSUES/YR: 6

Journal of Extra-Corporeal Technology
ISSN: 0022-1058
J. Extra-Corpor. Technol.
American Society of Extra-Corporeal Technology, P.O. Box 11086, Richmond, VA 23230, United States, Tel: +1(703)4358556, Fax: +1(703)4350056 ISSUES/YR: 4

Journal of Facial and Somato Prosthetics
ISSN: 1082-1821
J. Facial Somato Prosthet.
ABI Professional Publications, P.O. Box 17446, Clearwater, FL 33762, United States, Tel: +1(727)5560950, Fax: +1(727)5562560 ISSUES/YR: 2

Journal of Family Planning and Reproductive Health Care
ISSN: 1471-1893
J. Fam. Plann. Reprod. Health Care
Royal College of Obstetricians and Gynaecologists, 27 Sussex Place, Regent's Park, NW1 4RG, London, United Kingdom, Tel: +44(207)7242441, Fax: +44(207)7230575 ISSUES/YR: 4

Journal of Family Practice
ISSN: 0094-3509
J. Fam. Pract.
Dowden Health Media,Inc, 110 Summit Avenue, Montvale, NJ 07645, United States, Tel: +1(201)7825714, Fax: +1(201)5055890 ISSUES/YR: 12

Journal of Family Psychotherapy
ISSN: 0897-5353
J. Fam. Psychother.
Haworth Press Inc., 10 Alice Street, Binghamton, NY 13904-1580, United States, Tel: +1(607)7225857, Fax: +1(607)7710012 ISSUES/YR: 4

Journal of Fluency Disorders
ISSN: 0094-730X
J. Fluency Disord.
Elsevier Inc., 360 Park Avenue South, New York, NY 10010, United States, Tel: +1(212)6333950, Fax: +1(212)6333990 ISSUES/YR: 4

Journal of Food and Drug Analysis
ISSN: 1021-9498
J. Food Drug Anal.
National Laboratories of Foods and Drugs, 161-2 Kuen Yang Street, Nankang, Taipei, Taiwan, Tel: +886(2)7858283, Fax: +886(2)7856793 ISSUES/YR: 4

Journal of Foot and Ankle Surgery
ISSN: 1067-2516
J. Foot Ankle Surg.
Academic Press Inc., 525 B Street, Suite 1900, San Diego, CA 92101-4495, United States ISSUES/YR: 6

Journal of Forensic Identification
ISSN: 0895-173X
J. Forensic Identif.
International Association for Identification, 2535 Pilot Knob Road, Suite 117, Mendota Heights, MN 55120-1120, United States ISSUES/YR: 6

Journal of Forensic Medicine and Toxicology
ISSN: 0971-1929
J. Forensic Med. Toxicol.
Medicolegal Society, Old Operation Theatre Building, 110029, New Delhi, India, Tel: +91(11)6964851/EXT:3274, Fax: +91(11)6862663 ISSUES/YR: 2

Journal of Forensic Neuropsychology
ISSN: 1521-1029
J. Forensic Neuropsychol.
Haworth Press Inc., 10 Alice Street, Binghamton, NY 13904-1580, United States, Tel: +1(607)7225857 ISSUES/YR: 4

● **Journal of Forensic Odonto-Stomatology**
ISSN: 0258-414X
J. Forensic Odonto-Stomatol.
International Organisation for Forensic Odonto-Stomatology, SA 5005, Adelaide, Australia, Tel: +61(8)3035431 ISSUES/YR: 2

● **Journal of Forensic Psychiatry and Psychology**
ISSN: 1478-9949
J. Forensic Psychiatry Psychol.
Routledge, 4 Park Square, Milton Park, OX14 4RN, Abingdon, Oxfordshire, United Kingdom, Tel: +44(1235)828600, Fax: +44(1235)829000 ISSUES/YR: 4

Journal of Forensic Psychology Practice
ISSN: 1522-8932
J. For. Psychol. Prac.
Haworth Press Inc., 10 Alice Street, Binghamton, NY 13904-1580, United States, Tel: +1(607)7225857, Fax: +1(607)7221424 ISSUES/YR: 4

● **Journal of Forensic Sciences**
ISSN: 0022-1198
J. Forensic Sci.
Blackwell Publishing Inc., 350 Main Street, Malden, MA 02148, United States, Tel: +1(781)3888200, Fax: +1(781)3888210 ISSUES/YR: 6

● **Journal of Gastroenterology**
ISSN: 0944-1174
J. Gastroenterol.
Springer-Verlag Tokyo, 3-13, Hongo 3-chome, Bunkyo-ku, 113-8480, Tokyo, Japan, Tel: +81(3)38180861, Fax: +81(3)38180864 ISSUES/YR: 12

● **Journal of Gastroenterology and Hepatology**
ISSN: 0815-9319
J. Gastroenterol. Hepatol.
Blackwell Publishing, 550 Swanston Street, VIC 3053, Carlton South, Australia, Tel: +61(3)8359 1011, Fax: +61(3)8359 1120 ISSUES/YR: 12

Journal of Gastrointestinal and Liver Diseases
ISSN: 1841-8724
J. Gastrointest. Liver Dis.
Romanian Society of Gastroenterology, Str. Croitorilor 19-21, 3400, Cluj-Napoca, Romania ISSUES/YR: 4

● **Journal of Gene Medicine**
ISSN: 1099-498X
J. Gene Med.
John Wiley and Sons Ltd, Southern Gate, PO19 8SQ, Chichester, West Sussex, United Kingdom, Tel: +44(1243)779777, Fax: +44(1243)775878 ISSUES/YR: 12

Journal of General and Applied Microbiology
ISSN: 0022-1260
J. Gen. Appl. Microbiol.
Microbiology Research Foundation, 4-16 Yayoi 2-chome, Bunkyo-ku, 113-0032, Tokyo, Japan, Tel: +81(3)38175821, Fax: +81(3)58205830 ISSUES/YR: 6

Journal of General Internal Medicine
ISSN: 0884-8734
J. Gen. Intern. Med.
Blackwell Publishing Inc., 350 Main Street, Malden, MA 02148, United States, Tel: +1(781)3888200, Fax: +1(781)3888210 ISSUES/YR: 12

Journal of General Physiology
ISSN: 0022-1295
J. Gen. Physiol.
Rockefeller University Press, 1114 First Avenue, 4th Floor, New York, NY 10021-8325, United States, Tel: +1(212)3278572, Fax: +1(212)3277944 ISSUES/YR: 12

● **Journal of General Virology**
ISSN: 0022-1317
J. Gen. Virol.
Society for General Microbiology, Basingstoke Road, Spencers Wood, RG7 1AE, Reading, Berkshire, United Kingdom, Tel: +44(118)9885577, Fax: +44(118)9885656 ISSUES/YR: 12

Journal of Generic Medicines
ISSN: 1741-1343
J. Generic Med.
Palgrave Macmillan Ltd., Houndmills, RG21 6XS, Basingstoke, Hants., United Kingdom, Tel: +44(1256)329242, Fax: +44(1256)810526 ISSUES/YR: 4

Journal of Genetic Counseling
ISSN: 1059-7700
J. Genet. Couns.
Kluwer Academic/Human Sciences Press Inc., 233 Spring Street, New York, NY 10013-1578, United States, Tel: +1(212)6208000, Fax: +1(212)4630742 ISSUES/YR: 6

Journal of Geriatric Cardiology
ISSN: 1671-5411
J. Geriatr. Cardiol.
Institute of Geriatric Cardiology, 28 Fuxing Road, 100853, Beijing, China, Tel: +86-10 88270497, Fax: +86-10 88270497 ISSUES/YR: 4

Journal of Geriatric Physical Therapy
ISSN: 1539-8412
J. Geriatr. Phys. Ther
American Physical Therapy Association, 1111 North Fairfax Street, Alexandria, VA 22314, United States, Tel: +1(703)6842782 ISSUES/YR: 3

• **Journal of Geriatric Psychiatry and Neurology**
ISSN: 0891-9887
J. Geriatr. Psychiatry Neurol.
SAGE Publications Inc., 2455 Teller Road, Thousand Oaks, CA 91320, United States, Tel: +1(805)4990721, Fax: +1(805)4990871 ISSUES/YR: 4

• **Journal of Glaucoma**
ISSN: 1057-0829
J. Glaucoma
Lippincott Williams and Wilkins, 530 Walnut Street, Philadelphia, PA 19106-3621, United States, Tel: +1(215)5218342, Fax: +1(215)5218326 ISSUES/YR: 6

• **Journal of Gynecologic Surgery**
ISSN: 1042-4067
J. Gynecol. Surg.
Mary Ann Liebert Inc., 140 Huguenot Street, New Rochelle, NY 10801-5215, United States, Tel: +1(914)7402100, Fax: +1(914)7402101 ISSUES/YR: 4

• **Journal of Hand Surgery**
ISSN: 0363-5023
J. Hand Surg. (USA)
W.B. Saunders, Independence Square West, Philadelphia, PA 19106-3399, United States, Tel: +1(215)2387800, Fax: +1(215)2386445 ISSUES/YR: 6

Journal of Hand Surgery
ISSN: 0266-7681
J. Hand Surg. (GBR)
Churchill Livingstone, 1-3 Baxter's Place, Leith Walk, EH1 3AF, Edinburgh, United Kingdom, Tel: +44(131)5562424, Fax: +44(131)5581278 ISSUES/YR: 6

Journal of Hand Therapy
ISSN: 0894-1130
J. Hand Ther.
Hanley and Belfus Inc., 210 South 13th Street, Philadelphia, PA 19107, United States, Tel: +1(215)5467293 ISSUES/YR: 4

Journal of Hazardous Materials
ISSN: 0304-3894
J. Hazard. Mater.
Elsevier, P.O. Box 211, 1000 AE, Amsterdam, Netherlands, Tel: +31(20)4853757, Fax: +31(20)4853432 ISSUES/YR: 33

Journal of Head Trauma Rehabilitation
ISSN: 0885-9701
J. Head Trauma Rehabil.
Lippincott Williams and Wilkins, 351 West Camden Street, Baltimore, MD 21201-2436, United States, Tel: +1(410)5284000 ISSUES/YR: 6

• **Journal of Headache and Pain**
ISSN: 1129-2369
J. Headache Pain
Springer-Verlag Italia s.r.l, Via Podgora 4, I-20122, Milano, Italy, Tel: +39(2)5425971, Fax: +39(2)55193360 ISSUES/YR: 3

Journal of Health and Human Services Administration
ISSN: 1079-3739
J. Health Hum. Serv. Adm.
Southern Public Administration Education Foundation Inc., 2103 Fairway Lane, Harrisburg, PA 17112, United States, Tel: +1(717)5405477, Fax: +1(215)8931763 ISSUES/YR: 4

Journal of Health and Social Policy
ISSN: 0897-7186
J. Health Soc. Policy
Haworth Press Inc., 10 Alice Street, Binghamton, NY 13904-1580, United States, Tel: +1(607)7225857, Fax: +1(607)7710012 ISSUES/YR: 4

Journal of Health Care Finance
ISSN: 1078-6767
J. Health Care Financ.
Aspen Publishers Inc., 1185 Avenue of the Americas, New York, NY 10036, United States, Tel: +1(212)5970200, Fax: +1(212)5970338 ISSUES/YR: 4

Journal of Health Economics
ISSN: 0167-6296
J. Health Econ.
Elsevier, P.O. Box 211, 1000 AE, Amsterdam, Netherlands, Tel: +31(20)4853757, Fax: +31(20)4853432 ISSUES/YR: 6

Journal of Health Politics, Policy and Law
ISSN: 0361-6878
J. Health Polit. Policy Law
Duke University Press, Box 90660, Durham, NC 27708-0660, United States, Tel: +1(919)6873653 ISSUES/YR: 6

• **Journal of Health Psychology**
ISSN: 1359-1053
J. Health Psychol.
SAGE Publications Ltd, 55 City Road, EC1Y 1SP, London, United Kingdom, Tel: +44(207)3248500, Fax: +44(207)3248600 ISSUES/YR: 6

Journal of Health Science
ISSN: 1344-9702
J. Health Sci.
Pharmaceutical Society of Japan, 2-12-15-201, Shibuya, Shibuya-ku, 150, Tokyo, Japan, Tel: +81(3)34063321, Fax: +81(3)34981835 ISSUES/YR: 6

• **Journal of Health Services Research and Policy**
ISSN: 1355-8196
J. Health Serv. Res. Policy
Royal Society of Medicine Press Ltd, P.O. Box 9002, W1A 0ZA, London, United Kingdom, Tel: +44(207)2902927, Fax: +44(207)2902929 ISSUES/YR: 4

Journal of Health, Population and Nutrition
ISSN: 1606-0997
J. Health Popul. Nutr.
International Center for Diarrhoeal Disease Research, GPO Box 128, 1000, Dacca, Bangladesh, Tel: +880(2)600171 ISSUES/YR: 4

Journal of Healthcare Management
ISSN: 1096-9012
J. Healthc. Manage.
Foundation of the American College of Healthcare Executives, One N. Franklin Street, Suite 1700, Chicago, IL 60606-3491, United States, Tel: +1(312)8459869, Fax: +1(312)4240014 ISSUES/YR: 6

• **Journal of Heart and Lung Transplantation**
ISSN: 1053-2498
J. Heart Lung Transplant.
Elsevier USA, 6277 Sea Harbor Drive, Orlando, FL 32862 8239, United States, Tel: +1-407 345 4000, Fax: +1-407 363 9661 ISSUES/YR: 12

• **Journal of Heart Valve Disease**
ISSN: 0966-8519
J. Heart Valve Dis.
ICR Publishers Ltd, 12/A South Approach, Moor Park, HA6 2ET, Northwood, United Kingdom, Tel: +44(1923)836871 ISSUES/YR: 6

Journal of Helminthology
ISSN: 0022-149X
J. Helminthol.
CAB International, Wallingford, OX10 8DE, Oxon, United Kingdom, Tel: +44(1491)832111, Fax: +44(1491)826090 ISSUES/YR: 4

Journal of Hepato-Biliary-Pancreatic Surgery
ISSN: 0944-1166
J. Hepato-Biliary-Pancreatic Surg.
Springer-Verlag Tokyo, 3-13, Hongo 3-chome, Bunkyo-ku, 113-8480, Tokyo, Japan, Tel: +81(3)38180861, Fax: +81(3)38180864 ISSUES/YR: 6

• **Journal of Hepatology**
ISSN: 0168-8278
J. Hepatol.
Elsevier, P.O. Box 211, 1000 AE, Amsterdam, Netherlands, Tel: +31(20)4853757, Fax: +31(20)4853432 ISSUES/YR: 12

Current

- **Journal of Herbal Pharmacotherapy**
  ISSN: 1522-8940
  J. Herbal Pharmacother.
  Haworth Press Inc., 10 Alice Street, Binghamton, NY 13904-1580, United States, Tel: +1(607)7225857, Fax: +1(607)7221424 ISSUES/YR: 4

Journal of Herbs, Spices and Medicinal Plants
ISSN: 1049-6475
J. Herbs Spices Med. Plants
Haworth Press Inc., 10 Alice Street, Binghampton, NY 13904-1580, United States, Tel: +1(607)7225857, Fax: +1(607)7226362 ISSUES/YR: 4

- **Journal of Heredity**
  ISSN: 0022-1503
  J. Hered.
  Oxford University Press, 2001 Evans Road, Cary, NC 27513, United States, Tel: +1(919)6770977, Fax: +1(919)6771714 ISSUES/YR: 6

Journal of Heterocyclic Chemistry
ISSN: 0022-152X
J. Heterocycl. Chem.
HeteroCorporation, PO Box 170, Provo, UTAH 84603-0170, United States ISSUES/YR: 6

- **Journal of Histochemistry and Cytochemistry**
  ISSN: 0022-1554
  J. Histochem. Cytochem.
  Histochemical Society Inc., P.O. Box 357420, Seattle, WA 98195-9420, United States ISSUES/YR: 12

Journal of Histotechnology
ISSN: 0147-8885
J. Histotechnol.
National Society for Histotechnology, 4201 Northview Drive, Suite 502, Bowie, MD 20716-1073, United States, Tel: +1(301)2626221 ISSUES/YR: 4

Journal of HIV Therapy
ISSN: 1462-0308
J. HIV Ther.
Mediscript Ltd, 1 Mountview Court,310 Friern Barnet Lande, N20 0LD, London, United Kingdom, Tel: +44-208 446 8898, Fax: +44-208 446 9194 ISSUES/YR: 4

Journal of HIV/AIDS Prevention in Children and Youth
ISSN: 1553-8346
J. HIV AIDS Prev. Child. Youth
Haworth Press Inc., 10 Alice Street, Binghamton, NY 13904-1580, United States, Tel: +1(607)7225857, Fax: +1(607)7710012 ISSUES/YR: 4

Journal of Homosexuality
ISSN: 0091-8369
J. Homosex.
Haworth Press Inc., 10 Alice Street, Binghamton, NY 13904-1580, United States, Tel: +1(607)7225857, Fax: +1(607)7710012 ISSUES/YR: 4

Journal of Hospital Infection
ISSN: 0195-6701
J. Hosp. Infect.
W.B. Saunders Ltd, 32 Jamestown Road, NW1 7BY, London, United Kingdom, Tel: +44(20)74244200, Fax: +44(20)74854752 ISSUES/YR: 12

- **Journal of Hospital Marketing and Public Relations**
  ISSN: 1539-0942
  J. Hosp. Mark. Public Relat.
  Haworth Press Inc., 10 Alice Street, Binghamton, NY 13904-1580, United States, Tel: +1(607)7225857, Fax: +1(607)7710012 ISSUES/YR: 2

Journal of Human Genetics
ISSN: 1434-5161
J. Hum. Genet.
Springer-Verlag Tokyo, 3-13, Hongo 3-chome, Bunkyo-ku, 113-8480, Tokyo, Japan, Tel: +81(3)38180861, Fax: +81(3)38180864 ISSUES/YR: 4

Journal of Human Hypertension
ISSN: 0950-9240
J. Hum. Hypertens.
Nature Publishing Group, Houndmills, RG21 6XS, Basingstoke, Hampshire, United Kingdom, Tel: +44(1256)329242, Fax: +44(1256)810526 ISSUES/YR: 12

Journal of Human Lactation
ISSN: 0890-3344
J. Human Lactation
SAGE Publications Inc., 2455 Teller Road, Thousand Oaks, CA 91320, United States, Tel: +1(805)4990721, Fax: +1(805)4990871 ISSUES/YR: 4

Journal of Human Nutrition and Dietetics
ISSN: 0952-3871
J. Hum. Nutr. Diet.
Blackwell Publishing Ltd, 9600 Garsington Road, OX4 2XG, Oxford, United Kingdom, Tel: +44(1865)776868, Fax: +44(1865)714591 ISSUES/YR: 6

- **Journal of Hypertension**
  ISSN: 0263-6352
  J. Hypertens.
  Lippincott Williams and Wilkins, 241 Borough High Street, SE1 1GB, London, United Kingdom, Tel: +1(207)9407500, Fax: +1(207)9407575 ISSUES/YR: 12
- **Journal of Hypertension, Supplement**
  ISSN: 0952-1178
  J. Hypertens. Suppl.
  Lippincott Williams and Wilkins, 241 Borough High Street, SE1 1GB, London, United Kingdom, Tel: +44(207)9407500, Fax: +44(207)9407575 ISSUES/YR: 1
- **Journal of Immigrant and Minority Health**
  ISSN: 1557-1912
  J. Immigr. Minor. Health
  Springer Science and Business Media Deutschland GmbH, Heidelberger Platz 3, 14197, Berlin, Germany, Tel: +1-30 827875282, Fax: +1-30 827875707 ISSUES/YR: 4

Journal of Immune Based Therapies and Vaccines
ISSN: 1476-8518
J. Immune Based Ther. Vaccines
BioMed Central Ltd., 34 - 42 Cleveland Street, W1T 4LB, London, United Kingdom, Tel: +44(207)3230323, Fax: +44(207)6319926 ISSUES/YR: 1

- **Journal of Immunoassay and Immunochemistry**
  ISSN: 1532-1819
  J. Immunoassay Immunochem.
  Taylor and Francis Inc., 325 Chestnut St, Suite 800, Philadelphia PA, PA 19106, United States, Tel: +1(800)3541420, Fax: +1(215)6258914 ISSUES/YR: 4
- **Journal of Immunological Methods**
  ISSN: 0022-1759
  J. Immunol. Methods
  Elsevier, P.O. Box 211, 1000 AE, Amsterdam, Netherlands, Tel: +31(20)4853757, Fax: +31(20)4853432 ISSUES/YR: 24
- **Journal of Immunology**
  ISSN: 0022-1767
  J. Immunol.
  American Association of Immunologists, 9650 Rockville Pike, Bethesda, MD 20814, United States, Tel: +1(301)5711792 ISSUES/YR: 24
- **Journal of Immunotherapy**
  ISSN: 1524-9557
  J. Immunother.
  Lippincott Williams and Wilkins, 530 Walnut Street, Philadelphia, PA 19106-3621, United States, Tel: +1(215)5218342, Fax: +1(215)5218326 ISSUES/YR: 6
- **Journal of Immunotoxicology**
  ISSN: 1547-691X
  J. Immunotoxicol.
  Taylor and Francis Inc., 325 Chestnut St, Suite 800, Philadelphia PA, PA 19106, United States, Tel: +1(800)3541420, Fax: +1(215)6258914 ISSUES/YR: 4

Journal of Indian Association for Child and Adolescent Mental Health
ISSN: 0973-1342
J. Indian Assoc. Child Adolesc. Ment. Health
Indian Adolescent for Child and Adolescent Mental Health, N8-180/118 Rajendra Vihar, Newada, PO Sunderpur, 221005, Varanasi (UP), India, Tel: +91(542)2316694, Fax: +91(542)2316694 ISSUES/YR: 1

Journal of Industrial Microbiology and Biotechnology
ISSN: 1367-5435
J. Ind. Microbiol. Biotechnol.
Springer Verlag, Tiergartenstrasse 17, D-69121, Heidelberg, Germany, Tel: +49(6221)4870, Fax: +49(6221)487366 ISSUES/YR: 12

Journal of Infection
ISSN: 0163-4453
J. Infect.
W.B. Saunders Ltd, 32 Jamestown Road, NW1 7BY, London, United Kingdom, Tel: +44(20)74244200, Fax: +44(20)74854752 ISSUES/YR: 8

Journal of Infection and Chemotherapy
ISSN: 1341-321X
J. Infect. Chemother.
Springer-Verlag Tokyo, 3-13, Hongo 3-chome, Bunkyo-ku, 113-8480, Tokyo, Japan, Tel: +81(3)38180861, Fax: +81(3)38180864 ISSUES/YR: 6

• **Journal of Infectious Diseases**
ISSN: 0022-1899
J. Infect. Dis.
University of Chicago Press, 1427 E. 60th Street, Chicago, IL 60637-2954, United States, Tel: +1(773)7027600, Fax: +1(773)7020694 ISSUES/YR: 24

Journal of Inflammation
ISSN: 1476-9255
J. Inflamm.
BioMed Central Ltd., 34 - 42 Cleveland Street, W1T 4LB, London, United Kingdom, Tel: +44(207)3230323, Fax: +44(207)6319926 ISSUES/YR: 1

• **Journal of Infusion Nursing**
ISSN: 1533-1458
J. Infus. Nurs.
Lippincott Williams and Wilkins, 530 Walnut Street, Philadelphia, PA 19106-3621, United States, Tel: +1(215)5218342, Fax: +1(215)5218326 ISSUES/YR: 6

Journal of Inherited Metabolic Disease
ISSN: 0141-8955
J. Inherit. Metab. Dis.
Kluwer Academic Publishers, Van Godewijckstraat 30, 3311 GZ, Dordrecht, Netherlands, Tel: +31(78)6576000, Fax: +31(78)6576254 ISSUES/YR: 6

Journal of Inorganic Biochemistry
ISSN: 0162-0134
J. Inorg. Biochem.
Elsevier Inc., 360 Park Avenue South, New York, NY 10010, United States, Tel: +1(212)6333950, Fax: +1(212)6333990 ISSUES/YR: 12

Journal of Integrated Care Pathways
ISSN: 1473-2297
J. Integr. Care Pathways
Royal Society of Medicine Press Ltd, P.O. Box 9002, W1A 0ZA, London, United Kingdom, Tel: +44(207)2902927, Fax: +44(207)2902929 ISSUES/YR: 3

Journal of Integrative Neuroscience
ISSN: 0219-6352
J. Integr. Neurosci.
World Scientific Publishing Co. Pte Ltd, 5 Toh Tuck Link, 596224, Singapore, Singapore, Tel: +65(6466)5775, Fax: +65(6467)7667 ISSUES/YR: 4

Journal of Intellectual and Developmental Disability
ISSN: 1366-8250
J. Intellect. Dev. Disabil.
Taylor and Francis Ltd., 4 Park Square, Milton Park, OX14 4RN, Abingdon, Oxfordshire, United Kingdom, Tel: +44(20)70175541, Fax: +44(20)70175198 ISSUES/YR: 4

• **Journal of Intellectual Disabilities**
ISSN: 1744-6295
J. Intellect. Disabil.
SAGE Publications Ltd, 55 City Road, EC1Y 1SP, London, United Kingdom, Tel: +44(207)3248500, Fax: +44(207)3248600 ISSUES/YR: 4

Journal of Intellectual Disability Research
ISSN: 0964-2633
J. Intellect. Disabil. Res.
Blackwell Publishing Ltd, 9600 Garsington Road, OX4 2XG, Oxford, United Kingdom, Tel: +44(1865)776868, Fax: +44(1865)714591 ISSUES/YR: 8

Journal of Intellectual Disability Research, Supplement
ISSN: 1351-0886
J. Intellect. Disabil. Res. Suppl.
Blackwell Publishing Ltd, 9600 Garsington Road, OX4 2XG, Oxford, United Kingdom, Tel: +44(1865)776868, Fax: +44(1865)714591 ISSUES/YR: 1

• **Journal of Intensive Care Medicine**
ISSN: 0885-0666
J. Intensive Care Med.
SAGE Publications Inc., 2455 Teller Road, Thousand Oaks, CA 91320, United States, Tel: +1(805)4990721, Fax: +1(805)4990871 ISSUES/YR: 6

• **Journal of Interferon and Cytokine Research**
ISSN: 1079-9907
J. Interferon Cytokine Res.
Mary Ann Liebert Inc., 140 Huguenot Street, New Rochelle, NY 10801-5215, United States, Tel: +1(914)7402100, Fax: +1(914)7402101 ISSUES/YR: 12

• **Journal of Internal Medicine**
ISSN: 0954-6820
J. Intern. Med. (GBR)
Blackwell Publishing Ltd, 9600 Garsington Road, OX4 2XG, Oxford, United Kingdom, Tel: +44(1865)776868, Fax: +44(1865)714591 ISSUES/YR: 12

Journal of Internal Medicine of India
ISSN: 0972-1096
J. Intern. Med. India
BRD Medical College, 273 013 UP, Gorakhpur, India, Tel: +91(522)364043 ISSUES/YR: 4

Journal of Internal Medicine of Taiwan
ISSN: 1016-7390
J. Intern. Med. Taiwan
Society of Internal Medicine of Taiwan, Zhong Xiao Xi Lu, Sec.1, no.50, 22nd Fl, Rm 26, Taipei, Taiwan, Tel: +886(2)23758068, Fax: +886(2)23758072 ISSUES/YR: 4

Journal of International Medical Research
ISSN: 0300-0605
J. Int. Med. Res.
Cambridge Medical Publications, High Street, BN11 1DJ, Worthing, West Sussex, United Kingdom, Tel: +44(1903)205884, Fax: +44(1903)234862 ISSUES/YR: 6

Journal of Interprofessional Care
ISSN: 1356-1820
J. Interprof. Care
Taylor and Francis Ltd., 4 Park Square, Milton Park, OX14 4RN, Abingdon, Oxfordshire, United Kingdom, Tel: +44(20)70175541, Fax: +44(20)70175198 ISSUES/YR: 6

• **Journal of Interventional Cardiac Electrophysiology**
ISSN: 1383-875X
J. Intervent. Card. Electrophysiol.
Kluwer Academic Publishers, Van Godewijckstraat 30, 3311 GZ, Dordrecht, Netherlands, Tel: +31(78)6576000, Fax: +31(78)6576254 ISSUES/YR: 6

• **Journal of Interventional Cardiology**
ISSN: 0896-4327
J. Intervent. Cardiol.
Blackwell Publishing Inc., 350 Main Street, Malden, MA 02148, United States, Tel: +1(781)3888200, Fax: +1(781)3888210 ISSUES/YR: 6

Journal of Interventional Radiology
ISSN: 1008-794X
J. Intervent. Radiol.
Publishing House of the Journal of Interventional, No 17 Lane 1220 Huashan Road, 200052, Shanghai, China ISSUES/YR: 6

Journal of Intravenous Nursing
ISSN: 0896-5846
J. Intraven. Nurs.
Lippincott Williams and Wilkins, 530 Walnut Street, Philadelphia, PA 19106-3621, United States, Tel: +1(215)5218342, Fax: +1(215)5218326 ISSUES/YR: 6

Journal of Invasive Cardiology
ISSN: 1042-3931
J. Invasive Cadiol.
HMP Communications, 83 General Warren Blvd. Suite 100, Malven, PA 19355, United States, Tel: +1-610 237 7285, Fax: +1(610)560 0501 ISSUES/YR: 12

Journal of Investigational Allergology and Clinical Immunology
ISSN: 1018-9068
J. Invest. Allergol. Clin. Immunol.
ESMON Publicidad S.A., C/Balmes,209,number 3 2a, 08006, Barcelona, Spain, Tel: +34-93 215 90 34, Fax: +34-93 487 40 64 ISSUES/YR: 4

- **Journal of Investigative Dermatology**
ISSN: 0022-202X
J. Invest. Dermatol.
Nature Publishing Group, Houndmills, RG21 6XS, Basingstoke, Hampshire, United Kingdom, Tel: +1(1256)329242, Fax: +1(1256)810526 ISSUES/YR: 12

Journal of Investigative Dermatology Symposium Proceedings
ISSN: 0022-202X
J. Invest. Dermatol. Symp. Proc.
Nature Publishing Group, Houndmills, RG21 6XS, Basingstoke, Hampshire, United Kingdom, Tel: +44(1256)329242, Fax: +44(1256)810526 ISSUES/YR: 1

Journal of Investigative Medicine
ISSN: 1081-5589
J. Invest. Med.
BC Decker Inc., P.O. Box 620, LCD 1, L8N 3K7, Hamilton, Ontario, Canada, Tel: +1(905)522-7017, Fax: +1(905)522-7839 ISSUES/YR: 8

- **Journal of Investigative Surgery**
ISSN: 0894-1939
J. Invest. Surg.
Taylor and Francis Ltd., 4 Park Square, Milton Park, OX14 4RN, Abingdon, Oxfordshire, United Kingdom, Tel: +44(20)70175541, Fax: +44(20)70175198 ISSUES/YR: 6

Journal of Japanese Dental Society of Anesthesiology
ISSN: 0386-5835
J. Jpn. Dent. Soc. Anesthesiol.
Japanese Dental Society of Anesthesiology, 1-44-2 Komagome, Toshimaku, 170, Tokyo, Japan, Tel: +81(3)39478891, Fax: +81(3)39478341 ISSUES/YR: 5

Journal of JASTRO
ISSN: 1040-9564
J. JASTRO
Editorial Board of JASTRO, Sanwa Building, 3-39-1 Hongo, Bunkyo-ku, 113, Tokyo, Japan ISSUES/YR: 4

Journal of Jilin University Medicine Edition
ISSN: 1671-587X
J. Jilin Univ. Med. Ed.
Jilin University Press, 8 Xinmin St., 130023, Changchun, China, Tel: +86(431)892 3189 2876, Fax: +86(431)892 3907 ISSUES/YR: 6

Journal of Korean Society for Clinical Pharmacology and Therapeutics
ISSN: 1225-5467
J. Korean Soc. Clin. Pharmacol. Ther.
Korean Society Clinical Pharmacology and Therapeutics, 28 Yongon-dong, Chongno-gu, 110-744, Seoul, South Korea, Tel: +82(2)7602302, Fax: +82(2)7429252 ISSUES/YR: 2

Journal of Labelled Compounds and Radiopharmaceuticals
ISSN: 0362-4803
J. Label. Compd. Radiopharm.
John Wiley and Sons Ltd, Southern Gate, PO19 8SQ, Chichester, West Sussex, United Kingdom, Tel: +44(1243)779777, Fax: +44(1243)775878 ISSUES/YR: 14

Journal of Laboratory and Clinical Medicine
ISSN: 0022-2143
J. Lab. Clin. Med.
Mosby Inc., 11830 Westline Industrial Drive, St. Louis, MO 63146, United States, Tel: +1(314)8728370, Fax: +1(314)4321380 ISSUES/YR: 12

- **Journal of Laparoendoscopic and Advanced Surgical Techniques - Part A**
ISSN: 1092-6429
J. Laparoendosc. Adv. Surg. Techn. Part A
Mary Ann Liebert Inc., 140 Huguenot Street, New Rochelle, NY 10801-5215, United States, Tel: +1(914)7402100, Fax: +1(914)7402101 ISSUES/YR: 6

Journal of Laryngology and Otology
ISSN: 0022-2151
J. Laryngol. Otol.
Royal Society of Medicine Press Ltd, P.O. Box 9002, W1A 0ZA, London, United Kingdom, Tel: +44(207)2902927, Fax: +44(207)2902929 ISSUES/YR: 12

Journal of Legal Medicine
ISSN: 0194-7648
J. Leg. Med.
Taylor and Francis Ltd., 4 Park Square, Milton Park, OX14 4RN, Abingdon, Oxfordshire, United Kingdom, Tel: +44(20)70175541, Fax: +44(20)70175198 ISSUES/YR: 4

- **Journal of Leukocyte Biology**
ISSN: 0741-5400
J. Leukocyte Biol.
Federation of American Societies for Experimental Biology, 9650 Rockville Pike, Bethesda, MD 20814-3998, United States, Tel: +1(301)5307027 ISSUES/YR: 12

Journal of LGBT Health Research
ISSN: 1557-4091
J. LGBT Health Res.
Haworth Press Inc., 10 Alice Street, Binghamton, NY 13904-1580, United States, Tel: +1(607)7225857 ISSUES/YR: 4

Journal of Liaquat University of Medical and Health Sciences
ISSN: 1729-0341
J. Liaquat Univ. Med. Health Sci.
Liaquat University of Medical and Health Sciences, Jamshoro, Sindh, Pakistan, Tel: +92(221)77123940, Fax: +92-221 771 ISSUES/YR: 3

- **Journal of Lipid Research**
ISSN: 0022-2275
J. Lipid Res.
Lipid Research Inc., 9650 Rockville Pike, Bethesda, MD 20814-3998, United States, Tel: +1(301)5307027 ISSUES/YR: 12

- **Journal of Liposome Research**
ISSN: 0898-2104
J. Liposome Res.
Taylor and Francis Inc., 325 Chestnut St, Suite 800, Philadelphia PA, PA 19106, United States, Tel: +1(800)3541420, Fax: +1(215)6258914 ISSUES/YR: 4

Journal of Liquid Chromatography and Related Technologies
ISSN: 1082-6076
J. Liq. Chromatogr. Relat. Technol.
Taylor and Francis Inc., 325 Chestnut St, Suite 800, Philadelphia PA, PA 19106, United States, Tel: +1(800)3541420, Fax: +1(215)6258914 ISSUES/YR: 20

Journal of Long-Term Effects of Medical Implants
ISSN: 1050-6934
J. Long-Term Eff. Med. Implants
Begell House Inc., 145 Madison Avenue 6th Floor, New York, NY 10016, United States, Tel: +1(212)7251999, Fax: +1(212)2138368 ISSUES/YR: 4

Journal of Lower Genital Tract Disease
ISSN: 1089-2591
J. Lower Genital Tract Dis.
Lippincott Williams and Wilkins, 530 Walnut Street, Philadelphia, PA 19106-3621, United States, Tel: +1(215)5218342, Fax: +1(215)5218326 ISSUES/YR: 4

- **Journal of Magnetic Resonance Imaging**
ISSN: 1053-1807
J. Magn. Reson. Imaging
John Wiley and Sons Inc., P.O.Box 18667, Newark, NJ 07191-8667, United States, Tel: +1(212)8506000, Fax: +1(212)8506012 ISSUES/YR: 12

Journal of Mammary Gland Biology and Neoplasia
ISSN: 1083-3021
J. Mammary Gland Biol. Neoplasia
Springer Science and Business Media Deutschland GmbH, Heidelberger Platz 3, 14197, Berlin, Germany, Tel: +1-30 827875282, Fax: +1-30 827875707 ISSUES/YR: 4

Journal of Manipulative and Physiological Therapeutics
ISSN: 0161-4754
J. Manip. Physiol. Ther.
Mosby Inc., 11830 Westline Industrial Drive, St. Louis, MO 63146, United States, Tel: +1(314)8728370, Fax: +1(314)4321380 ISSUES/YR: 9

Journal of Manual and Manipulative Therapy
ISSN: 1066-9817
J. Man. Manip. Ther.
Journal of Manual and Manipulative Therapy Inc., P.O. Box 713, Forest Grove, OR 97116, United States, Tel: +1(503)3592322 ISSUES/YR: 4

● **Journal of Mass Spectrometry**
ISSN: 1076-5174
J. Mass Spectrom.
John Wiley and Sons Ltd, Southern Gate, PO19 8SQ, Chichester, West Sussex, United Kingdom, Tel: +44(1243)779777, Fax: +44(1243)775878 ISSUES/YR: 12

● **Journal of Materials Science: Materials in Medicine**
ISSN: 0957-4530
J. Mater. Sci. Mater. Med.
Kluwer Academic Publishers, Van Godewijckstraat 30, 3311 GZ, Dordrecht, Netherlands, Tel: +31(78)6576000, Fax: +31(78)6576254 ISSUES/YR: 12

● **Journal of Maternal-Fetal and Neonatal Medicine**
ISSN: 1476-7058
J. Matern.-Fetal Neonatal Med.
Taylor and Francis Ltd., 4 Park Square, Milton Park, OX14 4RN, Abingdon, Oxfordshire, United Kingdom, Tel: +44(20)70175541, Fax: +44(20)70175198 ISSUES/YR: 12

Journal of Medical and Biological Engineering
ISSN: 1609-0985
J. Med. Biol. Eng.
Biomedical Engineering Society, 1-1 Jen-Ai Road, Taipei, Taiwan, Tel: +886(2)3970800 EXT. 1444, Fax: +886(2)3940049 ISSUES/YR: 4

Journal of Medical Economics
ISSN: 1369-698X
J. Med. Econ.
PJB Publications, Sheepen Place, C03 3LP, Colchester, United Kingdom, Tel: +44-20 70175540 ISSUES/YR: 1

Journal of Medical Engineering and Technology
ISSN: 0309-1902
J. Med. Eng. Technol.
Taylor and Francis Ltd., 4 Park Square, Milton Park, OX14 4RN, Abingdon, Oxfordshire, United Kingdom, Tel: +44(20)70175541, Fax: +44(20)70175198 ISSUES/YR: 6

Journal of Medical Ethics
ISSN: 0306-6800
J. Med. Ethics
BMJ Publishing Group, Tavistock Square, WC1H 9JR, London, United Kingdom, Tel: +44(207)3874499, Fax: +44(207)3836668 ISSUES/YR: 6

● **Journal of Medical Genetics**
ISSN: 0022-2593
J. Med. Genet.
BMJ Publishing Group, Tavistock Square, WC1H 9JR, London, United Kingdom, Tel: +44(207)3874499, Fax: +44(207)3836668 ISSUES/YR: 12

Journal of Medical Internet Research
ISSN: 1438-8871
J. Med. Internet Res.
Journal of Medical Internet Research, R. Fraser Elliott Bldg, 4th Fl, 190 Elizabeth St, ON M5G 2C4, Toronto, Canada ISSUES/YR: 5

Journal of Medical Investigation
ISSN: 1343-1420
J. Med. Invest.
University of Tokushima, 18-15 Kuramoto-cho 3-chome, 770 8503, Tokushima, Japan, Tel: +81-88 633 7104, Fax: +81-88 633 7115 ISSUES/YR: 4

● **Journal of Medical Microbiology**
ISSN: 0022-2615
J. Med. Microbiol.
Society for General Microbiology, Basingstoke Road, Spencers Wood, RG7 1AE, Reading, Berkshire, United Kingdom, Tel: +44(118)9885577, Fax: +44(118)9885656 ISSUES/YR: 12

Journal of Medical Practice Management
ISSN: 8755-0229
J. Med. Pract. Manage.
Greenbranch Publishing LLC, P.O. Box 208, Phoenix, MD 21131-0208, United States, Tel: +1(410)3299788, Fax: +1(410)3291510 ISSUES/YR: 6

● **Journal of Medical Primatology**
ISSN: 0047-2565
J. Med. Primatol.
Blackwell Munksgaard, 1 Rosenorns Alle, P.O. Box 227, DK-1502, Copenhagen V, Denmark, Tel: +45(77)333333, Fax: +45(77)333377 ISSUES/YR: 6

Journal of Medical Sciences
ISSN: 1011-4564
J. Med. Sci.(Taiwan)
National Defense Medical Center, Min-Chuan East Road Section 6, 114, Taipei, Taiwan, Tel: +886(2)87923125, Fax: +886(2)87923111 ISSUES/YR: 6

Journal of Medical Sciences
ISSN: 1682-4474
J. Med. Sci. (Pakistan)
Asian Network for Scientific Information, 308-Lasani Town, Sargodha Road, Faisalabad, Pakistan, Tel: +92(41)2001145-46, Fax: +92(21)5206036 ISSUES/YR: 4

Journal of Medical Screening
ISSN: 0969-1413
J. Med. Screen.
Royal Society of Medicine Press Ltd, P.O. Box 9002, W1A 0ZA, London, United Kingdom, Tel: +44(207)2902927, Fax: +44(207)2902929 ISSUES/YR: 4

● **Journal of Medical Speech-Language Pathology**
ISSN: 1065-1438
J. Med. Speech-Lang. Pathol.
Delmar Learning, 5 Maxwell Drive, P.O. Box 8007, Clifton Park, NY 12065-8007, United States, Tel: +1(888)4275900, Fax: +1(570)3479072 ISSUES/YR: 4

Journal of Medical Systems
ISSN: 0148-5598
J. Med. Syst.
Springer Science and Business Media Deutschland GmbH, Heidelberger Platz 3, 14197, Berlin, Germany, Tel: +1-30 827875282, Fax: +1-30 827875707 ISSUES/YR: 6

Journal of Medical Teachers Federation
ISSN: 1560-9006
J. Med. Teach. Fed.
Journal of Medical Teachers Federation, Shahbag Avenue, 1000, Dhaka, Bangladesh, Tel: +880(2)8619115, Fax: +880(2)8826497 ISSUES/YR: 2

Journal of Medical Ultrasonics
ISSN: 1346-4523
J. Med. Ultrason.
Japan Society of Ultrasonics Medicine, Hongo 323-1 Crocevia Hongo 3F, Bunkyo-ku, 113, Tokyo, Japan, Tel: +81(3)38135540 ISSUES/YR: 4

Journal of Medical Ultrasound
ISSN: 0929-6441
J. Med. Ultrasound
Elsevier (Singapore) Pte Ltd, 77 Leighton Road, Causeway Bay, Hong Kong, China, Hong Kong, Tel: +86(852)29651300, Fax: +86(852)29760778 ISSUES/YR: 4

Journal of Medical Virology
ISSN: 0146-6615
J. Med. Virol.
Wiley-Liss Inc., 111 River Street, Hoboken, NJ 07030-5774, United States, Tel: +1(201)7486645, Fax: +1(201)7486021 ISSUES/YR: 12

Journal of Medicinal Chemistry
ISSN: 0022-2623
J. Med. Chem.
American Chemical Society, 2540 Olentangy River Road, P.O. Box 3337, Columbus, OH 43210-3337, United States, Tel: +1(614)4473776, Fax: +1(614)4473671 ISSUES/YR: 26

- **Journal of Medicinal Food**
  ISSN: 1096-620X
  J. Med. Food
  Mary Ann Liebert Inc., 140 Huguenot Street, New Rochelle, NY 10801-5215, United States, Tel: +1(914)7402100, Fax: +1(914)7402101 ISSUES/YR: 4

Journal of Medicinal Plants
ISSN: 1684-0240
J. Med. Plants
Institute of Medicinal Plants, P.O. Box. 13145-1446, No.97 Bozorgmehr St., Qods St.,, Enghelab Ave. Tehran, Iran, Tel: +98-9821 6950447, Fax: +98-9821 6463334 ISSUES/YR: 4

Journal of Medicine
ISSN: 0025-7850
J. Med.
PJD Publications Ltd, P.O. Box 966, Westbury, NY 11590-0966, United States, Tel: +1(516)6260650, Fax: +1(516)6265546 ISSUES/YR: 6

Journal of Medicine and Philosophy
ISSN: 0360-5310
J. Med. Philos.
Taylor and Francis Ltd., 4 Park Square, Milton Park, OX14 4RN, Abingdon, Oxfordshire, United Kingdom, Tel: +44(20)70175541, Fax: +44(20)70175198 ISSUES/YR: 6

Journal of Membrane Biology
ISSN: 0022-2631
J. Membr. Biol.
Springer New York LLC, 233 Springer Street, New York, 10013-1578, United States, Tel: +1-212 460 1501, Fax: +1-212 460 1595 ISSUES/YR: 18

- **Journal of Membrane Science**
  ISSN: 0376-7388
  J. Membr. Sci.
  Elsevier, P.O. Box 211, 1000 AE, Amsterdam, Netherlands, Tel: +31(20)4853757, Fax: +31(20)4853432 ISSUES/YR: 36

- **Journal of Men's Health and Gender**
  ISSN: 1571-8913
  J. Men's Health Gender
  Elsevier, P.O. Box 211, 1000 AE, Amsterdam, Netherlands, Tel: +31(20)4853757, Fax: +31(20)4853432 ISSUES/YR: 4

Journal of Mental Health
ISSN: 0963-8237
J. Ment. Health
Routledge, 4 Park Square, Milton Park, OX14 4RN, Abingdon, Oxfordshire, United Kingdom, Tel: +44(1235)828600, Fax: +44(1235)829000 ISSUES/YR: 6

Journal of Mental Health and Aging
ISSN: 1078-4470
J. Ment. Health Aging
Springer Publishing Company, 11 West 42nd Street, 15th Floor, New York, NY 10036, United States, Tel: +1(212)4314370, Fax: +1(212)9417842 ISSUES/YR: 4

Journal of Mental Health Policy and Economics
ISSN: 1091-4358
J. Mental Health Policy Econ.
ICMPE, Via Daniele Crespi 7, 20123, Milano, Italy, Fax: +39-02 5810 6901 ISSUES/YR: 4

- **Journal of Microbiological Methods**
  ISSN: 0167-7012
  J. Microbiol. Methods
  Elsevier, P.O. Box 211, 1000 AE, Amsterdam, Netherlands, Tel: +31(20)4853757, Fax: +31(20)4853432 ISSUES/YR: 12

Journal of Microbiology and Biotechnology
ISSN: 1017-7825
J. Microbiol. Biotechnol.
Korean Society for Microbiolog and Biotechnology, 635-4 Yeogsam-Dong, Kangnam-Gu, 135-703, Seoul, South Korea, Tel: +82(2)5524733, Fax: +82(2)5524732 ISSUES/YR: 6

Journal of Microbiology, Immunology and Infection
ISSN: 1684-1182
J. Microbiol. Immunol. Infect.
Chinese Society of Microbiology and Immunology, No. 1 Section 1, Jen-Ai Road, 100, Taipei, Taiwan, Tel: +886(2)23562220, Fax: +886(2)23915180 ISSUES/YR: 4

Journal of Microencapsulation
ISSN: 0265-2048
J. Microencapsulation
Taylor and Francis Ltd., 4 Park Square, Milton Park, OX14 4RN, Abingdon, Oxfordshire, United Kingdom, Tel: +44(20)70175541, Fax: +44(20)70175198 ISSUES/YR: 8

- **Journal of Microscopy**
  ISSN: 0022-2720
  J. Microsc.
  Blackwell Publishing Ltd, 9600 Garsington Road, OX4 2XG, Oxford, United Kingdom, Tel: +44(1865)776868, Fax: +44(1865)714591 ISSUES/YR: 12

- **Journal of Midwifery and Women's Health**
  ISSN: 1526-9523
  J. Midwifery Women's Health
  Elsevier Inc., 360 Park Avenue South, New York, NY 10010, United States, Tel: +1(212)6333950, Fax: +1(212)6333990 ISSUES/YR: 6

Journal of Mind and Behavior
ISSN: 0271-0137
J. Mind Behav.
Institute of Mind and Behavior Inc., P.O. Box 522, New York, NY 10014, United States, Tel: +1(212)5954853 ISSUES/YR: 4

Journal of Minimally Invasive Gynecology
ISSN: 1553-4650
J. Minimally Invasive Gynecol.
Elsevier, P.O. Box 211, 1000 AE, Amsterdam, Netherlands, Tel: +31(20)4853757, Fax: +31(20)4853432 ISSUES/YR: 6

- **Journal of Molecular and Cellular Cardiology**
  ISSN: 0022-2828
  J. Mol. Cell. Cardiol.
  Academic Press, 24-28 Oval Road, NW1 7DX, London, United Kingdom, Tel: +44(207)2674466 ISSUES/YR: 12

- **Journal of Molecular Biology**
  ISSN: 0022-2836
  J. Mol. Biol.
  Academic Press, 24-28 Oval Road, NW1 7DX, London, United Kingdom, Tel: +44(207)2674466 ISSUES/YR: 50

Journal of Molecular Catalysis B: Enzymatic
ISSN: 1381-1177
J. Mol. Catal. B Enzym.
Elsevier, P.O. Box 211, 1000 AE, Amsterdam, Netherlands, Tel: +31(20)4853757, Fax: +31(20)4853432 ISSUES/YR: 12

Journal of Molecular Diagnostics
ISSN: 1525-1578
J. Mol. Diagn.
Association of Molecular Pathology, 9650 Rockville Pike, Bethesda, MD 20814-3993, United States, Tel: +1(301)5710107, Fax: +1(301)5710108 ISSUES/YR: 4

- **Journal of Molecular Endocrinology**
  ISSN: 0952-5041
  J. Mol. Endocrinol.
  Society for Endocrinology, 22 Apex Court, Woodlands, Bradley Stoke, BS32 4JT, Bristol, United Kingdom, Tel: +44(1454)642220, Fax: +44(1454)642222 ISSUES/YR: 6

Journal of Molecular Evolution
ISSN: 0022-2844
J. Mol. Evol.
Springer New York LLC, 233 Springer Street, New York, 10013-1578, United States, Tel: +1-212 460 1501, Fax: +1-212 460 1595 ISSUES/YR: 12

- **Journal of Molecular Graphics and Modelling**
  ISSN: 1093-3263
  J. Mol. Graph. Model.
  Elsevier Inc., 360 Park Avenue South, New York, NY 10010, United States, Tel: +1(212)6333950, Fax: +1(212)6333990 ISSUES/YR: 6

- **Journal of Molecular Histology**
  ISSN: 1567-2379
  J. Mol. Histol.
  Kluwer Academic Publishers, Van Godewijckstraat 30, 3311 GZ, Dordrecht, Netherlands, Tel: +31(78)6576000, Fax: +31(78)6576254 ISSUES/YR: 9

Journal of Molecular Medicine
ISSN: 0946-2716
J. Mol. Med.
Springer Verlag, Tiergartenstrasse 17, D-69121, Heidelberg, Germany, Tel: +49(6221)4870, Fax: +49(6221)487366 ISSUES/YR: 12

Journal of Molecular Microbiology and Biotechnology
ISSN: 1464-1801
J. Mol. Microbiol. Biotechnol.
S. Karger AG, Allschwilerstrasse 10, P.O. Box, CH-4009, Basel, Switzerland, Tel: +41(61)3061111, Fax: +41(61)3061234 ISSUES/YR: 8

• **Journal of Molecular Modeling**
ISSN: 1610-2940
J. Mol. Model.
Springer Verlag, Tiergartenstrasse 17, D-69121, Heidelberg, Germany, Tel: +49(6221)4870, Fax: +49(6221)487366 ISSUES/YR: 6

Journal of Molecular Neuroscience
ISSN: 0895-8696
J. Mol. Neurosci.
Humana Press Inc., 999 Riverview Drive, Suite 208, Totowa, NJ 07512-1165, United States, Tel: +1(973)2561699, Fax: +1(973)2568341 ISSUES/YR: 9

Journal of Motor Behavior
ISSN: 0022-2895
J. Mot. Behav.
Heldref Publications, 1319 Eighteenth Street NW, Washington, DC 20036-1802, United States, Tel: +1(202)2966267, Fax: +1(202)2936130 ISSUES/YR: 6

• **Journal of Muscle Research and Cell Motility**
ISSN: 0142-4319
J. Muscle Res. Cell Motil.
Kluwer Academic Publishers, Van Godewijckstraat 30, 3311 GZ, Dordrecht, Netherlands, Tel: +31(78)6576000, Fax: +31(78)6576254 ISSUES/YR: 8

Journal of Musculoskeletal Neuronal Interactions
ISSN: 1108-7161
J. Musculoskelet. Neuronal Interact.
International Society of Musculoskeletal and Neuronal Interactions, 2 Thrakis St., 15124, Maroussi, Greece, Tel: +30-210 6120 382, Fax: +30-2752 099 343 ISSUES/YR: 4

Journal of Musculoskeletal Pain
ISSN: 1058-2452
J. Musculoskelet. Pain
Haworth Press Inc., 10 Alice Street, Binghamton, NY 13904-1580, United States, Tel: +1(607)7225857 ISSUES/YR: 4

Journal of Musculoskeletal Research
ISSN: 0218-9577
J. Musculoskelet. Res.
World Scientific Publishing Co. Pte Ltd, 5 Toh Tuck Link, 596224, Singapore, Singapore, Tel: +65(6466)5775, Fax: +65(6467)7667 ISSUES/YR: 4

Journal of Nanobiotechnology
ISSN: 1477-3155
J. Nanobiotechnology
BioMed Central Ltd., 34 - 42 Cleveland Street, W1T 4LB, London, United Kingdom, Tel: +44(207)3230323, Fax: +44(207)6319926 ISSUES/YR: 1

Journal of Nara Medical Association
ISSN: 1345-0069
J. Nara Med. Assoc.
Nara Medical Association, Kashihara, 634, Nara, Japan, Tel: +81(7442)23051 ISSUES/YR: 6

Journal of Natural Products
ISSN: 0163-3864
J. Nat. Prod.
American Chemical Society, 2540 Olentangy River Road, P.O. Box 3337, Columbus, OH 43210-3337, United States, Tel: +1(614)4473776, Fax: +1(614)4473671 ISSUES/YR: 12

Journal of Natural Remedies
ISSN: 0972-5547
J. Nat. Rem.
Natural Remedies Private Limited, Plot 5B, Veerasandra Indl. Area, 19th KM Stone, Hosur Road, 561229, Bangalore, India, Tel: +91(8110)225633092, Fax: +91(8110)226534369 ISSUES/YR: 2

Journal of Negative Results in BioMedicine
ISSN: 1477-5751
J. Negat. Results Biomed.
BioMed Central Ltd., 34 - 42 Cleveland Street, W1T 4LB, London, United Kingdom, Tel: +44(207)3230323, Fax: +44(207)6319926 ISSUES/YR: 1

Journal of Neonatology
ISSN: 0973-2179
J. Neonatal.
National Neonatology Forum, Sir Ganga Ram Hospital,Rajindera Nagar, 110060, New Dehli, India, Tel: +91-11 55 443838 ISSUES/YR: 4

Journal of Nephrology
ISSN: 1121-8428
J. Nephrol.
Wichtig Editore s.r.l., Via Friuli 72/74, 20135, Milan, Italy, Tel: +39(02)55195443, Fax: +39(02)55195971 ISSUES/YR: 6

Journal of Nephrology, Urology and Transplantation
ISSN: 1606-9331
J. Nephrol. Urol. Transplant.
Sindh Institute for Urology and Transplantation, 74200, Karachi, Pakistan, Tel: +92(21)7730351, Fax: +92(21)7767301 ISSUES/YR: 4

Journal of Nervous and Mental Disease
ISSN: 0022-3018
J. Nerv. Ment. Dis.
Lippincott Williams and Wilkins, 351 West Camden Street, Baltimore, MD 21201-2436, United States, Tel: +1(410)5284000 ISSUES/YR: 12

• **Journal of Neural Engineering**
ISSN: 1741-2560
J. Neural Eng.
Institute of Physics Publishing, Temple Back, BS1 6BE, Bristol, United Kingdom, Tel: +44(117)9297481, Fax: +44(117)9294318 ISSUES/YR: 4

• **Journal of Neural Transmission**
ISSN: 0300-9564
J. Neural Transm.
Springer-Verlag Wien, Sachsenplatz 4-6, P.O. Box 89, A-1201, Vienna, Austria, Tel: +43(1)3302415, Fax: +43(1)3302426 ISSUES/YR: 12

• **Journal of Neural Transmission, Supplement**
ISSN: 0303-6995
J. Neural Transm. Suppl.
Springer-Verlag Wien, Sachsenplatz 4-6, P.O. Box 89, A-1201, Vienna, Austria, Tel: +43(1)3302415, Fax: +43(1)3302426 ISSUES/YR: 1

Journal of Neuro-Oncology
ISSN: 0167-594X
J. Neuro-Oncol.
Kluwer Academic Publishers, 101 Philip Drive, Assinippi Park, Norwell, MA 02061, United States, Tel: +1(781)8716600, Fax: +1(781)8716528 ISSUES/YR: 15

• **Journal of Neuro-Ophthalmology**
ISSN: 1070-8022
J. Neuro-Ophthalmol.
Lippincott Williams and Wilkins, 530 Walnut Street, Philadelphia, PA 19106-3621, United States, Tel: +1(215)5218342, Fax: +1(215)5218326 ISSUES/YR: 4

• **Journal of Neurobiology**
ISSN: 0022-3034
J. Neurobiol.
John Wiley and Sons Inc., P.O.Box 18667, Newark, NJ 07191-8667, United States, Tel: +1(212)8506000, Fax: +1(212)8506012 ISSUES/YR: 16

• **Journal of Neurochemistry**
ISSN: 0022-3042
J. Neurochem.
Blackwell Publishing Ltd, 9600 Garsington Road, OX4 2XG, Oxford, United Kingdom, Tel: +44(1865)776868, Fax: +44(1865)714591 ISSUES/YR: 24

• **Journal of Neurocytology**
ISSN: 0300-4864
J. Neurocytol.
Kluwer Academic Publishers, Van Godewijckstraat 30, 3311 GZ, Dordrecht, Netherlands, Tel: +31(78)6576000, Fax: +31(78)6576254 ISSUES/YR: 9

- **Journal of Neuroendocrinology**
  ISSN: 0953-8194
  J. Neuroendocrinol.
  Blackwell Publishing Ltd, 9600 Garsington Road, OX4 2XG, Oxford, United Kingdom, Tel: +44(1865)776868, Fax: +44(1865)714591 ISSUES/YR: 12

Journal of NeuroEngineering and Rehabilitation
  ISSN: 1743-0003
  J. NeuroEng. Rehabil.
  BioMed Central Ltd., 34 - 42 Cleveland Street, W1T 4LB, London, United Kingdom, Tel: +44(207)3230323, Fax: +44(207)6319926 ISSUES/YR: 1

- **Journal of Neurogenetics**
  ISSN: 0167-7063
  J. Neurogenet.
  Taylor and Francis Ltd., 4 Park Square, Milton Park, OX14 4RN, Abingdon, Oxfordshire, United Kingdom, Tel: +44(20)70175541, Fax: +44(20)70175198 ISSUES/YR: 4

Journal of Neuroimaging
  ISSN: 1051-2284
  J. Neuroimaging
  Blackwell Publishing Inc., 350 Main Street, Malden, MA 02148, United States, Tel: +1(781)3888200, Fax: +1(781)3888210 ISSUES/YR: 4

- **Journal of Neuroimmunology**
  ISSN: 0165-5728
  J. Neuroimmunol.
  Elsevier, P.O. Box 211, 1000 AE, Amsterdam, Netherlands, Tel: +31(20)4853757, Fax: +31(20)4853432 ISSUES/YR: 24

Journal of Neuroinflammation
  ISSN: 1742-2094
  J. Neuroinflamm.
  BioMed Central Ltd., 34 - 42 Cleveland Street, W1T 4LB, London, United Kingdom, Tel: +44(207)3230323, Fax: +44(207)6319926 ISSUES/YR: 1

- **Journal of Neurolinguistics**
  ISSN: 0911-6044
  J. Neurolinguist.
  Elsevier Ltd, Langford Lane, Kidlington, OX5 1GB, Oxford, United Kingdom, Tel: +44(1865)843000, Fax: +44(1865)843010 ISSUES/YR: 6

- **Journal of Neurology**
  ISSN: 0340-5354
  J. Neurol.
  Dr. Dietrich Steinkopff Verlag GmbH and Co. KG, P.O. Box 100462, D-64204, Darmstadt, Germany, Tel: +49(6151)828990, Fax: +49(6151)8289940 ISSUES/YR: 12

- **Journal of Neurology, Neurosurgery and Psychiatry**
  ISSN: 0022-3050
  J. Neurol. Neurosurg. Psychiatry
  BMJ Publishing Group, Tavistock Square, WC1H 9JR, London, United Kingdom, Tel: +44(207)3874499, Fax: +44(207)3836668 ISSUES/YR: 12

- **Journal of Neurology, Supplement**
  ISSN: 0939-1517
  J. Neurol. Suppl.
  Dr. Dietrich Steinkopff Verlag GmbH and Co. KG, P.O. Box 100462, D-64204, Darmstadt, Germany, Tel: +49(6151)828990, Fax: +49(6151)8289940 ISSUES/YR: 1

- **Journal of Neuropathology and Experimental Neurology**
  ISSN: 0022-3069
  J. Neuropathol. Exp. Neurol.
  Lippincott Williams and Wilkins, 241 Borough High Street, SE1 1GB, London, United Kingdom, Tel: +1(207)9407500, Fax: +1(207)9407575 ISSUES/YR: 12

- **Journal of Neurophysiology**
  ISSN: 0022-3077
  J. Neurophysiol.
  American Physiological Society, 9650 Rockville Pike, Bethesda, MD 20814-3991, United States, Tel: +1(301)6347180, Fax: +1(301)6347241 ISSUES/YR: 12

- **Journal of Neuropsychiatry and Clinical Neurosciences**
  ISSN: 0895-0172
  J. Neuropsychiatry Clin. Neurosci.
  American Psychiatric Publishing Inc., 1000 Wilson Blvd, Suite 1825, VA 22209, United States, Tel: +1(703)907-7322, Fax: +1(703)90-1091 ISSUES/YR: 4

Journal of Neuroradiology
  ISSN: 0150-9861
  J. Neuroradiol.
  Masson, 21 Rue Camille Desmoulins, 92789, Issy les Moulineaux Cedex 9, France, Tel: +33(1)73281634, Fax: +33(1)73281649 ISSUES/YR: 5

- **Journal of Neuroscience**
  ISSN: 0270-6474
  J. Neurosci.
  Society for Neuroscience, 11 Dupont Circle NW, Suite 500, Washington, DC 20036, United States, Tel: +1(202)4626688 ISSUES/YR: 50

- **Journal of Neuroscience Methods**
  ISSN: 0165-0270
  J. Neurosci. Methods
  Elsevier, P.O. Box 211, 1000 AE, Amsterdam, Netherlands, Tel: +31(20)4853757, Fax: +31(20)4853432 ISSUES/YR: 16

- **Journal of Neuroscience Research**
  ISSN: 0360-4012
  J. Neurosci. Res.
  Wiley-Liss Inc., 111 River Street, Hoboken, NJ 07030-5774, United States, Tel: +1(201)7486645, Fax: +1(201)7486021 ISSUES/YR: 18

- **Journal of Neurosurgery**
  ISSN: 0022-3085
  J. Neurosurg.
  American Association of Neurological Surgeons, 1224 West Main Street Suite 450, Charlottesville, VA 22903, United States, Tel: +1-434 924 5503, Fax: +1-434 924 2702 ISSUES/YR: 12

- **Journal of Neurosurgical Anesthesiology**
  ISSN: 0898-4921
  J. Neurosurg. Anesthesiol.
  Lippincott Williams and Wilkins, 530 Walnut Street, Philadelphia, PA 19106-3621, United States, Tel: +1(215)5218342, Fax: +1(215)5218326 ISSUES/YR: 4

Journal of Neurosurgical Sciences
  ISSN: 0026-4881
  J. Neurosurg. Sci.
  Edizioni Minerva Medica S.p.A., Corso Bramante 83-85, 10126, Turin, Italy, Tel: +39(011)678282, Fax: +39(011)674502 ISSUES/YR: 4

- **Journal of Neurotherapy**
  ISSN: 1087-4208
  J. Neurother.
  Haworth Press Inc., 10 Alice Street, Binghamton, NY 13904-1580, United States, Tel: +1(607)7225857 ISSUES/YR: 4

Journal of Neurotrauma
  ISSN: 0897-7151
  J. Neurotrauma
  Mary Ann Liebert Inc., 140 Huguenot Street, New Rochelle, NY 10801-5215, United States, Tel: +1(914)7402100, Fax: +1(914)7402101 ISSUES/YR: 12

- **Journal of NeuroVirology**
  ISSN: 1355-0284
  J. Neurovirol.
  Taylor and Francis Inc., 325 Chestnut St, Suite 800, Philadelphia PA, PA 19106, United States, Tel: +1(800)3541420, Fax: +1(215)6258914 ISSUES/YR: 6

Journal of Nippon Medical School
  ISSN: 1345-4676
  J. Nippon Med. Sch.
  Medical Association of Nippon Medical School, 1-1-5 Sendagi, Bunkyo-ku, 113-8602, Tokyo, Japan, Tel: +81(3)38222131, Fax: +81(3)38223759 ISSUES/YR: 6

- **Journal of Nuclear Cardiology**
  ISSN: 1071-3581
  J. Nucl. Cardiol.
  Mosby Inc., 11830 Westline Industrial Drive, St. Louis, MO 63146, United States, Tel: +1(314)8728370, Fax: +1(314)4321380 ISSUES/YR: 6

- **Journal of Nuclear Medicine**
  ISSN: 0161-5505
  J. Nucl. Med.
  Society of Nuclear Medicine Inc., 1850 Samuel Morse Drive, Reston, VA 22090-5316, United States, Tel: +1(703)7089000 ISSUES/YR: 12

Journal of Nuclear Medicine Technology
ISSN: 0091-4916
J. Nucl. Med. Technol.
Society of Nuclear Medicine Inc., 1850 Samuel Morse Drive, Reston, VA 22090-5316, United States, Tel: +1(703)7089000 ISSUES/YR: 4

Journal of Nursing Administration
ISSN: 0002-0443
J. Nurs. Adm.
Lippincott Williams and Wilkins, 530 Walnut Street, Philadelphia, PA 19106-3621, United States, Tel: +1(215)5218342, Fax: +1(215)5218326 ISSUES/YR: 12

Journal of Nutraceuticals, Functional and Medical Foods
ISSN: 1089-4179
J. Neutraceuticals Funct. Med. Foods
Haworth Press Inc., 10 Alice Street, Binghamton, NY 13904-1580, United States, Tel: +1(607)7225857, Fax: +1(607)7221424 ISSUES/YR: 4

Journal of Nutrition
ISSN: 0022-3166
J. Nutr.
American Institute of Nutrition, 9650 Rockville Pike, Bethesda, MD 20814-3990, United States, Tel: +1(301)5307050 ISSUES/YR: 12

● **Journal of Nutrition, Health and Aging**
ISSN: 1279-7707
J. Nutr. Health Aging
Serdi Publishing Company, 320 rue Saint-Honore, 75001, Paris, France, Tel: +33(1)44542796, Fax: +33(5)61751128 ISSUES/YR: 6

● **Journal of Nutritional and Environmental Medicine**
ISSN: 1359-0847
J. Nutr. Environ. Med.
Routledge, 4 Park Square, Milton Park, OX14 4RN, Abingdon, Oxfordshire, United Kingdom, Tel: +44(1235)828600, Fax: +44(1235)829000 ISSUES/YR: 4

Journal of Nutritional Biochemistry
ISSN: 0955-2863
J. Nutr. Biochem.
Elsevier Inc., 360 Park Avenue South, New York, NY 10010, United States, Tel: +1(212)6333950, Fax: +1(212)6333990 ISSUES/YR: 12

Journal of Nutritional Science and Vitaminology
ISSN: 0301-4800
J. Nutr. Sci. Vitaminol.
Center for Academic Publications Japan, 4-16, Yayoi 2-chome, Bunkyo-ku, 113-0032, Tokyo, Japan, Tel: +81(3)38175830, Fax: +81(3)58175820 ISSUES/YR: 6

● **Journal of Obstetrics and Gynaecology**
ISSN: 0144-3615
J. Obstet. Gynaecol.
Taylor and Francis Ltd., 4 Park Square, Milton Park, OX14 4RN, Abingdon, Oxfordshire, United Kingdom, Tel: +44(20)70175541, Fax: +44(20)70175198 ISSUES/YR: 8

Journal of Obstetrics and Gynaecology Research
ISSN: 1341-8076
J. Obstet. Gynaecol. Res.
Blackwell Publishing, 550 Swanston Street, VIC 3053, Carlton South, Australia, Tel: +81(3)8359 1011, Fax: +81(3)8359 1120 ISSUES/YR: 6

● **Journal of Occupational and Environmental Hygiene**
ISSN: 1545-9624
J. Occup. Environ. Hyg.
Taylor and Francis Ltd., 4 Park Square, Milton Park, OX14 4RN, Abingdon, Oxfordshire, United Kingdom, Tel: +44(20)70175541, Fax: +44(20)70175198 ISSUES/YR: 12

Journal of Occupational and Environmental Medicine
ISSN: 1076-2752
J. Occup. Environ. Med.
Lippincott Williams and Wilkins, 351 West Camden Street, Baltimore, MD 21201-2436, United States, Tel: +1(410)5284000 ISSUES/YR: 12

Journal of Occupational Health
ISSN: 1341-9145
J. Occup. Health
Japan Society for Occupational Health, 29-8, Shinjuku, 1-chome, Shinjuku-ku, 160-0022, Tokyo, Japan, Tel: +81(3)33561536, Fax: +81(3)53623746 ISSUES/YR: 6

Journal of Occupational Health and Safety - Australia and New Zealand
ISSN: 0815-6409
J. Occup. Health Saf. Aust. New Zealand
CCH Australia Limited, G.P.O. Box 4072, NSW 2001, Sydney, Australia, Tel: +61(2)98571300, Fax: +61(2)98571600 ISSUES/YR: 6

● **Journal of Occupational Health Psychology**
ISSN: 1076-8998
J. Occup. Health Psychol.
Educational Publishing Foundation, 750 First Street NE, Washington, DC 20002-4242, United States, Tel: +1(202)3365540, Fax: +1(202)3365549 ISSUES/YR: 4

Journal of Occupational Rehabilitation
ISSN: 1053-0487
J. Occup. Rehabil.
Springer Science and Business Media Deutschland GmbH, Heidelberger Platz 3, 14197, Berlin, Germany, Tel: +1-30 827875282, Fax: +1-30 827875707 ISSUES/YR: 4

Journal of Ocular Pharmacology and Therapeutics
ISSN: 1080-7683
J. Ocul. Pharmacol. Ther.
Mary Ann Liebert Inc., 140 Huguenot Street, New Rochelle, NY 10801-5215, United States, Tel: +1(914)7402100, Fax: +1(914)7402101 ISSUES/YR: 6

● **Journal of Oncology Pharmacy Practice**
ISSN: 1078-1552
J. Oncol. Pharm. Pract.
Arnold, 338 Euston Road, NW1 3BH, London, United Kingdom, Tel: +1(20)78736000, Fax: +1(20)78736376 ISSUES/YR: 4

Journal of Oral and Maxillofacial Surgery
ISSN: 0278-2391
J. Oral Maxillofac. Surg.
W.B. Saunders, Independence Square West, Philadelphia, PA 19106-3399, United States, Tel: +1(215)2387800, Fax: +1(215)2386445 ISSUES/YR: 12

● **Journal of Oral Pathology and Medicine**
ISSN: 0904-2512
J. Oral Pathol. Med.
Blackwell Publishing Ltd, 9600 Garsington Road, OX4 2XG, Oxford, United Kingdom, Tel: +44(1865)776868, Fax: +44(1865)714591 ISSUES/YR: 10

● **Journal of Organ Dysfunction**
ISSN: 1747-1060
J. Organ Dysfunct.
Taylor and Francis Ltd., 4 Park Square, Milton Park, OX14 4RN, Abingdon, Oxfordshire, United Kingdom, Tel: +44(20)70175541, Fax: +44(20)70175198 ISSUES/YR: 4

Journal of Organic Chemistry
ISSN: 0022-3263
J. Org. Chem.
American Chemical Society, 2540 Olentangy River Road, P.O. Box 3337, Columbus, OH 43210-3337, United States, Tel: +1(614)4473776, Fax: +1(614)4473671 ISSUES/YR: 26

● **Journal of Orofacial Orthopedics**
ISSN: 1434-5293
J. Orofacial Orthop.
Urban und Vogel GmbH, Neumarkter Strasse 43, D-81673, Munchen, Germany, Tel: +49(89)43721300, Fax: +49(89)43721399 ISSUES/YR: 6

Journal of Orthomolecular Medicine
ISSN: 0317-0209
J. Orthomol. Med.
International Society for Orthomolecular Medicine, 16 Florence Avenue, ONT M2N 1E9, Toronto, Canada, Tel: +1(416)7332117 ISSUES/YR: 4

Journal of Orthopaedic and Sports Physical Therapy
ISSN: 0190-6011
J. Orthop. Sports Phys. Ther.
JOSPT, 1111 North Fairfax Street, Suite 100, VA 22314-1436, United States, Tel: +1-877 776 3450, Fax: +1-703 836 2210 ISSUES/YR: 12

- **Journal of Orthopaedic Research**
  ISSN: 0736-0266
  J. Orthop. Res.
  John Wiley and Sons Inc., P.O.Box 18667, Newark, NJ 07191-8667, United States, Tel: +1(212)8506000, Fax: +1(212)8506012 ISSUES/YR: 6

Journal of Orthopaedic Science
ISSN: 0949-2658
J. Ortop. Sci.
Springer-Verlag Tokyo, 3-13, Hongo 3-chome, Bunkyo-ku, 113-8480, Tokyo, Japan, Tel: +81(3)38180861, Fax: +81(3)38180864 ISSUES/YR: 6

- **Journal of Orthopaedic Trauma**
  ISSN: 0890-5339
  J. Orthop. Trauma
  Lippincott Williams and Wilkins, 530 Walnut Street, Philadelphia, PA 19106-3621, United States, Tel: +1(215)5218342, Fax: +1(215)5218326 ISSUES/YR: 10
- **Journal of Orthopaedics and Traumatology**
  ISSN: 1590-9921
  J. Orthop. Traumatol.
  Springer-Verlag Italia s.r.l, Via Podgora 4, I-20122, Milano, Italy, Tel: +39(2)5425971, Fax: +39(2)55193360 ISSUES/YR: 3
- **Journal of Otolaryngology**
  ISSN: 0381-6605
  J. Otolaryngol.
  BC Decker Inc., P.O. Box 620, LCD 1, L8N 3K7, Hamilton, Ontario, Canada, Tel: +1(905)522-7017, Fax: +1(905)522-7839 ISSUES/YR: 6

Journal of Otolaryngology of Japan
ISSN: 0030-6622
J. Otolaryngol. Jpn.
Oto-Rhino-Laryngological Society of Japan Inc., 23-14, 3-Chome Takanawa, Minato-ku, 108, Tokyo, Japan, Tel: +81(3)34433085 ISSUES/YR: 12

Journal of Outcomes Research
ISSN: 1369-698X
J. Outcomes Res.
PJB Publications, Sheepen Place, C03 3LP, Colchester, United Kingdom, Tel: +44-20 70175540 ISSUES/YR: 1

- **Journal of Paediatrics and Child Health**
  ISSN: 1034-4810
  J. Paediatr. Child Health
  Blackwell Publishing, 550 Swanston Street, VIC 3053, Carlton South, Australia, Tel: +61(3)8359 1011, Fax: +61(3)8359 1120 ISSUES/YR: 12

Journal of Pain
ISSN: 1526-5900
J. Pain
Churchill Livingstone Inc., 650 Avenue of the Americas, New York, NY 10011, United States, Tel: +1(212)2065000, Fax: +1(212)7277808 ISSUES/YR: 9

Journal of Pain and Palliative Care Pharmacotherapy
ISSN: 1536-0288
J. Pain Palliative Care Pharmacother.
Haworth Press Inc., 10 Alice Street, Binghamton, NY 13904-1580, United States, Tel: +1(607)7225857, Fax: +1(607)7221424 ISSUES/YR: 4

Journal of Pain and Symptom Management
ISSN: 0885-3924
J. Pain Symptom Manage.
Elsevier Inc., 360 Park Avenue South, New York, NY 10010, United States, Tel: +1(212)6333950, Fax: +1(212)6333990 ISSUES/YR: 12

Journal of Pakistan Association of Dermatologists
ISSN: 1560-9014
J. Pak. Assoc. Dermatol.
Pakistan Association of Dermatologists, 17/II-B, Khayaban-E-Janbaz, Phase V, D.H.A., Karachi, Pakistan, Tel: +92-5854535 ISSUES/YR: 4

Journal of Palliative Medicine
ISSN: 1096-6218
J. Palliative Med.
Mary Ann Liebert Inc., 140 Huguenot Street, New Rochelle, NY 10801-5215, United States, Tel: +1(914)7402100, Fax: +1(914)7402101 ISSUES/YR: 6

Journal of Parasitology
ISSN: 0022-3395
J. Parasitol.
American Society of Parasitologists, P.O. Box 1897, Lawrence, KS 66044-8897, United States, Tel: +1(913)8431221, Fax: +1(913)8431274 ISSUES/YR: 6

- **Journal of Parenteral and Enteral Nutrition**
  ISSN: 0148-6071
  J. Parenter. Enter. Nutr.
  American Society for Parenteral and Enteral Nutrition, 8630 Fenton Steet, Suite 412, Silver Spring, MD 20910-3805, United States, Tel: +1(301)5876315 ISSUES/YR: 6
- **Journal of Pathology**
  ISSN: 0022-3417
  J. Pathol.
  John Wiley and Sons Ltd, Southern Gate, PO19 8SQ, Chichester, West Sussex, United Kingdom, Tel: +44(1243)779777, Fax: +44(1243)775878 ISSUES/YR: 12
- **Journal of Pediatric and Adolescent Gynecology**
  ISSN: 1083-3188
  J. Pediatr. Adolesc. Gynecol.
  Elsevier USA, 6277 Sea Harbor Drive, Orlando, FL 32862 8239, United States, Tel: +1-407 345 4000, Fax: +1-407 363 9661 ISSUES/YR: 6

Journal of Pediatric Endocrinology and Metabolism
ISSN: 0334-018X
J. Pediatr. Endocrinol. Metab.
Freund Publishing House Ltd, 61 Nachmani Street, P.O. Box 35010, 61350, Tel Aviv, Israel, Tel: +972(3)5628540, Fax: +972(3)5628538 ISSUES/YR: 12

- **Journal of Pediatric Gastroenterology and Nutrition**
  ISSN: 0277-2116
  J. Pediatr. Gastroenterol. Nutr.
  Lippincott Williams and Wilkins, 530 Walnut Street, Philadelphia, PA 19106-3621, United States, Tel: +1(215)5218342, Fax: +1(215)5218326 ISSUES/YR: 10
- **Journal of Pediatric Hematology/Oncology**
  ISSN: 1077-4114
  J. Pediatr. Hematol. Oncol.
  Lippincott Williams and Wilkins, 530 Walnut Street, Philadelphia, PA 19106-3621, United States, Tel: +1(215)5218342, Fax: +1(215)5218326 ISSUES/YR: 12

Journal of Pediatric Neurology
ISSN: 1304-2580
J. Pediatr. Neurol.
Yuzuncu Yil University, PO Box 2, 65100 Van, Turkey, Tel: +90(432)159164, Fax: +90(432)2155281 ISSUES/YR: 4

Journal of Pediatric Neurosciences
ISSN: 1817-1745
J. Pediatr. Neurosci.
Medknow Publication, 12 Manisha Plaza, MN Rd, 400 070 MUMBAI, Kurla, India, Tel: +91(22)25032970, Fax: +91(22)25032398 ISSUES/YR: 4

Journal of Pediatric Ophthalmology and Strabismus
ISSN: 0191-3913
J. Pediatr. Ophthalmol. Strabismus
Slack Incorporated, 6900 Grove Road, Thorofare, NJ 08086-9447, United States, Tel: +1(856)8481000, Fax: +1(856)8535991 ISSUES/YR: 6

- **Journal of Pediatric Orthopaedics**
  ISSN: 0271-6798
  J. Pediatr. Orthop.
  Lippincott Williams and Wilkins, 530 Walnut Street, Philadelphia, PA 19106-3621, United States, Tel: +1(215)5218342, Fax: +1(215)5218326 ISSUES/YR: 6
- **Journal of Pediatric Orthopaedics Part B**
  ISSN: 1060-152X
  J. Pediatr. Orthop. Part B
  Lippincott Williams and Wilkins, 530 Walnut Street, Philadelphia, PA 19106-3621, United States, Tel: +1(215)5218342, Fax: +1(215)5218326 ISSUES/YR: 6

Journal of Pediatric Psychology
ISSN: 0146-8693
J. Pediatr. Psychol.
Oxford University Press, Great Clarendon Street, OX2 6DP, Oxford, United Kingdom, Tel: +1(1865)353907, Fax: +1(1865)353486 ISSUES/YR: 8

- **Journal of Pediatric Surgery**
  ISSN: 0022-3468
  J. Pediatr. Surg.
  W.B. Saunders, Independence Square West, Philadelphia, PA 19106-3399, United States, Tel: +1(215)2387800, Fax: +1(215)2386445 ISSUES/YR: 12
- **Journal of Pediatric Urology**
  ISSN: 1477-5131
  J. Pediatr. Urol.
  Elsevier Ltd, Langford Lane, Kidlington, OX5 1GB, Oxford, United Kingdom, Tel: +44(1865)843000, Fax: +44(1865)843010 ISSUES/YR: 6
- **Journal of Pediatrics**
  ISSN: 0022-3476
  J. Pediatr.
  Mosby Inc., 11830 Westline Industrial Drive, St. Louis, MO 63146, United States, Tel: +1(314)8728370, Fax: +1(314)4321380 ISSUES/YR: 12
- **Journal of Pelvic Medicine and Surgery**
  ISSN: 1542-5983
  J. Pelvic Med. Surg.
  Lippincott Williams and Wilkins, 530 Walnut Street, Philadelphia, PA 19106-3621, United States, Tel: +1(215)5218342, Fax: +1(215)5218326 ISSUES/YR: 6
- **Journal of Peptide Science**
  ISSN: 1075-2617
  J. Pept. Sci.
  John Wiley and Sons Ltd, Southern Gate, PO19 8SQ, Chichester, West Sussex, United Kingdom, Tel: +44(1243)779777, Fax: +44(1243)775878 ISSUES/YR: 12
- **Journal of Perinatal Medicine**
  ISSN: 0300-5577
  J. Perinat. Med.
  Walter de Gruyter and Co., Genthiner Strasse 13, D-10785, Berlin, Germany, Tel: +49(30)260-050, Fax: +49(30)260-05251 ISSUES/YR: 6

Journal of Perinatology
ISSN: 0743-8346
J. Perinatol.
Nature Publishing Group, Houndmills, RG21 6XS, Basingstoke, Hampshire, United Kingdom, Tel: +44(1256)329242, Fax: +44(1256)810526 ISSUES/YR: 8

Journal of Periodontology
ISSN: 0022-3492
J. Periodontol.
American Academy of Periodontology, 737 North Michigan Avenue, Suite 800, Chicago, IL 60611-2690, United States, Tel: +1(312)7875518 ISSUES/YR: 12

Journal of Personality Assessment
ISSN: 0022-3891
J. Pers. Assess.
Lawrence Erlbaum Associates Inc., 10 Industrial Avenue, Mahwah, NJ 07430-2262, United States, Tel: +1(201)2582200, Fax: +1(201)2360072 ISSUES/YR: 6

Journal of Personality Disorders
ISSN: 0885-579X
J. Pers. Disord.
Guilford Publications, 72 Spring Street, New York, NY 10012, United States, Tel: +1(212)4319800 ISSUES/YR: 6

- **Journal of Pharmaceutical and Biomedical Analysis**
  ISSN: 0731-7085
  J. Pharm. Biomed. Anal.
  Elsevier, P.O. Box 211, 1000 AE, Amsterdam, Netherlands, Tel: +31(20)4853757, Fax: +31(20)4853432 ISSUES/YR: 15
- **Journal of Pharmaceutical Finance, Economics and Policy**
  ISSN: 1538-5698
  J. Pharm. Financ. Econ. Policy
  Haworth Press Inc., 10 Alice Street, Binghamton, NY 13904-1580, United States, Tel: +1(607)7225857, Fax: +1(607)7221424 ISSUES/YR: 4

Journal of Pharmaceutical Sciences
ISSN: 0022-3549
J. Pharm. Sci.
John Wiley and Sons Inc., P.O.Box 18667, Newark, NJ 07191-8667, United States, Tel: +1(212)8506000, Fax: +1(212)8506012 ISSUES/YR: 12

- **Journal of Pharmacokinetics and Pharmacodynamics**
  ISSN: 1567-567X
  J. Pharmacokinet. Pharmacodyn.
  Springer Science and Business Media Deutschland GmbH, Heidelberger Platz 3, 14197, Berlin, Germany, Tel: +1-30 827875282, Fax: +1-30 827875707 ISSUES/YR: 6

Journal of Pharmacological and Toxicological Methods
ISSN: 1056-8719
J. Pharmacol. Toxicol. Methods
Elsevier Inc., 360 Park Avenue South, New York, NY 10010, United States, Tel: +1(212)6333950, Fax: +1(212)6333990 ISSUES/YR: 6

Journal of Pharmacological Sciences
ISSN: 1347-8613
J. Pharmacol. Sci.
Japanese Pharmacological Society, Gokomachi-Ebisugawa, Nakagyo-ku, 604-0982, Kyoto, Japan, Tel: +81(75)2524641, Fax: +81(75)2524618 ISSUES/YR: 12

- **Journal of Pharmacology and Experimental Therapeutics**
  ISSN: 0022-3565
  J. Pharmacol. Exp. Ther.
  American Society for Pharmacology and Experimental Therapy, 9650 Rockville Pike, Bethesda, MD 20814, United States, Tel: +1(301)5307024, Fax: +1(301)5715728 ISSUES/YR: 12

Journal of Pharmacy and Pharmaceutical Sciences
ISSN: 1482-1826
J. Pharm. Pharm. Sci.
Canadian Society for Pharmaceutical Sciences, 3118 Dentistry/Pharmacy Centre, Univ. of Alberta Campus, AL T6G 2N8, Edmonton, Canada, Tel: +1(780)4920950, Fax: +1(780)4920951 ISSUES/YR: 3

Journal of Pharmacy and Pharmacology
ISSN: 0022-3573
J. Pharm. Pharmacol.
Pharmaceutical Press, 1 Lambeth High Street, SE1 7JN, London, United Kingdom, Tel: +44(207)7359141, Fax: +44(207)7355085 ISSUES/YR: 12

Journal of Pharmacy of Istanbul University
ISSN: 0367-7524
J. Pharm. Istanb. Univ.
University of Istanbul, Beyasit, 34452, Istanbul, Turkey, Tel: +90(212)5282384, Fax: +90(212)4400252 ISSUES/YR: 2

Journal of Pharmacy Practice
ISSN: 0897-1900
J. Pharm. Pract.
SAGE Publications Inc., 2455 Teller Road, Thousand Oaks, CA 91320, United States, Tel: +1(805)4990721, Fax: +1(805)4990871 ISSUES/YR: 6

Journal of Pharmacy Practice and Research
ISSN: 1445-937X
J. Pharm. Res.
Society of Hospital Pharmacists of Australia, 27-33 Raglan Street, Suite 3, VIC 3205, South Melbourne, Australia, Tel: +61(39)96906733, Fax: +61(39)96967634 ISSUES/YR: 4

Journal of Pharmacy Technology
ISSN: 8755-1225
J. Pharm. Technol.
Harvey Whitney Books Company, 8044 Montgomery Road, Suite 415, Cincinnati, OH 45236, United States, Tel: +1(513)7933555, Fax: +1(513)7933600 ISSUES/YR: 6

- **Journal of Photochemistry and Photobiology B: Biology**
  ISSN: 1011-1344
  J. Photochem. Photobiol. B Biol.
  Elsevier, P.O. Box 211, 1000 AE, Amsterdam, Netherlands, Tel: +31(20)4853757, Fax: +31(20)4853432 ISSUES/YR: 15
- **Journal of Physiology**
  ISSN: 0022-3751
  J. Physiol.
  Blackwell Publishing Ltd, 9600 Garsington Road, OX4 2XG, Oxford, United Kingdom, Tel: +44(1865)776868, Fax: +44(1865)714591 ISSUES/YR: 24

Current

Journal of Physiology and Biochemistry
ISSN: 1138-7548
J. Physiol. Biochem.
Servicio de Publicaciones de la Universidad de Navarra, P.O. Box 177, 31080, Pamplona, Spain, Tel: +34(948)425600/6259 AND 6337, Fax: +34(948)425649 ISSUES/YR: 4

Journal of Physiology and Pharmacology
ISSN: 0867-5910
J. Physiol. Pharmacol.
Polish Physiological Society, ul. Grzegorzecka 16, 31-531, Krakow, Poland, Tel: +48(12)64211006, Fax: +48(12)64211578 ISSUES/YR: 4

Journal of Physiology Paris
ISSN: 0928-4257
J. Physiol. Paris
Editions Scientifiques et Medicales Elsevier, 23 rue Linois, 75724, Paris Cedex 15, France, Tel: +33(1)45589826, Fax: +33(1)45589428 ISSUES/YR: 6

Journal of Phytomedicine and Therapeutics
ISSN: 1118-1028
J. Phytomed. Ther.
Nat. Inst. for Pharmaceutical Research and Development, P.M.B. 21, Abuja, Nigeria, Tel: +234(9)5231602, Fax: +234(9)5231043 ISSUES/YR: 2

Journal of Pineal Research
ISSN: 0742-3098
J. Pineal Res.
Blackwell Publishing Ltd, 9600 Garsington Road, OX4 2XG, Oxford, United Kingdom, Tel: +44(1865)776868, Fax: +44(1865)714591 ISSUES/YR: 8

Journal of Planar Chromatography - Modern TLC
ISSN: 0933-4173
J. Planar Chromatogr. Mod. TLC
Research Institute for Medicinal Plants, Lupaszigeti St. 4, P.O. Box 11, H-2011, Budakalasz, Hungary, Tel: +36(26)340203, Fax: +36(26)340426 ISSUES/YR: 6

Journal of Plastic Dermatology
ISSN: not available yet
J. Plast. Dermatol.
Edizioni Scripta Manent s.n.c., Via Bassini 41, 20133, Milano, Italy, Tel: +39(02)70608060, Fax: +39(02)70606917 ISSUES/YR: 3

• **Journal of Plastic, Reconstructive and Aesthetic Surgery**
ISSN: 1748-6815
J. Plast. Reconstr. Aesthetic Surg.
Churchill Livingstone, 1-3 Baxter's Place, Leith Walk, EH1 3AF, Edinburgh, United Kingdom, Tel: +44(131)5562424, Fax: +44(131)5581278 ISSUES/YR: 8

Journal of Postgraduate Medical Institute
ISSN: 1013-5472
J. Postgrad. Med. Inst.
Postgraduate Medical Institute, Lady Reading Hospital, Peshawar, Pakistan ISSUES/YR: 4

Journal of Postgraduate Medicine
ISSN: 0022-3859
J. Postgrad. Med.
Medknow Publication, 12 Manisha Plaza, MN Rd, 400 070 MUMBAI, Kurla, India, Tel: +91(22)25032970, Fax: +91(22)25032398 ISSUES/YR: 4

Journal of Prevention and Intervention in the Community
ISSN: 1085-2352
J. Prev. Intervention Community
Haworth Press Inc., 10 Alice Street, Binghamton, NY 13904-1580, United States, Tel: +1(607)7225857, Fax: +1(607)7710012 ISSUES/YR: 4

Journal of Preventive Medicine and Hygiene
ISSN: 1121-2233
J. Prev. Med. Hyg.
Pacini Editore s.r.l., Via A. Gherardesca 1, 56121, Ospedaletto (Pisa), Italy, Tel: +39(050)313011, Fax: +39(050)3130300 ISSUES/YR: 4

Journal of Primary Prevention
ISSN: 0278-095X
J. Prim. Prev.
Kluwer Academic/Human Sciences Press Inc., 233 Spring Street, New York, NY 10013-1578, United States, Tel: +1(212)6208000, Fax: +1(212)4630742 ISSUES/YR: 4

• **Journal of Proteome Research**
ISSN: 1535-3893
J. Proteome Res.
American Chemical Society, 2540 Olentangy River Road, P.O. Box 3337, Columbus, OH 43210-3337, United States, Tel: +1(614)4473776, Fax: +1(614)4473671 ISSUES/YR: 6

Journal of Psychiatric Practice
ISSN: 1527-4160
J. Psychiatr. Pract.
Lippincott Williams and Wilkins, 530 Walnut Street, Philadelphia, PA 19106-3621, United States, Tel: +1(215)5218342, Fax: +1(215)5218326 ISSUES/YR: 6

• **Journal of Psychiatric Research**
ISSN: 0022-3956
J. Psychiatr. Res.
Elsevier Ltd, Langford Lane, Kidlington, OX5 1GB, Oxford, United Kingdom, Tel: +44(1865)843000, Fax: +44(1865)843010 ISSUES/YR: 6

Journal of Psychiatry and Law
ISSN: 0093-1853
J. Psychiatry Law
Federal Legal Publications Inc., 39 Lakeview Rd, Carmel, NY 10512, United States, Tel: +1-845 228 5086, Fax: +1-845 225 2686 ISSUES/YR: 4

Journal of Psychiatry and Neuroscience
ISSN: 1180-4882
J. Psychiatry Neurosci.
Canadian Medical Association, 1740 boul.Cote-Vertu Blvd, H4L 2A4, St-Laurent, Canada, Tel: +1-514 738 3111, Fax: +1-514 738 5199 ISSUES/YR: 6

Journal of Psychoactive Drugs
ISSN: 0279-1072
J. Psychoact. Drugs
Haight-Ashbury Publications, 612 Clayton Street, San Francisco, CA 94117-2958, United States, Tel: +1(415)5651904 ISSUES/YR: 4

Journal of Psychopathology and Behavioral Assessment
ISSN: 0882-2689
J. Psychopathol. Behav. Assess.
Springer Science and Business Media Deutschland GmbH, Heidelberger Platz 3, 14197, Berlin, Germany, Tel: +1-30 827875282, Fax: +1-30 827875707 ISSUES/YR: 4

• **Journal of Psychopharmacology**
ISSN: 0269-8811
J. Psychopharmacol.
SAGE Publications Ltd, 55 City Road, EC1Y 1SP, London, United Kingdom, Tel: +44(207)3248500, Fax: +44(207)3248600 ISSUES/YR: 4

Journal of Psychophysiology
ISSN: 0269-8803
J. Psychophysiol.
Hogrefe and Huber Publishers, Rohnsweg 25, D-37085, Gottingen, Germany, Tel: +49(551)49609-0, Fax: +49(551)49609-88 ISSUES/YR: 4

Journal of Psychosocial Oncology
ISSN: 0734-7332
J. Psychosoc. Oncol.
Haworth Press Inc., 10 Alice Street, Binghamton, NY 13904-1580, United States, Tel: +1(607)7225857 ISSUES/YR: 4

Journal of Psychosomatic Obstetrics and Gynecology
ISSN: 0167-482X
J. Psychosom. Obstet. Gynecol.
Taylor and Francis Ltd., 4 Park Square, Milton Park, OX14 4RN, Abingdon, Oxfordshire, United Kingdom, Tel: +44(20)70175541, Fax: +44(20)70175198 ISSUES/YR: 4

Journal of Psychosomatic Research
ISSN: 0022-3999
J. Psychosom. Res.
Elsevier Inc., 360 Park Avenue South, New York, NY 10010, United States, Tel: +1(212)6333950, Fax: +1(212)6333990 ISSUES/YR: 12

Journal of Psychotherapy Integration
ISSN: 1053-0479
J. Psychother. Integr.
American Psychological Association Inc., 750 First Street NE, Washington, DC 20002-4242, United States, Tel: +1(202)3365600, Fax: +1(202)3365568 ISSUES/YR: 4

- **Journal of Public Health**
  ISSN: 1741-3842
  J. Public Health (United Kingdom)
  Oxford University Press, Great Clarendon Street, OX2 6DP, Oxford, United Kingdom, Tel: +44(1865)353907, Fax: +44(1865)353486 ISSUES/YR: 4

  Journal of Public Health
  ISSN: 0943-1853
  J. Public Health (Germany)
  Springer Verlag, Tiergartenstrasse 17, D-69121, Heidelberg, Germany, Tel: +49(6221)4870, Fax: +49(6221)487366 ISSUES/YR: 4
- **Journal of Public Health Management and Practice**
  ISSN: 1078-4659
  J. Public Health Manage. Pract.
  Lippincott Williams and Wilkins, 530 Walnut Street, Philadelphia, PA 19106-3621, United States, Tel: +1(215)5218342, Fax: +1(215)5218326 ISSUES/YR: 6

  Journal of Public Health Policy
  ISSN: 0197-5897
  J. Public Health Policy
  Journal of Public Health Policy Inc., 213 West Canton Street, Boston, MA 02116, United States, Tel: +1(617)5366903, Fax: +1(617)5365190 ISSUES/YR: 4

  Journal of Radioanalytical and Nuclear Chemistry
  ISSN: 0236-5731
  J. Radioanal. Nucl. Chem.
  Kluwer Academic Publishers, Van Godewijckstraat 30, 3311 GZ, Dordrecht, Netherlands, Tel: +31(78)6576000, Fax: +31(78)6576254 ISSUES/YR: 12
- **Journal of Radiological Protection**
  ISSN: 0952-4746
  J. Radiol. Prot.
  Institute of Physics Publishing, Temple Back, BS1 6BE, Bristol, United Kingdom, Tel: +44(117)9297481, Fax: +44(117)9294318 ISSUES/YR: 4
- **Journal of Radiology Nursing**
  ISSN: 1546-0843
  J. Radiol. Nurs.
  Elsevier Inc., 360 Park Avenue South, New York, NY 10010, United States, Tel: +1(212)6333950, Fax: +1(212)6333990 ISSUES/YR: 4

  Journal of Radiotherapy in Practice
  ISSN: 1460-3969
  J. Radiother. Pract.
  Cambridge University Press, Shaftesbury Road, CB2 2RU, Cambridge, United Kingdom, Tel: +44(1223)312393, Fax: +44(1223)315052 ISSUES/YR: 4

  Journal of Receptor and Signal Transduction
  ISSN: 1079-9893
  J. Recept. Signal Transduction
  Taylor and Francis Inc., 325 Chestnut St, Suite 800, Philadelphia PA, PA 19106, United States, Tel: +1(800)3541420, Fax: +1(215)6258914 ISSUES/YR: 6
- **Journal of Reconstructive Microsurgery**
  ISSN: 0743-684X
  J. Reconstr. Microsurg.
  Thieme Medical Publishers, Inc., 333 7th Avenue, New York, NY 10001-5004, United States, Tel: +1(212)7600888, Fax: +1(212)7799020 ISSUES/YR: 8
- **Journal of Refractive Surgery**
  ISSN: 1081-597X
  J. Refractive Surg.
  Slack Incorporated, 6900 Grove Road, Thorofare, NJ 08086-9447, United States, Tel: +1(856)8481000, Fax: +1(856)8535991 ISSUES/YR: 9

  Journal of Rehabilitation Medicine
  ISSN: 1650-1977
  J. Rehabil. Med.
  Taylor and Francis Ltd., 4 Park Square, Milton Park, OX14 4RN, Abingdon, Oxfordshire, United Kingdom, Tel: +44(20)70175541, Fax: +44(20)70175198 ISSUES/YR: 6

  Journal of Rehabilitation Medicine, Supplement
  ISSN: 1650-1969
  J. Rehabil. Med. Suppl.
  Taylor and Francis Ltd., 4 Park Square, Milton Park, OX14 4RN, Abingdon, Oxfordshire, United Kingdom, Tel: +44(20)70175541, Fax: +44(20)70175198 ISSUES/YR: 1
- **Journal of Rehabilitation Research and Development**
  ISSN: 0748-7711
  J. Rehabil. Res. Dev.
  Rehabilitation Research and Development Service, 103 South Gay Street, Baltimore, MD 21202-4051, United States, Tel: +1(410)9621800 ISSUES/YR: 6

  Journal of Reproductive and Infant Psychology
  ISSN: 0264-6838
  J. Reprod. Infant Psychol.
  Routledge, 4 Park Square, Milton Park, OX14 4RN, Abingdon, Oxfordshire, United Kingdom, Tel: +44(1235)828600, Fax: +44(1235)829000 ISSUES/YR: 4
- **Journal of Reproductive Immunology**
  ISSN: 0165-0378
  J. Reprod. Immunol.
  Elsevier Ireland Ltd, P.O. Box 85, Limerick, Ireland, Tel: +353(61)471944, Fax: +353(61)472144 ISSUES/YR: 8
- **Journal of Reproductive Medicine for the Obstetrician and Gynecologist**
  ISSN: 0024-7758
  J. Reprod. Med. Obstet. Gynecol.
  Journal of Reproductive Medicine Inc., 8342 Olive Boulevard, P.O. Box 12425, St. Louis, MO 63132-2814, United States, Tel: +1(314)9914440 ISSUES/YR: 12

  Journal of Research in Crime and Delinquency
  ISSN: 0022-4278
  J. Res. Crime Delinq.
  SAGE Publications Inc., 2455 Teller Road, Thousand Oaks, CA 91320, United States, Tel: +1(805)4990721, Fax: +1(805)4990871 ISSUES/YR: 4

  Journal of Research in Medical Sciences
  ISSN: 1735-1995
  J. Res. Med. Sci.
  Isfahan University of Medical Sciences(IUMS), Hezar Jerib Avenue, P.O. Box 81745-319, Isfahan, Iran, Tel: +98-0098 792 3065, Fax: +98-0098 668 7898 ISSUES/YR: 6

  Journal of Respiratory Diseases
  ISSN: 0194-259X
  J. Respir. Dis.
  Cliggott Publishing Co., 330 Boston Post Road, Box 4027, Darien, CT 06820-4027, United States, Tel: +1(203)6626400 ISSUES/YR: 12
- **Journal of Rheumatology**
  ISSN: 0315-162X
  J. Rheumatol.
  Journal of Rheumatology, 365 Bloor Street East, Suit 901, ON M4W 3L4, Toronto, Canada, Tel: +1-416 967 5155 EXT: 26, Fax: +1-416 976 7556 ISSUES/YR: 12

  Journal of Rheumatology and Medical Rehabilitation
  ISSN: 1300-0691
  J. Rheumatol. Med. Rehabil.
  Hacettepe Universitesi Tip Fakultesi, 06100, Ankara, Turkey, Tel: +90(312)3103545 ISSUES/YR: 4

  Journal of Rural Health
  ISSN: 0890-765X
  J. Rural Health
  Blackwell Publishing Ltd, 9600 Garsington Road, OX4 2XG, Oxford, United Kingdom, Tel: +44(1865)776868, Fax: +44(1865)714591 ISSUES/YR: 4

  Journal of Science and Medicine in Sport
  ISSN: 1440-2440
  J. Sci. Med. Sport
  Thomson Scientific Ltd., 14 Great Queen Street, WC2B 5DF, London, United Kingdom, Tel: +44-20 7344 2800, Fax: +44-20 7344 2900 ISSUES/YR: 4
- **Journal of Separation Science**
  ISSN: 1615-9306
  J. Sep. Sci.
  Wiley-VCH Verlag, P.O. Box 101161, D-69451, Weinheim, Germany, Tel: +49(6201)606147, Fax: +49(6201)606117 ISSUES/YR: 18

  Journal of Sex and Marital Therapy
  ISSN: 0092-623X
  J. Sex Marital Ther.
  Routledge, 4 Park Square, Milton Park, OX14 4RN, Abingdon, Oxfordshire, United Kingdom, Tel: +44(1235)828600, Fax: +44(1235)829000 ISSUES/YR: 5

Journal of Sex Education and Therapy
ISSN: 0161-4576
J. Sex Educ. Ther.
American Assoc. of Sex Educators, Counselors and Therapists, P.O. Box 5488, Richmond, VA 23220-0488, United States, Tel: +1(804)6443288, Fax: +1(804)6443290 ISSUES/YR: 4

Journal of Sex Research
ISSN: 0022-4499
J. Sex Res.
Society for the Scientific Study of Sex Inc., P.O. Box 416, Allentown, PA 18105-0416, United States, Tel: +1(610)5302483, Fax: +1(610)5302485 ISSUES/YR: 4

Journal of Sexual and Reproductive Medicine
ISSN: 1488-5069
J. Sex. Reprod. Med.
Pulsus Group Inc., 2902 South Sheridan Way, ONT L6J 7L6, Oakville, Canada, Tel: +1(905)8294770, Fax: +1(905)8294799 ISSUES/YR: 3

• **Journal of Sexual Medicine**
ISSN: 1743-6095
J. Sex. Med.
Blackwell Publishing Ltd, 9600 Garsington Road, OX4 2XG, Oxford, United Kingdom, Tel: +44(1865)776868, Fax: +44(1865)714591 ISSUES/YR: 6

• **Journal of Shoulder and Elbow Surgery**
ISSN: 1058-2746
J. Shoulder Elbow Surg.
Mosby Inc., 11830 Westline Industrial Drive, St. Louis, MO 63146, United States, Tel: +1(314)8728370, Fax: +1(314)4321380 ISSUES/YR: 6

Journal of Sichuan University (Medical Science Edition)
ISSN: 1672-173X
J. Sichuan Univ. Med. Sci. Ed.
Sichuan University, 17, Section 3, Renmin Nanlu, Chengdu, 610041, Sichuan, China ISSUES/YR: 6

• **Journal of Sleep Research**
ISSN: 0962-1105
J. Sleep Res.
Blackwell Publishing Ltd, 9600 Garsington Road, OX4 2XG, Oxford, United Kingdom, Tel: +44(1865)776868, Fax: +44(1865)714591 ISSUES/YR: 4

• **Journal of Sleep Research, Supplement**
ISSN: 0966-6826
J. Sleep Res. Suppl.
Blackwell Publishing Ltd, 9600 Garsington Road, OX4 2XG, Oxford, United Kingdom, Tel: +44(1865)776868, Fax: +44(1865)714591 ISSUES/YR: 1

Journal of Smooth Muscle Research
ISSN: 0916-8737
J. Smooth Muscle Res.
Japan Society of Smooth Muscle Research, 40-206 Hasekura-cho 4-chome, Aoba-ku, Sendai-shi, 980-0824, Miyagi-ken, Japan, Tel: +81(22)7177587, Fax: +81(22)7177508 ISSUES/YR: 6

Journal of Speech, Language, and Hearing Research
ISSN: 1092-4388
J. Speech Lang. Hear. Res.
American Speech-Language-Hearing Association, 10801 Rockville Pike, Rockville, MD 20852-3279, United States, Tel: +1(301)8975700 ISSUES/YR: 6

Journal of Spinal Cord Medicine
ISSN: 1079-0268
J. Spinal Cord Med.
American Paraplegia Society, 75-20 Astoria Boulevard, Jackson Heights, NJ 11370-1177, United States ISSUES/YR: 4

• **Journal of Spinal Disorders and Techniques**
ISSN: 1536-0652
J. Spinal Disord. Tech.
Lippincott Williams and Wilkins, 530 Walnut Street, Philadelphia, PA 19106-3621, United States, Tel: +1(215)5218342, Fax: +1(215)5218326 ISSUES/YR: 6

• **Journal of Spirochetal and Tick-borne Diseases**
ISSN: 1060-0051
J. Spirochetal Tick-borne Dis.
Slack Incorporated, 6900 Grove Road, Thorofare, NJ 08086-9447, United States, Tel: +1(856)8481000, Fax: +1(856)8535991 ISSUES/YR: 2

Journal of Sport Rehabilitation
ISSN: 1056-6716
J. Sport Rehabil.
Human Kinetics Publishers Inc., 1607 N. Market Street, Champaign, IL 61820-2200, United States, Tel: +1(217)3515076, Fax: +1(217)3511549 ISSUES/YR: 4

Journal of Sports Medicine and Physical Fitness
ISSN: 0022-4707
J. Sports Med. Phys. Fitness
Edizioni Minerva Medica S.p.A., Corso Bramante 83-85, 10126, Turin, Italy, Tel: +39(011)678282, Fax: +39(011)674502 ISSUES/YR: 4

Journal of Sports Sciences
ISSN: 0264-0414
J. Sports Sci.
Routledge, 4 Park Square, Milton Park, OX14 4RN, Abingdon, Oxfordshire, United Kingdom, Tel: +44(1235)828600, Fax: +44(1235)829000 ISSUES/YR: 12

Journal of Steroid Biochemistry and Molecular Biology
ISSN: 0960-0760
J. Steroid Biochem. Mol. Biol.
Elsevier Ltd, Langford Lane, Kidlington, OX5 1GB, Oxford, United Kingdom, Tel: +44(1865)843000, Fax: +44(1865)843010 ISSUES/YR: 20

• **Journal of Stroke and Cerebrovascular Diseases**
ISSN: 1052-3057
J. Stroke Cerebrovasc. Dis.
W.B. Saunders, Independence Square West, Philadelphia, PA 19106-3399, United States, Tel: +1(215)2387800, Fax: +1(215)2386445 ISSUES/YR: 6

• **Journal of Structural and Functional Genomics**
ISSN: 1345-711X
J. Struc. Funct. Genomics
Kluwer Academic Publishers, Van Godewijckstraat 30, 3311 GZ, Dordrecht, Netherlands, Tel: +31(78)6576000, Fax: +31(78)6576254 ISSUES/YR: 4

• **Journal of Structural Biology**
ISSN: 1047-8477
J. Struct. Biol.
Academic Press Inc., 6277 Sea Harbor Drive, Orlando, FL 32887-4900, United States, Tel: +1(407)3454040, Fax: +1(407)3639661 ISSUES/YR: 12

Journal of Studies on Alcohol
ISSN: 0096-882X
J. Stud. Alcohol
Alcohol Research Documentation Inc., New Brunswick, NJ 08903, United States ISSUES/YR: 6

• **Journal of Substance Abuse Treatment**
ISSN: 0740-5472
J. Subst. Abuse Treat.
Elsevier Inc., 360 Park Avenue South, New York, NY 10010, United States, Tel: +1(212)6333950, Fax: +1(212)6333990 ISSUES/YR: 8

• **Journal of Substance Use**
ISSN: 1465-9891
J. Subst. Use
Taylor and Francis Ltd., 4 Park Square, Milton Park, OX14 4RN, Abingdon, Oxfordshire, United Kingdom, Tel: +44(20)70175541, Fax: +44(20)70175198 ISSUES/YR: 6

Journal of Supportive Oncology
ISSN: 1544-6794
J. Supportive Oncol.
Biolink Communications, 46 Green Street, 2nd Floor, Huntington, NY 11743, United States, Tel: +1(631)4248900, Fax: +1(631)4248905 ISSUES/YR: 6

• **Journal of Surgical Oncology**
ISSN: 0022-4790
J. Surg. Oncol.
Wiley-Liss Inc., 111 River Street, Hoboken, NJ 07030-5774, United States, Tel: +1(201)7486645, Fax: +1(201)7486021 ISSUES/YR: 16

- **Journal of Surgical Research**
  ISSN: 0022-4804
  J. Surg. Res.
  Academic Press Inc., 6277 Sea Harbor Drive, Orlando, FL 32887-4900, United States, Tel: +1(407)3454040, Fax: +1(407)3639661 ISSUES/YR: 14

Journal of Taiwan Otolaryngology - Head and Neck Surgery
ISSN: 1019-6102
J. Taiwan Otolaryngol. Head Neck Surg.
Taiwan Otolaryngological Society, 87 Tung-Teh Road, 115, Nankang, Taipei, Taiwan, Tel: +886(2)27834768, Fax: +886(2)27834763 ISSUES/YR: 6

Journal of Terminal Oncology
ISSN: 1538-8573
J. Terminal Oncol.
Prime National Publishing Corp., 470 Boston Post Road, Weston, MA 02193-1529, United States, Tel: +1(781)8992702, Fax: +1(781)8994900 ISSUES/YR: 4

- **Journal of the Acoustical Society of America**
  ISSN: 0001-4966
  J. Acoust. Soc. Am.
  Acoustical Society of America, 2 Huntington Quadrangle, Ste 1NO1, Melville, NY 11747-4502, United States, Tel: +1(800)8746383, Fax: +1(516)5762604 ISSUES/YR: 12

- **Journal of the Air and Waste Management Association**
  ISSN: 1047-3289
  J. Air Waste Manage. Assoc.
  Air and Waste Management Association, 420 Ft. Duquesne Blvd., Pittsburgh, PA 15222, United States, Tel: +1(412)2323444, Fax: +1(412)2323450 ISSUES/YR: 12

- **Journal of the American Academy of Audiology**
  ISSN: 1050-0545
  J. Am. Acad. Audiol.
  American Academy of Audiology, 11730 Plaza America Drive, Suite 300 Reston, VA 20190, United States, Tel: +1-703 790 8466, Fax: +1-703 790 8631 ISSUES/YR: 10

- **Journal of the American Academy of Child and Adolescent Psychiatry**
  ISSN: 0890-8567
  J. Am. Acad. Child Adolesc. Psychiatry
  Lippincott Williams and Wilkins, 351 West Camden Street, Baltimore, MD 21201-2436, United States, Tel: +1(410)5284000 ISSUES/YR: 12

- **Journal of the American Academy of Dermatology**
  ISSN: 0190-9622
  J. Am. Acad. Dermatol.
  Mosby Inc., 11830 Westline Industrial Drive, St. Louis, MO 63146, United States, Tel: +1(314)8728370, Fax: +1(314)4321380 ISSUES/YR: 12

Journal of the American Academy of Psychiatry and the Law
ISSN: 1093-6793
J. Am. Acad. Psychiatry Law
American Academy of Psychiatry and the Law, 1 Regency Drive, P.O. Box 30, Bloomfield, CT 06002, United States ISSUES/YR: 4

Journal of the American Academy of Psychoanalysis and Dynamic Psychiatry
ISSN: 1546-0371
J. Am. Acad. Psychoanal. Dyn. Psychiatry
Guilford Publications, 72 Spring Street, New York, NY 10012, United States, Tel: +1(212)4319800 ISSUES/YR: 4

Journal of the American Aging Association
ISSN: 0161-9152
J. Am. Aging Assoc.
American Aging Association (AGE), 110 Chesley Drive, Media, PA 19063, United States, Tel: +1(610)6272626, Fax: +1(610)5659747 ISSUES/YR: 4

Journal of the American Animal Hospital Association
ISSN: 1547-3317
J. Am. Anim. Hosp. Assoc.
American Animal Hospital Association, 12575 West Bayaud Avenue, Lakewood, CO 80228, United States, Tel: +1(303)9862800, Fax: +1(303)9861700 ISSUES/YR: 6

Journal of the American Board of Family Practice
ISSN: 0893-8652
J. Am. Board Fam. Pract.
American Board of Family Medicine, 2228 Young Drive, Lexington, KY 40505-4294, United States, Tel: +1(206)6853993, Fax: +1(206)5438911 ISSUES/YR: 6

Journal of the American Chemical Society
ISSN: 0002-7863
J. Am. Chem. Soc.
American Chemical Society, 2540 Olentangy River Road, P.O. Box 3337, Columbus, OH 43210-3337, United States, Tel: +1(614)4473776, Fax: +1(614)4473671 ISSUES/YR: 51

- **Journal of the American College of Cardiology**
  ISSN: 0735-1097
  J. Am. Coll. Cardiol.
  Elsevier USA, 6277 Sea Harbor Drive, Orlando, FL 32862 8239, United States, Tel: +1-407 345 4000, Fax: +1-407 363 9661 ISSUES/YR: 24

Journal of the American College of Nutrition
ISSN: 0731-5724
J. Am. Coll. Nutr.
American College of Nutrition, 300 South Duncan Avenue, Clearwater, FL 33755, United States, Tel: +1(727)446-6086, Fax: +1(727)446-6202 ISSUES/YR: 6

- **Journal of the American College of Surgeons**
  ISSN: 1072-7515
  J. Am. Coll. Surg.
  Elsevier Inc., 360 Park Avenue South, New York, NY 10010, United States, Tel: +1(212)6333950, Fax: +1(212)6333990 ISSUES/YR: 12

Journal of the American Geriatrics Society
ISSN: 0002-8614
J. Am. Geriatr. Soc.
Blackwell Publishing Inc., 350 Main Street, Malden, MA 02148, United States, Tel: +1(781)3888200, Fax: +1(781)3888210 ISSUES/YR: 12

Journal of the American Health Information Management Association
ISSN: 1060-5487
J. Am. Health Inf. Manage. Assoc.
American Health Information Management Association, 233 N. Michigan Ave Ste 2150, Chicago, IL 60601, United States, Tel: +1(312)7872672 ISSUES/YR: 10

- **Journal of the American Medical Association**
  ISSN: 0098-7484
  J. Am. Med. Assoc.
  American Medical Association, 515 North State Street, Chicago, IL 60610, United States, Tel: +1(312)4645000 ISSUES/YR: 48

Journal of the American Medical Directors Association
ISSN: 1525-8610
J. Am. Med. Dir. Assoc.
Elsevier, P.O. Box 211, 1000 AE, Amsterdam, Netherlands, Tel: +1(20)4853757, Fax: +1(20)4853432 ISSUES/YR: 6

Journal of the American Medical Informatics Association
ISSN: 1067-5027
J. Am. Med. Informatics Assoc.
Hanley and Belfus Inc., 210 South 13th Street, Philadelphia, PA 19107, United States, Tel: +1(215)5467293 ISSUES/YR: 6

Journal of the American Osteopathic Association
ISSN: 0098-6151
J. Am. Osteopath. Assoc.
American Osteopathic Association, 142 East Ontario Street, Chicago, IL 60611-2864, United States, Tel: +1(312)2805800 ISSUES/YR: 12

Journal of the American Pharmacists Association
ISSN: 1544-3191
J. Am. Pharm. Assoc.
American Pharmacists Association, 2215 Constitution Avenue NW, Washington, DC 20037, United States, Tel: +1(202)6284410 ISSUES/YR: 6

Current

Journal of the American Society for Mass Spectrometry
ISSN: 1044-0305
J. Am. Soc. Mass Spectrom.
Elsevier Inc., 360 Park Avenue South, New York, NY 10010, United States, Tel: +1(212)6333950, Fax: +1(212)6333990 ISSUES/YR: 12

Journal of the American Society of Echocardiography
ISSN: 0894-7317
J. Am. Soc. Echocardiogr.
Mosby Inc., 11830 Westline Industrial Drive, St. Louis, MO 63146, United States, Tel: +1(314)8728370, Fax: +1(314)4321380 ISSUES/YR: 12

• **Journal of the American Society of Nephrology**
ISSN: 1046-6673
J. Am. Soc. Nephrol.
American Society of Nephrology, 1725 I Street NW,510, Washington, DC 20006, United States, Tel: +1-202 659 0599, Fax: +1-202 659 0709 ISSUES/YR: 12

Journal of the American Veterinary Medical Association
ISSN: 0003-1488
J. Am. Vet. Med. Assoc.
American Veterinary Medical Association, 1931 N. Meacham Road, Suite 100, Schaumburg, IL 60173-4360, United States, Tel: +1(847)9258070, Fax: +1(847)9251329 ISSUES/YR: 24

Journal of the Bahrain Medical Society
ISSN: 1015-6321
J. Bahrain Med. Soc.
Bahrain Medical Society, P.O. Box 26136, Manama, Bahrain, Tel: +973-742666 ISSUES/YR: 4

Journal of the British Menopause Society
ISSN: 1362-1807
J. Br. Menopause Soc.
Royal Society of Medicine Press Ltd, P.O. Box 9002, W1A 0ZA, London, United Kingdom, Tel: +44(207)2902927, Fax: +44(207)2902929 ISSUES/YR: 4

Journal of the Canadian Academy of Child and Adolescent Psychiatry
ISSN: 1209-7268
J. Can. Acad. Child Adolesc. Psychiatry
Canadian Academy of Child and Adolescent Psychiatry, 701-141 Laurier Ave. West, ON KIP 5J3, Ottawa, Canada, Tel: +1(613)288-0408, Fax: +1(613)234-9857 ISSUES/YR: 4

Journal of the Canadian Society of Forensic Science
ISSN: 0008-5030
J. Can. Soc. Forensic Sci.
Canadian Society of Forensic Science, P.O.Box 37040, K1V 0W0, 3332 McCarthy Road Ottawa Ontario, Canada, Tel: +1(613)7380001, Fax: +1-613 7381987 ISSUES/YR: 4

• **Journal of the Chartered Institution of Water and Environmental Management**
ISSN: 0951-7359
J. Chart. Inst. Water Environ. Manage.
Leadmedia, 6 Hertforde C. John Tate Road, SG13 7NW, Hertforde, United Kingdom, Tel: +44(1992)535323, Fax: +44(1992)535324 ISSUES/YR: 4

Journal of the Chinese Medical Association
ISSN: 1726-4901
J. Chin. Med. Assoc.
Chinese Medical Association - Taipei, 201 Shih-Pai Road, Section 2, P.O. Box 3043, 11217, Taipei, Taiwan, Tel: +886(2)28712121 EXT.3043, Fax: +886(2)28771873 ISSUES/YR: 12

Journal of the College of Physicians and Surgeons Pakistan
ISSN: 1022-386X
J. Coll. Phys. Surg. Pak.
College of Physicians and Surgeons Pakistan, 7th Central Street, 755000, Karachi, Pakistan, Tel: +92(21)5892801, Fax: +92(21)5887513 ISSUES/YR: 12

Journal of the Diabetic Association of India
ISSN: 0970-4035
J. Diabetic Assoc. India
Diabetic Association of India, Raheja Hospital Road, Mahim, 400 016, Mumbai, India, Tel: +91-4467569 ISSUES/YR: 4

• **Journal of the European Academy of Dermatology and Venereology**
ISSN: 0926-9959
J. Eur. Acad. Dermatol. Venereol.
Blackwell Publishing Ltd, 9600 Garsington Road, OX4 2XG, Oxford, United Kingdom, Tel: +44(1865)776868, Fax: +44(1865)714591 ISSUES/YR: 6

Journal of the Food Hygienic Society of Japan
ISSN: 0015-6426
J. Food Hyg. Soc. Jpn.
Food Hygienic Society of Japan, 6-1 Jingu-Mae 2-chome, Shibuya-ku, 150-0001, Tokyo, Japan, Tel: +81(3)34702933, Fax: +81(3)34702975 ISSUES/YR: 6

Journal of the Formosan Medical Association
ISSN: 0929-6646
J. Formos. Med. Assoc.
Scientific Communications International Ltd, Suite C, 10/F Wo On Building, 10 Wo On Lane, Central, Hong Kong, Tel: +852-28 68 9171, Fax: +852-28 68 9269 ISSUES/YR: 12

Journal of the History of the Neurosciences
ISSN: 0964-704X
J. Hist. Neurosci.
Taylor and Francis Ltd., 4 Park Square, Milton Park, OX14 4RN, Abingdon, Oxfordshire, United Kingdom, Tel: +44(20)70175541, Fax: +44(20)70175198 ISSUES/YR: 4

Journal of the Hoffman Heart Institute of Connecticut
ISSN: 1091-630X
J. Hoffman Heart Inst. Conn.
Saint Francis Hospital and Medical Center, 114 Woodland Street, Hartford, CT 06105-1299, United States, Tel: +1(203)5484019 ISSUES/YR: 2

Journal of the Hong Kong College of Radiologists
ISSN: 1029-5097
J. HK Coll. Radiol.
Scientific Communications International Ltd, Suite C, 10/F Wo On Building, 10 Wo On Lane, Central, Hong Kong, Tel: +852-28 68 9171, Fax: +852-28 68 9269 ISSUES/YR: 4

Journal of the ICRU
ISSN: 1473-6691
J. ICRU
Oxford University Press, 2001 Evans Road, Cary, NC 27513, United States, Tel: +1(919)6770977, Fax: +1(919)6771714 ISSUES/YR: 2

Journal of the Indian Chemical Society
ISSN: 0019-4522
J. Indian Chem. Soc.
Indian Chemical Society, 92 Acharya Prafulla Chandra Road, 700 009, Calcutta, India, Tel: +91(33)353478 ISSUES/YR: 12

Journal of the Indian Medical Association
ISSN: 0019-5847
J. Indian Med. Assoc.
Indian Medical Association, 53 Creek Row, 700014, Calcutta, India, Tel: +91(33)263598, Fax: +91(33)266437 ISSUES/YR: 12

• **Journal of the International Neuropsychological Society**
ISSN: 1355-6177
J. Int. Neuropsychol. Soc.
Cambridge University Press, Shaftesbury Road, CB2 2RU, Cambridge, United Kingdom, Tel: +44(1223)312393, Fax: +44(1223)315052 ISSUES/YR: 7

Journal of the Japan Diabetes Society
ISSN: 0021-437X
J. Jpn. Diabetes Soc.
Japan Diabetes Society, 3-38-11 Hongo, Bunkyo-ku, 113, Tokyo, Japan, Tel: +81(3)38154364 ISSUES/YR: 12

Journal of the Japan Epilepsy Society
ISSN: 0912-0890
J. Jpn. Epilepsy Soc.
Japan Epilepsy Society, 886 Urushiyama, 420-8688, Shizuoka, Japan, Tel: +81(54)2489077, Fax: +81(54)2489077 ISSUES/YR: 3

Journal of the Medical Association of Thailand
ISSN: 0125-2208
J. Med. Assoc. Thailand
Medical Association of Thailand, 4th Royal Golden Jubilee Building, 2 Soi Soonvijai, 10310, New Petch Huayakwang Bangkok, Thailand, Tel: +66(2)3144333, Fax: +66(2)3146305 ISSUES/YR: 12

Journal of the Medical Library Association
ISSN: 1536-5050
J. Med. Libr. Assoc.
Medical Library Association, 65 East Wacker Place, suite 1900, Chicago, IL 60601-7298, United States, Tel: +1(312)4199094, Fax: +1(312)4198950 ISSUES/YR: 4

Journal of the Medical Society of Toho University
ISSN: 0040-8670
J. Med. Soc. Toho Univ.
Medical Society of Toho University, 5-21-16 Omori Nishi, Ota-ku, 143, Tokyo, Japan, Tel: +81(3)37624151 ISSUES/YR: 6

● **Journal of the National Cancer Institute**
ISSN: 0027-8874
J. Natl. Cancer Inst.
Oxford University Press, Great Clarendon Street, OX2 6DP, Oxford, United Kingdom, Tel: +44(1865)353907, Fax: +44(1865)353486 ISSUES/YR: 24

● **Journal of the National Medical Association**
ISSN: 0027-9684
J. Natl. Med. Assoc.
National Medical Association, 1012 Tenth Street NW, Washington, DC 20001, United States, Tel: +1(202)3471895, Fax: +1(202)8423293 ISSUES/YR: 12

Journal of the Nepal Medical Association
ISSN: 0028-2715
J. Nepal Med. Assoc.
Nepal Medical Association, Exhibition Road, post box 189, Kathmandu, Nepal, Tel: +977-1-4225860 ISSUES/YR: 4

● **Journal of the Neurological Sciences**
ISSN: 0022-510X
J. Neurol. Sci.
Elsevier, P.O. Box 211, 1000 AE, Amsterdam, Netherlands, Tel: +31(20)4853757, Fax: +31(20)4853432 ISSUES/YR: 22

Journal of the Osaka City Medical Center
ISSN: 0386-4103
J. Osaka City Med. Cent.
Osaka City Medical Center, 4-54 Asahimachi 1-chome, Abeno-ku, 545, Osaka, Japan, Tel: +81(6)6452121 ISSUES/YR: 4

Journal of the Pakistan Medical Association
ISSN: 0030-9982
J. Pak. Med. Assoc.
Pakistan Medical Association, Garden Road, Karachi - 3, Pakistan, Tel: +92(21)7214632, Fax: +92(21)7226443 ISSUES/YR: 12

Journal of the Pancreas
ISSN: 1590-8577
J. Pancreas
E.S. Burioni Ricerche Bibliografiche, Corso Firenze 41/2, 16136, Genova, Italy, Fax: +39(10)2722913 ISSUES/YR: 6

● **Journal of the Peripheral Nervous System**
ISSN: 1085-9489
J. Peripher. Nerv. Syst.
Blackwell Publishing Inc., 350 Main Street, Malden, MA 02148, United States, Tel: +1(781)3888200, Fax: +1(781)3888210 ISSUES/YR: 4

Journal of The Royal Society for the Promotion of Health
ISSN: 1466-4240
J. R. Soc. Promot. Health
Royal Society for the Promotion of Health, 38A St. George's Drive, SW1V 4BH, London, United Kingdom, Tel: +44(207)6300121, Fax: +44(207)9766847 ISSUES/YR: 4

Journal of the Royal Society Interface
ISSN: 1742-5689
J. R. Soc. Interface
Royal Society of London, 6 Carlton House Terrace, SW1Y 5AG, London, United Kingdom, Tel: +44(207)8395561, Fax: +44(207)9761837 ISSUES/YR: 6

Journal of the Royal Society of Medicine
ISSN: 0141-0768
J. R. Soc. Med.
Royal Society of Medicine Press Ltd, P.O. Box 9002, W1A 0ZA, London, United Kingdom, Tel: +44(207)2902927, Fax: +44(207)2902929 ISSUES/YR: 12

Journal of the Royal Society of Medicine, Supplement
ISSN: 0267-5331
J. R. Soc. Med. Suppl.
Royal Society of Medicine Press Ltd, P.O. Box 9002, W1A 0ZA, London, United Kingdom, Tel: +44(207)2902927, Fax: +44(207)2902929 ISSUES/YR: 1

Journal of the Showa Medical Association
ISSN: 0037-4342
J. Showa Med. Assoc.
Showa Medical Association, 5-8 Hatanodai, 1-chome, Shinagawa-ku, 142, Tokyo, Japan, Tel: +81(3)37848042 ISSUES/YR: 6

● **Journal of the Society for Gynecologic Investigation**
ISSN: 1071-5576
J. Soc. Gynecol. Invest.
Elsevier USA, 6277 Sea Harbor Drive, Orlando, FL 32862 8239, United States, Tel: +1-407 345 4000, Fax: +1-407 363 9661 ISSUES/YR: 8

Journal of the South African Veterinary Association
ISSN: 0038-2809
J. S. Afr. Vet. Assoc.
South African Veterinary Association, Monument Park, P.O. Box 25033, 0105, Pretoria, South Africa, Tel: +27(12)3461150 ISSUES/YR: 4

Journal of the Turkish German Gynecology Association Artemis
ISSN: 1303-9695
J. Turkish German Gynecol. Assoc. Artemis
Turgut Publishing, Bomonti Firin sk No:61/2, 80260, Istanbul, Turkey, Tel: +90(0212)2330223, Fax: +90(0212)2336545 ISSUES/YR: 4

Journal of the Wakayama Medical Society
ISSN: 0043-0013
J. Wakayama Med. Soc.
Wakayama Medical Society, 9-27 Kyubancho, 640, Wakayama City, Japan, Tel: +81(734)268319 ISSUES/YR: 4

● **Journal of Theoretical Biology**
ISSN: 0022-5193
J. Theor. Biol.
Academic Press, 24-28 Oval Road, NW1 7DX, London, United Kingdom, Tel: +44(207)2674466 ISSUES/YR: 24

Journal of Theoretical Medicine
ISSN: 1027-3662
J. Theor. Med.
Taylor and Francis Ltd., 4 Park Square, Milton Park, OX14 4RN, Abingdon, Oxfordshire, United Kingdom, Tel: +44(20)70175541, Fax: +44(20)70175198 ISSUES/YR: 4

Journal of Thermal Biology
ISSN: 0306-4565
J. Therm. Biol.
Elsevier Ltd, Langford Lane, Kidlington, OX5 1GB, Oxford, United Kingdom, Tel: +44(1865)843000, Fax: +44(1865)843010 ISSUES/YR: 8

● **Journal of Thoracic and Cardiovascular Surgery**
ISSN: 0022-5223
J. Thorac. Cardiovasc. Surg.
Mosby Inc., 11830 Westline Industrial Drive, St. Louis, MO 63146, United States, Tel: +1(314)8728370, Fax: +1(314)4321380 ISSUES/YR: 12

Journal of Thoracic Imaging
ISSN: 0883-5993
J. Thorac. Imaging
Lippincott Williams and Wilkins, 530 Walnut Street, Philadelphia, PA 19106-3621, United States, Tel: +1(215)5218342, Fax: +1(215)5218326 ISSUES/YR: 4

● **Journal of Thrombosis and Haemostasis**
ISSN: 1538-7933
J. Thromb. Haemost.
Blackwell Publishing Ltd, 9600 Garsington Road, OX4 2XG, Oxford, United Kingdom, Tel: +44(1865)776868, Fax: +44(1865)714591 ISSUES/YR: 12

- **Journal of Thrombosis and Thrombolysis**
  ISSN: 0929-5305
  J. Thromb. Trombolysis
  Kluwer Academic Publishers, Van Godewijckstraat 30, 3311 GZ, Dordrecht, Netherlands, Tel: +31(78)6576000, Fax: +31(78)6576254 ISSUES/YR: 6

Journal of Tokyo Medical University
ISSN: 0040-8905
J. Tokyo Med. Univ.
Medical Society of Tokyo Medical College, 1-1 Shinjuku 6-chome, Shinjuku-ku, 160, Tokyo, Japan, Tel: +81(3)33516141 ISSUES/YR: 6

Journal of Toxicological Sciences
ISSN: 0388-1350
J. Toxicol. Sci.
Japanese Society of Toxicology, 35 Shinanomachi, Shinjyuku-ku, 160-0016, Tokyo, Japan, Tel: +81(3)53617075, Fax: +81(3)53617091 ISSUES/YR: 5

- **Journal of Toxicology and Environmental Health - Part A**
  ISSN: 1528-7394
  J. Toxicol. Environ. Health Part A
  Taylor and Francis Inc., 325 Chestnut St, Suite 800, Philadelphia PA, PA 19106, United States, Tel: +44(800)3541420, Fax: +44(215)6258914 ISSUES/YR: 24

Journal of Toxicology and Environmental Health - Part B: Critical Reviews
ISSN: 1093-7404
J. Toxicol. Environ. Health Part B
Taylor and Francis Inc., 325 Chestnut St, Suite 800, Philadelphia PA, PA 19106, United States, Tel: +1(800)3541420, Fax: +1(215)6258914 ISSUES/YR: 6

- **Journal of Trace Elements in Medicine and Biology**
  ISSN: 0946-672X
  J. Trace Elem. Med. Biol.
  Urban und Fischer Verlag GmbH und Co. KG, P.O. Box 201930, 80019, Munchen, Germany, Tel: +49(89)53830, Fax: +49(89)5383221 ISSUES/YR: 4

Journal of Traditional Chinese Medicine
ISSN: 0255-2922
J. Trad. Chin. Med.
Journal of Traditional Chinese Medicine, 18 Beixincang, Dongzhimen Nei, 100700, Beijing, China, Fax: +86(10)86472013 ISSUES/YR: 4

Journal of Translational Medicine
ISSN: 1479-5876
J. Transl. Med.
BioMed Central Ltd., 34 - 42 Cleveland Street, W1T 4LB, London, United Kingdom, Tel: +44(207)3230323, Fax: +44(207)6319926 ISSUES/YR: 1

- **Journal of Trauma - Injury, Infection and Critical Care**
  ISSN: 0022-5282
  J. Trauma Inj. Infect. Crit. Care
  Lippincott Williams and Wilkins, 351 West Camden Street, Baltimore, MD 21201-2436, United States, Tel: +1(410)5284000 ISSUES/YR: 12

- **Journal of Trauma and Dissociation**
  ISSN: 1529-9732
  J. Trauma Dissociation
  Haworth Press Inc., 10 Alice Street, Binghamton, NY 13904-1580, United States, Tel: +1(607)7225857 ISSUES/YR: 4

Journal of Traumatic Stress
ISSN: 0894-9867
J. Trauma. Stress
Springer Science and Business Media Deutschland GmbH, Heidelberger Platz 3, 14197, Berlin, Germany, Tel: +1-30 827875282, Fax: +1-30 827875707 ISSUES/YR: 6

Journal of Travel Medicine
ISSN: 1195-1982
J. Travel Med.
Blackwell Publishing Ltd, 9600 Garsington Road, OX4 2XG, Oxford, United Kingdom, Tel: +44(1865)776868, Fax: +44(1865)714591 ISSUES/YR: 6

Journal of Tropical Pediatrics
ISSN: 0142-6338
J. Trop. Pediatr.
Oxford University Press, Great Clarendon Street, OX2 6DP, Oxford, United Kingdom, Tel: +44(1865)353907, Fax: +44(1865)353486 ISSUES/YR: 6

Journal of Ultrasound in Medicine
ISSN: 0278-4297
J. Ultrasound Med.
American Institute of Ultrasound in Medicine, 14750 Sweitzer Lane, Suite 100, Laurel, MD 20707-5906, United States, Tel: +1(301)4984100, Fax: +1(301)4984450 ISSUES/YR: 12

Journal of UOEH
ISSN: 0387-821X
J. UOEH
University of Occupational and Environmental Health, Iseigaoka 1-1, Yahata-Nishiku, Kitakyushu, 807, Fukuoka, Japan, Tel: +81(93)6031611 ISSUES/YR: 4

- **Journal of Urban Health**
  ISSN: 1099-3460
  Urban Health
  Springer Science and Business Media Deutschland GmbH, Heidelberger Platz 3, 14197, Berlin, Germany, Tel: +49-30 827875282, Fax: +49-30 827875707 ISSUES/YR: 5

- **Journal of Urology**
  ISSN: 0022-5347
  J. Urol.
  Elsevier Inc., 360 Park Avenue South, New York, NY 10010, United States, Tel: +1(212)6333950, Fax: +1(212)6333990 ISSUES/YR: 12

Journal of Vascular Access
ISSN: 1129-7298
J. Vasc. Access
Wichtig Editore s.r.l., Via Friuli 72/74, 20135, Milan, Italy, Tel: +39(02)55195443, Fax: +39(02)55195971 ISSUES/YR: 4

- **Journal of Vascular and Interventional Radiology**
  ISSN: 1051-0443
  J. Vasc. Intervent. Radiol.
  Lippincott Williams and Wilkins, 530 Walnut Street, Philadelphia, PA 19106-3621, United States, Tel: +1(215)5218342, Fax: +1(215)5218326 ISSUES/YR: 12

- **Journal of Vascular Research**
  ISSN: 1018-1172
  J. Vasc. Res.
  S. Karger AG, Allschwilerstrasse 10, P.O. Box, CH-4009, Basel, Switzerland, Tel: +41(61)3061111, Fax: +41(61)3061234 ISSUES/YR: 6

- **Journal of Vascular Surgery**
  ISSN: 0741-5214
  J. Vasc. Surg.
  Mosby Inc., 11830 Westline Industrial Drive, St. Louis, MO 63146, United States, Tel: +1(314)8728370, Fax: +1(314)4321380 ISSUES/YR: 12

Journal of Vector Borne Diseases
ISSN: 0972-9062
J. Vector Borne Dis.
Malaria Research Center, 22-Sham Nath Marg., 110 054, Delhi, India, Tel: +91(11)22533201, Fax: +91(11)22515086 ISSUES/YR: 4

- **Journal of Vestibular Research: Equilibrium and Orientation**
  ISSN: 0957-4271
  J. Vestibular Res. Equilib. Orientat.
  IOS Press, Nieuwe Hemweg 6B, 1013 BG, Amsterdam, Netherlands, Tel: +31(20)6883355, Fax: +31(20)6203419 ISSUES/YR: 6

- **Journal of Veterinary Cardiology**
  ISSN: 1760-2734
  J. Vet. Cardiol.
  Elsevier, P.O. Box 211, 1000 AE, Amsterdam, Netherlands, Tel: +31(20)4853757, Fax: +31(20)4853432 ISSUES/YR: 2

- **Journal of Veterinary Pharmacology and Therapeutics**
  ISSN: 0140-7783
  J. Vet. Pharmacol. Ther.
  Blackwell Publishing Ltd, 9600 Garsington Road, OX4 2XG, Oxford, United Kingdom, Tel: +44(1865)776868, Fax: +44(1865)714591 ISSUES/YR: 6

- **Journal of Viral Hepatitis**
ISSN: 1352-0504
J. Viral Hepatitis
Blackwell Publishing Ltd, 9600 Garsington Road, OX4 2XG, Oxford, United Kingdom, Tel: +44(1865)776868, Fax: +44(1865)714591 ISSUES/YR: 6
- **Journal of Virological Methods**
ISSN: 0166-0934
J. Virol. Methods
Elsevier, P.O. Box 211, 1000 AE, Amsterdam, Netherlands, Tel: +31(20)4853757, Fax: +31(20)4853432 ISSUES/YR: 14
- **Journal of Virology**
ISSN: 0022-538X
J. Virol.
American Society for Microbiology, 1752 N Street N.W., Washington, DC 20036-2904, United States, Tel: +1(202)942-9244, Fax: +1(202)942-9355 ISSUES/YR: 24

Journal of Vision
ISSN: 1534-7362
J. Vis.
Association for Research in Vision and Opthalmology Inc., 9650 Rockville Pike - Suite 1610, Bethesda, MD 20814-3998, United States, Tel: +1(301)5711844 ISSUES/YR: 12

Journal of Visual Impairment and Blindness
ISSN: 0145-482X
J. Vis. Impairm. Blindn.
AFB Press, 11 Penn Plaza, Suite 300, New York, NY 10001, United States, Tel: +1(212)5027774, Fax: +1(212)5027647 ISSUES/YR: 12

Journal of Vocational Rehabilitation
ISSN: 1052-2263
J. Vocat. Rehabil.
IOS Press, Nieuwe Hemweg 6B, 1013 BG, Amsterdam, Netherlands, Tel: +31(20)6883355, Fax: +31(20)6203419 ISSUES/YR: 6

Journal of Voice
ISSN: 0892-1997
J. Voice
Mosby Inc., 11830 Westline Industrial Drive, St. Louis, MO 63146, United States, Tel: +1(314)8728370, Fax: +1(314)4321380 ISSUES/YR: 4

Journal of Water and Health
ISSN: 1477-8920
J. Water Health
IWA Publishing, 12 Caxton Street, SW1H 0QS, London, United Kingdom, Tel: +44(207)6545500, Fax: +44(207)6545555 ISSUES/YR: 4

- **Journal of Whiplash and Related Disorders**
ISSN: 1533-2888
J. Whiplash Rel. Disord.
Haworth Press Inc., 10 Alice Street, Binghamton, NY 13904-1580, United States, Tel: +1(607)7225857 ISSUES/YR: 2

Journal of Women's Cancer
ISSN: 1529-0719
J. Women's Cancer
Smith-Edward-Dunlap Company, 2867 East Allegheny Avenue, Philadelphia, PA 19134, United States, Tel: +1(215)4258800, Fax: +1(215)4259715 ISSUES/YR: 4

- **Journal of Women's Health**
ISSN: 1540-9996
J. Women's Health
Mary Ann Liebert Inc., 140 Huguenot Street, New Rochelle, NY 10801-5215, United States, Tel: +1(914)7402100, Fax: +1(914)7402101 ISSUES/YR: 10
- **Journal of Women's Imaging**
ISSN: 1084-824X
J. Women's Imaging
Lippincott Williams and Wilkins, 530 Walnut Street, Philadelphia, PA 19106-3621, United States, Tel: +1(215)5218342, Fax: +1(215)5218326 ISSUES/YR: 4

Journal of Xi'an Jiaotong University (Medical Sciences)
ISSN: 1671-8259
J. Xi'An Jiaotong Univ. Med. Sci.
Xi'an Medical University, 16 Scarlet-Bird Avenue, 710061, Xi'an, China, Tel: +86(29)711621 ISSUES/YR: 6

Journal on Information Technology in Healthcare
ISSN: 1479-649X
J. Inf. Technol. Healthc.
Health Technology Press, 72 Churston Drive, SM4 4JQ, Morden, Surrey, United Kingdom ISSUES/YR: 6

Journal, Indian Academy of Clinical Medicine
ISSN: 0972-3560
J. Indian Acad. Clin. Med.
Indian Academy of Clinical Medicine, D II/65, Kaka Nagar, 110 003, New Delhi, India, Tel: +91(011)2435 6446, Fax: +91(011)2435 9445 ISSUES/YR: 4

- **Journals of Gerontology - Series A Biological Sciences and Medical Sciences**
ISSN: 1079-5006
J. Gerontol. Ser. A Biol. Sci. Med. Sci.
Gerontological Society of America, 1030 15th Street, NW Suite 250, Washington, DC 20005-1503, United States, Tel: +1(202)4083375, Fax: +1(202)8421150 ISSUES/YR: 12
- **Journals of Gerontology - Series B Psychological Sciences and Social Sciences**
ISSN: 1079-5014
J. Gerontol. Ser. B Psychol. Sci. Soc. Sci.
Gerontological Society of America, 1030 15th Street, NW Suite 250, Washington, DC 20005-1503, United States, Tel: +1(202)4083375, Fax: +1(202)8421150 ISSUES/YR: 6

JRAAS - Journal of the Renin-Angiotensin-Aldosterone System
ISSN: 1470-3203
JRAAS J. Renin-Angiotensin-Aldosterone Syst.
JRAAS Limited, Edgbaston House, 3 Duchess Place, Edgbaston, B16 8NH, Birmingham, United Kingdom, Tel: +44(121)4544114, Fax: +44(121)4541190 ISSUES/YR: 4

Jugoslovenska Medicinska Biohemija
ISSN: 0354-3447
Jugosl. Med. Biohem.
Society of Medical Biochemists of Serbia and Montenegro, Vojvode Stepe 450, P.O. Box 146, 11000, Belgrade, Serbia and Montenegro, Tel: +381(11)684283, Fax: +381(11)3615631 ISSUES/YR: 4

Kakuigaku
ISSN: 0022-7854
Kakuigaku
Japanese Society of Nuclear Medicine, 28-45 Hon-Komagome 2-chome, Bunkyo-ku, 113-0021, Tokyo, Japan, Tel: +81(3)39470976, Fax: +81(3)39472535 ISSUES/YR: 4

Kaohsiung Journal of Medical Sciences
ISSN: 0257-5655
Kaohsiung J. Med. Sci.
Elsevier (Singapore) Pte Ltd, 77 Leighton Road, Causeway Bay, Hong Kong, China, Hong Kong, Tel: +852-29651300, Fax: +852-29760778 ISSUES/YR: 12

Kardiologia
ISSN: 1210-0048
Kardiologia
SYMEKARD s.r.o., Pri Viohradoch 15, P.O. Box 29, 830 05, Bratislava 35, Slovakia, Tel: +421(2)4488 5817, Fax: +421(2)4487 2486 ISSUES/YR: 6

Kardiologia Polska
ISSN: 0022-9032
Kardiol. Pol.
Klinika Kardiologii CMKP, ul. Grenadierow 51/59, 04-073, Warsaw, Poland, Tel: +48(22)101738 ISSUES/YR: 12

Kardiotechnik
ISSN: 0941-2670
Kardiotechnik
Deutsche Gesellschaft fur Kardiotechnik e.V., Herderstrasse 67, 50931, Koln, Germany ISSUES/YR: 4

Kathmandu University Medical Journal
ISSN: 1812-2027
Kathmandu Univ. Med. J.
Kathmandu University, Sinamangal, P. O. Box 21266, Katmandu, Nepal ISSUES/YR: 4

Keio Journal of Medicine
ISSN: 0022-9717
Keio J. Med.
Keio University School of Medicine, 35 Shinanomachi, Shinjuku-ku, 160-8582, Tokyo, Japan, Tel: +81(3)33531211 EXT: 64023, Fax: +81(3)32665748 ISSUES/YR: 4

Kekkaku
ISSN: 0022-9776
Kekkaku
Japanese Society for Tuberculosis, 3-1-24 Matsuyama, Kiyose-shi, 204, Tokyo, Japan, Tel: +81(424)922091 ISSUES/YR: 12

- **Kidney and Blood Pressure Research**
ISSN: 1420-4096
Kidney Blood Press. Res.
S. Karger AG, Allschwilerstrasse 10, P.O. Box, CH-4009, Basel, Switzerland, Tel: +41(61)3061111, Fax: +41(61)3061234 ISSUES/YR: 6

- **Kidney International**
ISSN: 0085-2538
Kidney Int.
Nature Publishing Group, Houndmills, RG21 6XS, Basingstoke, Hampshire, United Kingdom, Tel: +44(1256)329242, Fax: +44(1256)810526 ISSUES/YR: 12

Kinderanalyse
ISSN: 0942-6051
Kinderanalyse
J.C. Cotta'sche Buchhandlung Nachvolger GmbH, P.O. Box 106016, 70049, Stuttgart, Germany, Tel: +49(711)66720, Fax: +49(711)66722030 ISSUES/YR: 4

Kinetoplastid Biology and Disease
ISSN: 1475-9292
Kinetoplastid Biol. Dis.
BioMed Central Ltd., 34 - 42 Cleveland Street, W1T 4LB, London, United Kingdom, Tel: +44(207)3230323, Fax: +44(207)6319926 ISSUES/YR: 1

Kitakanto Medical Journal
ISSN: 1343-2826
Kitakanto Med. J.
Kitakanto Medical Society, 3-39-22 Showamachi, 371, Maebashi, Japan, Tel: +81(272)207111 ISSUES/YR: 4

Klinicka Biochemie a Metabolismus
ISSN: 1210-7921
Klin. Biochem. Metab.
Czech Medical Association J.E. Purkyne, Sokolska 31, P.O. Box 88, 120 26, Praha 2, Czech Republic, Tel: +420(2)24266201-4, Fax: +420(2)24266212 ISSUES/YR: 4

Klinicka Imunologia a Alergologia
ISSN: 1335-0013
Klin. Imunol. Alergol.
Oddelenie Klinickej Imunologie FN, Americke nam. 3, SK-813 69, Bratislava, Slovakia, Tel: +421(7)54774144, Fax: +421(7)55565622 ISSUES/YR: 4

Klinicka Mikrobiologie a Infekcni Lekarstvi
ISSN: 1211-264X
Klin. Mikrobiol. Infekcni Lek.
Trios spol. s.r.o., Zakourilova 142, 149 00, Praha 4, Czech Republic, Tel: +420(267)912030, Fax: +420(267)915563 ISSUES/YR: 6

Klinicka Onkologie
ISSN: 0862-495X
Klin. Onkol.
Czech Medical Association J.E. Purkyne, Sokolska 31, P.O. Box 88, 120 26, Praha 2, Czech Republic, Tel: +420(2)24266201-4, Fax: +420(2)24266212 ISSUES/YR: 6

Klinik Psikiyatri Dergisi
ISSN: 1302-0099
Klin. Psikiyatr. Derg.
Klinik Psikiyatri Dergisi, Alper Sokak No: 6/5, Cankaya / Ankara, Turkey, Tel: +90-312 442 67 92, Fax: +90-312 439 65 87 ISSUES/YR: 4

Klinik Psikofarmakoloji Bulteni
ISSN: 1017-7833
Klin. Psikofarmakol. Bul.
Cukurova Univ Tip Fakultesi Psikiyatri Anabilim Dali, 01330, Balcali, Adana, Turkey, Tel: +90(542)4171769, Fax: +90(322)3386505 ISSUES/YR: 4

Klinika Oczna
ISSN: 0023-2157
Klin. Oczna
Volumed s.c., ul. Olsztynska 3, 51-423, Wroclaw, Poland, Tel: +48(71)3253561, Fax: +48(71)3253561 ISSUES/YR: 6

Klinikarzt
ISSN: 0341-2350
Klinikartz
Karl Demeter Verlag GmbH, Rudigerstr. 14, D-70469, Stuttgart, Germany, Tel: +49(0711)89310, Fax: +49(0711)8931536 ISSUES/YR: 12

Klinische Monatsblatter fur Augenheilkunde
ISSN: 0023-2165
Klin. Monatsbl. Augenheilkd.
Georg Thieme Verlag, Rudigerstrasse 14, D-70469, Stuttgart, Germany, Tel: +49(711)8931-321, Fax: +49(711)8931-422 ISSUES/YR: 12

Klinische Neurophysiologie
ISSN: 1434-0275
Klin. Neurophysiol.
Georg Thieme Verlag, Rudigerstrasse 14, D-70469, Stuttgart, Germany, Tel: +49(711)8931-321, Fax: +49(711)8931-422 ISSUES/YR: 4

Klinische Neuroradiologie
ISSN: 0939-7116
Klin. Neuroradiol.
Urban und Vogel GmbH, Neumarkter Strasse 43, D-81673, Munchen, Germany, Tel: +49(89)43721300, Fax: +49(89)43721399 ISSUES/YR: 4

Klinische Padiatrie
ISSN: 0300-8630
Klin. Padiatr.
Georg Thieme Verlag, Rudigerstrasse 14, D-70469, Stuttgart, Germany, Tel: +49(711)8931-321, Fax: +49(711)8931-422 ISSUES/YR: 6

- **Knee**
ISSN: 0968-0160
Knee
Elsevier, P.O. Box 211, 1000 AE, Amsterdam, Netherlands, Tel: +31(20)4853757, Fax: +31(20)4853432 ISSUES/YR: 6

Kobe Journal of Medical Sciences
ISSN: 0023-2513
Kobe J. Med. Sci.
Kobe University School of Medicine, 5-1, Kusunokicho 7-chome, Chuo-ku, 650-0017, Kobe, Japan, Tel: +81(78)382-5200 ISSUES/YR: 6

Koomesh
ISSN: 1608-7046
Koomesh
Semnan University of Medical Sciences, Bolvar Basij, 35195-163, Semnan, Iran, Tel: +98-2314441420, Fax: +98-2314440225 ISSUES/YR: 4

Korean Circulation Journal
ISSN: 1738-5520
Korean Circ. J.
Korean Society of Circulation, 553 Dowha-dong, Mapo-gu, 121-040, Seoul, South Korea, Tel: +82(2)32755258, Fax: +82(2)32755259 ISSUES/YR: 12

Korean Journal of Cardiovascular Diseases
ISSN: 1229-974X
Korean J. Cardiovasc. Dis.
Korean Society of Circulation, 553 Dowha-dong, Mapo-gu, 121-040, Seoul, South Korea, Tel: +82(2)32755258, Fax: +82(2)32755259 ISSUES/YR: 2

Korean Journal of Dermatology
ISSN: 0494-4739
Korean J. Dermatol.
Korean Dermatological Association, 1330-16 Seocho-2-dong, Seocho-ku, 137-072, Seoel, South Korea, Tel: +82(2)34730284 ISSUES/YR: 12

Korean Journal of Internal Medicine
ISSN: 1226-3303
Korean J. Intern. Med.
Korean Association of Internal Medicine, Korean Medical Association Building, 140-031, Seoul, South Korea ISSUES/YR: 4

Korean Journal of Medical Mycology
ISSN: 1226-4709
Korean J. Med. Mycol.
Korean Society for Medical Mycology, 388-1 Poongnap Dong, Songpa-Ku, 138-736, Seoul, South Korea, Tel: +82(2)22243300, Fax: +82(2)22246970 ISSUES/YR: 4

Korean Journal of Microbiology and Biotechnology
ISSN: 1598-642X
Korean J. Microbiol. Biotechnol.
Korean Society for Microbiolog and Biotechnology, 635-4 Yeogsam-Dong, Kangnam-Gu, 135-703, Seoul, South Korea, Tel: +82(2)5524733, Fax: +82(2)5524732 ISSUES/YR: 4

Korean Journal of Pharmacognosy
ISSN: 0253-3073
Korean J. Pharmacogn.
Korean Society of Pharmacognosy, 28 Yungun-Dong, Jongro-ku, 110-460, Seoul, South Korea, Tel: +82(2)7408925 ISSUES/YR: 4

Korean Journal of Physiology and Pharmacology
ISSN: 1226-4512
Korean J. Physiol. Pharmacol.
Korean Physiological Soc. and Korean Soc. of Pharmacology, 28 Yongno-dong, Chongno-Ku, 110-799, Seoul, South Korea ISSUES/YR: 6

Korean Journal of Urology
ISSN: 0494-4747
Korean J. Urol.
Korean Urological Association, Room 419, 4F, Jungmyung Honeyville Plaza # 965, 135-270, Kangnnam-gu, Seoul, South Korea, Tel: +82-2-573-8190, Fax: +82-2-573-8192 ISSUES/YR: 12

Kosmetische Medizin
ISSN: 1430-4031
Kosmet. Med.
BMV Berliner Medizinische Verlagsanstalt GmbH, Lietzenburger Strasse 97, 10719, Berlin, Germany, Tel: +49(30)8867490, Fax: +49(30)88674999 ISSUES/YR: 6

Krankenhaushygiene und Infektionsverhutung
ISSN: 0720-3373
Krankenh.hyg. Infektionsverhutung
Urban und Fischer Verlag Jena, P.O. Box 100537, 07705, Jena, Germany, Tel: +49(3641)6263, Fax: +49(3641)626500 ISSUES/YR: 6

Krankenhauspharmazie
ISSN: 0173-7597
Krankenhauspharmazie
Deutscher Apotheker Verlag, P.O. Box 10 10 61, D-70 009, Stuttgart, Germany, Tel: +49(711)25820 ISSUES/YR: 12

Krankenhauspsychiatrie
ISSN: 0937-289X
Krankenhauspsychiatrie
Georg Thieme Verlag, Rudigerstrasse 14, D-70469, Stuttgart, Germany, Tel: +49(711)8931-321, Fax: +49(711)8931-422 ISSUES/YR: 4

Krankenhauspsychiatrie, Supplement
ISSN: 1615-6625
Krankenhauspsychiatr. Suppl.
Georg Thieme Verlag, Rudigerstrasse 14, D-70469, Stuttgart, Germany, Tel: +49(711)8931-321, Fax: +49(711)8931-422 ISSUES/YR: 1

Kumamoto Medical Journal
ISSN: 0023-5326
Kumamoto Med. J.
Kumamoto University School of Medicine, Honjo 2-2-1, 860, Kumamoto, Japan, Tel: +81(96)3442111 ISSUES/YR: 4

Kurume Medical Journal
ISSN: 0023-5679
Kurume Med. J.
Kurume University School of Medicine, 67 Asahi-machi, 830, Kurume, Japan, Tel: +81(942)353311 ISSUES/YR: 4

● **Lab on a Chip - Miniaturisation for Chemistry and Biology**
ISSN: 1473-0197
Lab Chip Miniaturisation Chem. Biol.
Royal Society of Chemistry, Milton Road, CB4 OWF, Cambridge, United Kingdom, Tel: +44(1223)420066, Fax: +44(1223)420247 ISSUES/YR: 12

LaboratoriumsMedizin
ISSN: 0342-3026
LaboratoriumsMedizin
Walter de Gruyter and Co., Genthiner Strasse 13, D-10785, Berlin, Germany, Tel: +49(30)260-050, Fax: +49(30)260-05251 ISSUES/YR: 6

Laboratory Animals
ISSN: 0023-6772
Lab. Anim.
Royal Society of Medicine Press Ltd, P.O. Box 9002, W1A 0ZA, London, United Kingdom, Tel: +44(207)2902927, Fax: +44(207)2902929 ISSUES/YR: 4

Laboratory Hematology
ISSN: 1080-2924
Lab. Hematol.
Carden Jennings Publishing Co. Ltd, 375 Greenbrier Drive #100, Charlottesville, VA 22901-1618, United States, Tel: +1(434)8172000, Fax: +1(434)8172020 ISSUES/YR: 4

● **Laboratory Investigation**
ISSN: 0023-6837
Lab. Invest.
Nature Publishing Group, Houndmills, RG21 6XS, Basingstoke, Hampshire, United Kingdom, Tel: +44(1256)329242, Fax: +44(1256)810526 ISSUES/YR: 12

Laboratory Medicine
ISSN: 0007-5027
Lab. Med.
American Society of Clinical Pathologists, 2100 West Harrison Street, Chicago, IL 60612-3798, United States, Tel: +1(312)7381336 ISSUES/YR: 12

● **Lancet**
ISSN: 0140-6736
Lancet
Elsevier Limited, 32 Jamestown Road, NW1 7BY, London, United Kingdom ISSUES/YR: 52

● **Lancet Infectious Diseases**
ISSN: 1473-3099
Lancet Infect. Dis.
Lancet Publishing Group, 84 Theobald's Road, WC1X 8RR, London, United Kingdom, Tel: +44(20)76114100, Fax: +44(20)76114479 ISSUES/YR: 12

● **Lancet Neurology**
ISSN: 1474-4422
Lancet Neurol.
Lancet Publishing Group, 84 Theobald's Road, WC1X 8RR, London, United Kingdom, Tel: +44(20)76114100, Fax: +44(20)76114479 ISSUES/YR: 12

● **Lancet Oncology**
ISSN: 1470-2045
Lancet Oncol.
Lancet Publishing Group, 650 Avenue of the Americas, New York, NY 10010, United States, Tel: +1(212)6333800, Fax: +1(212)6333850 ISSUES/YR: 12

● **Langenbeck's Archives of Surgery**
ISSN: 1435-2443
Langenbeck's Arch. Surg.
Springer Verlag, Tiergartenstrasse 17, D-69121, Heidelberg, Germany, Tel: +49(6221)4870, Fax: +49(6221)487366 ISSUES/YR: 6

● **Language and Cognitive Processes**
ISSN: 0169-0965
Lang. Cogn. Proc.
Psychology Press Ltd, 4 Park Square, Milton Park, OX14 4RN, Abingdon, Oxfordshire, United Kingdom, Tel: +44(1235)828600, Fax: +44(1235)829000 ISSUES/YR: 8

● **Laryngo- Rhino- Otologie**
ISSN: 0935-8943
Laryngo- Rhino- Otol.
Georg Thieme Verlag, Rudigerstrasse 14, D-70469, Stuttgart, Germany, Tel: +49(711)8931-321, Fax: +49(711)8931-422 ISSUES/YR: 12

● **Laryngoscope**
ISSN: 0023-852X
Laryngoscope
Lippincott Williams and Wilkins, 530 Walnut Street, Philadelphia, PA 19106-3621, United States, Tel: +1(215)5218342, Fax: +1(215)5218326 ISSUES/YR: 12

● **Lasers in Medical Science**
ISSN: 0268-8921
Lasers Med. Sci.
Springer-Verlag London Ltd, Catteshall Road, Farncombe, GU7 3DJ, Godalming, Surrey, United Kingdom, Tel: +44(1483)414142, Fax: +44(1483)415144 ISSUES/YR: 4

Current

- **Lasers in Surgery and Medicine**
  ISSN: 0196-8092
  Lasers Surg. Med.
  Wiley-Liss Inc., 111 River Street, Hoboken, NJ 07030-5774, United States, Tel: +1(201)7486645, Fax: +1(201)7486021 ISSUES/YR: 10

Laterality
ISSN: 1357-650X
Laterality
Psychology Press Ltd, 4 Park Square, Milton Park, OX14 4RN, Abingdon, Oxfordshire, United Kingdom, Tel: +44(1235)828600, Fax: +44(1235)829000 ISSUES/YR: 6

Law and Human Behavior
ISSN: 0147-7307
Law Hum. Behav.
Springer Science and Business Media Deutschland GmbH, Heidelberger Platz 3, 14197, Berlin, Germany, Tel: +1-30 827875282, Fax: +1-30 827875707 ISSUES/YR: 6

Learning and Behavior
ISSN: 1543-4494
Learn. Behav.
Psychonomic Society Inc., 1710 Fortview Road, Austin, TX 78704, United States, Tel: +1(512)4622442 ISSUES/YR: 4

- **Learning and Memory**
  ISSN: 1072-0502
  Learn. Mem.
  Cold Spring Harbor Laboratory Press, 500 Sunnyside Boulevard, Woodbury New York, NY 11797, United States, Tel: +1(516)3491930, Fax: +1(516)3491946 ISSUES/YR: 6

- **Legal and Criminological Psychology**
  ISSN: 1355-3259
  Leg. Criminol. Psychol.
  British Psychological Society, 48 Princess Road East, LE1 7DR, Leicester, United Kingdom, Tel: +44(116)2549568, Fax: +44(116)2470787 ISSUES/YR: 2

Legal Medicine
ISSN: 1344-6223
Legal Med.
Elsevier, P.O. Box 211, 1000 AE, Amsterdam, Netherlands, Tel: +31(20)4853757, Fax: +31(20)4853432 ISSUES/YR: 4

Lege Artis Medicine
ISSN: 0866-4811
Lege Artis Med.
Literatura Medica Publishing House, P.O. Box 603, 1539, Budapest, Hungary, Tel: +36(1)3164556, Fax: +36(1)3169600 ISSUES/YR: 12

Lekarsky Obzor
ISSN: 0457-4214
Lek. Obz.
Vydavatel'stvo HERBA spol. s.r.o., P.O. Box 53, 830 07, Bratislava, Slovakia, Tel: +421(7)376683, Fax: +421(7)376683 ISSUES/YR: 12

Leprosy Review
ISSN: 0305-7518
Lepr. Rev.
British Leprosy Relief Association, Fairfax House, Causton Road, CO1 7PU, Colchester, United Kingdom ISSUES/YR: 4

Letters in Applied Microbiology
ISSN: 0266-8254
Lett. Appl. Microbiol.
Blackwell Publishing Ltd, 9600 Garsington Road, OX4 2XG, Oxford, United Kingdom, Tel: +44(1865)776868, Fax: +44(1865)714591 ISSUES/YR: 12

- **Letters in Drug Design and Discovery**
  ISSN: 1570-1808
  Lett. Drug. Des. Discov.
  Bentham Science Publishers B.V., P.O. Box 294, 1400 AG, Bussum, Netherlands, Tel: +31-356257290, Fax: +31-847511310 ISSUES/YR: 8

- **Leukemia**
  ISSN: 0887-6924
  Leukemia
  Nature Publishing Group, Houndmills, RG21 6XS, Basingstoke, Hampshire, United Kingdom, Tel: +44(1256)329242, Fax: +44(1256)810526 ISSUES/YR: 12

- **Leukemia and Lymphoma**
  ISSN: 1042-8194
  Leuk. Lymphoma
  Taylor and Francis Ltd., 4 Park Square, Milton Park, OX14 4RN, Abingdon, Oxfordshire, United Kingdom, Tel: +44(20)70175541, Fax: +44(20)70175198 ISSUES/YR: 12

- **Leukemia Research**
  ISSN: 0145-2126
  Leuk. Res.
  Elsevier Ltd, Langford Lane, Kidlington, OX5 1GB, Oxford, United Kingdom, Tel: +44(1865)843000, Fax: +44(1865)843010 ISSUES/YR: 12

Libri Oncologici
ISSN: 0300-8142
Libri Oncol.
University Hospital for Tumors, Ilica 197, HR-10000, Zagreb, Croatia, Tel: +385(41)572111 ISSUES/YR: 3

Life Sciences
ISSN: 0024-3205
Life Sci.
Elsevier Inc., 360 Park Avenue South, New York, NY 10010, United States, Tel: +1(212)6333950, Fax: +1(212)6333990 ISSUES/YR: 52

Lijecnicki Vjesnik
ISSN: 0024-3477
Lijec. Vjesn.
Urednistvu Lijecnickog Vjesnika, Subiceva ulica 9, HR-41000, Zagreb, Croatia, Tel: +385(41)440621 ISSUES/YR: 12

Lijecnicki Vjesnik, Supplement
ISSN: 1330-4917
Lijec. Vjesn. Suppl.
Urednistvu Lijecnickog Vjesnika, Subiceva ulica 9, HR-41000, Zagreb, Croatia, Tel: +385(41)440621 ISSUES/YR: 1

Lipids
ISSN: 0024-4201
Lipids
American Oil Chemists Society, P.O. Box 3489, Champaign, IL 61826-3489, United States, Tel: +1(217)3592344, Fax: +1(217)3518091 ISSUES/YR: 12

Lipids in Health and Disease
ISSN: 1476-511X
Lipids Health Dis.
BioMed Central Ltd., 34 - 42 Cleveland Street, W1T 4LB, London, United Kingdom, Tel: +44(207)3230323, Fax: +44(207)6319926 ISSUES/YR: 1

Liver International
ISSN: 1478-3223
Liver Int.
Blackwell Publishing Ltd, 9600 Garsington Road, OX4 2XG, Oxford, United Kingdom, Tel: +44(1865)776868, Fax: +44(1865)714591 ISSUES/YR: 6

- **Liver Transplantation**
  ISSN: 1527-6465
  Liver Transplant.
  John Wiley and Sons Ltd, Southern Gate, PO19 8SQ, Chichester, West Sussex, United Kingdom, Tel: +44(1243)779777, Fax: +44(1243)775878 ISSUES/YR: 12

- **Logopedics Phoniatrics Vocology**
  ISSN: 1401-5439
  Logop. Phoniatr. Vocology
  Taylor and Francis Ltd., 4 Park Square, Milton Park, OX14 4RN, Abingdon, Oxfordshire, United Kingdom, Tel: +44(20)70175541, Fax: +44(20)70175198 ISSUES/YR: 4

- **Logopedics Phoniatrics Vocology, Supplement**
  ISSN: 1401-5455
  Logop. Phoniatr. Vocology Suppl.
  Taylor and Francis Ltd., 4 Park Square, Milton Park, OX14 4RN, Abingdon, Oxfordshire, United Kingdom, Tel: +44(20)70175541, Fax: +44(20)70175198 ISSUES/YR: 1

Lotta Contro la Tuberculosi e le Malattie Polmonari Sociali
ISSN: 0368-7546
Lotta Contro Tuberc. Mal. Polm. Soc.
Federazione Italiana Contro Tubercol. Malat. Polmon. Sociali, Via Frua 15, 20148 Milano, Italy, Tel: +39-02 439 82 610 ISSUES/YR: 4

Current

Louvain Medical
ISSN: 0024-6956
Louvain Med.
Louvain Medical asbl, Avenue E. Mounier 52.65, B-1200, Brussels, Belgium, Tel: +32(2)7645265, Fax: +32(2)7645280 ISSUES/YR: 10

- **Lung**
ISSN: 0341-2040
Lung
Springer New York LLC, 233 Springer Street, New York, 10013-1578, United States, Tel: +1-212 460 1501, Fax: +1-212 460 1595 ISSUES/YR: 6

- **Lung Cancer**
ISSN: 0169-5002
Lung Cancer
Elsevier Ireland Ltd, P.O. Box 85, Limerick, Ireland, Tel: +353(61)471944, Fax: +353(61)472144 ISSUES/YR: 12

- **Lupus**
ISSN: 0961-2033
Lupus
Arnold, 338 Euston Road, NW1 3BH, London, United Kingdom, Tel: +44(20)78736000, Fax: +44(20)78736376 ISSUES/YR: 12

- **Lymphatic Research and Biology**
ISSN: 1539-6851
Lymphatic Res. Biol.
Mary Ann Liebert Inc., 140 Huguenot Street, New Rochelle, NY 10801-5215, United States, Tel: +1(914)7402100, Fax: +1(914)7402101 ISSUES/YR: 4

Lymphologie in Forschung und Praxis
ISSN: 1433-5255
Lymphol. Forsch. Prax.
Viavital Verlag, Theodor Althoff Strasse 39, 45133, Essen, Germany, Tel: +49-201-87126830, Fax: +49-201-87126840 ISSUES/YR: 2

- **Lymphology**
ISSN: 0024-7766
Lymphology
International Society of Lymphology, P.O. Box 245063, Tucson, AZ 85724-5063, United States, Tel: +1(520)626-6118, Fax: +1(520)626-0822 ISSUES/YR: 4

Macromolecular Bioscience
ISSN: 1616-5187
Macromol. Biosci.
Wiley-VCH Verlag, P.O. Box 101161, D-69451, Weinheim, Germany, Tel: +49(6201)606147, Fax: +49(6201)606117 ISSUES/YR: 12

Maghreb Medical
ISSN: 0330-258X
Maghreb Med.
Societe d'Edition du Maghreb Medical, 18 rue El Gorjani, 1008, Montfleury, Tunisia, Tel: +216(1)259448, Fax: +216(1)344077 ISSUES/YR: 6

- **Magnetic Resonance Imaging**
ISSN: 0730-725X
Magn. Reson. Imaging
Elsevier Inc., 360 Park Avenue South, New York, NY 10010, United States, Tel: +1(212)6333950, Fax: +1(212)6333990 ISSUES/YR: 10

Magnetic Resonance Imaging Clinics of North America
ISSN: 1064-9689
Magn. Reson. Imaging Clin. North Am.
W.B. Saunders, Independence Square West, Philadelphia, PA 19106-3399, United States, Tel: +1(215)2387800, Fax: +1(215)2386445 ISSUES/YR: 4

Magnetic Resonance in Medicine
ISSN: 0740-3194
Magn. Reson. Med.
John Wiley and Sons Inc., P.O.Box 18667, Newark, NJ 07191-8667, United States, Tel: +1(212)8506000, Fax: +1(212)8506012 ISSUES/YR: 12

- **Magnetic Resonance Materials in Physics, Biology and Medicine**
ISSN: 0968-5243
Magn. Reson. Mater. Phys. Biol. Med.
Springer Verlag, Tiergartenstrasse 17, D-69121, Heidelberg, Germany, Tel: +49(6221)4870, Fax: +49(6221)487366 ISSUES/YR: 6

Magyar Noorvosok Lapja
ISSN: 0025-021X
Magy. Noorv. Lapja
Magyar Noovos Tarsasag, Szabolcs u. 35, P.O. Box 112, H-1135, Budapest, Hungary, Tel: +36(1)1296726 ISSUES/YR: 6

Magyar Urologia
ISSN: 0864-8921
Magy. Urol.
Society of Hungarian Urological Surgeons, Nagyerdei krt. 98, P.O. Box 29, H-4032, Debrecen, Hungary, Tel: +36(52)413279 ISSUES/YR: 4

Malaria Journal
ISSN: 1475-2875
Malaria J.
BioMed Central Ltd., 34 - 42 Cleveland Street, W1T 4LB, London, United Kingdom, Tel: +44(207)3230323, Fax: +44(207)6319926 ISSUES/YR: 1

Malaysian Journal of Medical Sciences
ISSN: 1394-195X
Malays. J. Med. Sci.
Malaysian Journal of Medical Sciences, 16150 Kubang Kerian, Kentan, Malaysia, Tel: +60(9)7651700, Fax: +60(9)7656532 ISSUES/YR: 2

Malignant Hyperthermia
ISSN: 1609-7041
Malignant Hyperthermia
VICER Publishing, P.O. Box 14, A-1097, Vienna, Austria, Tel: +43(676)9568085, Fax: +43(676)9568086 ISSUES/YR: 4

Mammalian Genome
ISSN: 0938-8990
Mamm. Genome
Springer New York LLC, 233 Springer Street, New York, 10013-1578, United States, Tel: +1-212 460 1501, Fax: +1-212 460 1595 ISSUES/YR: 12

Managed Care Interface
ISSN: 1096-5645
Managed Care Interface
Medicom International Inc., 66 Palmer Avenue, Suite 49, Bronxville, NY 10708, United States, Tel: +1(914)3377878, Fax: +1(914)3375023 ISSUES/YR: 12

Managed Care Pharmacy Update
ISSN: 1540-2355
Managed Care Pharm. Update
Excerpta Medica Inc., 105 Raider Blvd, Suite 101, Hillsborough, NJ 08844, United States, Tel: +1(908)8748550, Fax: +1(908)8745633 ISSUES/YR: 4

- **Manual Therapy**
ISSN: 1356-689X
Man. Ther.
Churchill Livingstone, 1-3 Baxter's Place, Leith Walk, EH1 3AF, Edinburgh, United Kingdom, Tel: +44(131)5562424, Fax: +44(131)5581278 ISSUES/YR: 4

Mapfre Medicina
ISSN: 1130-5665
Mapfre Med.
Fundacion Mapfre Medicina, Ctra. Pozuelo a Majadahonda, s/n, 28220, Majadahonda (Madrid), Spain, Tel: +34(91)6265704, Fax: +34(91)6265825 ISSUES/YR: 4

Marine Drugs
ISSN: 1660-3397
Mar. Drugs
Molecular Diversity Preservation International, Matthaeusstrasse 11, CH-4057, Basel, Switzerland, Tel: +41-79 322 3379, Fax: +41-61 302 8918 ISSUES/YR: 4

Marine Environmental Research
ISSN: 0141-1136
Mar. Environ. Res.
Elsevier Ltd, Langford Lane, Kidlington, OX5 1GB, Oxford, United Kingdom, Tel: +44(1865)843000 ISSUES/YR: 10

Marine Pollution Bulletin
ISSN: 0025-326X
Mar. Pollut. Bull.
Elsevier Ltd, Langford Lane, Kidlington, OX5 1GB, Oxford, United Kingdom, Tel: +44(1865)843000, Fax: +44(1865)843010 ISSUES/YR: 24

Current

Marketing Health Services
ISSN: 1094-1304
Mark. Health Serv.
American Marketing Association, 311 S. Wacker Drive, Suite 5800, Chicago, IL 60606, United States, Tel: +1(312)5429000, Fax: +1(312)5429001 ISSUES/YR: 4

Marmara Medical Journal
ISSN: 1019-1941
Marmara Med. J.
Marmara University, Haydarpasa, Istanbul, Turkey ISSUES/YR: 4

• **Maternal and Child Health Journal**
ISSN: 1092-7875
Matern. Child Health J.
Kluwer Academic/Plenum Press New York, 233 Spring Street, New York, NY 10013, United States ISSUES/YR: 4

• **Maternal and Child Nutrition**
ISSN: 1740-8695
Matern. Child Nutr.
Blackwell Publishing Ltd, 9600 Garsington Road, OX4 2XG, Oxford, United Kingdom, Tel: +44(1865)776868, Fax: +44(1865)714591 ISSUES/YR: 4

Mathematical Biosciences
ISSN: 0025-5564
Math. Biosci.
Elsevier Inc., 360 Park Avenue South, New York, NY 10010, United States, Tel: +1(212)6333950, Fax: +1(212)6333990 ISSUES/YR: 10

Mathematical Medicine and Biology
ISSN: 1477-8599
Math. Med. Biol.
Oxford University Press, Great Clarendon Street, OX2 6DP, Oxford, United Kingdom, Tel: +44(1865)353907, Fax: +44(1865)353486 ISSUES/YR: 4

• **Matrix Biology**
ISSN: 0945-053X
Matrix Biol.
Elsevier, P.O. Box 211, 1000 AE, Amsterdam, Netherlands, Tel: +31(20)4853757, Fax: +31(20)4853432 ISSUES/YR: 8

Matronas Profesion
ISSN: 1578-0740
Matronas Profesion
Ediciones Mayo S.A., Aribau 185-187 2e planta, 08021, Barcelona, Spain, Tel: +34(93)2090255, Fax: +34(93)2020643 ISSUES/YR: 4

Maturitas
ISSN: 0378-5122
Maturitas
Elsevier Ireland Ltd, P.O. Box 85, Limerick, Ireland, Tel: +353(61)471944, Fax: +353(61)472144 ISSUES/YR: 12

Mayo Clinic Proceedings
ISSN: 0025-6196
Mayo Clin. Proc.
Mayo Medical Ventures, 200 First Street Southwest, Rochester, MN 55905, United States, Tel: +1(507)2842094 ISSUES/YR: 12

McGill Journal of Medicine
ISSN: 1201-026X
McGill J. Med.
McGill University, School fof Nursing, 3655 Promenade Sir William Osler, QUE H3G 1Y6, Montreal, Canada, Tel: +1(514)3986987, Fax: +1(514)3983595 ISSUES/YR: 2

Measurement in Physical Education and Exercise Science
ISSN: 1091-367X
Meas. Phys. Educ. Exerc. Sci.
Lawrence Erlbaum Associates Inc., 10 Industrial Avenue, Mahwah, NJ 07430-2262, United States, Tel: +1(201)2582200, Fax: +1(201)2360072 ISSUES/YR: 4

• **Mechanisms of Ageing and Development**
ISSN: 0047-6374
Mech. Ageing Dev.
Elsevier Ireland Ltd, P.O. Box 85, Limerick, Ireland, Tel: +353(61)471944, Fax: +353(61)472144 ISSUES/YR: 12

• **Mechanisms of Development**
ISSN: 0925-4773
Mech. Dev.
Elsevier Ireland Ltd, P.O. Box 85, Limerick, Ireland, Tel: +353(61)471944, Fax: +353(61)472144 ISSUES/YR: 12

Medecine et Chirurgie du Pied
ISSN: 0759-2280
Med. Chir. Pied
Springer-Verlag France, 1 rue Paul Cezanne, 75008, Paris, France, Tel: +33(1)53933628, Fax: +33(1)53933683 ISSUES/YR: 4

Medecine et Droit
ISSN: 1246-7391
Med. Droit
Editions Scientifiques et Medicales Elsevier, 23 rue Linois, 75724, Paris Cedex 15, France, Tel: +33(1)45589826, Fax: +33(1)45589428 ISSUES/YR: 6

Medecine et Hygiene
ISSN: 0025-6749
Med. Hyg.
Editions Medecine et Hygiene, 7, Avenue de la Roseraie, P.O. Box 456, CH-1211, Geneva 4, Switzerland, Tel: +41(22)7029311, Fax: +41(22)7029355 ISSUES/YR: 52

Medecine et Maladies Infectieuses
ISSN: 0399-077X
Med. Mal. Infect.
Editions Scientifiques et Medicales Elsevier, 23 rue Linois, 75724, Paris Cedex 15, France, Tel: +33(1)45589826, Fax: +33(1)45589428 ISSUES/YR: 12

Medecine Nucleaire
ISSN: 0928-1258
Med. Nucl.
SFBMN - Diffusion, 45 Rue des Saints Peres, 75270, Paris Cedex 06, France, Tel: +33(1)42860377, Fax: +33(1)42860278 ISSUES/YR: 12

Medecine Therapeutique
ISSN: 1264-6520
Med. Ther.
John Libbey Eurotext, 127, avenue de la Republique, 92120, Montrouge, France, Tel: +33(01)46730660, Fax: +33(01)40840999 ISSUES/YR: 4

Medecine Therapeutique - Cardio
ISSN: 1774-8747
Med. Ther. Cardio
John Libbey Eurotext, 127, avenue de la Republique, 92120, Montrouge, France, Tel: +33(01)46730660, Fax: +33(01)40840999 ISSUES/YR: 6

Medecine Therapeutique Endocrinologie et Reproduction
ISSN: 1763-3850
Med. Ther. Endocrinol. Reprod.
John Libbey Eurotext, 127, avenue de la Republique, 92120, Montrouge, France, Tel: +33(01)46730660, Fax: +33(01)40840999 ISSUES/YR: 6

Medecine Therapeutique Pediatrie
ISSN: 1286-5494
Med. Ther. Pediatr.
John Libbey Eurotext, 127, avenue de la Republique, 92120, Montrouge, France, Tel: +33(01)46730660, Fax: +33(01)40840999 ISSUES/YR: 6

Medecine Tropicale
ISSN: 0025-682X
Med. Trop.
Institut de Medecine Tropicale, BP 46, 13998, Marseille - Armees, France, Tel: +33(04)91523568, Fax: +33(04)91594477 ISSUES/YR: 4

Medecine/Sciences
ISSN: 0767-0974
Med. Sci.
Editions EDK, 33 Rue Bezout, 75014, Paris, France, Tel: +33-1 5391 0606, Fax: +33-1 5391 0607 ISSUES/YR: 12

MedGenMed Medscape General Medicine
ISSN: 1531-0132
Medgenmed Medscape Gen. Med.
Medscape Health Network, 224 W. 30th Street, New York, NY 10001-5399, United States, Tel: +1(888)5066098 ISSUES/YR: 1

• **Mediators of Inflammation**
ISSN: 0962-9351
Mediators Inflamm.
Hindawi Publishing Corporation, 410 Park Avenue, 15th Floor, 287 pmb, New York, NY 10022, United States, Tel: +1(20)2-2706352, Fax: +1(20)2-2706351 ISSUES/YR: 6

Medica Jadertina
ISSN: 0351-0093
Med. Jadertina
Opca Bolnica Zadar, Boze Pericica 5, 23000, Zadar, Croatia, Tel: +385(23)25440, Fax: +385(23)25440 ISSUES/YR: 4

Medical and Biological Engineering and Computing
ISSN: 0140-0118
Med. Biol. Eng. Comput.
Springer Verlag, Tiergartenstrasse 17, D-69121, Heidelberg, Germany, Tel: +49(6221)4870, Fax: +49(6221)487366 ISSUES/YR: 6

Medical Care Research and Review
ISSN: 1077-5587
Med. Care Res. Rev.
SAGE Publications Inc., 2455 Teller Road, Thousand Oaks, CA 91320, United States, Tel: +1(805)4990721, Fax: +1(805)4990871 ISSUES/YR: 4

• **Medical Clinics of North America**
ISSN: 0025-7125
Med. Clin. North Am.
W.B. Saunders, Independence Square West, Philadelphia, PA 19106-3399, United States, Tel: +1(215)2387800, Fax: +1(215)2386445 ISSUES/YR: 6

Medical Crossfire
ISSN: 1525-9404
Med. Crossfire
Liberty Communications Network Inc., 50 Millstone Road, Building 400, Suite 110, East Windsor, NJ 08520, United States, Tel: +1(609)9189400, Fax: +1(609)9189411 ISSUES/YR: 6

Medical Decision Making
ISSN: 0272-989X
Med. Decis. Mak.
SAGE Publications Inc., 2455 Teller Road, Thousand Oaks, CA 91320, United States, Tel: +1(805)4990721, Fax: +1(805)4990871 ISSUES/YR: 6

• **Medical Dosimetry**
ISSN: 0958-3947
Med. Dosim.
Elsevier Inc., 360 Park Avenue South, New York, NY 10010, United States, Tel: +1(212)6333950, Fax: +1(212)6333990 ISSUES/YR: 4

Medical Education
ISSN: 0308-0110
Med. Educ.
Blackwell Publishing Ltd, 9600 Garsington Road, OX4 2XG, Oxford, United Kingdom, Tel: +44(1865)776868, Fax: +44(1865)714591 ISSUES/YR: 12

Medical Education, Supplement
ISSN: 1352-3929
Med. Educ. Suppl.
Blackwell Publishing Ltd, 9600 Garsington Road, OX4 2XG, Oxford, United Kingdom, Tel: +44(1865)776868, Fax: +44(1865)714591 ISSUES/YR: 1

• **Medical Electron Microscopy**
ISSN: 0918-4287
Med. Electron Microsc.
Springer-Verlag Tokyo, 3-13, Hongo 3-chome, Bunkyo-ku, 113-8480, Tokyo, Japan, Tel: +81(3)38180861, Fax: +81(3)38180864 ISSUES/YR: 4

• **Medical Engineering and Physics**
ISSN: 1350-4533
Med. Eng. Phys.
Elsevier Ltd, Langford Lane, Kidlington, OX5 1GB, Oxford, United Kingdom, Tel: +44(1865)843000, Fax: +44(1865)843010 ISSUES/YR: 10

Medical Forum Monthly
ISSN: 1029-385X
Med. Forum Monthly
Medical Forum Monthly, Gujjar Singh,, 5460, Lahore, Pakistan, Tel: +92(42)6361436, Fax: +92(42)6366809 ISSUES/YR: 12

Medical Group Management Journal
ISSN: 0899-8949
Med. Group Manage. J.
Medical Group Management Association, 104 Inverness Terrace East, Englewood, CO 80112-5306, United States, Tel: +1(303)7991111 ISSUES/YR: 6

Medical Humanities
ISSN: 1468-215X
Med. Humanit.
BMJ Publishing Group, Tavistock Square, WC1H 9JR, London, United Kingdom, Tel: +44(207)3874499, Fax: +44(207)3836668 ISSUES/YR: 2

• **Medical Hypotheses**
ISSN: 0306-9877
Med. Hypotheses
Churchill Livingstone, 1-3 Baxter's Place, Leith Walk, EH1 3AF, Edinburgh, United Kingdom, Tel: +44(131)5562424, Fax: +44(131)5581278 ISSUES/YR: 12

• **Medical Image Analysis**
ISSN: 1361-8415
Med. Image Anal.
Elsevier, P.O. Box 211, 1000 AE, Amsterdam, Netherlands, Tel: +31(20)4853757, Fax: +31(20)4853432 ISSUES/YR: 6

Medical Immunology
ISSN: 1476-9433
Med. Immunol.
BioMed Central Ltd., 34 - 42 Cleveland Street, W1T 4LB, London, United Kingdom, Tel: +44(207)3230323, Fax: +44(207)6319926 ISSUES/YR: 1

Medical Informatics and the Internet in Medicine
ISSN: 1463-9238
Med. Informatics Internet Med.
Taylor and Francis Ltd., 4 Park Square, Milton Park, OX14 4RN, Abingdon, Oxfordshire, United Kingdom, Tel: +44(20)70175541, Fax: +44(20)70175198 ISSUES/YR: 4

Medical Journal Armed Forces India
ISSN: 0377-1237
Med. J. Armed Forces India
Medical Journal Armed Forces India, 411040, Pune, India, Tel: +91(20)6815647, Fax: +91(20)6815647 ISSUES/YR: 4

Medical Journal of Australia
ISSN: 0025-729X
Med. J. Aust.
Australasian Medical Publishing Co. Ltd, Level 2, 26-32 Pyrmont Bridge Road, NSW 2009, Pyrmont, Australia, Tel: +61(2)99548666, Fax: +61(2)99548699 ISSUES/YR: 24

Medical Journal of Kobe University
ISSN: 0075-6431
Med. J. Kobe Univ.
Kobe University School of Medicine, 5-1 Kusunokicho 7 chome, Chuo-ku, 650, Kobe, Japan, Tel: +81(78)3417451 ISSUES/YR: 4

Medical Journal of Malaysia
ISSN: 0300-5283
Med. J. Malays.
Malaysian Medical Association, 124 Jalan Pahang, P.O. Box S-20, 53000, Kuala Lumpur, Malaysia, Tel: +60(3)4420617, Fax: +60(3)4418187 ISSUES/YR: 4

Medical Journal of Minami Osaka Hospital
ISSN: 0540-1259
Med. J. Minami Osaka Hosp.
Minami Osaka Hospital, 18-18 Higashi-Kagaya 1-chome, Suminoe-ku, 559, Osaka, Japan, Tel: +81(6)6850221 ISSUES/YR: 3

Medical Journal of the Islamic Republic of Iran
ISSN: 1016-1430
Med. J. Islam. Repub. Iran
National Center for Scientific Research, P.O. Box 13145-478, 13158, Teheran, Iran, Tel: +98(21)6462778, Fax: +98(21)6462254 ISSUES/YR: 4

Medical Laser Application
ISSN: 1615-1615
Med. Laser Appl.
Urban und Fischer Verlag Jena, P.O. Box 100537, 07705, Jena, Germany, Tel: +49(3641)6263, Fax: +49(3641)626500 ISSUES/YR: 4

Medical Law International
ISSN: 0968-5332
Med. Law Int.
A B Academic Publishers, P.O. Box 42, OX26 6NW, Bicester, Oxfordshire, United Kingdom, Tel: +44(1869)320949 ISSUES/YR: 4

Medical Letter on Drugs and Therapeutics
ISSN: 0025-732X
Med. Lett. Drugs Ther.
Medical Letter Inc., 1000 Main Street, New Rochelle, NY 10801-7537, United States, Tel: +1(343)2585250 ISSUES/YR: 26

• **Medical Microbiology and Immunology**
ISSN: 0300-8584
Med. Microbiol. Immunol.
Springer Verlag, Tiergartenstrasse 17, D-69121, Heidelberg, Germany, Tel: +49(6221)4870, Fax: +49(6221)487366 ISSUES/YR: 4

Medical Mycology
ISSN: 1369-3786
Med. Mycol.
Taylor and Francis Ltd., 4 Park Square, Milton Park, OX14 4RN, Abingdon, Oxfordshire, United Kingdom, Tel: +44(20)70175541, Fax: +44(20)70175198 ISSUES/YR: 8

• **Medical Oncology**
ISSN: 1357-0560
Med. Oncol.
Humana Press Inc., 999 Riverview Drive, Suite 208, Totowa, NJ 07512-1165, United States, Tel: +1(973)2561699, Fax: +1(973)2568341 ISSUES/YR: 4

• **Medical Physics**
ISSN: 0094-2405
Med. Phys.
AAPM - American Association of Physicists in Medicine, One Physics Ellipse, College Park, MD 20740, United States, Tel: +1-3012093350, Fax: +1-3012090862 ISSUES/YR: 12

Medical Principles and Practice
ISSN: 1011-7571
Med. Princ. Pract.
S. Karger AG, Allschwilerstrasse 10, P.O. Box, CH-4009, Basel, Switzerland, Tel: +41(61)3061111, Fax: +41(61)3061234 ISSUES/YR: 6

Medical Problems of Performing Artists
ISSN: 0885-1158
Med. Probl. Perform. Artists
Science and Medicine Inc., P.O.Box 313, Narberth, PA 19072, United States, Tel: +1-610 660 9187, Fax: +1-610 660 0348 ISSUES/YR: 4

Medical Quality
ISSN: 1682-4059
Med. Qual.
VICER Publishing, P.O. Box 14, A-1097, Vienna, Austria, Tel: +43(676)9568085, Fax: +43(676)9568086 ISSUES/YR: 4

Medical Science Monitor
ISSN: 1234-1010
Med. Sci. Monit.
International Scientific Literature Inc., 1125 Willis Ave., Albertson, NY 11507, United States ISSUES/YR: 12

Medical Teacher
ISSN: 0142-159X
Med. Teach.
Taylor and Francis Ltd., 4 Park Square, Milton Park, OX14 4RN, Abingdon, Oxfordshire, United Kingdom, Tel: +44(20)70175541, Fax: +44(20)70175198 ISSUES/YR: 8

MedicaMundi
ISSN: 0025-7664
MedicaMundi
Philips Medical Systems Nederland, P.O. Box 10000, 5680 DA, Best, Netherlands, Tel: +31(3140)762972 ISSUES/YR: 3

Medicina
ISSN: 0025-7680
Medicina (Argentina)
Instituto de Investigaciones Medicas, Combatientes de Malvinas 3150, 1427, Buenos Aires, Argentina, Tel: +54(1)5236619, Fax: +54(1)5236619 ISSUES/YR: 6

Medicina
ISSN: 0025-7729
Medicina (Croatia)
Hrvatski lijecnicki zbor - Podruznica Rijeka, Kresimirova 52a/V, p.p. 227, 51001, Rijeka, Croatia, Tel: +385(51)334542, Fax: +385(51)334542 ISSUES/YR: 4

Medicina
ISSN: 0076-6046
Medicina (Brazil)
Faculdade de Medicina de Ribeirao Preto - U.S.P., Monte Alegre, CEP 14048-900, Ribeirao Preto, Brazil, Tel: +55(16)6331000, Fax: +55(16)6331144 ISSUES/YR: 4

Medicina Clinica
ISSN: 0025-7753
Med. Clin.
Ediciones Doyma, S.L., Travesera de Gracia 17-21, 08021, Barcelona, Spain, Tel: +34(93)2000711, Fax: +34(93)2091136 ISSUES/YR: 40

Medicina Clinica e Termale
ISSN: 1123-9395
Med. Clin. Termal.
Edizione Selecta Medica, Via Indipendenza 27, 27100, Pavia, Italy, Tel: +39(0382)539672 ISSUES/YR: 4

Medicina Cutanea Ibero-Latino-Americana
ISSN: 0210-5187
Med. Cutanea Ibero-Lat.-Am.
Catedra Universitario, Facultad de Medicina, Campus Universitairo Teatinos s/n, 29071, Malaga, Spain, Tel: +34-952131557, Fax: +34-952131512 ISSUES/YR: 6

Medicina del Lavoro
ISSN: 0025-7818
Med. Lav.
Mattioli 1885 S.p.A., Via Coduro 1/b, 43036, Fidenza (PR), Italy, Tel: +39(524)84547, Fax: +39(524)84751 ISSUES/YR: 6

Medicina dello Sport
ISSN: 0025-7826
Med. Sport
Edizioni Minerva Medica S.p.A., Corso Bramante 83-85, 10126, Turin, Italy, Tel: +39(011)678282, Fax: +39(011)674502 ISSUES/YR: 4

Medicina Intensiva
ISSN: 0210-5691
Med. Intensiva
Ediciones Doyma, S.L., Travesera de Gracia 17-21, 08021, Barcelona, Spain, Tel: +34(93)2000711, Fax: +34(93)2091136 ISSUES/YR: 9

Medicina Interna de Mexico
ISSN: 0186-4866
Med. Interna Mex.
Comunicaciones Cientificas Mexicanas S.A. de C.V., Calz. Gen. Anaya 330, Col. de Carmen Coyoacan, CP 04100, Mexico D.F., Mexico, Tel: +52(55)5-659-3863, Fax: +52(55)5-554-6054 ISSUES/YR: 6

Medicina Oral, Patologia Oral y Cirugia Bucal
ISSN: 1698-4447
Med. Oral Patol. Oral Cir. Bucal
Medicina Oral, Patologia Oral y Cirugia Bucal, C/ Daniel Balaciart No. 4 pta.17, 46020, Valencia, Spain, Fax: +34(6)3941854 ISSUES/YR: 5

Medicina Paliativa
ISSN: 1134-248X
Med. Paliativa
ARAN Ediciones S.A., Castello 128 - 1o, 28006, Madrid, Spain, Tel: +34(91)7820035, Fax: +34(91)5615787 ISSUES/YR: 4

Medicina Preventiva
ISSN: 1135-2841
Med. Prev.
Editores Medicos S.A. (EDIMSA), Gabriela Mistral 2, 28035, Madrid, Spain, Tel: +34(91)3860033, Fax: +34(91)3860366 ISSUES/YR: 3

• **Medicinal Chemistry**
ISSN: 1573-4064
Med. Chem.
Bentham Science Publishers B.V., P.O. Box 294, 1400 AG, Bussum, Netherlands, Tel: +31-356257290, Fax: +31-847511310 ISSUES/YR: 6

Medicinal Chemistry Research
ISSN: 1054-2523
Med. Chem. Res.
Birkhauser Boston, 675 Massachusettes Avenue, Cambridge, MA 02139, United States, Tel: +1(617)8762333 ISSUES/YR: 9

Medicinal Research Reviews
ISSN: 0198-6325
Med. Res. Rev.
John Wiley and Sons Inc., P.O.Box 18667, Newark, NJ 07191-8667, United States, Tel: +1(212)8506000, Fax: +1(212)8506012 ISSUES/YR: 6

● **Medicine**
ISSN: 0025-7974
Medicine (USA)
Lippincott Williams and Wilkins, 351 West Camden Street, Baltimore, MD 21201-2436, United States, Tel: +1(410)5284000 ISSUES/YR: 6

Medicine and Law
ISSN: 0723-1393
Med. Law
Yozmot Ltd, P.O. Box 56055, 61560, Tel Aviv, Israel, Tel: +972(3)5284851, Fax: +972(3)5285397 ISSUES/YR: 4

Medicine and Science in Sports and Exercise
ISSN: 0195-9131
Med. Sci. Sports Exerc.
American College of Sports Medicine, 401 West Michigan Street, Indianapolis, IN 46202, United States, Tel: +1(317)6379200 ISSUES/YR: 12

Medicine Today
ISSN: 1443-430X
Med. Today
Medicine Today Pty Ltd, P.O. Box 1473, NSW 2089, Neutral Bay, Australia, Tel: +61(2)99088577, Fax: +61(2)99087488 ISSUES/YR: 12

Medicinski Glasnik
ISSN: 1840-0132
Med. Glas.
Medical Association of Zenica-Doboj Canton, Bul.Kralja Tyrtka 4/11, 72000, Zenica, Bosnia and Herzegovina, Tel: +387-32414756, Fax: +387-32414756 ISSUES/YR: 2

Medico e Bambino
ISSN: 1591-3090
Med. Bambino
Edifarm S.p.A., Via P. Palagi 4, 20129, Milano, Italy, Tel: +39(02)29520315, Fax: +39(02)29520449 ISSUES/YR: 10

Medico-Legal Update
ISSN: 0971-720X
Med.-Leg. Update
World Informations Syndicate, 3/258-259 POBOX 9108, 110091, Trilok Puri Delhi, India, Tel: +91-11 22720006, Fax: +91-11 22720006 ISSUES/YR: 4

Medicus
ISSN: 1330-013X
Medicus
PLIVA d.d., Ulica grada Vukovara 49, 10000, Zagreb, Croatia, Tel: +385(1)6120837, Fax: +385(1)6111835 ISSUES/YR: 2

Medicus
ISSN: 1450-7994
Medicus
University of Kragujevac, Faculty of Science, P.O. Box 60, 34 000, Kragujevac, Serbia and Montenegro, Tel: +381-34-331876, Fax: +381-34-335040 ISSUES/YR: 2

Mediterranean Journal of Pacing and Electrophysiology
ISSN: 1128-4293
Mediterr. J. Pacing Electrophysiol.
Edizioni Luigi Pozzi S.r.l., Via Panama 68, 00198, Roma, Italy, Tel: +39(06)8553548, Fax: +39(06)8554105 ISSUES/YR: 4

Medizinische Ausbildung
ISSN: 1435-5116
Med. Ausbildung
Georg Thieme Verlag, Rudigerstrasse 14, D-70469, Stuttgart, Germany, Tel: +49(711)8931-321, Fax: +49(711)8931-422 ISSUES/YR: 2

Medizinische Genetik
ISSN: 0936-5931
Med. Genet.
Berufsverband Medizinische Genetik e.V., Goethestrasse 29, 80336, Munchen, Germany, Tel: +49(89)532516, Fax: +49(89)5438382 ISSUES/YR: 4

Medizinische Klinik
ISSN: 0723-5003
Med. Klin.
Urban und Vogel GmbH, Neumarkter Strasse 43, D-81673, Munchen, Germany, Tel: +49(89)43721300, Fax: +49(89)43721399 ISSUES/YR: 12

Medizinische Monatsschrift fur Pharmazeuten
ISSN: 0342-9601
Med. Monatsschr. Pharm.
Wissenschaftliche Verlagsgesellschaft mbH, P.O. Box 10 10 61, D-70009, Stuttgart, Germany, Tel: +49(711)2582234, Fax: +49(711)2582283 ISSUES/YR: 12

Medizinische Welt
ISSN: 0025-8512
Med. Welt
F.K. Schattauer Verlagsgesellschaft mbH, P.O. Box 10 45 45, D-70040, Stuttgart, Germany, Tel: +49(711)229870, Fax: +49(711)2298750 ISSUES/YR: 12

Medizintechnik
ISSN: 0344-9416
Medizintechnik
TUV-Verlag GmbH, Am Grauen Stein, D-51105, Koln, Germany, Tel: +49(221)8063511, Fax: +49(221)8063510 ISSUES/YR: 6

● **Melanoma Research**
ISSN: 0960-8931
Melanoma Res.
Lippincott Williams and Wilkins, 241 Borough High Street, SE1 1GB, London, United Kingdom, Tel: +44(207)9407500, Fax: +44(207)9407575 ISSUES/YR: 6

● **Memory**
ISSN: 0965-8211
Memory
Psychology Press Ltd, 4 Park Square, Milton Park, OX14 4RN, Abingdon, Oxfordshire, United Kingdom, Tel: +44(1235)828600, Fax: +44(1235)829000 ISSUES/YR: 8

Memory and Cognition
ISSN: 0090-502X
Mem. Cognit.
Psychonomic Society Inc., 1710 Fortview Road, Austin, TX 78704, United States, Tel: +1(512)4622442 ISSUES/YR: 8

Men's Health Journal
ISSN: 1473-3021
Men's Health J.
Medical Education Partnership, 53 Hargrave Rd, N19 5SH, London, United Kingdom, Tel: +44-207 561 5400, Fax: +44-207 7561 5401 ISSUES/YR: 4

Menopause
ISSN: 1072-3714
Menopause
Lippincott Williams and Wilkins, 530 Walnut Street, Philadelphia, PA 19106-3621, United States, Tel: +1(215)5218342, Fax: +1(215)5218326 ISSUES/YR: 6

Menopause Review
ISSN: 1272-9868
Menopause Rev.
Editions ESKA, 12 rue du 4 Septembre, 75002, Paris, France, Tel: +33(01)42865600, Fax: +33(01)42603269 ISSUES/YR: 4

Mental Health Aspects of Developmental Disabilities
ISSN: 1057-3291
Ment. Health Asp. Dev. Disabil.
Psych-Media Inc., P.O. Box 57, Bear Creek, NC 27207-0057, United States, Tel: +1(336)5813700, Fax: +1(336)5813766 ISSUES/YR: 4

Mental Health Services Research
ISSN: 1522-3434
Ment. Health Serv. Res.
Kluwer Academic Publishers, Van Godewijckstraat 30, 3311 GZ, Dordrecht, Netherlands, Tel: +31(78)6576000, Fax: +31(78)6576254 ISSUES/YR: 4

Mental Health, Religion and Culture
ISSN: 1367-4676
Ment. Health Relig. Cult.
Routledge, 4 Park Square, Milton Park, OX14 4RN, Abingdon, Oxfordshire, United Kingdom, Tel: +44(1235)828600, Fax: +44(1235)829000 ISSUES/YR: 5

- **Mental Retardation and Developmental Disabilities Research Reviews**
ISSN: 1080-4013
Ment. Retard. Dev. Disabil. Res. Rev.
Wiley-Liss Inc., 111 River Street, Hoboken, NJ 07030-5774, United States, Tel: +1(201)7486645, Fax: +1(201)7486021 ISSUES/YR: 4

MeReC Briefing
ISSN: not available yet
MeReC Brief.
National Prescribing Centre, 70 Pembroke Place, L69 3GF, Liverpool, United Kingdom, Tel: +44(151)7948146, Fax: +44(151)7948139 ISSUES/YR: 1

MeReC Bulletin
ISSN: 1465-5659
MeReC Bull.
National Prescribing Centre, 70 Pembroke Place, L69 3GF, Liverpool, United Kingdom, Tel: +44(151)7948146, Fax: +44(151)7948139 ISSUES/YR: 6

MeReC Extra
ISSN: not available yet
MeReC Extra
National Prescribing Centre, 70 Pembroke Place, L69 3GF, Liverpool, United Kingdom, Tel: +44(151)7948146, Fax: +44(151)7948139 ISSUES/YR: 1

Metabolic Brain Disease
ISSN: 0885-7490
Metab. Brain Dis.
Springer Science and Business Media Deutschland GmbH, Heidelberger Platz 3, 14197, Berlin, Germany, Tel: +1-30 827875282, Fax: +1-30 827875707 ISSUES/YR: 4

- **Metabolic Engineering**
ISSN: 1096-7176
Metab. Eng.
Academic Press Inc., 6277 Sea Harbor Drive, Orlando, FL 32887-4900, United States, Tel: +1(407)3454040, Fax: +1(407)3639661 ISSUES/YR: 6

- **Metabolic Syndrome and Related Disorders**
ISSN: 1540-4196
Metab. Syndr. Relat. Disord.
Mary Ann Liebert Inc., 140 Huguenot Street, New Rochelle, NY 10801-5215, United States, Tel: +1(914)7402100, Fax: +1(914)7402101 ISSUES/YR: 4

- **Metabolism: Clinical and Experimental**
ISSN: 0026-0495
Metab. Clin. Exp.
W.B. Saunders, Independence Square West, Philadelphia, PA 19106-3399, United States, Tel: +1(215)2387800, Fax: +1(215)2386445 ISSUES/YR: 12

- **Methods**
ISSN: 1046-2023
Methods
Academic Press Inc., 6277 Sea Harbor Drive, Orlando, FL 32887-4900, United States, Tel: +1(407)3454040, Fax: +1(407)3639661 ISSUES/YR: 12

Methods and Findings in Experimental and Clinical Pharmacology
ISSN: 0379-0355
Methods Find. Exp. Clin. Pharmacol.
Prous Science, P.O. Box 540, 08080, Barcelona, Spain, Tel: +34(93)4592220, Fax: +34(93)4581535 ISSUES/YR: 10

Methods in Cell Science
ISSN: 1381-5741
Methods Cell Sci.
Kluwer Academic Publishers, Van Godewijckstraat 30, 3311 GZ, Dordrecht, Netherlands, Tel: +31(78)6576000, Fax: +31(78)6576254 ISSUES/YR: 4

- **Methods in Enzymology**
ISSN: 0076-6879
Methods Enzymol.
Academic Press Inc., 525 B Street, Suite 1900, San Diego, CA 92101-4495, United States ISSUES/YR: 1

- **Methods of Information in Medicine**
ISSN: 0026-1270
Methods Inf. Med.
F.K. Schattauer Verlagsgesellschaft mbH, P.O. Box 10 45 45, D-70040, Stuttgart, Germany, Tel: +49(711)229870, Fax: +49(711)2298750 ISSUES/YR: 5

- **Microbes and Infection**
ISSN: 1286-4579
Microbes Infect.
Editions Scientifiques et Medicales Elsevier, 23 rue Linois, 75724, Paris Cedex 15, France, Tel: +33(1)45589826, Fax: +33(1)45589428 ISSUES/YR: 15

Microbial Cell Factories
ISSN: 1475-2859
Microb. Cell Fact.
BioMed Central Ltd., 34 - 42 Cleveland Street, W1T 4LB, London, United Kingdom, Tel: +44(207)3230323, Fax: +44(207)6319926 ISSUES/YR: 1

- **Microbial Drug Resistance**
ISSN: 1076-6294
Microb. Drug Resist.
Mary Ann Liebert Inc., 140 Huguenot Street, New Rochelle, NY 10801-5215, United States, Tel: +1(914)7402100, Fax: +1(914)7402101 ISSUES/YR: 4

- **Microbial Ecology in Health and Disease**
ISSN: 0891-060X
Microb. Ecol. Health Dis.
Taylor and Francis Ltd., 4 Park Square, Milton Park, OX14 4RN, Abingdon, Oxfordshire, United Kingdom, Tel: +44(20)70175541, Fax: +44(20)70175198 ISSUES/YR: 4

- **Microbial Pathogenesis**
ISSN: 0882-4010
Microb. Pathog.
Academic Press, 24-28 Oval Road, NW1 7DX, London, United Kingdom, Tel: +44(207)2674466 ISSUES/YR: 12

- **Microbiology**
ISSN: 1350-0872
Microbiology
Society for General Microbiology, Basingstoke Road, Spencers Wood, RG7 1AE, Reading, Berkshire, United Kingdom, Tel: +44(118)9885577, Fax: +44(118)9885656 ISSUES/YR: 12

Microbiology and Immunology
ISSN: 0385-5600
Microbiol. Immunol.
Center for Academic Publications Japan, 4-16, Yayoi 2-chome, Bunkyo-ku, 113-0032, Tokyo, Japan, Tel: +81(3)38175830, Fax: +81(3)58175820 ISSUES/YR: 12

Microbiology and Molecular Biology Reviews
ISSN: 1092-2172
Microbiol. Mol. Biol. Rev.
American Society for Microbiology, 1752 N Street N.W., Washington, DC 20036-2904, United States, Tel: +1(202)942-9244, Fax: +1(202)942-9355 ISSUES/YR: 4

- **Microcirculation**
ISSN: 1073-9688
Microcirculation
Taylor and Francis Inc., 325 Chestnut St, Suite 800, Philadelphia PA, PA 19106, United States, Tel: +1(800)3541420, Fax: +1(215)6258914 ISSUES/YR: 8

- **Microscopy Research and Technique**
ISSN: 1059-910X
Microsc. Res. Tech.
Wiley-Liss Inc., 111 River Street, Hoboken, NJ 07030-5774, United States, Tel: +1(201)7486645, Fax: +1(201)7486021 ISSUES/YR: 18

- **Microsurgery**
ISSN: 0738-1085
Microsurgery
Wiley-Liss Inc., 111 River Street, Hoboken, NJ 07030-5774, United States, Tel: +1(201)7486645, Fax: +1(201)7486021 ISSUES/YR: 8

- **Microvascular Research**
ISSN: 0026-2862
Microvasc. Res.
Academic Press Inc., 6277 Sea Harbor Drive, Orlando, FL 32887-4900, United States, Tel: +1(407)3454040, Fax: +1(407)3639661 ISSUES/YR: 6

Mid-Taiwan Journal of Medicine
ISSN: 1029-3507
Mid-Taiwan J. Med.
Mid-Taiwan Journal of Medicine, No. 91, Hsueh-Shih Road, 404, Taichung, Taiwan, Tel: +886(4)2205366 5901/5902, Fax: +886(4)22035718 ISSUES/YR: 4

Middle East Fertility Society Journal
ISSN: 1110-5690
Middle East Fertil. Soc. J.
Middle East Fertility Society, 10 Geziret El Arab Street, Mohandessin, 12411, Cairo, Egypt, Tel: +20(2)3498488, Fax: +20(2)3516649 ISSUES/YR: 3

Middle East Journal of Emergency Medicine
ISSN: 1729-6455
Middle East J. Emerg. Med.
Ministry of Public Health and Hamad Medical, PO Box 3050, Doha, Qatar, Tel: +974-4392050, Fax: +974-4392179 ISSUES/YR: 2

Middle East Paediatrics
ISSN: 1368-5937
Middle East Paediatr.
PHARMedia International, 6 tobin Close Livingstone Park, KT19 8AE, Epsom, Surrey, United Kingdom, Tel: +44-1372745187 ISSUES/YR: 4

Mie Medical Journal
ISSN: 0026-3532
Mie Med. J.
Library of Mie University, 2-174 Edobashi, 514, Tsu City, Japan, Tel: +81(592)321211 ISSUES/YR: 3

Mikologia Lekarska
ISSN: 1232-986X
Mikol. Lek.
Cornetis, ul. Dlugosza 2-6, 51-162, Wroclaw, Poland, Tel: +48-071 326 3101, Fax: +48-071 325 2803 ISSUES/YR: 4

Military Medicine
ISSN: 0026-4075
Mil. Med.
Association of Military Surgeons of the US, 9320 Old Georgetown Road, Bethesda, MD 20814, United States, Tel: +1(301)8978800, Fax: +1(301)5305446 ISSUES/YR: 12

Minerva Anestesiologica
ISSN: 0375-9393
Minerva Anestesiol.
Edizioni Minerva Medica S.p.A., Corso Bramante 83-85, 10126, Turin, Italy, Tel: +39(011)678282, Fax: +39(011)674502 ISSUES/YR: 12

Minerva Biotecnologica
ISSN: 1120-4826
Minerva Biotecnol.
Edizioni Minerva Medica S.p.A., Corso Bramante 83-85, 10126, Turin, Italy, Tel: +39(011)678282, Fax: +39(011)674502 ISSUES/YR: 4

Minerva Cardioangiologica
ISSN: 0026-4725
Minerva Cardioangiol.
Edizioni Minerva Medica S.p.A., Corso Bramante 83-85, 10126, Turin, Italy, Tel: +39(011)678282, Fax: +39(011)674502 ISSUES/YR: 6

Minerva Chirurgica
ISSN: 0026-4733
Minerva Chir.
Edizioni Minerva Medica S.p.A., Corso Bramante 83-85, 10126, Turin, Italy, Tel: +39(011)678282, Fax: +39(011)674502 ISSUES/YR: 6

Minerva Gastroenterologica e Dietologica
ISSN: 0026-4776
Minerva Gastroenterol. Dietol.
Edizioni Minerva Medica S.p.A., Corso Bramante 83-85, 10126, Turin, Italy, Tel: +39(011)678282, Fax: +39(011)674502 ISSUES/YR: 4

Minerva Ginecologica
ISSN: 0026-4784
Minerva Ginecol.
Edizioni Minerva Medica S.p.A., Corso Bramante 83-85, 10126, Turin, Italy, Tel: +39(011)678282, Fax: +39(011)674502 ISSUES/YR: 6

Minerva Medica
ISSN: 0026-4806
Minerva Med.
Edizioni Minerva Medica S.p.A., Corso Bramante 83-85, 10126, Turin, Italy, Tel: +39(011)678282, Fax: +39(011)674502 ISSUES/YR: 6

Minerva Ortopedica e Traumatologica
ISSN: 0026-4911
Minerva Ortop. Traumatol.
Edizioni Minerva Medica S.p.A., Corso Bramante 83-85, 10126, Turin, Italy, Tel: +39(011)678282, Fax: +39(011)674502 ISSUES/YR: 6

Minerva Pediatrica
ISSN: 0026-4946
Minerva Pediatr.
Edizioni Minerva Medica S.p.A., Corso Bramante 83-85, 10126, Turin, Italy, Tel: +39(011)678282, Fax: +39(011)674502 ISSUES/YR: 6

Minerva Pneumologica
ISSN: 0026-4954
Minerva Pneumol.
Edizioni Minerva Medica S.p.A., Corso Bramante 83-85, 10126, Turin, Italy, Tel: +39(011)678282, Fax: +39(011)674502 ISSUES/YR: 4

Minerva Psichiatrica
ISSN: 0374-9320
Minerva Psichiatr.
Edizioni Minerva Medica S.p.A., Corso Bramante 83-85, 10126, Turin, Italy, Tel: +39(011)678282, Fax: +39(011)674502 ISSUES/YR: 4

Minerva Urologica e Nefrologica
ISSN: 0393-2249
Minerva Urol. Nefrol.
Edizioni Minerva Medica S.p.A., Corso Bramante 83-85, 10126, Turin, Italy, Tel: +39(011)678282, Fax: +39(011)674502 ISSUES/YR: 4

Mini-Reviews in Medicinal Chemistry
ISSN: 1389-5575
Mini-Rev. Med. Chem.
Bentham Science Publishers B.V., P.O. Box 294, 1400 AG, Bussum, Netherlands, Tel: +31-356257290, Fax: +31-847511310 ISSUES/YR: 12

Minimal Invasive Chirurgie
ISSN: 0941-455X
Minimal Invasive Chir.
Urban und Fischer Verlag Jena, P.O. Box 100537, 07705, Jena, Germany, Tel: +49(3641)6263, Fax: +49(3641)626500 ISSUES/YR: 4

Minimally Invasive Neurosurgery
ISSN: 0946-7211
Minimally Invasive Neurosurg.
Georg Thieme Verlag, Rudigerstrasse 14, D-70469, Stuttgart, Germany, Tel: +49(711)8931-321, Fax: +49(711)8931-422 ISSUES/YR: 4

• **Minimally Invasive Therapy and Allied Technologies**
ISSN: 1364-5706
Minimally Invasive Ther. Allied Technol.
Taylor and Francis Ltd., 4 Park Square, Milton Park, OX14 4RN, Abingdon, Oxfordshire, United Kingdom, Tel: +44(20)70175541, Fax: +44(20)70175198 ISSUES/YR: 6

• **Mitochondrion**
ISSN: 1567-7249
Mitochondrion
Elsevier, P.O. Box 211, 1000 AE, Amsterdam, Netherlands, Tel: +31(20)4853757, Fax: +31(20)4853432 ISSUES/YR: 6

MMW-Fortschritte der Medizin
ISSN: 1438-3276
MMW-Fortschr. Med.
Urban und Vogel GmbH, Neumarkter Strasse 43, D-81673, Munchen, Germany, Tel: +49(89)43721300, Fax: +49(89)43721399 ISSUES/YR: 52

Modern Aspects in Sepsis
ISSN: not available yet
Mod. Asp. Sepsis
VICER Publishing, P.O. Box 14, A-1097, Vienna, Austria, Tel: +43(676)9568085, Fax: +43(676)9568086 ISSUES/YR: 4

Modern Aspects of Immunobiology
ISSN: not available yet
Mod. Asp. Immunobiol.
IA-Verlag, St.-Martins-Gasse 5, 79108, Freiburg, Germany, Tel: +49(7665)912691, Fax: +49(7665)912691 ISSUES/YR: 4

Current

Modern Medicine
ISSN: 0759-240X
Mod. Med.
HCP-Healthcare Publications Ltd, 11 Boumboulinas straat, Nicosia, Cyprus, Tel: +357(2 245)8500, Fax: +357(2 275)1648 ISSUES/YR: 8

• **Modern Pathology**
ISSN: 0893-3952
Mod. Pathol.
Nature Publishing Group, Houndmills, RG21 6XS, Basingstoke, Hampshire, United Kingdom, Tel: +44(1256)329242, Fax: +44(1256)810526 ISSUES/YR: 12

• **Modern Rheumatology**
ISSN: 1439-7595
Mod. Rheumatol.
Springer-Verlag Tokyo, 3-13, Hongo 3-chome, Bunkyo-ku, 113-8480, Tokyo, Japan, Tel: +81(3)38180861, Fax: +81(3)38180864 ISSUES/YR: 4

• **Molecular and Biochemical Parasitology**
ISSN: 0166-6851
Mol. Biochem. Parasitol.
Elsevier, P.O. Box 211, 1000 AE, Amsterdam, Netherlands, Tel: +31(20)4853757, Fax: +31(20)4853432 ISSUES/YR: 12

Molecular and Cellular Biochemistry
ISSN: 0300-8177
Mol. Cell. Biochem.
Kluwer Academic Publishers, Van Godewijckstraat 30, 3311 GZ, Dordrecht, Netherlands, Tel: +31(78)6576000, Fax: +31(78)6576254 ISSUES/YR: 24

• **Molecular and Cellular Biology**
ISSN: 0270-7306
Mol. Cell. Biol.
American Society for Microbiology, 1752 N Street N.W., Washington, DC 20036-2904, United States, Tel: +1(202)942-9244, Fax: +1(202)942-9355 ISSUES/YR: 24

• **Molecular and Cellular Endocrinology**
ISSN: 0303-7207
Mol. Cell. Endocrinol.
Elsevier Ireland Ltd, P.O. Box 85, Limerick, Ireland, Tel: +353(61)471944, Fax: +353(61)472144 ISSUES/YR: 20

• **Molecular and Cellular Neuroscience**
ISSN: 1044-7431
Mol. Cell. Neurosci.
Academic Press Inc., 6277 Sea Harbor Drive, Orlando, FL 32887-4900, United States, Tel: +1(407)3454040, Fax: +1(407)3639661 ISSUES/YR: 12

• **Molecular and Cellular Probes**
ISSN: 0890-8508
Mol. Cell. Probes
Academic Press, 24-28 Oval Road, NW1 7DX, London, United Kingdom, Tel: +44(207)2674466 ISSUES/YR: 6

• **Molecular and Cellular Proteomics**
ISSN: 1535-9476
Mol. Cell. Proteomics
American Society for Biochemistry and Molecular Biology Inc., 9650 Rockville Pike, Bethesda, MD 20814, United States, Tel: +1-301 530 7145, Fax: +1-301 571 1824 ISSUES/YR: 12

Molecular Aspects of Medicine
ISSN: 0098-2997
Mol. Asp. Med.
Elsevier Ltd, Langford Lane, Kidlington, OX5 1GB, Oxford, United Kingdom, Tel: +44(1865)843000, Fax: +44(1865)843010 ISSUES/YR: 6

Molecular Biology and Evolution
ISSN: 0737-4038
Mol. Biol. Evol.
Oxford University Press, 2001 Evans Road, Cary, NC 27513, United States, Tel: +1(919)6770977, Fax: +1(919)6771714 ISSUES/YR: 12

• **Molecular Biology of the Cell**
ISSN: 1059-1524
Mol. Biol. Cell
American Society for Cell Biology, 8120 Woodmont Avenue, Suite 750, Bethesda, MD 20814-2755, United States, Tel: +1(301)3479300, Fax: +1(301)3479310 ISSUES/YR: 12

Molecular Biology Reports
ISSN: 0301-4851
Mol. Biol. Rep.
Kluwer Academic Publishers, Van Godewijckstraat 30, 3311 GZ, Dordrecht, Netherlands, Tel: +31(78)6576000, Fax: +31(78)6576254 ISSUES/YR: 4

Molecular Biotechnology
ISSN: 1073-6085
Mol. Biotechnol.
Humana Press Inc., 999 Riverview Drive, Suite 208, Totowa, NJ 07512-1165, United States, Tel: +1(973)2561699, Fax: +1(973)2568341 ISSUES/YR: 9

• **Molecular Brain Research**
ISSN: 0169-328X
Mol. Brain Res.
Elsevier, P.O. Box 211, 1000 AE, Amsterdam, Netherlands, Tel: +31(20)4853757, Fax: +31(20)4853432 ISSUES/YR: 24

Molecular Cancer
ISSN: 1476-4598
Mol. Cancer
BioMed Central Ltd., 34 - 42 Cleveland Street, W1T 4LB, London, United Kingdom, Tel: +44(207)3230323, Fax: +44(207)6319926 ISSUES/YR: 1

• **Molecular Cancer Research**
ISSN: 1541-7786
Mol. Cancer Res.
American Association for Cancer Research Inc., 150 South Independence Mall West, Philadelphia, PA 19106-3483, United States, Tel: +1(215)4409300, Fax: +1(215)4409354 ISSUES/YR: 12

• **Molecular Cancer Therapeutics**
ISSN: 1535-7163
Mol. Cancer Ther.
American Association for Cancer Research Inc., 150 South Independence Mall West, Philadelphia, PA 19106-3483, United States, Tel: +1(215)4409300, Fax: +1(215)4409354 ISSUES/YR: 12

• **Molecular Carcinogenesis**
ISSN: 0899-1987
Mol. Carcinog.
Wiley-Liss Inc., 111 River Street, Hoboken, NJ 07030-5774, United States, Tel: +1(201)7486645, Fax: +1(201)7486021 ISSUES/YR: 12

Molecular Cell
ISSN: 1097-2765
Mol. Cell
Cell Press, 1100 Massachusetts Avenue, Cambridge, MA 02138-9957, United States, Tel: +1(617)6617057, Fax: +1(617)6617061 ISSUES/YR: 24

• **Molecular Diagnosis and Therapy**
ISSN: 1177-1062
Mol. Diagn. Ther.
Adis International Ltd, 41 Centorian Drive, Private Bag 65901, Mairangi Bay, 1311, Auckland 10, New Zealand, Tel: +64(9)4770700, Fax: +64(9)4770766 ISSUES/YR: 6

• **Molecular Diversity**
ISSN: 1381-1991
Mol. Diversity
Kluwer Academic Publishers, Van Godewijckstraat 30, 3311 GZ, Dordrecht, Netherlands, Tel: +31(78)6576000, Fax: +31(78)6576254 ISSUES/YR: 4

• **Molecular Endocrinology**
ISSN: 0888-8809
Mol. Endocrinol.
Endocrine Society, 8401 Connecticut Ave. Suite 900, Chevy Chase, MD 20815, United States, Tel: +1-888-363-6274 ISSUES/YR: 12

• **Molecular Genetics and Genomics**
ISSN: 1617-4615
Mol. Genet. Genomics
Springer Verlag, Tiergartenstrasse 17, D-69121, Heidelberg, Germany, Tel: +49(6221)4870, Fax: +49(6221)487366 ISSUES/YR: 12

- **Molecular Genetics and Metabolism**
ISSN: 1096-7192
Mol. Genet. Metab.
Academic Press Inc., 6277 Sea Harbor Drive, Orlando, FL 32887-4900, United States, Tel: +1(407)3454040, Fax: +1(407)3639661 ISSUES/YR: 12
- **Molecular Human Reproduction**
ISSN: 1360-9947
Mol. Hum. Reprod.
Oxford University Press, Great Clarendon Street, OX2 6DP, Oxford, United Kingdom, Tel: +44(1865)353907, Fax: +44(1865)353486 ISSUES/YR: 12
- **Molecular Imaging and Biology**
ISSN: 1536-1632
Mol. Imaging Biol.
Springer Verlag, Tiergartenstrasse 17, D-69121, Heidelberg, Germany, Tel: +1(6221)4870, Fax: +1(6221)487366 ISSUES/YR: 6
- **Molecular Immunology**
ISSN: 0161-5890
Mol. Immunol.
Elsevier Ltd, Langford Lane, Kidlington, OX5 1GB, Oxford, United Kingdom, Tel: +44(1865)843000, Fax: +44(1865)843010 ISSUES/YR: 12

Molecular Interventions
ISSN: 1534-0384
Mol. Interventions
American Society for Pharmacology and Experimental Therapy, 9650 Rockville Pike, Bethesda, MD 20814, United States, Tel: +1(301)5307024, Fax: +1(301)5715728 ISSUES/YR: 6
- **Molecular Medicine**
ISSN: 1076-1551
Mol. Med.
Uninversity of Michigan, 1150 West Medical Center Drive, Ann Arbor, MI 48109-0640, United States, Tel: +1-734 764 0030, Fax: +1-734 764 0101 ISSUES/YR: 12
- **Molecular Membrane Biology**
ISSN: 0968-7688
Mol. Membr. Biol.
Taylor and Francis Ltd., 4 Park Square, Milton Park, OX14 4RN, Abingdon, Oxfordshire, United Kingdom, Tel: +44(20)70175541, Fax: +44(20)70175198 ISSUES/YR: 6
- **Molecular Microbiology**
ISSN: 0950-382X
Mol. Microbiol.
Blackwell Publishing Ltd, 9600 Garsington Road, OX4 2XG, Oxford, United Kingdom, Tel: +44(1865)776868, Fax: +44(1865)714591 ISSUES/YR: 23

Molecular Neurobiology
ISSN: 0893-7648
Mol. Neurobiol.
Humana Press Inc., 999 Riverview Drive, Suite 208, Totowa, NJ 07512-1165, United States, Tel: +1(973)2561699, Fax: +1(973)2568341 ISSUES/YR: 6
- **Molecular Pain**
ISSN: 1744-8069
Mol. Pain
BioMed Central Ltd., 34 - 42 Cleveland Street, W1T 4LB, London, United Kingdom, Tel: +44(207)3230323, Fax: +44(207)6319926 ISSUES/YR: 1
- **Molecular Pharmaceutics**
ISSN: 1543-8384
Mol. Pharm.
American Chemical Society, 2540 Olentangy River Road, P.O. Box 3337, Columbus, OH 43210-3337, United States, Tel: +1(614)4473776, Fax: +1(614)4473671 ISSUES/YR: 6
- **Molecular Pharmacology**
ISSN: 0026-895X
Mol. Pharmacol.
American Society for Pharmacology and Experimental Therapy, 9650 Rockville Pike, Bethesda, MD 20814, United States, Tel: +1(301)5307024, Fax: +1(301)5715728 ISSUES/YR: 12
- **Molecular Psychiatry**
ISSN: 1359-4184
Mol. Psychiatry
Nature Publishing Group, Houndmills, RG21 6XS, Basingstoke, Hampshire, United Kingdom, Tel: +44(1256)329242, Fax: +44(1256)810526 ISSUES/YR: 12
- **Molecular Reproduction and Development**
ISSN: 1040-452X
Mol. Reprod. Dev.
Wiley-Liss Inc., 111 River Street, Hoboken, NJ 07030-5774, United States, Tel: +1(201)7486645, Fax: +1(201)7486021 ISSUES/YR: 12
- **Molecular Systems Biology**
ISSN: 1744-4292
Mol. Syst. Biol.
Nature Publishing Group, Houndmills, RG21 6XS, Basingstoke, Hampshire, United Kingdom, Tel: +44(1256)329242, Fax: +44(1256)810526 ISSUES/YR: 1

Molecular Therapy
ISSN: 1525-0016
Mol. Ther.
Academic Press Inc., 6277 Sea Harbor Drive, Orlando, FL 32887-4900, United States, Tel: +1(407)3454040, Fax: +1(407)3639661 ISSUES/YR: 12
- **Molecular Vision**
ISSN: 1090-0535
Mol. Vision
Molecular Vision, 1327 Clifton Road, N.E., Atlanta, GA 30322, United States ISSUES/YR: 1

Molecules and Cells
ISSN: 1016-8478
Mol. Cells
Korean Society for Molecular and Cellular Biology, 635-4 Yeogsam-dong,Kangnam-gu, 135-703, Seoul, South Korea, Tel: +82-2 568 4490, Fax: +82-2 568 4544 ISSUES/YR: 6

Monaldi Archives for Chest Disease - Cardiac Series
ISSN: 1122-0643
Monaldi Arch. Chest Dis. Card. Ser.
Fondazione Salvatore Maugeri, Via Ferrata 8, 27100, Pavia, Italy, Tel: +39-0382 592 990, Fax: +39-0382 592 514 ISSUES/YR: 4

Monaldi Archives for Chest Disease - Pulmonary Series
ISSN: 1122-0643
Monaldi Arch. Chest Dis. Pulm. Ser.
PI-ME Tipografia Editrice S.r.l., Via Vigentina 136, 27100, Pavia, Italy, Tel: +39(0382)572169, Fax: +39(0382)572102 ISSUES/YR: 4

Monatsschrift fur Kinderheilkunde
ISSN: 0026-9298
Monatsschr. Kinderheilkd.
Springer Verlag, Tiergartenstrasse 17, D-69121, Heidelberg, Germany, Tel: +49(6221)4870, Fax: +49(6221)487366 ISSUES/YR: 12

Morphologie
ISSN: 1286-0115
Morphologie
Masson, 21 Rue Camille Desmoulins, 92789, Issy les Moulineaux Cedex 9, France, Tel: +33(1)73281634, Fax: +33(1)73281649 ISSUES/YR: 4

Motor Control
ISSN: 1087-1640
Mot. Control
Human Kinetics Publishers Inc., 1607 N. Market Street, Champaign, IL 61820-2200, United States, Tel: +1(217)3515076, Fax: +1(217)3511549 ISSUES/YR: 4

Mount Sinai Journal of Medicine
ISSN: 0027-2507
Mt. Sinai J. Med.
Mount Sinai Medical Center, One Gustave L. Levy Place, Box 1094, New York, NY 10029-6574, United States, Tel: +1(212)2416108, Fax: +1(212)7226386 ISSUES/YR: 6
- **Movement Disorders**
ISSN: 0885-3185
Mov. Disord.
John Wiley and Sons Inc., P.O.Box 18667, Newark, NJ 07191-8667, United States, Tel: +1(212)8506000, Fax: +1(212)8506012 ISSUES/YR: 12

Multiple Sclerosis
ISSN: 1352-4585
Mult. Scler.
Arnold, 338 Euston Road, NW1 3BH, London, United Kingdom, Tel: +44(20)78736000, Fax: +44(20)78736376
ISSUES/YR: 6

- **Muscle and Nerve**
ISSN: 0148-639X
Muscle Nerve
John Wiley and Sons Inc., P.O.Box 18667, Newark, NJ 07191-8667, United States, Tel: +1(212)8506000, Fax: +1(212)8506012 ISSUES/YR: 12

- **Mutagenesis**
ISSN: 0267-8357
Mutagenesis
Oxford University Press, Great Clarendon Street, OX2 6DP, Oxford, United Kingdom, Tel: +44(1865)353907, Fax: +44(1865)353486 ISSUES/YR: 6

- **Mutation Research - Fundamental and Molecular Mechanisms of Mutagenesis**
ISSN: 0027-5107
Mutat. Res. Fundam. Mol. Mech. Mutagen.
Elsevier, P.O. Box 211, 1000 AE, Amsterdam, Netherlands, Tel: +31(20)4853757, Fax: +31(20)4853432 ISSUES/YR: 24

- **Mutation Research - Genetic Toxicology and Environmental Mutagenesis**
ISSN: 1383-5718
Mutat. Res. Genet. Toxicol. Environ. Mutagen.
Elsevier, P.O. Box 211, 1000 AE, Amsterdam, Netherlands, Tel: +31(20)4853757, Fax: +31(20)4853432 ISSUES/YR: 18

- **Mutation Research - Reviews in Mutation Research**
ISSN: 1383-5742
Mutat. Res. Rev. Mutat. Res.
Elsevier, P.O. Box 211, 1000 AE, Amsterdam, Netherlands, Tel: +31(20)4853757, Fax: +31(20)4853432 ISSUES/YR: 6

- **Mycoses**
ISSN: 0933-7407
Mycoses
Blackwell Publishing Ltd, 9600 Garsington Road, OX4 2XG, Oxford, United Kingdom, Tel: +44(1865)776868, Fax: +44(1865)714591 ISSUES/YR: 6

- **Mycoses, Supplement**
ISSN: 0943-7312
Mycoses Suppl.
Blackwell Publishing Ltd, 9600 Garsington Road, OX4 2XG, Oxford, United Kingdom, Tel: +44(1865)776868, Fax: +44(1865)714591 ISSUES/YR: 1

NAD Publication
ISSN: 0359-7024
NAD Publ.
Nordic Council for Alcohol and Drug Research (NAD), Annankatu 29 A 23, SF-00100, Helsinki, Finland, Tel: +358(0)6948082, Fax: +358(0)6949081 ISSUES/YR: 1

Nadcisnienie Tetnicze
ISSN: 1428-5851
Nadcisnienie Tetnicze
Via Medica, Ul. Swietokrzyska 73, 80-180, Gdansk, Poland, Tel: +48(58)3209494, Fax: +48(58)3209460 ISSUES/YR: 4

Nano Today
ISSN: 1748-0132
Nano Today
Elsevier, P.O. Box 211, 1000 AE, Amsterdam, Netherlands, Tel: +31(20)4853757, Fax: +31(20)4853432 ISSUES/YR: 4

Nanomedicine: Nanotechnology, Biology, and Medicine
ISSN: 1549-9634
Nanomed. Nanotechnol. Biol. Med.
Elsevier Inc., 360 Park Avenue South, New York, NY 10010, United States, Tel: +1(212)6333950, Fax: +1(212)6333990
ISSUES/YR: 4

- **Nanotoxicology**
ISSN: 1743-5390
Nanotoxicology
Taylor and Francis Ltd., 4 Park Square, Milton Park, OX14 4RN, Abingdon, Oxfordshire, United Kingdom, Tel: +44(20)70175541, Fax: +44(20)70175198 ISSUES/YR: 4

Nascer e Crescer
ISSN: 0872-0754
Nascer Crescer
Hospital Central Especializado de Crincias Maria Pia, Rua da Boavista 827, 4000, Porto, Portugal, Tel: +351(2)6099861
ISSUES/YR: 4

National Medical Journal of China
ISSN: 0376-2491
Nat. Med. J. China
Chinese Medical Association, 42 Dongsi Xidajie, P.O. Box 2258, 100710, Beijing, China, Tel: +86(10)65244339, Fax: +86(10)65134886 ISSUES/YR: 49

National Medical Journal of India
ISSN: 0970-258X
Natl. Med. J. India
All India Institute of Medical Sciences, Ansari Nagar, 110 029, New Delhi, India, Tel: +91(11)6863002, Fax: +91(11)6862663 ISSUES/YR: 6

Natural Computing
ISSN: 1567-7818
Nat. Comput.
Springer Science and Business Media Deutschland GmbH, Heidelberger Platz 3, 14197, Berlin, Germany, Tel: +1-30 827875282, Fax: +1-30 827875707 ISSUES/YR: 4

Natural Medicines
ISSN: 1340-3443
Nat. Med.
Springer-Verlag Tokyo, 3-13, Hongo 3-chome, Bunkyo-ku, 113-8480, Tokyo, Japan, Tel: +81(3)38180861, Fax: +81(3)38180864 ISSUES/YR: 6

Natural Product Research
ISSN: 1478-6427
Nat. Prod. Res.
Taylor and Francis Ltd., 4 Park Square, Milton Park, OX14 4RN, Abingdon, Oxfordshire, United Kingdom, Tel: +44(20)70175541, Fax: +44(20)70175198 ISSUES/YR: 14

Natural Product Sciences
ISSN: 1226-3907
Nat. Prod. Sci.
Korean Society of Pharmacognosy, 28 Yeongun-dong, Jongro-ku, 110-460, Seoul, South Korea, Tel: +82(2)7408923, Fax: +82(2)7667818 ISSUES/YR: 6

- **Nature**
ISSN: 0028-0836
Nature
Nature Publishing Group, Houndmills, RG21 6XS, Basingstoke, Hampshire, United Kingdom, Tel: +44(1256)329242, Fax: +44(1256)810526 ISSUES/YR: 51

- **Nature Biotechnology**
ISSN: 1087-0156
Nat. Biotechnol.
Nature Publishing Group, 345 Park Avenue South, New York, NY 10010-1707, United States, Tel: +1(212)7269200, Fax: +1(212)6960052 ISSUES/YR: 12

- **Nature Cell Biology**
ISSN: 1465-7392
Nature Cell Biol.
Nature Publishing Group, Houndmills, RG21 6XS, Basingstoke, Hampshire, United Kingdom, Tel: +44(1256)329242, Fax: +44(1256)810526 ISSUES/YR: 12

- **Nature Chemical Biology**
ISSN: 1552-4450
Nat. Chem. Biol.
Nature Publishing Group, Houndmills, RG21 6XS, Basingstoke, Hampshire, United Kingdom, Tel: +44(1256)329242, Fax: +44(1256)810526 ISSUES/YR: 12

- **Nature Clinical Practice Cardiovascular Medicine**
ISSN: 1743-4297
Nat. Clin. Pract. Cardiovasc. Med.
Nature Publishing Group, Houndmills, RG21 6XS, Basingstoke, Hampshire, United Kingdom, Tel: +44(1256)329242, Fax: +44(1256)810526 ISSUES/YR: 12

- **Nature Clinical Practice Endocrinology and Metabolism**
ISSN: 1745-8366
Nat. Clin. Pract. Endocrinol. Metab.
Nature Publishing Group, Houndmills, RG21 6XS, Basingstoke, Hampshire, United Kingdom, Tel: +44(1256)329242, Fax: +44(1256)810526 ISSUES/YR: 12

- **Nature Clinical Practice Gastroenterology and Hepatology**
  ISSN: 1743-4378
  Nat. Clin. Pract. Gastroenterol. Hepatol.
  Nature Publishing Group, Houndmills, RG21 6XS, Basingstoke, Hampshire, United Kingdom, Tel: +44(1256)329242, Fax: +44(1256)810526 ISSUES/YR: 12
- **Nature Clinical Practice Nephrology**
  ISSN: 1745-8323
  Nat. Clin. Pract. Nephrol.
  Nature Publishing Group, Houndmills, RG21 6XS, Basingstoke, Hampshire, United Kingdom, Tel: +44(1256)329242, Fax: +44(1256)810526 ISSUES/YR: 12
- **Nature Clinical Practice Neurology**
  ISSN: 1745-834X
  Nat. Clin. Pract. Neurol.
  Nature Publishing Group, Houndmills, RG21 6XS, Basingstoke, Hampshire, United Kingdom, Tel: +44(1256)329242, Fax: +44(1256)810526 ISSUES/YR: 12
- **Nature Clinical Practice Oncology**
  ISSN: 1743-4254
  Nat. Clin. Pract. Oncol.
  Nature Publishing Group, Houndmills, RG21 6XS, Basingstoke, Hampshire, United Kingdom, Tel: +44(1256)329242, Fax: +44(1256)810526 ISSUES/YR: 12
- **Nature Clinical Practice Rheumatology**
  ISSN: 1745-8382
  Nat. Clin. Pract. Rheumatol.
  Nature Publishing Group, Houndmills, RG21 6XS, Basingstoke, Hampshire, United Kingdom, Tel: +44(1256)329242, Fax: +44(1256)810526 ISSUES/YR: 12
- **Nature Clinical Practice Urology**
  ISSN: 1743-4270
  Nat. Clin. Pract. Urol.
  Nature Publishing Group, Houndmills, RG21 6XS, Basingstoke, Hampshire, United Kingdom, Tel: +44(1256)329242, Fax: +44(1256)810526 ISSUES/YR: 12
- **Nature Genetics**
  ISSN: 1061-4036
  Nat. Genet.
  Nature Publishing Group, 345 Park Avenue South, New York, NY 10010-1707, United States, Tel: +1(212)7269200, Fax: +1(212)6960052 ISSUES/YR: 12
- **Nature Immunology**
  ISSN: 1529-2908
  Nat. Immunol.
  Nature Publishing Group, Houndmills, RG21 6XS, Basingstoke, Hampshire, United Kingdom, Tel: +44(1256)329242, Fax: +44(1256)810526 ISSUES/YR: 12
- **Nature Medicine**
  ISSN: 1078-8956
  Nat. Med.
  Nature Publishing Group, Houndmills, RG21 6XS, Basingstoke, Hampshire, United Kingdom, Tel: +44(1256)329242, Fax: +44(1256)810526 ISSUES/YR: 12
- **Nature Methods**
  ISSN: 1548-7091
  Nat. Methods
  Nature Publishing Group, Houndmills, RG21 6XS, Basingstoke, Hampshire, United Kingdom, Tel: +44(1256)329242, Fax: +44(1256)810526 ISSUES/YR: 12
- **Nature Neuroscience**
  ISSN: 1097-6256
  Nat. Neurosci.
  Nature Publishing Group, 345 Park Avenue South, New York, NY 10010-1707, United States, Tel: +1(212)7269200, Fax: +1(212)6960052 ISSUES/YR: 12
- **Nature Reviews Cancer**
  ISSN: 1474-175X
  Nat. Rev. Cancer
  Nature Publishing Group, Houndmills, RG21 6XS, Basingstoke, Hampshire, United Kingdom, Tel: +44(1256)329242, Fax: +44(1256)810526 ISSUES/YR: 12
- **Nature Reviews Drug Discovery**
  ISSN: 1474-1776
  Nat. Rev. Drug Discov.
  Nature Publishing Group, Houndmills, RG21 6XS, Basingstoke, Hampshire, United Kingdom, Tel: +44(1256)329242, Fax: +44(1256)810526 ISSUES/YR: 12
- **Nature Reviews Genetics**
  ISSN: 1471-0056
  Nat. Rev. Gen.
  Nature Publishing Group, Houndmills, RG21 6XS, Basingstoke, Hampshire, United Kingdom, Tel: +44(1256)329242, Fax: +44(1256)810526 ISSUES/YR: 12
- **Nature Reviews Immunology**
  ISSN: 1474-1733
  Nat. Rev. Immunol.
  Nature Publishing Group, Houndmills, RG21 6XS, Basingstoke, Hampshire, United Kingdom, Tel: +44(1256)329242, Fax: +44(1256)810526 ISSUES/YR: 12
- **Nature Reviews Microbiology**
  ISSN: 1740-1526
  Nat. Rev. Microbiol.
  Nature Publishing Group, Houndmills, RG21 6XS, Basingstoke, Hampshire, United Kingdom, Tel: +44(1256)329242, Fax: +44(1256)810526 ISSUES/YR: 12
- **Nature Reviews Molecular Cell Biology**
  ISSN: 1471-0072
  Nat. Rev. Mol. Cell Biol.
  Nature Publishing Group, Houndmills, RG21 6XS, Basingstoke, Hampshire, United Kingdom, Tel: +44(1256)329242, Fax: +44(1256)810526 ISSUES/YR: 12
- **Nature Reviews Neuroscience**
  ISSN: 1471-003X
  Nat. Rev. Neurosci.
  Nature Publishing Group, Houndmills, RG21 6XS, Basingstoke, Hampshire, United Kingdom, Tel: +44(1256)329242, Fax: +44(1256)810526 ISSUES/YR: 12
- **Nature Structural and Molecular Biology**
  ISSN: 1545-9993
  Nat. Struct. Mol. Biol.
  Nature Publishing Group, 345 Park Avenue South, New York, NY 10010-1707, United States, Tel: +1(212)7269200, Fax: +1(212)6960052 ISSUES/YR: 12

Naturwissenschaften
ISSN: 0028-1042
Naturwissenschaften
Springer Verlag, Tiergartenstrasse 17, D-69121, Heidelberg, Germany, Tel: +49(6221)4870, Fax: +49(6221)487366 ISSUES/YR: 12

Naunyn-Schmiedeberg's Archives of Pharmacology
ISSN: 0028-1298
Naunyn-Schmiedeberg's Arch. Pharmacol.
Springer Verlag, Tiergartenstrasse 17, D-69121, Heidelberg, Germany, Tel: +49(6221)4870, Fax: +49(6221)487366 ISSUES/YR: 12

NCRP Report
ISSN: 0083-209X
NCRP Rep.
National Council Radiation Protection and Measurements, 7910 Woodmont Avenue, Bethesda, MD 20814-3095, United States, Tel: +1(301)6572652 ISSUES/YR: 1

Nederlands Tijdschrift voor Anesthesiologie
ISSN: 0921-8769
Ned. Tijdschr. Anesthesiol.
Elsevier Bedrijfsinformatie bv, P.O. Box 1110, 3600 BC, Maarssen, Netherlands, Tel: +31(346)577577, Fax: +31(346)554287 ISSUES/YR: 4

Nederlands Tijdschrift voor Dermatologie en Venereologie
ISSN: 0925-8604
Ned. Tijdschr. Dermatol. Venereol.
Elsevier Bedrijfsinformatie bv, P.O. Box 1110, 3600 BC, Maarssen, Netherlands, Tel: +31(346)577577, Fax: +31(346)554287 ISSUES/YR: 10

Nederlands Tijdschrift voor Geneeskunde
ISSN: 0028-2162
Ned. Tijdschr. Geneeskd.
Bohn Stafleu Van Loghum bv, P.O. Box 246, 3990 GA, Houten, Netherlands, Tel: +31(030)6383830, Fax: +31(030)6383999 ISSUES/YR: 52

Nederlands Tijdschrift voor Keel- Neus- Oorheelkunde
ISSN: 1381-6683
Ned. Tijds. Keel- Neus- Oorheelkunde
Secretariaat KNO-vereniging, Jan van Goyenkade 11, 1075 HP, Amsterdam, Netherlands, Tel: +31(20)6793826, Fax: +31(20)6751274 ISSUES/YR: 4

Nederlands Tijdschrift voor Klinische Chemie en Laboratoriumgeneeskunde
ISSN: 1570-8306
Ned. Tijdschr. Klin. Chem. Lab. Geneeskd.
Nederlandse Vereniging voor Klinische Chemie, Catharijnesingel 49 A, 3511 GC, Utrecht, Netherlands, Tel: +31(30)2328623, Fax: +31(30)2311178 ISSUES/YR: 6

Nederlands Tijdschrift voor Obstetrie en Gynaecologie
ISSN: 0921-4011
Ned. Tijdschr. Obstet. Gynaecol.
Elsevier Bedrijfsinformatie bv, P.O. Box 1110, 3600 BC, Maarssen, Netherlands, Tel: +31(346)577577, Fax: +31(346)554287 ISSUES/YR: 10

Nederlands Tijdschrift voor Urologie
ISSN: 0929-0184
Ned. Tijdschr. Urol.
Elsevier Bedrijfsinformatie bv, P.O. Box 1110, 3600 BC, Maarssen, Netherlands, Tel: +31(346)577577, Fax: +31(346)554287 ISSUES/YR: 8

• **Neoplasia**
ISSN: 1522-8002
Neoplasia
Neoplasia, 1331 East Ann Street, Room 4131A, Ann Arbor, MI 48109-0582, United States, Tel: +1-734 647 6679, Fax: +1-734 615 5669 ISSUES/YR: 12

Neoplasma
ISSN: 0028-2685
Neoplasma
SAP - Slovak Academic Press, spol. s.r.o., Nam. Slobody 6, P.O. Box 57, 81005, Bratislava, Slovakia, Tel: +421(7)211729 ISSUES/YR: 6

Nephrologie et Therapeutique
ISSN: 1769-7255
Nephrol. Ther.
Editions Scientifiques et Medicales Elsevier, 23 rue Linois, 75724, Paris Cedex 15, France, Tel: +33(1)45589826, Fax: +33(1)45589428 ISSUES/YR: 6

• **Nephrology**
ISSN: 1320-5358
Nephrology
Blackwell Publishing, 550 Swanston Street, VIC 3053, Carlton South, Australia, Tel: +61(3)8359 1011, Fax: +61(3)8359 1120 ISSUES/YR: 6

• **Nephrology Dialysis Transplantation**
ISSN: 0931-0509
Nephrol. Dial. Transplant.
Oxford University Press, Great Clarendon Street, OX2 6DP, Oxford, United Kingdom, Tel: +44(1865)353907, Fax: +44(1865)353486 ISSUES/YR: 12

Nephrology, Dialysis and Transplantation
ISSN: 1312-5257
Nephrol. Dial. Transplant.
Medical Information Center, 1 Sv. G. Sofijski str., 1431, Sofia, Bulgaria, Tel: +359-522342, Fax: +359-522393 ISSUES/YR: 4

• **Nephron - Clinical Practice**
ISSN: 1660-2110
Nephron Clin. Pract.
S. Karger AG, Allschwilerstrasse 10, P.O. Box, CH-4009, Basel, Switzerland, Tel: +41(61)3061111, Fax: +41(61)3061234 ISSUES/YR: 12

• **Nephron - Experimental Nephrology**
ISSN: 1660-2129
Nephron Exp. Nephrol.
S. Karger AG, Allschwilerstrasse 10, P.O. Box, CH-4009, Basel, Switzerland, Tel: +41(61)3061111, Fax: +41(61)3061234 ISSUES/YR: 12

• **Nephron - Physiology**
ISSN: 1660-2137
Nephron Physiol.
S. Karger AG, Allschwilerstrasse 10, P.O. Box, CH-4009, Basel, Switzerland, Tel: +41(61)3061111, Fax: +41(61)3061234 ISSUES/YR: 12

Nervenarzt
ISSN: 0028-2804
Nervenarzt
Springer Verlag, Tiergartenstrasse 17, D-69121, Heidelberg, Germany, Tel: +49(6221)4870, Fax: +49(6221)487366 ISSUES/YR: 12

Nervenheilkunde
ISSN: 0722-1541
Nervenheilkunde
F.K. Schattauer Verlagsgesellschaft mbH, P.O. Box 10 45 45, D-70040, Stuttgart, Germany, Tel: +49(711)229870, Fax: +49(711)2298750 ISSUES/YR: 10

Netherlands Heart Journal
ISSN: 0929-7456
Neth. Heart J.
Bohn Stafleu Van Loghum bv, P.O. Box 246, 3990 GA, Houten, Netherlands, Tel: +31(030)6383830, Fax: +31(030)6383999 ISSUES/YR: 11

Netherlands Journal of Medicine
ISSN: 0300-2977
Neth. J. Med.
Van Zuiden Communications BV, Postbus 2122, 2400 CC, Alphen aan de Rijn, Netherlands, Tel: +31-172 471 882, Fax: +31-172 476 191 ISSUES/YR: 11

• **Neural Networks**
ISSN: 0893-6080
Neural Netw.
Elsevier Ltd, Langford Lane, Kidlington, OX5 1GB, Oxford, United Kingdom, Tel: +44(1865)843000, Fax: +44(1865)843010 ISSUES/YR: 10

Neural Plasticity
ISSN: 0792-8483
Neural Plast.
Hindawi Publishing Corporation, 410 Park Avenue, 15th Floor, 287 pmb, New York, NY 10022, United States, Tel: +1(20)2-2706352, Fax: +1(20)2-2706351 ISSUES/YR: 4

Neuro-Oncology
ISSN: 1522-8517
Neuro-Oncology
Duke University Press, Box 90660, Durham, NC 27708-0660, United States, Tel: +1(919)6873653 ISSUES/YR: 4

Neuro-Ophthalmology
ISSN: 0165-8107
Neuro-Ophthalmology
Taylor and Francis Ltd., 4 Park Square, Milton Park, OX14 4RN, Abingdon, Oxfordshire, United Kingdom, Tel: +44(20)70175541, Fax: +44(20)70175198 ISSUES/YR: 6

Neuro-Ophthalmology Japan
ISSN: 0289-7024
Neuro-Ophthalmol. Jpn.
Neuro-Ophthalmology Society of Japan, 15-1-1 Kitasato, Sagamihara, 228, Kanagawa, Japan, Tel: +81(427)788464 ISSUES/YR: 4

Neuro-Psy News
ISSN: 1633-5767
Neuro-Psy News
Interligne, Levallois-Perret Cedex, 92593, Paris, France, Tel: +33(1)4140 7663, Fax: +33(1)4140 7529 ISSUES/YR: 5

Neuroanatomy
ISSN: 1303-1783
Neuroanatomy
M. Mustafa Aldur, 06100, Ankara, Turkey, Tel: +90(312)3052466, Fax: +90(312)4785200 ISSUES/YR: 1

• **Neurobiology of Aging**
ISSN: 0197-4580
Neurobiol. Aging
Elsevier Inc., 360 Park Avenue South, New York, NY 10010, United States, Tel: +1(212)6333950, Fax: +1(212)6333990 ISSUES/YR: 10

• **Neurobiology of Disease**
ISSN: 0969-9961
Neurobiol. Dis.
Academic Press Inc., 6277 Sea Harbor Drive, Orlando, FL 32887-4900, United States, Tel: +1(407)3454040, Fax: +1(407)3639661 ISSUES/YR: 9

Neurobiology of Learning and Memory
ISSN: 1074-7427
Neurobiol. Learn. Mem.
Academic Press Inc., 6277 Sea Harbor Drive, Orlando, FL 32887-4900, United States, Tel: +1(407)3454040, Fax: +1(407)3639661 ISSUES/YR: 6

Neurocase
ISSN: 1355-4794
Neurocase
Psychology Press Ltd, 4 Park Square, Milton Park, OX14 4RN, Abingdon, Oxfordshire, United Kingdom, Tel: +44(1235)828600, Fax: +44(1235)829000 ISSUES/YR: 6

• **Neurochemical Research**
ISSN: 0364-3190
Neurochem. Res.
Springer Science and Business Media Deutschland GmbH, Heidelberger Platz 3, 14197, Berlin, Germany, Tel: +1-30 827875282, Fax: +1-30 827875707 ISSUES/YR: 12

• **Neurochemistry International**
ISSN: 0197-0186
Neurochem. Int.
Elsevier Ltd, Langford Lane, Kidlington, OX5 1GB, Oxford, United Kingdom, Tel: +44(1865)843000, Fax: +44(1865)843010 ISSUES/YR: 16

Neurochirurgie
ISSN: 0028-3770
Neurochirurgie
Masson, 21 Rue Camille Desmoulins, 92789, Issy les Moulineaux Cedex 9, France, Tel: +33(1)73281634, Fax: +33(1)73281649 ISSUES/YR: 6

Neurocirugia
ISSN: 1130-1473
Neurocirugia
Neurocirugia, Avda. Castaneda 37, C.10o, 39005, Santander, Spain, Tel: +34(942)202701, Fax: +34(942)202655 ISSUES/YR: 6

• **Neurocomputing**
ISSN: 0925-2312
Neurocomputing
Elsevier, P.O. Box 211, 1000 AE, Amsterdam, Netherlands, Tel: +31(20)4853757, Fax: +31(20)4853432 ISSUES/YR: 24

• **Neurocritical Care**
ISSN: 1541-6933
Neurocrit. Care
Humana Press Inc., 999 Riverview Drive, Suite 208, Totowa, NJ 07512-1165, United States, Tel: +1(973)2561699, Fax: +1(973)2568341 ISSUES/YR: 6

• **Neurodegenerative Diseases**
ISSN: 1660-2854
Neurodegenerative Dis.
S. Karger AG, Allschwilerstrasse 10, P.O. Box, CH-4009, Basel, Switzerland, Tel: +41(61)3061111, Fax: +41(61)3061234 ISSUES/YR: 6

• **Neuroembryology and Aging**
ISSN: 1661-3406
Neuroembryology Aging
S. Karger AG, Allschwilerstrasse 10, P.O. Box, CH-4009, Basel, Switzerland, Tel: +41(61)3061111, Fax: +41(61)3061234 ISSUES/YR: 4

• **Neuroendocrinology**
ISSN: 0028-3835
Neuroendocrinology
S. Karger AG, Allschwilerstrasse 10, P.O. Box, CH-4009, Basel, Switzerland, Tel: +41(61)3061111, Fax: +41(61)3061234 ISSUES/YR: 12

Neuroendocrinology Letters
ISSN: 0172-780X
Neuroendocrinol. Lett.
Maghira and Maas Publications, P.O. Box 26132, S-100 41, Stockholm, Sweden, Tel: +46(8)6115565, Fax: +46(8)6780747 ISSUES/YR: 6

Neuroepidemiology
ISSN: 0251-5350
Neuroepidemiology
S. Karger AG, Allschwilerstrasse 10, P.O. Box, CH-4009, Basel, Switzerland, Tel: +41(61)3061111, Fax: +41(61)3061234 ISSUES/YR: 4

Neuroforum
ISSN: 0947-0875
Neuroforum
Elsevier GmbH, Slevogtstr. 3-5, 69126, Heidelberg, Germany, Tel: +49(6221)9126-200, Fax: +49(6221)9126-370 ISSUES/YR: 4

• **Neurogastroenterology and Motility**
ISSN: 1350-1925
Neurogastroenterol. Motil.
Blackwell Publishing Ltd, 9600 Garsington Road, OX4 2XG, Oxford, United Kingdom, Tel: +44(1865)776868, Fax: +44(1865)714591 ISSUES/YR: 6

• **Neurogenetics**
ISSN: 1364-6745
Neurogenetics
Springer Verlag, Tiergartenstrasse 17, D-69121, Heidelberg, Germany, Tel: +49(6221)4870, Fax: +49(6221)487366 ISSUES/YR: 4

• **NeuroImage**
ISSN: 1053-8119
NeuroImage
Academic Press Inc., 6277 Sea Harbor Drive, Orlando, FL 32887-4900, United States, Tel: +1(407)3454040, Fax: +1(407)3639661 ISSUES/YR: 12

• **Neuroimaging Clinics of North America**
ISSN: 1052-5149
Neuroimaging Clin. North Am.
W.B. Saunders, Independence Square West, Philadelphia, PA 19106-3399, United States, Tel: +1(215)2387800, Fax: +1(215)2386445 ISSUES/YR: 4

• **NeuroImmunoModulation**
ISSN: 1021-7401
NeuroImmunomodulation
S. Karger AG, Allschwilerstrasse 10, P.O. Box, CH-4009, Basel, Switzerland, Tel: +41(61)3061111, Fax: +41(61)3061234 ISSUES/YR: 8

• **Neuroinformatics**
ISSN: 1539-2791
Neuroinformatics
Humana Press Inc., 999 Riverview Drive, Suite 208, Totowa, NJ 07512-1165, United States, Tel: +1(973)2561699, Fax: +1(973)2568341 ISSUES/YR: 4

Neurologia
ISSN: 0213-4853
Neurologia
STM Editores S.A, Principe de Vergara 211, 28002, Esc Izda 60D, Madrid, Spain ISSUES/YR: 10

Neurologia Croatica
ISSN: 0353-8842
Neurol. Croatica
University Hospital School of Medicine Zagreb, Kispaticeva 12, 41000, Zagreb, Croatia, Tel: +385(41)222706, Fax: +385(41)222706 ISSUES/YR: 4

Neurologia i Neurochirurgia Polska
ISSN: 0028-3843
Neurol. Neurochir. Pol.
Termedia Publishing House Ltd., ul Kleberga 8,, 61-615, Poznan, Poland, Tel: +48-61 822 77 81, Fax: +48-61 822 7781 ISSUES/YR: 6

Neurologia Medico-Chirurgica
ISSN: 0470-8105
Neurol. Med.-Chir.
Japan Neurosurgical Society, 5-25-16 Hongo, Bunkyo-ku, 113-0033, Tokyo, Japan, Tel: +81(3)38128092, Fax: +81(3)38128092 ISSUES/YR: 12

Neurologia Suplementos
ISSN: 1695-5374
Neurol. Supl.
STM Editores S.A, Principe de Vergara 211, 28002, Esc Izda 60D, Madrid, Spain ISSUES/YR: 1

• **Neurologic Clinics**
ISSN: 0733-8619
Neurol. Clin.
W.B. Saunders, Independence Square West, Philadelphia, PA 19106-3399, United States, Tel: +1(215)2387800, Fax: +1(215)2386445 ISSUES/YR: 4

Neurological Research
ISSN: 0161-6412
Neurol. Res.
Maney Publishing, Hudson Road, LS9 7DL, Leeds, United Kingdom, Tel: +1(113)2497481, Fax: +1(113)2486983 ISSUES/YR: 8

Neurological Sciences
ISSN: 1590-1874
Neurol. Sci.
Springer-Verlag Italia s.r.l, Via Podgora 4, I-20122, Milano, Italy, Tel: +39(2)5425971, Fax: +39(2)55193360 ISSUES/YR: 6

Neurological Surgery
ISSN: 0301-2603
Neurol. Surg.
Igaku-Shoin Ltd, 5-24-3 Hongo, Bunkyo-ku, 113-91, Tokyo, Japan, Tel: +81(3)38175600 ISSUES/YR: 12

Neurologie und Rehabilitation
ISSN: 0947-2177
Neurol. Rehabil.
Hippocampus Verlag, Bismarckstr. 8, 53604, Bad Honnef, Germany, Tel: +49(2224)940757 ISSUES/YR: 6

• **Neurologist**
ISSN: 1074-7931
Neurologist
Lippincott Williams and Wilkins, 351 West Camden Street, Baltimore, MD 21201-2436, United States, Tel: +1(410)5284000 ISSUES/YR: 6

• **Neurology**
ISSN: 0028-3878
Neurology
Lippincott Williams and Wilkins, 530 Walnut Street, Philadelphia, PA 19106-3621, United States, Tel: +1(215)5218342, Fax: +1(215)5218326 ISSUES/YR: 24

• **Neurology and Clinical Neurophysiology**
ISSN: 1526-8748
Neurol. Clin. Neurophysiol.
American Academy of Clinical Neurophysiology, 104 13-Street Hudson, WI 54016, United States, Tel: +1-715 381 3440, Fax: +1-715 381 3442 ISSUES/YR: 1

• **Neurology in Practice**
ISSN: 1473-7086
Neurol. Practice
BMJ Publishing Group, Tavistock Square, WC1H 9JR, London, United Kingdom, Tel: +44(207)3874499, Fax: +44(207)3836668 ISSUES/YR: 1

Neurology India
ISSN: 0028-3886
Neurol. India
Medknow Publication, 12 Manisha Plaza, MN Rd, 400 070 MUMBAI, Kurla, India, Tel: +91(22)25032970, Fax: +91(22)25032398 ISSUES/YR: 4

• **Neurology Psychiatry and Brain Research**
ISSN: 0941-9500
Neurol. Psychiatry Brain Res.
Universitatsverlag Ulm GmbH, Bahnhofstrasse 20, D-89073, Ulm, Germany, Tel: +49(731)152860, Fax: +49(731) 152862 ISSUES/YR: 4

Neuromodulation
ISSN: 1094-7159
Neuromodulation
Blackwell Publishing Inc., 350 Main Street, Malden, MA 02148, United States, Tel: +1(781)3888200, Fax: +1(781)3888210 ISSUES/YR: 4

• **NeuroMolecular Medicine**
ISSN: 1535-1084
NeuroMol. Med.
Humana Press Inc., 999 Riverview Drive, Suite 208, Totowa, NJ 07512-1165, United States, Tel: +1(973)2561699, Fax: +1(973)2568341 ISSUES/YR: 6

• **Neuromuscular Disorders**
ISSN: 0960-8966
Neuromuscular Disord.
Elsevier Ltd, Langford Lane, Kidlington, OX5 1GB, Oxford, United Kingdom, Tel: +44(1865)843000, Fax: +44(1865)843010 ISSUES/YR: 12

• **Neuron**
ISSN: 0896-6273
Neuron
Cell Press, 1100 Massachusetts Avenue, Cambridge, MA 02138-9957, United States, Tel: +1(617)6617057, Fax: +1(617)6617061 ISSUES/YR: 24

• **Neuron Glia Biology**
ISSN: 1740-925X
Neuron Glia Biol.
Cambridge University Press, Shaftesbury Road, CB2 2RU, Cambridge, United Kingdom, Tel: +44(1223)312393, Fax: +44(1223)315052 ISSUES/YR: 4

• **Neuropathology**
ISSN: 0919-6544
Neuropathology
Blackwell Publishing, 550 Swanston Street, VIC 3053, Carlton South, Australia, Tel: +61(3)8359 1011, Fax: +61(3)8359 1120 ISSUES/YR: 4

• **Neuropathology and Applied Neurobiology**
ISSN: 0305-1846
Neuropathol. Appl. Neurobiol.
Blackwell Publishing Ltd, 9600 Garsington Road, OX4 2XG, Oxford, United Kingdom, Tel: +44(1865)776868, Fax: +44(1865)714591 ISSUES/YR: 6

• **Neuropediatrics**
ISSN: 0174-304X
Neuropediatrics
Hippokrates Verlag GmbH, Rudigerstrasse 14, D-70469, Stuttgart, Germany, Tel: +49(711)8931-321, Fax: +49(711)8931-422 ISSUES/YR: 6

• **Neuropeptides**
ISSN: 0143-4179
Neuropeptides
Churchill Livingstone, 1-3 Baxter's Place, Leith Walk, EH1 3AF, Edinburgh, United Kingdom, Tel: +44(131)5562424, Fax: +44(131)5581278 ISSUES/YR: 6

• **Neuropharmacology**
ISSN: 0028-3908
Neuropharmacology
Elsevier Ltd, Langford Lane, Kidlington, OX5 1GB, Oxford, United Kingdom, Tel: +44(1865)843000, Fax: +44(1865)843010 ISSUES/YR: 16

Neurophysiologie Clinique
ISSN: 0987-7053
Neurophysiol. Clin.
Editions Scientifiques et Medicales Elsevier, 23 rue Linois, 75724, Paris Cedex 15, France, Tel: +33(1)45589826, Fax: +33(1)45589428 ISSUES/YR: 6

Neurophysiologie-Labor
ISSN: 1439-4847
Neurophysiol.-Labor
Urban und Fischer Verlag Jena, P.O. Box 100537, 07705, Jena, Germany, Tel: +49(3641)6263, Fax: +49(3641)626500 ISSUES/YR: 4

Neurophysiology
ISSN: 0090-2977
Neurophysiology
Springer Science and Business Media Deutschland GmbH, Heidelberger Platz 3, 14197, Berlin, Germany, Tel: +1-30 827875282, Fax: +1-30 827875707 ISSUES/YR: 6

Neuropsychiatric Disease and Treatment
ISSN: 1176-6328
Neuropsychiatr. Dis. Treat.
Dove Press Ltd., PO Box 300-008, Albany, Auckland, New Zealand, Tel: +64(9)4485032, Fax: +64(9)4485034 ISSUES/YR: 4

Neuropsychiatrie
ISSN: 0948-6259
Neuropsychiatrie
Dustri-Verlag Dr. Karl Feistle, P.O. Box 1351, D-82032, Deisenhofen/Munchen, Germany, Tel: +49(89)6138610, Fax: +49(89)6135412 ISSUES/YR: 4

Neuropsychiatrie de l'Enfance et de l'Adolescence
ISSN: 0222-9617
Neuropsychiatr. Enfance Adolesc.
Editions Scientifiques et Medicales Elsevier, 23 rue Linois, 75724, Paris Cedex 15, France, Tel: +33(1)45589826, Fax: +33(1)45589428 ISSUES/YR: 8

Current

• **Neuropsychobiology**
ISSN: 0302-282X
Neuropsychobiology
S. Karger AG, Allschwilerstrasse 10, P.O. Box, CH-4009, Basel, Switzerland, Tel: +41(61)3061111, Fax: +41(61)3061234 ISSUES/YR: 8

Neuropsychologia
ISSN: 0028-3932
Neuropsychologia
Elsevier Ltd, Langford Lane, Kidlington, OX5 1GB, Oxford, United Kingdom, Tel: +44(1865)843000, Fax: +44(1865)843010 ISSUES/YR: 14

Neuropsychological Rehabilitation
ISSN: 0960-2011
Neuropsychol. Rehabil.
Psychology Press Ltd, 4 Park Square, Milton Park, OX14 4RN, Abingdon, Oxfordshire, United Kingdom, Tel: +44(1235)828600, Fax: +44(1235)829000 ISSUES/YR: 6

• **Neuropsychology**
ISSN: 0894-4105
Neuropsychology
American Psychological Association Inc., 750 First Street NE, Washington, DC 20002-4242, United States, Tel: +1(202)3365600, Fax: +1(202)3365568 ISSUES/YR: 6

Neuropsychology Review
ISSN: 1040-7308
Neuropsychol. Rev.
Springer Science and Business Media Deutschland GmbH, Heidelberger Platz 3, 14197, Berlin, Germany, Tel: +1-30 827875282, Fax: +1-30 827875707 ISSUES/YR: 4

• **Neuropsychopharmacology**
ISSN: 0893-133X
Neuropsychopharmacology
Nature Publishing Group, Houndmills, RG21 6XS, Basingstoke, Hampshire, United Kingdom, Tel: +44(1256)329242, Fax: +44(1256)810526 ISSUES/YR: 12

• **Neuroradiology**
ISSN: 0028-3940
Neuroradiology
Springer Verlag, Tiergartenstrasse 17, D-69121, Heidelberg, Germany, Tel: +49(6221)4870, Fax: +49(6221)487366 ISSUES/YR: 12

NeuroRehabilitation
ISSN: 1053-8135
NeuroRehabilitation
IOS Press, Nieuwe Hemweg 6B, 1013 BG, Amsterdam, Netherlands, Tel: +31(20)6883355, Fax: +31(20)6203419 ISSUES/YR: 4

Neurorehabilitation and Neural Repair
ISSN: 1545-9683
Neurorehabil. Neural Repair
SAGE Publications Inc., 2455 Teller Road, Thousand Oaks, CA 91320, United States, Tel: +1(805)4990721, Fax: +1(805)4990871 ISSUES/YR: 4

• **NeuroReport**
ISSN: 0959-4965
NeuroReport
Lippincott Williams and Wilkins, 241 Borough High Street, SE1 1GB, London, United Kingdom, Tel: +44(207)9407500, Fax: +44(207)9407575 ISSUES/YR: 18

NeuroRx
ISSN: 1545-5343
NeuroRx
Elsevier Inc., 360 Park Avenue South, New York, NY 10010, United States, Tel: +1(212)6333950, Fax: +1(212)6333990 ISSUES/YR: 4

• **Neuroscience**
ISSN: 0306-4522
Neuroscience
Elsevier Ltd, Langford Lane, Kidlington, OX5 1GB, Oxford, United Kingdom, Tel: +44(1865)843000, Fax: +44(1865)843010 ISSUES/YR: 28

Neuroscience and Behavioral Physiology
ISSN: 0097-0549
Neurosci. Behav. Physiol.
Springer Science and Business Media Deutschland GmbH, Heidelberger Platz 3, 14197, Berlin, Germany, Tel: +1-30 827875282, Fax: +1-30 827875707 ISSUES/YR: 9

• **Neuroscience and Biobehavioral Reviews**
ISSN: 0149-7634
Neurosci. Biobehav. Rev.
Elsevier Ltd, Langford Lane, Kidlington, OX5 1GB, Oxford, United Kingdom, Tel: +44(1865)843000, Fax: +44(1865)843010 ISSUES/YR: 8

• **Neuroscience Letters**
ISSN: 0304-3940
Neurosci. Lett.
Elsevier Ireland Ltd, P.O. Box 85, Limerick, Ireland, Tel: +353(61)471944, Fax: +353(61)472144 ISSUES/YR: 57

• **Neuroscience Research**
ISSN: 0168-0102
Neurosci. Res.
Elsevier Ireland Ltd, P.O. Box 85, Limerick, Ireland, Tel: +353(61)471944, Fax: +353(61)472144 ISSUES/YR: 12

Neurosciences
ISSN: 1319-6138
Neurosciences
Saudi Arabian Armed Forces Hospital, P.O. Box 7897, 11159, Riyadh, Saudi Arabia, Tel: +966(1)4777714 EXT.6570, Fax: +966(1)4761810 ISSUES/YR: 4

• **Neuroscientist**
ISSN: 1073-8584
Neuroscientist
SAGE Publications Inc., 2455 Teller Road, Thousand Oaks, CA 91320, United States, Tel: +1(805)4990721, Fax: +1(805)4990871 ISSUES/YR: 6

NeuroSignals
ISSN: 1424-862X
Neurosignals
S. Karger AG, Allschwilerstrasse 10, P.O. Box, CH-4009, Basel, Switzerland, Tel: +41(61)3061111, Fax: +41(61)3061234 ISSUES/YR: 6

• **Neurosurgery**
ISSN: 0148-396X
Neurosurgery
Lippincott Williams and Wilkins, 351 West Camden Street, Baltimore, MD 21201-2436, United States, Tel: +1(410)5284000 ISSUES/YR: 12

• **Neurosurgery Clinics of North America**
ISSN: 1042-3680
Neurosurg. Clin. North Am.
W.B. Saunders, Independence Square West, Philadelphia, PA 19106-3399, United States, Tel: +1(215)2387800, Fax: +1(215)2386445 ISSUES/YR: 4

• **Neurosurgery Quarterly**
ISSN: 1050-6438
Neurosurg. Q.
Lippincott Williams and Wilkins, 530 Walnut Street, Philadelphia, PA 19106-3621, United States, Tel: +1(215)5218342, Fax: +1(215)5218326 ISSUES/YR: 4

• **Neurosurgical Review**
ISSN: 0344-5607
Neurosurg. Rev.
Springer Verlag, Tiergartenstrasse 17, D-69121, Heidelberg, Germany, Tel: +49(6221)4870, Fax: +49(6221)487366 ISSUES/YR: 4

• **NeuroToxicology**
ISSN: 0161-813X
NeuroToxicology
Elsevier, P.O. Box 211, 1000 AE, Amsterdam, Netherlands, Tel: +31(20)4853757, Fax: +31(20)4853432 ISSUES/YR: 6

• **Neurotoxicology and Teratology**
ISSN: 0892-0362
Neurotoxicol. Teratol.
Elsevier Inc., 360 Park Avenue South, New York, NY 10010, United States, Tel: +1(212)6333950, Fax: +1(212)6333990 ISSUES/YR: 6

Neurourology and Urodynamics
ISSN: 0733-2467
Neurourol. Urodyn.
Wiley-Liss Inc., 111 River Street, Hoboken, NJ 07030-5774, United States, Tel: +1(201)7486645, Fax: +1(201)7486021 ISSUES/YR: 7

- **New England Journal of Medicine**
  ISSN: 0028-4793
  New Engl. J. Med.
  Massachussetts Medical Society, 860 Winter Street, Waltham, MA 02451-1413, United States, Tel: +1(781)8933800, Fax: +1(781)8930413 ISSUES/YR: 52
- **New Genetics and Society**
  ISSN: 1463-6778
  New Genet. Soc.
  Routledge, 4 Park Square, Milton Park, OX14 4RN, Abingdon, Oxfordshire, United Kingdom, Tel: +44(1235)828600, Fax: +44(1235)829000 ISSUES/YR: 3
- **New Journal of Chemistry**
  ISSN: 1144-0546
  New J. Chem.
  Royal Society of Chemistry, Milton Road, CB4 OWF, Cambridge, United Kingdom, Tel: +44(1223)420066, Fax: +44(1223)420247 ISSUES/YR: 12

New Medicine
ISSN: 1427-0994
New Med.
Borgis Ltd, ul. E. Ciolka 11A, 01-445, Warsaw, Poland, Tel: +48(22)8369669, Fax: +48(22)8369669 ISSUES/YR: 1

New Zealand Journal of Medical Laboratory Science
ISSN: 1171-0195
New Zealand J. Med. Lab. Sci.
New Zealand Institute of Medical Laboratory Science Inc., P.O. Box 505, 8254, Rangiora, New Zealand, Tel: +64(3)313 4761, Fax: +64(3)313 2098 ISSUES/YR: 3

New Zealand Medical Journal
ISSN: 1175-8716
New Zealand Med. J.
New Zealand Medical Association, 26 The Terrace, P.O. Box 156, Wellington, New Zealand, Tel: +64(4)4724741, Fax: +64(4)4710838 ISSUES/YR: 20

New Zealand Public Health Surveillance Report
ISSN: 1176-2888
New Zealand Public Health Surveill. Rep.
Institute of Environmental Science and Research Ltd, Kenepuru Drive, P.O. Box 50-348, Porirua, Wellington, New Zealand, Tel: +64(4)2370149, Fax: +64(4)2372370 ISSUES/YR: 4

News in Medical Products
ISSN: 1683-6316
News Med. Prod.
VICER Publishing, P.O. Box 14, A-1097, Vienna, Austria, Tel: +43(676)9568085, Fax: +43(676)9568086 ISSUES/YR: 4

- **Nicotine and Tobacco Research**
  ISSN: 1462-2203
  Nicotine Tob. Res.
  Routledge, 4 Park Square, Milton Park, OX14 4RN, Abingdon, Oxfordshire, United Kingdom, Tel: +44(1235)828600, Fax: +44(1235)829000 ISSUES/YR: 6

NIDA Research Monograph Series
ISSN: 1046-9516
NIDA Res. Monogr. Ser.
National Institute on Drug Abuse, 6001 Executive Boulevard. Room 5213, Bethesda, MD 20892-9561, United States, Tel: +1(301)443 1124 ISSUES/YR: 1

Nieren- und Hochdruckkrankheiten
ISSN: 0300-5224
Nieren- Hochdruckkr.
Dustri-Verlag Dr. Karl Feistle, P.O. Box 1351, D-82032, Deisenhofen/Munchen, Germany, Tel: +49(89)6138610, Fax: +49(89)6135412 ISSUES/YR: 12

Nigerian Journal of Surgical Research
ISSN: 1595-1103
Niger. J. Surg. Res.
Surgical Research Society, P.O. Box 76, 810001, Zaria, Nigeria ISSUES/YR: 4

Nishinihon Journal of Dermatology
ISSN: 0386-9784
Nishinihon J. Dermatol.
Kyushu University, 1-1 Maidashi 3-chome, Higashi-ku, 812, Fukuoka, Japan, Tel: +81(92)6411151, Fax: +81(92)6510165 ISSUES/YR: 6

Nishinihon Journal of Urology
ISSN: 0029-0726
Nishinihon J. Urol.
Nishinihon Section of Japan Urological Association, 1-1 Maidashi 3-chome, Higashi-ku, 812, Fukuoka, Japan, Tel: +81(92)6411151 ISSUES/YR: 12

- **Nitric Oxide - Biology and Chemistry**
  ISSN: 1089-8603
  Nitric Oxide Biol. Chem.
  Academic Press Inc., 6277 Sea Harbor Drive, Orlando, FL 32887-4900, United States, Tel: +1(407)3454040, Fax: +1(407)3639661 ISSUES/YR: 8
- **NMR in Biomedicine**
  ISSN: 0952-3480
  NMR Biomed.
  John Wiley and Sons Ltd, Southern Gate, PO19 8SQ, Chichester, West Sussex, United Kingdom, Tel: +44(1243)779777, Fax: +44(1243)775878 ISSUES/YR: 8

No To Hattatsu
ISSN: 0029-0831
No To Hattatsu
Japanese Society of Child Neurology, 6-13 Wakamatsu-cho, Shinjuku-ku, 162, Tokyo, Japan, Tel: +81(3)33514125 ISSUES/YR: 6

Nobel Medicus
ISSN: 1305-2381
Nobel Med.
Nobelmedicus, Barbaros Bulvari 76-78, 34353, Besiktas Istanbul, Turkey, Tel: +90(212)2597490 EXT 371, Fax: +90(212)2588789 ISSUES/YR: 3

Noise and Health
ISSN: 1463-1741
Noise and Health
NRN Publications, 330 Gray's Inn Road, WC1X 8EE, London, United Kingdom, Tel: +44(20)69151590, Fax: +44(20)72788041 ISSUES/YR: 4

Nordic Journal of Psychiatry
ISSN: 0803-9488
Nord. J. Psychiatry
Taylor and Francis Ltd., P.O. Box 2562 Solli, 0202, Oslo, Norway, Tel: +47(22)129880, Fax: +47(22)129890 ISSUES/YR: 6

Nordic Journal of Psychiatry, Supplement
ISSN: 0803-9496
Nord. J. Psychiatry Suppl.
Taylor and Francis A.S., P.O. Box 2562 Solli, 0202, Oslo, Norway, Tel: +47(22)129880, Fax: +47(22)129890 ISSUES/YR: 1

Nordisk Tidsskrift for Biologisk Medisin
ISSN: 1502-1475
Nord. Tidsskr. Biol. Med.
Vital Press, Toften 24, 8610, Grubhei, Norway, Tel: +47(75)130371 ISSUES/YR: 4

Norsk Epidemiologi
ISSN: 0803-2491
Nor. Epidemiol.
Norwegian Epidemiological Society, NTNU, 7491, Trondheim, Norway, Tel: +47-+47 73591806 ISSUES/YR: 1

North and West London Journal of General Practice
ISSN: 1475-1763
North West London J. Gen. Pract.
Kingscos Medical Publishers, P.O. Box 18265, EC2A 3TT, London, United Kingdom, Tel: +44(20)77497245, Fax: +44(20)77398683 ISSUES/YR: 3

Notarzt
ISSN: 0177-2309
Notarzt
Georg Thieme Verlag, Rudigerstrasse 14, D-70469, Stuttgart, Germany, Tel: +49(711)8931-321, Fax: +49(711)8931-422 ISSUES/YR: 6

Notfall und Hausarztmedizin
ISSN: 1612-8591
Notf. Hausarztmed.
Georg Thieme Verlag, Rudigerstrasse 14, D-70469, Stuttgart, Germany, Tel: +49(711)8931-321, Fax: +49(711)8931-422 ISSUES/YR: 12

Nouvelles Dermatologiques
ISSN: 0752-5370
Nouv. Dermatol.
Initiatives Sante/Groupe Liaisons, 1 Avenue Edouard-Berlin, 92856, Rueil Malmaison Cedex, France, Tel: +33-3 88 60 46 00, Fax: +33-3 88 60 46 70 ISSUES/YR: 10

Nowotwory
ISSN: 0029-540X
Nowotwory
Centrum Onkologii, ul. W.K. Roentgena 5, P.O. Box 47, 02-781, Warsaw, Poland, Tel: +48(22)6445024 ISSUES/YR: 6

Nuclear Medicine and Biology
ISSN: 0969-8051
Nucl. Med. Biol.
Elsevier Inc., 360 Park Avenue South, New York, NY 10010, United States, Tel: +1(212)6333950, Fax: +1(212)6333990 ISSUES/YR: 8

Nuclear Medicine Communications
ISSN: 0143-3636
Nucl. Med. Commun.
Lippincott Williams and Wilkins, 241 Borough High Street, SE1 1GB, London, United Kingdom, Tel: +44(207)9407500, Fax: +44(207)9407575 ISSUES/YR: 12

Nuclear Medicine Review
ISSN: 1506-9680
Nucl. Med. Rev.
Via Medica, Ul. Swietokrzyska 73, 80-180, Gdansk, Poland, Tel: +48(58)3209494, Fax: +48(58)3209460 ISSUES/YR: 2

Nuclear Receptor
ISSN: 1478-1336
Nucl. Recept.
BioMed Central Ltd., 34 - 42 Cleveland Street, W1T 4LB, London, United Kingdom, Tel: +44(207)3230323, Fax: +44(207)6319926 ISSUES/YR: 1

Current

• **Nucleic Acids Research**
ISSN: 0305-1048
Nucleic Acids Res.
Oxford University Press, Great Clarendon Street, OX2 6DP, Oxford, United Kingdom, Tel: +44(1865)353907, Fax: +44(1865)353486 ISSUES/YR: 22

Nucleosides, Nucleotides and Nucleic Acids
ISSN: 1525-7770
Nucleosides Nucleotides Nucleic Acids
Taylor and Francis Inc., 325 Chestnut St, Suite 800, Philadelphia PA, PA 19106, United States, Tel: +1(800)3541420, Fax: +1(215)6258914 ISSUES/YR: 12

NuklearMedizin
ISSN: 0029-5566
NuklearMedizin
F.K. Schattauer Verlagsgesellschaft mbH, P.O. Box 10 45 45, D-70040, Stuttgart, Germany, Tel: +49(711)229870, Fax: +49(711)2298750 ISSUES/YR: 6

• **Nursing and Health Sciences**
ISSN: 1441-0745
Nurs. Health Sci.
Blackwell Publishing, 550 Swanston Street, VIC 3053, Carlton South, Australia, Tel: +61(3)8359 1011, Fax: +61(3)8359 1120 ISSUES/YR: 4

Nutricion Clinica y Dietetica Hospitalaria
ISSN: 0211-6057
Nutr. Clin. Diet. Hosp.
Alpe Editores S.A., Pedro Rico 27, oficinas 11-12, 28029, Madrid, Spain, Tel: +34(91)7338811, Fax: +34(91)3159652 ISSUES/YR: 6

Nutricion Hospitalaria
ISSN: 0212-1611
Nutr. Hosp.
Grupo Aula Medica S.A., C.I. Venecia 2, Alfa III, 28050, Madrid, Spain, Tel: +34(91)3588757, Fax: +34(91)3589067 ISSUES/YR: 6

• **Nutrition**
ISSN: 0899-9007
Nutrition
Elsevier Inc., 360 Park Avenue South, New York, NY 10010, United States, Tel: +1(212)6333950, Fax: +1(212)6333990 ISSUES/YR: 12

Nutrition and Cancer
ISSN: 0163-5581
Nutr. Cancer
Lawrence Erlbaum Associates Inc., 10 Industrial Avenue, Mahwah, NJ 07430-2262, United States, Tel: +1(201)2582200, Fax: +1(201)2360072 ISSUES/YR: 6

Nutrition and Metabolic Disorders in HIV Infection
ISSN: 1579-7376
Nutr. Metab. Disord. HIV Infect.
AAMIC, C/Ecuador 14, Bajo B, 28220, Majadhonda,Madrid, Spain, Tel: +34-916343103, Fax: +34-916343103 ISSUES/YR: 4

Nutrition and Metabolism
ISSN: 1743-7075
Nutr. Metab.
BioMed Central Ltd., 34 - 42 Cleveland Street, W1T 4LB, London, United Kingdom, Tel: +44(207)3230323, Fax: +44(207)6319926 ISSUES/YR: 1

• **Nutrition Bulletin**
ISSN: 1471-9827
Nutr. Bull.
Blackwell Publishing Ltd, 9600 Garsington Road, OX4 2XG, Oxford, United Kingdom, Tel: +44(1865)776868, Fax: +44(1865)714591 ISSUES/YR: 4

Nutrition Clinique et Metabolisme
ISSN: 0985-0562
Nutr. Clin. Metab.
Editions Scientifiques et Medicales Elsevier, 23 rue Linois, 75724, Paris Cedex 15, France, Tel: +33(1)45589826, Fax: +33(1)45589428 ISSUES/YR: 4

Nutrition in Clinical Care
ISSN: 1096-6781
Nutr. Clin. Care
International Life Sciences Institute, 810 East 10th Street, Lawrence, KS 66044-8897, United States, Tel: +1(913)8431221 ISSUES/YR: 4

Nutrition in Clinical Practice
ISSN: 0884-5336
Nutr. Clin. Prac.
American Society for Parenteral and Enteral Nutrition, 8630 Fenton Steet, Suite 412, Silver Spring, MD 20910-3805, United States, Tel: +1(301)5876315 ISSUES/YR: 6

Nutrition Journal
ISSN: 1475-2891
Nutr. J.
BioMed Central Ltd., 34 - 42 Cleveland Street, W1T 4LB, London, United Kingdom, Tel: +44(207)3230323, Fax: +44(207)6319926 ISSUES/YR: 1

• **Nutrition Research**
ISSN: 0271-5317
Nutr. Res.
Elsevier Inc., 360 Park Avenue South, New York, NY 10010, United States, Tel: +1(212)6333950, Fax: +1(212)6333990 ISSUES/YR: 12

Nutrition Research Reviews
ISSN: 0954-4224
Nut. Res. Rev.
CAB International, Wallingford, OX10 8DE, Oxon, United Kingdom, Tel: +44(1491)832111, Fax: +44(1491)826090 ISSUES/YR: 2

Nutrition Reviews
ISSN: 0029-6643
Nutr. Rev.
International Life Sciences Institute, 810 East 10th Street, Lawrence, KS 66044-8897, United States, Tel: +1(913)8431221 ISSUES/YR: 12

• **Nutritional Neuroscience**
ISSN: 1028-415X
Nutr. Neurosci.
Taylor and Francis Ltd., 4 Park Square, Milton Park, OX14 4RN, Abingdon, Oxfordshire, United Kingdom, Tel: +44(20)70175541, Fax: +44(20)70175198 ISSUES/YR: 6

Nutritional Sciences Journal
ISSN: 1011-6958
Nutr. Sci. J.
Nutrition Society in Taipei, Fu Jen University, 24205, Taipei, Taiwan, Tel: +886(8862)9028725 ISSUES/YR: 4

Cardiometabolic Risk and Weight Management
ISSN: 1750-0621
Cardiometab. Risk Weight Manage.
Medical Education Partnership, 53 Hargrave Rd, N19 5SH, London, United Kingdom, Tel: +44-207 561 5400, Fax: +44-207 7561 5401 ISSUES/YR: 4

Obesity Management
ISSN: 1545-1712
Obes. Manage.
Mary Ann Liebert Inc., 140 Huguenot Street, New Rochelle, NY 10801-5215, United States, Tel: +1(914)7402100, Fax: +1(914)7402101 ISSUES/YR: 6

Obesity Reviews
ISSN: 1467-7881
Obes. Rev.
Blackwell Publishing Ltd, 9600 Garsington Road, OX4 2XG, Oxford, United Kingdom, Tel: +44(1865)776868, Fax: +44(1865)714591 ISSUES/YR: 4

Obesity Reviews, Supplement
ISSN: 1743-7008
Obes. Rev. Suppl.
Blackwell Publishing Ltd, 9600 Garsington Road, OX4 2XG, Oxford, United Kingdom, Tel: +44(1865)776868, Fax: +44(1865)714591 ISSUES/YR: 1

• **Obesity Surgery**
ISSN: 0960-8923
Obes. Surg.
FD Communications Inc., 1448 Lawrence Avenue East, Suite 302, ONT M4A 2V6, Toronto, Canada, Tel: +1(416)7509608, Fax: +1(416)2245455 ISSUES/YR: 10

Obstetrical and Gynecological Survey
ISSN: 0029-7828
Obstet. Gynecol. Surv.
Lippincott Williams and Wilkins, 351 West Camden Street, Baltimore, MD 21201-2436, United States, Tel: +1(410)5284000 ISSUES/YR: 12

• **Obstetrics and Gynecology**
ISSN: 0029-7844
Obstet. Gynecol.
Lippincott Williams and Wilkins, 241 Borough High Street, SE1 1GB, London, United Kingdom, Tel: +1(207)9407500, Fax: +1(207)9407575 ISSUES/YR: 12

• **Obstetrics and Gynecology Clinics of North America**
ISSN: 0889-8545
Obstet. Gynecol. Clin. North Am.
W.B. Saunders, Independence Square West, Philadelphia, PA 19106-3399, United States, Tel: +1(215)2387800, Fax: +1(215)2386445 ISSUES/YR: 4

• **Occupational and Environmental Medicine**
ISSN: 1351-0711
Occup. Environ. Med.
BMJ Publishing Group, Tavistock Square, WC1H 9JR, London, United Kingdom, Tel: +44(207)3874499, Fax: +44(207)3836668 ISSUES/YR: 12

Occupational Health
ISSN: 0029-7917
Occup. Health
Reed Business Information Ltd, Brighton Road, SM2 5AS, Sutton, Surrey, United Kingdom, Tel: +44-20 8652 3500, Fax: +44-20 8652 8932 ISSUES/YR: 12

• **Occupational Medicine**
ISSN: 0962-7480
Occup. Med.
Oxford University Press, Great Clarendon Street, OX2 6DP, Oxford, United Kingdom, Tel: +44(1865)353907, Fax: +44(1865)353486 ISSUES/YR: 8

Occupational Therapy in Health Care
ISSN: 0738-0577
Occup. Ther. Health Care
Haworth Press Inc., 10 Alice Street, Binghamton, NY 13904-1580, United States, Tel: +1(607)7225857, Fax: +1(607)7710012 ISSUES/YR: 4

Occupational Therapy in Mental Health
ISSN: 0164-212X
Occup. Ther. Ment. Health
Haworth Press Inc., 10 Alice Street, Binghamton, NY 13904-1580, United States, Tel: +1(607)7225857, Fax: +1(607)7710012 ISSUES/YR: 4

Occupational Therapy International
ISSN: 0966-7903
Occup. Ther. Int.
John Wiley and Sons Ltd, Southern Gate, PO19 8SQ, Chichester, West Sussex, United Kingdom, Tel: +44(1243)779777, Fax: +44(1243)775878 ISSUES/YR: 4

Ochsner Journal
ISSN: 1524-5012
Ochsner J.
Ochsner Clinic, 1514 Jefferson Highway, New Orleans, LA 70121, United States, Tel: +1(504)8426096, Fax: +1(504)8422013 ISSUES/YR: 4

Ocular Immunology and Inflammation
ISSN: 0927-3948
Ocul. Immunol. Inflamm.
Taylor and Francis Ltd., 4 Park Square, Milton Park, OX14 4RN, Abingdon, Oxfordshire, United Kingdom, Tel: +44(20)70175541, Fax: +44(20)70175198 ISSUES/YR: 6

Ocular Surface
ISSN: 1542-0124
Ocul. Surf.
ETHIS COMMUNICATIONS, INC., 75 Maiden Lane, Suite 907, New York, 10038, United States, Tel: +1(212)791-1440, Fax: +1(212)791-4980 ISSUES/YR: 4

• **Oligonucleotides**
ISSN: 1545-4576
Oligonucleotides
Mary Ann Liebert Inc., 140 Huguenot Street, New Rochelle, NY 10801-5215, United States, Tel: +1(914)7402100, Fax: +1(914)7402101 ISSUES/YR: 4

• **OMICS A Journal of Integrative Biology**
ISSN: 1536-2310
OMICS J. Integr. Biol.
Mary Ann Liebert Inc., 140 Huguenot Street, New Rochelle, NY 10801-5215, United States, Tel: +1(914)7402100, Fax: +1(914)7402101 ISSUES/YR: 4

• **Oncogene**
ISSN: 0950-9232
Oncogene
Nature Publishing Group, Houndmills, RG21 6XS, Basingstoke, Hampshire, United Kingdom, Tel: +44(1256)329242, Fax: +44(1256)810526 ISSUES/YR: 56

Oncologia
ISSN: 0378-4835
Oncologia (Spain)
Alpe Editores S.A., Pedro Rico 27, oficinas 11-12, 28029, Madrid, Spain, Tel: +34(91)7338811, Fax: +34(91)3159652 ISSUES/YR: 10

Oncologie
ISSN: 1292-3818
Oncologie
Springer-Verlag France, 1 rue Paul Cezanne, 75008, Paris, France, Tel: +33(1)53933628, Fax: +33(1)53933683 ISSUES/YR: 8

• **Oncologist**
ISSN: 1083-7159
Oncologist
AlphaMed Press, 318 Blackwell St. Suite 260, Durham, NC 27701-2884, United States, Tel: +1(937)2912355, Fax: +1(937)2914229 ISSUES/YR: 10

• **Oncology**
ISSN: 0030-2414
Oncology (Switzerland)
S. Karger AG, Allschwilerstrasse 10, P.O. Box, CH-4009, Basel, Switzerland, Tel: +41(61)3061111, Fax: +41(61)3061234 ISSUES/YR: 8

ONCOLOGY
ISSN: 0890-9091
ONCOLOGY (USA)
PRR Inc., 48 South Service Road, Melville, NY 11474-2335, United States, Tel: +1(516)4248900 ISSUES/YR: 12

Oncology Forum
ISSN: 1027-779X
Oncol. Forum
Scientific Communications International Ltd, Suite C, 10/F Wo On Building, 10 Wo On Lane, Central, Hong Kong, Tel: +852-28 68 9171, Fax: +852-28 68 9269 ISSUES/YR: 4

Current

Oncology Report
ISSN: 1548-5323
Oncol. Rep.
Biolink Communications, 46 Green Street, 2nd Floor, Huntington, NY 11743, United States, Tel: +1(631)4248900, Fax: +1(631)4248905 ISSUES/YR: 2

- **Oncology Research**
ISSN: 0965-0407
Oncol. Res.
Cognizant Communication Corporation, 3 Hartsdale Road, Elmsford, NY 10523-3701, United States, Tel: +1(914)5927720, Fax: +1(914)5928981 ISSUES/YR: 12

Ondokuz Mayis Universitesi Tip Dergisi
ISSN: 1300-2996
Ondokuz Mayis Univ. Tip Derg.
Ondokuz Mayis Universitesi, Kurupelit, 55139, Samsun, Turkey, Tel: +90(362)4576000 ISSUES/YR: 4

Onkologe
ISSN: 0947-8965
Onkologe
Springer Verlag, Tiergartenstrasse 17, D-69121, Heidelberg, Germany, Tel: +49(6221)4870, Fax: +49(6221)487366 ISSUES/YR: 12

Onkologia Polska
ISSN: 1505-6732
Onkol. Pols.
LEXiMED, ul. Moniuszki 7, 48-300, Nysa, Poland, Tel: +48-077 433 41 88, Fax: +48-077 433 41 88 ISSUES/YR: 4

Onkologie
ISSN: 0378-584X
Onkologie
S. Karger AG, Allschwilerstrasse 10, P.O. Box, CH-4009, Basel, Switzerland, Tel: +49(61)3061111, Fax: +49(61)3061234 ISSUES/YR: 6

Online Brazilian Journal of Nursing
ISSN: 1676-4285
Online Braz. J. Nurs.
Fluminense Federal University, Rua Dr. Celestino, 74, 24020-091, Niteroi, Brazil, Tel: +55(51)21-27194411, Fax: +55(51)21-27198073 ISSUES/YR: 3

Online Journal of Health and Allied Sciences
ISSN: 0972-5997
Online J. Health Allied Sci.
Online Journal of Health and Allied Sciences, Light House Hill Road, 575001, Mangalore, India ISSUES/YR: 1

Online Journal of Issues in Nursing
ISSN: 1091-3734
Online J. Issues Nurs.
American Nurses Association, 600 Maryland Ave S W, Ste 100 W, Washington, DC 20024-2571, United States, Tel: +1(202)6517000, Fax: +1(202)6517005 ISSUES/YR: 3

Online Journal of Knowledge Synthesis for Nursing
ISSN: 1072-7639
Online J. Knowl. Synth. Nurs.
Sigma Theta Tau International Honor Society of Nursing, 550 W North Street, Indianapolis, IN 46202, United States, Tel: +1(317)6348171, Fax: +1(317)6348188 ISSUES/YR: 2

Online Journal of Nursing Informatics
ISSN: 1089-9758
Online J. Nurs. Inform.
OJNI Corp., RD 7, Box 44, Kittanning, PA 16201, United States ISSUES/YR: 3

Online Publishing in Medicine
ISSN: 1609-2775
Online Publ. Med.
VICER Publishing, P.O. Box 14, A-1097, Vienna, Austria, Tel: +43(676)9568085, Fax: +43(676)9568086 ISSUES/YR: 4

- **Operative Orthopadie und Traumatologie**
ISSN: 0934-6694
Oper. Orthop. Traumatol.
Urban und Vogel GmbH, Neumarkter Strasse 43, D-81673, Munchen, Germany, Tel: +49(89)43721300, Fax: +49(89)43721399 ISSUES/YR: 4

- **Operative Techniques in General Surgery**
ISSN: 1524-153X
Oper. Tech. Gen. Surg.
W.B. Saunders, Independence Square West, Philadelphia, PA 19106-3399, United States, Tel: +1(215)2387800, Fax: +1(215)2386445 ISSUES/YR: 4

- **Operative Techniques in Orthopaedics**
ISSN: 1048-6666
Oper. Tech. Orthop.
W.B. Saunders, Independence Square West, Philadelphia, PA 19106-3399, United States, Tel: +1(215)2387800, Fax: +1(215)2386445 ISSUES/YR: 4

- **Operative Techniques in Otolaryngology - Head and Neck Surgery**
ISSN: 1043-1810
Oper. Tech. Otolaryngol. Head Neck Surg.
W.B. Saunders, Independence Square West, Philadelphia, PA 19106-3399, United States, Tel: +1(215)2387800, Fax: +1(215)2386445 ISSUES/YR: 4

- **Operative Techniques in Plastic and Reconstructive Surgery**
ISSN: 1071-0949
Oper. Tech. Plast. Reconstr. Surg.
W.B. Saunders, Independence Square West, Philadelphia, PA 19106-3399, United States, Tel: +1(215)2387800, Fax: +1(215)2386445 ISSUES/YR: 4

- **Operative Techniques in Sports Medicine**
ISSN: 1060-1872
Oper. Tech. Sports Med.
W.B. Saunders, Independence Square West, Philadelphia, PA 19106-3399, United States, Tel: +1(215)2387800, Fax: +1(215)2386445 ISSUES/YR: 4

- **Operative Techniques in Thoracic and Cardiovascular Surgery**
ISSN: 1522-2942
Oper. Tech. Thorac. Cardiovasc. Surg.
W.B. Saunders, Independence Square West, Philadelphia, PA 19106-3399, United States, Tel: +1(215)2387800, Fax: +1(215)2386445 ISSUES/YR: 4

- **Ophthalmic and Physiological Optics**
ISSN: 0275-5408
Ophthalmic Physiol. Opt.
Blackwell Publishing Ltd, 9600 Garsington Road, OX4 2XG, Oxford, United Kingdom, Tel: +44(1865)776868, Fax: +44(1865)714591 ISSUES/YR: 6

- **Ophthalmic Epidemiology**
ISSN: 0928-6586
Ophthalmic Epidemiol.
Taylor and Francis Ltd., 4 Park Square, Milton Park, OX14 4RN, Abingdon, Oxfordshire, United Kingdom, Tel: +44(20)70175541, Fax: +44(20)70175198 ISSUES/YR: 6

- **Ophthalmic Genetics**
ISSN: 1381-6810
Ophthalmic Genet.
Taylor and Francis Ltd., 4 Park Square, Milton Park, OX14 4RN, Abingdon, Oxfordshire, United Kingdom, Tel: +44(20)70175541, Fax: +44(20)70175198 ISSUES/YR: 4

- **Ophthalmic Plastic and Reconstructive Surgery**
ISSN: 0740-9303
Ophthalmic Plast. Reconstr. Surg.
Lippincott Williams and Wilkins, 530 Walnut Street, Philadelphia, PA 19106-3621, United States, Tel: +1(215)5218342, Fax: +1(215)5218326 ISSUES/YR: 6

- **Ophthalmic Research**
ISSN: 0030-3747
Ophthalmic Res.
S. Karger AG, Allschwilerstrasse 10, P.O. Box, CH-4009, Basel, Switzerland, Tel: +41(61)3061111, Fax: +41(61)3061234 ISSUES/YR: 6

- **Ophthalmic Surgery Lasers and Imaging**
ISSN: 1542-8877
Ophthalmic Surg. Lasers Imaging
Slack Incorporated, 6900 Grove Road, Thorofare, NJ 08086-9447, United States, Tel: +1(856)8481000, Fax: +1(856)8535991 ISSUES/YR: 6

Ophthalmologe
ISSN: 0941-293X
Ophthalmologe
Springer Verlag, Tiergartenstrasse 17, D-69121, Heidelberg, Germany, Tel: +49(6221)4870, Fax: +49(6221)487366 ISSUES/YR: 12

Ophthalmologica
ISSN: 0030-3755
Ophthalmologica
S. Karger AG, Allschwilerstrasse 10, P.O. Box, CH-4009, Basel, Switzerland, Tel: +41(61)3061111, Fax: +41(61)3061234 ISSUES/YR: 6

• **Ophthalmology**
ISSN: 0161-6420
Ophthalmology
Elsevier Inc., 360 Park Avenue South, New York, NY 10010, United States, Tel: +1(212)6333950, Fax: +1(212)6333990 ISSUES/YR: 12

• **Ophthalmology Clinics of North America**
ISSN: 0896-1549
Ophthalmol. Clin. North Am.
W.B. Saunders, Independence Square West, Philadelphia, PA 19106-3399, United States, Tel: +1(215)2387800, Fax: +1(215)2386445 ISSUES/YR: 4

Ophthalmology in China
ISSN: 1004-4469
Ophthalmol. China
Editorial Board of Ophthalmology in China, 17, Chongnei Dajie Hougouhutong, 100005, Beijing, China, Tel: +86(10)58265902 ISSUES/YR: 6

• **Optometry and Vision Science**
ISSN: 1040-5488
Optom. Vis. Sci.
Lippincott Williams and Wilkins, 351 West Camden Street, Baltimore, MD 21201-2436, United States, Tel: +1(410)5284000 ISSUES/YR: 12

Oral and Maxillofacial Surgery Clinics of North America
ISSN: 1042-3699
Oral Maxillofac. Surg. Clin. North Am.
W.B. Saunders, Independence Square West, Philadelphia, PA 19106-3399, United States, Tel: +1(215)2387800, Fax: +1(215)2386445 ISSUES/YR: 4

• **Oral Diseases**
ISSN: 1354-523X
Oral Dis.
Blackwell Munksgaard, 1 Rosenorns Alle, P.O. Box 227, DK-1502, Copenhagen V, Denmark, Tel: +45(77)333333, Fax: +45(77)333377 ISSUES/YR: 6

Oral Microbiology and Immunology
ISSN: 0902-0055
Oral Microbiol. Immunol.
Blackwell Publishing Ltd, 9600 Garsington Road, OX4 2XG, Oxford, United Kingdom, Tel: +44(1865)776868, Fax: +44(1865)714591 ISSUES/YR: 6

• **Oral Oncology**
ISSN: 1368-8375
Oral Oncol.
Elsevier Ltd, Langford Lane, Kidlington, OX5 1GB, Oxford, United Kingdom, Tel: +44(1865)843000, Fax: +44(1865)843010 ISSUES/YR: 10

Oral Radiology
ISSN: 0911-6028
Oral Radiol.
Japanese Soc. for Oral and Maxillofacial Radiology, 1 - 8 Yamadaoka, Suita, 565, Osaka, Japan, Tel: +81(6)8765711 ISSUES/YR: 2

Oral Therapeutics and Pharmacology
ISSN: 0288-1012
Oral Ther. Pharmacol.
Japanese Society of Oral Therapeutics and Pharmacology, Fukagawa 2-Chome, TOKYO -135-0033, Koto-ku, Japan, Tel: +81-03 3441 7262, Fax: +81-03 3441 3472 ISSUES/YR: 3

Orbit
ISSN: 0167-6830
Orbit
Taylor and Francis Ltd., 4 Park Square, Milton Park, OX14 4RN, Abingdon, Oxfordshire, United Kingdom, Tel: +44(20)70175541, Fax: +44(20)70175198 ISSUES/YR: 4

Organogenesis
ISSN: 1547-6278
Organogenesis
Landes Bioscience, 810 S Church Street, Georgetown, 78626, United States, Tel: +1-512.863.7762, Fax: +1-512.863.0081 ISSUES/YR: 4

• **ORL**
ISSN: 0301-1569
ORL
S. Karger AG, Allschwilerstrasse 10, P.O. Box, CH-4009, Basel, Switzerland, Tel: +41(61)3061111, Fax: +41(61)3061234 ISSUES/YR: 6

ORL-DIPS
ISSN: 0210-7309
ORL DIPS
Nexus Medica, c/Caracas 33, 08030, Barcelona, Spain, Tel: +34(93)4593492, Fax: +34(93)4590283 ISSUES/YR: 4

Orthopade
ISSN: 0085-4530
Orthopade
Springer Verlag, Tiergartenstrasse 17, D-69121, Heidelberg, Germany, Tel: +49(6221)4870, Fax: +49(6221)487366 ISSUES/YR: 12

• **Orthopedic Clinics of North America**
ISSN: 0030-5898
Orthop. Clin. North Am.
W.B. Saunders, Independence Square West, Philadelphia, PA 19106-3399, United States, Tel: +1(215)2387800, Fax: +1(215)2386445 ISSUES/YR: 4

Orthopedics
ISSN: 0147-7447
Orthopedics
Slack Incorporated, 6900 Grove Road, Thorofare, NJ 08086-9447, United States, Tel: +1(856)8481000, Fax: +1(856)8535991 ISSUES/YR: 12

Ortopedia Traumatologia Rehabilitacja
ISSN: 1509-3492
Ortop. Traumatol. Rehab.
Agencja Wydawnicza Medsportpress, ul. Marymoncka 34 (skr. 23), 01-813, Warsaw, Poland, Tel: +48(22)8346772, Fax: +48(22)8346772 ISSUES/YR: 4

Orvosi Hetilap
ISSN: 0030-6002
Orvosi Hetil.
Medicina Kiado RT, Zoltan u. 8., 1054, Budapest, Hungary ISSUES/YR: 52

Orvoskepzes
ISSN: 0030-6037
Orvoskepzes
Orvoskepzes, Szabolcs utca 33-35, H-1389, Budapest, Hungary, Tel: +36(1)1208588 ISSUES/YR: 6

Ospedale Maggiore
ISSN: 0369-7843
Osp. Magg.
Masson SpA, Via F.lli Bressan 2, 20126, Milan, Italy, Tel: +39(02)270741, Fax: +39(02)27074211 ISSUES/YR: 4

• **Osteoarthritis and Cartilage**
ISSN: 1063-4584
Osteoarthritis Cartilage
W.B. Saunders Ltd, 32 Jamestown Road, NW1 7BY, London, United Kingdom, Tel: +44(20)74244200, Fax: +44(20)74854752 ISSUES/YR: 12

Osteologicky Bulletin
ISSN: 1211-3778
Osteol. Bull.
Trios spol. s.r.o., Zakourilova 142, 149 00, Praha 4, Czech Republic, Tel: +420(267)912030, Fax: +420(267)915563 ISSUES/YR: 4

Osteologie
ISSN: 1019-1291
Osteologie
Verlag Hans Huber AG, Langgassstrasse 76, CH-3000, Bern, Switzerland, Tel: +41(31)3004500, Fax: +41(31)3004590 ISSUES/YR: 4

Osteopathische Medizin
ISSN: 1615-9071
Osteopath. Med.
Urban und Fischer Verlag Jena, P.O. Box 100537, 07705, Jena, Germany, Tel: +49(3641)6263, Fax: +49(3641)626500 ISSUES/YR: 4

• **Osteoporosis International**
ISSN: 0937-941X
Osteoporosis Int.
Springer-Verlag London Ltd, Catteshall Road, Farncombe, GU7 3DJ, Godalming, Surrey, United Kingdom, Tel: +44(1483)414142, Fax: +44(1483)415144 ISSUES/YR: 12

OTJR Occupation, Participation and Health
ISSN: 1539-4492
OTJR Occup. Part. Health
Slack Incorporated, 6900 Grove Road, Thorofare, NJ 08086-9447, United States, Tel: +1(856)8481000, Fax: +1(856)8535991 ISSUES/YR: 4

Oto-Rhino-Laryngology Tokyo
ISSN: 0386-9687
Oto-Rhino-Laryngol. Tokyo
Society of Oto-Rhino-Laryngology Tokyo, 25-8 Nishishinbashi 3-chome, Minato-ku, 105, Tokyo, Japan, Tel: +81(3)34331111 ISSUES/YR: 6

Otolaryngologic Clinics of North America
ISSN: 0030-6665
Otolaryngol. Clin. North Am.
W.B. Saunders, Independence Square West, Philadelphia, PA 19106-3399, United States, Tel: +1(215)2387800, Fax: +1(215)2386445 ISSUES/YR: 6

Otolaryngology - Head and Neck Surgery
ISSN: 0194-5998
Otolaryngol. Head Neck Surg.
Mosby Inc., 11830 Westline Industrial Drive, St. Louis, MO 63146, United States, Tel: +1(314)8728370, Fax: +1(314)4321380 ISSUES/YR: 12

Otolaryngology - Head and Neck Surgery (Tokyo)
ISSN: 0914-3491
Otolaryngol. Head Neck Surg. Tokyo
Igaku-Shoin Ltd, 5-24-3 Hongo, Bunkyo-ku, 113-91, Tokyo, Japan, Tel: +81(3)38175600 ISSUES/YR: 13

• **Otology and Neurotology**
ISSN: 1531-7129
Otol. Neurotol.
Lippincott Williams and Wilkins, 530 Walnut Street, Philadelphia, PA 19106-3621, United States, Tel: +1(215)5218342, Fax: +1(215)5218326 ISSUES/YR: 6

Otorinolaringologia
ISSN: 0026-4938
Otorinolaringologia
Edizioni Minerva Medica S.p.A., Corso Bramante 83-85, 10126, Turin, Italy, Tel: +39(011)678282, Fax: +39(011)674502 ISSUES/YR: 4

Otorinolaringologia Pediatrica
ISSN: 1120-3455
Otorinolaringol. Pediatr.
CIC Edizioni Internazionali s.r.l., Corso Trieste 42, 00198, Roma, Italy, Tel: +39(06)8412673, Fax: +39(06)8412688 ISSUES/YR: 4

Otorinolaryngologie a Foniatrie
ISSN: 1210-7867
Otorinolaryngol. Foniatr.
Czech Medical Association J.E. Purkyne, Sokolska 31, P.O. Box 88, 120 26, Praha 2, Czech Republic, Tel: +420(2)24266201-4, Fax: +420(2)24266212 ISSUES/YR: 4

P and T
ISSN: 1052-1372
P T
Medi Media USA Inc, 780 Township Line Road, Yardley, PA 19067, United States, Tel: +1-267 685 2779, Fax: +1-267 685 2966 ISSUES/YR: 12

PACE - Pacing and Clinical Electrophysiology
ISSN: 0147-8389
PACE Pacing Clin. Electrophysiol.
Blackwell Publishing Inc., 350 Main Street, Malden, MA 02148, United States, Tel: +1(781)3888200, Fax: +1(781)3888210 ISSUES/YR: 12

Padiatrische Praxis
ISSN: 0030-9346
Padiatr. Prax.
Hans Marseille Verlag GmbH, P.O. Box 22 13 41, 80503, Munchen, Germany, Tel: +49(89)227988, Fax: +49(89)2904643 ISSUES/YR: 4

Paediatria Croatica
ISSN: 1330-1403
Paediatr. Croat.
Children's Hospital Zagreb, Klaiceva 16, 10 000, Zagreb, Croatia, Tel: +385(1) 46 00 111 ISSUES/YR: 4

• **Paediatric Anaesthesia**
ISSN: 1155-5645
Paediatr. Anaesth.
Blackwell Publishing Ltd, 9600 Garsington Road, OX4 2XG, Oxford, United Kingdom, Tel: +44(1865)776868, Fax: +44(1865)714591 ISSUES/YR: 12

Paediatric and Perinatal Drug Therapy
ISSN: 1463-0095
Paediatr. Perinat. Drug Ther.
Librapharm, Venture West, New Greenham Park, RG19 6HX, Newbury, Berks., United Kingdom, Tel: +44(1635)522651, Fax: +44(1635)36294 ISSUES/YR: 4

Paediatric and Perinatal Epidemiology
ISSN: 0269-5022
Paediatr. Perinat. Epidemiol.
Blackwell Publishing Ltd, 9600 Garsington Road, OX4 2XG, Oxford, United Kingdom, Tel: +44(1865)776868, Fax: +44(1865)714591 ISSUES/YR: 6

• **Paediatric Respiratory Reviews**
ISSN: 1526-0542
Paediatr. Respir. Rev.
W.B. Saunders Ltd, 32 Jamestown Road, NW1 7BY, London, United Kingdom, Tel: +44(20)74244200, Fax: +44(20)74854752 ISSUES/YR: 4

Paediatrics and Child Health
ISSN: 1205-7088
Paediatr. Child Health
Pulsus Group Inc., 2902 South Sheridan Way, ONT L6J 7L6, Oakville, Canada, Tel: +1(905)8294770, Fax: +1(905)8294799 ISSUES/YR: 10

• **Pain**
ISSN: 0304-3959
Pain
Elsevier, P.O. Box 211, 1000 AE, Amsterdam, Netherlands, Tel: +31(20)4853757, Fax: +31(20)4853432 ISSUES/YR: 15

Pain Clinic
ISSN: 0169-1112
Pain Clinic
VSP BV, P.O.Box 9000, 2300 PA, Leiden, Netherlands, Tel: +31-71 535 3500, Fax: +31-71 531 7532 ISSUES/YR: 4

Pain Digest
ISSN: 0938-9016
Pain Dig.
Springer New York LLC, 233 Springer Street, New York, 10013-1578, United States, Tel: +1-212 460 1501, Fax: +1-212 460 1595 ISSUES/YR: 6

Pain Medicine
ISSN: 1526-2375
Pain Med. (USA)
Blackwell Publishing Inc., 350 Main Street, Malden, MA 02148, United States, Tel: +1(781)3888200, Fax: +1(781)3888210 ISSUES/YR: 6

Pain Medicine
ISSN: 1681-1232
Pain Med. (Austria)
VICER Publishing, P.O. Box 14, A-1097, Vienna, Austria, Tel: +43(676)9568085, Fax: +43(676)9568086 ISSUES/YR: 4

Current

Pain Physician
ISSN: 1533-3159
Pain Phys.
Association of Pain Management Anesthesiologists, 2831 Lone Oak Road, Paducah, KT 42003, United States, Tel: +1(270)5549412, Fax: +1(270)5548987 ISSUES/YR: 4

Pain Practice
ISSN: 1530-7085
Pain Pract.
Blackwell Publishing Inc., 350 Main Street, Malden, MA 02148, United States, Tel: +1(781)3888200, Fax: +1(781)3888210 ISSUES/YR: 4

Pain Research and Management
ISSN: 1203-6765
Pain Res. Manage.
Pulsus Group Inc., 2902 South Sheridan Way, ONT L6J 7L6, Oakville, Canada, Tel: +1(905)8294770, Fax: +1(905)8294799 ISSUES/YR: 4

Pakistan Journal of Medical Sciences
ISSN: 1682-024X
Pak. J. Med. Sci.
Professional Medical Publications, Raja Ghazanfar Ali Road, Saddar, Karachi, Pakistan, Tel: +92(21)5688791, Fax: +92(21)5689860 ISSUES/YR: 4

Pakistan Journal of Nutrition
ISSN: 1680-5194
Pak. J. Nutr.
Asian Network for Scientific Information, 308-Lasani Town, Sargodha Road, Faisalabad, Pakistan, Tel: +92(41)2001145-46, Fax: +92(21)5206036 ISSUES/YR: 6

Palliative Medicine
ISSN: 0269-2163
Palliative Med.
Arnold, 338 Euston Road, NW1 3BH, London, United Kingdom, Tel: +44(20)78736000, Fax: +44(20)78736376 ISSUES/YR: 8

Pan Arab Journal of Neurosurgery
ISSN: 1319-6995
Pan Arab J. Neurosurg.
Pan Arab Neurosurgical Society, PO BOX 7897, 11159, Riyadh, Saudi Arabia, Tel: +966-4777714, Fax: +966-4784057 ISSUES/YR: 2

• **Pancreas**
ISSN: 0885-3177
Pancreas
Lippincott Williams and Wilkins, 530 Walnut Street, Philadelphia, PA 19106-3621, United States, Tel: +1(215)5218342, Fax: +1(215)5218326 ISSUES/YR: 8

• **Pancreatology**
ISSN: 1424-3903
Pancreatology
S. Karger AG, Allschwilerstrasse 10, P.O. Box, CH-4009, Basel, Switzerland, Tel: +41(61)3061111, Fax: +41(61)3061234 ISSUES/YR: 6

• **Parasite Immunology**
ISSN: 0141-9838
Parasite Immunol.
Blackwell Publishing Ltd, 9600 Garsington Road, OX4 2XG, Oxford, United Kingdom, Tel: +44(1865)776868, Fax: +44(1865)714591 ISSUES/YR: 12

• **Parasitology**
ISSN: 0031-1820
Parasitology
Cambridge University Press, Shaftesbury Road, CB2 2RU, Cambridge, United Kingdom, Tel: +44(1223)312393, Fax: +44(1223)315052 ISSUES/YR: 12

• **Parasitology International**
ISSN: 1383-5769
Parasitol. Int.
Elsevier Ireland Ltd, P.O. Box 85, Limerick, Ireland, Tel: +353(61)471944, Fax: +353(61)472144 ISSUES/YR: 4

• **Parasitology Research**
ISSN: 0932-0113
Parasitol. Res.
Springer Verlag, Tiergartenstrasse 17, D-69121, Heidelberg, Germany, Tel: +49(6221)4870, Fax: +49(6221)487366 ISSUES/YR: 12

• **Parkinsonism and Related Disorders**
ISSN: 1353-8020
Parkinsonism Relat. Disord.
Elsevier Ltd, Langford Lane, Kidlington, OX5 1GB, Oxford, United Kingdom, Tel: +44(1865)843000, Fax: +44(1865)843010 ISSUES/YR: 8

• **Particle and Fibre Toxicology**
ISSN: 1743-8977
Part. Fibre Toxicol.
BioMed Central Ltd., 34 - 42 Cleveland Street, W1T 4LB, London, United Kingdom, Tel: +44(207)3230323, Fax: +44(207)6319926 ISSUES/YR: 1

• **Particulate Science and Technology**
ISSN: 0272-6351
Part. Sci. Technol.
Taylor and Francis Inc., 325 Chestnut St, Suite 800, Philadelphia PA, PA 19106, United States, Tel: +1(800)3541420, Fax: +1(215)6258914 ISSUES/YR: 4

PAT
ISSN: 1549-4446
PAT
Russell Publishing LLC, 8888 Keystone Crossing, Suite 1420, Indianapolis, IN 46240-2182, United States, Tel: +1(317)8168787, Fax: +1(317)8168788 ISSUES/YR: 6

• **Pathobiology**
ISSN: 1015-2008
Pathobiology
S. Karger AG, Allschwilerstrasse 10, P.O. Box, CH-4009, Basel, Switzerland, Tel: +41(61)3061111, Fax: +41(61)3061234 ISSUES/YR: 6

Pathologe
ISSN: 0172-8113
Pathologe
Springer Verlag, Tiergartenstrasse 17, D-69121, Heidelberg, Germany, Tel: +49(6221)4870, Fax: +49(6221)487366 ISSUES/YR: 6

Pathologica
ISSN: 0031-2983
Pathologica
Pacini Editore s.r.l., Via A. Gherardesca 1, 56121, Ospedaletto (Pisa), Italy, Tel: +39(050)313011, Fax: +39(050)3130300 ISSUES/YR: 6

Pathologie Biologie
ISSN: 0369-8114
Pathol. Biol.
Editions Scientifiques et Medicales Elsevier, 23 rue Linois, 75724, Paris Cedex 15, France, Tel: +33(1)45589826, Fax: +33(1)45589428 ISSUES/YR: 10

Pathology
ISSN: 0031-3025
Pathology
Taylor and Francis Ltd., 4 Park Square, Milton Park, OX14 4RN, Abingdon, Oxfordshire, United Kingdom, Tel: +44(20)70175541, Fax: +44(20)70175198 ISSUES/YR: 6

• **Pathology Case Reviews**
ISSN: 1082-9784
Pathol. Case Rev.
Lippincott Williams and Wilkins, 530 Walnut Street, Philadelphia, PA 19106-3621, United States, Tel: +1(215)5218342, Fax: +1(215)5218326 ISSUES/YR: 6

• **Pathology International**
ISSN: 1320-5463
Pathol. Int.
Blackwell Publishing, 550 Swanston Street, VIC 3053, Carlton South, Australia, Tel: +61(3)8359 1011, Fax: +61(3)8359 1120 ISSUES/YR: 12

Pathology Oncology Research
ISSN: 1219-4956
Pathol. Oncol. Res.
1st Institute of Pathology and Experimental Cancer Research, Ullio ut 26, 1085, Budapest, Hungary, Tel: +36-1 317 0891, Fax: +36-1 317 0891 ISSUES/YR: 4

Pathology Research and Practice
ISSN: 0344-0338
Pathol. Res. Pract.
Urban und Fischer Verlag GmbH und Co. KG, P.O. Box 201930, 80019, Munchen, Germany, Tel: +49(89)53830, Fax: +49(89)5383221 ISSUES/YR: 12

Current

- **Pathophysiology**
  ISSN: 0928-4680
  Pathophysiology
  Elsevier, P.O. Box 211, 1000 AE, Amsterdam, Netherlands, Tel: +31(20)4853757, Fax: +31(20)4853432 ISSUES/YR: 4
- **Pathophysiology of Haemostasis and Thrombosis**
  ISSN: 1424-8832
  Pathophysiol. Haemost. Thromb.
  S. Karger AG, Allschwilerstrasse 10, P.O. Box, CH-4009, Basel, Switzerland, Tel: +41(61)3061111, Fax: +41(61)3061234 ISSUES/YR: 6
- **Patient Education and Counseling**
  ISSN: 0738-3991
  Patient Educ. Couns.
  Elsevier Ireland Ltd, P.O. Box 85, Limerick, Ireland, Tel: +353(61)471944, Fax: +353(61)472144 ISSUES/YR: 12

Patologia del Aparato Locomotor
ISSN: 1696-5981
Patol. Aparato Locomotor
Fundacion Mapfre Medicina, Ctra. Pozuelo a Majadahonda, s/n, 28220, Majadahonda (Madrid), Spain, Tel: +34(91)6265704, Fax: +34(91)6265825 ISSUES/YR: 4

Patoloji Bulteni
ISSN: not available yet
Patol. Bul.
Ankara Society of Pathology, Sihhiye, 06100, Ankara, Turkey, Tel: +90(312)3242348, Fax: +90(312)4689420 ISSUES/YR: 3

PDA Journal of Pharmaceutical Science and Technology
ISSN: 1079-7440
PDA J. Pharm. Sci. Technol.
Parenteral Drug Association Inc., 3Bethesda MetroCenter, Bethesda, MD 20814, United States, Tel: +1-301 656 5900, Fax: +1-301 986 1093 ISSUES/YR: 6

Pediatria Catalana
ISSN: 1135-8831
Pediatr. Catalana
Fundacio Catalana de Pediatria, Passeig de la Bonanova 47, 2n, 08017, Barcelona, Spain, Tel: +34(93)2123895 ISSUES/YR: 6

Pediatria i Medycyna Rodzinna
ISSN: 1734-1531
Pediatr. Med. Rodzinna
Medical Communications, Ul.Ojcowska 11, 02-918, Warszawa, Poland, Tel: +48-226519783, Fax: +48-8425363 ISSUES/YR: 4

Pediatria Integral
ISSN: 1135-4542
Pediatr. Integr.
Ediciones Ergon SA, C/Arboleda 1, 28220, Majadahonda (Madrid), Spain, Tel: +34(91)6362937, Fax: +34(91)6362930 ISSUES/YR: 4

Pediatria Medica e Chirurgica
ISSN: 0391-5387
Pediatr. Med. Chir.
Ospedale Civile, 36100, Vicenza, Italy, Tel: +39(0444)993887 ISSUES/YR: 6

Pediatria Oggi Medica e Chirurgica
ISSN: 0391-898X
Pediatr. Oggi Med. Chir.
CIC Edizioni Internazionali s.r.l., Corso Trieste 42, 00198, Roma, Italy, Tel: +39(06)8412673, Fax: +39(06)8412688 ISSUES/YR: 4

Pediatria Polska
ISSN: 0031-3939
Pediatr. Pol.
Urban and Partner, ul. M. Sklodowskiej-Curie 55/61, 50-950, Wroclaw, Poland, Tel: +48(71)3285487, Fax: +48(71)3284391 ISSUES/YR: 12

Pediatria Wspolczesna
ISSN: 1507-5532
Pediatr. Wspolczesna
LEXiMED, ul. Moniuszki 7, 48-300, Nysa, Poland, Tel: +48-077 433 41 88, Fax: +48-077 433 41 88 ISSUES/YR: 4

- **Pediatric Allergy and Immunology**
  ISSN: 0905-6157
  Pediatr. Allergy Immunol.
  Blackwell Munksgaard, 1 Rosenorns Alle, P.O. Box 227, DK-1502, Copenhagen V, Denmark, Tel: +45(77)333333, Fax: +45(77)333377 ISSUES/YR: 8
- **Pediatric and Developmental Pathology**
  ISSN: 1093-5266
  Pediatr. Dev. Pathol.
  Society for Pediatric Pathology, P. O. Box 7075, Lawrence, KS 66044, United States, Tel: +1-785 843 1235 EXT: 297, Fax: +1(785)843 1274 ISSUES/YR: 6

Pediatric Annals
ISSN: 0090-4481
Pediatr. Ann.
Slack Incorporated, 6900 Grove Road, Thorofare, NJ 08086-9447, United States, Tel: +1(856)8481000, Fax: +1(856)8535991 ISSUES/YR: 12

- **Pediatric Asthma, Allergy and Immunology**
  ISSN: 0883-1874
  Pediatr. Asthma Allergy Immunol.
  Mary Ann Liebert Inc., 140 Huguenot Street, New Rochelle, NY 10801-5215, United States, Tel: +1(914)7402100, Fax: +1(914)7402101 ISSUES/YR: 4
- **Pediatric Blood and Cancer**
  ISSN: 1545-5009
  Pediatr. Blood Cancer
  Wiley-Liss Inc., 111 River Street, Hoboken, NJ 07030-5774, United States, Tel: +1(201)7486645, Fax: +1(201)7486021 ISSUES/YR: 14

Pediatric Cardiac Surgery Annual
ISSN: 1092-9126
Pediatr. Card. Surg. Annu.
W.B. Saunders, Independence Square West, Philadelphia, PA 19106-3399, United States, Tel: +1(215)2387800, Fax: +1(215)2386445 ISSUES/YR: 1

Pediatric Cardiology
ISSN: 0172-0643
Pediatr. Cardiol.
Springer New York LLC, 233 Springer Street, New York, 10013-1578, United States, Tel: +1-212 460 1501, Fax: +1-212 460 1595 ISSUES/YR: 6

Pediatric Case Reviews
ISSN: 1532-9798
Pediatr. Case Rev.
Lippincott Williams and Wilkins, 530 Walnut Street, Philadelphia, PA 19106-3621, United States, Tel: +1(215)5218342, Fax: +1(215)5218326 ISSUES/YR: 4

Pediatric Clinics of North America
ISSN: 0031-3955
Pediatr. Clin. North Am.
W.B. Saunders, Independence Square West, Philadelphia, PA 19106-3399, United States, Tel: +1(215)2387800, Fax: +1(215)2386445 ISSUES/YR: 6

- **Pediatric Critical Care Medicine**
  ISSN: 1529-7535
  Pediatr. Crit. Care Med.
  Lippincott Williams and Wilkins, 530 Walnut Street, Philadelphia, PA 19106-3621, United States, Tel: +1(215)5218342, Fax: +1(215)5218326 ISSUES/YR: 6
- **Pediatric Dermatology**
  ISSN: 0736-8046
  Pediatr. Dermatol.
  Blackwell Publishing Inc., 350 Main Street, Malden, MA 02148, United States, Tel: +1(781)3888200, Fax: +1(781)3888210 ISSUES/YR: 6
- **Pediatric Diabetes**
  ISSN: 1399-543X
  Pediatr. Diabetes
  Blackwell Munksgaard, 1 Rosenorns Alle, P.O. Box 227, DK-1502, Copenhagen V, Denmark, Tel: +45(77)333333, Fax: +45(77)333377 ISSUES/YR: 4
- **Pediatric Diabetes, Supplement**
  ISSN: 1745-1426
  Pediatr. Diabetes Suppl.
  Blackwell Munksgaard, 1 Rosenorns Alle, P.O. Box 227, DK-1502, Copenhagen V, Denmark, Tel: +45(77)333333, Fax: +45(77)333377 ISSUES/YR: 1

Current

- **Pediatric Drugs**
ISSN: 1174-5878
Pediatr. Drugs
Adis International Ltd, 41 Centorian Drive, Private Bag 65901, Mairangi Bay, 1311, Auckland 10, New Zealand, Tel: +64(9)4770700, Fax: +64(9)4770766 ISSUES/YR: 6

Pediatric Emergency Care
ISSN: 0749-5161
Pediatr. Emerg. Care
Lippincott Williams and Wilkins, 351 West Camden Street, Baltimore, MD 21201-2436, United States, Tel: +1(410)5284000 ISSUES/YR: 12

Pediatric Endocrinology Reviews
ISSN: 1565-4753
Pediatr. Endocrinol. Rev.
YS Medical Media Ltd., P.O. Box 8214, Kiryat-Nordau Industrial Area, 42504, Netanya, Israel, Tel: +972(9)8656599, Fax: +972(9)8656965 ISSUES/YR: 8

- **Pediatric Endosurgery and Innovative Techniques**
ISSN: 1092-6410
Pediatr. Endosurg. Innovative Techn.
Mary Ann Liebert Inc., 140 Huguenot Street, New Rochelle, NY 10801-5215, United States, Tel: +1(914)7402100, Fax: +1(914)7402101 ISSUES/YR: 4

Pediatric Exercise Science
ISSN: 0899-8493
Pediatr. Exerc. Sci.
Human Kinetics Publishers Inc., 1607 N. Market Street, Champaign, IL 61820-2200, United States, Tel: +1(217)3515076, Fax: +1(217)3511549 ISSUES/YR: 4

Pediatric Hematology and Oncology
ISSN: 0888-0018
Pediatr. Hematol. Oncol.
Taylor and Francis Ltd., 4 Park Square, Milton Park, OX14 4RN, Abingdon, Oxfordshire, United Kingdom, Tel: +44(20)70175541, Fax: +44(20)70175198 ISSUES/YR: 8

- **Pediatric Infectious Disease Journal**
ISSN: 0891-3668
Pediatr. Infect. Dis. J.
Lippincott Williams and Wilkins, 351 West Camden Street, Baltimore, MD 21201-2436, United States, Tel: +1(410)5284000 ISSUES/YR: 12

- **Pediatric Nephrology**
ISSN: 0931-041X
Pediatr. Nephrol.
Springer Verlag, Tiergartenstrasse 17, D-69121, Heidelberg, Germany, Tel: +49(6221)4870, Fax: +49(6221)487366 ISSUES/YR: 12

- **Pediatric Neurology**
ISSN: 0887-8994
Pediatr. Neurol.
Elsevier Inc., 360 Park Avenue South, New York, NY 10010, United States, Tel: +1(212)6333950, Fax: +1(212)6333990 ISSUES/YR: 10

- **Pediatric Neurosurgery**
ISSN: 1016-2291
Pediatr. Neurosurg.
S. Karger AG, Allschwilerstrasse 10, P.O. Box, CH-4009, Basel, Switzerland, Tel: +41(61)3061111, Fax: +41(61)3061234 ISSUES/YR: 6

Pediatric Physical Therapy
ISSN: 0898-5669
Pediatr. Phys. Ther.
Lippincott Williams and Wilkins, 351 West Camden Street, Baltimore, MD 21201-2436, United States, Tel: +1(410)5284000 ISSUES/YR: 4

- **Pediatric Pulmonology**
ISSN: 8755-6863
Pediatr. Pulmonol.
Wiley-Liss Inc., 111 River Street, Hoboken, NJ 07030-5774, United States, Tel: +1(201)7486645, Fax: +1(201)7486021 ISSUES/YR: 12

- **Pediatric Radiology**
ISSN: 0301-0449
Pediatr. Radiol.
Springer Verlag, Tiergartenstrasse 17, D-69121, Heidelberg, Germany, Tel: +49(6221)4870, Fax: +49(6221)487366 ISSUES/YR: 12

- **Pediatric Rehabilitation**
ISSN: 1363-8491
Pediatr. Rehab.
Taylor and Francis Ltd., 4 Park Square, Milton Park, OX14 4RN, Abingdon, Oxfordshire, United Kingdom, Tel: +44(20)70175541, Fax: +44(20)70175198 ISSUES/YR: 4

- **Pediatric Research**
ISSN: 0031-3998
Pediatr. Res.
Lippincott Williams and Wilkins, 351 West Camden Street, Baltimore, MD 21201-2436, United States, Tel: +1(410)5284000 ISSUES/YR: 12

- **Pediatric Surgery International**
ISSN: 0179-0358
Pediatr. Surg. Int.
Springer Verlag, Tiergartenstrasse 17, D-69121, Heidelberg, Germany, Tel: +49(6221)4870, Fax: +49(6221)487366 ISSUES/YR: 8

Pediatric Transplantation
ISSN: 1397-3142
Pediatr. Transplant.
Blackwell Publishing Ltd, 9600 Garsington Road, OX4 2XG, Oxford, United Kingdom, Tel: +44(1865)776868, Fax: +44(1865)714591 ISSUES/YR: 6

- **Pediatrics**
ISSN: 0031-4005
Pediatrics
American Academy of Pediatrics, 141 Northwest Point Blvd, P.O. Box 927, Elk Grove Village, IL 60007-1098, United States, Tel: +1(847)2285005, Fax: +1(847)2285097 ISSUES/YR: 12

Pediatrics in Review
ISSN: 0191-9601
Pediatr. Rev.
American Academy of Pediatrics, 141 Northwest Point Blvd, P.O. Box 927, Elk Grove Village, IL 60007-1098, United States, Tel: +1(847)2285005, Fax: +1(847)2285097 ISSUES/YR: 12

- **Pediatrics International**
ISSN: 1328-8067
Pediatr. Int.
Blackwell Publishing, 550 Swanston Street, VIC 3053, Carlton South, Australia, Tel: +61(3)8359 1011, Fax: +61(3)8359 1120 ISSUES/YR: 6

Pediatrika
ISSN: 0211-3465
Pediatrika
Alpe Editores S.A., Pedro Rico 27, oficinas 11-12, 28029, Madrid, Spain, Tel: +34(91)7338811, Fax: +34(91)3159652 ISSUES/YR: 10

Pediatriya
ISSN: 0479-7876
Pediatriya
Academy of Medicine, 1 G. Sofiiski Boul., 1431, Sofia, Bulgaria, Tel: +359(2)523171, Fax: +359(2)522393 ISSUES/YR: 4

- **Peptides**
ISSN: 0196-9781
Peptides
Elsevier Inc., 360 Park Avenue South, New York, NY 10010, United States, Tel: +1(212)6333950, Fax: +1(212)6333990 ISSUES/YR: 12

- **Perfusion**
ISSN: 0267-6591
Perfusion
Arnold, 338 Euston Road, NW1 3BH, London, United Kingdom, Tel: +44(20)78736000, Fax: +44(20)78736376 ISSUES/YR: 6

Perinatology
ISSN: 0972-2408
Perinatology
Prism Books Pvt Ltd, 1865, 32nd Cross, 10th Main, BSK 2nd Stage, 560070, Bangalore, India, Tel: +91-6713979, Fax: +91-6713991 ISSUES/YR: 6

Current

Periodicum Biologorum
ISSN: 0031-5362
Period. Biol.
Croatian Society of Natural Sciences, Ilica 16/III, HR-41000, Zagreb, Croatia, Tel: +385(41)425288 ISSUES/YR: 6

- **Peritoneal Dialysis International**
ISSN: 0896-8608
Peritoneal Dial. Int.
Multimed Inc., 66 Martin Street, ONT L9T 2R2, Milton, Canada, Tel: +1(905)8752456, Fax: +1(905)8752864 ISSUES/YR: 6

- **Personalized Medicine**
ISSN: 1741-0541
Personalized Med.
Future Medicine Ltd., 2nd Albert Place, Finchley Central, N3 1QB, London, United Kingdom, Tel: +44-208 343 3883, Fax: +44-208 343 2313 ISSUES/YR: 4

- **Perspectives in Biology and Medicine**
ISSN: 0031-5982
Perspect. Biol. Med.
Johns Hopkins University Press, 2715 N. Charles Street, Baltimore, MD 21218-4363, United States, Tel: +1(410)5166987, Fax: +1(410)5166968 ISSUES/YR: 4

Perspectives on Sexual and Reproductive Health
ISSN: 1538-6341
Perspect. Sex. Reprod. Health
Alan Guttmacher Institute, 120 Wall Street, New York, NY 10005, United States, Tel: +1(212)2481111 ISSUES/YR: 6

- **Pflugers Archiv European Journal of Physiology**
ISSN: 0031-6768
Pflug. Arch. Eur. J. Physiol.
Springer Verlag, Tiergartenstrasse 17, D-69121, Heidelberg, Germany, Tel: +49(6221)4870, Fax: +49(6221)487366 ISSUES/YR: 12

Pharma Fokus Herz-Kreislauf
ISSN: 1860-4730
Pharm. Fokus Herz Kreisl.
Georg Thieme Verlag, Rudigerstrasse 14, D-70469, Stuttgart, Germany, Tel: +49(711)8931-321, Fax: +49(711)8931-422 ISSUES/YR: 1

Pharma-Kritik
ISSN: 1010-5409
Pharma-Krit.
Infomed-Verlags-AG, P.O. Box 528, 9501, Wil, Switzerland, Tel: +41(71)9100866, Fax: +41(71)9100877 ISSUES/YR: 20

Pharmaca
ISSN: 0031-6857
Pharmaca
Udruzenje Organizacija Zdravstva Hrvatske, Savska cesta 41/VII, P.O. Box 878, HR-41000, Zagreb, Croatia, Tel: +385(41)539011 ISSUES/YR: 4

Pharmaceutical Biology
ISSN: 1388-0209
Pharm. Biol.
Taylor and Francis Ltd., 4 Park Square, Milton Park, OX14 4RN, Abingdon, Oxfordshire, United Kingdom, Tel: +44(20)70175541, Fax: +44(20)70175198 ISSUES/YR: 9

Pharmaceutical Care and Research
ISSN: 1671-2838
Pharm. Care Res.
Publishing House of Pharmaceutical Care and Research, 174 Changhai Road, 200433, Shanghai, China, Tel: +86(21)65519829, Fax: +86(21)65519829 ISSUES/YR: 4

Pharmaceutical Care Espana
ISSN: 1139-6202
Pharm. Care Espana
SANED S.L. (Sanidad y Ediciones), Capitan Haya 60, 28020, Madrid, Spain, Tel: +34(91)7499506, Fax: +34(91)7499507 ISSUES/YR: 6

Pharmaceutical Chemistry Journal
ISSN: 0091-150X
Pharm. Chem. J.
Springer Science and Business Media Deutschland GmbH, Heidelberger Platz 3, 14197, Berlin, Germany, Tel: +1-30 827875282, Fax: +1-30 827875707 ISSUES/YR: 12

- **Pharmaceutical Development and Technology**
ISSN: 1083-7450
Pharm. Dev. Technol.
Taylor and Francis Inc., 325 Chestnut St, Suite 800, Philadelphia PA, PA 19106, United States, Tel: +1(800)3541420, Fax: +1(215)6258914 ISSUES/YR: 4

Pharmaceutical Engineering
ISSN: 0273-8139
Pharm. Eng.
Int. Society for Pharmaceutical Engineering, Inc., 3109 W. Dr. Martin Luther King Jr. Blvd. Suite 250, Tampa, FL 33607, United States, Tel: +1(813)9602105, Fax: +1(813)2642816 ISSUES/YR: 6

Pharmaceutical Journal
ISSN: 0031-6873
Pharm. J.
Pharmaceutical Press, 1 Lambeth High Street, SE1 7JN, London, United Kingdom, Tel: +44(207)7359141, Fax: +44(207)7355085 ISSUES/YR: 56

- **Pharmaceutical Research**
ISSN: 0724-8741
Pharm. Res.
Springer Science and Business Media Deutschland GmbH, Heidelberger Platz 3, 14197, Berlin, Germany, Tel: +1-30 827875282, Fax: +1-30 827875707 ISSUES/YR: 12

Pharmaceutical Statistics
ISSN: 1539-1604
Pharm. Stat.
John Wiley and Sons Ltd, Southern Gate, PO19 8SQ, Chichester, West Sussex, United Kingdom, Tel: +44(1243)779777, Fax: +44(1243)775878 ISSUES/YR: 4

Pharmaceutical Technology
ISSN: 1543-2521
Pharm. Technol.
Advanstar Communications, 131 West First Street, Duluth, MN 55802-2065, United States, Tel: +1(218)7239200, Fax: +1(218)7239437 ISSUES/YR: 13

Pharmaceutical Technology Europe
ISSN: 0164-6826
Pharm. Technol. Eur.
Advanstar Communications, 131 West First Street, Duluth, MN 55802-2065, United States, Tel: +44(218)7239200, Fax: +44(218)7239437 ISSUES/YR: 12

Pharmaceutical Visions
ISSN: 0966-1719
Pharm. Vis.
Highbury House Communications PLC, 1-3 Highbury Station Road, Islington, N1 1SE, London, United Kingdom, Tel: +44(207)2262222, Fax: +44(207)7041021 ISSUES/YR: 4

Pharmaceuticals
ISSN: 1424-8247
Pharmaceuticals
Molecular Diversity Preservation International, Matthaeusstrasse 11, CH-4057, Basel, Switzerland, Tel: +41-79 322 3379, Fax: +41-61 302 8918 ISSUES/YR: 4

Pharmaceutisch Weekblad
ISSN: 0031-6911
Pharm. Weekbl.
Kon. Ned. Mij. ter Bevordering der Pharmacie (KNMP), P.O. Box 30460, 2500 GL, The Hague, Netherlands, Tel: +31(70)3624111 ISSUES/YR: 52

Pharmacia
ISSN: 0428-0296
Pharmacia
Medical University Faculty of Pharmacia, 2 Dunav Str., 1000, Sofia, Bulgaria ISSUES/YR: 4

Pharmacien Hospitalier
ISSN: 0768-9179
Pharm. Hosp.
Masson, 21 Rue Camille Desmoulins, 92789, Issy les Moulineaux Cedex 9, France, Tel: +33(1)73281634, Fax: +33(1)73281649 ISSUES/YR: 4

- **PharmacoEconomics**
ISSN: 1170-7690
Pharmacoeconomics
Adis International Ltd, 41 Centorian Drive, Private Bag 65901, Mairangi Bay, 1311, Auckland 10, New Zealand, Tel: +64(9)4770700, Fax: +64(9)4770766 ISSUES/YR: 18

PharmacoEconomics - Italian Research Articles
ISSN: 1590-9158
PharmacoEcon. Ital. Res. Artic.
Adis International Ltd, Via Lanino 5, 20144, Milan, Italy, Tel: +39-024 23 45 62, Fax: +39-024 22 12 00 ISSUES/YR: 2

- **Pharmacoepidemiology and Drug Safety**
ISSN: 1053-8569
Pharmacoepidemiol. Drug Saf.
John Wiley and Sons Ltd, Southern Gate, PO19 8SQ, Chichester, West Sussex, United Kingdom, Tel: +44(1243)779777, Fax: +44(1243)775878 ISSUES/YR: 7

- **Pharmacogenetics and Genomics**
ISSN: 1744-6872
Pharmacogenet. Genomics
Lippincott Williams and Wilkins, 241 Borough High Street, SE1 1GB, London, United Kingdom, Tel: +44(207)9407500, Fax: +44(207)9407575 ISSUES/YR: 12

Pharmacogenomics
ISSN: 1462-2416
Pharmacogenomics
Future Drugs Ltd, 2 Albert Place, Finchley Central, N3 1QB, London, United Kingdom, Tel: +44(20)83492033, Fax: +44(20)83432313 ISSUES/YR: 8

- **Pharmacogenomics Journal**
ISSN: 1470-269X
Pharmacogenomics J.
Nature Publishing Group, Houndmills, RG21 6XS, Basingstoke, Hampshire, United Kingdom, Tel: +44(1256)329242, Fax: +44(1256)810526 ISSUES/YR: 6

Pharmacological Reports
ISSN: 1734-1140
Pharmacol. Rep.
Polish Academy of Sciences, 12 Smetna Street, 31-343, Krakow, Poland, Tel: +48(12)6374022, Fax: +48(12)6374500 ISSUES/YR: 6

- **Pharmacological Research**
ISSN: 1043-6618
Pharmacol. Res.
Academic Press, 24-28 Oval Road, NW1 7DX, London, United Kingdom, Tel: +44(207)2674466 ISSUES/YR: 12

- **Pharmacological Reviews**
ISSN: 0031-6997
Pharmacol. Rev.
American Society for Pharmacology and Experimental Therapy, 9650 Rockville Pike, Bethesda, MD 20814, United States, Tel: +1(301)5307024, Fax: +1(301)5715728 ISSUES/YR: 4

- **Pharmacology**
ISSN: 0031-7012
Pharmacology
S. Karger AG, Allschwilerstrasse 10, P.O. Box, CH-4009, Basel, Switzerland, Tel: +41(61)3061111, Fax: +41(61)3061234 ISSUES/YR: 12

- **Pharmacology and Therapeutics**
ISSN: 0163-7258
Pharmacol. Ther.
Elsevier Inc., 360 Park Avenue South, New York, NY 10010, United States, Tel: +1(212)6333950, Fax: +1(212)6333990 ISSUES/YR: 12

- **Pharmacology Biochemistry and Behavior**
ISSN: 0091-3057
Pharmacol. Biochem. Behav.
Elsevier Inc., 360 Park Avenue South, New York, NY 10010, United States, Tel: +1(212)6333950, Fax: +1(212)6333990 ISSUES/YR: 12

Pharmacometrics
ISSN: 0300-8533
Pharmacometrics
Japanese Society of Pharmacometrics, Sendai Chuo, P.O. Box 180, 980-91, Sendai, Japan, Tel: +81(22)2673810 ISSUES/YR: 6

- **Pharmacopsychiatry**
ISSN: 0176-3679
Pharmacopsychiatry
Georg Thieme Verlag, Rudigerstrasse 14, D-70469, Stuttgart, Germany, Tel: +49(711)8931-321, Fax: +49(711)8931-422 ISSUES/YR: 6

Pharmacotherapy
ISSN: 0277-0008
Pharmacotherapy
Pharmacotherapy Publications Inc., 750 Washington Street, Boston, MA 02111, United States, Tel: +1(617)9565390 ISSUES/YR: 12

- **Pharmacy Education**
ISSN: 1560-2214
Pharm. Educ.
Taylor and Francis Ltd., 4 Park Square, Milton Park, OX14 4RN, Abingdon, Oxfordshire, United Kingdom, Tel: +44(20)70175541, Fax: +44(20)70175198 ISSUES/YR: 4

Pharmacy in Practice
ISSN: 1358-1538
Pharm. Pract.
Medicom Group, Thameside House, Hurst Road,, KT8 9AY, Hampton Court, Surrey, United Kingdom, Tel: +44-208 481 8100, Fax: +44-208 481 8105 ISSUES/YR: 9

Pharmacy World and Science
ISSN: 0928-1231
Pharm. World Sci.
Kluwer Academic Publishers, Van Godewijckstraat 30, 3311 GZ, Dordrecht, Netherlands, Tel: +31(78)6576000, Fax: +31(78)6576254 ISSUES/YR: 6

Pharmazeutische Industrie
ISSN: 0031-711X
Pharm. Ind.
Editio Cantor Verlag GmbH, P.O. Box 1255, D-88322, Aulendorf, Germany, Tel: +49(7525)9400, Fax: +49(7525)940180 ISSUES/YR: 12

Pharmazeutische Zeitung
ISSN: 0031-7136
Pharm. Ztg.
Govi-Verlag Pharmazeutischer Verlag GmbH, P.O. Box 5360, D-65728, Eschborn, Germany, Tel: +49(6196)9280, Fax: +49(6196)928203 ISSUES/YR: 52

Pharmazie
ISSN: 0031-7144
Pharmazie
Govi-Verlag Pharmazeutischer Verlag GmbH, P.O. Box 5360, D-65728, Eschborn, Germany, Tel: +49(6196)9280, Fax: +49(6196)928203 ISSUES/YR: 12

Pharmazie in Unserer Zeit
ISSN: 0048-3664
Pharm. Unserer Zeit
Wiley-VCH Verlag, P.O. Box 101161, D-69451, Weinheim, Germany, Tel: +49(6201)606147, Fax: +49(6201)606117 ISSUES/YR: 6

Phlebologie
ISSN: 0939-978X
Phlebologie
F.K. Schattauer Verlagsgesellschaft mbH, P.O. Box 10 45 45, D-70040, Stuttgart, Germany, Tel: +49(711)229870, Fax: +49(711)2298750 ISSUES/YR: 6

- **Phlebology**
ISSN: 0268-3555
Phlebology
Royal Society of Medicine Press Ltd, P.O. Box 9002, W1A 0ZA, London, United Kingdom, Tel: +44(207)2902927, Fax: +44(207)2902929 ISSUES/YR: 4

Phlebolymphology
ISSN: 1286-0107
Phlebolymphology
Servier International, 31 rue du Pont, 92200, Neuilly-sur-Seine, France, Tel: +33(1)55726000, Fax: +33(1)55726995 ISSUES/YR: 1

- **Photochemical and Photobiological Sciences**
ISSN: 1474-905X
Photochem. Photobiol. Sci.
Royal Society of Chemistry, Milton Road, CB4 0WF, Cambridge, United Kingdom, Tel: +44(1223)420066, Fax: +44(1223)420247 ISSUES/YR: 12

- **Photodermatology Photoimmunology and Photomedicine**
ISSN: 0905-4383
Photodermatol. Photoimmunol. Photomed.
Blackwell Munksgaard, 1 Rosenorns Alle, P.O. Box 227, DK-1502, Copenhagen V, Denmark, Tel: +45(77)333333, Fax: +45(77)333377 ISSUES/YR: 6
- **Photodiagnosis and Photodynamic Therapy**
ISSN: 1572-1000
Photodiagn. Photodyn. Ther.
Elsevier, P.O. Box 211, 1000 AE, Amsterdam, Netherlands, Tel: +31(20)4853757, Fax: +31(20)4853432 ISSUES/YR: 4

Photomedicine and Laser Surgery
ISSN: 1549-5418
Photomed. Laser Surg.
Mary Ann Liebert Inc., 140 Huguenot Street, New Rochelle, NY 10801-5215, United States, Tel: +1(914)7402100, Fax: +1(914)7402101 ISSUES/YR: 6

Physica Medica
ISSN: 1120-1797
Phys. Med.
Giardini Editore e Stampatori in Pisa S.r.l., via delle Sorgenti 23, Agnano Pisano, 56010, Pisa, Italy ISSUES/YR: 4

Physical and Occupational Therapy in Geriatrics
ISSN: 0270-3181
Phys. Occup. Ther. Geriatr.
Haworth Press Inc., 10 Alice Street, Binghamton, NY 13904-1580, United States, Tel: +1(607)7225857, Fax: +1(607)7710012 ISSUES/YR: 4

Physical and Occupational Therapy in Pediatrics
ISSN: 0194-2638
Phys. Occup. Ther. Pediatr.
Haworth Press Inc., 10 Alice Street, Binghamton, NY 13904-1580, United States, Tel: +1(607)7225857, Fax: +1(607)7710012 ISSUES/YR: 4

- **Physical Biology**
ISSN: 1478-3975
Phys. Biol.
Institute of Physics Publishing, Temple Back, BS1 6BE, Bristol, United Kingdom, Tel: +44(117)9297481, Fax: +44(117)9294318 ISSUES/YR: 4
- **Physical Medicine and Rehabilitation Clinics of North America**
ISSN: 1047-9651
Phys. Med. Rehabil. Clin. North Am.
W.B. Saunders, Independence Square West, Philadelphia, PA 19106-3399, United States, Tel: +1(215)2387800, Fax: +1(215)2386445 ISSUES/YR: 4

Physical Therapy
ISSN: 0031-9023
Phys. Ther.
American Physical Therapy Association, 1111 North Fairfax Street, Alexandria, VA 22314, United States, Tel: +1(703)6842782 ISSUES/YR: 12

- **Physical Therapy in Sport**
ISSN: 1466-853X
Phys. Ther. Sport
Churchill Livingstone, 1-3 Baxter's Place, Leith Walk, EH1 3AF, Edinburgh, United Kingdom, Tel: +44(131)5562424, Fax: +44(131)5581278 ISSUES/YR: 4

Physician and Sportsmedicine
ISSN: 0091-3847
Phys. Sportsmed.
McGraw-Hill Companies, 1221 Avenue of the Americas, New York, NY 10020, United States, Tel: +1(212 512)2000, Fax: +1-212 512 6111 ISSUES/YR: 12

- **Physics in Medicine and Biology**
ISSN: 0031-9155
Phys. Med. Biol.
Institute of Physics Publishing, Temple Back, BS1 6BE, Bristol, United Kingdom, Tel: +44(117)9297481, Fax: +44(117)9294318 ISSUES/YR: 24

Physikalische Medizin Rehabilitationsmedizin Kurortmedizin
ISSN: 0940-6689
Phys. Med. Rehabil. Kurortmed.
Georg Thieme Verlag, Rudigerstrasse 14, D-70469, Stuttgart, Germany, Tel: +49(711)8931-321, Fax: +49(711)8931-422 ISSUES/YR: 6

- **Physiological Genomics**
ISSN: 1094-8341
Physiol. Genomics
American Physiological Society, 9650 Rockville Pike, Bethesda, MD 20814-3991, United States, Tel: +1(301)6347180, Fax: +1(301)6347241 ISSUES/YR: 4
- **Physiological Measurement**
ISSN: 0967-3334
Physiol. Meas.
Institute of Physics Publishing, Temple Back, BS1 6BE, Bristol, United Kingdom, Tel: +44(117)9297481, Fax: +44(117)9294318 ISSUES/YR: 6

Physiological Research
ISSN: 0862-8408
Physiol. Res.
Czech Academy of Sciences, Videnska 1083, 142 20, Prague 4, Czech Republic ISSUES/YR: 6

- **Physiological Reviews**
ISSN: 0031-9333
Physiol. Rev.
American Physiological Society, 9650 Rockville Pike, Bethesda, MD 20814-3991, United States, Tel: +1(301)6347180, Fax: +1(301)6347241 ISSUES/YR: 4
- **Physiology and Behavior**
ISSN: 0031-9384
Physiol. Behav.
Elsevier Inc., 360 Park Avenue South, New York, NY 10010, United States, Tel: +1(212)6333950, Fax: +1(212)6333990 ISSUES/YR: 15
- **Physiotherapy**
ISSN: 0031-9406
Physiotherapy
Elsevier Ltd, Langford Lane, Kidlington, OX5 1GB, Oxford, United Kingdom, Tel: +44(1865)843000 ISSUES/YR: 12

Physiotherapy Singapore
ISSN: 0219-1091
Physiother. Singap.
Singapore Physiotherapy Association, Tanglin P.O. Box 442, 912415, Singapore, Singapore ISSUES/YR: 4

Physiotherapy Theory and Practice
ISSN: 0959-3985
Physiother. Theory Pract.
Taylor and Francis Ltd., 4 Park Square, Milton Park, OX14 4RN, Abingdon, Oxfordshire, United Kingdom, Tel: +44(20)70175541, Fax: +44(20)70175198 ISSUES/YR: 4

Phytochemical Analysis
ISSN: 0958-0344
Phytochem. Anal.
John Wiley and Sons Ltd, Southern Gate, PO19 8SQ, Chichester, West Sussex, United Kingdom, Tel: +44(1243)779777, Fax: +44(1243)775878 ISSUES/YR: 6

- **Phytochemistry Reviews**
ISSN: 1568-7767
Phytochem. Rev.
Kluwer Academic Publishers, Van Godewijckstraat 30, 3311 GZ, Dordrecht, Netherlands, Tel: +31(78)6576000, Fax: +31(78)6576254 ISSUES/YR: 3

Phytomedica
ISSN: 0972-3293
Phytomedica
Society of Phytomedica, Sharda Nagar, 247 001, Saharanpur, Uttar Pradesh, India, Tel: +91(132)725044, Fax: +91(132)726288 ISSUES/YR: 1

- **Phytomedicine**
ISSN: 0944-7113
Phytomedicine
Urban und Fischer Verlag Jena, P.O. Box 100537, 07705, Jena, Germany, Tel: +49(3641)6263, Fax: +49(3641)626500 ISSUES/YR: 10

Phytotherapie
ISSN: 1624-8597
Phytotherapie
Springer-Verlag France, 1 rue Paul Cezanne, 75008, Paris, France, Tel: +33(1)53933628, Fax: +33(1)53933683 ISSUES/YR: 6

Phytotherapie Europeenne
ISSN: 1628-6847
Phytother. Eur.
Meditions Carline, 1-3 rue du Depart, 75014, Paris, France, Tel: +33(1)40640075, Fax: +33(1)43222699 ISSUES/YR: 6

Phytotherapy Research
ISSN: 0951-418X
Phytother. Res.
John Wiley and Sons Ltd, Southern Gate, PO19 8SQ, Chichester, West Sussex, United Kingdom, Tel: +44(1243)779777, Fax: +44(1243)775878 ISSUES/YR: 10

Pigment Cell Research
ISSN: 0893-5785
Pigm. Cell Res.
Blackwell Publishing Ltd, 9600 Garsington Road, OX4 2XG, Oxford, United Kingdom, Tel: +44(1865)776868, Fax: +44(1865)714591 ISSUES/YR: 6

• **Pituitary**
ISSN: 1386-341X
Pituitary
Kluwer Academic Publishers, 101 Philip Drive, Assinippi Park, Norwell, MA 02061, United States, Tel: +1(781)8716600, Fax: +1(781)8716528 ISSUES/YR: 4

Placebo
ISSN: 1608-5523
Placebo
VICER Publishing, P.O. Box 14, A-1097, Vienna, Austria, Tel: +43(676)9568085, Fax: +43(676)9568086 ISSUES/YR: 4

• **Placenta**
ISSN: 0143-4004
Placenta
W.B. Saunders Ltd, 32 Jamestown Road, NW1 7BY, London, United Kingdom, Tel: +44(20)74244200, Fax: +44(20)74854752 ISSUES/YR: 10

Planta Medica
ISSN: 0032-0943
Planta Med.
Georg Thieme Verlag, Rudigerstrasse 14, D-70469, Stuttgart, Germany, Tel: +49(711)8931-321, Fax: +49(711)8931-422 ISSUES/YR: 12

Plasmid
ISSN: 0147-619X
Plasmid
Academic Press Inc., 6277 Sea Harbor Drive, Orlando, FL 32887-4900, United States, Tel: +1(407)3454040, Fax: +1(407)3639661 ISSUES/YR: 6

• **Plastic and Reconstructive Surgery**
ISSN: 0032-1052
Plast. Reconstr. Surg.
Lippincott Williams and Wilkins, 351 West Camden Street, Baltimore, MD 21201-2436, United States, Tel: +1(410)5284000 ISSUES/YR: 14

• **Platelets**
ISSN: 0953-7104
Platelets
Taylor and Francis Ltd., 4 Park Square, Milton Park, OX14 4RN, Abingdon, Oxfordshire, United Kingdom, Tel: +44(20)70175541, Fax: +44(20)70175198 ISSUES/YR: 8

PLoS Biology
ISSN: 1545-7885
PloS Biol.
Public Library of Science, 185 Berry Street, Suite 1300, San Francisco, CA 94107, United States, Tel: +1(415)6241200, Fax: +1(415)5464090 ISSUES/YR: 12

PLoS Computational Biology
ISSN: 1553-734X
PLoS Comput. Biol.
Public Library of Science, 185 Berry Street, Suite 1300, San Francisco, CA 94107, United States, Tel: +1(415)6241200, Fax: +1(415)5464090 ISSUES/YR: 4

PLoS Genetics
ISSN: 1553-7390
PLoS Genet.
Public Library of Science, 185 Berry Street, Suite 1300, San Francisco, CA 94107, United States, Tel: +1(415)6241200, Fax: +1(415)5464090 ISSUES/YR: 2

PLoS Medicine
ISSN: 1549-1277
PLoS Med.
Public Library of Science, 185 Berry Street, Suite 1300, San Francisco, CA 94107, United States, Tel: +1(415)6241200, Fax: +1(415)5464090 ISSUES/YR: 12

PLoS Pathogens
ISSN: 1553-7366
PLoS Pathog.
Public Library of Science, 185 Berry Street, Suite 1300, San Francisco, CA 94107, United States, Tel: +1(415)6241200, Fax: +1(415)5464090 ISSUES/YR: 4

• **Pneumologie**
ISSN: 0934-8387
Pneumologie
Georg Thieme Verlag, Rudigerstrasse 14, D-70469, Stuttgart, Germany, Tel: +49(711)8931-321, Fax: +49(711)8931-422 ISSUES/YR: 12

Pneumonologia i Alergologia Polska
ISSN: 0867-7077
Pneumonol. Alergol. Pol.
Instytut Gruzlicy i Chorob Pluc, ul. Plocka 26, 01-138, Warsaw, Poland, Tel: +48(22)324451 ISSUES/YR: 12

Policlinico - Sezione Medica
ISSN: 0048-4717
Policlin. Sez. Med.
Edizioni Luigi Pozzi S.r.l., Via Panama 68, 00198, Roma, Italy, Tel: +39(06)8553548, Fax: +39(06)8554105 ISSUES/YR: 4

Polish Journal of Gynaecological Investigations
ISSN: 1506-4794
Pol. J. Gynaecol. Invest.
Medical Science International Ltd., ul. Ustrzycka 11, 02-141 Warsaw, Poland, Tel: +48(22)8681265, Fax: +48(22)7813232 ISSUES/YR: 4

Polish Journal of Microbiology
ISSN: 1733-1331
Pol. J. Microbiol.
Polish Society of Microbiologists, Ul. Chocimska 24, 00-791, Warsaw, Poland, Tel: +48(22)8497781, Fax: +48(22)8497484 ISSUES/YR: 4

Polish Journal of Pathology
ISSN: 1233-9687
Pol. J. Pathol.
Versalius University Medical Publisher, Wislisko 1, 31-538, Krakow, Poland, Tel: +48(12)6213387 ISSUES/YR: 4

Polish Journal of Radiology
ISSN: 0137-7183
Pol. J. Radiol.
Medical Science International Ltd., ul. Ustrzycka 11, 02-141 Warsaw, Poland, Tel: +48(22)8681265, Fax: +48(22)7813232 ISSUES/YR: 4

Polska Medycyna Paliatywna
ISSN: 1644-115X
Polska Medycyna Paliatywna
Via Medica, Ul. Swietokrzyska 73, 80-180, Gdansk, Poland, Tel: +48(58)3209494, Fax: +48(58)3209460 ISSUES/YR: 4

Polski Merkuriusz Lekarski
ISSN: 1426-9686
Pol. Merkuriusz Lek.
Medpress, ul. Slonki 67, 02-814, Warsaw, Poland, Tel: +48(22)6444820, Fax: +48(22)6444820 ISSUES/YR: 12

Polski Przeglad Chirurgiczny
ISSN: 0032-373X
Pol. Prz. Chir.
Infor-Press, ul. Prosta 2, 05-092, Lomianki, Poland, Tel: +48(22)7514409, Fax: +48(22)7510921 ISSUES/YR: 12

Polski Przeglad Kardiologiczny
ISSN: 1507-5540
Pol. Prz. Kardiol.
LEXiMED, ul. Moniuszki 7, 48-300, Nysa, Poland, Tel: +48-077 433 41 88, Fax: +48-077 433 41 88 ISSUES/YR: 4

Polskie Archiwum Medycyny Wewnetrznej
ISSN: 0032-3772
Pol. Arch. Med. Wewn.
III Oddzial Wewnetrzny Szpitala Praskiego, Pl. Weteranow 4, 03-701, Warsaw, Poland, Tel: +48(22)6190897, Fax: +48(22)6189666 ISSUES/YR: 12

Population Health Metrics
ISSN: 1478-7954
Popul. Health Metr.
BioMed Central Ltd., 34 - 42 Cleveland Street, W1T 4LB, London, United Kingdom, Tel: +44(207)3230323, Fax: +44(207)6319926 ISSUES/YR: 1

Postepy Dermatologii I Alergologii
ISSN: 1642-395X
Postepy Dermatol. Alergol.
Termedia Publishing House Ltd., ul Kleberga 8,, 61-615, Poznan, Poland, Tel: +48-61 822 77 81, Fax: +48-61 822 7781 ISSUES/YR: 6

Postepy Mikrobiologii
ISSN: 0079-4252
Postepy Mikrobiol.
Polish Society of Microbiologists, Ul. Chocimska 24, 00-791, Warsaw, Poland, Tel: +48(22)8497781, Fax: +48(22)8497484 ISSUES/YR: 4

Postepy Rehabilitacji
ISSN: 0860-6161
Postepy Rehabilitacji
Polish Scientific Publishers PWN Ltd, Ul. Miodowa 10, 00-251, Warszawa, Poland, Tel: +48-22-6954181, Fax: +48-22-6954288 ISSUES/YR: 4

Postgraduate Medical Journal
ISSN: 0032-5473
Postgrad. Med. J.
BMJ Publishing Group, Tavistock Square, WC1H 9JR, London, United Kingdom, Tel: +44(207)3874499, Fax: +44(207)3836668 ISSUES/YR: 12

Postgraduate Medicine
ISSN: 0032-5481
Postgrad. Med.
McGraw-Hill Companies, 1221 Avenue of the Americas, New York, NY 10020, United States, Tel: +1(212 512)2000, Fax: +1-212 512 6111 ISSUES/YR: 12

Powder Technology
ISSN: 0032-5910
Powder Technol.
Elsevier, P.O. Box 211, 1000 AE, Amsterdam, Netherlands, Tel: +31(20)4853757, Fax: +31(20)4853432 ISSUES/YR: 15

• **PPmP Psychotherapie Psychosomatik Medizinische Psychologie**
ISSN: 0937-2032
PPmP Psychother. Psychosom. Med. Psychol.
Georg Thieme Verlag, Rudigerstrasse 14, D-70469, Stuttgart, Germany, Tel: +49(711)8931-321, Fax: +49(711)8931-422 ISSUES/YR: 12

Pracovni Lekarstvi
ISSN: 0032-6291
Prac. Lek.
Czech Medical Association J.E. Purkyne, Sokolska 31, P.O. Box 88, 120 26, Praha 2, Czech Republic, Tel: +420(2)24266201-4, Fax: +420(2)24266212 ISSUES/YR: 4

Practica Oto-Rhino-Laryngologica
ISSN: 0032-6313
Pract. Oto-Rhino-Laryngol.
Society of Practical Otolaryngology, 39 Kawabata-Higashi, Marutamachi-Dohri, Sakyo-ku, 606-8395, Kyoto, Japan, Tel: +81(75)7712301 ISSUES/YR: 12

Practica Otologica, Supplement
ISSN: 0912-1870
Pract. Otol. Suppl.
Society of Practical Otolaryngology, 39 Kawabata-Higashi, Marutamachi-Dohri, Sakyo-ku, 606-8395, Kyoto, Japan, Tel: +81(75)7712301 ISSUES/YR: 1

Practica Pediatrica
ISSN: 1405-1168
Pract. Pediatr.
Mundo Medico S.A. de C.V., Ejercito Nacional 381, Col. Granada, 11520, Mexico D.F., Mexico, Tel: +52(55)5-5451798, Fax: +52(55)5-2036418 ISSUES/YR: 12

Practical Cardiovascular Risk Management
ISSN: 1478-6575
Pract. Cardiovasc. Risk Manage.
Medical Education Partnership, 53 Hargrave Rd, N19 5SH, London, United Kingdom, Tel: +44-207 561 5400, Fax: +44-207 7561 5401 ISSUES/YR: 4

Practical Diabetes International
ISSN: 1357-8170
Pract. Diabetes Int.
John Wiley and Sons Ltd, Southern Gate, PO19 8SQ, Chichester, West Sussex, United Kingdom, Tel: +44(1243)779777, Fax: +44(1243)775878 ISSUES/YR: 9

Practical Gastroenterology
ISSN: 0277-4208
Pract. Gastroenterol.
Shugar Publishing Inc., 99B Main Street, Westhampton Beach, NY 11978, United States, Tel: +1(516)2884404, Fax: +1(516)2884435 ISSUES/YR: 12

Practical Neurology
ISSN: 1474-7758
Pract. Neurol.
BMJ Publishing Group, Tavistock Square, WC1H 9JR, London, United Kingdom, Tel: +44(207)3874499, Fax: +44(207)3836668 ISSUES/YR: 6

Prakticky Lekar
ISSN: 0032-6739
Prakt. Lek.
Czech Medical Association J.E. Purkyne, Sokolska 31, P.O. Box 88, 120 26, Praha 2, Czech Republic, Tel: +420(2)24266201-4, Fax: +420(2)24266212 ISSUES/YR: 12

Pravention und Rehabilitation
ISSN: 0937-552X
Pravent. Rehabil.
Dustri-Verlag Dr. Karl Feistle, P.O. Box 1351, D-82032, Deisenhofen/Munchen, Germany, Tel: +49(89)6138610, Fax: +49(89)6135412 ISSUES/YR: 4

Prehospital Emergency Care
ISSN: 1090-3127
Prehosp. Emerg. Care
Taylor and Francis Inc., 325 Chestnut St, Suite 800, Philadelphia PA, PA 19106, United States, Tel: +1(800)3541420, Fax: +1(215)6258914 ISSUES/YR: 4

• **Prenatal Diagnosis**
ISSN: 0197-3851
Prenat. Diagn.
John Wiley and Sons Ltd, Southern Gate, PO19 8SQ, Chichester, West Sussex, United Kingdom, Tel: +44(1243)779777, Fax: +44(1243)775878 ISSUES/YR: 13

Prensa Medica Argentina
ISSN: 0032-745X
Prensa Med. Argent.
Prensa Medica Argentina, Junin 845, 1113, Buenos Aires, Argentina, Tel: +54(1)9619793 ISSUES/YR: 10

Prescrire International
ISSN: 1167-7422
Prescrire Int.
Association Mieux Prescrire, 83, boulevard Voltaire, 75011, Paris, France, Tel: +33(1)49237293, Fax: +33(1)47004101 ISSUES/YR: 6

Presse Medicale
ISSN: 0755-4982
Presse Med.
Masson, 21 Rue Camille Desmoulins, 92789, Issy les Moulineaux Cedex 9, France, Tel: +33(1)73281634, Fax: +33(1)73281649 ISSUES/YR: 22

• **Prevention and Control**
ISSN: 1573-2088
Prev. Control
Elsevier, P.O. Box 211, 1000 AE, Amsterdam, Netherlands, Tel: +31(20)4853757, Fax: +31(20)4853432 ISSUES/YR: 4

• **Preventive Medicine**
ISSN: 0091-7435
Prev. Med.
Academic Press Inc., 6277 Sea Harbor Drive, Orlando, FL 32887-4900, United States, Tel: +1(407)3454040, Fax: +1(407)3639661 ISSUES/YR: 12

- **Primary Care - Clinics in Office Practice**
  ISSN: 0095-4543
  Prim. Care Clin. Off. Pract.
  W.B. Saunders, Independence Square West, Philadelphia, PA 19106-3399, United States, Tel: +1(215)2387800, Fax: +1(215)2386445 ISSUES/YR: 4
- **Primary Care and Community Psychiatry**
  ISSN: 1746-8841
  Prim. Care Community Psychiatry
  Librapharm, Venture West, New Greenham Park, RG19 6HX, Newbury, Berks., United Kingdom, Tel: +44(1635)522651, Fax: +44(1635)36294 ISSUES/YR: 4

Primary Care Case Reviews
ISSN: 1096-8954
Prim. Care Case Rev.
Lippincott Williams and Wilkins, 530 Walnut Street, Philadelphia, PA 19106-3621, United States, Tel: +1(215)5218342, Fax: +1(215)5218326 ISSUES/YR: 4

Primary Care Companion to the Journal of Clinical Psychiatry
ISSN: 1523-5998
Prim. Care Companion J. Clin. Psych.
Physicians Postgraduate Press Inc., P.O. Box 752870, Memphis, TN 38119-2870, United States, Tel: +1(901)7513800 ISSUES/YR: 2

Primary Care Mental Health
ISSN: 1476-4717
Prim. Care Ment. Health
Radcliffe Publishing Ltd, 18 Marcham Road, OX14 1AA, Abingdon, Oxon, United Kingdom, Tel: +44(1235)528820, Fax: +44(1235)528830 ISSUES/YR: 4

Primary Care Respiratory Journal
ISSN: 1471-4418
Prim. Care Respir. J.
Elsevier, P.O. Box 211, 1000 AE, Amsterdam, Netherlands, Tel: +31(20)4853757, Fax: +31(20)4853432 ISSUES/YR: 4

Primary Psychiatry
ISSN: 1082-6319
Prim. Psychiatry
MBL Communications, 333 Hudson St. 7th Floor, NY 10013, United States, Tel: +1-212 328 0800 EXT: 222, Fax: +1-212 328 0600 ISSUES/YR: 12

Problems of Infectious and Parasitic Diseases
ISSN: 0204-9155
Probl. Infect. Parasit. Dis.
Izdatelstvo Medicina i Fizkultura, 11 Slaveikov Square, 1000, Sofia, Bulgaria, Tel: +359(2)879111 ISSUES/YR: 2

Problemy HIV i AIDS
ISSN: 1234-9496
Probl. HIV AIDS
Polskie Towarzystwo Naukowe AIDS, ul.Koszarowa 5, 51-149, Wroclaw, Poland, Tel: +48-71 326 13 25, Fax: +48(71 325)52-42 ISSUES/YR: 4

Proceedings of the American Thoracic Society
ISSN: 1546-3222
Proc. Am. Thorac. Soc.
American Thoracic Society, 61 Broadway 4th Floor, New York, 10006 - 2755, United States, Tel: +1(212)315 8600, Fax: +1(212)315 8689 ISSUES/YR: 1

Proceedings of the Controlled Release Society
ISSN: 1022-0178
Proc. Control. Release Soc.
Controlled Release Society, 13355 Tenth Avenue North, Suite 108, Minneapolis, MN 55441-5554, United States, Tel: +1(763)5120909, Fax: +1(763)7652329 ISSUES/YR: 1

- **Proceedings of the National Academy of Sciences of the United States of America**
  ISSN: 0027-8424
  Proc. Natl. Acad. Sci. U. S. A.
  National Academy of Sciences, 2101 Constitution Avenue NW, Washington, DC 20418, United States, Tel: +1(202)3342525 ISSUES/YR: 52

Proceedings of the Nutrition Society
ISSN: 0029-6651
Proc. Nutr. Soc.
CAB International, Wallingford, OX10 8DE, Oxon, United Kingdom, Tel: +44(1491)832111, Fax: +44(1491)826090 ISSUES/YR: 4

- **Proceedings of the Royal Society - Biological Sciences (Series B)**
  ISSN: 0962-8452
  Proc. R. Soc. Biol. Sci. Ser. B
  Royal Society of London, 6 Carlton House Terrace, SW1Y 5AG, London, United Kingdom, Tel: +44(207)8395561, Fax: +44(207)9761837 ISSUES/YR: 24

Proceedings of the Western Pharmacology Society
ISSN: 0083-8969
Proc. West. Pharmacol. Soc.
Western Pharmacology Society, P.O Box 245050, Tucson, AZ 85724-5050, United States, Tel: +1(520)6267843, Fax: +1(520)6266883 ISSUES/YR: 1

- **Progress in Brain Research**
  ISSN: 0079-6123
  Prog. Brain Res.
  Elsevier, P.O. Box 211, 1000 AE, Amsterdam, Netherlands, Tel: +31(20)4853757, Fax: +31(20)4853432 ISSUES/YR: 1

Progress in Cardiovascular Diseases
ISSN: 0033-0620
Prog. Cardiovasc. Dis.
W.B. Saunders, Independence Square West, Philadelphia, PA 19106-3399, United States, Tel: +1(215)2387800, Fax: +1(215)2386445 ISSUES/YR: 6

- **Progress in Drug Research**
  ISSN: 0071-786X
  Prog. Drug Res.
  Birkhauser Verlag AG, Klosterberg 23, P.O. Box 133, CH-4010, Basel, Switzerland, Tel: +41(61)2717400, Fax: +41(61)2050791 ISSUES/YR: 1

Progress in Histochemistry and Cytochemistry
ISSN: 0079-6336
Progr. Histochem. Cytochem.
Urban und Fischer Verlag Jena, P.O. Box 100537, 07705, Jena, Germany, Tel: +49(3641)6263, Fax: +49(3641)626500 ISSUES/YR: 4

- **Progress in Lipid Research**
  ISSN: 0163-7827
  Prog. Lipid Res.
  Elsevier Ltd, Langford Lane, Kidlington, OX5 1GB, Oxford, United Kingdom, Tel: +44(1865)843000, Fax: +44(1865)843010 ISSUES/YR: 6

Progress in Neuro-Psychopharmacology and Biological Psychiatry
ISSN: 0278-5846
Prog. Neuro-Psychopharmacol. Biol. Psychiatry
Elsevier Inc., 360 Park Avenue South, New York, NY 10010, United States, Tel: +1(212)6333950, Fax: +1(212)6333990 ISSUES/YR: 8

- **Progress in Neurobiology**
  ISSN: 0301-0082
  Prog. Neurobiol.
  Elsevier Ltd, Langford Lane, Kidlington, OX5 1GB, Oxford, United Kingdom, Tel: +44(1865)843000, Fax: +44(1865)843010 ISSUES/YR: 18

Progress in Nutrition
ISSN: 1129-8723
Prog. Nutr.
Mattioli 1885 S.p.A., Via Coduro 1/b, 43036, Fidenza (PR), Italy, Tel: +39(524)84547, Fax: +39(524)84751 ISSUES/YR: 4

Progress in Palliative Care
ISSN: 0969-9260
Prog. Palliative Care
Maney Publishing, Hudson Road, LS9 7DL, Leeds, United Kingdom, Tel: +44(113)2497481, Fax: +44(113)2486983 ISSUES/YR: 6

- **Progress in Pediatric Cardiology**
  ISSN: 1058-9813
  Prog. Pediatr. Cardiol.
  Elsevier Ireland Ltd, P.O. Box 85, Limerick, Ireland, Tel: +353(61)471944, Fax: +353(61)472144 ISSUES/YR: 2
- **Progress in Retinal and Eye Research**
  ISSN: 1350-9462
  Prog. Retinal Eye Res.
  Elsevier Ltd, Langford Lane, Kidlington, OX5 1GB, Oxford, United Kingdom, Tel: +44(1865)843000, Fax: +44(1865)843010 ISSUES/YR: 6

Progress in Transplantation
ISSN: 1526-9248
Prog. Transplant.
InnoVision Communications, 169 Saxony Road, Suite 103, Encinitas, CA 92024, United States, Tel: +1(760)633 3910, Fax: +1(760)633 3918 ISSUES/YR: 4

Progressi in Reumatologia
ISSN: 1129-8758
Progressi Reumatol.
Mattioli 1885 S.p.A., Via Coduro 1/b, 43036, Fidenza (PR), Italy, Tel: +39(524)84547, Fax: +39(524)84751 ISSUES/YR: 4

Proktologia
ISSN: 1640-5382
Proktologia
Medical Communications, Ul.Ojcowska 11, 02-918, Warszawa, Poland, Tel: +48-226519783, Fax: +48-8425363 ISSUES/YR: 4

Prostaglandins and Other Lipid Mediators
ISSN: 1098-8823
Prostaglandins Other Lipid Mediators
Elsevier Inc., 360 Park Avenue South, New York, NY 10010, United States, Tel: +1(212)6333950, Fax: +1(212)6333990 ISSUES/YR: 12

- **Prostaglandins Leukotrienes and Essential Fatty Acids**
ISSN: 0952-3278
Prostaglandins Leukotrienes Essent. Fatty Acids
Churchill Livingstone, 1-3 Baxter's Place, Leith Walk, EH1 3AF, Edinburgh, United Kingdom, Tel: +44(131)5562424, Fax: +44(131)5581278 ISSUES/YR: 12

- **Prostate**
ISSN: 0270-4137
Prostate
Wiley-Liss Inc., 111 River Street, Hoboken, NJ 07030-5774, United States, Tel: +1(201)7486645, Fax: +1(201)7486021 ISSUES/YR: 16

- **Prostate Cancer and Prostatic Diseases**
ISSN: 1365-7852
Prostate Cancer Prostatic Dis.
Nature Publishing Group, Houndmills, RG21 6XS, Basingstoke, Hampshire, United Kingdom, Tel: +44(1256)329242, Fax: +44(1256)810526 ISSUES/YR: 4

Prosthetics and Orthotics International
ISSN: 0309-3646
Prosthet. Orthot. Int.
Taylor and Francis Ltd., 4 Park Square, Milton Park, OX14 4RN, Abingdon, Oxfordshire, United Kingdom, Tel: +44(20)70175541, Fax: +44(20)70175198 ISSUES/YR: 3

- **Protein Engineering, Design and Selection**
ISSN: 1741-0126
Protein Eng. Des. Sel.
Oxford University Press, Great Clarendon Street, OX2 6DP, Oxford, United Kingdom, Tel: +44(1865)353907, Fax: +44(1865)353486 ISSUES/YR: 12

Protein Journal
ISSN: 1572-3887
Protein J.
Springer Science and Business Media Deutschland GmbH, Heidelberger Platz 3, 14197, Berlin, Germany, Tel: +1-30 827875282, Fax: +1-30 827875707 ISSUES/YR: 8

- **Protein Science**
ISSN: 0961-8368
Protein Sci.
Cold Spring Harbor Laboratory Press, 500 Sunnyside Boulevard, Woodbury New York, NY 11797, United States, Tel: +1(516)3491930, Fax: +1(516)3491946 ISSUES/YR: 12

- **Proteins: Structure, Function and Genetics**
ISSN: 0887-3585
Proteins Struct. Funct. Genet.
Wiley-Liss Inc., 111 River Street, Hoboken, NJ 07030-5774, United States, Tel: +1(201)7486645, Fax: +1(201)7486021 ISSUES/YR: 12

Proteome Science
ISSN: 1477-5956
Proteome Sci.
BioMed Central Ltd., 34 - 42 Cleveland Street, W1T 4LB, London, United Kingdom, Tel: +44(207)3230323, Fax: +44(207)6319926 ISSUES/YR: 1

- **Proteomics**
ISSN: 1615-9853
Proteomics
Wiley-VCH Verlag, P.O. Box 101161, D-69451, Weinheim, Germany, Tel: +49(6201)606147, Fax: +49(6201)606117 ISSUES/YR: 18

Przeglad Dermatologiczny
ISSN: 0033-2526
Prz. Dermatol.
Wydawnictwo Czelej Sp. z.o.o., ul. Boczna Lubomelskiej 4, 20-070, Lublin, Poland, Tel: +48(81)7437766, Fax: +48(81)5347788 ISSUES/YR: 6

Przeglad Menopauzalny
ISSN: 1643-8876
Prz. Menopauzalny
Termedia Publishing House Ltd., ul Kleberga 8,, 61-615, Poznan, Poland, Tel: +48-61 822 77 81, Fax: +48-61 822 7781 ISSUES/YR: 6

Przeglad Pediatryczny
ISSN: 0137-723X
Prz. Pediatr.
Cornetis, ul. Dlugosza 2-6, 51-162, Wroclaw, Poland, Tel: +48-071 326 3101, Fax: +48-071 325 2803 ISSUES/YR: 4

Psicologia Conductual
ISSN: 1132-9483
Psicol. Conductual
Fundacion para el Avance de la Psicologia Clinica Conductual, P.O. Box 1245, 18080, Granada, Spain ISSUES/YR: 3

Psicoterapia Cognitiva e Comportamentale
ISSN: 1126-1072
Psicoter. Cogn. Comport.
Tipografia PI-ME Editrice Srl, Via Vigentina 136, 27100, Pavia, Italy, Tel: +39(382)572169, Fax: +39(382)572102 ISSUES/YR: 4

Psicoterapia e Scienze Umane
ISSN: 0394-2864
Psicoter. Sci. Um.
Franco Angeli Edizioni, Viale Monza 106, 20127, 20127 Milano, Italy, Tel: +39-02 2837141, Fax: +39-02 26144793 ISSUES/YR: 4

Psihoterapija
ISSN: 0350-3186
Psihoterapija
Klinika za Psiholosku Medicinu, Kispaticeva 12, 10000, Zagreb, Croatia, Tel: +385(1)2332825, Fax: +385(1)2335818 ISSUES/YR: 4

Psiquiatria Biologica
ISSN: 1134-5934
Psiquiatr. Biol.
Ediciones Doyma, S.L., Travesera de Gracia 17-21, 08021, Barcelona, Spain, Tel: +34(93)2000711, Fax: +34(93)2091136 ISSUES/YR: 6

Psiquis
ISSN: 0210-8348
Psiquis
Alpe Editores S.A., Pedro Rico 27, oficinas 11-12, 28029, Madrid, Spain, Tel: +34(91)7338811, Fax: +34(91)3159652 ISSUES/YR: 6

Psyche
ISSN: 0033-2623
Psyche (Germany)
J.C. Cotta'sche Buchhandlung Nachvolger GmbH, P.O. Box 106016, 70049, Stuttgart, Germany, Tel: +49(711)66720, Fax: +49(711)66722030 ISSUES/YR: 12

Psychiatria Danubina
ISSN: 0353-5053
Psychiatr. Danub.
Medicinska Naklada Zagreb, Vlaska 69, 10000, Zagreb, Croatia ISSUES/YR: 4

Psychiatria i Psychologia Kliniczna
ISSN: 1644-6313
Psychiatr. Psychol. Klin.
Medical Communications, Ul.Ojcowska 11, 02-918, Warszawa, Poland, Tel: +48-226519783, Fax: +48-8425363 ISSUES/YR: 4

Psychiatria Polska
ISSN: 0033-2674
Psychiatr. Pol.
Polish Psychiatric Association, Lenartowicza 14, 31-138, Cracow, Poland, Tel: +48(12)6331203, Fax: +48(12)6334067 ISSUES/YR: 6

Psychiatric Bulletin
ISSN: 0955-6036
Psychiatr. Bull.
Royal College of Psychiatrists, 17 Belgrave Square, SW1X 8PG, London, United Kingdom, Tel: +44(20)72352351, Fax: +44(20)72451231 ISSUES/YR: 12

• **Psychiatric Clinics of North America**
ISSN: 0193-953X
Psychiatr. Clin. North Am.
W.B. Saunders, Independence Square West, Philadelphia, PA 19106-3399, United States, Tel: +1(215)2387800, Fax: +1(215)2386445 ISSUES/YR: 4

• **Psychiatric Genetics**
ISSN: 0955-8829
Psychiatr. Genet.
Lippincott Williams and Wilkins, 241 Borough High Street, SE1 1GB, London, United Kingdom, Tel: +44(207)9407500, Fax: +44(207)9407575 ISSUES/YR: 4

Psychiatric Quarterly
ISSN: 0033-2720
Psychiatr. Q.
Kluwer Academic/Human Sciences Press Inc., 233 Spring Street, New York, NY 10013-1578, United States, Tel: +1(212)6208000, Fax: +1(212)4630742 ISSUES/YR: 4

Psychiatric Rehabilitation Journal
ISSN: 1095-158X
Psychiatr. Rehab. J.
Boston University, 940 Commonwealth Avenue West, Boston, MA 02215, United States, Tel: +1(617)3533549, Fax: +1(617)3539209 ISSUES/YR: 4

Psychiatric Services
ISSN: 1075-2730
Psychiatr. Serv.
American Psychiatric Association, 1000 Wilson Boulevard, Suite 1825, Arlington, VA 22209-3901, United States, Tel: +1(703)9077300 ISSUES/YR: 12

Psychiatrie
ISSN: 1211-7579
Psychiatrie
Tigis s.r.o., Na Habrove 17, 152 00, Prague 5, Czech Republic, Tel: +420-251 813 1924, Fax: +420-251 681 217 ISSUES/YR: 4

Psychiatrische Praxis
ISSN: 0303-4259
Psychiatr. Prax.
Georg Thieme Verlag, Rudigerstrasse 14, D-70469, Stuttgart, Germany, Tel: +49(711)8931-321, Fax: +49(711)8931-422 ISSUES/YR: 8

Psychiatrische Praxis, Supplement
ISSN: 0934-3008
Psychiatr. Prax. Suppl.
Georg Thieme Verlag, Rudigerstrasse 14, D-70469, Stuttgart, Germany, Tel: +49(711)8931-321, Fax: +49(711)8931-422 ISSUES/YR: 1

Psychiatry
ISSN: 0033-2747
Psychiatry
Guilford Publications, 72 Spring Street, New York, NY 10012, United States, Tel: +1(212)4319800 ISSUES/YR: 4

Psychiatry and Clinical Neurosciences
ISSN: 1323-1316
Psychiatry Clin. Neurosci.
Blackwell Publishing Ltd, 9600 Garsington Road, OX4 2XG, Oxford, United Kingdom, Tel: +44(1865)776868, Fax: +44(1865)714591 ISSUES/YR: 6

Psychiatry Investigation
ISSN: 1738-3684
Psychiatr. Invest.
Korean Neuropsychiatric Association, RN 522, G-five Central Plaza 1685-8, 137-882, Seocho-Dong, Seocho-Gu, Seoul, South Korea, Tel: +82-25376171, Fax: +82-25376174 ISSUES/YR: 2

• **Psychiatry Research**
ISSN: 0165-1781
Psychiatry Res.
Elsevier Ireland Ltd, P.O. Box 85, Limerick, Ireland, Tel: +353(61)471944, Fax: +353(61)472144 ISSUES/YR: 15

• **Psychiatry Research - Neuroimaging**
ISSN: 0925-4927
Psychiatry Res. Neuroimaging
Elsevier Ireland Ltd, P.O. Box 85, Limerick, Ireland, Tel: +353(61)471944, Fax: +353(61)472144 ISSUES/YR: 9

Psychiatry, Psychology and Law
ISSN: 1321-8719
Psychiatry Psychol. Law
Australian Academic Press Pty. Ltd, 32 Jeays Street, QLD 4006, Bowen Hills, Australia, Tel: +61(7)32571176, Fax: +61(7)32525908 ISSUES/YR: 2

Psycho-Oncology
ISSN: 1057-9249
Psycho-Oncology
John Wiley and Sons Ltd, Southern Gate, PO19 8SQ, Chichester, West Sussex, United Kingdom, Tel: +44(1243)779777, Fax: +44(1243)775878 ISSUES/YR: 8

Psychoanalysis and Psychotherapy
ISSN: 0736-508X
Psychoanal. Psychother.
International Universities Press Inc., P.O. Box 389, Guilford, CT 0643-0389, United States, Tel: +1(203)2454000, Fax: +1(203)2450775 ISSUES/YR: 2

Psychoanalytic Psychology
ISSN: 0736-9735
Psychoanal. Psychol.
American Psychological Association Inc., 750 First Street NE, Washington, DC 20002-4242, United States, Tel: +1(202)3365600, Fax: +1(202)3365568 ISSUES/YR: 4

• **Psychoanalytic Psychotherapy**
ISSN: 0266-8734
Psychoanal. Psychother.
Routledge, 4 Park Square, Milton Park, OX14 4RN, Abingdon, Oxfordshire, United Kingdom, Tel: +44(1235)828600, Fax: +44(1235)829000 ISSUES/YR: 4

Psychoanalytic Quarterly
ISSN: 0033-2828
Psychoanal. Q.
Psychoanalytic Quarterly Inc., West Coast Office, 670 Berry Ave. Los Altos, CA 94024, United States, Tel: +1-650 941 5420, Fax: +1-650 941 5420 ISSUES/YR: 4

Psychodynamic Practice
ISSN: 1475-3634
Psychodyn. Pract.
Routledge, 4 Park Square, Milton Park, OX14 4RN, Abingdon, Oxfordshire, United Kingdom, Tel: +44(1235)828600, Fax: +44(1235)829000 ISSUES/YR: 4

Psychogeriatria Polska
ISSN: 1732-2642
Psychogeriatr. Pol.
Psychogeriatria Polska, ul. Nozownicza 4/8, 50-119, Wroclaw, Poland, Tel: +48(71)3432152, Fax: +48(71)3432155 ISSUES/YR: 4

• **Psychogeriatrics**
ISSN: 1346-3500
Psychogeriatrics
Blackwell Publishing, 550 Swanston Street, VIC 3053, Carlton South, Australia, Tel: +61(3)8359 1011, Fax: +61(3)8359 1120 ISSUES/YR: 4

Psychological Assessment
ISSN: 1040-3590
Psychol. Assess.
American Psychological Association Inc., 750 First Street NE, Washington, DC 20002-4242, United States, Tel: +1(202)3365600, Fax: +1(202)3365568 ISSUES/YR: 4

Current

Psychological Medicine
ISSN: 0033-2917
Psychol. Med.
Cambridge University Press, Shaftesbury Road, CB2 2RU, Cambridge, United Kingdom, Tel: +44(1223)312393, Fax: +44(1223)315052 ISSUES/YR: 8

Psychologie et NeuroPsychiatrie du Vieillissement
ISSN: 1760-1703
Psychol. Neuropsychiatr. Vieil.
John Libbey Eurotext, 127, avenue de la Republique, 92120, Montrouge, France, Tel: +33(01)46730660, Fax: +33(01)40840999 ISSUES/YR: 4

Psychology and Aging
ISSN: 0882-7974
Psychol. Aging
American Psychological Association Inc., 750 First Street NE, Washington, DC 20002-4242, United States, Tel: +1(202)3365600, Fax: +1(202)3365568 ISSUES/YR: 4

Psychology and Psychotherapy: Theory, Research and Practice
ISSN: 1476-0835
Psychol. Psychother. Theory Res. Pract.
British Psychological Society, 48 Princess Road East, LE1 7DR, Leicester, United Kingdom, Tel: +44(116)2549568, Fax: +44(116)2470787 ISSUES/YR: 4

Psychology of Addictive Behaviors
ISSN: 0893-164X
Psychol. Addict. Behav.
Educational Publishing Foundation, 750 First Street NE, Washington, DC 20002-4242, United States, Tel: +1(202)3365540, Fax: +1(202)3365549 ISSUES/YR: 4

Current

• **Psychology, Health and Medicine**
ISSN: 1354-8506
Psychol. Health Med.
Routledge, 4 Park Square, Milton Park, OX14 4RN, Abingdon, Oxfordshire, United Kingdom, Tel: +44(1235)828600, Fax: +44(1235)829000 ISSUES/YR: 4

PsychoNeuro
ISSN: 1611-9991
PsychoNeuro
Karl Demeter Verlag GmbH, Rudigerstr. 14, D-70469, Stuttgart, Germany, Tel: +49(0711)89310, Fax: +49(0711)8931536 ISSUES/YR: 12

• **Psychoneuroendocrinology**
ISSN: 0306-4530
Psychoneuroendocrinology
Elsevier Ltd, Langford Lane, Kidlington, OX5 1GB, Oxford, United Kingdom, Tel: +44(1865)843000, Fax: +44(1865)843010 ISSUES/YR: 10

Psychopathology
ISSN: 0254-4962
Psychopathology
S. Karger AG, Allschwilerstrasse 10, P.O. Box, CH-4009, Basel, Switzerland, Tel: +41(61)3061111, Fax: +41(61)3061234 ISSUES/YR: 6

• **Psychopharmacology**
ISSN: 0033-3158
Psychopharmacology
Springer Verlag, Tiergartenstrasse 17, D-69121, Heidelberg, Germany, Tel: +49(6221)4870, Fax: +49(6221)487366 ISSUES/YR: 24

Psychopharmakotherapie
ISSN: 0944-6877
Psychopharmakotherapie
Wissenschaftliche Verlagsgesellschaft mbH, Birkenwaldstrasse 44, D-70191, Stuttgart, Germany, Tel: +49(711)25820 ISSUES/YR: 6

Psychopharmakotherapie, Supplement
ISSN: 0946-5146
Psychopharmakother. Suppl.
Wissenschaftliche Verlagsgesellschaft mbH, Birkenwaldstrasse 44, D-70191, Stuttgart, Germany, Tel: +49(711)25820 ISSUES/YR: 1

• **Psychophysiology**
ISSN: 0048-5772
Psychophysiology
Blackwell Publishing Inc., 350 Main Street, Malden, MA 02148, United States, Tel: +1(781)3888200, Fax: +1(781)3888210 ISSUES/YR: 6

• **Psychosomatic Medicine**
ISSN: 0033-3174
Psychosom. Med.
Lippincott Williams and Wilkins, 351 West Camden Street, Baltimore, MD 21201-2436, United States, Tel: +1(410)5284000 ISSUES/YR: 6

Psychosomatics
ISSN: 0033-3182
Psychosomatics
American Psychiatric Publishing Inc., 1000 Wilson Blvd, Suite 1825, VA 22209, United States, Tel: +1(703)907-7322, Fax: +1(703)90-1091 ISSUES/YR: 6

Psychoterapia
ISSN: 0239-4170
Psychoterapia
Polish Psychiatric Association, Lenartowicza 14, 31-138, Cracow, Poland, Tel: +48(12)6331203, Fax: +48(12)6334067 ISSUES/YR: 4

Psychotherapeut
ISSN: 0935-6185
Psychotherapeut
Springer Verlag, Tiergartenstrasse 17, D-69121, Heidelberg, Germany, Tel: +49(6221)4870, Fax: +49(6221)487366 ISSUES/YR: 6

Psychotherapies
ISSN: 0251-737X
Psychotherapies
Editions Medecine et Hygiene, 7, Avenue de la Roseraie, P.O. Box 456, CH-1211, Geneva 4, Switzerland, Tel: +41(22)7029311, Fax: +41(22)7029355 ISSUES/YR: 4

Psychotherapy
ISSN: 0033-3204
Psychotherapy
American Psychological Association, 6557 E. Riverdale St., Mesa, AZ 85215, United States, Tel: +1(602)3639211, Fax: +1(480)854-8966 ISSUES/YR: 4

• **Psychotherapy and Psychosomatics**
ISSN: 0033-3190
Psychother. Psychosom.
S. Karger AG, Allschwilerstrasse 10, P.O. Box, CH-4009, Basel, Switzerland, Tel: +41(61)3061111, Fax: +41(61)3061234 ISSUES/YR: 6

• **Psychotherapy Research**
ISSN: 1050-3307
Psychother. Res.
Routledge, 4 Park Square, Milton Park, OX14 4RN, Abingdon, Oxfordshire, United Kingdom, Tel: +44(1235)828600, Fax: +44(1235)829000 ISSUES/YR: 5

• **Pteridines**
ISSN: 0933-4807
Pteridines
International Society of Pteridinology, PO3 6AD, Portsmouth, United Kingdom, Tel: +44(23)92822331 ISSUES/YR: 4

Public Health
ISSN: 0033-3506
Public Health
Elsevier, P.O. Box 211, 1000 AE, Amsterdam, Netherlands, Tel: +31(20)4853757, Fax: +31(20)4853432 ISSUES/YR: 6

• **Public Health Medicine**
ISSN: 1465-1505
Public Health Med.
Rila Publications Ltd, 73 Newman Street, W1A 4PG, London, United Kingdom, Tel: +44(207)6311299, Fax: +44(207)5807166 ISSUES/YR: 3

• **Public Health Nutrition**
ISSN: 1368-9800
Public Health Nutr.
CAB International, Wallingford, OX10 8DE, Oxon, United Kingdom, Tel: +44(1491)832111, Fax: +44(1491)826090 ISSUES/YR: 8

• **Public Health Reports**
ISSN: 0033-3549
Public Health Rep.
Association of Schools of Public Health, 1101 15th Street NW Suite 910, Washington, DC 20005, United States, Tel: +1(202)2961099, Fax: +1(202)2961252 ISSUES/YR: 6

• **Public Health Reviews**
ISSN: 0301-0422
Public Health Rev.
Technosdar Ltd, 5 Levontin Street, P.O. Box 31684, Tel Aviv, Israel, Tel: +972(3)622418 ISSUES/YR: 4

Publications of Takeda Research Laboratories
ISSN: 1346-7271
Publ. Takeda Res. Lab.
Takeda Pharmaceutical Co. Ltd., 12-10 Nihonbashi 2-chome, Chuo-ku, 103-8668, Tokyo, Japan, Tel: +81(3)32782111, Fax: +81(3)32782000 ISSUES/YR: 1

Puesta al Dia en Urgencias, Emergencias y Catastrofes
ISSN: 1576-0316
Puesta Dia Urgencias Emerg. Catastr.
ARAN Ediciones S.A., Castello 128 - 1o, 28006, Madrid, Spain, Tel: +34(91)7820035, Fax: +34(91)5615787 ISSUES/YR: 4

• **Pulmonary Pharmacology and Therapeutics**
ISSN: 1094-5539
Pulm. Pharmacol. Ther.
Academic Press, 24-28 Oval Road, NW1 7DX, London, United Kingdom, Tel: +44(207)2674466 ISSUES/YR: 6

PZ Prisma
ISSN: 0945-5566
PZ Prisma
Govi-Verlag Pharmazeutischer Verlag GmbH, P.O. Box 5360, D-65728, Eschborn, Germany, Tel: +49(6196)9280, Fax: +49(6196)928203 ISSUES/YR: 4

Qatar Medical Journal
ISSN: 0253-8253
Qatar Med. J.
Hamad Medical Corporation, P.O. Box 3050, Doha, Qatar, Tel: +974-392132, Fax: +974-314564 ISSUES/YR: 2

• **QJM - Monthly Journal of the Association of Physicians**
ISSN: 1460-2725
QJM Mon. J. Assoc. Phys.
Oxford University Press, Great Clarendon Street, OX2 6DP, Oxford, United Kingdom, Tel: +44(1865)353907, Fax: +44(1865)353486 ISSUES/YR: 12

• **QSAR and Combinatorial Science**
ISSN: 1611-020X
QSAR Comb. Sci.
Wiley-VCH Verlag, P.O. Box 101161, D-69451, Weinheim, Germany, Tel: +49(6201)606147, Fax: +49(6201)606117 ISSUES/YR: 10

Quaderni ACP
ISSN: not available yet
Quad. ACP
Associazione Culturale Pediatri, Corso U. Comandini 10, 47023, Cesena, Italy, Tel: +39-054729304, Fax: +39-054729304 ISSUES/YR: 6

• **Qualitative Health Research**
ISSN: 1049-7323
Qual. Health Res.
SAGE Publications Inc., 2455 Teller Road, Thousand Oaks, CA 91320, United States, Tel: +1(805)4990721, Fax: +1(805)4990871 ISSUES/YR: 10

Quality and Safety in Health Care
ISSN: 1475-3898
Qual. Saf. Health Care
BMJ Publishing Group, Tavistock Square, WC1H 9JR, London, United Kingdom, Tel: +44(207)3874499, Fax: +44(207)3836668 ISSUES/YR: 6

• **Quality Assurance Journal**
ISSN: 1087-8378
Qual. Assur. J.
John Wiley and Sons Ltd, Southern Gate, PO19 8SQ, Chichester, West Sussex, United Kingdom, Tel: +44(1243)779777, Fax: +44(1243)775878 ISSUES/YR: 4

Quality in Primary Care
ISSN: 1479-1072
Qual. Prim. Care
University of Leicester, Gwendolen Road, LE5 4PW, Leicester, United Kingdom, Tel: +44(116)2584873, Fax: +44(116)2584982 ISSUES/YR: 4

• **Quality of Life Research**
ISSN: 0962-9343
Qual. Life Res.
Kluwer Academic Publishers, Van Godewijckstraat 30, 3311 GZ, Dordrecht, Netherlands, Tel: +31(78)6576000, Fax: +31(78)6576254 ISSUES/YR: 8

Quarterly Journal of Nuclear Medicine and Molecular Imaging
ISSN: 1125-0135
Q. J. Nucl. Med. Mol. Imaging
Edizioni Minerva Medica S.p.A., Corso Bramante 83-85, 10126, Turin, Italy, Tel: +39(011)678282, Fax: +39(011)674502 ISSUES/YR: 4

• **Quarterly Reviews of Biophysics**
ISSN: 0033-5835
Q. Rev. Biophys.
Cambridge University Press, Shaftesbury Road, CB2 2RU, Cambridge, United Kingdom, Tel: +44(1223)312393, Fax: +44(1223)315052 ISSUES/YR: 4

Quimica Clinica
ISSN: 1139-2436
Quim. Clin.
Ediciones Mayo S.A., Aribau 185-187 2e planta, 08021, Barcelona, Spain, Tel: +34(93)2090255, Fax: +34(93)2020643 ISSUES/YR: 6

Radiation and Environmental Biophysics
ISSN: 0301-634X
Radiat. Environ. Biophys.
Springer New York LLC, 233 Springer Street, New York, 10013-1578, United States, Tel: +1-212 460 1501, Fax: +1-212 460 1595 ISSUES/YR: 4

Radiation Medicine - Medical Imaging and Radiation Oncology
ISSN: 0288-2043
Radiat. Med. Med. Imaging Radiat. Oncol.
Keio University, 35 Shinano-machi, 160-8582, Shinjuku-ku, Tokyo, Japan, Tel: +81(3)5363-3824, Fax: +81(3)3353-0677 ISSUES/YR: 8

Radiation Physics and Chemistry
ISSN: 0969-806X
Radiat. Phys. Chem.
Elsevier Ltd, Langford Lane, Kidlington, OX5 1GB, Oxford, United Kingdom, Tel: +44(1865)843000, Fax: +44(1865)843010 ISSUES/YR: 12

Radiation Protection Dosimetry
ISSN: 0144-8420
Radiat. Prot. Dosim.
Oxford University Press, Great Clarendon Street, OX2 6DP, Oxford, United Kingdom, Tel: +44(1865)353907, Fax: +44(1865)353486 ISSUES/YR: 20

• **Radiation Research**
ISSN: 0033-7587
Radiat. Res.
Radiation Research Society, 10105 Cottesmore Court, Great Falls, VA 22066-3540, United States, Tel: +1-703 757 4585, Fax: +1-703 757 0454 ISSUES/YR: 12

• **Radiography**
ISSN: 1078-8174
Radiography
W.B. Saunders Ltd, 32 Jamestown Road, NW1 7BY, London, United Kingdom, Tel: +44(20)74244200, Fax: +44(20)74854752 ISSUES/YR: 4

Radiologe
ISSN: 0033-832X
Radiologe
Springer Verlag, Tiergartenstrasse 17, D-69121, Heidelberg, Germany, Tel: +49(6221)4870, Fax: +49(6221)487366 ISSUES/YR: 12

Radiologia
ISSN: 0033-8338
Radiologia
Ediciones Doyma, S.L., Travesera de Gracia 17-21, 08021, Barcelona, Spain, Tel: +34(93)2000711, Fax: +34(93)2091136 ISSUES/YR: 6

Radiologia Medica
ISSN: 0033-8362
Radiol. Med.
Edizioni Minerva Medica S.p.A., Corso Bramante 83-85, 10126, Turin, Italy, Tel: +39(011)678282, Fax: +39(011)674502 ISSUES/YR: 12

● **Radiologic Clinics of North America**
ISSN: 0033-8389
Radiol. Clin. North Am.
W.B. Saunders, Independence Square West, Philadelphia, PA 19106-3399, United States, Tel: +1(215)2387800, Fax: +1(215)2386445 ISSUES/YR: 6

● **Radiology**
ISSN: 0033-8419
Radiology
Radiological Society of North America Inc., 820 Jorie Boulevard, Oak Brook, IL 60523-2251, United States, Tel: +1(630)5712670, Fax: +1(630)5717837 ISSUES/YR: 12

Radiology and Oncology
ISSN: 1318-2099
Radiol. Oncol.
Slovenian Medical Society / Croatian Medical Association, Vrazov trg 4, 61105, Ljubljana, Slovenia, Tel: +386(61)1320068 ISSUES/YR: 4

Radiology Now
ISSN: 1461-4650
Radiol. Now
Discovery London, Avonmore Road, W14 8DG, London, United Kingdom, Tel: +44-20 7173 4020, Fax: +44-20 7173 4101 ISSUES/YR: 3

● **Radiotherapy and Oncology**
ISSN: 0167-8140
Radiother. Oncol.
Elsevier Ireland Ltd, P.O. Box 85, Limerick, Ireland, Tel: +353(61)471944, Fax: +353(61)472144 ISSUES/YR: 12

RAS - Radioloski Arhiv Srbije
ISSN: 0354-1452
RAS. Radioloski. Arhiv. Srbije
Institute of Radiology, UMC Nis, Bui. Cara Konstantina 48, 18000, Nis, Serbia and Montenegro ISSUES/YR: 2

Rassegna di Patologia dell'Apparato Respiratorio
ISSN: 0033-9563
Rass. Patol. Appar. Respir.
EdiAipo Scientifica s.c.a.r.l, A.Gherardesca, 56121, Ospedaletto, Italy, Tel: +39-50 3130260, Fax: +39-50 3130290 ISSUES/YR: 4

Rays - International Journal of Radiological Sciences
ISSN: 0390-7740
Rays Int. J. Radiol. Sci.
Ecoedizioni International srl, Via Cola di Rienzo 28, 00192, Roma, Italy, Tel: +39-6 631 350, Fax: +39-6 631 350 ISSUES/YR: 4

Reanimation
ISSN: 1624-0693
Reanimation
Editions Scientifiques et Medicales Elsevier, 23 rue Linois, 75724, Paris Cedex 15, France, Tel: +33(1)45589826, Fax: +33(1)45589428 ISSUES/YR: 8

Recenti Progressi in Medicina
ISSN: 0034-1193
Recenti Prog. Med.
Il Pensiero Scientifico Editore s.r.l., Via Bradano 3/C, 00199, Roma, Italy, Tel: +39(06)86207158, Fax: +39(06)86207160 ISSUES/YR: 12

Recht und Psychiatrie
ISSN: 0724-2247
Recht Psychiatrie
Psychiatrie-Verlag GmbH, Thomas-Mann-Strasse 49a, 53111, Bonn, Germany ISSUES/YR: 4

Rechtsmedizin
ISSN: 0937-9819
Rechtsmedizin
Springer Verlag, Tiergartenstrasse 17, D-69121, Heidelberg, Germany, Tel: +49(6221)4870, Fax: +49(6221)487366 ISSUES/YR: 6

● **Redox Report**
ISSN: 1351-0002
Redox Rep.
Maney Publishing, Hudson Road, LS9 7DL, Leeds, United Kingdom, Tel: +44(113)2497481, Fax: +44(113)2486983 ISSUES/YR: 6

References en Gynecologie Obstetrique
ISSN: 1244-8168
Ref. Gynecol. Obstet.
General Sciences Publications International (GSP), 53 Rue de Bellevue, F-92100, Boulogne, France, Tel: +33-1 55 38 05 80, Fax: +33-1 55 38 05 86 ISSUES/YR: 6

● **Regional Anesthesia and Pain Medicine**
ISSN: 1098-7339
Reg. Anesth. Pain Med.
W.B. Saunders, Independence Square West, Philadelphia, PA 19106-3399, United States, Tel: +1(215)2387800, Fax: +1(215)2386445 ISSUES/YR: 6

● **Regulatory Peptides**
ISSN: 0167-0115
Regul. Pept.
Elsevier, P.O. Box 211, 1000 AE, Amsterdam, Netherlands, Tel: +31(20)4853757, Fax: +31(20)4853432 ISSUES/YR: 21

● **Regulatory Toxicology and Pharmacology**
ISSN: 0273-2300
Regul. Toxicol. Pharmacol.
Academic Press Inc., 6277 Sea Harbor Drive, Orlando, FL 32887-4900, United States, Tel: +1(407)3454040, Fax: +1(407)3639661 ISSUES/YR: 6

Rehabilitace a Fyzikalni Lekarstvi
ISSN: 1211-2658
Rehabil. Fyz. Lek.
Czech Medical Association J.E. Purkyne, Sokolska 31, P.O. Box 88, 120 26, Praha 2, Czech Republic, Tel: +420(2)24266201-4, Fax: +420(2)24266212 ISSUES/YR: 4

Rehabilitacia
ISSN: 0375-0922
Rehabilitacia
LIECREH GUTH, P.O. Box 77, 830 03, Bratislava, Slovakia, Tel: +421(7)54772323, Fax: +421(7)54414700 ISSUES/YR: 4

Rehabilitacja Medyczna
ISSN: 1427-9622
Rehabilitacja Medyczna
Rehabilitacja Medyczna, Al. Jana Pawla II 82, 31-579, Krakow, Poland, Tel: +48(12)6416749 ISSUES/YR: 1

Rehabilitation
ISSN: 0034-3536
Rehabilitation
Georg Thieme Verlag, Rudigerstrasse 14, D-70469, Stuttgart, Germany, Tel: +49(711)8931-321, Fax: +49(711)8931-422 ISSUES/YR: 6

Rehabilitation Psychology
ISSN: 0090-5550
Rehabil. Psychol.
American Psychological Association Inc., 750 First Street NE, Washington, DC 20002-4242, United States, Tel: +1(202)3365600, Fax: +1(202)3365568 ISSUES/YR: 4

● **Rejuvenation Research**
ISSN: 1549-1684
Rejuvenation Res.
Mary Ann Liebert Inc., 140 Huguenot Street, New Rochelle, NY 10801-5215, United States, Tel: +1(914)7402100, Fax: +1(914)7402101 ISSUES/YR: 4

Remedia - Klinicka Mikrobiologie
ISSN: 1211-7684
Rem. Klin. Mikrobiol.
Panax Publishing House, Sumavska 5, 120 00, Prague 2, Czech Republic, Tel: +420(2)24256947, Fax: +420(2)24255217 ISSUES/YR: 12

Current

- **Renal Failure**
ISSN: 0886-022X
Renal Fail.
Taylor and Francis Inc., 325 Chestnut St, Suite 800, Philadelphia PA, PA 19106, United States, Tel: +1(800)3541420, Fax: +1(215)6258914 ISSUES/YR: 8

Rentgenologiya i Radiologiya
ISSN: 0486-400X
Rentgenol. Radiol.
Izdatelstvo Medicina i Fizkultura, 11 Slaveikov Square, 1000, Sofia, Bulgaria, Tel: +359(2)879111 ISSUES/YR: 4

Reports of Practical Oncology and Radiotherapy
ISSN: 1507-1367
Rep. Pract. Oncol. Radiother.
Great Poland Cancer Center, 15 Garbary Street, 61-866, Poznan, Poland, Tel: +48(61)8540700, Fax: +48(61)8521948 ISSUES/YR: 4

Reproducao e Climaterio
ISSN: 1413-2087
Reprod. Climaterio
Reproducao e Climaterio, Campus Universitario, 14049-900, Ribeirao Preto, Sao Paulo, Brazil, Tel: +55(16)6022821, Fax: +55(16)6022821 ISSUES/YR: 4

- **Reproduction**
ISSN: 1470-1626
Reproduction
BioScientifica Ltd., 22 Apex Court, Woodlands, Bradley Stoke, BS32 4JT, Bristol, United Kingdom, Tel: +44(1454)642240, Fax: +44(1454)642222 ISSUES/YR: 12

Reproduction Humaine et Hormones
ISSN: 0994-3919
Reprod. Hum. Horm.
Editions ESKA, 12 rue du 4 Septembre, 75002, Paris, France, Tel: +33(01)42865600, Fax: +33(01)42603269 ISSUES/YR: 6

Reproduction Nutrition Development
ISSN: 0926-5287
Reprod. Nutr. Dev.
EDP Sciences, 17 Avenue du Hoggar - BP 112, F-91944, Les Ulis Cedex A, France, Tel: +33(1)69187575, Fax: +33(1)69288491 ISSUES/YR: 6

Reproduction, Fertility and Development
ISSN: 1031-3613
Reprod. Fertil. Dev.
CSIRO, P.O. Box 1139, VIC 3066, Collingwood, Australia, Tel: +61(39)6627686, Fax: +61(39)6627555 ISSUES/YR: 8

Reproductive Biology and Endocrinology
ISSN: 1477-7827
Reprod. Biol. Endocrinol.
BioMed Central Ltd., 34 - 42 Cleveland Street, W1T 4LB, London, United Kingdom, Tel: +44(207)3230323, Fax: +44(207)6319926 ISSUES/YR: 1

Reproductive BioMedicine Online
ISSN: 1472-6483
Reprod. BioMed. Online
Reproductive Healthcare Ltd, Duck End Farm, Park Lane, Dry Drayton, CB3 8DB, Cambridge, United Kingdom, Tel: +44(1954)781812, Fax: +44(1954)781816 ISSUES/YR: 12

Reproductive Health
ISSN: 1742-4755
Reprod. Health
BioMed Central Ltd., 34 - 42 Cleveland Street, W1T 4LB, London, United Kingdom, Tel: +44(207)3230323, Fax: +44(207)6319926 ISSUES/YR: 1

- **Reproductive Health Matters**
ISSN: 0968-8080
Reprod. Health Matters
Elsevier, P.O. Box 211, 1000 AE, Amsterdam, Netherlands, Tel: +44(20)4853757, Fax: +44(20)4853432 ISSUES/YR: 2

- **Reproductive Medicine and Biology**
ISSN: 1445-5781
Reprod. Med. Biol.
Blackwell Publishing, 550 Swanston Street, VIC 3053, Carlton South, Australia, Tel: +61(3)8359 1011, Fax: +61(3)8359 1120 ISSUES/YR: 4

- **Reproductive Medicine Review**
ISSN: 0962-2799
Reprod. Med. Rev.
Cambridge University Press, Shaftesbury Road, CB2 2RU, Cambridge, United Kingdom, Tel: +44(1223)312393, Fax: +44(1223)315052 ISSUES/YR: 3

Reproductive Toxicology
ISSN: 0890-6238
Reprod. Toxicol.
Elsevier Inc., 360 Park Avenue South, New York, NY 10010, United States, Tel: +1(212)6333950, Fax: +1(212)6333990 ISSUES/YR: 8

Research and Clinical Forums
ISSN: 0143-3083
Res. Clin. Forums
Wells Medical Ltd, Speldhurst Road, TN4 0JB, Royal Tunbridge Wells, Kent, United Kingdom, Tel: +44(1892)511600, Fax: +44(1892)511400 ISSUES/YR: 1

Research and Practice in Alzheimer's Disease
ISSN: 1284-8360
Res. Pract. Alzheimer's Dis.
Serdi Publishing Company, 320 rue Saint-Honore, 75001, Paris, France, Tel: +33(1)44542796, Fax: +33(5)61751128 ISSUES/YR: 1

Research Communications in Alcohol and Substances of Abuse
ISSN: 1080-8388
Res. Commun. Alcohol Subst. Abuse
PJD Publications Ltd, P.O. Box 966, Westbury, NY 11590-0966, United States, Tel: +1(516)6260650, Fax: +1(516)6265546 ISSUES/YR: 4

Research Communications in Biochemistry and Cell and Molecular Biology
ISSN: 1087-111X
Res. Commun. Biochem. Cell Mol. Biol.
PJD Publications Ltd, P.O. Box 966, Westbury, NY 11590-0966, United States, Tel: +1(516)6260650, Fax: +1(516)6265546 ISSUES/YR: 4

- **Research Communications in Biological Psychology and Psychiatry**
ISSN: 1087-0695
Res. Commun. Biol. Psychol. Psychiatry
PJD Publications Ltd, P.O. Box 966, Westbury, NY 11590-0966, United States, Tel: +1(516)6260650, Fax: +1(516)6265546 ISSUES/YR: 4

- **Research Communications in Molecular Pathology and Pharmacology**
ISSN: 1078-0297
Res. Commun. Mol. Pathol. Pharmacol.
PJD Publications Ltd, P.O. Box 966, Westbury, NY 11590-0966, United States, Tel: +1(516)6260650, Fax: +1(516)6265546 ISSUES/YR: 12

Research Communications in Pharmacology and Toxicology
ISSN: 1087-1101
Res. Commun. Pharmacol. Toxicol.
PJD Publications Ltd, P.O. Box 966, Westbury, NY 11590-0966, United States, Tel: +1(516)6260650, Fax: +1(516)6265546 ISSUES/YR: 4

Research in Developmental Disabilities
ISSN: 0891-4222
Res. Dev. Disabil.
Elsevier Inc., 360 Park Avenue South, New York, NY 10010, United States, Tel: +1(212)6333950, Fax: +1(212)6333990 ISSUES/YR: 6

Research in Healthcare Financial Management
ISSN: 1524-1521
Res. Healthc. Fin. Manage.
Int. Soc. for Res. in Healthcare Financial Management, 6600 York Road, Suite 107, Baltimore, MD 21212, United States, Tel: +1-305 348 2861, Fax: +1(810)8155378 ISSUES/YR: 1

- **Research in Microbiology**
ISSN: 0923-2508
Res. Microbiol.
Editions Scientifiques et Medicales Elsevier, 23 rue Linois, 75724, Paris Cedex 15, France, Tel: +33(1)45589826, Fax: +33(1)45589428 ISSUES/YR: 10

Research in Sports Medicine
ISSN: 1543-8627
Res. Sports Med.
Taylor and Francis Ltd., 4 Park Square, Milton Park, OX14 4RN, Abingdon, Oxfordshire, United Kingdom, Tel: +44(20)70175541, Fax: +44(20)70175198 ISSUES/YR: 4

Research in Veterinary Science
ISSN: 0034-5288
Res. Vet. Sci.
Elsevier, P.O. Box 211, 1000 AE, Amsterdam, Netherlands, Tel: +44(20)4853757, Fax: +44(20)4853432 ISSUES/YR: 6

Research on Aging
ISSN: 0164-0275
Res. Aging
SAGE Publications Inc., 2455 Teller Road, Thousand Oaks, CA 91320, United States, Tel: +1(805)4990721, Fax: +1(805)4990871 ISSUES/YR: 6

Resources, Conservation and Recycling
ISSN: 0921-3449
Resour. Conserv. Recycl.
Elsevier, P.O. Box 211, 1000 AE, Amsterdam, Netherlands, Tel: +31(20)4853757, Fax: +31(20)4853432 ISSUES/YR: 12

● **Respiration**
ISSN: 0025-7931
Respiration
S. Karger AG, Allschwilerstrasse 10, P.O. Box, CH-4009, Basel, Switzerland, Tel: +41(61)3061111, Fax: +41(61)3061234 ISSUES/YR: 6

Respiration and Circulation
ISSN: 0452-3458
Respir. Circ.
Igaku-Shoin Ltd, 5-24-3 Hongo, Bunkyo-ku, 113-91, Tokyo, Japan, Tel: +81(3)38175600 ISSUES/YR: 12

Respiratory Care
ISSN: 0020-1324
Respir. Care
Daedalus Enterprises Inc., 9425 North MacArthur Blvd, Suite 100, Irving, TX 75063, United States, Tel: +1-972 2432272, Fax: +1-972 4842720 ISSUES/YR: 12

Respiratory Care Clinics of North America
ISSN: 1078-5337
Respir. Care Clin. North Am.
W.B. Saunders, Independence Square West, Philadelphia, PA 19106-3399, United States, Tel: +1(215)2387800, Fax: +1(215)2386445 ISSUES/YR: 4

● **Respiratory Medicine**
ISSN: 0954-6111
Respir. Med.
W.B. Saunders Ltd, 32 Jamestown Road, NW1 7BY, London, United Kingdom, Tel: +44(20)74244200, Fax: +44(20)74854752 ISSUES/YR: 12

● **Respiratory Medicine Extra**
ISSN: 1744-9049
Respir. Med. Extra
W.B. Saunders Ltd, 32 Jamestown Road, NW1 7BY, London, United Kingdom, Tel: +44(20)74244200, Fax: +44(20)74854752 ISSUES/YR: 4

● **Respiratory Medicine: COPD Update**
ISSN: 1745-0454
Respir. Med. COPD Update
W.B. Saunders Ltd, 32 Jamestown Road, NW1 7BY, London, United Kingdom, Tel: +44(20)74244200, Fax: +44(20)74854752 ISSUES/YR: 4

● **Respiratory Physiology and Neurobiology**
ISSN: 1569-9048
Respir. Physiol. Neurobiol.
Elsevier, P.O. Box 211, 1000 AE, Amsterdam, Netherlands, Tel: +31(20)4853757, Fax: +31(20)4853432 ISSUES/YR: 15

Respiratory Research
ISSN: 1465-993X
Respir. Res.
BioMed Central Ltd., 34 - 42 Cleveland Street, W1T 4LB, London, United Kingdom, Tel: +44(207)3230323, Fax: +44(207)6319926 ISSUES/YR: 6

● **Respirology**
ISSN: 1323-7799
Respirology
Blackwell Publishing, 550 Swanston Street, VIC 3053, Carlton South, Australia, Tel: +61(3)8359 1011, Fax: +61(3)8359 1120 ISSUES/YR: 4

● **Restorative Neurology and Neuroscience**
ISSN: 0922-6028
Restor. Neurol. Neurosci.
IOS Press, Nieuwe Hemweg 6B, 1013 BG, Amsterdam, Netherlands, Tel: +31(20)6883355, Fax: +31(20)6203419 ISSUES/YR: 6

● **Resuscitation**
ISSN: 0300-9572
Resuscitation
Elsevier Ireland Ltd, P.O. Box 85, Limerick, Ireland, Tel: +353(61)471944, Fax: +353(61)472144 ISSUES/YR: 12

Retina
ISSN: 0275-004X
Retina
Lippincott Williams and Wilkins, 530 Walnut Street, Philadelphia, PA 19106-3621, United States, Tel: +1(215)5218342, Fax: +1(215)5218326 ISSUES/YR: 6

Retrovirology
ISSN: 1742-4690
Retrovirology
BioMed Central Ltd., 34 - 42 Cleveland Street, W1T 4LB, London, United Kingdom, Tel: +44(207)3230323, Fax: +44(207)6319926 ISSUES/YR: 1

Reumatismo
ISSN: 0048-7449
Reumatismo
Societa Italiana di Reumatologia, Via Giustinaiani, 2, 35128, Padova, Italy, Tel: +39(049)8212190, Fax: +39(049)8763852 ISSUES/YR: 4

Reumatologia
ISSN: 0034-6233
Reumatologia
Instytut Reumatologiczny, ul. Spartanska 1, 02-637, Warsaw, Poland, Tel: +48(22)444241 ISSUES/YR: 4

Reumatologia Clinica
ISSN: 1699-258X
Reumatol. Clin.
Ediciones Doyma, S.L., Travesera de Gracia 17-21, 08021, Barcelona, Spain, Tel: +34(93)2000711, Fax: +34(93)2091136 ISSUES/YR: 10

Review of Clinical Pharmacology and Pharmacokinetics, International Edition
ISSN: 1011-6583
Rev. Clin. Pharmacol. Pharmacokinet. Int. Ed.
PHARMAKON-Press, 20 Daskalaki Str., GR-115 26, Athens, Greece, Tel: +30(210)7778101 ISSUES/YR: 3

Reviews in Analgesia
ISSN: 1542-961X
Rev. Analg.
Cognizant Communication Corporation, 3 Hartsdale Road, Elmsford, NY 10523-3701, United States, Tel: +1(914)5927720, Fax: +1(914)5928981 ISSUES/YR: 2

Reviews in Cardiovascular Medicine
ISSN: 1530-6550
Rev. Cardiovasc. Med.
MedReviews LLC, 1333 Broadway, Suite 1120, New York, NY 10018, United States, Tel: +1(212)9714032, Fax: +1(212)9714047 ISSUES/YR: 4

Reviews in Clinical and Experimental Hematology
ISSN: 1825-151X
Rev. Clin. Exp. Hematol.
Forum Service Editore, Via G. Mameli 3/14, 16122, Genova, Italy, Tel: +39-010-8390832, Fax: +39-010-8390860 ISSUES/YR: 4

Reviews in Clinical Gerontology
ISSN: 0959-2598
Rev. Clin. Gerontol.
Cambridge University Press, Shaftesbury Road, CB2 2RU, Cambridge, United Kingdom, Tel: +44(1223)312393, Fax: +44(1223)315052 ISSUES/YR: 4

Reviews in Contemporary Pharmacotherapy
ISSN: 0954-8602
Rev. Contemp. Pharmacother.
Marius Press, P.O. Box 15, LA6 1HW, Carnforth, Lancashire, United Kingdom, Tel: +44(1542)733027 ISSUES/YR: 6

Reviews in Endocrine and Metabolic Disorders
ISSN: 1389-9155
Rev. Endocr. Metab. Disord.
Kluwer Academic Publishers, Van Godewijckstraat 30, 3311 GZ, Dordrecht, Netherlands, Tel: +31(78)6576000, Fax: +31(78)6576254 ISSUES/YR: 4

Reviews in Gastroenterological Disorders
ISSN: 1533-001X
Rev. Gastroenterol. Disord.
MedReviews LLC, 1333 Broadway, Suite 1120, New York, NY 10018, United States, Tel: +1(212)9714032, Fax: +1(212)9714047 ISSUES/YR: 4

• **Reviews in Gynaecological and Perinatal Practice**
ISSN: 1871-2320
Rev. Gynaecol. Perinat. Pract.
Elsevier, P.O. Box 211, 1000 AE, Amsterdam, Netherlands, Tel: +31(20)4853757, Fax: +31(20)4853432 ISSUES/YR: 4

• **Reviews in Medical Microbiology**
ISSN: 0954-139X
Rev. Med. Microbiol.
Lippincott Williams and Wilkins, 241 Borough High Street, SE1 1GB, London, United Kingdom, Tel: +44(207)9407500, Fax: +44(207)9407575 ISSUES/YR: 4

Reviews in Medical Virology
ISSN: 1052-9276
Rev. Med. Virol.
John Wiley and Sons Ltd, Southern Gate, PO19 8SQ, Chichester, West Sussex, United Kingdom, Tel: +44(1243)779777, Fax: +44(1243)775878 ISSUES/YR: 6

Reviews in Neurological Diseases
ISSN: 1545-2913
Rev. Neurol. Dis.
MedReviews LLC, 1333 Broadway, Suite 1120, New York, NY 10018, United States, Tel: +1(212)9714032, Fax: +1(212)9714047 ISSUES/YR: 4

• **Reviews in the Neurosciences**
ISSN: 0334-1763
Rev. Neurosci.
Freund Publishing House Ltd, 61 Nachmani Street, P.O. Box 35010, 61350, Tel Aviv, Israel, Tel: +972(3)5628540, Fax: +972(3)5628538 ISSUES/YR: 6

Reviews in Toxicology
ISSN: 1382-6980
Rev. Toxicol.
IOS Press, Nieuwe Hemweg 6B, 1013 BG, Amsterdam, Netherlands, Tel: +31(20)6883355, Fax: +31(20)6203419 ISSUES/YR: 2

Reviews on Environmental Health
ISSN: 0048-7554
Rev. Environ. Health
Freund Publishing House Ltd, 61 Nachmani Street, P.O. Box 35010, 61350, Tel Aviv, Israel, Tel: +972(3)5628540, Fax: +972(3)5628538 ISSUES/YR: 4

Revisiones en Cancer
ISSN: 0213-8573
Revis. Cancer
ARAN Ediciones S.A., Castello 128 - 1o, 28006, Madrid, Spain, Tel: +34(91)7820035, Fax: +34(91)5615787 ISSUES/YR: 6

Revista Argentina de Dermatologia
ISSN: 0325-2787
Rev. Argent. Dermatol.
Asociation Argentina de Dermatologia, Mexico 1720, 1100, Buenos Aires, Argentina, Tel: +54(1)3812737 ISSUES/YR: 4

Revista Argentina de Endocrinologia y Metabolismo
ISSN: 0326-4610
Rev. Argent. Endocrinol. Metab.
Sociedad Argentina de Endocrinologia y Metabolismo, Av. Diaz Velez 3889/91, 1200, Buenos Aires, Argentina, Tel: +54(1)9838178, Fax: +54(1)9838178 ISSUES/YR: 4

Revista Argentina de Microbiologia
ISSN: 0325-7541
Rev. Argent. Microbiol.
Asociacion Argentina de Microbiologia, Bulnes 44, Planta Baja 'B', 1176, Buenos Aires, Argentina, Tel: +54(1)9584888, Fax: +54(1)9828557 ISSUES/YR: 4

Revista Brasileira de Anestesiologia
ISSN: 0034-7094
Rev. Bras. Anestesiol.
Sociedade Brasileira de Anestesiologia, R. Prof. Alfredo Gomes 36, 22251-080, Rio de Janeiro, Brazil, Tel: +55(21)2666324 ISSUES/YR: 6

Revista Brasileira de Ciencias Farmaceuticas/Brazilian Journal of Pharmaceutical Sciences
ISSN: 1516-9332
Rev. Bras. Cienc. Farm.
Faculdade de Ciencias Farmaceuticas (Biblioteca), P.O. Box 66083, 05389-970, Sao Paulo, Brazil, Tel: +55(11)8137251 ISSUES/YR: 3

Revista Brasileira de Medicina
ISSN: 0034-7264
Rev. Bras. Med.
Moreira Junior Editora Ltda., Rua Henrique Martins 493, 04504, Sao Paulo, Brazil, Tel: +55(11)8849911 ISSUES/YR: 11

Revista Brasileira de Medicina do Esporte
ISSN: 1517-8692
Rev. Bras. Med. Esporte
Redprint Editora Ltda, Rua Domingos de Morais 2.777 - 13.o, CEP 04035-001, Sao Paulo, Brazil, Tel: +55(11)55724813, Fax: +55(11)55711719 ISSUES/YR: 6

Revista Brasileira de Neurologia
ISSN: 0101-8469
Rev. Bras. Neurol.
Editora Cientifica Nacional Ltda, Rua da Gloria 366, 3o andar, P.O. Box 590, 20241-180, Rio de Janeiro, Brazil, Tel: +55(21)2213235, Fax: +55(21)2521691 ISSUES/YR: 6

Revista Brasileira de Neurologia e Psiquiatria
ISSN: 1414-0365
Rev. Bras. Neurol. Psiquiatr.
Editora Cientifica Nacional Ltda, Rua da Gloria 366 - 3o andar, 20241-180 RJ, Rio de Janeiro, Brazil ISSUES/YR: 3

Revista Brasileira de Otorrinolaringologia
ISSN: 0034-7299
Rev. Bras. Otorrinolaringol.
Sociedade Brasileira de Otorrinolaringologia, Av. Indianopolis 740, Moema, Sao Paulo SP, Brazil, Tel: +55(11)50529515, Fax: +55(11)50521025 ISSUES/YR: 6

Revista Brasileira de Plantas Medicinais
ISSN: 1516-0572
Rev. Bras. Plantas Med.
Instituto de Biociencias, Distrito de Rubiao Jr., Campus da UNESP, 18.618 - 000, Botucatu, SP, Brazil, Tel: +55(14)8212121, Fax: +55(14)8213744 ISSUES/YR: 2

Revista Brasileira de Psiquiatria
ISSN: 1516-4446
Rev. Bras. Psiquiatr.
Associacao Brasileira de Psiquiatria, Rua Pedro de Toledo 967 - casa 01, CEP 04039-032, Sao Paulo - SP, Brazil, Tel: +55(11)5496699, Fax: +55(11)5706210 ISSUES/YR: 4

Revista Brasileira de Reumatologia
ISSN: 0482-5004
Rev. Bras. Reumatol.
Redprint Editora Ltda, Rua Domingos de Morais 2.777 - 13.o, CEP 04035-001, Sao Paulo, Brazil, Tel: +55(11)55724813, Fax: +55(11)55711719 ISSUES/YR: 6

Revista Brasileira de Toxicologia
ISSN: 1415-2983
Rev. Bras. Toxicol.
Sociedade Brasileira de Toxicologia, Rua Pamplona 788, Sala 32 - 3.o andar, CEP 01405-001, Sao Paulo, SP, Brazil, Tel: +55(11)32849568, Fax: +55(11)32532848 ISSUES/YR: 2

Revista Chilena de Neuro-Psiquiatria
ISSN: 0034-7388
Rev. Chil. Neuro-Psiquiatr.
Sociedad de Neurologia Psiquiatria y Neurocirugia, Carlos Silva 1292, Depto. 22, Providencia, Santiago, Chile, Tel: +56(2)2319247, Fax: +56(2)2319287 ISSUES/YR: 4

Revista Chilena de Pediatria
ISSN: 0370-4106
Rev. Chil. Pediatr.
Sociedad Chilena de Pediatria, Alcalde Eduardo Castillo Velasco 1838, Nunoa, Santiago, Chile, Tel: +56(2)2371598, Fax: +56(2)2380046 ISSUES/YR: 6

Revista Clinica Espanola
ISSN: 0014-2565
Rev. Clin. Esp.
Ediciones Doyma, S.L., Travesera de Gracia 17-21, 08021, Barcelona, Spain, Tel: +34(93)2000711, Fax: +34(93)2091136 ISSUES/YR: 12

Revista da Associacao Brasileira de Medicina Psicossomatica
ISSN: 1518-7365
Rev. Assoc. Bras. Med. Psicossomatica
Editora Cientifica Nacional Ltda, Rua da Gloria 366 - 3o andar, 20241-180 RJ, Rio de Janeiro, Brazil ISSUES/YR: 4

Revista da Sociedade Brasileira de Cancerologia
ISSN: 1415-6725
Rev. Soc. Bras. Cancerol.
Sociedad Brasileira Cancerologia, R.Para 197 Pituba, 41927-000, Salvador, Brazil, Tel: +55-7132404868, Fax: +55-7132489134 ISSUES/YR: 3

Revista da Sociedade Brasileira de Medicina Tropical
ISSN: 0037-8682
Rev. Soc. Bras. Med. Trop.
Sociedade Brasileira de Medicina Tropical, P.O. Box 04-671, 70919-970, Brasilia, Distrito Federal, Brazil, Tel: +55(61)2740022 ISSUES/YR: 6

Revista de Calidad Asistencial
ISSN: 1134-282X
Rev. Calidad Asist.
Ediciones Doyma, S.L., Travesera de Gracia 17-21, 08021, Barcelona, Spain, Tel: +34(93)2000711, Fax: +34(93)2091136 ISSUES/YR: 8

Revista de Ciencias Farmaceuticas
ISSN: 0101-3793
Rev. Cienc. Farm.
Universidade Estadual Paulista (UNESP), Rodovia Araraquara - Jau km 1, P.O. Box 502, 14801 902, Araraquara, Brazil ISSUES/YR: 2

Revista de Investigacion Clinica
ISSN: 0034-8376
Rev. Invest. Clin.
Instituto Nacional de la Nutricion Salvador Zubiran, Vasco de Quiroga No. 15, Delegacion Tlalpan, 14000, Mexico D.F., Mexico ISSUES/YR: 6

Revista de la Asociacion Espanola de Especialistas en Medicina del Trabajo
ISSN: 1132-6255
Rev. Asoc. Esp. Espec. Med. Trab.
Accion Medica S.A., c/Fernandez de la Hoz, 61, Entreplanta, 28003, Madrid, Spain, Tel: +34(91)5360814, Fax: +34(91)5360607 ISSUES/YR: 4

Revista de la Sociedad Espanola del Dolor
ISSN: 1134-8046
Rev. Soc. Esp. Dolor
ARAN Ediciones S.A., Castello 128 - 1o, 28006, Madrid, Spain, Tel: +34(91)7820035, Fax: +34(91)5615787 ISSUES/YR: 8

Revista de Logopedia, Foniatria y Audiologia
ISSN: 0214-4603
Rev. Logop. Foniatr. Audiol.
Grupo Ars XXI de Communicacion, S.A., Paseo de Gracia, 25, 3., 08007, Barcelona, Spain, Tel: +34-34 93 2721750, Fax: +34-34 93 4881193 ISSUES/YR: 4

Revista de Nefrologia, Dialisis y Transplante
ISSN: 0326-3428
Rev. Nefrol. Dial. Transplant.
Asociacion Regional de Dialisi y Transplantes Renales, Carlos Pellegrini 887 9 piso, C1009ABQ, Buenos Aires, Argentina, Tel: +54(11)5555-1700, Fax: +54(11)5555-1700 ISSUES/YR: 4

Revista de Neurologia
ISSN: 0210-0010
Rev. Neurol.
Revista de Neurologia, P.O. Box 94121, 08080, Barcelona, Spain, Tel: +34(93)2478188, Fax: +34(93)2317250 ISSUES/YR: 24

Revista de Nutricao
ISSN: 1415-5273
Rev. Nutr.
Revista de Nutricao, Av. John B. Dunlop s/n, Bloco C, Jd. Ipaussurama, 13020-904, Campinas SP, Brazil, Tel: +55(19)37298576, Fax: +55(19)37296875 ISSUES/YR: 6

Revista de Ortopedia y Traumatologia
ISSN: 0482-5985
Rev. Ortop. Traumatol.
Ediciones Doyma, S.L., Travesera de Gracia 17-21, 08021, Barcelona, Spain, Tel: +34(93)2000711, Fax: +34(93)2091136 ISSUES/YR: 6

Revista de Psiquiatria Clinica
ISSN: 0044-5967
Rev. Psiquiatr. Clin.
Lemos Editorial e Graficos Ltda, Rua Rui Barbosa 70, Bela Vista, 01326-010, Sao Paulo SP, Brazil, Tel: +55(11)33711855, Fax: +55(11)33711855 ISSUES/YR: 6

Revista de Psiquiatria Consiliar e de Ligacao
ISSN: 0873-612X
Rev. Psiquiatr. Consiliar Ligacao
Revista de Psiquiatria Consiliar e de Ligacao, MEDISA - Centro Empresarial MAPRE, 347 -s/217, 4000-270, Porto, Portugal, Tel: +351(22)2001479, Fax: +351(22)2001490 ISSUES/YR: 2

Revista de Psiquiatria de la Facultad de Medicina de Barcelona
ISSN: 0213-7429
Rev. Psiquiatr. Fac. Med. Barc.
Nexus Medica, c/Caracas 33, 08030, Barcelona, Spain, Tel: +34(93)4593492, Fax: +34(93)4590283 ISSUES/YR: 6

Revista de Psiquiatria do Rio Grande do Sul
ISSN: 0101-8108
Rev. Psiqiatr. Rio Grande Sul
Sociedade de Psiquiatria do Rio Grande do Sul, Av. Ipiranga, 5311 Sala 202, RS, Porto Alegre, Brazil, Tel: +55-33364846, Fax: +55-33396277 ISSUES/YR: 3

Revista de Saude Publica
ISSN: 0034-8910
Rev. Saude Publica
Universidade de Sao Paulo, Av. Dr. Arnaldo 715, 01246-904, Sao Paulo, Brazil, Tel: +55(11)2803233 EXT.260 ISSUES/YR: 6

Revista de Toxicologia
ISSN: 0212-7113
Rev. Toxicol.
Asociacion Espanola de Toxicologia, C/Serrano, 115 dpdo., E-28006, Madrid, Spain ISSUES/YR: 3

Revista del Hospital Psiquiatrico de la Habana
ISSN: 0138-7103
Rev. Hosp. Psiquiatr. Habana
Ministerio de Salud Publica, Aven. de la Independencia 26520, Mazorra, Boyeros, Habana, Cuba, Tel: +53(7)451868 ISSUES/YR: 3

Revista del Instituto Nacional de Enfermedades Respiratorias
ISSN: 0187-7585
Rev. Inst. Nac. Enferm. Respir.
Instituto Nacional de Enfermedades Respiratorias, Calzada de Tlalpan 4502, CP 14080, Mexico D.F., Mexico, Tel: +52(55)5-6664539 EXT.145 ISSUES/YR: 4

Revista do Instituto de Medicina Tropical de Sao Paulo
ISSN: 0036-4665
Rev. Inst. Med. Trop. Sao Paulo
Instituto de Medicina Tropical de Sao Paulo, Av. Dr. Eneas de Carvalho Aguiar 470, 05403-000, Sao Paulo, Brazil, Tel: +55(11)8522174, Fax: +55(11)8522174 ISSUES/YR: 6

Revista Ecuatoriana de Neurologia
ISSN: 1019-8113
Rev. Ecuat. Neurol.
Sociedad Ecuatoriana de Neurologia, P.O. Box (09-01) 3734, Guayaquil, Ecuador ISSUES/YR: 3

Revista Espanola de Cardiologia
ISSN: 0300-8932
Rev. Esp. Cardiol.
Ediciones Doyma, S.L., Travesera de Gracia 17-21, 08021, Barcelona, Spain, Tel: +34(93)2000711, Fax: +34(93)2091136 ISSUES/YR: 12

Revista Espanola de Enfermedades Digestivas
ISSN: 1130-0108
Rev. Esp. Enferm. Dig.
ARAN Ediciones S.A., Castello 128 - 1o, 28006, Madrid, Spain, Tel: +34(91)7820035, Fax: +34(91)5615787 ISSUES/YR: 12

Revista Espanola de Enfermedades Metabolicas Oseas
ISSN: 1132-8460
Rev. Esp. Enferm. Metab. Oseas
Ediciones Doyma, S.L., Travesera de Gracia 17-21, 08021, Barcelona, Spain, Tel: +34(93)2000711, Fax: +34(93)2091136 ISSUES/YR: 6

Revista Espanola de Geriatria y Gerontologia
ISSN: 0211-139X
Rev. Esp. Geriatr. Gerontol.
Ediciones Doyma, S.L., Travesera de Gracia 17-21, 08021, Barcelona, Spain, Tel: +34(93)2000711, Fax: +34(93)2091136 ISSUES/YR: 6

Revista Espanola de Medicina Nuclear
ISSN: 0212-6982
Rev. Esp. Med. Nucl.
Ediciones Mayo S.A., Aribau 185-187 2e planta, 08021, Barcelona, Spain, Tel: +34(93)2090255, Fax: +34(93)2020643 ISSUES/YR: 6

Revista Espanola de Nutricion Comunitaria
ISSN: 1135-3074
Rev. Esp. Nutr. Comunitaria
Nexus Medica, c/Caracas 33, 08030, Barcelona, Spain, Tel: +34(93)4593492, Fax: +34(93)4590283 ISSUES/YR: 4

Revista Espanola de Obesidad
ISSN: 1696-6112
Rev. Esp. Obesidad
Spanish Publishers Associate, Edificio Vertice - Antonio Lopez 249- 1 piso, 28041, Madrid, Spain, Tel: +34-91500 2075, Fax: +34-91500 2075 ISSUES/YR: 6

Revista Espanola de Pediatria
ISSN: 0034-947X
Rev. Esp. Pediatr.
Ediciones Ergon SA, C/Arboleda 1, 28220, Majadahonda (Madrid), Spain, Tel: +34(91)6362937, Fax: +34(91)6362930 ISSUES/YR: 6

Revista Espanola de Quimioterapia
ISSN: 0214-3429
Rev. Esp. Quimioter.
Prous Science, P.O. Box 540, 08080, Barcelona, Spain, Tel: +34(93)4592220, Fax: +34(93)4581535 ISSUES/YR: 4

Revista Espanola de Salud Publica
ISSN: 1135-5727
Rev. Esp. Salud Publica
Ministerio de Sanidad y Consumo, Paseo del Prado 18-20, 28071, Madrid, Spain, Tel: +34(91)5964175, Fax: +34(91)5964195 ISSUES/YR: 6

Revista Iberoamericana de Fertilidad y Reproduccion Humana
ISSN: 1132-0249
Rev. Iberoam. Fertil. Reprod. Humana
Sociedad Espanola de Fertilidad, C/Gamonal 5, 5a Planta, No 9, Edificio Valencia, 28031, Madrid, Spain, Tel: +34(91)7775360, Fax: +34(91)7775360 ISSUES/YR: 6

Revista Iberoamericana de Micologia
ISSN: 1130-1406
Rev. Iberoam. Micol.
Asociacion Espanola de Micologia, Aptdo. 699, 48080, Bilbao, Spain, Tel: +34(4)4648800 EXT.2708, Fax: +34(4)4649266 ISSUES/YR: 4

Revista Iberoamericana de Revisiones en Menopausia
ISSN: 1139-4773
Rev. Iberoam. Revis. Menopausia
Drug Farma S.L., Antonio Lopez 249, 1o, 28041, Madrid, Spain, Tel: +34(91)5002077, Fax: +34(91)5002075 ISSUES/YR: 4

Revista Latina de Cardiologia
ISSN: 1578-2700
Rev. Lat. Cardiol.
Ediciones Doyma, S.L., Travesera de Gracia 17-21, 08021, Barcelona, Spain, Tel: +34(93)2000711, Fax: +34(93)2091136 ISSUES/YR: 6

Revista Latinoamericana de Microbiologia
ISSN: 0034-9771
Rev. Latinoam. Microbiol.
Asociacion Mexicana de Microbiologia, Ave.Universidad 2001 Col.Chamilpa, CP 62210, Cuernavaca Morelos, Mexico, Tel: +52-777 329 1627, Fax: +52-777 313 8673 ISSUES/YR: 4

Revista Medica de Aguascalientes
ISSN: 1405-1648
Rev. Med. Aguascalientes
ISEA, Instituto de Salud del Estado de Aguascalientes, Margil de Jesus 1501, Fracc. Arboledas, 20020, Aguascalientes, Ags., Mexico ISSUES/YR: 4

Revista Medica de Rosario
ISSN: 0327-5019
Rev. Med. Rosario
Circulo Medico de Rosario, Argentina, Tel: +54(0341)4210120, Fax: +54(0341)4259089 ISSUES/YR: 3

Revista Mexicana de Anestesiologia
ISSN: 0185-1012
Rev. Mex. Anestesiol.
Colegio Mexicano de Anestesiologia A.C., CP 03810, Mexico D.F., Mexico, Tel: +52(555)6691457, Fax: +52(555)6691659 ISSUES/YR: 4

Revista Mexicana de Ciencias Farmaceuticas
ISSN: 1027-3956
Rev. Mex. Ciencia. Farm.
Asociacion Farmaceutica Mexicana A.C., Adolfo Prieto 1649-6o piso, 03100, Mexico D.F., Mexico, Tel: +52(55)5-5240993, Fax: +52(55)5-5345098 ISSUES/YR: 4

Revista Mexicana de Oftalmologia
ISSN: 0187-4519
Rev. Mex. Oftalmol.
Sociedad Mexicana de Oftalmologia S.C., Boston 99, Col. Nochebuena, 03720, Mexico D.F., Mexico, Tel: +52(55)5-5639393, Fax: +52(55)5-6111343 ISSUES/YR: 6

Revista Multidisciplinar de Gerontologia
ISSN: 1139-0921
Rev. Multidiscip. Gerontol.
Nexus Medica, c/Caracas 33, 08030, Barcelona, Spain, Tel: +34(93)4593492, Fax: +34(93)4590283 ISSUES/YR: 4

Revista Neurologica Argentina
ISSN: 0325-0938
Rev. Neurol. Argent.
Argentine Neurological Society, Thames 2127, 4773-5850, Ciudad de Buenos Aires, Argentina, Tel: +54(4773)5850 ISSUES/YR: 4

Revista Panamericana de Salud Publica/Pan American Journal of Public Health
ISSN: 1020-4989
Rev. Panam. Salud Publica Pan Am. J. Public Health
Pan American Health Organization, 525 Twenty-Third Street NW, Washington, DC 20037-2895, United States, Tel: +1(202)9743000, Fax: +1(202)2235971 ISSUES/YR: 12

Revista Portuguesa de Cardiologia
ISSN: 0304-4750
Rev. Port. Cardiol.
Sociedade Portuguesa de Cardiologia, Campo Grande 28, 4o-C., 1700, Lisbon, Portugal, Tel: +351(1)7970685 ISSUES/YR: 12

Revista Portuguesa de Farmacia
ISSN: 0484-811X
Rev. Port. Farm.
Ordem dos Farmaceuticos, Rua da Sociedado Farmaceutica 18, 1150, Lisbon, Portugal, Tel: +351(1)521424 ISSUES/YR: 4

Revista Portuguesa de Pneumologia
ISSN: 0873-2159
Rev. Port. Pneumol.
Sociedade Portuguesa de Pneumologia, Rua Rodrigues Sampaio 112, 2o-Dto.B, 1150-281, Lisbon, Portugal, Tel: +351(21)3523735, Fax: +351(21)3523735 ISSUES/YR: 6

Revue d'Epidemiologie et de Sante Publique
ISSN: 0398-7620
Rev. Epidemiol. Sante Publ.
Masson, 21 Rue Camille Desmoulins, 92789, Issy les Moulineaux Cedex 9, France, Tel: +33(1)73281634, Fax: +33(1)73281649 ISSUES/YR: 6

Revue de Chirurgie Orthopedique et Reparatrice de l'Appareil Moteur
ISSN: 0035-1040
Rev. Chir. Orthop. Repar. Appar. Mot.
Masson, 21 Rue Camille Desmoulins, 92789, Issy les Moulineaux Cedex 9, France, Tel: +33(1)73281634, Fax: +33(1)73281649 ISSUES/YR: 8

Revue de Geriatrie
ISSN: 0397-7927
Rev. Geriatr.
EDIMEDICA, 146, boulevard Voltaire, 92600, Asnieres, France, Tel: +33(1)47935603, Fax: +33(1)47934052 ISSUES/YR: 10

Revue de Laryngologie Otologie Rhinologie
ISSN: 0035-1334
Rev. Laryngol. Otol. Rhinol.
Revue de Laryngologie Otologie Rhinologie, 114 Avenue d'Ares, 33074, Bordeaux, France, Tel: +33(05)56243015 ISSUES/YR: 5

Revue de Medecine Interne
ISSN: 0248-8663
Rev. Med. Interne
Editions Scientifiques et Medicales Elsevier, 23 rue Linois, 75724, Paris Cedex 15, France, Tel: +33(1)45589826, Fax: +33(1)45589428 ISSUES/YR: 12

Revue de Pneumologie Clinique
ISSN: 0761-8417
Rev. Pneumol. Clin.
Masson, 21 Rue Camille Desmoulins, 92789, Issy les Moulineaux Cedex 9, France, Tel: +33(1)73281634, Fax: +33(1)73281649 ISSUES/YR: 6

Revue des Maladies Respiratoires
ISSN: 0761-8425
Rev. Mal. Respir.
Masson, 21 Rue Camille Desmoulins, 92789, Issy les Moulineaux Cedex 9, France, Tel: +33(1)73281634, Fax: +33(1)73281649 ISSUES/YR: 6

Revue du Praticien
ISSN: 0035-2640
Rev. Prat.
Revue du Praticien, 2 cite Paradis, 75010, Paris, France, Tel: +33(01)55336800, Fax: +33(01)55336815 ISSUES/YR: 20

Revue du Praticien - Gynecologie et Obstetrique
ISSN: 1280-1437
Rev. Prat. Gynecol. Obstet.
J.B. Bailliere Editeur, 2 cite Paradis, 75010, Paris, France, Tel: +33(01)55 336800, Fax: +33(01)55 336815 ISSUES/YR: 10

Revue Francaise d'Allergologie et d'Immunologie Clinique
ISSN: 0335-7457
Rev. Fr. Allergol. Immunol. Clin.
Editions Scientifiques et Medicales Elsevier, 23 rue Linois, 75724, Paris Cedex 15, France, Tel: +33(1)45589826, Fax: +33(1)45589428 ISSUES/YR: 8

Revue Francophone de Psycho-Oncologie
ISSN: 1760-4133
Rev. Francophone Psycho-Oncologie
Springer-Verlag France, 1 rue Paul Cezanne, 75008, Paris, France, Tel: +33(1)53933628, Fax: +33(1)53933683 ISSUES/YR: 4

Revue Internationale de Pediatrie
ISSN: 0048-8135
Rev. Int. Pediatr.
Meditions Carline, 1-3 rue du Depart, 75014, Paris, France, Tel: +33(1)40640075, Fax: +33(1)43222699 ISSUES/YR: 6

Revue Medicale de Bruxelles
ISSN: 0035-3639
Rev. Med. Brux.
Bureau d'Information Medicale (A.M.U.B.) asbl, Route de Lennik 808, bte 612, P.O. Box 612, 1070, Brussels, Belgium, Tel: +32(2)5556062 ISSUES/YR: 6

Revue Medicale de Liege
ISSN: 0035-3663
Rev. Med. Liege
Revue Medicale de Liege, Sart Tilman, B-4000, Liege, Belgium, Tel: +32(4)366219192 ISSUES/YR: 12

Revue Neurologique
ISSN: 0035-3787
Rev. Neurol.
Masson, 21 Rue Camille Desmoulins, 92789, Issy les Moulineaux Cedex 9, France, Tel: +33(1)73281634, Fax: +33(1)73281649 ISSUES/YR: 12

• **Rheumatic Disease Clinics of North America**
ISSN: 0889-857X
Rheum. Dis. Clin. North Am.
W.B. Saunders, Independence Square West, Philadelphia, PA 19106-3399, United States, Tel: +1(215)2387800, Fax: +1(215)2386445 ISSUES/YR: 4

Rheumatologia
ISSN: 1210-1931
Rheumatologia
SAP - Slovak Academic Press, spol. s.r.o., Nam. Slobody 6, P.O. Box 57, 81005, Bratislava, Slovakia, Tel: +421(7)211729 ISSUES/YR: 4

Rheumatology
ISSN: 1310-0505
Rheumatology (Bulgaria)
Medical Information Center, 1 Sv. G. Sofijski str., 1431, Sofia, Bulgaria, Tel: +359-522342, Fax: +359-522393 ISSUES/YR: 4

• **Rheumatology**
ISSN: 1462-0324
Rheumatology (UK)
Oxford University Press, Great Clarendon Street, OX2 6DP, Oxford, United Kingdom, Tel: +44(1865)353907, Fax: +44(1865)353486 ISSUES/YR: 12

• **Rheumatology International**
ISSN: 0172-8172
Rheumatol. Int.
Springer Verlag, Tiergartenstrasse 17, D-69121, Heidelberg, Germany, Tel: +49(6221)4870, Fax: +49(6221)487366 ISSUES/YR: 6

Rhinology
ISSN: 0300-0729
Rhinology
University Hospital Utrecht, P.O. Box 85500, 3508 GA, Utrecht, Netherlands, Tel: +31(030)2509111 ISSUES/YR: 4

Ricerca e Pratica
ISSN: 1120-379X
Ric. Prat.
Il Pensiero Scientifico Editore s.r.l., Via Bradano 3/C, 00199, Roma, Italy, Tel: +39(06)86207158, Fax: +39(06)86207160 ISSUES/YR: 6

Ricerca in Psicoterapia
ISSN: 1592-8543
Ric. Psicoter.
Universita degli Studi di Milano, Via Ippocrate 45, 20161, Milan, Italy, Tel: +39(2)64445194, Fax: +39(2)64445222 ISSUES/YR: 3

Ricerche in Psichiatria
ISSN: 1724-8922
Ric. Psichiatr.
CIC Edizioni Internazionali s.r.l., Corso Trieste 42, 00198, Roma, Italy, Tel: +39(06)8412673, Fax: +39(06)8412688 ISSUES/YR: 3

Risk Analysis
ISSN: 0272-4332
Risk Anal.
Blackwell Publishing Inc., 350 Main Street, Malden, MA 02148, United States, Tel: +1(781)3888200, Fax: +1(781)3888210 ISSUES/YR: 6

Rivista di Medicina di Laboratorio
ISSN: 1592-3479
Riv. Med. Labor.
SIMeL Societa Italiana di Medicina di Laboratorio, Via Ospedale 18, 31033, Castelfranco Veneto (TV), Italy, Tel: +39(0423)722201, Fax: +39(02)700508828 ISSUES/YR: 4

Rivista di Neurobiologia
ISSN: 0035-6336
Riv. Neurobiol.
Editrice Pisani s.a.s., via Napoli 85, 03036, Isola di Liri (FR), Italy, Tel: +39(0776)808036 ISSUES/YR: 6

Rivista di Neuroradiologia
ISSN: 1120-9976
Riv. Neuroradiol.
Edizioni del Centauro, Via del Pratello, 8, 40122, Bologna, Italy, Tel: +39(0)51227634, Fax: +39(0)51220099 ISSUES/YR: 6

Rivista di Ostetricia e Ginecologia
ISSN: 0394-977X
Riv. Ostet. Ginecol.
Servizi Editoriali S.p.A., Via Porrettana 111, 40135, Bologna, Italy ISSUES/YR: 4

Rivista di Psichiatria
ISSN: 0035-6484
Riv. Psichiatr.
Il Pensiero Scientifico Editore s.r.l., Via Bradano 3/C, 00199, Roma, Italy, Tel: +39(06)86207158, Fax: +39(06)86207160 ISSUES/YR: 6

Rivista Italiana di Chirurgia Plastica
ISSN: 0391-2221
Riv. Ital. Chir. Plast.
Editrice La Garangola, via E. dalla Costa, 6, 35129, Padova, Italy, Tel: +39(049)8750550, Fax: +39(049)7806580 ISSUES/YR: 4

Rivista Italiana di Medicina dell'Adolescenza
ISSN: not available yet
Riv. Ital. Med. Adolesc.
Edizioni Scripta Manent s.n.c., Via Bassini 41, 20133, Milano, Italy, Tel: +39(02)70608060, Fax: +39(02)70606917 ISSUES/YR: 3

Rivista Italiana di Nutrizione Parenterale ed Enterale
ISSN: 0393-5582
Riv. Ital. Nutr. Parenter. Enter.
Wichtig Editore s.r.l., Via Friuli 72/74, 20135, Milan, Italy, Tel: +39(02)55195443, Fax: +39(02)55195971 ISSUES/YR: 4

Rivista Italiana di Ossigeno-Ozonoterapia
ISSN: 1720-6561
Rev. Ital. Ossigeno-Ozonter.
Edizioni del Centauro, Via del Pratello, 8, 40122, Bologna, Italy, Tel: +39(0)51227634, Fax: +39(0)51220099 ISSUES/YR: 2

Rivista Italiana di Otorinolaringologia Audiologia e Foniatria
ISSN: 0392-1360
Riv. Ital. Otorinolaringol. Audiol. Foniatr.
CIC Edizioni Internazionali s.r.l., Corso Trieste 42, 00198, Roma, Italy, Tel: +39(06)8412673, Fax: +39(06)8412688 ISSUES/YR: 4

Rivista Medica
ISSN: 1127-6339
Riv. Med.
New Magazine Edizioni S.r.l., Viale Rovereto 51, 38100, Trento, Italy, Tel: +39(0461)925007 ISSUES/YR: 4

• **RNA**
ISSN: 1355-8382
RNA
Cold Spring Harbor Laboratory Press, 500 Sunnyside Boulevard, Woodbury New York, NY 11797, United States, Tel: +1(516)3491930, Fax: +1(516)3491946 ISSUES/YR: 12

RNA Biology
ISSN: 1547-6286
RNA Biology
Landes Bioscience, 810 S Church Street, Georgetown, 78626, United States, Tel: +1-512.863.7762, Fax: +1-512.863.0081 ISSUES/YR: 6

• **RoFo Fortschritte auf dem Gebiet der Rontgenstrahlen und der Bildgebenden Verfahren**
ISSN: 1438-9029
RoFo Fortschr. Geb. Rontgenstr. Bildgebenden Verfahren
Georg Thieme Verlag, Rudigerstrasse 14, D-70469, Stuttgart, Germany, Tel: +49(711)8931-321, Fax: +49(711)8931-422 ISSUES/YR: 12

Romanian Journal of Endocrinology
ISSN: 1221-356X
Rom. J. Endocrinol.
Editura Academiei Romane, Calea 13 Septembrie No. 13, sector 5, R-76117, Bucharest, Romania, Tel: +40(1)410 3200, Fax: +40(1)410 3983 ISSUES/YR: 4

Romanian Journal of Legal Medicine
ISSN: 1221-8618
Rom. J. Leg. Med.
Romanian Legal Medicine Society, Sos. Vitan-Birzesti 9, 75669, Bucharest, Romania, Tel: +40(1)6343890 ISSUES/YR: 4

Rontgenpraxis
ISSN: 0035-7820
Rontgenpraxis
Urban und Fischer Verlag Jena, P.O. Box 100537, 07705, Jena, Germany, Tel: +49(3641)6263, Fax: +49(3641)626500 ISSUES/YR: 6

Round Table Series - Royal Society of Medicine
ISSN: 0268-3091
Round Table Ser. R. Soc. Med.
Royal Society of Medicine Press Ltd, P.O. Box 9002, W1A 0ZA, London, United Kingdom, Tel: +44(207)2902927, Fax: +44(207)2902929 ISSUES/YR: 3

Russian Journal of Immunology
ISSN: 1028-7221
Russ. J. Immunol.
Russian Society of Immunology, P.O. Box 28 Mocscow, Russian Federation, Tel: +7-095 4344565 ISSUES/YR: 24

S.T.P. Pharma Pratiques
ISSN: 1157-1497
STP Pharma Prat.
Editions de Sante, 49 rue Galilee, 75116, Paris, France, Tel: +33(1)40701615, Fax: +33(1)40701614 ISSUES/YR: 6

• **Safety Science**
ISSN: 0925-7535
Saf. Sci.
Elsevier, P.O. Box 211, 1000 AE, Amsterdam, Netherlands, Tel: +31(20)4853757, Fax: +31(20)4853432 ISSUES/YR: 10

SAGGI - Child Development and Disabilities
ISSN: 0390-5179
SAGGI Child Dev. Disabil.
Associazione La Nostra Famiglia, Via Don Luigi Monza 20, 23842, Bosilio Parini LC, Italy, Tel: +39(031)877111, Fax: +39(031)877499 ISSUES/YR: 4

Salud (i) Sciencia
ISSN: 1667-8982
Salud Sci.
Sociedad Iberoamericana de Informacion Cientifica, Av.Belgrano 430, C1092AAR, Buenos Aires, Argentina, Tel: +54-1143424901, Fax: +54-1143313305 ISSUES/YR: 6

Salud Mental
ISSN: 0185-3325
Salud Ment.
Instituto Mexicano de Psiquiatria, Calz. Mexico-Xochimilco 101 San Lorenzo Huipulco, 14370, Mexico D.F., Del. Tlalpan, Mexico, Tel: +52(55)56552811 ISSUES/YR: 6

Salud Publica de Mexico
ISSN: 0036-3634
Salud Publica Mex.
Instituto Nacional de Salud Publica, Av. Universidad 655, Col. Santa Maria Ahuacatitlan, 62508, Cuernavaca, Morelos, Mexico, Tel: +52(73)175745 ISSUES/YR: 6

Saludarte
ISSN: 1657-4400
Saludarte
Colsubsidio, Calle 67, #10-67, Bogota, Colombia, Tel: +57(1)2171859, Fax: +57(1)2352865 ISSUES/YR: 1

Salus
ISSN: 1316-7138
Salus
Revista Salus, Psiquiatrico de Barbula, Modulo 13, Valencia, Edo Carabobo, Venezuela, Tel: +58(241)8666259 EXT.229-230 ISSUES/YR: 3

Sang Thrombose Vaisseaux
ISSN: 0999-7385
Sang Thromb. Vaiss.
John Libbey Eurotext, 127, avenue de la Republique, 92120, Montrouge, France, Tel: +33(01)46730660, Fax: +33(01)40840999 ISSUES/YR: 10

Sangre
ISSN: 0036-4355
Sangre
Sangre, P.O. Box 687, 50080, Zaragoza, Spain, Tel: +34(76)222638, Fax: +34(76)222638 ISSUES/YR: 6

Sante Publique
ISSN: 0995-3914
Sante Publique
Societe Francaise de Sante Publique, 2 Avenue du Doyen J. Parisot, P.O. Box 7, 54501, Vandoeuvre-les-Nancy, France, Tel: +33(03)83448747, Fax: +33(03)83443776 ISSUES/YR: 4

● **Santo Tomas Journal of Medicine**
ISSN: 0115-1126
Santo Tomas J. Med.
Elsevier (Singapore) Pte Ltd, 77 Leighton Road, Causeway Bay, Hong Kong, China, Hong Kong, Tel: +852-29651300, Fax: +852-29760778 ISSUES/YR: 4

Sao Paulo Medical Journal
ISSN: 1516-3180
Sao Paulo Med. J.
Associacao Paulista de Medicina, Av. Brig. Luiz Antonio 278 - 7o andar, 01318-901, Sao Paulo, Brazil, Tel: +55(11)2323141, Fax: +55(11)31077979 ISSUES/YR: 6

Sapporo Medical Journal
ISSN: 0036-472X
Sapporo Med. J.
Sapporo Medical University, S.1, W.17, Chuo-ku, 060, Sapporo, Japan ISSUES/YR: 6

● **Sarcoidosis Vasculitis and Diffuse Lung Diseases**
ISSN: 1124-0490
Sarcoidosis Vasc. Diffuse Lung Dis.
Pacini Editore s.r.l., Via A. Gherardesca 1, 56121, Ospedaletto (Pisa), Italy, Tel: +39(050)313011, Fax: +39(050)3130300 ISSUES/YR: 3

● **Sarcoma**
ISSN: 1357-714X
Sarcoma
Hindawi Publishing Corporation, 410 Park Avenue, 15th Floor, 287 pmb, New York, NY 10022, United States, Tel: +1(20)2-2706352, Fax: +1(20)2-2706351 ISSUES/YR: 4

Saudi Medical Journal
ISSN: 0379-5284
Saudi Med. J.
Saudi Arabian Armed Forces Hospital, P.O. Box 7897, 11159, Riyadh, Saudi Arabia, Tel: +966(1)4777714 EXT.6570, Fax: +966(1)4761810 ISSUES/YR: 12

Saudi Pharmaceutical Journal
ISSN: 1319-0164
Saudi Pharm. J.
Saudi Pharmaceutical Society, P.O. Box 2457, 11451, Riyadh, Saudi Arabia, Tel: +966(1)4675575, Fax: +966(1)4676789 ISSUES/YR: 4

● **Scandinavian Cardiovascular Journal**
ISSN: 1401-7431
Scand. Cardiovasc. J.
Taylor and Francis A.S., P.O. Box 2562 Solli, 0202, Oslo, Norway, Tel: +47(22)129880, Fax: +47(22)129890 ISSUES/YR: 6

● **Scandinavian Cardiovascular Journal, Supplement**
ISSN: 1401-7458
Scand. Cardiovasc. J. Suppl.
Taylor and Francis A.S., P.O. Box 2562 Solli, 0202, Oslo, Norway, Tel: +47(22)129880, Fax: +47(22)129890 ISSUES/YR: 1

Scandinavian Journal of Caring Sciences
ISSN: 0283-9318
Scand. J. Caring Sci.
Blackwell Publishing Ltd, 9600 Garsington Road, OX4 2XG, Oxford, United Kingdom, Tel: +44(1865)776868, Fax: +44(1865)714591 ISSUES/YR: 4

● **Scandinavian Journal of Clinical and Laboratory Investigation**
ISSN: 0036-5513
Scand. J. Clin. Lab. Invest.
Taylor and Francis A.S., P.O. Box 2562 Solli, 0202, Oslo, Norway, Tel: +47(22)129880, Fax: +47(22)129890 ISSUES/YR: 8

● **Scandinavian Journal of Clinical and Laboratory Investigation, Supplement**
ISSN: 0085-591X
Scand. J. Clin. Lab. Invest. Suppl.
Taylor and Francis A.S., P.O. Box 2562 Solli, 0202, Oslo, Norway, Tel: +47(22)129880, Fax: +47(22)129890 ISSUES/YR: 1

Scandinavian Journal of Food and Nutrition
ISSN: 1748-2976
Scand. J. Food Nutr.
Taylor and Francis AS, Hollandargatan 20, SE-10365, P.O. Box 3255, Stockholm, Sweden, Tel: +46-8 440 8040, Fax: +46-8 440 8050 ISSUES/YR: 4

● **Scandinavian Journal of Gastroenterology**
ISSN: 0036-5521
Scand. J. Gastroenterol.
Taylor and Francis A.S., P.O. Box 2562 Solli, 0202, Oslo, Norway, Tel: +47(22)129880, Fax: +47(22)129890 ISSUES/YR: 12

● **Scandinavian Journal of Gastroenterology, Supplement**
ISSN: 0085-5928
Scand. J. Gastroenterol. Suppl.
Taylor and Francis A.S., P.O. Box 2562 Solli, 0202, Oslo, Norway, Tel: +47(22)129880, Fax: +47(22)129890 ISSUES/YR: 1

● **Scandinavian Journal of Immunology**
ISSN: 0300-9475
Scand. J. Immunol.
Blackwell Publishing Ltd, 9600 Garsington Road, OX4 2XG, Oxford, United Kingdom, Tel: +44(1865)776868, Fax: +44(1865)714591 ISSUES/YR: 12

● **Scandinavian Journal of Immunology, Supplement**
ISSN: 0301-6323
Scand. J. Immunol. Suppl.
Blackwell Publishing Ltd., 9600 Garsington Road, OX4 2XG, Oxford, United Kingdom, Tel: +44(1865)776868, Fax: +44(1865)714591 ISSUES/YR: 1

Scandinavian Journal of Infectious Diseases
ISSN: 0036-5548
Scand. J. Infect. Dis.
Taylor and Francis A.S., P.O. Box 2562 Solli, 0202, Oslo, Norway, Tel: +46(22)129880, Fax: +46(22)129890 ISSUES/YR: 12

Scandinavian Journal of Infectious Diseases, Supplement
ISSN: 0300-8878
Scand. J. Infect. Dis. Suppl.
Taylor and Francis A.S., P.O. Box 2562 Solli, 0202, Oslo, Norway, Tel: +47(22)129880, Fax: +47(22)129890 ISSUES/YR: 1

Scandinavian Journal of Laboratory Animal Science
ISSN: 0901-3393
Scand. J. Lab. Anim. Sci.
Swedish Research Council, SE-10378, Stockholm, Sweden, Tel: +45(8)54644281, Fax: +45(8)54644275 ISSUES/YR: 4

Scandinavian Journal of Nutrition/Naringsforskning
ISSN: 1102-6480
Scand. J. Nutr. Naringsforsk.
Taylor and Francis AS, Hollandargatan 20, SE-10365, P.O. Box 3255, Stockholm, Sweden, Tel: +46-8 440 8040, Fax: +46-8 440 8050 ISSUES/YR: 4

Scandinavian Journal of Occupational Therapy
ISSN: 1103-8128
Scand. J. Occup. Ther.
Taylor and Francis A.S., P.O. Box 2562 Solli, 0202, Oslo, Norway, Tel: +47(22)129880, Fax: +47(22)129890 ISSUES/YR: 4

Scandinavian Journal of Plastic and Reconstructive Surgery and Hand Surgery
ISSN: 0284-4311
Scand. J. Plast. Reconstr. Surg. Hand Surg.
Taylor and Francis A.S., P.O. Box 2562 Solli, 0202, Oslo, Norway, Tel: +47(22)129880, Fax: +47(22)129890 ISSUES/YR: 6

Scandinavian Journal of Plastic and Reconstructive Surgery and Hand Surgery, Supplement
ISSN: 1235-662X
Scand. J. Plast. Reconstr. Surg. Hand Surg. Suppl.
Taylor and Francis A.S., P.O. Box 2562 Solli, 0202, Oslo, Norway, Tel: +47(22)129880, Fax: +47(22)129890 ISSUES/YR: 1

Scandinavian Journal of Primary Health Care
ISSN: 0281-3432
Scand. J. Prim. Health Care
Taylor and Francis A.S., P.O. Box 2562 Solli, 0202, Oslo, Norway, Tel: +47(22)129880, Fax: +47(22)129890 ISSUES/YR: 4

Scandinavian Journal of Primary Health Care, Supplement
ISSN: 0284-6020
Scand. J. Prim. Health Care Suppl.
Taylor and Francis A.S., P.O. Box 2562 Solli, 0202, Oslo, Norway, Tel: +47(22)129880, Fax: +47(22)129890 ISSUES/YR: 1

• **Scandinavian Journal of Public Health**
ISSN: 1403-4948
Scand. J. Public Health
Taylor and Francis A.S., P.O. Box 2562 Solli, 0202, Oslo, Norway, Tel: +47(22)129880, Fax: +47(22)129890 ISSUES/YR: 6

• **Scandinavian Journal of Public Health, Supplement**
ISSN: 1403-4956
Scand. J. Public Health Suppl.
Taylor and Francis A.S., P.O. Box 2562 Solli, 0202, Oslo, Norway, Tel: +47(22)129880, Fax: +47(22)129890 ISSUES/YR: 1

• **Scandinavian Journal of Rheumatology**
ISSN: 0300-9742
Scand. J. Rheumatol.
Taylor and Francis A.S., P.O. Box 2562 Solli, 0202, Oslo, Norway, Tel: +47(22)129880, Fax: +47(22)129890 ISSUES/YR: 6

• **Scandinavian Journal of Rheumatology, Supplement**
ISSN: 0301-3847
Scand. J. Rheumatol. Suppl.
Taylor and Francis A.S., P.O. Box 2562 Solli, 0202, Oslo, Norway, Tel: +47(22)129880, Fax: +47(22)129890 ISSUES/YR: 1

Scandinavian Journal of Surgery
ISSN: 1457-4969
Scand. J. Surg.
Finnish Surgical Society, Makelankatu 2A, SF-00500, Helsinki, Finland, Tel: +358(0)393091 ISSUES/YR: 4

• **Scandinavian Journal of Urology and Nephrology**
ISSN: 0036-5599
Scand. J. Urol. Nephrol.
Taylor and Francis A.S., P.O. Box 2562 Solli, 0202, Oslo, Norway, Tel: +47(22)129880, Fax: +47(22)129890 ISSUES/YR: 6

• **Scandinavian Journal of Urology and Nephrology, Supplement**
ISSN: 0300-8886
Scand. J. Urol. Nephrol. Suppl.
Taylor and Francis A.S., P.O. Box 2562 Solli, 0202, Oslo, Norway, Tel: +47(22)129880, Fax: +47(22)129890 ISSUES/YR: 1

• **Scandinavian Journal of Work, Environment and Health**
ISSN: 0355-3140
Scand. J. Work Environ. Health
Finnish Institute of Occupational Health, Topeliuksenkatu 41 aA, 00250, Helsinki, Finland, Tel: +358(0)94747694, Fax: +358(0)98783326 ISSUES/YR: 6

Scandinavian Psychoanalytic Review
ISSN: 0106-2301
Scan. Psychoanal. Rev.
University Press of Southern Denmark, Campusvej 55, 5230, Odense M, Denmark, Tel: +45-6615 7999 ISSUES/YR: 2

• **Scanning**
ISSN: 0161-0457
Scanning
Foundation for Advances in Medicine and Science Inc., P.O. Box 832, Mahwah, NJ 07430-0832, United States, Tel: +1(201)8181010, Fax: +1(201)8180086 ISSUES/YR: 6

• **Schizophrenia Bulletin**
ISSN: 0586-7614
Schizophr. Bull.
Oxford University Press, Great Clarendon Street, OX2 6DP, Oxford, United Kingdom, Tel: +44(1865)353907, Fax: +44(1865)353486 ISSUES/YR: 4

• **Schizophrenia Research**
ISSN: 0920-9964
Schizophr. Res.
Elsevier, P.O. Box 211, 1000 AE, Amsterdam, Netherlands, Tel: +31(20)4853757, Fax: +31(20)4853432 ISSUES/YR: 18

Schmerz
ISSN: 0932-433X
Schmerz
Springer Verlag, Tiergartenstrasse 17, D-69121, Heidelberg, Germany, Tel: +49(6221)4870, Fax: +49(6221)487366 ISSUES/YR: 6

Schweizer Archiv fur Neurologie und Psychiatrie
ISSN: 0258-7661
Schweiz. Arch. Neurol. Psychiatr.
Schwabe A.G. Verlag, Steinentorstrasse 13, CH-4010, Basel, Switzerland, Tel: +41(61)4678555, Fax: +41(61)4678556 ISSUES/YR: 8

Schweizerische Rundschau fur Medizin - Praxis
ISSN: 1013-2058
Schweiz. Rundsch. Med. Prax.
Verlag Hans Huber AG, Langgassstrasse 76, CH-3000, Bern, Switzerland, Tel: +41(31)3004500, Fax: +41(31)3004590 ISSUES/YR: 52

Schweizerische Zeitschrift fur GanzheitsMedizin
ISSN: 1015-0684
Schweiz. Z. GanzheitsMed.
Dr.Becker Media Consulting, Peter Merian-Strasse 58, 4002, Basel, Switzerland, Tel: +41(61)2729009, Fax: +41(61)2729008 ISSUES/YR: 8

Schweizerische Zeitschrift fur Sportmedizin und Sporttraumatologie
ISSN: 1422-0644
Schweiz. Z. Sportmed. Sporttraumatol.
Rub Graf-Lehmann AG, Murtenstrasse 40, CH-3001, Bern, Switzerland, Tel: +41-31 380 1490, Fax: +41-31 380 1491 ISSUES/YR: 4

• **Science**
ISSN: 0036-8075
Science
American Association for the Advancement of Science, 1200 New York Avenue NW, Washington, DC 20005, United States, Tel: +1(202)3266430, Fax: +1(202)8421065 ISSUES/YR: 51

Science and Justice - Journal of the Forensic Science Society
ISSN: 1355-0306
Sci. Justice J. Forensic Sci. Soc.
Forensic Science Society, 18A Mount Parade, HG1 1BX, Harrogate, North Yorkshire, United Kingdom, Tel: +44(1423)566973 ISSUES/YR: 4

• **Science and Sports**
ISSN: 0765-1597
Sci. Sports
Editions Scientifiques et Medicales Elsevier, 23 rue Linois, 75724, Paris Cedex 15, France, Tel: +33(1)45589826, Fax: +33(1)45589428 ISSUES/YR: 6

• **Science of the Total Environment**
ISSN: 0048-9697
Sci. Total Environ.
Elsevier, P.O. Box 211, 1000 AE, Amsterdam, Netherlands, Tel: +31(20)4853757, Fax: +31(20)4853432 ISSUES/YR: 54

Scientia Pharmaceutica
ISSN: 0036-8709
Sci. Pharm.
Osterreichische Apotheker-Verlagsgesellschaft m.b.H., Spitalgasse 31, A-1094, Vienna, Austria, Tel: +43(1)4023588 ISSUES/YR: 4

Scientific Review of Alternative Medicine
ISSN: 1095-0656
Sci. Rev. Altern. Med.
Prometheus Books, 59 John Glenn Drive, Amherst, NY 14228-2197, United States, Tel: +1(716)6910133, Fax: +1(716)6910137 ISSUES/YR: 2

Scottish Medical Journal
ISSN: 0036-9330
Scott. Med. J.
Hermiston Publications Ltd, 9 Stonelaws, EH40 3DX, Whitekirk, East Lothian, United Kingdom ISSUES/YR: 6

Scripta Medica Facultatis Medicae Universitatis Brunensis Masarykianae
ISSN: 1211-3395
Scr. Med. Fac. Med. Univ. Brun. Masarykianae
Universitatis Brunensis Masarykianae, Scripta Medica, 662 43, Jostova 10, Czech Republic, Tel: +420(5)42126268, Fax: +420(5)42126200 ISSUES/YR: 6

SD Revista Medica Internacional sobre el Sindrome de Down
ISSN: 1138-011X
SD Rev. Med. Int. Sindrome Down
Fundacio Catalana Sindrome de Down, Comte Borrell 201-203, entresuelo, 08029, Barcelona, Spain, Tel: +34(93)2157423, Fax: +34(93)2157699 ISSUES/YR: 3

Seguimiento Farmacoterapeutico
ISSN: 1885-642X
Seguimiento Farmacoter.
Grupo de Investigacion en Atencion Farmaceutica, Campus de la Cartuja, 18071, Granada, Spain, Fax: +34(986)401889 ISSUES/YR: 4

• **Seizure**
ISSN: 1059-1311
Seizure
W.B. Saunders Ltd, 32 Jamestown Road, NW1 7BY, London, United Kingdom, Tel: +44(20)74244200, Fax: +44(20)74854752 ISSUES/YR: 8

Seksuologia Polska
ISSN: 1731-6677
Seksuologia Pol.
Via Medica Medical Publishers, ul. Swietokrzyska 73, 80-180 Gdansk, Poland, Tel: +48-58 3209494, Fax: +48-58 3209460 ISSUES/YR: 2

Seleccion
ISSN: 0214-8927
Seleccion
Alpe Editores S.A., Pedro Rico 27, oficinas 11-12, 28029, Madrid, Spain, Tel: +34(91)7338811, Fax: +34(91)3159652 ISSUES/YR: 2

Seminarios de la Fundacion Espanola de Reumatologia
ISSN: 1577-3566
Semin. Fund. Esp. Reumatol.
ARAN Ediciones S.A., Castello 128 - 1o, 28006, Madrid, Spain, Tel: +34(91)7820035, Fax: +34(91)5615787 ISSUES/YR: 4

• **Seminars in Anesthesia, Perioperative Medicine and Pain**
ISSN: 0277-0326
Semin. Anesth. Perioper. Med. Pain
W.B. Saunders, Independence Square West, Philadelphia, PA 19106-3399, United States, Tel: +1(215)2387800, Fax: +1(215)2386445 ISSUES/YR: 4

• **Seminars in Arthritis and Rheumatism**
ISSN: 0049-0172
Semin. Arthritis Rheum.
W.B. Saunders, Independence Square West, Philadelphia, PA 19106-3399, United States, Tel: +1(215)2387800, Fax: +1(215)2386445 ISSUES/YR: 6

Seminars in Arthroplasty
ISSN: 1045-4527
Semin. Arthroplasty
W.B. Saunders, Independence Square West, Philadelphia, PA 19106-3399, United States, Tel: +1(215)2387800, Fax: +1(215)2386445 ISSUES/YR: 4

• **Seminars in Breast Disease**
ISSN: 1092-4450
Semin. Breast Dis.
W.B. Saunders, Independence Square West, Philadelphia, PA 19106-3399, United States, Tel: +1(215)2387800, Fax: +1(215)2386445 ISSUES/YR: 4

Seminars in Cancer Biology
ISSN: 1044-579X
Semin. Cancer Biol.
Academic Press, 24-28 Oval Road, NW1 7DX, London, United Kingdom, Tel: +44(207)2674466 ISSUES/YR: 6

Seminars in Cardiothoracic and Vascular Anesthesia
ISSN: 1089-2532
Semin. Cardiothorac. Vasc. Anesth.
Westminster Publications Inc., 708 Glen Cove Avenue, Glen Head, NY 11545, United States, Tel: +1(516)7590025, Fax: +1(516)7595524 ISSUES/YR: 4

Seminars in Cell and Developmental Biology
ISSN: 1084-9521
Semin. Cell. Dev. Biol.
Academic Press, 24-28 Oval Road, NW1 7DX, London, United Kingdom, Tel: +44(207)2674466 ISSUES/YR: 6

• **Seminars in Cerebrovascular Diseases and Stroke**
ISSN: 1528-9931
Semin. Cerebrovasc. Dis. Stroke
W.B. Saunders, Independence Square West, Philadelphia, PA 19106-3399, United States, Tel: +1(215)2387800, Fax: +1(215)2386445 ISSUES/YR: 4

Seminars in Colon and Rectal Surgery
ISSN: 1043-1489
Semin. Colon Rectal Surg.
W.B. Saunders, Independence Square West, Philadelphia, PA 19106-3399, United States, Tel: +1(215)2387800, Fax: +1(215)2386445 ISSUES/YR: 4

Seminars in Cutaneous Medicine and Surgery
ISSN: 1085-5629
Semin. Cutaneous Med. Surg.
W.B. Saunders, Independence Square West, Philadelphia, PA 19106-3399, United States, Tel: +1(215)2387800, Fax: +1(215)2386445 ISSUES/YR: 4

• **Seminars in Diagnostic Pathology**
ISSN: 0740-2570
Semin. Diagn. Pathol.
W.B. Saunders, Independence Square West, Philadelphia, PA 19106-3399, United States, Tel: +1(215)2387800, Fax: +1(215)2386445 ISSUES/YR: 4

Seminars in Fetal and Neonatal Medicine
ISSN: 1744-165X
Semin. Fetal Neonatal Med.
W.B. Saunders Ltd, 32 Jamestown Road, NW1 7BY, London, United Kingdom, Tel: +44(20)74244200, Fax: +44(20)74854752 ISSUES/YR: 6

Current

Seminars in Hearing
ISSN: 0734-0451
Semin. Hear.
Thieme Medical Publishers, Inc., 333 7th Avenue, New York, NY 10001-5004, United States, Tel: +1(212)7600888, Fax: +1(212)7799020 ISSUES/YR: 4

• **Seminars in Hematology**
ISSN: 0037-1963
Semin. Hematol.
W.B. Saunders, Independence Square West, Philadelphia, PA 19106-3399, United States, Tel: +1(215)2387800, Fax: +1(215)2386445 ISSUES/YR: 4

Seminars in Immunology
ISSN: 1044-5323
Semin. Immunol.
Academic Press, 24-28 Oval Road, NW1 7DX, London, United Kingdom, Tel: +44(207)2674466 ISSUES/YR: 6

Seminars in Integrative Medicine
ISSN: 1543-1150
Semin. Integr. Med.
Elsevier Inc., 360 Park Avenue South, New York, NY 10010, United States, Tel: +1(212)6333950, Fax: +1(212)6333990 ISSUES/YR: 4

Seminars in Interventional Radiology
ISSN: 0739-9529
Semin. Intervent. Radiol.
Thieme Medical Publishers, Inc., 333 7th Avenue, New York, NY 10001-5004, United States, Tel: +1(212)7600888, Fax: +1(212)7799020 ISSUES/YR: 4

• **Seminars in Liver Disease**
ISSN: 0272-8087
Semin. Liver Dis.
Thieme Medical Publishers, Inc., 333 7th Avenue, New York, NY 10001-5004, United States, Tel: +1(212)7600888, Fax: +1(212)7799020 ISSUES/YR: 4

Seminars in Musculoskeletal Radiology
ISSN: 1089-7860
Semin. Muscoloskelet. Radiol.
Thieme Medical Publishers, Inc., 333 7th Avenue, New York, NY 10001-5004, United States, Tel: +1(212)7600888, Fax: +1(212)7799020 ISSUES/YR: 4

• **Seminars in Nephrology**
ISSN: 0270-9295
Semin. Nephrol.
W.B. Saunders, Independence Square West, Philadelphia, PA 19106-3399, United States, Tel: +1(215)2387800, Fax: +1(215)2386445 ISSUES/YR: 6

• **Seminars in Neurology**
ISSN: 0271-8235
Semin. Neurol.
Thieme Medical Publishers, Inc., 333 7th Avenue, New York, NY 10001-5004, United States, Tel: +1(212)7600888, Fax: +1(212)7799020 ISSUES/YR: 4

Seminars in Nuclear Medicine
ISSN: 0001-2998
Semin. Nucl. Med.
W.B. Saunders, Independence Square West, Philadelphia, PA 19106-3399, United States, Tel: +1(215)2387800, Fax: +1(215)2386445 ISSUES/YR: 4

• **Seminars in Oncology**
ISSN: 0093-7754
Semin. Oncol.
W.B. Saunders, Independence Square West, Philadelphia, PA 19106-3399, United States, Tel: +1(215)2387800, Fax: +1(215)2386445 ISSUES/YR: 6

• **Seminars in Ophthalmology**
ISSN: 0882-0538
Semin. Ophthalmol.
Taylor and Francis Ltd., 4 Park Square, Milton Park, OX14 4RN, Abingdon, Oxfordshire, United Kingdom, Tel: +44(20)70175541, Fax: +44(20)70175198 ISSUES/YR: 4

• **Seminars in Pain Medicine**
ISSN: 1537-5897
Semin. Pain Med.
W.B. Saunders, Independence Square West, Philadelphia, PA 19106-3399, United States, Tel: +1(215)2387800, Fax: +1(215)2386445 ISSUES/YR: 4

Seminars in Pediatric Infectious Diseases
ISSN: 1045-1870
Semin. Pediatr. Infect. Dis.
W.B. Saunders, Independence Square West, Philadelphia, PA 19106-3399, United States, Tel: +1(215)2387800, Fax: +1(215)2386445 ISSUES/YR: 4

Seminars in Pediatric Neurology
ISSN: 1071-9091
Semin. Pediatr. Neurol.
W.B. Saunders, Independence Square West, Philadelphia, PA 19106-3399, United States, Tel: +1(215)2387800, Fax: +1(215)2386445 ISSUES/YR: 4

• **Seminars in Pediatric Surgery**
ISSN: 1055-8586
Semin. Pediatr. Surg.
W.B. Saunders, Independence Square West, Philadelphia, PA 19106-3399, United States, Tel: +1(215)2387800, Fax: +1(215)2386445 ISSUES/YR: 4

• **Seminars in Perinatology**
ISSN: 0146-0005
Semin. Perinatol.
W.B. Saunders, Independence Square West, Philadelphia, PA 19106-3399, United States, Tel: +1(215)2387800, Fax: +1(215)2386445 ISSUES/YR: 6

• **Seminars in Radiation Oncology**
ISSN: 1053-4296
Semin. Radiat. Oncol.
W.B. Saunders, Independence Square West, Philadelphia, PA 19106-3399, United States, Tel: +1(215)2387800, Fax: +1(215)2386445 ISSUES/YR: 4

Seminars in Reproductive Medicine
ISSN: 1526-8004
Semin. Reprod. Med.
Thieme Medical Publishers, Inc., 333 7th Avenue, New York, NY 10001-5004, United States, Tel: +1(212)7600888, Fax: +1(212)7799020 ISSUES/YR: 4

• **Seminars in Respiratory and Critical Care Medicine**
ISSN: 1069-3424
Semin. Respir. Crit. Care Med.
Thieme Medical Publishers, Inc., 333 7th Avenue, New York, NY 10001-5004, United States, Tel: +1(212)7600888, Fax: +1(212)7799020 ISSUES/YR: 6

• **Seminars in Roentgenology**
ISSN: 0037-198X
Semin. Roentgenol.
W.B. Saunders, Independence Square West, Philadelphia, PA 19106-3399, United States, Tel: +1(215)2387800, Fax: +1(215)2386445 ISSUES/YR: 4

Seminars in Speech and Language
ISSN: 0734-0478
Semin. Speech Lang.
Thieme Medical Publishers, Inc., 333 7th Avenue, New York, NY 10001-5004, United States, Tel: +1(212)7600888, Fax: +1(212)7799020 ISSUES/YR: 4

• **Seminars in Spine Surgery**
ISSN: 1040-7383
Semin. Spine Surg.
W.B. Saunders, Independence Square West, Philadelphia, PA 19106-3399, United States, Tel: +1(215)2387800, Fax: +1(215)2386445 ISSUES/YR: 4

• **Seminars in Surgical Oncology**
ISSN: 8756-0437
Semin. Surg. Oncol.
Wiley-Liss Inc., 111 River Street, Hoboken, NJ 07030-5774, United States, Tel: +1(201)7486645, Fax: +1(201)7486021 ISSUES/YR: 4

Seminars in Thoracic and Cardiovascular Surgery
ISSN: 1043-0679
Semin. Thorac. Cardiovasc. Surg.
W.B. Saunders, Independence Square West, Philadelphia, PA 19106-3399, United States, Tel: +1(215)2387800, Fax: +1(215)2386445 ISSUES/YR: 4

• **Seminars in Thrombosis and Hemostasis**
ISSN: 0094-6176
Semin. Thromb. Hemost.
Thieme Medical Publishers, Inc., 333 7th Avenue, New York, NY 10001-5004, United States, Tel: +1(212)7600888, Fax: +1(212)7799020 ISSUES/YR: 6

- **Seminars in Ultrasound, CT and MRI**
  ISSN: 0887-2171
  Semin. Ultrasound CT MRI
  W.B. Saunders, Independence Square West, Philadelphia, PA 19106-3399, United States, Tel: +1(215)2387800, Fax: +1(215)2386445 ISSUES/YR: 6

Seminars in Vascular Medicine
ISSN: 1528-9648
Semin. Vasc. Med.
Thieme Medical Publishers, Inc., 333 7th Avenue, New York, NY 10001-5004, United States, Tel: +1(212)7600888, Fax: +1(212)7799020 ISSUES/YR: 4

- **Seminars in Vascular Surgery**
  ISSN: 0895-7967
  Semin. Vasc. Surg.
  W.B. Saunders, Independence Square West, Philadelphia, PA 19106-3399, United States, Tel: +1(215)2387800, Fax: +1(215)2386445 ISSUES/YR: 4

Sen
ISSN: 1641-6007
Sen
Via Medica Medical Publishers, ul. Swietokrzyska 73, 80-180 Gdansk, Poland, Tel: +48-58 3209494, Fax: +48-58 3209460 ISSUES/YR: 2

SENDROM
ISSN: 1016-5134
SENDROM
Turkish Anaesthesiology and Intensive Care Society, Inonu Cad. 7/4 Zambak apt., Gumussuyu Taksim. Istanbul, Turkey ISSUES/YR: 12

Sexual Addiction and Compulsivity
ISSN: 1072-0162
Sex. Addict. Compulsivity
Routledge, 4 Park Square, Milton Park, OX14 4RN, Abingdon, Oxfordshire, United Kingdom, Tel: +44(1235)828600, Fax: +44(1235)829000 ISSUES/YR: 4

Sexual and Relationship Therapy
ISSN: 1468-1994
Sex. Relatsh. Ther.
Routledge, 4 Park Square, Milton Park, OX14 4RN, Abingdon, Oxfordshire, United Kingdom, Tel: +44(1235)828600, Fax: +44(1235)829000 ISSUES/YR: 4

Sexual Health
ISSN: 1448-5028
Sex. Health
CSIRO, P.O. Box 1139, VIC 3066, Collingwood, Australia, Tel: +61(39)6627686, Fax: +61(39)6627555 ISSUES/YR: 4

- **Sexualities, Evolution and Gender**
  ISSN: 1479-2508
  Sex. Evol. Gender
  Routledge, 4 Park Square, Milton Park, OX14 4RN, Abingdon, Oxfordshire, United Kingdom, Tel: +44(1235)828600, Fax: +44(1235)829000 ISSUES/YR: 3

Sexuality and Disability
ISSN: 0146-1044
Sex. Disabil.
Kluwer Academic/Human Sciences Press Inc., 233 Spring Street, New York, NY 10013-1578, United States, Tel: +1(212)6208000, Fax: +1(212)4630742 ISSUES/YR: 4

- **Sexuality, Reproduction and Menopause**
  ISSN: 1546-2501
  Sex. Reprod. Menopause
  Elsevier Inc., 360 Park Avenue South, New York, NY 10010, United States, Tel: +1(212)6333950, Fax: +1(212)6333990 ISSUES/YR: 4

Sexually Transmitted Diseases
ISSN: 0148-5717
Sex. Transm. Dis.
Lippincott Williams and Wilkins, 530 Walnut Street, Philadelphia, PA 19106-3621, United States, Tel: +1(215)5218342, Fax: +1(215)5218326 ISSUES/YR: 12

- **Sexually Transmitted Infections**
  ISSN: 1368-4973
  Sex. Transm. Infect.
  BMJ Publishing Group, Tavistock Square, WC1H 9JR, London, United Kingdom, Tel: +44(207)3874499, Fax: +44(207)3836668 ISSUES/YR: 6

Sexuologie
ISSN: 0944-7105
Sexuologie
Urban und Fischer Verlag Jena, P.O. Box 100537, 07705, Jena, Germany, Tel: +49(3641)6263, Fax: +49(3641)626500 ISSUES/YR: 4

Shock
ISSN: 1073-2322
Shock
Lippincott Williams and Wilkins, 530 Walnut Street, Philadelphia, PA 19106-3621, United States, Tel: +1(215)5218342, Fax: +1(215)5218326 ISSUES/YR: 12

Signal
ISSN: 1471-9967
Signal
ADIS International Ltd, Chowley Oak Lane, Tattenhall, CH3 9GA, Chester, United Kingdom, Tel: +44(1829)771155, Fax: +44(1829)770330 ISSUES/YR: 4

- **Signal Transduction**
  ISSN: 1615-4053
  Signal Transduct.
  Wiley-VCH Verlag, P.O. Box 101161, D-69451, Weinheim, Germany, Tel: +49(6201)606147, Fax: +49(6201)606117 ISSUES/YR: 6

Sinapse
ISSN: 1645-281X
Sinapse
Sociedade Portuguesa de Neurologia, Rua D. Manuel II, 33 - Sala 41, 4050-345, Porto, Portugal, Tel: +351-226 001 553 ISSUES/YR: 2

Singapore General Hospital Proceedings
ISSN: 0218-3048
Singapore Gen. Hosp. Proc.
SGH-PGMI Press, Block 6 Level 1, Singapore General Hospital, Outram Road, Singapore 169608, Singapore, Tel: +65-6326 6691, Fax: +65-6226 0356 ISSUES/YR: 3

Singapore Medical Journal
ISSN: 0037-5675
Singapore Med. J.
Singapore Medical Association, 2 College Road, Level 2, 0316, Singapore, Singapore, Tel: +65-62231264 ISSUES/YR: 12

Singapore Paediatric Journal
ISSN: 0218-9941
Singapore Paediatr. J.
Singapore Paediatric Society, Lower Kent Ridge Road, 119074, Singapore, Singapore, Tel: +65-67724112 ISSUES/YR: 2

- **Skeletal Radiology**
  ISSN: 0364-2348
  Skelet. Radiol.
  Springer Verlag, Tiergartenstrasse 17, D-69121, Heidelberg, Germany, Tel: +49(6221)4870, Fax: +49(6221)487366 ISSUES/YR: 12

Skin Cancer
ISSN: 0871-2549
Skin Cancer
Revismedica - Revistas Medicas e Congressos Lda., Av. Marques de Tomar 102, 2-Esq., 1050, Lisbon, Portugal, Tel: +351(1)7970143, Fax: +351(1)7969961 ISSUES/YR: 4

- **Skin Pharmacology and Physiology**
  ISSN: 1660-5527
  Skin Pharmacol. Physiol.
  S. Karger AG, Allschwilerstrasse 10, P.O. Box, CH-4009, Basel, Switzerland, Tel: +41(61)3061111, Fax: +41(61)3061234 ISSUES/YR: 6

Skin Research
ISSN: 1347-1813
Skin Res.
Osaka University Medical School, 2-2 Yamadaoka, Suita-shi, 553, Osaka, Japan, Tel: +81(6)8793037 ISSUES/YR: 6

Skin Research and Technology
ISSN: 0909-752X
Skin Res. Technol.
Blackwell Publishing Ltd, 9600 Garsington Road, OX4 2XG, Oxford, United Kingdom, Tel: +44(1865)776868, Fax: +44(1865)714591 ISSUES/YR: 4

- **Skull Base**
  ISSN: 1531-5010
  Skull Base Surg.
  Thieme Medical Publishers, Inc., 333 7th Avenue, New York, NY 10001-5004, United States, Tel: +1(212)7600888, Fax: +1(212)7799020 ISSUES/YR: 4
- **Sleep**
  ISSN: 0161-8105
  Sleep
  American Academy of Sleep Medicine, 1 Westbrook Corporate Center, Suite 920, Westchester, IL 60154, United States, Tel: +1(708)4920930, Fax: +1(708)4920943 ISSUES/YR: 8
- **Sleep and Biological Rhythms**
  ISSN: 1446-9235
  Sleep Biol. Rhythms
  Blackwell Publishing, 550 Swanston Street, VIC 3053, Carlton South, Australia, Tel: +61(3)8359 1011, Fax: +61(3)8359 1120 ISSUES/YR: 3
- **Sleep and Breathing**
  ISSN: 1520-9512
  Sleep Breathing
  Springer Verlag, Tiergartenstrasse 17, D-69121, Heidelberg, Germany, Tel: +49(6221)4870, Fax: +49(6221)487366 ISSUES/YR: 4

Sleep and Hypnosis
ISSN: 1302-1192
Sleep Hypn.
Kure Iletisim Grubu A.S., Siracevizler Cad. 43/3, 34381, Sisli, Istanbul, Turkey, Tel: +90(212)219-4900, Fax: +90(212)230-5009 ISSUES/YR: 4

- **Sleep Medicine**
  ISSN: 1389-9457
  Sleep Med.
  Elsevier, P.O. Box 211, 1000 AE, Amsterdam, Netherlands, Tel: +31(20)4853757, Fax: +31(20)4853432 ISSUES/YR: 6

Sleep Medicine Reviews
ISSN: 1087-0792
Sleep Med. Rev.
W.B. Saunders Ltd, 32 Jamestown Road, NW1 7BY, London, United Kingdom, Tel: +44(20)74244200, Fax: +44(20)74854752 ISSUES/YR: 6

Sleep Research Online
ISSN: 1096-214X
Sleep Res. Online
Sleep Research Online, 1263 Westwood Blvd. Suite 200, Los Angeles, CA 90024, United States, Tel: +1(310)4786648, Fax: +1(310)2352067 ISSUES/YR: 4

Social Psychiatry and Psychiatric Epidemiology
ISSN: 0933-7954
Soc. Psychiatry Psychiatr. Epidemiol.
Dr. Dietrich Steinkopff Verlag GmbH and Co. KG, P.O. Box 100462, D-64204, Darmstadt, Germany, Tel: +49(6151)828990, Fax: +49(6151)8289940 ISSUES/YR: 12

Social Science and Medicine
ISSN: 0277-9536
Soc. Sci. Med.
Elsevier Ltd, Langford Lane, Kidlington, OX5 1GB, Oxford, United Kingdom, Tel: +44(1865)843000, Fax: +44(1865)843010 ISSUES/YR: 24

Social Work in Health Care
ISSN: 0098-1389
Soc. Work Health Care
Haworth Press Inc., 10 Alice Street, Binghamton, NY 13904-1580, United States, Tel: +1(607)7225857, Fax: +1(607)7710012 ISSUES/YR: 8

Socijalna Psihijatrija
ISSN: 0303-7908
Socijalna Psihijatr.
Croatian Academy for Medical Sciences, KBC Rebro, Kispaticeva 12, 10000, Zagreb, Croatia, Tel: +385(1)2333043 ISSUES/YR: 4

- **Somatic Cell and Molecular Genetics**
  ISSN: 0740-7750
  Somatic Cell Mol. Genet.
  Springer Science and Business Media Deutschland GmbH, Heidelberger Platz 3, 14197, Berlin, Germany, Tel: +1-30 827875282, Fax: +1-30 827875707 ISSUES/YR: 6
- **Somatosensory and Motor Research**
  ISSN: 0899-0220
  Somatosens. Mot. Res.
  Taylor and Francis Ltd., 4 Park Square, Milton Park, OX14 4RN, Abingdon, Oxfordshire, United Kingdom, Tel: +44(20)70175541, Fax: +44(20)70175198 ISSUES/YR: 4

Somnologie
ISSN: 1432-9123
Somnologie
Blackwell Publishing Ltd, 9600 Garsington Road, OX4 2XG, Oxford, United Kingdom, Tel: +44(1865)776868, Fax: +44(1865)714591 ISSUES/YR: 4

South African Family Practice
ISSN: 1726-426X
S. Afr. Fam. Pract.
Medpharm Publications, Medpharm Publications (Pty) Ltd, 0157, PO Box 14804, Lyttelton, South Africa, Tel: +27(012) 664 7460, Fax: +27(012) 664 6276 ISSUES/YR: 10

South African Journal of Clinical Nutrition
ISSN: 1011-8578
S. Afr. J. Clin. Nutr.
South African Medical Association, Private Bag X1, 7430, Pinelands, South Africa, Tel: +27(21)5313081, Fax: +27(21)5314126 ISSUES/YR: 3

South African Journal of Obstetrics and Gynaecology
ISSN: 0038-2329
S. Afr. J. Obs. Gyn.
South African Medical Association, Private Bag X1, 7430, Pinelands, South Africa, Tel: +27(21)5313081, Fax: +27(21)5314126 ISSUES/YR: 3

South African Journal of Psychiatry
ISSN: 1608-9685
S. Afr. J. Psychiatry
South African Medical Association, Private Bag X1, 7430, Pinelands, South Africa, Tel: +27(21)5313081, Fax: +27(21)5314126 ISSUES/YR: 3

South African Journal of Surgery
ISSN: 0038-2361
S. Afr. J. Surg.
South African Medical Association, Private Bag X1, 7430, Pinelands, South Africa, Tel: +27(21)5313081, Fax: +27(21)5314126 ISSUES/YR: 4

South African Medical Journal
ISSN: 0256-9574
S. Afr. Med. J.
South African Medical Association, Private Bag X1, 7430, Pinelands, South Africa, Tel: +27(21)5313081, Fax: +27(21)5314126 ISSUES/YR: 12

South African Psychiatry Review
ISSN: 1811-7805
S. Afr. Psychiatry Rev.
In House Publications, P.O. Box 412748, Craighall, 2024, Johannesburg, South Africa, Tel: +27(11)6466129, Fax: +27(11)6467558 ISSUES/YR: 4

Southern African Journal of Critical Care
ISSN: 1562-8264
South. Afr. J. Crit. Care
South African Medical Association, Private Bag X1, 7430, Pinelands, South Africa, Tel: +27(21)5313081, Fax: +27(21)5314126 ISSUES/YR: 2

Southern African Journal of HIV Medicine
ISSN: 1608-9693
South. Afr. J. HIV Med.
South African Medical Association, Private Bag X1, 7430, Pinelands, South Africa, Tel: +27(21)5313081, Fax: +27(21)5314126 ISSUES/YR: 3

Southern Medical Journal
ISSN: 0038-4348
South. Med. J.
Lippincott Williams and Wilkins, 530 Walnut Street, Philadelphia, PA 19106-3621, United States, Tel: +1(215)5218342, Fax: +1(215)5218326 ISSUES/YR: 12

Sozial- und Praventivmedizin
ISSN: 0303-8408
Soz.- Praventivmed.
Birkhauser Verlag AG, Klosterberg 23, P.O. Box 133, CH-4010, Basel, Switzerland, Tel: +41(61)2717400, Fax: +41(61)2050791 ISSUES/YR: 6

Spectroscopy
ISSN: 0712-4813
Spectroscopy
IOS Press, Nieuwe Hemweg 6B, 1013 BG, Amsterdam, Netherlands, Tel: +31(20)6883355, Fax: +31(20)6203419 ISSUES/YR: 4

Spektrum der Augenheilkunde
ISSN: 0930-4282
Spektrum Augenheilkd.
Springer-Verlag Wien, Sachsenplatz 4-6, P.O. Box 89, A-1201, Vienna, Austria, Tel: +43(1)3302415, Fax: +43(1)3302426 ISSUES/YR: 6

• **Spinal Cord**
ISSN: 1362-4393
Spinal Cord
Nature Publishing Group, Houndmills, RG21 6XS, Basingstoke, Hampshire, United Kingdom, Tel: +44(1256)329242, Fax: +44(1256)810526 ISSUES/YR: 12

• **Spine**
ISSN: 0362-2436
Spine
Lippincott Williams and Wilkins, 530 Walnut Street, Philadelphia, PA 19106-3621, United States, Tel: +1(215)5218342, Fax: +1(215)5218326 ISSUES/YR: 24

• **Spine Journal**
ISSN: 1529-9430
Spine J.
Elsevier Inc., 360 Park Avenue South, New York, NY 10010, United States, Tel: +1(212)6333950, Fax: +1(212)6333990 ISSUES/YR: 6

Sports Medicine
ISSN: 0112-1642
Sports Med.
Adis International Ltd, 41 Centorian Drive, Private Bag 65901, Mairangi Bay, 1311, Auckland 10, New Zealand, Tel: +64(9)4770700, Fax: +64(9)4770766 ISSUES/YR: 15

• **Sports Medicine and Arthroscopy Review**
ISSN: 1062-8592
Sports Med. Arthroscopy Rev.
Lippincott Williams and Wilkins, 530 Walnut Street, Philadelphia, PA 19106-3621, United States, Tel: +1(215)5218342, Fax: +1(215)5218326 ISSUES/YR: 4

Sprache Stimme Gehor
ISSN: 0342-0477
Sprache Stimme Gehor
Georg Thieme Verlag, Rudigerstrasse 14, D-70469, Stuttgart, Germany, Tel: +49(711)8931-321, Fax: +49(711)8931-422 ISSUES/YR: 4

• **Springer Seminars in Immunopathology**
ISSN: 0344-4325
Springer Semin. Immunopathol.
Springer Verlag, Tiergartenstrasse 17, D-69121, Heidelberg, Germany, Tel: +49(6221)4870, Fax: +49(6221)487366 ISSUES/YR: 4

Statistical Applications in Genetics and Molecular Biology
ISSN: 1544-6115
Stat. Appl. Genet. Mol. Biol.
Berkeley Electronic Press, 2809 Telegraph Avenue, Suite 202, Berkeley, CA 94705, United States, Tel: +1(510)6651200, Fax: +1(510)6651201 ISSUES/YR: 1

Statistical Methods in Medical Research
ISSN: 0962-2802
Stat. Methods Med. Res.
Arnold, 338 Euston Road, NW1 3BH, London, United Kingdom, Tel: +44(20)78736000, Fax: +44(20)78736376 ISSUES/YR: 6

Statistics in Medicine
ISSN: 0277-6715
Stat. Med.
John Wiley and Sons Ltd, Southern Gate, PO19 8SQ, Chichester, West Sussex, United Kingdom, Tel: +44(1243)779777, Fax: +44(1243)775878 ISSUES/YR: 24

Stem Cells
ISSN: 1066-5099
Stem Cells
AlphaMed Press, 318 Blackwell St. Suite 260, Durham, NC 27701-2884, United States, Tel: +1(937)2912355, Fax: +1(937)2914229 ISSUES/YR: 10

• **Stem Cells and Development**
ISSN: 1547-3287
Stem Cells Dev.
Mary Ann Liebert Inc., 140 Huguenot Street, New Rochelle, NY 10801-5215, United States, Tel: +1(914)7402100, Fax: +1(914)7402101 ISSUES/YR: 6

Stereotactic and Functional Neurosurgery
ISSN: 1011-6125
Stereotactic Funct. Neurosurg.
S. Karger AG, Allschwilerstrasse 10, P.O. Box, CH-4009, Basel, Switzerland, Tel: +41(61)3061111, Fax: +41(61)3061234 ISSUES/YR: 8

Steroids
ISSN: 0039-128X
Steroids
Elsevier Inc., 360 Park Avenue South, New York, NY 10010, United States, Tel: +1(212)6333950, Fax: +1(212)6333990 ISSUES/YR: 14

• **Strabismus**
ISSN: 0927-3972
Strabismus
Taylor and Francis Ltd., 4 Park Square, Milton Park, OX14 4RN, Abingdon, Oxfordshire, United Kingdom, Tel: +44(20)70175541, Fax: +44(20)70175198 ISSUES/YR: 4

Strahlentherapie und Onkologie
ISSN: 0179-7158
Strahlenther. Onkol.
Urban und Vogel GmbH, Neumarkter Strasse 43, D-81673, Munchen, Germany, Tel: +49(89)43721300, Fax: +49(89)43721399 ISSUES/YR: 12

• **Stress**
ISSN: 1025-3890
Stress
Taylor and Francis Ltd., 4 Park Square, Milton Park, OX14 4RN, Abingdon, Oxfordshire, United Kingdom, Tel: +44(20)70175541, Fax: +44(20)70175198 ISSUES/YR: 4

Stress and Health
ISSN: 1532-3005
Stress Health
John Wiley and Sons Ltd, Southern Gate, PO19 8SQ, Chichester, West Sussex, United Kingdom, Tel: +44(1243)779777, Fax: +44(1243)775878 ISSUES/YR: 5

• **Stroke**
ISSN: 0039-2499
Stroke
Lippincott Williams and Wilkins, 530 Walnut Street, Philadelphia, PA 19106-3621, United States, Tel: +1(215)5218342, Fax: +1(215)5218326 ISSUES/YR: 12

• **Structure**
ISSN: 0969-2126
Structure
Cell Press, 1100 Massachusetts Avenue, Cambridge, MA 02138-9957, United States, Tel: +1(617)6617057, Fax: +1(617)6617061 ISSUES/YR: 12

Studia Pneumologica et Phthiseologica
ISSN: 1213-810X
Stud. Pneumol. Phtiseol.
Trios spol. s.r.o., Zakourilova 142, 149 00, Praha 4, Czech Republic, Tel: +420(267)912030, Fax: +420(267)915563 ISSUES/YR: 6

Substance Abuse
ISSN: 0889-7077
Subst. Abuse
Haworth Press Inc., 10 Alice Street, Binghamton, NY 13904-1580, United States, Tel: +1(607)7225857, Fax: +1(607)7710012 ISSUES/YR: 4

• **Substance Use and Misuse**
ISSN: 1082-6084
Subst. Use Misuse
Taylor and Francis Inc., 325 Chestnut St, Suite 800, Philadelphia PA, PA 19106, United States, Tel: +1(800)3541420, Fax: +1(215)6258914 ISSUES/YR: 14

Sucht
ISSN: 0939-5911
Sucht
Neuland-Verlagsgesellschaft mbH, P.O. Box 1422, D-21496, Geesthacht, Germany, Tel: +49(4152)81342, Fax: +49(4152)81343 ISSUES/YR: 6

Suchtmedizin in Forschung und Praxis
ISSN: 1437-5567
Suchtmed. Forsch. Prax.
Ecomed Publishers, Rudolf-Diesel-Strasse 3, D-86899, Landsberg, Germany, Tel: +49(8191)125802, Fax: +49(8191)125492 ISSUES/YR: 4

Suicide and Life-Threatening Behavior
ISSN: 0363-0234
Suicide Life-Threat. Behav.
Guilford Publications, 72 Spring Street, New York, NY 10012, United States, Tel: +1(212)4319800 ISSUES/YR: 4

Supportive Cancer Therapy
ISSN: 1543-2912
Supportive Cancer Ther.
Cancer Information Group, LP, 3535 Worth Street, Dallas, TX 75246, United States, Tel: +1(214)8207126, Fax: +1(214)8208844 ISSUES/YR: 4

• **Supportive Care in Cancer**
ISSN: 0941-4355
Supportive Care Cancer
Springer Verlag, Tiergartenstrasse 17, D-69121, Heidelberg, Germany, Tel: +49(6221)4870, Fax: +49(6221)487366 ISSUES/YR: 12

Surgeon
ISSN: 1479-666X
Surgeon
Edinburgh University Press, 22 George Square, EH8 9LF, Edinburgh, United Kingdom, Tel: +44(131)6504218, Fax: +44(131)6620053 ISSUES/YR: 6

• **Surgery**
ISSN: 0039-6060
Surgery
Mosby Inc., 11830 Westline Industrial Drive, St. Louis, MO 63146, United States, Tel: +1(314)8728370, Fax: +1(314)4321380 ISSUES/YR: 12

• **Surgery for Obesity and Related Diseases**
ISSN: 1550-7289
Surg. Obes. Relat. Dis.
Elsevier Inc., 360 Park Avenue South, New York, NY 10010, United States, Tel: +1(212)6333950, Fax: +1(212)6333990 ISSUES/YR: 6

Surgery Today
ISSN: 0941-1291
Surg. Today
Springer-Verlag Tokyo, 3-13, Hongo 3-chome, Bunkyo-ku, 113-8480, Tokyo, Japan, Tel: +81(3)38180861, Fax: +81(3)38180864 ISSUES/YR: 12

• **Surgical and Radiologic Anatomy**
ISSN: 0930-1038
Surg. Radiol. Anat.
Springer-Verlag France, 1 rue Paul Cezanne, 75008, Paris, France, Tel: +33(1)53933628, Fax: +33(1)53933683 ISSUES/YR: 6

Surgical Chronicles
ISSN: 1108-5002
Surg. Chronicles
Surgical Society of Northern Greece, P.O. Box 1514, 54006, Thessaloniki, Greece, Tel: +30(2310)994640, Fax: +30(2310)994791 ISSUES/YR: 4

• **Surgical Clinics of North America**
ISSN: 0039-6109
Surg. Clin. North Am.
W.B. Saunders, Independence Square West, Philadelphia, PA 19106-3399, United States, Tel: +1(215)2387800, Fax: +1(215)2386445 ISSUES/YR: 6

• **Surgical Endoscopy**
ISSN: 0930-2794
Surg. Endosc.
Springer New York LLC, 233 Springer Street, New York, 10013-1578, United States, Tel: +1-212 460 1501, Fax: +1-212 460 1595 ISSUES/YR: 12

• **Surgical Infections**
ISSN: 1096-2964
Surg. Infect.
Mary Ann Liebert Inc., 140 Huguenot Street, New Rochelle, NY 10801-5215, United States, Tel: +1(914)7402100, Fax: +1(914)7402101 ISSUES/YR: 4

Surgical Innovation
ISSN: 1553-3506
Surg. Innov.
Westminster Publications Inc., 708 Glen Cove Avenue, Glen Head, NY 11545, United States, Tel: +1(516)7590025, Fax: +1(516)7595524 ISSUES/YR: 4

• **Surgical Laparoscopy, Endoscopy and Percutaneous Techniques**
ISSN: 1530-4515
Surg. Laparoscopy Endosc. Percutaneous Tech.
Lippincott Williams and Wilkins, 530 Walnut Street, Philadelphia, PA 19106-3621, United States, Tel: +1(215)5218342, Fax: +1(215)5218326 ISSUES/YR: 6

Surgical Neurology
ISSN: 0090-3019
Surg. Neurol.
Elsevier Inc., 360 Park Avenue South, New York, NY 10010, United States, Tel: +1(212)6333950, Fax: +1(212)6333990 ISSUES/YR: 12

• **Surgical Oncology**
ISSN: 0960-7404
Surg. Oncol.
Elsevier Ltd, Langford Lane, Kidlington, OX5 1GB, Oxford, United Kingdom, Tel: +44(1865)843000, Fax: +44(1865)843010 ISSUES/YR: 4

Surgical Oncology Clinics of North America
ISSN: 1055-3207
Surg. Oncol. Clin. North Am.
W.B. Saunders, Independence Square West, Philadelphia, PA 19106-3399, United States, Tel: +1(215)2387800, Fax: +1(215)2386445 ISSUES/YR: 4

Surgical Practice
ISSN: 1744-1625
Surg. Pract.
Blackwell Publishing Ltd, 9600 Garsington Road, OX4 2XG, Oxford, United Kingdom, Tel: +1(1865)776868, Fax: +1(1865)714591 ISSUES/YR: 4

• **Survey of Ophthalmology**
ISSN: 0039-6257
Surv. Ophthalmol.
Elsevier USA, 6277 Sea Harbor Drive, Orlando, FL 32862 8239, United States, Tel: +1-407 345 4000, Fax: +1-407 363 9661 ISSUES/YR: 6

Swiss Medical Weekly
ISSN: 1424-7860
Swiss Med. Wkly
EMH Swiss Medical Publishers Ltd., Steinentorstrasse 13, CH-4010, Basel, Switzerland, Tel: +41(61)4678555, Fax: +41(61)4678556 ISSUES/YR: 46

• **Synapse**
ISSN: 0887-4476
Synapse
Wiley-Liss Inc., 111 River Street, Hoboken, NJ 07030-5774, United States, Tel: +1(201)7486645, Fax: +1(201)7486021 ISSUES/YR: 16

Synlett
ISSN: 0936-5214
Synlett
Georg Thieme Verlag, Rudigerstrasse 14, D-70469, Stuttgart, Germany, Tel: +49(711)8931-321, Fax: +49(711)8931-422 ISSUES/YR: 20

Synthesis
ISSN: 0039-7881
Synthesis
Georg Thieme Verlag, Rudigerstrasse 14, D-70469, Stuttgart, Germany, Tel: +49(711)8931-321, Fax: +49(711)8931-422 ISSUES/YR: 20

Synthetic Communications
ISSN: 0039-7911
Synth. Commun.
Taylor and Francis Inc., 325 Chestnut St, Suite 800, Philadelphia PA, PA 19106, United States, Tel: +1(800)3541420, Fax: +1(215)6258914 ISSUES/YR: 24

Current

- **Systematic and Applied Microbiology**
ISSN: 0723-2020
Syst. Appl. Microbiol.
Urban und Fischer Verlag Jena, P.O. Box 100537, 07705, Jena, Germany, Tel: +49(3641)6263, Fax: +49(3641)626500 ISSUES/YR: 8

Tagliche Praxis
ISSN: 0494-464X
Tagl. Prax.
Hans Marseille Verlag GmbH, P.O. Box 22 13 41, 80503, Munchen, Germany, Tel: +49(89)227988, Fax: +49(89)2904643 ISSUES/YR: 4

Taiwan Journal of Public Health
ISSN: 1023-2141
Taiwain J. Public Health
Chinese Public Health Association of Taiwan, No. 19, Hsu-Chow Rd,, 100, Taipei, Taiwan, Tel: +886(2)8267051, Fax: +886(2)8228026 ISSUES/YR: 6

Taiwanese Journal of Obstetrics and Gynecology
ISSN: 1028-4559
Taiwanese J. Obstet. Gynecol.
Elsevier (Singapore) Pte Ltd, 77 Leighton Road, Causeway Bay, Hong Kong, China, Hong Kong, Tel: +852-29651300, Fax: +852-29760778 ISSUES/YR: 4

Techniques in Coloproctology
ISSN: 1123-6337
Tech. Coloproctol.
Springer-Verlag Italia s.r.l, Via Podgora 4, I-20122, Milano, Italy, Tel: +39(2)5425971, Fax: +39(2)55193360 ISSUES/YR: 3

- **Techniques in Foot and Ankle Surgery**
ISSN: 1536-0644
Tech. Foot Ankle Surg.
Lippincott Williams and Wilkins, 530 Walnut Street, Philadelphia, PA 19106-3621, United States, Tel: +1(215)5218342, Fax: +1(215)5218326 ISSUES/YR: 4

Techniques in Gastrointestinal Endoscopy
ISSN: 1096-2883
Tech. Gastrointest. Endosc.
W.B. Saunders, Independence Square West, Philadelphia, PA 19106-3399, United States, Tel: +1(215)2387800, Fax: +1(215)2386445 ISSUES/YR: 4

- **Techniques in Hand and Upper Extremity Surgery**
ISSN: 1089-3393
Tech. Hand Upper Extremity Surg.
Lippincott Williams and Wilkins, 530 Walnut Street, Philadelphia, PA 19106-3621, United States, Tel: +1(215)5218342, Fax: +1(215)5218326 ISSUES/YR: 4

- **Techniques in Knee Surgery**
ISSN: 1536-0636
Tech. Knee Surg.
Lippincott Williams and Wilkins, 530 Walnut Street, Philadelphia, PA 19106-3621, United States, Tel: +1(215)5218342, Fax: +1(215)5218326 ISSUES/YR: 4

- **Techniques in Ophthalmology**
ISSN: 1542-1929
Tech. Ophthalmol.
Lippincott Williams and Wilkins, 530 Walnut Street, Philadelphia, PA 19106-3621, United States, Tel: +1(215)5218342, Fax: +1(215)5218326 ISSUES/YR: 4

Techniques in Orthopaedics
ISSN: 0885-9698
Tech. Orthop.
Lippincott Williams and Wilkins, 530 Walnut Street, Philadelphia, PA 19106-3621, United States, Tel: +1(215)5218342, Fax: +1(215)5218326 ISSUES/YR: 4

Techniques in Regional Anesthesia and Pain Management
ISSN: 1084-208X
Tech. Reg. Anesth. Pain Manage.
W.B. Saunders, Independence Square West, Philadelphia, PA 19106-3399, United States, Tel: +1(215)2387800, Fax: +1(215)2386445 ISSUES/YR: 4

Techniques in Shoulder and Elbow Surgery
ISSN: 1523-9896
Techn. Shoulder Elbow Surg.
Lippincott Williams and Wilkins, 530 Walnut Street, Philadelphia, PA 19106-3621, United States, Tel: +1(215)5218342, Fax: +1(215)5218326 ISSUES/YR: 4

Techniques in Vascular and Interventional Radiology
ISSN: 1089-2516
Tech. Vasc. Intervent. Radiol.
W.B. Saunders, Independence Square West, Philadelphia, PA 19106-3399, United States, Tel: +1(215)2387800, Fax: +1(215)2386445 ISSUES/YR: 4

Technology and Disability
ISSN: 1055-4181
Technol. Disabil.
IOS Press, Nieuwe Hemweg 6B, 1013 BG, Amsterdam, Netherlands, Tel: +31(20)6883355, Fax: +31(20)6203419 ISSUES/YR: 4

- **Technology and Health Care**
ISSN: 0928-7329
Technol. Health Care
IOS Press, Nieuwe Hemweg 6B, 1013 BG, Amsterdam, Netherlands, Tel: +31(20)6883355, Fax: +31(20)6203419 ISSUES/YR: 6

Technology in Cancer Research and Treatment
ISSN: 1533-0346
Technol. Cancer Res. Treat.
Adenine Press, 2066 Central Avenue, Schenectady, NY 12304, United States, Tel: +1(518)4560784, Fax: +1(518)4524955 ISSUES/YR: 6

Tecniche Chirurgiche in Ortopedia e Traumatologia
ISSN: 1724-269X
Tec. Chir. Ortop. Traumatol.
CIC Edizioni Internazionali s.r.l., Corso Trieste 42, 00198, Roma, Italy, Tel: +39(06)8412673, Fax: +39(06)8412688 ISSUES/YR: 3

Teikyo Medical Journal
ISSN: 0387-5547
Teikyo Med. J.
Teikyo University School of Medicine, 11-1 Kaga 2-chome, Itabashi-ku, 173, Tokyo, Japan, Tel: +81(3)39641211 EXT.1755 ISSUES/YR: 6

- **Telemedicine Journal and e-Health**
ISSN: 1530-5627
Telemedicine J. e-Health
Mary Ann Liebert Inc., 140 Huguenot Street, New Rochelle, NY 10801-5215, United States, Tel: +1(914)7402100, Fax: +1(914)7402101 ISSUES/YR: 4

TEN Trends in Evidence-Based Neuropsychiatry
ISSN: 1527-0815
TEN Trends Evid.-based Neuropsychiatr.
MedWorks Media LLC, 333 Hudson Street, 7th Floor, New York, NY 10013, United States, Tel: +1(212)3280800, Fax: +1(212)3280600 ISSUES/YR: 12

- **Tetrahedron**
ISSN: 0040-4020
Tetrahedron
Elsevier Ltd, Langford Lane, Kidlington, OX5 1GB, Oxford, United Kingdom, Tel: +44(1865)843000, Fax: +44(1865)843010 ISSUES/YR: 52

- **Tetrahedron Asymmetry**
ISSN: 0957-4166
Tetrahedron Asymmetry
Elsevier Ltd, Langford Lane, Kidlington, OX5 1GB, Oxford, United Kingdom, Tel: +44(1865)843000, Fax: +44(1865)843010 ISSUES/YR: 24

Tetrahedron Letters
ISSN: 0040-4039
Tetrahedron Lett.
Elsevier Ltd, Langford Lane, Kidlington, OX5 1GB, Oxford, United Kingdom, Tel: +44(1865)843000, Fax: +44(1865)843010 ISSUES/YR: 52

Texas Heart Institute Journal
ISSN: 0730-2347
Tex. Heart Inst. J.
Texas Heart Institute, 1101 Bates Street, P-514, P.O. Box 20345, Houston, TX 77030, United States, Tel: +1(713)794-6630 ISSUES/YR: 4

- **Thalamus and Related Systems**
ISSN: 1472-9288
Thalamus Relat. Syst.
Cambridge University Press, Shaftesbury Road, CB2 2RU, Cambridge, United Kingdom, Tel: +44(1223)312393, Fax: +44(1223)315052 ISSUES/YR: 4

Theoretical Biology and Medical Modelling
ISSN: 1742-4682
Theor. Biol. Med. Model.
BioMed Central Ltd., 34 - 42 Cleveland Street, W1T 4LB, London, United Kingdom, Tel: +44(207)3230323, Fax: +44(207)6319926 ISSUES/YR: 1

Theoretical Medicine and Bioethics
ISSN: 1386-7415
Theor. Med. Bioethics
Kluwer Academic Publishers, Van Godewijckstraat 30, 3311 GZ, Dordrecht, Netherlands, Tel: +31(78)6576000, Fax: +31(78)6576254 ISSUES/YR: 6

Theory in Biosciences
ISSN: 1431-7613
Theory Biosci.
Urban und Fischer Verlag Jena, P.O. Box 100537, 07705, Jena, Germany, Tel: +49(3641)6263, Fax: +49(3641)626500 ISSUES/YR: 4

• **Therapeutic Apheresis and Dialysis**
ISSN: 1744-9979
Ther. Apheresis Dial.
Blackwell Publishing Inc., 350 Main Street, Malden, MA 02148, United States, Tel: +1(781)3888200, Fax: +1(781)3888210 ISSUES/YR: 6

Therapeutic Communities: the International Journal for Therapeutic and Supportive Organizations
ISSN: 0964-1866
Ther. Communities Int. J. Ther. Supportive Organ.
Association of Therapeutic Communities, 13-15 Pine Street, EC1R 0JH, London, United Kingdom, Tel: +44(208)9509557 ISSUES/YR: 4

• **Therapeutic Drug Monitoring**
ISSN: 0163-4356
Ther. Drug Monit.
Lippincott Williams and Wilkins, 530 Walnut Street, Philadelphia, PA 19106-3621, United States, Tel: +1(215)5218342, Fax: +1(215)5218326 ISSUES/YR: 6

Therapeutic Research
ISSN: 0289-8020
Ther. Res.
Life Science Publishing Co. Ltd, 11-7 Nihonbashi, Kobunacho, Chuo-ku, 103, Tokyo, Japan, Tel: +81(3)36647750, Fax: +81(3)36647734 ISSUES/YR: 12

Therapeutics and Clinical Risk Management
ISSN: 1176-6336
Ther. Clin. Risk Manage.
Dove Press Ltd., PO Box 300-008, Albany, Auckland, New Zealand, Tel: +64(9)4485032, Fax: +64(9)4485034 ISSUES/YR: 4

Therapeutische Umschau
ISSN: 0040-5930
Ther. Umsch.
Verlag Hans Huber AG, Langgassstrasse 76, CH-3000, Bern, Switzerland, Tel: +41(31)3004500, Fax: +41(31)3004590 ISSUES/YR: 12

• **Therapie**
ISSN: 0040-5957
Therapie
ADIS International Ltd, Chowley Oak Lane, Tattenhall, CH3 9GA, Chester, United Kingdom, Tel: +64(1829)771155, Fax: +64(1829)770330 ISSUES/YR: 6

Therapy
ISSN: 1475-0708
Therapy
Future Drugs Ltd, 2 Albert Place, Finchley Central, N3 1QB, London, United Kingdom, Tel: +44(20)83492033, Fax: +44(20)83432313 ISSUES/YR: 6

Thermology International
ISSN: 1560-604X
Thermology Int.
Ludwig Boltzmann Forsch. fur Physikalishe Diagnostik, Heinrich Collinstrasse 30, 1140, Vienna, Austria, Tel: +43(1)9149701, Fax: +43(1)9149264 ISSUES/YR: 4

TheScientificWorldJournal
ISSN: 1537-744X
TheScientificWorldJournal
ScientificWorld Ltd., Venture West, New Greenham Park,, RG19 6HX, Newbury, Berkshire, United Kingdom, Tel: +44-1635584185, Fax: +44-163536294 ISSUES/YR: 1

• **Thoracic and Cardiovascular Surgeon**
ISSN: 0171-6425
Thorac. Cardiovasc. Surg.
Georg Thieme Verlag, Rudigerstrasse 14, D-70469, Stuttgart, Germany, Tel: +49(711)8931-321, Fax: +49(711)8931-422 ISSUES/YR: 6

• **Thoracic and Cardiovascular Surgeon, Supplement**
ISSN: 0946-4778
Thorac. Cardiovasc. Surg. Suppl.
Georg Thieme Verlag, Rudigerstrasse 14, D-70469, Stuttgart, Germany, Tel: +49(711)8931-321, Fax: +49(711)8931-422 ISSUES/YR: 1

• **Thoracic Surgery Clinics**
ISSN: 1547-4127
Thorac. Surg. Clin.
W.B. Saunders, Independence Square West, Philadelphia, PA 19106-3399, United States, Tel: +1(215)2387800, Fax: +1(215)2386445 ISSUES/YR: 4

• **Thorax**
ISSN: 0040-6376
Thorax
BMJ Publishing Group, Tavistock Square, WC1H 9JR, London, United Kingdom, Tel: +44(207)3874499, Fax: +44(207)3836668 ISSUES/YR: 12

• **Thrombosis and Haemostasis**
ISSN: 0340-6245
Thromb. Haemost.
F.K. Schattauer Verlagsgesellschaft mbH, P.O. Box 10 45 45, D-70040, Stuttgart, Germany, Tel: +49(711)229870, Fax: +49(711)2298750 ISSUES/YR: 12

Thrombosis Journal
ISSN: 1477-9560
Thromb. J.
BioMed Central Ltd., 34 - 42 Cleveland Street, W1T 4LB, London, United Kingdom, Tel: +44(207)3230323, Fax: +44(207)6319926 ISSUES/YR: 1

• **Thrombosis Research**
ISSN: 0049-3848
Thromb. Res.
Elsevier Ltd, Langford Lane, Kidlington, OX5 1GB, Oxford, United Kingdom, Tel: +44(1865)843000, Fax: +44(1865)843010 ISSUES/YR: 24

• **Thyroid**
ISSN: 1050-7256
Thyroid
Mary Ann Liebert Inc., 140 Huguenot Street, New Rochelle, NY 10801-5215, United States, Tel: +1(914)7402100, Fax: +1(914)7402101 ISSUES/YR: 12

Thyroidology Clinical and Experimental
ISSN: 1121-7596
Thyroidology Clin. Exp.
Pacini Editore s.r.l., Via A. Gherardesca 1, 56121, Ospedaletto (Pisa), Italy, Tel: +39(050)313011, Fax: +39(050)3130300 ISSUES/YR: 3

Tidsskrift for den Norske Laegeforening
ISSN: 0029-2001
Tidsskr. Nor. Laegeforen.
Norwegian Medical Association, P.O. Box 1152 Sentrum, N-0107, Oslo, Norway, Tel: +47(2)67124600 ISSUES/YR: 24

Tijdschrift voor Fertiliteitsonderzoek
ISSN: 0921-7304
Tijdschr. Fertil.
Excerpta Medica, Radarweg 29, 1043 NX, Amsterdam, Netherlands ISSUES/YR: 4

Tijdschrift voor Geneeskunde
ISSN: 0371-683X
Tijdschr. Geneeskd.
Tijdschrift voor Geneeskunde, De Pintelaan 185, B-9000, Gent, Belgium, Tel: +32(9)2403330, Fax: +32(9)2403390 ISSUES/YR: 24

Tijdschrift voor Gerontologie en Geriatrie
ISSN: 0167-9228
Tijdschr. Gerontol. Geriatr.
Nederlands Instituut voor Gerontologie (NIG), Christiaan Krammlaan 8, 3571 AX, Utrecht, Netherlands, Tel: +31(030)2769960, Fax: +31(030)2762880 ISSUES/YR: 6

Tijdschrift voor Kindergeneeskunde
ISSN: 0376-7442
Tijdschr. Kindergeneeskd.
Bohn Stafleu Van Loghum bv, P.O. Box 246, 3990 GA, Houten, Netherlands, Tel: +31(030)6383830, Fax: +31(030)6383999 ISSUES/YR: 6

Tijdschrift voor Psychiatrie
ISSN: 0303-7339
Tijdschr. Psychiatr.
Uitgeverij Boom, P.O. Box 400, 7940 AK, Meppel, Netherlands, Tel: +31(522)257012 ISSUES/YR: 12

Tijdschrift voor Verpleeghuisgeneeskunde
ISSN: 1386-1514
Tijdschr. Verpleeghuisgeneeskd.
NVVA - Nederlandse Vereniging van Verpleeghuisartsen, Postbus 20069, 3502 LB, Utrecht, Netherlands, Tel: +31(030)2803622 ISSUES/YR: 6

Tissue and Cell
ISSN: 0040-8166
Tissue Cell
Churchill Livingstone, 1-3 Baxter's Place, Leith Walk, EH1 3AF, Edinburgh, United Kingdom, Tel: +44(131)5562424, Fax: +44(131)5581278 ISSUES/YR: 6

• **Tissue Antigens**
ISSN: 0001-2815
Tissue Antigens
Blackwell Publishing Ltd, 9600 Garsington Road, OX4 2XG, Oxford, United Kingdom, Tel: +44(1865)776868, Fax: +44(1865)714591 ISSUES/YR: 12

Current

• **Tissue Engineering**
ISSN: 1076-3279
Tissue Eng.
Mary Ann Liebert Inc., 140 Huguenot Street, New Rochelle, NY 10801-5215, United States, Tel: +1(914)7402100, Fax: +1(914)7402101 ISSUES/YR: 12

Today's Life Science
ISSN: 1033-6893
Today's Life Sci.
Reed Business Publishing Pty. Ltd, 475 Victoria Avenue, NSW 2067, Chatswood, Australia, Tel: +61(2)94222999, Fax: +61(2)94222922 ISSUES/YR: 6

Today's Therapeutic Trends
ISSN: 0741-2320
Today's Ther. Trends
Communications Media for Education Inc., P.O. Box 712, Princeton Junction, NJ 08550, United States, Tel: +1(609)7992300 ISSUES/YR: 4

Tohoku Journal of Experimental Medicine
ISSN: 0040-8727
Tohoku J. Exp. Med.
Tohoku University Medical Press, 2-1 Seiryomachi, Aoba-ku, 980-8575, Sendai, Japan, Fax: +81(22)717-8185 ISSUES/YR: 12

Tokai Journal of Experimental and Clinical Medicine
ISSN: 0385-0005
Tokai J. Exp. Clin. Med.
Tokai University School of Medicine, 2-28-4 Tomigaya, Shibuya-ku, 151, Tokyo, Japan ISSUES/YR: 4

Tokyo Jikeikai Medical Journal
ISSN: 0375-9172
Tokyo Jikeikai Med. J.
Jikei University School of Medicine, 25-8 Nishi-Shinbashi 3-chome, Minato-ku, 105, Tokyo, Japan, Tel: +81(3)34331111 ISSUES/YR: 6

Topics in Advanced Practice Nursing
ISSN: 1535-2250
Top. Adv. Pract. Nurs.
Medscape Health Network, 224 W. 30th Street, New York, NY 10001-5399, United States, Tel: +1(888)5066098 ISSUES/YR: 4

• **Topics in Clinical Nutrition**
ISSN: 0883-5691
Top. Clin. Nutr.
Lippincott Williams and Wilkins, 530 Walnut Street, Philadelphia, PA 19106-3621, United States, Tel: +1(215)5218342, Fax: +1(215)5218326 ISSUES/YR: 4

• **Topics in Emergency Medicine**
ISSN: 0164-2340
Top. Emerg. Med.
Lippincott Williams and Wilkins, 530 Walnut Street, Philadelphia, PA 19106-3621, United States, Tel: +1(215)5218342, Fax: +1(215)5218326 ISSUES/YR: 4

Topics in Geriatric Rehabilitation
ISSN: 0882-7524
Top. Geriatr. Rehabil.
Lippincott Williams and Wilkins, 530 Walnut Street, Philadelphia, PA 19106-3621, United States, Tel: +1(215)5218342, Fax: +1(215)5218326 ISSUES/YR: 4

• **Topics in Magnetic Resonance Imaging**
ISSN: 0899-3459
Top. Magn. Reson. Imaging
Lippincott Williams and Wilkins, 530 Walnut Street, Philadelphia, PA 19106-3621, United States, Tel: +1(215)5218342, Fax: +1(215)5218326 ISSUES/YR: 6

Topics in Spinal Cord Injury Rehabilitation
ISSN: 1082-0744
Top. Spinal Cord Inj. Rehabil.
Thomas Land Publishers Inc., 255 Jefferson Road, St. Louis, MO 63119, United States, Tel: +1(314)9637445, Fax: +1(314)9639345 ISSUES/YR: 4

Topics in Stroke Rehabilitation
ISSN: 1074-9357
Top. Stroke Rehabil.
Thomas Land Publishers Inc., 255 Jefferson Road, St. Louis, MO 63119, United States, Tel: +1(314)9637445, Fax: +1(314)9639345 ISSUES/YR: 4

• **Toxicologic Pathology**
ISSN: 0192-6233
Toxicol. Pathol.
Taylor and Francis Ltd., 4 Park Square, Milton Park, OX14 4RN, Abingdon, Oxfordshire, United Kingdom, Tel: +44(20)70175541, Fax: +44(20)70175198 ISSUES/YR: 6

Toxicological Reviews
ISSN: 1176-2551
Toxicol. Rev.
ADIS International Ltd, Chowley Oak Lane, Tattenhall, CH3 9GA, Chester, United Kingdom, Tel: +64(1829)771155, Fax: +64(1829)770330 ISSUES/YR: 4

Toxicological Sciences
ISSN: 1096-6080
Toxicol. Sci.
Oxford University Press, 2001 Evans Road, Cary, NC 27513, United States, Tel: +1(919)6770977, Fax: +1(919)6771714 ISSUES/YR: 12

• **Toxicology**
ISSN: 0300-483X
Toxicology
Elsevier Ireland Ltd, P.O. Box 85, Limerick, Ireland, Tel: +353(61)471944, Fax: +353(61)472144 ISSUES/YR: 33

Toxicology and Applied Pharmacology
ISSN: 0041-008X
Toxicol. Appl. Pharmacol.
Academic Press Inc., 6277 Sea Harbor Drive, Orlando, FL 32887-4900, United States, Tel: +1(407)3454040, Fax: +1(407)3639661 ISSUES/YR: 24

Toxicology and Industrial Health
ISSN: 0748-2337
Toxicol. Ind. Health
Arnold, 338 Euston Road, NW1 3BH, London, United Kingdom, Tel: +44(20)78736000, Fax: +44(20)78736376 ISSUES/YR: 10

Toxicology in Vitro
ISSN: 0887-2333
Toxicol. Vitro
Elsevier Ltd, Langford Lane, Kidlington, OX5 1GB, Oxford, United Kingdom, Tel: +44(1865)843000, Fax: +44(1865)843010 ISSUES/YR: 6

Toxicology International
ISSN: 0971-6580
Toxicol. Int.
Society of Toxicology India, C-44 Rajendra Nagar, 243122, Bareilly, India, Tel: +91(581)2300628 ISSUES/YR: 2

- **Toxicology Letters**
ISSN: 0378-4274
Toxicol. Lett.
Elsevier Ireland Ltd, P.O. Box 85, Limerick, Ireland, Tel: +353(61)471944, Fax: +353(61)472144 ISSUES/YR: 24

- **Toxicology Mechanisms and Methods**
ISSN: 1537-6516
Toxicol. Mechan. Methods
Taylor and Francis Ltd., 4 Park Square, Milton Park, OX14 4RN, Abingdon, Oxfordshire, United Kingdom, Tel: +44(20)70175541, Fax: +44(20)70175198 ISSUES/YR: 9

- **Toxicon**
ISSN: 0041-0101
Toxicon
Elsevier Ltd, Langford Lane, Kidlington, OX5 1GB, Oxford, United Kingdom, Tel: +44(1865)843000, Fax: +44(1865)843010 ISSUES/YR: 16

Toxin Reviews
ISSN: 1556-9543
Toxin Rev.
Taylor and Francis Inc., 325 Chestnut St, Suite 800, Philadelphia PA, PA 19106, United States, Tel: +1(800)3541420, Fax: +1(215)6258914 ISSUES/YR: 4

- **TrAC - Trends in Analytical Chemistry**
ISSN: 0165-9936
TrAC Trends Anal. Chem.
Elsevier, P.O. Box 211, 1000 AE, Amsterdam, Netherlands, Tel: +31(20)4853757, Fax: +31(20)4853432 ISSUES/YR: 11

- **Trace Elements and Electrolytes**
ISSN: 0946-2104
Trace Elem. Electrolytes
Dustri-Verlag Dr. Karl Feistle, P.O. Box 1351, D-82032, Deisenhofen/Munchen, Germany, Tel: +49(89)6138610, Fax: +49(89)6135412 ISSUES/YR: 4

- **Traffic**
ISSN: 1398-9219
Traffic
Blackwell Munksgaard, 1 Rosenorns Alle, P.O. Box 227, DK-1502, Copenhagen V, Denmark, Tel: +45(77)333333, Fax: +45(77)333377 ISSUES/YR: 12

Traffic Injury Prevention
ISSN: 1538-9588
Traffic Inj. Prev.
Taylor and Francis Inc., 325 Chestnut St, Suite 800, Philadelphia PA, PA 19106, United States, Tel: +1(800)3541420, Fax: +1(215)6258914 ISSUES/YR: 4

- **Transactions of the American Ophthalmological Society**
ISSN: 1545-6110
Trans. Am. Ophthalmol. Soc.
American Ophthalmological Society, P.O. Box 193940, San Francisco, CA 94119, United States, Tel: +1-415 561 8578, Fax: +1-415 561 8531 ISSUES/YR: 1

Transactions of the Royal Society of Tropical Medicine and Hygiene
ISSN: 0035-9203
Trans. R. Soc. Trop. Med. Hyg.
Elsevier, P.O. Box 211, 1000 AE, Amsterdam, Netherlands, Tel: +31(20)4853757, Fax: +31(20)4853432 ISSUES/YR: 12

Transcultural Psychiatry
ISSN: 1363-4615
Transcult. Psychiatry
SAGE Publications Ltd, 55 City Road, EC1Y 1SP, London, United Kingdom, Tel: +44(207)3248500, Fax: +44(207)3248600 ISSUES/YR: 4

Transfusion
ISSN: 0041-1132
Transfusion
Blackwell Publishing Inc., 350 Main Street, Malden, MA 02148, United States, Tel: +1(781)3888200, Fax: +1(781)3888210 ISSUES/YR: 12

Transfusion and Apheresis Science
ISSN: 1473-0502
Transfus. Apheresis Sci.
Elsevier Ltd, Langford Lane, Kidlington, OX5 1GB, Oxford, United Kingdom, Tel: +44(1865)843000, Fax: +44(1865)843010 ISSUES/YR: 6

Transfusion Clinique et Biologique
ISSN: 1246-7820
Transfus. Clin. Biol.
Editions Scientifiques et Medicales Elsevier, 23 rue Linois, 75724, Paris Cedex 15, France, Tel: +33(1)45589826, Fax: +33(1)45589428 ISSUES/YR: 6

Transfusion Medicine
ISSN: 0958-7578
Transfus. Med.
Blackwell Publishing Ltd, 9600 Garsington Road, OX4 2XG, Oxford, United Kingdom, Tel: +44(1865)776868, Fax: +44(1865)714591 ISSUES/YR: 6

Transfusion Medicine and Hemotherapy
ISSN: 1660-3796
Transfus. Med. Hemotherapy
S. Karger AG, Allschwilerstrasse 10, P.O. Box, CH-4009, Basel, Switzerland, Tel: +41(61)3061111, Fax: +41(61)3061234 ISSUES/YR: 6

Transfusion Medicine Reviews
ISSN: 0887-7963
Transfus. Med. Rev.
W.B. Saunders, Independence Square West, Philadelphia, PA 19106-3399, United States, Tel: +1(215)2387800, Fax: +1(215)2386445 ISSUES/YR: 4

Transfuze a Hematologie Dnes
ISSN: 1213-5763
Transfuze Hematol. Dnes
Czech Medical Association J.E. Purkyne, Sokolska 31, P.O. Box 88, 120 26, Praha 2, Czech Republic, Tel: +420(2)24266201-4, Fax: +420(2)24266212 ISSUES/YR: 4

TransPharma
ISSN: 1473-5792
TransPharma
Libra pharm, Venture West, New Greenham Park, RG19 6HX, Newbury, Berks., United Kingdom, Tel: +44(1635)522651, Fax: +44(1635)36294 ISSUES/YR: 6

- **Transplant Immunology**
ISSN: 0966-3274
Transplant Immunol.
Elsevier, P.O. Box 211, 1000 AE, Amsterdam, Netherlands, Tel: +31(20)4853757, Fax: +31(20)4853432 ISSUES/YR: 4

- **Transplant Infectious Disease**
ISSN: 1398-2273
Transplant Infect. Dis.
Blackwell Publishing Ltd, 9600 Garsington Road, OX4 2XG, Oxford, United Kingdom, Tel: +44(1865)776868, Fax: +44(1865)714591 ISSUES/YR: 4

- **Transplant International**
ISSN: 0934-0874
Transplant Int.
Blackwell Publishing Ltd, 9600 Garsington Road, OX4 2XG, Oxford, United Kingdom, Tel: +44(1865)776868, Fax: +44(1865)714591 ISSUES/YR: 12

- **Transplantation**
ISSN: 0041-1337
Transplantation
Lippincott Williams and Wilkins, 351 West Camden Street, Baltimore, MD 21201-2436, United States, Tel: +1(410)5284000 ISSUES/YR: 24

- **Transplantation Proceedings**
ISSN: 0041-1345
Transplant. Proc.
Elsevier USA, 6277 Sea Harbor Drive, Orlando, FL 32862 8239, United States, Tel: +1-407 345 4000, Fax: +1-407 363 9661 ISSUES/YR: 10

Transplantation Reviews
ISSN: 0955-470X
Transpl. Rev.
W.B. Saunders, Independence Square West, Philadelphia, PA 19106-3399, United States, Tel: +1(215)2387800, Fax: +1(215)2386445 ISSUES/YR: 4

Transplantationsmedizin: Organ der Deutschen Transplantationsgesellschaft
ISSN: 0946-9648
Transplantationsmed. Organ Dtsch. Transplantationsges.
Pabst Science Publishers, Eichengrund 28, D-49525, Lengerich, Germany, Tel: +49(5484)308, Fax: +49(5484)550 ISSUES/YR: 2

TransplantLinc
ISSN: 1439-0027
TransplantLinc
Pabst Science Publishers, Eichengrund 28, D-49525, Lengerich, Germany, Tel: +49(5484)308, Fax: +49(5484)550 ISSUES/YR: 1

Trasfusione del Sangue
ISSN: 0041-1787
Trasfus. Sangue
SIMTI SrL, Viale Beatrice d'Este, 5, 20122, Milano, Italy, Tel: +39(06)4452751, Fax: +39(06)4441439 ISSUES/YR: 6

Trauma
ISSN: 0564-1470
Trauma (USA)
Matthew Bender and Co. Inc., 136 Carlin Road, Conklin, NY 13748-1531, United States ISSUES/YR: 6

Trauma
ISSN: 1460-4086
Trauma (GBR)
Arnold, 338 Euston Road, NW1 3BH, London, United Kingdom, Tel: +44(20)78736000, Fax: +44(20)78736376 ISSUES/YR: 4

• **Travel Medicine and Infectious Disease**
ISSN: 1477-8939
Travel Med. Infect. Dis.
Elsevier USA, 6277 Sea Harbor Drive, Orlando, FL 32862 8239, United States, Tel: +1-407 345 4000, Fax: +1-407 363 9661 ISSUES/YR: 4

• **Treatments in Endocrinology**
ISSN: 1175-6349
Treat. Endocrinol.
Adis International Ltd, 41 Centorian Drive, Private Bag 65901, Mairangi Bay, 1311, Auckland 10, New Zealand, Tel: +64(9)4770700, Fax: +64(9)4770766 ISSUES/YR: 6

• **Treatments in Respiratory Medicine**
ISSN: 1176-3450
Treat. Respir. Med.
Adis International Ltd, 41 Centorian Drive, Private Bag 65901, Mairangi Bay, 1311, Auckland 10, New Zealand, Tel: +64(9)4770700, Fax: +64(9)4770766 ISSUES/YR: 6

Trends in Amplification
ISSN: 1084-7138
Trends Amplif.
Westminster Publications Inc., 708 Glen Cove Avenue, Glen Head, NY 11545, United States, Tel: +1(516)7590025, Fax: +1(516)7595524 ISSUES/YR: 4

• **Trends in Biochemical Sciences**
ISSN: 0968-0004
Trends Biochem. Sci.
Elsevier Ltd, Langford Lane, Kidlington, OX5 1GB, Oxford, United Kingdom, Tel: +44(1865)843300 ISSUES/YR: 12

• **Trends in Biotechnology**
ISSN: 0167-7799
Trends Biotechnol.
Elsevier Ltd, Langford Lane, Kidlington, OX5 1GB, Oxford, United Kingdom, Tel: +44(1865)843300 ISSUES/YR: 12

• **Trends in Cardiovascular Medicine**
ISSN: 1050-1738
Trends Cardiovasc. Med.
Elsevier Inc., 360 Park Avenue South, New York, NY 10010, United States, Tel: +1(212)6333950, Fax: +1(212)6333990 ISSUES/YR: 8

• **Trends in Cell Biology**
ISSN: 0962-8924
Trends Cell Biol.
Elsevier Ltd, Langford Lane, Kidlington, OX5 1GB, Oxford, United Kingdom, Tel: +44(1865)843300 ISSUES/YR: 12

Trends in Cognitive Sciences
ISSN: 1364-6613
Trends Cogn. Sci.
Elsevier Ltd, Langford Lane, Kidlington, OX5 1GB, Oxford, United Kingdom, Tel: +44(1865)843300 ISSUES/YR: 12

• **Trends in Endocrinology and Metabolism**
ISSN: 1043-2760
Trends Endocrinol. Metab.
Elsevier Inc., 360 Park Avenue South, New York, NY 10010, United States, Tel: +1(212)6333950, Fax: +1(212)6333990 ISSUES/YR: 10

• **Trends in Genetics**
ISSN: 0168-9525
Trends Genet.
Elsevier Ltd, Langford Lane, Kidlington, OX5 1GB, Oxford, United Kingdom, Tel: +44(1865)843300 ISSUES/YR: 12

Trends in Immunology
ISSN: 1471-4906
Trends Immunol.
Elsevier Ltd, Langford Lane, Kidlington, OX5 1GB, Oxford, United Kingdom, Tel: +44(1865)843300 ISSUES/YR: 12

Trends in Medicine
ISSN: 1594-2848
Trends Med.
Pharma Project Group Edizioni Scientifiche (PPG), Viale Rimembranze 43/A, 21047, Saronno (VA), Italy, Tel: +39-0296702708, Fax: +39-0296702677 ISSUES/YR: 4

• **Trends in Microbiology**
ISSN: 0966-842X
Trends Microbiol.
Elsevier Ltd, Langford Lane, Kidlington, OX5 1GB, Oxford, United Kingdom, Tel: +44(1865)843300 ISSUES/YR: 12

Trends in Molecular Medicine
ISSN: 1471-4914
Trends Mol. Med.
Elsevier Ltd, Langford Lane, Kidlington, OX5 1GB, Oxford, United Kingdom, Tel: +44(1865)843300 ISSUES/YR: 12

• **Trends in Neurosciences**
ISSN: 0166-2236
Trends Neurosci.
Elsevier Ltd, Langford Lane, Kidlington, OX5 1GB, Oxford, United Kingdom, Tel: +44(1865)843300 ISSUES/YR: 12

Trends in Parasitology
ISSN: 1471-4922
Trends Parasitol.
Elsevier Ltd, Langford Lane, Kidlington, OX5 1GB, Oxford, United Kingdom, Tel: +44(1865)843300 ISSUES/YR: 12

• **Trends in Pharmacological Sciences**
ISSN: 0165-6147
Trends Pharmacol. Sci.
Elsevier Ltd, Langford Lane, Kidlington, OX5 1GB, Oxford, United Kingdom, Tel: +44(1865)843300 ISSUES/YR: 12

Tropical Doctor
ISSN: 0049-4755
Trop. Doct.
Royal Society of Medicine Press Ltd, P.O. Box 9002, W1A 0ZA, London, United Kingdom, Tel: +44(207)2902927, Fax: +44(207)2902929 ISSUES/YR: 4

Tropical Medicine and International Health
ISSN: 1360-2276
Trop. Med. Int. Health
Blackwell Publishing Ltd, 9600 Garsington Road, OX4 2XG, Oxford, United Kingdom, Tel: +44(1865)776868, Fax: +44(1865)714591 ISSUES/YR: 12

• **Tuberculosis**
ISSN: 1472-9792
Tuberculosis
Churchill Livingstone, 1-3 Baxter's Place, Leith Walk, EH1 3AF, Edinburgh, United Kingdom, Tel: +44(131)5562424, Fax: +44(131)5581278 ISSUES/YR: 6

Tuberculosis and Respiratory Diseases
ISSN: 1738-3536
Tuberc. Respir. Dis.
Korean National Tuberculosis Association, 121-150 Dangsandong-6-ga Yeoungdeungpogu, 150-046, Seoul, South Korea, Tel: +82(2)26339461, Fax: +82(2)6776673 ISSUES/YR: 12

Tuberkuloz ve Toraks
ISSN: 0494-1373
Tuberk. Toraks
Ankara University, C/o Dr. Doganay Alper, 06100, Cebeci, Ankara, Turkey, Tel: +90-312 3190027, Fax: +90-312 3190046 ISSUES/YR: 4

• **Tumor Biology**
ISSN: 1010-4283
Tumor Biol.
S. Karger AG, Allschwilerstrasse 10, P.O. Box, CH-4009, Basel, Switzerland, Tel: +41(61)3061111, Fax: +41(61)3061234 ISSUES/YR: 6

Tumor Diagnostik und Therapie
ISSN: 0722-219X
Tumor Diagn. Ther.
Georg Thieme Verlag, Rudigerstrasse 14, D-70469, Stuttgart, Germany, Tel: +49(711)8931-321, Fax: +49(711)8931-422 ISSUES/YR: 6

Tumor Research
ISSN: 0041-4093
Tumor Res.
Cancer Research Institute, South 1, West 17, Chuo-ku, 060, Sapporo, Japan, Tel: +81(11)6112111 ISSUES/YR: 1

Tumori
ISSN: 0300-8916
Tumori
Il Pensiero Scientifico Editore s.r.l., Via Bradano 3/C, 00199, Roma, Italy, Tel: +39(06)86207158, Fax: +39(06)86207160 ISSUES/YR: 6

Tunisie Medicale
ISSN: 0041-4131
Tunis. Med.
Maison du Medicine, 16 rue de Touraine, 1002, Tunis Belvedere, Tunisia, Tel: +216(71)790924, Fax: +216(71)796602 ISSUES/YR: 12

Turk Anesteziyoloji ve Reanimasyon Dernegi Dergisi
ISSN: 1304-0871
Turk Anesteziyol. Reanim. Dernegi Derg.
Turkish Anaesthesiology and Intensive Care Society, Inonu Cad. 7/4 Zambak apt., Gumussuyu Taksim. Istanbul, Turkey ISSUES/YR: 6

Turk Geriatri Dergisi
ISSN: 1304-2947
Turk Geriatr. Derg.
Geriatrics Society, Kultur Mahallesi Icel Sokak 3/23, Kizilay, Ankara, Turkey ISSUES/YR: 4

Turk Kardiyoloji Dernegi Arsivi
ISSN: 1016-5169
Turk Kardiyol. Dernegi Ars.
Turkish Anaesthesiology and Intensive Care Society, Inonu Cad. 7/4 Zambak apt., Gumussuyu Taksim. Istanbul, Turkey ISSUES/YR: 9

Turk Pediatri Arsivi
ISSN: 1016-5150
Turk Pediatr. Ars.
Turkish Medical Association, Turkey, Tel: +90-212 6328633, Fax: +90-212 6328633 ISSUES/YR: 4

Turk Serebrovaskuler Hastaliklar Dergisi
ISSN: 1301-1375
Turk Serebrovaskuler Hast. Derg.
Turkish Society of Cerebrovascular Diseases, Sair Fuzuli Cad. Kidan Ishani 20/2, 26010, Eskisehir, Turkey, Tel: +90(222)2309696, Fax: +90(222)2309696 ISSUES/YR: 3

Turk Uroloji Dergisi
ISSN: 1300-5804
Turk Urol. Derg.
Turk Uroloji Dernegi, Sokak Lale Palas Apt 18/2, 80260, Sisli Istanbul, Turkey, Tel: +90(212)2324689, Fax: +90(212)2339804 ISSUES/YR: 4

Turkish Journal of Cancer
ISSN: 1019-3103
Turk. J. Cancer
Turkish Association for Cancer Research and Control, Atac Sokak 21, Yenisehir 06420, Ankara, Turkey, Tel: +90(312)3092908, Fax: +90(312)4313958 ISSUES/YR: 4

Turkish Journal of Gastroenterology
ISSN: 1300-4948
Turk. J. Gastroenterol.
Turkish Society of Gastroenterology, Bayindir Sokak 17/7, 06420, Kizilay Ankara, Turkey, Tel: +90(312)4354373 ISSUES/YR: 4

Turkish Journal of Haematology
ISSN: 1300-7777
Turk. J. Haematol.
Turkish Society of Haematology, P.O. Box 25, Samanpazari, 06100, Ankara, Turkey, Fax: +90(312)3115474 ISSUES/YR: 4

Turkish Journal of Immunology
ISSN: 1301-109X
Turk. J. Immunol.
Turkish Society of Immunology, 1 Cerrahpasa 34303, Istanbul, Turkey, Tel: +90(212)5302095, Fax: +90(212)5861595 ISSUES/YR: 2

Turkish Journal of Medical Sciences
ISSN: 1300-0144
Turk. J. Med. Sci.
Turkiye Klinikleri Journal of Medical Sciences, Talapapa Bulvary no. 102, 1 06230, Hamammonu, Turkey, Tel: +90-312 309 36 66, Fax: +90-312 312 67 41 ISSUES/YR: 6

Turkish Journal of Pediatrics
ISSN: 0041-4301
Turk. J. Pediatr.
Turkish Journal of Pediatrics, Samanpazan, P.O. Box 66, 06240, Ankara, Turkey, Tel: +90(312)3244291, Fax: +90(312)3243284 ISSUES/YR: 4

Turkish Journal of Pharmaceutical Sciences
ISSN: 1304-530X
Turk. J. Pharm. Sci.
Turkish Pharmacists Association, Willy Brandt Str. 9, TR06690, Cankaya Ankara, Turkey, Tel: +90-3124098100, Fax: +90-3124098109 ISSUES/YR: 3

Turkish Journal of Surgery
ISSN: 1300-0705
Turk. J. Surg.
Turkish Surgical Society, Gape, Istanbul, Turkey, Tel: +90(212)5323246, Fax: +90(212)5348078 ISSUES/YR: 4

Turkish Neurosurgery
ISSN: 1019-5149
Turk. Neurosurg.
Turkish Neurological Society, Taskent Caddesi 13/4, Bahcelievler, 06500, Ankara, Turkey, Tel: +90(312)2126408, Fax: +90(312)2154626 ISSUES/YR: 4

Turkiye Fiziksel Tip ve Rehabilitasyon Dergisi
ISSN: 1302-0234
Turk. Fiz. Tip Rehab. Derg.
Turkish Society of Physical Medicine and Rehabilitation, c/o Barbaros Bulvari Salim Aktas Is Hani 79, D.3, K.3, Besiktas - Istanbul, Turkey, Tel: +90(212)2369052, Fax: +90(212)2369053 ISSUES/YR: 6

Turkiye Klinikleri Journal of Medical Sciences
ISSN: 1300-0292
Turk. Klinikleri J. Med. Sci.
Turkiye Klinikleri Journal of Medical Sciences, Talapapa Bulvary no. 102, 1 06230, Hamammonu, Turkey, Tel: +90-312 309 36 66, Fax: +90-312 312 67 41 ISSUES/YR: 6

• **Twin Research and Human Genetics**
ISSN: 1832-4274
Twin Res. Hum. Genet.
Australian Academic Press Pty. Ltd, 32 Jeays Street, QLD 4006, Bowen Hills, Australia, Tel: +61(7)32571176, Fax: +61(7)32525908 ISSUES/YR: 6

Tzu Chi Medical Journal
ISSN: 1016-3190
Tzu Chi Med. J.
Buddhist Compassion Relief Tzu Chi Foundation, 21 Kanglo Village, Shinchen Hsien, Hualien County, Taiwan ISSUES/YR: 6

Udar Mozgu - Problemy Interdyscyplinarne
ISSN: 1505-6740
Udar Mozgu Problemy Interdyscyplinarne
Via Medica Medical Publishers, ul. Swietokrzyska 73, 80-180 Gdansk, Poland, Tel: +48-58 3209494, Fax: +48-58 3209460 ISSUES/YR: 2

Ugeskrift for Laeger
ISSN: 0041-5782
Ugeskr. Laeg.
Almindelige Danske Laegeforening, Tromdhjemsgade 9, DK-2100, Copenhagen, Denmark, Tel: +45-35 44 8500, Fax: +45-35 44 8370 ISSUES/YR: 50

UHOD - Uluslararasi Hematoloji-Onkoloji Dergisi
ISSN: 1306-133X
UHOD Uluslararasi Hematol.-Onkol. Derg.
UHOD - Uluslararasi Hematoloji Onkoloji Dergisi, 4 Cadde, 56 Sokak, No: 49-E Hilal, 06550, Cankaya, Ankara, Turkey, Tel: +90(312)4402028, Fax: +90(312)4404749 ISSUES/YR: 4

Ulster Medical Journal
ISSN: 0041-6193
Ulster Med. J.
Ulster Medical Society, 97 Lisburn Road, BT9 7BL, Belfast, United Kingdom ISSUES/YR: 2

Ultramicroscopy
ISSN: 0304-3991
Ultramicroscopy
Elsevier, P.O. Box 211, 1000 AE, Amsterdam, Netherlands, Tel: +31(20)4853757, Fax: +31(20)4853432 ISSUES/YR: 20

Ultraschall in der Medizin
ISSN: 0172-4614
Ultraschall Med.
Georg Thieme Verlag, Rudigerstrasse 14, D-70469, Stuttgart, Germany, Tel: +49(711)8931-321, Fax: +49(711)8931-422 ISSUES/YR: 6

Ultraschall in der Medizin, Supplement
ISSN: 1431-4894
Ultraschall Med. Suppl.
Georg Thieme Verlag, Rudigerstrasse 14, D-70469, Stuttgart, Germany, Tel: +49(711)8931-321, Fax: +49(711)8931-422 ISSUES/YR: 1

Ultrasonic Imaging
ISSN: 0161-7346
Ultrason. Imaging
Dynamedia Inc., 2 Fulham Court, Silver Spring, MD 20902, United States ISSUES/YR: 4

• **Ultrasonics Sonochemistry**
ISSN: 1350-4177
Ultrason. Sonochem.
Elsevier, P.O. Box 211, 1000 AE, Amsterdam, Netherlands, Tel: +31(20)4853757, Fax: +31(20)4853432 ISSUES/YR: 6

• **Ultrasound in Medicine and Biology**
ISSN: 0301-5629
Ultrasound Med. Biol.
Elsevier USA, 6277 Sea Harbor Drive, Orlando, FL 32862 8239, United States, Tel: +1-407 345 4000, Fax: +1-407 363 9661 ISSUES/YR: 12

• **Ultrasound in Obstetrics and Gynecology**
ISSN: 0960-7692
Ultrasound Obstet. Gynecol.
John Wiley and Sons Ltd, Southern Gate, PO19 8SQ, Chichester, West Sussex, United Kingdom, Tel: +44(1243)779777, Fax: +44(1243)775878 ISSUES/YR: 12

• **Ultrasound International**
ISSN: 0971-6874
Ultrasound Int.
Ultrasound International, 5A/157, W.E.A., Karol Bagh, 110 005, New Dehli, India, Tel: +91(11)5726124, Fax: +91(11)5753040 ISSUES/YR: 4

• **Ultrasound Quarterly**
ISSN: 0894-8771
Ultrasound Q.
Lippincott Williams and Wilkins, 530 Walnut Street, Philadelphia, PA 19106-3621, United States, Tel: +1(215)5218342, Fax: +1(215)5218326 ISSUES/YR: 4

Ultrasound Review of Obstetrics and Gynecology
ISSN: 1472-2240
Ultrasound Rev. Obstet. Gynecol.
Taylor and Francis Ltd., 4 Park Square, Milton Park, OX14 4RN, Abingdon, Oxfordshire, United Kingdom, Tel: +44(20)70175541, Fax: +44(20)70175198 ISSUES/YR: 4

• **Ultrastructural Pathology**
ISSN: 0191-3123
Ultrastruct. Pathol.
Taylor and Francis Ltd., 4 Park Square, Milton Park, OX14 4RN, Abingdon, Oxfordshire, United Kingdom, Tel: +44(20)70175541, Fax: +44(20)70175198 ISSUES/YR: 6

Ulusal Travma ve Acil Cerrahi Dergisi
ISSN: 1300-6738
Ulusal Travma Acil Cerrahi Derg.
Turkish Association of Trauma and Emergency Surgery, Erayh Mah Borekci Veli Sok, SEHREMINI 34280, Istanbul, Turkey, Tel: +90(0212)5886246, Fax: +90(0212)5331882 ISSUES/YR: 4

Umweltmedizin in Forschung und Praxis
ISSN: 1430-8681
Umweltmed. Forsch. Prax.
Ecomed Publishers, Rudolf-Diesel-Strasse 3, D-86899, Landsberg, Germany, Tel: +49(8191)125802, Fax: +49(8191)125492 ISSUES/YR: 6

Umweltwissenschaften und Schadstoff-Forschung
ISSN: 0934-3504
Umweltwiss. Schadstoff-Forsch.
Ecomed Publishers, Rudolf-Diesel-Strasse 3, D-86899, Landsberg, Germany, Tel: +49(8191)125802, Fax: +49(8191)125492 ISSUES/YR: 4

Unfallchirurg
ISSN: 0177-5537
Unfallchirurg
Springer Verlag, Tiergartenstrasse 17, D-69121, Heidelberg, Germany, Tel: +49(6221)4870, Fax: +49(6221)487366 ISSUES/YR: 12

Update in Anaesthesia
ISSN: 1353-4882
Update Anaesth.
Roayal Devon and Exeter Healthcare NHS Trust, Barrack Road, EX2 5DW, Exter, United Kingdom, Tel: +44(1392)411611 ISSUES/YR: 1

Urodinamica
ISSN: 1120-5989
Urodinamica
Editrice Kurtis s.r.l., Via Luigi Zoja 30, 20153, Milan, Italy, Tel: +39(02)48202740, Fax: +39(02)48201219 ISSUES/YR: 4

Urogynaecologia International Journal
ISSN: 1121-3086
Urogynaecol. Int. J.
Associato All'USPI - Unione Stampa Periodica Italia, Pallonetto S. Chiara 8, 80134, Napoli, Italy, Tel: +39(081)446146 ISSUES/YR: 3

Urologe - Ausgabe A
ISSN: 0340-2592
Urologe Ausg. A
Springer Verlag, Tiergartenstrasse 17, D-69121, Heidelberg, Germany, Tel: +49(6221)4870, Fax: +49(6221)487366 ISSUES/YR: 12

• **Urologia Internationalis**
ISSN: 0042-1138
Urol. Int.
S. Karger AG, Allschwilerstrasse 10, P.O. Box, CH-4009, Basel, Switzerland, Tel: +41(61)3061111, Fax: +41(61)3061234 ISSUES/YR: 8

• **Urologic Clinics of North America**
ISSN: 0094-0143
Urol. Clin. North Am.
W.B. Saunders, Independence Square West, Philadelphia, PA 19106-3399, United States, Tel: +1(215)2387800, Fax: +1(215)2386445 ISSUES/YR: 4

• **Urologic Oncology: Seminars and Original Investigations**
ISSN: 1078-1439
Urol. Oncol. Semin. Orig. Invest.
Elsevier Inc., 360 Park Avenue South, New York, NY 10010, United States, Tel: +1(212)6333950, Fax: +1(212)6333990 ISSUES/YR: 6

- **Urological Research**
  ISSN: 0300-5623
  Urol. Res.
  Springer Verlag, Tiergartenstrasse 17, D-69121, Heidelberg, Germany, Tel: +49(6221)4870, Fax: +49(6221)487366 ISSUES/YR: 6
- **Urology**
  ISSN: 0090-4295
  Urology
  Elsevier Inc., 360 Park Avenue South, New York, NY 10010, United States, Tel: +1(212)6333950, Fax: +1(212)6333990 ISSUES/YR: 12
- **Vaccine**
  ISSN: 0264-410X
  Vaccine
  Elsevier Ltd, Langford Lane, Kidlington, OX5 1GB, Oxford, United Kingdom, Tel: +44(1865)843000, Fax: +44(1865)843010 ISSUES/YR: 32
- **Vaccines: Children and Practice**
  ISSN: 1460-6887
  Vaccines Child. Pract.
  ADIS International Ltd, Chowley Oak Lane, Tattenhall, CH3 9GA, Chester, United Kingdom, Tel: +64(1829)771155, Fax: +64(1829)770330 ISSUES/YR: 2
- **Value in Health**
  ISSN: 1098-3015
  Value Health
  Blackwell Publishing Inc., 350 Main Street, Malden, MA 02148, United States, Tel: +1(781)3888200, Fax: +1(781)3888210 ISSUES/YR: 6

Vasa - Journal of Vascular Diseases
ISSN: 0301-1526
Vasa J. Vasc. Dis.
Verlag Hans Huber AG, Langgassstrasse 76, CH-3000, Bern, Switzerland, Tel: +41(31)3004500, Fax: +41(31)3004590 ISSUES/YR: 4

- **Vascular**
  ISSN: 1708-5381
  Vascular
  BC Decker Inc., P.O. Box 620, LCD 1, L8N 3K7, Hamilton, Ontario, Canada, Tel: +1(905)522-7017, Fax: +1(905)522-7839 ISSUES/YR: 6

Vascular and Endovascular Surgery
ISSN: 1538-5744
Vasc. Endovasc. Surg.
Westminster Publications Inc., 708 Glen Cove Avenue, Glen Head, NY 11545, United States, Tel: +1(516)7590025, Fax: +1(516)7595524 ISSUES/YR: 6

Vascular Disease Management
ISSN: 1553-8036
Vasc. Dis. Manage.
HMP Communications, 83 General Warren Blvd. Suite 100, Malven, PA 19355, United States, Tel: +1-610 237 7285, Fax: +1(610)560 0501 ISSUES/YR: 6

- **Vascular Disease Prevention**
  ISSN: 1567-2700
  Vasc. Dis. Prev.
  Bentham Science Publishers B.V., P.O. Box 294, 1400 AG, Bussum, Netherlands, Tel: +31-356257290, Fax: +31-847511310 ISSUES/YR: 4

Vascular Health and Risk Management
ISSN: 1176-6344
Vasc. Health Risk Manage.
Dove Press Ltd., PO Box 300-008, Albany, Auckland, New Zealand, Tel: +64(9)4485032, Fax: +64(9)4485034 ISSUES/YR: 4

- **Vascular Medicine**
  ISSN: 1358-863X
  Vasc. Med.
  Arnold, 338 Euston Road, NW1 3BH, London, United Kingdom, Tel: +44(20)78736000, Fax: +44(20)78736376 ISSUES/YR: 4
- **Vascular Pharmacology**
  ISSN: 1537-1891
  Vasc. Pharmacol.
  Elsevier Inc., 360 Park Avenue South, New York, NY 10010, United States, Tel: +1(212)6333950, Fax: +1(212)6333990 ISSUES/YR: 6

Vasomed
ISSN: 0942-1181
Vasomed
Viavital Verlag, Theodor Althoff Strasse 39, 45133, Essen, Germany, Tel: +49-201-87126830, Fax: +49-201-87126840 ISSUES/YR: 6

- **Vector-Borne and Zoonotic Diseases**
  ISSN: 1530-3667
  Vector Borne Zoonotic Dis.
  Mary Ann Liebert Inc., 140 Huguenot Street, New Rochelle, NY 10801-5215, United States, Tel: +1(914)7402100, Fax: +1(914)7402101 ISSUES/YR: 4

Verdauungskrankheiten
ISSN: 0174-738X
Verdauungskrankheiten
Dustri-Verlag Dr. Karl Feistle, P.O. Box 1351, D-82032, Deisenhofen/Munchen, Germany, Tel: +49(89)6138610, Fax: +49(89)6135412 ISSUES/YR: 6

Verhaltenstherapie
ISSN: 1016-6262
Verhaltenstherapie
S. Karger AG, Allschwilerstrasse 10, P.O. Box, CH-4009, Basel, Switzerland, Tel: +41(61)3061111, Fax: +41(61)3061234 ISSUES/YR: 4

Veterinary and Human Toxicology
ISSN: 0145-6296
Vet. Hum. Toxicol.
Comparative Toxicology Laboratories, Manhattan, KS 66506-5606, United States, Tel: +1(913)5324334 ISSUES/YR: 6

Veterinary Clinics of North America - Small Animal Practice
ISSN: 0195-5616
Vet. Clin. North Am. Small Anim. Pract.
W.B. Saunders, Independence Square West, Philadelphia, PA 19106-3399, United States, Tel: +1(215)2387800, Fax: +1(215)2386445 ISSUES/YR: 6

Veterinary Immunology and Immunopathology
ISSN: 0165-2427
Vet. Immunol. Immunopathol.
Elsevier, P.O. Box 211, 1000 AE, Amsterdam, Netherlands, Tel: +31(20)4853757, Fax: +31(20)4853432 ISSUES/YR: 24

Veterinary Journal
ISSN: 1090-0233
Vet. J.
Bailliere Tindall Ltd, 32 Jamestown Road, NW1 7BY, London, United Kingdom, Tel: +44(207)4244200, Fax: +44(207)4854752 ISSUES/YR: 6

Veterinary Microbiology
ISSN: 0378-1135
Vet. Microbiol.
Elsevier, P.O. Box 211, 1000 AE, Amsterdam, Netherlands, Tel: +31(20)4853757, Fax: +31(20)4853432 ISSUES/YR: 24

Veterinary Parasitology
ISSN: 0304-4017
Vet. Parasitol.
Elsevier, P.O. Box 211, 1000 AE, Amsterdam, Netherlands, Tel: +31(20)4853757, Fax: +31(20)4853432 ISSUES/YR: 28

Veterinary Quarterly
ISSN: 0165-2176
Vet. Q.
Euroscience, P.O. Box 408, 3720 AK, Bilthoven, Netherlands, Tel: +31(30)229-3906, Fax: +31(30)228-5772 ISSUES/YR: 4

Veterinary Record
ISSN: 0042-4900
Vet. Rec.
British Veterinary Association, 7 Mansfield Street, W1G 9NQ, London, United Kingdom, Tel: +44(20)76366541, Fax: +44(20)76370620 ISSUES/YR: 52

Veterinary Research
ISSN: 0928-4249
Vet. Res.
EDP Sciences, 17 Avenue du Hoggar - BP 112, F-91944, Les Ulis Cedex A, France, Tel: +33(1)69187575, Fax: +33(1)69288491 ISSUES/YR: 6

Veterinary Research Communications
ISSN: 0165-7380
Vet. Res. Commun.
Kluwer Academic Publishers, Van Godewijckstraat 30, 3311 GZ, Dordrecht, Netherlands, Tel: +31(78)6576000, Fax: +31(78)6576254 ISSUES/YR: 8

VHSJ - Veterans Health System Journal
ISSN: 1095-2586
VHSJ Veterans Health Syst. J.
Executive Health Systems Journals Inc., 5740 Executive Drive, Suite 208, Baltimore, MD 21228, United States, Tel: +1(410)7442120, Fax: +1(410)7442150 ISSUES/YR: 11

Violence and Victims
ISSN: 0886-6708
Violence Victims
Springer Publishing Company, 11 West 42nd Street, 15th Floor, New York, NY 10036, United States, Tel: +1(212)4314370, Fax: +1(212)9417842 ISSUES/YR: 6

Viral Immunology
ISSN: 0882-8245
Viral Immunol.
Mary Ann Liebert Inc., 140 Huguenot Street, New Rochelle, NY 10801-5215, United States, Tel: +1(914)7402100, Fax: +1(914)7402101 ISSUES/YR: 4

● **Virchows Archiv**
ISSN: 0945-6317
Virchows Arch.
Springer Verlag, Tiergartenstrasse 17, D-69121, Heidelberg, Germany, Tel: +49(6221)4870, Fax: +49(6221)487366 ISSUES/YR: 12

Virologica Sinica
ISSN: 1003-5125
Virol. Sin.
Science Press, 16 Donghuangchenggen North Street, 100717, Beijing, China ISSUES/YR: 6

Virologie
ISSN: 1267-8694
Virologie
John Libbey Eurotext, 127, avenue de la Republique, 92120, Montrouge, France, Tel: +33(01)46730660, Fax: +33(01)40840999 ISSUES/YR: 6

● **Virology**
ISSN: 0042-6822
Virology
Academic Press Inc., 6277 Sea Harbor Drive, Orlando, FL 32887-4900, United States, Tel: +1(407)3454040, Fax: +1(407)3639661 ISSUES/YR: 26

Virology Journal
ISSN: 1743-422X
Virol. J.
BioMed Central Ltd., 34 - 42 Cleveland Street, W1T 4LB, London, United Kingdom, Tel: +44(207)3230323, Fax: +44(207)6319926 ISSUES/YR: 1

● **Virus Genes**
ISSN: 0920-8569
Virus Genes
Kluwer Academic Publishers, Van Godewijckstraat 30, 3311 GZ, Dordrecht, Netherlands, Tel: +31(78)6576000, Fax: +31(78)6576254 ISSUES/YR: 6

● **Virus Research**
ISSN: 0168-1702
Virus Res.
Elsevier, P.O. Box 211, 1000 AE, Amsterdam, Netherlands, Tel: +31(20)4853757, Fax: +31(20)4853432 ISSUES/YR: 12

● **Vision Research**
ISSN: 0042-6989
Vis. Res.
Elsevier Ltd, Langford Lane, Kidlington, OX5 1GB, Oxford, United Kingdom, Tel: +44(1865)843000, Fax: +44(1865)843010 ISSUES/YR: 28

● **Visual Cognition**
ISSN: 1350-6285
Vis. Cogn.
Psychology Press Ltd, 4 Park Square, Milton Park, OX14 4RN, Abingdon, Oxfordshire, United Kingdom, Tel: +44(1235)828600, Fax: +44(1235)829000 ISSUES/YR: 8

● **Visual Impairment Research**
ISSN: 1388-235X
Vis. Impairment Res.
Taylor and Francis Ltd., 4 Park Square, Milton Park, OX14 4RN, Abingdon, Oxfordshire, United Kingdom, Tel: +44(20)70175541, Fax: +44(20)70175198 ISSUES/YR: 3

● **Visual Neuroscience**
ISSN: 0952-5238
Vis. Neurosci.
Cambridge University Press, 40 West 20th Street, New York, NY 10011-4211, United States, Tel: +1(212)9243900, Fax: +1(212)6913239 ISSUES/YR: 6

Viszeralchirurgie
ISSN: 1435-3067
Viszeralchirurgie
Georg Thieme Verlag, Rudigerstrasse 14, D-70469, Stuttgart, Germany, Tel: +49(711)8931-321, Fax: +49(711)8931-422 ISSUES/YR: 6

Vnitrni Lekarstvi
ISSN: 0042-773X
Vnitr. Lek.
Czech Medical Association J.E. Purkyne, Sokolska 31, P.O. Box 88, 120 26, Praha 2, Czech Republic, Tel: +420(2)24266201-4, Fax: +420(2)24266212 ISSUES/YR: 12

● **Vox Sanguinis**
ISSN: 0042-9007
Vox Sang.
Blackwell Publishing Ltd, 9600 Garsington Road, OX4 2XG, Oxford, United Kingdom, Tel: +44(1865)776868, Fax: +44(1865)714591 ISSUES/YR: 8

● **Vox Sanguinis, Supplement**
ISSN: 1741-6892
Vox Sang. Suppl.
Blackwell Publishing Ltd, 9600 Garsington Road, OX4 2XG, Oxford, United Kingdom, Tel: +44(1865)776868, Fax: +44(1865)714591 ISSUES/YR: 1

● **Waste Management**
ISSN: 0956-053X
Waste Manage.
Elsevier Ltd, Langford Lane, Kidlington, OX5 1GB, Oxford, United Kingdom, Tel: +44(1865)843000, Fax: +44(1865)843010 ISSUES/YR: 10

● **Waste Management and Research**
ISSN: 0734-242X
Waste Manage. Res.
SAGE Publications Ltd, 55 City Road, EC1Y 1SP, London, United Kingdom, Tel: +44(207)3248500, Fax: +44(207)3248600 ISSUES/YR: 6

● **Water Environment Research**
ISSN: 1061-4303
Water Environ. Res.
Water Environment Federation, 601 Wythe Street, Alexandria, VA 22314-1994, United States, Tel: +1(703)6842400, Fax: +1(703)6842492 ISSUES/YR: 6

● **Water Research**
ISSN: 0043-1354
Water Res.
Elsevier Ltd, Langford Lane, Kidlington, OX5 1GB, Oxford, United Kingdom, Tel: +44(1865)843000, Fax: +44(1865)843010 ISSUES/YR: 20

Water Science and Technology
ISSN: 0273-1223
Water Sci. Technol.
IWA Publishing, 12 Caxton Street, SW1H 0QS, London, United Kingdom, Tel: +44(207)6545500, Fax: +44(207)6545555 ISSUES/YR: 24

Current

Water Science and Technology: Water Supply
ISSN: 1606-9749
Water Sci. Technol. Water Supply
IWA Publishing, 12 Caxton Street, SW1H 0QS, London, United Kingdom, Tel: +44(207)6545500, Fax: +44(207)6545555 ISSUES/YR: 6
Water, Air, and Soil Pollution
ISSN: 0049-6979
Water Air Soil Pollut.
Kluwer Academic Publishers, Van Godewijckstraat 30, 3311 GZ, Dordrecht, Netherlands, Tel: +31(78)6576000, Fax: +31(78)6576254 ISSUES/YR: 32
West African Journal of Medicine
ISSN: 0189-160X
West Afr. J. Med.
West African Journal of Medicine, 6 Taylor Drive, P.M.B. 2023, Yaba, Lagos, Nigeria, Tel: +234-4710872 ISSUES/YR: 4
West Indian Medical Journal
ISSN: 0043-3144
West Indian Med. J.
University of the West Indies, Mona, Kingston 7, Jamaica, Tel: +1809(809)9271214 ISSUES/YR: 6
WHO Drug Information
ISSN: 1010-9609
WHO Drug Inf.
World Health Organization, 20 Ave. Appia, CH-1211, Geneva 27, Switzerland, Tel: +41(22)7912111, Fax: +41(22)7914857 ISSUES/YR: 4
Wiener Klinische Wochenschrift
ISSN: 0043-5325
Wien. Klin. Wochenschr.
Springer-Verlag Wien, Sachsenplatz 4-6, P.O. Box 89, A-1201, Vienna, Austria, Tel: +43(1)3302415, Fax: +43(1)3302426 ISSUES/YR: 24
Wiener Klinische Wochenschrift, Supplement
ISSN: 0300-5178
Wien. Klin. Wochenschr. Suppl.
Springer-Verlag Wien, Sachsenplatz 4-6, P.O. Box 89, A-1201, Vienna, Austria, Tel: +43(1)3302415, Fax: +43(1)3302426 ISSUES/YR: 1
Wiener Medizinische Wochenschrift
ISSN: 0043-5341
Wien. Med. Wochenschr.
Springer-Verlag Wien, Sachsenplatz 4-6, P.O. Box 89, A-1201, Vienna, Austria, Tel: +43(1)3302415, Fax: +43(1)3302426 ISSUES/YR: 24
Wilderness and Environmental Medicine
ISSN: 1080-6032
Wilderness Environ. Med.
Allen Press Inc., P.O. Box 7075, Lawrence, KS 66044-8897, United States, Tel: +1(785)8431234, Fax: +1(785)8431244 ISSUES/YR: 4
Wisconsin Medical Journal
ISSN: 1098-1861
Wisc. Med. J.
State Medical Society of Wisconsin, 330 E. Lakeside, P.O. Box 1109, Madison, WI 53701, United States, Tel: +1(608)2576781 ISSUES/YR: 8
Women and Health
ISSN: 0363-0242
Women Health
Haworth Press Inc., 10 Alice Street, Binghamton, NY 13904-1580, United States, Tel: +1(607)7225857 ISSUES/YR: 8
Women's Health
ISSN: 1745-5057
Women's Health
Future Medicine Ltd., 2nd Albert Place, Finchley Central, N3 1QB, London, United Kingdom, Tel: +44-208 343 3883, Fax: +44-208 343 2313 ISSUES/YR: 4
Women's Health Issues
ISSN: 1049-3867
Women's Health Issues
Elsevier USA, 6277 Sea Harbor Drive, Orlando, FL 32862 8239, United States, Tel: +1-407 345 4000, Fax: +1-407 363 9661 ISSUES/YR: 6
Women's Oncology Review
ISSN: 1473-3404
Women's Oncol. Rev.
Taylor and Francis Ltd., 4 Park Square, Milton Park, OX14 4RN, Abingdon, Oxfordshire, United Kingdom, Tel: +44(20)70175541, Fax: +44(20)70175198 ISSUES/YR: 4
Work
ISSN: 1051-9815
Work
IOS Press, Nieuwe Hemweg 6B, 1013 BG, Amsterdam, Netherlands, Tel: +31(20)6883355, Fax: +31(20)6203419 ISSUES/YR: 6
Work and Stress
ISSN: 0267-8373
Work Stress
Taylor and Francis Ltd., 4 Park Square, Milton Park, OX14 4RN, Abingdon, Oxfordshire, United Kingdom, Tel: +44(20)70175541, Fax: +44(20)70175198 ISSUES/YR: 4
Work Based Learning in Primary Care
ISSN: 1740-3715
Work Based Learn. Prim. Care
Radcliffe Publishing Ltd, 18 Marcham Road, OX14 1AA, Abingdon, Oxon, United Kingdom, Tel: +44(1235)528820, Fax: +44(1235)528830 ISSUES/YR: 4
World Chinese Journal of Digestology
ISSN: 1009-3079
World Chin. J. Dig.
WJC Press, 77 Shuangta Xijie, 030001, Taiyuan, Shanxi Province, China ISSUES/YR: 24
World Health Organization - Technical Report Series
ISSN: 0512-3054
WHO Tech. Rep. Ser.
World Health Organization, 20 Ave. Appia, CH-1211, Geneva 27, Switzerland, Tel: +41(22)7912111, Fax: +41(22)7914857 ISSUES/YR: 1
World Journal of Biological Psychiatry
ISSN: 1562-2975
World J. Biol. Psychiatry
Taylor and Francis A.S., P.O. Box 2562 Solli, 0202, Oslo, Norway, Tel: +47(22)129880, Fax: +47(22)129890 ISSUES/YR: 4
World Journal of Gastroenterology
ISSN: 1007-9327
World J. Gastroenterol.
WJG Press, P.O. Box 2345, 100023, Beijing, China, Tel: +86(10)65891893, Fax: +86(10)65891893 ISSUES/YR: 24
World Journal of Surgery
ISSN: 0364-2313
World J. Surg.
Springer New York LLC, 233 Springer Street, New York, 10013-1578, United States, Tel: +1-212 460 1501, Fax: +1-212 460 1595 ISSUES/YR: 12
World Journal of Surgical Oncology
ISSN: 1477-7819
World J. Surg. Oncol.
BioMed Central Ltd., 34 - 42 Cleveland Street, W1T 4LB, London, United Kingdom, Tel: +44(207)3230323, Fax: +44(207)6319926 ISSUES/YR: 1
World Wide Wounds
ISSN: 1369-2607
World Wide Wounds
Surgical Materials Testing Laboratory, Coity Road, Bridgend, CF31 1RQ, South Wales, United Kingdom, Tel: +44(1656)752820, Fax: +44(1656)752830 ISSUES/YR: 1
Worldviews on Evidence-Based Nursing
ISSN: 1545-102X
Worldviews Evid.-Based Nurs.
Blackwell Publishing Ltd, 9600 Garsington Road, OX4 2XG, Oxford, United Kingdom, Tel: +44(1865)776868, Fax: +44(1865)714591 ISSUES/YR: 4
Wound Ballistics Review
ISSN: 1055-0305
Wound Ballist. Rev.
International Wound Ballistics Association, P.O. Box 701, El Segundo, CA 90245, United States, Tel: +1(310)6406065 ISSUES/YR: 4

● **Wound Repair and Regeneration**
ISSN: 1067-1927
Wound Repair Regen.
Blackwell Publishing Inc., 350 Main Street, Malden, MA 02148, United States, Tel: +1(781)3888200, Fax: +1(781)3888210 ISSUES/YR: 6

Wounds UK
ISSN: 1746-6814
Wounds UK
Wounds UK, 36 Upperkirkgate 36, Suite 3.1, AB10 1BA, Aberdeen, United Kingdom, Tel: +44-01224637371 ISSUES/YR: 4

Wspolczesna Onkologia
ISSN: 1428-2526
Wspolczesna Onkol.
Termedia Publishing House Ltd., ul Kleberga 8,, 61-615, Poznan, Poland, Tel: +48-61 822 77 81, Fax: +48-61 822 7781 ISSUES/YR: 6

Xenobiotica
ISSN: 0049-8254
Xenobiotica
Taylor and Francis Ltd., 4 Park Square, Milton Park, OX14 4RN, Abingdon, Oxfordshire, United Kingdom, Tel: +44(20)70175541, Fax: +44(20)70175198 ISSUES/YR: 12

● **Xenotransplantation**
ISSN: 0908-665X
Xenotransplantation
Blackwell Publishing Ltd, 9600 Garsington Road, OX4 2XG, Oxford, United Kingdom, Tel: +44(1865)776868, Fax: +44(1865)714591 ISSUES/YR: 6

Yakhteh
ISSN: 1561-4921
Yakhteh
Royan Institute, P.O. Box 19395-4644, Tehran, Iran, Tel: +98(21)2413790, Fax: +98(21)2409314 ISSUES/YR: 4

Yakugaku Zasshi
ISSN: 0031-6903
Yakugaku Zasshi
Pharmaceutical Society of Japan, 2-12-15-201, Shibuya, Shibuya-ku, 150, Tokyo, Japan, Tel: +81(3)34063321, Fax: +81(3)34981835 ISSUES/YR: 12

Yale Journal of Biology and Medicine
ISSN: 0044-0086
Yale J. Biol. Med.
Yale Journal of Biology and Medicine Inc., 333 Cedar Street, P.O. Box 208000, New Haven, CT 06510-8000, United States, Tel: +1(203)7854251, Fax: +1(203)7854251 ISSUES/YR: 1

● **Yeast**
ISSN: 0749-503X
Yeast
John Wiley and Sons Ltd, Southern Gate, PO19 8SQ, Chichester, West Sussex, United Kingdom, Tel: +44(1243)779777, Fax: +44(1243)775878 ISSUES/YR: 16

Yeni Symposium
ISSN: 1300-8773
Yeni Symp.
Istanbul Universitesi, Aksaray, 34303, Istanbul, Turkey, Tel: +90(212)5884800 ISSUES/YR: 4

Yokohama Medical Journal
ISSN: 0372-7726
Yokohama Med. J.
Medical Association of Yokohama City University, 3-9 Fukuura, Kanazawa-ku, 236, Yokohama, Japan, Tel: +81(45)7872556 ISSUES/YR: 6

Yonago Acta Medica
ISSN: 0513-5710
Yonago Acta Med.
Tottori University Faculty of Medicine, 86 Nishimachi, 683, Yonago, Tottori, Japan, Tel: +81(859)348053 ISSUES/YR: 4

Yonsei Medical Journal
ISSN: 0513-5796
Yonsei Med. J.
Yonsei University College of Medicine, Shinchon dong 134, Seodaemoon ku, 120-752, Seoul, South Korea, Tel: +82(2)3925061, Fax: +82(2)3934945 ISSUES/YR: 6

Z Zagadnien Nauk Sadowych
ISSN: 1230-7483
Z Zagadnien Nauk Sadowych
Instytut Ekspertyz Sadowych, Ul. Westerplatte 9, 31-033, Krakow, Poland, Tel: +48(12)64228755, Fax: +48(12)64223850 ISSUES/YR: 1

Zeitschrift fur Allgemeinmedizin
ISSN: 0937-6801
Z. Allg.med.
Georg Thieme Verlag, Rudigerstrasse 14, D-70469, Stuttgart, Germany, Tel: +49(711)8931-321, Fax: +49(711)8931-422 ISSUES/YR: 12

Zeitschrift fur Arztliche Fortbildung und Qualitatssicherung
ISSN: 1431-7621
Z. Arztl. Fortbild. Qual.Sich.
Urban und Fischer Verlag Jena, P.O. Box 100537, 07705, Jena, Germany, Tel: +49(3641)6263, Fax: +49(3641)626500 ISSUES/YR: 10

Zeitschrift fur Epileptologie
ISSN: 1617-6782
Z. Epileptologie
Dr. Dietrich Steinkopff Verlag GmbH and Co. KG, P.O. Box 100462, D-64204, Darmstadt, Germany, Tel: +49(6151)828990, Fax: +49(6151)8289940 ISSUES/YR: 4

Zeitschrift fur Gastroenterologie
ISSN: 0044-2771
Z. Gastroenterol.
Karl Demeter Verlag GmbH, Rudigerstr. 14, D-70469, Stuttgart, Germany, Tel: +49(0711)89310, Fax: +49(0711)8931536 ISSUES/YR: 12

Zeitschrift fur Geburtshilfe und Neonatologie
ISSN: 0948-2393
Z. Geburtshilfe Neonatol.
Georg Thieme Verlag, Rudigerstrasse 14, D-70469, Stuttgart, Germany, Tel: +49(711)8931-321, Fax: +49(711)8931-422 ISSUES/YR: 6

Zeitschrift fur Geburtshilfe und Neonatologie, Supplement
ISSN: 1615-5300
Z. Geburtshilfe Neonatol.
Georg Thieme Verlag, Rudigerstrasse 14, D-70469, Stuttgart, Germany, Tel: +49(711)8931-321, Fax: +49(711)8931-422 ISSUES/YR: 1

Zeitschrift fur Gefassmedizin
ISSN: 1812-9501
Z. Gefassmed.
Krause und Pachernegg GmbH, Mozartgasse 10, A-3003, Gablitz, Austria, Tel: +43(2231)612580, Fax: +43(2231)6125810 ISSUES/YR: 4

Zeitschrift fur Gerontologie und Geriatrie
ISSN: 0948-6704
Z. Gerontol. Geriatr.
Dr. Dietrich Steinkopff Verlag GmbH and Co. KG, P.O. Box 100462, D-64204, Darmstadt, Germany, Tel: +49(6151)828990, Fax: +49(6151)8289940 ISSUES/YR: 6

Zeitschrift fur Gerontopsychologie und -psychiatrie
ISSN: 1011-6877
Z. Gerontopsychol. -psychiatr.
Verlag Hans Huber AG, Langgassstrasse 76, CH-3000, Bern, Switzerland, Tel: +41(31)3004500, Fax: +41(31)3004590 ISSUES/YR: 4

Zeitschrift fur Herz-, Thorax- und Gefasschirurgie
ISSN: 0930-9225
Z. Herz- Thorax- Gefasschir.
Dr. Dietrich Steinkopff Verlag GmbH and Co. KG, P.O. Box 100462, D-64204, Darmstadt, Germany, Tel: +49(6151)828990, Fax: +49(6151)8289940 ISSUES/YR: 6

Zeitschrift fur Kinder- und Jugendpsychiatrie und Psychotherapie
ISSN: 1422-4917
Z. Kinder- Jugendpsychiatr. Psychother.
Verlag Hans Huber AG, Langgassstrasse 76, CH-3000, Bern, Switzerland, Tel: +41(31)3004500, Fax: +41(31)3004590 ISSUES/YR: 4

Zeitschrift fur Klinische Psychologie und Psychotherapie
ISSN: 1616-3443
Z. Klin. Psychol. Psychother.
Hogrefe Verlag GmbH and Co. KG, Rohnsweg 25, D-37085, Gottingen, Germany, Tel: +49(551)496090, Fax: +49(551)4960988 ISSUES/YR: 4

Zeitschrift fur Larmbekampfung
ISSN: 0174-1098
Z. Larmbekampf.
Springer-VDI Verlag GmbH und Co. KG, Postfach 101022, 40001, Dusseldorf, Germany, Tel: +49(211)61030, Fax: +49-211 6103 414 ISSUES/YR: 6

Zeitschrift fur Medizinische Physik
ISSN: 0939-3889
Z. Med. Physik
Urban und Fischer Verlag Jena, P.O. Box 100537, 07705, Jena, Germany, Tel: +49(3641)6263, Fax: +49(3641)626500 ISSUES/YR: 4

Zeitschrift fur Orthopadie und Ihre Grenzgebiete
ISSN: 0044-3220
Z. Orthop. Ihre Grenzgeb.
Ferdinand Enke Verlag, Bludenzer Strasse 6, D-70469, Stuttgart, Germany, Tel: +49(711)1357980 ISSUES/YR: 6

Zeitschrift fur Phytotherapie
ISSN: 0722-348X
Z. Phytother.
Hippokrates Verlag GmbH, Rudigerstrasse 14, D-70469, Stuttgart, Germany, Tel: +49(711)89310, Fax: +49(711)8931453 ISSUES/YR: 6

Zeitschrift fur Psychiatrie, Psychologie und Psychotherapie
ISSN: 1661-4747
Z. Psych. Psychol. Psychother.
Verlag Hans Huber AG, Langgassstrasse 76, CH-3000, Bern, Switzerland, Tel: +41(31)3004500, Fax: +41(31)3004590 ISSUES/YR: 4

Zeitschrift fur Psychosomatische Medizin und Psychotherapie
ISSN: 1438-3608
Z. Psychosom. Med. Psychother.
Vandenhoeck and Ruprecht GmbH and Co. KG, Theaterstrasse 13, D-37073, Gottingen, Germany, Tel: +49(551)547820, Fax: +49(551)695917 ISSUES/YR: 4

Zeitschrift fur Rheumatologie
ISSN: 0340-1855
Z. Rheumatol.
Dr. Dietrich Steinkopff Verlag GmbH and Co. KG, P.O. Box 100462, D-64204, Darmstadt, Germany, Tel: +49(6151)828990, Fax: +49(6151)8289940 ISSUES/YR: 6

Zeitschrift fur Rheumatologie, Supplement
ISSN: 0941-8466
Z. Rheumatol. Suppl.
Dr. Dietrich Steinkopff Verlag GmbH and Co. KG, P.O. Box 100462, D-64204, Darmstadt, Germany, Tel: +49(6151)828990, Fax: +49(6151)8289940 ISSUES/YR: 1

Zeitschrift fur Umweltmedizin
ISSN: 1436-3208
Z. Umweltmed.
Promedico Verlag, P.O. Box 67 03 06, 22343, Hamburg, Germany, Tel: +49(40)6091540, Fax: +49(40)60915444 ISSUES/YR: 4

Zeitschrift fur Wundheilung
ISSN: 1439-670X
Z. Wundheilung
mhp-Verlag GmbH, Markplatz 13, D-65183, Wiesbaden, Germany, Tel: +49(611)505930, Fax: +49(611)5059311 ISSUES/YR: 6

Zentralblatt fur Arbeitsmedizin, Arbeitsschutz und Ergonomie
ISSN: 0944-2502
Zentralbl. Arbeitsmed. Arbeitssch. Ergonomie
Dr. Curt Haefner Verlag GmbH, Bachstrasse 14-16, D-69121, Heidelberg, Germany, Tel: +49(6221)49063 ISSUES/YR: 12

Zentralblatt fur Chirurgie
ISSN: 0044-409X
Zentralbl. Chir.
MVS Medizinverlage Stuttgart, Steiermarkerstr. 3-5, D-70469, Stuttgart, Germany, Tel: +49(711)8931-0, Fax: +49(711)8931706 ISSUES/YR: 6

Zentralblatt fur Chirurgie, Supplement
ISSN: 0233-1616
Zentralbl. Chir. Suppl.
MVS Medizinverlage Stuttgart, Steiermarkerstr. 3-5, D-70469, Stuttgart, Germany, Tel: +49(711)8931-0, Fax: +49(711)8931706 ISSUES/YR: 1

Zentralblatt fur Gynakologie
ISSN: 0044-4197
Zentralbl. Gynakol.
MVS Medizinverlage Stuttgart, Steiermarkerstr. 3-5, D-70469, Stuttgart, Germany, Tel: +49(711)8931-0, Fax: +49(711)8931706 ISSUES/YR: 6

Zentralblatt fur Neurochirurgie
ISSN: 0044-4251
Zentralbl. Neurochir.
MVS Medizinverlage Stuttgart, Steiermarkerstr. 3-5, D-70469, Stuttgart, Germany, Tel: +49(711)8931-0, Fax: +49(711)8931706 ISSUES/YR: 4

Zentralsterilisation - Central Service
ISSN: 0942-6086
Zentralsterilisation Cent. Serv.
mhp-Verlag GmbH, Markplatz 13, D-65183, Wiesbaden, Germany, Tel: +49(611)505930, Fax: +49(611)5059311 ISSUES/YR: 6

LIST OF PRIORITY JOURNALS

Abdominal Imaging
ISSN: 0942-8925

Academic Emergency Medicine
ISSN: 1069-6563

Academic Medicine
ISSN: 1040-2446

Academic Radiology
ISSN: 1076-6332

Acta Anaesthesiologica Scandinavica
ISSN: 0001-5172

Acta Anaesthesiologica Scandinavica, Supplement
ISSN: 0515-2720

Acta Biomaterialia
ISSN: 1742-7061

Acta Cytologica
ISSN: 0001-5547

Acta Diabetologica
ISSN: 0940-5429

Acta Haematologica
ISSN: 0001-5792

Acta Neurochirurgica
ISSN: 0001-6268

Acta Neuropathologica
ISSN: 0001-6322

Acta Obstetricia et Gynecologica Scandinavica
ISSN: 0001-6349

Acta Oncologica
ISSN: 0284-186X

Acta Oncologica, Supplement
ISSN: 1100-1704

Acta Ophthalmologica Scandinavica
ISSN: 1395-3907

Acta Ophthalmologica Scandinavica, Supplement
ISSN: 1395-3931

Acta Oto-Laryngologica
ISSN: 0001-6489

Acta Oto-Laryngologica, Supplement
ISSN: 0365-5237

Acta Paediatrica, International Journal of Paediatrics
ISSN: 0803-5253

Acta Paediatrica, International Journal of Paediatrics, Supplement
ISSN: 0803-5326

Acta Physiologica
ISSN: 1748-1708

Acta Psychiatrica Scandinavica
ISSN: 0001-690X

Acute Cardiac Care
ISSN: 1748-2941

Acute Pain
ISSN: 1366-0071

Addiction Biology
ISSN: 1355-6215

Addictive Disorders and their Treatment
ISSN: 1531-5754

Administration and Policy in Mental Health
ISSN: 0894-587X

Advanced Drug Delivery Reviews
ISSN: 0169-409X

Advances in Anatomic Pathology
ISSN: 1072-4109

Advances in Biophysics
ISSN: 0065-227X

Advances in Cancer Research
ISSN: 0065-230X

Advances in Carbohydrate Chemistry and Biochemistry
ISSN: 0065-2318

Advances in Chronic Kidney Disease
ISSN: 1548-5595

Advances in Environmental Research
ISSN: 1093-0191

Advances in Experimental Medicine and Biology
ISSN: 0065-2598

Advances in Immunology
ISSN: 0065-2776

Advances in Parasitology
ISSN: 0065-308X

Advances in Pharmacy
ISSN: 1532-3862

Advances in Protein Chemistry
ISSN: 0065-3233

Aerobiologia
ISSN: 0393-5965

Aerosol Science and Technology
ISSN: 0278-6826

Aesthetic Plastic Surgery
ISSN: 0364-216X

Aesthetic Surgery Journal
ISSN: 1090-820X

Age and Ageing
ISSN: 0002-0729

Aggression and Violent Behavior
ISSN: 1359-1789

Aging and Mental Health
ISSN: 1360-7863

Aging Cell
ISSN: 1474-9718

Aging Male
ISSN: 1368-5538

AIDS
ISSN: 0269-9370

AIDS Care - Psychological and Socio-Medical Aspects of AIDS/HIV
ISSN: 0954-0121

AIDS Research and Human Retroviruses
ISSN: 0889-2229

AIDS Research and Therapy
ISSN: 1742-6405

Air Medical Journal
ISSN: 1067-991X

Alcohol and Alcoholism
ISSN: 0735-0414

Alcoholism: Clinical and Experimental Research
ISSN: 0145-6008

Alimentary Pharmacology and Therapeutics
ISSN: 0269-2813

Alimentary Pharmacology and Therapeutics Symposium Series
ISSN: 1746-6334

Allergology International
ISSN: 1323-8930

Allergy: European Journal of Allergy and Clinical Immunology
ISSN: 0105-4538

Allergy: European Journal of Allergy and Clinical Immunology, Supplement
ISSN: 0108-1675

Alzheimer Disease and Associated Disorders
ISSN: 0893-0341

Alzheimer's and Dementia
ISSN: 1552-5260

American Heart Journal
ISSN: 0002-8703

American Journal Geriatric Pharmacotherapy
ISSN: 1543-5946

American Journal of Cancer
ISSN: 1175-6357

American Journal of Cardiology
ISSN: 0002-9149

American Journal of Cardiovascular Drugs
ISSN: 1175-3277

American Journal of Clinical Dermatology
ISSN: 1175-0561

American Journal of Clinical Pathology
ISSN: 0002-9173

American Journal of Dermatopathology
ISSN: 0193-1091

American Journal of Drug Delivery
ISSN: 1175-9038

American Journal of Emergency Medicine
ISSN: 0735-6757

American Journal of Gastroenterology
ISSN: 0002-9270

American Journal of Health-System Pharmacy
ISSN: 1079-2082

American Journal of Hematology
ISSN: 0361-8609

American Journal of Human Biology
ISSN: 1042-0533

American Journal of Human Genetics
ISSN: 0002-9297

American Journal of Hypertension
ISSN: 0895-7061

American Journal of Managed Care
ISSN: 1088-0224

American Journal of Medical Genetics
ISSN: 1552-4825

American Journal of Medical Genetics - Neuropsychiatric Genetics
ISSN: 1552-4841

American Journal of Medical Genetics - Seminars in Medical Genetics
ISSN: 1552-4868

American Journal of Medicine
ISSN: 0002-9343

American Journal of Nephrology
ISSN: 0250-8095

American Journal of Obstetrics and Gynecology
ISSN: 0002-9378

American Journal of Ophthalmology
ISSN: 0002-9394

American Journal of Otolaryngology - Head and Neck Medicine and Surgery
ISSN: 0196-0709

American Journal of Pathology
ISSN: 0002-9440

American Journal of Perinatology
ISSN: 0735-1631

American Journal of Physiology - Cell Physiology
ISSN: 0363-6143

American Journal of Physiology - Endocrinology and Metabolism
ISSN: 0193-1849

American Journal of Physiology - Gastrointestinal and Liver Physiology
ISSN: 0193-1857

American Journal of Physiology - Heart and Circulatory Physiology
ISSN: 0363-6135

American Journal of Physiology - Lung Cellular and Molecular Physiology
ISSN: 1040-0605

American Journal of Physiology - Regulatory Integrative and Comparative Physiology
ISSN: 0363-6119

American Journal of Physiology - Renal Physiology
ISSN: 0363-6127

American Journal of Psychiatric Rehabilitation
ISSN: 1548-7768

American Journal of Psychiatry
ISSN: 0002-953X

American Journal of Reproductive Immunology
ISSN: 1046-7408

American Journal of Respiratory and Critical Care Medicine
ISSN: 1073-449X

American Journal of Roentgenology
ISSN: 0361-803X

American Journal of Sports Medicine
ISSN: 0363-5465

American Journal of Surgery
ISSN: 0002-9610

American Journal of Therapeutics
ISSN: 1075-2765

American Journal of Transplantation
ISSN: 1600-6135

Amino Acids
ISSN: 0939-4451

Amyloid
ISSN: 1350-6129

Amyotrophic Lateral Sclerosis and Other Motor Neuron Disorders
ISSN: 1466-0822

Anaerobe
ISSN: 1075-9964

Analytica Chimica Acta
ISSN: 0003-2670

Analytical and Quantitative Cytology and Histology
ISSN: 0884-6812

Analytical Biochemistry
ISSN: 0003-2697

Anatomical Record - Part A Discoveries in Molecular, Cellular, and Evolutionary Biology
ISSN: 0003-276X

Anatomical Record - Part B New Anatomist
ISSN: 1552-4906

Anatomical Science International
ISSN: 1447-6959

Anatomy and Embryology
ISSN: 0340-2061

Anesthesia and Analgesia
ISSN: 0003-2999

Anesthesiology
ISSN: 0003-3022

Angiogenesis
ISSN: 0969-6970

Annals of Allergy, Asthma and Immunology
ISSN: 1081-1206

Annals of Clinical and Laboratory Science
ISSN: 0091-7370

Annals of Clinical Biochemistry
ISSN: 0004-5632

Annals of Clinical Psychiatry
ISSN: 1040-1237

Annals of Diagnostic Pathology
ISSN: 1092-9134

Annals of Emergency Medicine
ISSN: 0196-0644

Annals of Epidemiology
ISSN: 1047-2797

Annals of Hematology
ISSN: 0939-5555

Annals of Hematology, Supplement
ISSN: 0945-8077

Annals of Human Genetics
ISSN: 0003-4800

Annals of Internal Medicine
ISSN: 0003-4819

Annals of Medicine
ISSN: 0785-3890

Annals of Neurology
ISSN: 0364-5134

Annals of Noninvasive Electrocardiology
ISSN: 1082-720X

Annals of Nuclear Medicine
ISSN: 0914-7187

Annals of Nutrition and Metabolism
ISSN: 0250-6807

Annals of Occupational Hygiene
ISSN: 0003-4878

Annals of Oncology
ISSN: 0923-7534

Annals of Otology, Rhinology and Laryngology
ISSN: 0003-4894

Annals of Pharmacotherapy
ISSN: 1060-0280

Annals of Plastic Surgery
ISSN: 0148-7043

Annals of Saudi Medicine
ISSN: 1319-9226

Annals of Surgery
ISSN: 0003-4932

Annals of the Rheumatic Diseases
ISSN: 0003-4967

Annals of Thoracic Surgery
ISSN: 0003-4975

Annals of Vascular Surgery
ISSN: 0890-5096

Annual Review of Biochemistry
ISSN: 0066-4154

Annual Review of Biophysics and Biomolecular Structure
ISSN: 1056-8700

Annual Review of Cell and Developmental Biology
ISSN: 1081-0706

Annual Review of Genetics
ISSN: 0066-4197

Annual Review of Immunology
ISSN: 0732-0582

Annual Review of Medicine
ISSN: 0066-4219

Annual Review of Microbiology
ISSN: 0066-4227

Annual Review of Neuroscience
ISSN: 0147-006X

Annual Review of Nutrition
ISSN: 0199-9885

Annual Review of Pathology
ISSN: 0066-4286

Annual Review of Pharmacology and Toxicology
ISSN: 0362-1642

Annual Review of Physiology
ISSN: 0066-4278

Annual Review of Public Health
ISSN: 0163-7525

Anthropology and Medicine
ISSN: 1364-8470

Anti-Cancer Drugs
ISSN: 0959-4973

Anticancer Research
ISSN: 0250-7005

Antimicrobial Agents and Chemotherapy
ISSN: 0066-4804

Antioxidants and Redox Signaling
ISSN: 1523-0864

Antiviral Chemistry and Chemotherapy
ISSN: 0956-3202

Antiviral Research
ISSN: 0166-3542

Antiviral Therapy
ISSN: 1359-6535

Antonie van Leeuwenhoek, International Journal of General and Molecular Microbiology
ISSN: 0003-6072

APLAR Journal of Rheumatology
ISSN: 0219-0494

APMIS
ISSN: 0903-4641

APMIS, Supplement
ISSN: 0903-465X

Apoptosis
ISSN: 1360-8185

Applied Cardiopulmonary Pathophysiology
ISSN: 0920-5268

Applied Cognitive Psychology
ISSN: 0888-4080

Applied Immunohistochemistry and Molecular Morphology
ISSN: 1541-2016

Applied Radiation and Isotopes
ISSN: 0969-8043

Aquatic Toxicology
ISSN: 0166-445X

Archiv der Pharmazie
ISSN: 0365-6233

Archives of Andrology
ISSN: 0148-5016

Archives of Biochemistry and Biophysics
ISSN: 0003-9861

Archives of Dermatological Research
ISSN: 0340-3696

Archives of Dermatological Research, Supplement
ISSN: 0944-1948

Archives of Dermatology
ISSN: 0003-987X

Archives of Disease in Childhood
ISSN: 0003-9888

Archives of Disease in Childhood: Education and Practice Edition
ISSN: 1743-0585

Archives of Disease in Childhood: Fetal and Neonatal Edition
ISSN: 1359-2998

Archives of Environmental Contamination and Toxicology
ISSN: 0090-4341

Archives of Environmental Health
ISSN: 0003-9896

Archives of Gerontology and Geriatrics
ISSN: 0167-4943

Archives of Internal Medicine
ISSN: 0003-9926

Archives of Microbiology
ISSN: 0302-8933

Archives of Neurology
ISSN: 0003-9942

Archives of Ophthalmology
ISSN: 0003-9950

Archives of Orthopaedic and Trauma Surgery
ISSN: 0936-8051

Archives of Pediatrics and Adolescent Medicine
ISSN: 1072-4710

Archives of Surgery
ISSN: 0004-0010

Archives of Toxicology
ISSN: 0340-5761

Archives of Women's Mental Health
ISSN: 1434-1816

Arteriosclerosis, Thrombosis, and Vascular Biology
ISSN: 1079-5642

Arthritis and Rheumatism
ISSN: 0004-3591

Artificial Intelligence in Medicine
ISSN: 0933-3657

Artificial Organs
ISSN: 0160-564X

Arts in Psychotherapy
ISSN: 0197-4556

Asia Pacific Journal of Environmental Law
ISSN: 1385-2140

Asia-Pacific Journal of Clinical Oncology
ISSN: 1743-7555

Atherosclerosis
ISSN: 0021-9150

Atherosclerosis Supplements
ISSN: 1567-5688

ATLA Alternatives to Laboratory Animals
ISSN: 0261-1929

Atmospheric Environment
ISSN: 1352-2310

Attachment and Human Development
ISSN: 1461-6734

Audiological Medicine
ISSN: 1651-386X

Audiology and Neurotology
ISSN: 1420-3030

Australasian Radiology
ISSN: 0004-8461

Australia and New Zealand Health Policy
ISSN: 1743-8462

Australian and New Zealand Journal of Obstetrics and Gynaecology
ISSN: 0004-8666

Australian Journal of Rural Health
ISSN: 1038-5282

Autism
ISSN: 1362-3613

Autonomic Neuroscience: Basic and Clinical
ISSN: 1566-0702

Basic and Clinical Pharmacology and Toxicology
ISSN: 1742-7835

Basic and Clinical Pharmacology and Toxicology, Supplement
ISSN: 0901-9928

Behavioral and Brain Functions
ISSN: 1744-9081

Behavioural and Cognitive Psychotherapy
ISSN: 1352-4658

Behavioural Brain Research
ISSN: 0166-4328

Behavioural Neurology
ISSN: 0953-4180

Best Practice and Research in Clinical Anaesthesiology
ISSN: 1521-6896

Best Practice and Research in Clinical Haematology
ISSN: 1521-6926

Best Practice and Research in Clinical Obstetrics and Gynaecology
ISSN: 1521-6934

Best Practice and Research in Clinical Rheumatology
ISSN: 1521-6942

Binocular Vision and Strabismus Quarterly
ISSN: 1088-6281

Bio-Medical Materials and Engineering
ISSN: 0959-2989

Biochemical and Biophysical Research Communications
ISSN: 0006-291X

Biochemical Engineering Journal
ISSN: 1369-703X

Biochemical Journal
ISSN: 0264-6021

Biochemical Pharmacology
ISSN: 0006-2952

Biochemical Society Transactions
ISSN: 0300-5127

Biochemistry
ISSN: 0006-2960

Biochimica et Biophysica Acta - Bioenergetics
ISSN: 0005-2728

Biochimica et Biophysica Acta - Biomembranes
ISSN: 0005-2736

Biochimica et Biophysica Acta - Gene Structure and Expression
ISSN: 0167-4781

Biochimica et Biophysica Acta - General Subjects
ISSN: 0304-4165

Biochimica et Biophysica Acta - Molecular and Cell Biology of Lipids
ISSN: 1388-1981

Biochimica et Biophysica Acta - Molecular Basis of Disease
ISSN: 0925-4439

Biochimica et Biophysica Acta - Molecular Cell Research
ISSN: 0167-4889

Biochimica et Biophysica Acta - Proteins and Proteomics
ISSN: 1570-9639

Biochimica et Biophysica Acta - Reviews on Cancer
ISSN: 0304-419X

BioDrugs
ISSN: 1173-8804

Bioelectromagnetics
ISSN: 0197-8462

BioFactors
ISSN: 0951-6433

Biofilms
ISSN: 1479-0505

Biogerontology
ISSN: 1389-5729

Bioinformatics
ISSN: 1367-4803

Biological Chemistry
ISSN: 1431-6730

Biological Psychiatry
ISSN: 0006-3223

Biological Psychology
ISSN: 0301-0511

Biologicals
ISSN: 1045-1056

Biology Letters
ISSN: 1744-9561

Biology of Reproduction
ISSN: 0006-3363

Biology of the Cell
ISSN: 0248-4900

Biology of the Neonate
ISSN: 0006-3126

Biomacromolecules
ISSN: 1525-7797

Biomaterials
ISSN: 0142-9612

Biomedical Microdevices
ISSN: 1387-2176

Biomedical Signal Processing and Control
ISSN: 1746-8094

Biomedicine and Pharmacotherapy
ISSN: 0753-3322

Biomolecular Engineering
ISSN: 1389-0344

Bioorganic Chemistry
ISSN: 0045-2068

Biophysical Chemistry
ISSN: 0301-4622

Bioprocess and Biosystems Engineering
ISSN: 1615-7591

Bioresource Technology
ISSN: 0960-8524

Bipolar Disorders
ISSN: 1398-5647

Bipolar Disorders, Supplement
ISSN: 1399-2406

Birth
ISSN: 0730-7659

Birth Defects Research Part A - Clinical and Molecular Teratology
ISSN: 1542-0752

Birth Defects Research Part B - Developmental and Reproductive Toxicology
ISSN: 1542-9733

Birth Defects Research Part C - Embryo Today: Reviews
ISSN: 1542-975X

BJOG: An International Journal of Obstetrics and Gynaecology
ISSN: 1470-0328

BJU International
ISSN: 1464-4096

Blood
ISSN: 0006-4971

Blood Cells, Molecules, and Diseases
ISSN: 1079-9796

Blood Coagulation and Fibrinolysis
ISSN: 0957-5235

Blood Pressure
ISSN: 0803-7051

Blood Pressure Monitoring
ISSN: 1359-5237

Blood Pressure, Supplement
ISSN: 0803-8023

Blood Purification
ISSN: 0253-5068

Blood Reviews
ISSN: 0268-960X

BMC Cardiovascular Disorders
ISSN: 1471-2261

Bone Marrow Transplantation
ISSN: 0268-3369

Brachytherapy
ISSN: 1538-4721

Brain
ISSN: 0006-8950

Brain and Cognition
ISSN: 0278-2626

Brain Research
ISSN: 0006-8993

Brain Research Bulletin
ISSN: 0361-9230

Brain Research Reviews
ISSN: 0165-0173

Brain Topography
ISSN: 0896-0267

Brain, Behavior and Evolution
ISSN: 0006-8977

Brain, Behavior, and Immunity
ISSN: 0889-1591

Breast
ISSN: 0960-9776

Breast Cancer Online
ISSN: 1470-9031

Breast Cancer Research and Treatment
ISSN: 0167-6806

Breast Care
ISSN: 1661-3791

Breast Disease
ISSN: 0888-6008

British Journal of Anaesthesia
ISSN: 0007-0912

British Journal of Biomedical Science
ISSN: 0967-4845

British Journal of Cancer
ISSN: 0007-0920

British Journal of Clinical Pharmacology
ISSN: 0306-5251

British Journal of Clinical Pharmacology, Supplement
ISSN: 0264-3774

British Journal of Clinical Psychology
ISSN: 0144-6657

British Journal of Dermatology
ISSN: 0007-0963

British Journal of Haematology
ISSN: 0007-1048

British Journal of Health Psychology
ISSN: 1359-107X

British Journal of Neurosurgery
ISSN: 0268-8697

British Journal of Ophthalmology
ISSN: 0007-1161

British Journal of Pharmacology
ISSN: 0007-1188

British Journal of Surgery
ISSN: 0007-1323

British Medical Bulletin
ISSN: 0007-1420

British Medical Journal
ISSN: 0959-8146

Ca-A Cancer Journal for Clinicians
ISSN: 0007-9235

Calcified Tissue International
ISSN: 0171-967X

Canadian Journal of Anesthesia
ISSN: 0832-610X

Canadian Journal of Gastroenterology
ISSN: 0835-7900

Canadian Journal of Microbiology
ISSN: 0008-4166

Canadian Journal of Neurological Sciences
ISSN: 0317-1671

Canadian Journal of Physiology and Pharmacology
ISSN: 0008-4212

Canadian Respiratory Journal
ISSN: 1198-2241

Cancer
ISSN: 0008-543X

Cancer and Metastasis Reviews
ISSN: 0167-7659

Cancer Biotherapy and Radiopharmaceuticals
ISSN: 1084-9785

Cancer Causes and Control
ISSN: 0957-5243

Cancer Cell
ISSN: 1535-6108

Cancer Chemotherapy and Pharmacology
ISSN: 0344-5704

Cancer Chemotherapy and Pharmacology, Supplement
ISSN: 0943-9404

Cancer Detection and Prevention
ISSN: 0361-090X

Cancer Epidemiology Biomarkers and Prevention
ISSN: 1055-9965

Cancer Gene Therapy
ISSN: 0929-1903

Cancer Genetics and Cytogenetics
ISSN: 0165-4608

Cancer Immunology, Immunotherapy
ISSN: 0340-7004

Cancer Investigation
ISSN: 0735-7907

Cancer Journal
ISSN: 1528-9117

Cancer Letters
ISSN: 0304-3835

Cancer Nursing
ISSN: 0162-220X

Cancer Research
ISSN: 0008-5472

Cancer Science
ISSN: 1347-9032

Cancer/Radiotherapie
ISSN: 1278-3218

Carbohydrate Research
ISSN: 0008-6215

Carcinogenesis
ISSN: 0143-3334

Cardiology
ISSN: 0008-6312

Cardiology Clinics
ISSN: 0733-8651

CardioVascular and Interventional Radiology
ISSN: 0174-1551

Cardiovascular Drug Reviews
ISSN: 0897-5957

Cardiovascular Drugs and Therapy
ISSN: 0920-3206

Cardiovascular Pathology
ISSN: 1054-8807

Cardiovascular Radiation Medicine
ISSN: 1522-1865

Cardiovascular Research
ISSN: 0008-6363

Cardiovascular Revascularization Medicine
ISSN: 1553-8389

Cardiovascular Toxicology
ISSN: 1530-7905

Cell
ISSN: 0092-8674

Cell and Tissue Banking
ISSN: 1389-9333

Cell and Tissue Research
ISSN: 0302-766X

Cell Biochemistry and Function
ISSN: 0263-6484

Cell Biology and Toxicology
ISSN: 0742-2091

Cell Calcium
ISSN: 0143-4160

Cell Communication and Adhesion
ISSN: 1061-5385

Cell Communication and Signaling
ISSN: 1478-811X

Cell Death and Differentiation
ISSN: 1350-9047

Cell Metabolism
ISSN: 1550-4131

Cell Motility and the Cytoskeleton
ISSN: 0886-1544

Cell Preservation Technology
ISSN: 1538-344X

Cell Stress and Chaperones
ISSN: 1355-8145

Cell Structure and Function
ISSN: 0386-7196

Cell Transplantation
ISSN: 0963-6897

Cells and Materials
ISSN: 1051-6794

Cells Tissues Organs
ISSN: 1422-6405

Cellular and Molecular Neurobiology
ISSN: 0272-4340

Cellular Immunology
ISSN: 0008-8749

Cellular Microbiology
ISSN: 1462-5814

Cellular Oncology
ISSN: 1570-5870

Cellular Physiology and Biochemistry
ISSN: 1015-8987

Cellular Signalling
ISSN: 0898-6568

Cerebellum
ISSN: 1473-4222

Cerebral Cortex
ISSN: 1047-3211

Cerebrospinal Fluid Research
ISSN: 1743-8454

Cerebrovascular Diseases
ISSN: 1015-9770

ChemBioChem
ISSN: 1439-4227

Chemical Biology and Drug Design
ISSN: 1747-0277

Chemical Immunology and Allergy
ISSN: 1660-2242

Chemical Senses
ISSN: 0379-864X

Chemistry and Physics of Lipids
ISSN: 0009-3084

Chemometrics and Intelligent Laboratory Systems
ISSN: 0169-7439

Chemotherapy
ISSN: 0009-3157

Chest
ISSN: 0012-3692

Child and Adolescent Mental Health
ISSN: 1475-357X

Child and Adolescent Psychiatric Clinics of North America
ISSN: 1056-4993

Child Psychology and Psychiatry Review
ISSN: 1360-6417

Child's Nervous System
ISSN: 0256-7040

Child: Care, Health and Development
ISSN: 0305-1862

Chinese Journal of Digestive Diseases
ISSN: 1443-9611

Chinese Journal of Integrative Medicine
ISSN: 1672-0415

Chirality
ISSN: 0899-0042

Chiropractic and Osteopathy
ISSN: 1746-1340

Chromatographia
ISSN: 0009-5893

Chromosome Research
ISSN: 0967-3849

Chronic Respiratory Disease
ISSN: 1479-9723

Circulation
ISSN: 0009-7322

Circulation Research
ISSN: 0009-7330

Cleft Palate-Craniofacial Journal
ISSN: 1055-6656

Climacteric
ISSN: 1369-7137

Clinica Chimica Acta
ISSN: 0009-8981

Clinical Anatomy
ISSN: 0897-3806

Clinical and Applied Immunology Reviews
ISSN: 1529-1049

Clinical and Applied Thrombosis/ Hemostasis
ISSN: 1076-0296

Clinical and Developmental Immunology
ISSN: 1740-2522

Clinical and Diagnostic Laboratory Immunology
ISSN: 1071-412X

Clinical and Experimental Allergy
ISSN: 0954-7894

Clinical and Experimental Allergy Reviews
ISSN: 1472-9725

Clinical and Experimental Dermatology
ISSN: 0307-6938

Clinical and Experimental Dermatology, Supplement
ISSN: 0966-3487

Clinical and Experimental Immunology
ISSN: 0009-9104

Clinical and Experimental Medicine
ISSN: 1591-8890

Clinical and Experimental Rheumatology
ISSN: 0392-856X

Clinical and Investigative Medicine
ISSN: 0147-958X

Clinical and Laboratory Haematology
ISSN: 0141-9854

Clinical and Laboratory Haematology, Supplement
ISSN: 0960-3964

Clinical and Vaccine Immunology
ISSN: 1556-6811

Clinical Biochemistry
ISSN: 0009-9120

Clinical Biomechanics
ISSN: 0268-0033

Clinical Cancer Research
ISSN: 1078-0432

Clinical Chemistry and Laboratory Medicine
ISSN: 1434-6621

Clinical Drug Investigation
ISSN: 1173-2563

Clinical Dysmorphology
ISSN: 0962-8827

Clinical EEG and Neuroscience
ISSN: 0009-9155

Clinical Endocrinology
ISSN: 0300-0664

Clinical Genetics
ISSN: 0009-9163

Clinical Governance
ISSN: 1477-7274

Clinical Imaging
ISSN: 0899-7071

Clinical Immunology
ISSN: 1521-6616

Clinical Infectious Diseases
ISSN: 1058-4838

Clinical Journal of Pain
ISSN: 0749-8047

Clinical Journal of Sport Medicine
ISSN: 1050-642X

Clinical Linguistics and Phonetics
ISSN: 0269-9206

Clinical Microbiology and Infection
ISSN: 1198-743X

Clinical Microbiology and Infection, Supplement
ISSN: 1470-9465

Clinical Neuropathology
ISSN: 0722-5091

Clinical Neuropharmacology
ISSN: 0362-5664

Clinical Neurophysiology
ISSN: 1388-2457

Clinical Neuroscience Research
ISSN: 1566-2772

Clinical Oncology
ISSN: 0936-6555

Clinical Orthopaedics and Related Research
ISSN: 0009-921X

Clinical Otolaryngology
ISSN: 0307-7772

Clinical Pharmacokinetics
ISSN: 0312-5963

Clinical Pharmacology and Therapeutics
ISSN: 0009-9236

Clinical Physiology and Functional Imaging
ISSN: 1475-0961

Clinical Proteomics
ISSN: 1542-6416

Clinical Psychology and Psychotherapy
ISSN: 1063-3995

Clinical Radiology
ISSN: 0009-9260

Clinical Radiology Extra
ISSN: 1477-6804

Clinical Rheumatology
ISSN: 0770-3198

Clinical Science
ISSN: 0143-5221

Clinical Transplantation
ISSN: 0902-0063

Clinical Transplantation, Supplement
ISSN: 1399-6738

Clinical Trials
ISSN: 1740-7745

Clinician in Management
ISSN: 0965-5751

Clinics in Chest Medicine
ISSN: 0272-5231

Clinics in Dermatology
ISSN: 0738-081X

Clinics in Laboratory Medicine
ISSN: 0272-2712

Clinics in Perinatology
ISSN: 0095-5108

Clinics in Podiatric Medicine and Surgery
ISSN: 0891-8422

Cloning and Stem Cells
ISSN: 1536-2302

CNS Drugs
ISSN: 1172-7047

CNS Spectrums
ISSN: 1092-8529

Cognition
ISSN: 0010-0277

Cognition and Emotion
ISSN: 0269-9931

Cognitive and Behavioral Neurology
ISSN: 1543-3633

Cognitive and Behavioral Practice
ISSN: 1077-7229

Cognitive Neuropsychiatry
ISSN: 1354-6805

Cognitive Neuropsychology
ISSN: 0264-3294

Cognitive Processing
ISSN: 1612-4782

Cognitive Systems Research
ISSN: not available yet

Cold Spring Harbor Symposia on Quantitative Biology
ISSN: 0091-7451

Colloid and Polymer Science
ISSN: 0303-402X

Colloids and Surfaces A: Physicochemical and Engineering Aspects
ISSN: 0927-7757

Colloids and Surfaces B: Biointerfaces
ISSN: 0927-7765

Colorectal Disease
ISSN: 1462-8910

Combinatorial Chemistry and High Throughput Screening
ISSN: 1386-2073

Combustion and Flame
ISSN: 0010-2180

Community Genetics
ISSN: 1422-2795

Comparative and Functional Genomics
ISSN: 1531-6912

Comparative Biochemistry and Physiology - B Biochemistry and Molecular Biology
ISSN: 1096-4959

Comparative Biochemistry and Physiology - C Toxicology and Pharmacology
ISSN: 1532-0456

Comparative Biochemistry and Physiology - Part D: Genomics and Proteomics
ISSN: 1744-117X

Complexus
ISSN: 1424-8492

Computer Aided Surgery
ISSN: 1092-9088

Computerized Medical Imaging and Graphics
ISSN: 0895-6111

Computers in Biology and Medicine
ISSN: 0010-4825

Congenital Anomalies
ISSN: 0914-3505

Congenital Heart Disease
ISSN: 1747-079X

Consultant
ISSN: 0010-7069

Contact Lens and Anterior Eye
ISSN: 1367-0484

COPD: Journal of Chronic Obstructive Pulmonary Disease
ISSN: 1541-2555

Cornea
ISSN: 0277-3740

Coronary Artery Disease
ISSN: 0954-6928

Critical Care
ISSN: 1364-8535

Critical Care Clinics
ISSN: 0749-0704

Critical Care Medicine
ISSN: 0090-3493

Critical Pathways in Cardiology
ISSN: 1535-282X

Critical Reviews in Biochemistry and Molecular Biology
ISSN: 1040-9238

Critical Reviews in Biomedical Engineering
ISSN: 0278-940X

Critical Reviews in Biotechnology
ISSN: 0738-8551

Critical Reviews in Clinical Laboratory Sciences
ISSN: 1040-8363

Critical Reviews in Computed Tomography
ISSN: 1040-8371

Critical Reviews in Eukaryotic Gene Expression
ISSN: 1045-4403

Critical Reviews in Immunology
ISSN: 1040-8401

Critical Reviews in Microbiology
ISSN: 1040-841X

Critical Reviews in Oral Biology and Medicine
ISSN: 1045-4411

Critical Reviews in Physical and Rehabilitation Medicine
ISSN: 0896-2960

Cryobiology
ISSN: 0011-2240

Culture, Health and Sexuality
ISSN: 1369-1058

Current Alzheimer Research
ISSN: 1567-2050

Current Anaesthesia and Critical Care
ISSN: 0953-7112

Current Cancer Therapy Reviews
ISSN: 1573-3947

Current Cardiology Reviews
ISSN: 1573-3947

Current Diabetes Reviews
ISSN: 1573-3998

Current Drug Delivery
ISSN: 1567-2018

Current Drug Discovery Technologies
ISSN: 1570-1638

Current Eye Research
ISSN: 0271-3683

Current Genetics
ISSN: 0172-8083

Current Hypertension Reviews
ISSN: 1573-4021

Current Immunology Reviews
ISSN: 1573-3955

Current Medical Imaging Reviews
ISSN: 1573-4056

Current Microbiology
ISSN: 0343-8651

Current Nanoscience
ISSN: 1573-4137

Current Neurovascular Research
ISSN: 1567-2026

Current Nutrition and Food Science
ISSN: 1573-4013

Current Opinion in Allergy and Clinical Immunology
ISSN: 1528-4050

Current Opinion in Anaesthesiology
ISSN: 0952-7907

Current Opinion in Biotechnology
ISSN: 0958-1669

Current Opinion in Cardiology
ISSN: 0268-4705

Current Opinion in Cell Biology
ISSN: 0955-0674

Current Opinion in Genetics and Development
ISSN: 0959-437X

Current Opinion in Hematology
ISSN: 1065-6251

Current Opinion in Lipidology
ISSN: 0957-9672

Current Opinion in Nephrology and Hypertension
ISSN: 1062-4821

Current Opinion in Neurobiology
ISSN: 0959-4388

Current Opinion in Obstetrics and Gynecology
ISSN: 1040-872X

Current Opinion in Oncology
ISSN: 1040-8746

Current Opinion in Ophthalmology
ISSN: 1040-8738

Current Opinion in Orthopaedics
ISSN: 1041-9918

Current Opinion in Otolaryngology and Head and Neck Surgery
ISSN: 1068-9508

Current Opinion in Pediatrics
ISSN: 1040-8703

Current Opinion in Pharmacology
ISSN: 1471-4892

Current Opinion in Rheumatology
ISSN: 1040-8711

Current Opinion in Structural Biology
ISSN: 0959-440X

Current Opinion in Urology
ISSN: 0963-0643

Current Orthopaedics
ISSN: 0268-0890

Current Pediatric Reviews
ISSN: 1573-3963

Current Pharmaceutical Analysis
ISSN: 1573-4129

Current Pharmaceutical Design
ISSN: 1381-6128

Current Psychiatry Reviews
ISSN: 1573-4005

Current Respiratory Medicine Reviews
ISSN: 1573-398X

Current Rheumatology Reviews
ISSN: 1573-3971

Current Studies in Hematology and Blood Transfusion
ISSN: 0258-0330

Current Surgery
ISSN: 0149-7944

Current Therapeutic Research - Clinical and Experimental
ISSN: 0011-393X

Current Topics in Microbiology and Immunology
ISSN: 0070-217X

Current Women's Health Reviews
ISSN: 1573-4048

Cytogenetic and Genome Research
ISSN: 1424-8581

Cytokine
ISSN: 1043-4666

Cytokine and Growth Factor Reviews
ISSN: 1359-6101

Cytometry Part A
ISSN: 1552-4922

Cytometry Part B - Clinical Cytometry
ISSN: 1552-4949

Cytopathology
ISSN: 0956-5507

Cytotherapy
ISSN: 1465-3249

Dementia and Geriatric Cognitive Disorders
ISSN: 1420-8008

Depression and Anxiety
ISSN: 1091-4269

Dermatitis
ISSN: 1710-3568

Dermatologic Clinics
ISSN: 0733-8635

Dermatologic Surgery
ISSN: 1076-0512

Dermatology
ISSN: 1018-8665

Developing World Bioethics
ISSN: 1471-8731

Development
ISSN: 0950-1991

Development and Psychopathology
ISSN: 0954-5794

Development Genes and Evolution
ISSN: 0949-944X

Developmental and Comparative Immunology
ISSN: 0145-305X

Developmental Biology
ISSN: 0012-1606

Developmental Cell
ISSN: 1534-5807

Developmental Dynamics
ISSN: 1058-8388

Developmental Medicine and Child Neurology
ISSN: 0012-1622

Developmental Neuropsychology
ISSN: 8756-5641

Developmental Neuroscience
ISSN: 0378-5866

Diabetes
ISSN: 0012-1797

Diabetes Technology and Therapeutics
ISSN: 1520-9156

Diabetes, Nutrition and Metabolism - Clinical and Experimental
ISSN: 0394-3402

Diabetes/Metabolism Research and Reviews
ISSN: 1520-7552

Diabetologia
ISSN: 0012-186X

Diagnostic Cytopathology
ISSN: 8755-1039

Diagnostic Microbiology and Infectious Disease
ISSN: 0732-8893

Diagnostic Molecular Pathology
ISSN: 1052-9551

Dialysis and Transplantation
ISSN: 0090-2934

Differentiation
ISSN: 0301-4681

Digestion
ISSN: 0012-2823

Digestive Diseases
ISSN: 0257-2753

Digestive Diseases and Sciences
ISSN: 0163-2116

Disability and Rehabilitation
ISSN: 0379-0797

Disease Markers
ISSN: 0278-0240

Diseases of the Esophagus
ISSN: 1120-8694

DNA and Cell Biology
ISSN: 1044-5498

DNA Repair
ISSN: 1568-7864

DNA Sequence - Journal of DNA Sequencing and Mapping
ISSN: 1042-5179

Drug and Alcohol Dependence
ISSN: 0376-8716

Drug Delivery
ISSN: 1071-7544

Drug Design Reviews Online
ISSN: 1567-2697

Drug Information Journal
ISSN: 0092-8615

Drug Metabolism and Disposition
ISSN: 0090-9556

Drug Resistance Updates
ISSN: 1368-7646

Drug Safety
ISSN: 0114-5916

Drugs and Aging
ISSN: 1170-229X

Drugs in R and D
ISSN: 1174-5886

Dysphagia
ISSN: 0179-051X

Eating Behaviors
ISSN: 1471-0153

Echocardiography
ISSN: 0742-2822

EcoHealth
ISSN: 1612-9202

Ecotoxicology
ISSN: 0963-9292

Education for Health: Change in Learning and Practice
ISSN: 1357-6283

EMA - Emergency Medicine Australasia
ISSN: 1742-6731

EMBO Journal
ISSN: 0261-4189

EMBO Reports
ISSN: 1469-221X

Emergency Medicine Clinics of North America
ISSN: 0733-8627

Emergency Medicine Journal
ISSN: 1472-0205

Emergency Radiology
ISSN: 1070-3004

Emotional and Behavioural Difficulties
ISSN: 1363-2752

Endocrine
ISSN: 0969-711X

Endocrine Pathology
ISSN: 1046-3976

Endocrine Research
ISSN: 0743-5800

Endocrine Reviews
ISSN: 0163-769X

Endocrinology
ISSN: 0013-7227

Endocrinology and Metabolism Clinics of North America
ISSN: 0889-8529

Endoscopy
ISSN: 0013-726X

Endothelium: Journal of Endothelial Cell Research
ISSN: 1062-3329

Enhancer - Biotherapy of Cancer
ISSN: 1572-0004

Environment International
ISSN: 0160-4120

Environmental Health Perspectives
ISSN: 0091-6765

Environmental Research
ISSN: 0013-9351

Environmental Science and Policy
ISSN: 1462-9011

Environmental Toxicology
ISSN: 1520-4081

Environmental Toxicology and Chemistry
ISSN: 0730-7268

Environmental Toxicology and Pharmacology
ISSN: 1382-6689

Epidemiologic Reviews
ISSN: 0193-936X

Epidemiology
ISSN: 1044-3983

Epilepsia
ISSN: 0013-9580

Epilepsy Research
ISSN: 0920-1211

Epileptic Disorders
ISSN: 1294-9361

ESHRE Monographs
ISSN: 1477-741X

Esophagus
ISSN: 1612-9059

Ethical Human Psychology and Psychiatry
ISSN: 1523-150X

Europace
ISSN: 1099-5129

European Addiction Research
ISSN: 1022-6877

European Archives of Oto-Rhino-Laryngology
ISSN: 0937-4477

European Archives of Psychiatry and Clinical Neuroscience
ISSN: 0940-1334

European Heart Journal
ISSN: 0195-668X

European Heart Journal, Supplement
ISSN: 1520-765X

European Journal of Applied Physiology
ISSN: 1439-6319

European Journal of Cancer
ISSN: 0959-8049

European Journal of Cancer Prevention
ISSN: 0959-8278

European Journal of Cancer, Supplement
ISSN: 1359-6349

European Journal of Cardio-thoracic Surgery
ISSN: 1010-7940

European Journal of Cardiovascular Nursing
ISSN: 1474-5151

European Journal of Cardiovascular Prevention and Rehabilitation
ISSN: 1741-8267

European Journal of Cell Biology
ISSN: 0171-9335

European Journal of Clinical Investigation
ISSN: 0014-2972

European Journal of Clinical Microbiology and Infectious Diseases
ISSN: 0934-9723

European Journal of Clinical Pharmacology
ISSN: 0031-6970

European Journal of Cognitive Psychology
ISSN: 0954-1446

European Journal of Contraception and Reproductive Health Care
ISSN: 1362-5187

European Journal of Echocardiography
ISSN: 1525-2167

European Journal of Emergency Medicine
ISSN: 0969-9546

European Journal of Endocrinology
ISSN: 0804-4643

European Journal of Gastroenterology and Hepatology
ISSN: 0954-691X

European Journal of Haematology
ISSN: 0902-4441

European Journal of Haematology, Supplement
ISSN: 0902-4506

European Journal of Health Economics
ISSN: 1618-7598

European Journal of Heart Failure
ISSN: 1388-9842

European Journal of Heart Failure, Supplement
ISSN: 1567-4215

European Journal of Human Genetics
ISSN: 1018-4813

European Journal of Immunology
ISSN: 0014-2980

European Journal of Morphology
ISSN: 0924-3860

European Journal of Neurology
ISSN: 1351-5101

European Journal of Neurology, Supplement
ISSN: 1471-0552

European Journal of Neuroscience
ISSN: 0953-816X

European Journal of Obstetrics Gynecology and Reproductive Biology
ISSN: 0301-2115

European Journal of Ophthalmology
ISSN: 1120-6721

European Journal of Orthopaedic Surgery and Traumatology
ISSN: 0948-4817

European Journal of Paediatric Neurology
ISSN: 1090-3798

European Journal of Pain
ISSN: 1090-3801

European Journal of Pediatrics
ISSN: 0340-6199

European Journal of Pediatrics, Supplement
ISSN: 0943-9676

European Journal of Pharmaceutical Sciences
ISSN: 0928-0987

European Journal of Pharmacology
ISSN: 0014-2999

European Journal of Plastic Surgery
ISSN: 0930-343X

European Journal of Public Health
ISSN: 1101-1262

European Journal of Radiology
ISSN: 0720-048X

European Journal of Special Needs Education
ISSN: 0885-6257

European Journal of Surgical Oncology
ISSN: 0748-7983

European Journal of Vascular and Endovascular Surgery
ISSN: 1078-5884

European Neurology
ISSN: 0014-3022

European Neuropsychopharmacology
ISSN: 0924-977X

European Psychiatry
ISSN: 0924-9338

European Radiology
ISSN: 0938-7994

European Radiology, Supplement
ISSN: 1613-3749

European Respiratory Journal
ISSN: 0903-1936

European Respiratory Journal, Supplement
ISSN: 0904-1850

European Spine Journal
ISSN: 0940-6719

European Surgical Research
ISSN: 0014-312X

European Urology
ISSN: 0302-2838

European Urology, Supplements
ISSN: 1569-9056

Evidence-based Cardiovascular Medicine
ISSN: 1361-2611

Evidence-based Complementary and Alternative Medicine
ISSN: 1741-427X

Evidence-Based Healthcare and Public Health
ISSN: 1744-2249

Evidence-based Obstetrics and Gynecology
ISSN: 1361-259X

Evidence-Based Ophthalmology
ISSN: 1555-9203

Exercise and Sport Sciences Reviews
ISSN: 0091-6331

Exogenous Dermatology
ISSN: 1424-4616

Experimental and Clinical Endocrinology and Diabetes
ISSN: 0947-7349

Experimental Brain Research
ISSN: 0014-4819

Experimental Cell Research
ISSN: 0014-4827

Experimental Diabesity Research
ISSN: 1543-8600

Experimental Eye Research
ISSN: 0014-4835

Experimental Gerontology
ISSN: 0531-5565

Experimental Hematology
ISSN: 0301-472X

Experimental Lung Research
ISSN: 0190-2148

Experimental Neurology
ISSN: 0014-4886

Experimental Parasitology
ISSN: 0014-4894

Expert Review of Vaccines
ISSN: 1476-0584

Expert Reviews in Molecular Medicine
ISSN: 1462-3994

Explore: The Journal of Science and Healing
ISSN: 1550-8307

Eye and Contact Lens
ISSN: 1542-2321

Familial Cancer
ISSN: 1389-9600

FASEB Journal
ISSN: 0892-6638

FEBS Journal
ISSN: 1742-464X

FEBS Letters
ISSN: 0014-5793

FEMS Immunology and Medical Microbiology
ISSN: 0928-8244

FEMS Microbiology Ecology
ISSN: 0168-6496

FEMS Microbiology Letters
ISSN: 0378-1097

Fertility and Sterility
ISSN: 0015-0282

Fetal and Pediatric Pathology
ISSN: 1551-3815

Fetal Diagnosis and Therapy
ISSN: 1015-3837

Fitoterapia
ISSN: 0367-326X

Folia Phoniatrica et Logopaedica
ISSN: 1021-7762

Food Additives and Contaminants
ISSN: 0265-203X

Foodborne Pathogens and Disease
ISSN: 1535-3141

Foot
ISSN: 0958-2592

Foot and Ankle International
ISSN: 1071-1007

Foot and Ankle Surgery
ISSN: 1268-7731

Forensic Science International
ISSN: 0379-0738

Free Radical Biology and Medicine
ISSN: 0891-5849

Frontiers in Neuroendocrinology
ISSN: 0091-3022

Functional and Integrative Genomics
ISSN: 1438-793X

Fundamental and Clinical Pharmacology
ISSN: 0767-3981

Fungal Genetics and Biology
ISSN: 1087-1845

Future Cardiology
ISSN: 1479-6678

Future Oncology
ISSN: 1479-6694

Gait and Posture
ISSN: 0966-6362

Gann Monographs on Cancer Research
ISSN: 0072-0151

Gastric Cancer
ISSN: 1436-3291

Gastroenterology
ISSN: 0016-5085

Gastroenterology International
ISSN: 0950-5911

Gastrointestinal Endoscopy
ISSN: 0016-5107

Gastrointestinal Endoscopy Clinics of North America
ISSN: 1052-5157

Gender Medicine
ISSN: 1550-8579

Gene
ISSN: 0378-1119

Gene Expression Patterns
ISSN: 1567-133X

Gene Therapy
ISSN: 0969-7128

General and Comparative Endocrinology
ISSN: 0016-6480

Genes and Development
ISSN: 0890-9369

Genes and Immunity
ISSN: 1466-4879

Genes Chromosomes and Cancer
ISSN: 1045-2257

Genes to Cells
ISSN: 1356-9597

Genes, Brain and Behavior
ISSN: 1601-1848

Genesis
ISSN: 1526-954X

Genetics
ISSN: 0016-6731

Genome Research
ISSN: 1088-9051

Genomics
ISSN: 0888-7543

Gerontology
ISSN: 0304-324X

GLIA
ISSN: 0894-1491

Globalization and Health
ISSN: 1744-8603

Glycobiology
ISSN: 0959-6658

Glycoconjugate Journal
ISSN: 0282-0080

Graefe's Archive for Clinical and Experimental Ophthalmology
ISSN: 0721-832X

Growth Factors
ISSN: 0897-7194

Growth Hormone and IGF Research
ISSN: 1096-6374

Growth, Development and Aging
ISSN: 0017-4793

Gut
ISSN: 0017-5749

Gynecologic and Obstetric Investigation
ISSN: 0378-7346

Gynecologic Oncology
ISSN: 0090-8258

Gynecological Endocrinology
ISSN: 0951-3590

Haemophilia
ISSN: 1351-8216

Haemophilia, Supplement
ISSN: 1355-0691

Harvard Review of Psychiatry
ISSN: 1067-3229

Head and Neck
ISSN: 1043-3074

Headache
ISSN: 0017-8748

Headache and Pain: Diagnostic Challenges, Current Therapy
ISSN: 1545-8261

Headache Currents
ISSN: 1743-5005

Health and Place
ISSN: 1353-8292

Health Economics
ISSN: 1057-9230

Health Education Journal
ISSN: 0017-8969

Health Education Research
ISSN: 0268-1153

Health Expectations
ISSN: 1369-6513

Health Physics
ISSN: 0017-9078

Health Promotion International
ISSN: 0957-4824

Health Services and Outcomes Research Methodology
ISSN: 1387-3741

Health Services Management Research
ISSN: 0951-4848

Hearing Research
ISSN: 0378-5955

Heart
ISSN: 1355-6037

Heart and Lung: Journal of Acute and Critical Care
ISSN: 0147-9563

Heart and Metabolism
ISSN: 1566-0338

Heart and Vessels
ISSN: 0910-8327

Heart Lung and Circulation
ISSN: 1443-9506

Heart Rhythm
ISSN: 1547-5271

Heart Surgery Forum
ISSN: 1098-3511

Helicobacter
ISSN: 1083-4389

Helicobacter, Supplement
ISSN: 1478-4041

Helvetica Chimica Acta
ISSN: 0018-019X

Hematological Oncology
ISSN: 0278-0232

Hematology
ISSN: 1024-5332

Hematology/Oncology Clinics of North America
ISSN: 0889-8588

Hepato-Gastroenterology
ISSN: 0172-6390

Hepatology
ISSN: 0270-9139

Hepatology Research
ISSN: 1386-6346

Hernia
ISSN: 1265-4906

Heroin Addiction and Related Clinical Problems
ISSN: 1592-1638

Herpes
ISSN: 0969-7667

High Altitude Medicine and Biology
ISSN: 1527-0297

High Blood Pressure and Cardiovascular Prevention
ISSN: 1120-9879

Hippocampus
ISSN: 1050-9631

Hispanic Healthcare International
ISSN: 1540-4153

Histochemistry and Cell Biology
ISSN: 0948-6143

Histopathology
ISSN: 0309-0167

HIV Clinical Trials
ISSN: 1528-4336

HIV Medicine
ISSN: 1464-2662

Hong Kong Journal of Nephrology
ISSN: 1561-5413

Hong Kong Journal of Occupational Therapy
ISSN: 1569-1861

Hormone and Metabolic Research
ISSN: 0018-5043

Hormone and Metabolic Research, Supplement
ISSN: 0170-5903

Hormone Research
ISSN: 0301-0163

Hormones and Behavior
ISSN: 0018-506X

Human and Experimental Toxicology
ISSN: 0960-3271

Human Antibodies
ISSN: 1093-2607

Human Brain Mapping
ISSN: 1065-9471

Human Development
ISSN: 0018-716X

Human Genetics
ISSN: 0340-6717

Human Immunology
ISSN: 0198-8859

Human Molecular Genetics
ISSN: 0964-6906

Human Mutation
ISSN: 1059-7794

Human Psychopharmacology
ISSN: 0885-6222

Human Reproduction Update
ISSN: 1355-4786

Hybridoma
ISSN: 1554-0014

Hypertension
ISSN: 0194-911X

IARC Monographs on the Evaluation of Carcinogenic Risks to Humans
ISSN: 1017-1606

IEEE Transactions on Neural Systems and Rehabilitation Engineering
ISSN: 1534-4320

Imaging
ISSN: 0965-6812

Imaging Decisions MRI
ISSN: 1433-3317

Immunity
ISSN: 1074-7613

Immunity and Ageing
ISSN: 1742-4933

Immuno-Analyse et Biologie Specialisee
ISSN: 0923-2532

Immunobiology
ISSN: 0171-2985

Immunogenetics
ISSN: 0093-7711

Immunohematology
ISSN: 0894-203X

Immunologic Research
ISSN: 0257-277X

Immunological Investigations
ISSN: 0882-0139

Immunological Reviews
ISSN: 0105-2896

Immunology
ISSN: 0019-2805

Immunology and Allergy Clinics of North America
ISSN: 0889-8561

Immunology Letters
ISSN: 0165-2478

Immunology, Supplement
ISSN: 0953-4954

Immunopharmacology and Immunotoxicology
ISSN: 0892-3973

Indoor and Built Environment
ISSN: 1420-326X

Infant Behavior and Development
ISSN: 0163-6383

Infant Mental Health Journal
ISSN: 0163-9641

Infection and Immunity
ISSN: 0019-9567

Infection, Genetics and Evolution
ISSN: 1567-1348

Infectious Diseases in Obstetrics and Gynecology
ISSN: 1064-7449

Inflammatory Bowel Diseases
ISSN: 1078-0998

Inflammopharmacology
ISSN: 0925-4692

Inhalation Toxicology
ISSN: 0895-8378

Injury
ISSN: 0020-1383

Insulin
ISSN: 1557-0843

Integrative Cancer Therapies
ISSN: 1534-7354

Interactive Cardiovascular and Thoracic Surgery
ISSN: 1569-9293

Internal Medicine Journal
ISSN: 1444-0903

International Anesthesiology Clinics
ISSN: 0020-5907

International Archives of Allergy and Immunology
ISSN: 1018-2438

International Clinical Psychopharmacology
ISSN: 0268-1315

International Forum of Psychoanalysis
ISSN: 0803-706X

International Heart Journal
ISSN: 1349-2365

International Immunology
ISSN: 0953-8178

International Immunopharmacology
ISSN: 1567-5769

International Journal for Quality in Health Care
ISSN: 1353-4505

International Journal of Andrology
ISSN: 0105-6263

International Journal of Andrology, Supplement
ISSN: 1350-4908

International Journal of Angiology
ISSN: 1061-1711

International Journal of Antimicrobial Agents
ISSN: 0924-8579

International Journal of Audiology
ISSN: 1499-2027

International Journal of Cancer
ISSN: 0020-7136

International Journal of Cardiology
ISSN: 0167-5273

International Journal of Clinical Oncology
ISSN: 1341-9625

International Journal of Clinical Practice
ISSN: 1368-5031

International Journal of Clinical Practice, Supplement
ISSN: 1368-504X

International Journal of Colorectal Disease
ISSN: 0179-1958

International Journal of Dairy Technology
ISSN: 1364-727X

International Journal of Developmental Biology
ISSN: 0214-6282

International Journal of Developmental Neuroscience
ISSN: 0736-5748

International Journal of Disaster Medicine
ISSN: 1503-1438

International Journal of Drug Policy
ISSN: 0955-3959

International Journal of Eating Disorders
ISSN: 0276-3478

International Journal of Epidemiology
ISSN: 0300-5771

International Journal of Experimental Pathology
ISSN: 0959-9673

International Journal of Fertility and Women's Medicine
ISSN: 1069-3130

International Journal of Gastrointestinal Cancer
ISSN: 0169-4197

International Journal of Gynecological Cancer
ISSN: 1048-891X

International Journal of Gynecological Pathology
ISSN: 0277-1691

International Journal of Gynecology and Obstetrics
ISSN: 0020-7292

International Journal of Immunogenetics
ISSN: 1744-3121

International Journal of Immunopathology and Pharmacology
ISSN: 0394-6320

International Journal of Impotence Research
ISSN: 0955-9930

International Journal of Industrial Ergonomics
ISSN: 0169-8141

International Journal of Legal Medicine
ISSN: 0937-9827

International Journal of Life Cycle Assessment
ISSN: 0948-3349

International Journal of Lower Extremity Wounds
ISSN: 1534-7346

International Journal of Medical Informatics
ISSN: 1386-5056

International Journal of Men's Health
ISSN: 1532-6306

International Journal of Neuropsychopharmacology
ISSN: 1461-1457

International Journal of Obesity
ISSN: 0307-0565

International Journal of Occupational and Environmental Health
ISSN: 1077-3525

International Journal of Pediatric Otorhinolaryngology
ISSN: 0165-5876

International Journal of Pharmaceutical Medicine
ISSN: 1364-9027

International Journal of Pharmaceutics
ISSN: 0378-5173

International Journal of Pharmacy Practice
ISSN: 0961-7671

International Journal of Psychiatry in Clinical Practice
ISSN: 1365-1501

International Journal of Psychotherapy
ISSN: 1356-9082

International Journal of Radiation Biology
ISSN: 0955-3002

International Journal of Radiation Oncology Biology Physics
ISSN: 0360-3016

International Journal of Rehabilitation Research
ISSN: 0342-5282

International Journal of STD and AIDS
ISSN: 0956-4624

International Journal of Stroke
ISSN: 1747-4930

International Journal of Surgery
ISSN: 1743-9191

International Journal of Surgical Pathology
ISSN: 1066-8969

International Journal of Systematic and Evolutionary Microbiology
ISSN: 1466-5026

International Journal of Tuberculosis and Lung Disease
ISSN: 1027-3719

International Journal of Urology
ISSN: 0919-8172

International Ophthalmology Clinics
ISSN: 0020-8167

International Orthopaedics
ISSN: 0341-2695

International Review of Cytology
ISSN: 0074-7696

International Review of Psychiatry
ISSN: 0954-0261

International Reviews of Immunology
ISSN: 0883-0185

International Tinnitus Journal
ISSN: 0946-5448

International Urogynecology Journal and Pelvic Floor Dysfunction
ISSN: 0937-3462

International Wound Journal
ISSN: 1742-4801

Intervirology
ISSN: 0300-5526

Invertebrate Neuroscience
ISSN: 1354-2516

Investigational New Drugs
ISSN: 0167-6997

Investigative Ophthalmology and Visual Science
ISSN: 0146-0404

Investigative Radiology
ISSN: 0020-9996

Japanese Heart Journal
ISSN: 0021-4868

Japanese Journal of Ophthalmology
ISSN: 0021-5155

JARO - Journal of the Association for Research in Otolaryngology
ISSN: 1525-3961

JEUR
ISSN: 0993-9857

Journal Belge de Radiologie
ISSN: 0021-7646

Journal for Vascular Ultrasound
ISSN: 1544-3167

Journal of AAPOS
ISSN: 1091-8531

Journal of Acquired Immune Deficiency Syndromes
ISSN: 1525-4135

Journal of Adolescent Health
ISSN: 1054-139X

Journal of Aerosol Science
ISSN: 0021-8502

Journal of Affective Disorders
ISSN: 0165-0327

Journal of Allergy and Clinical Immunology
ISSN: 0091-6749

Journal of Alternative and Complementary Medicine
ISSN: 1075-5535

Journal of Alzheimer's Disease
ISSN: 1387-2877

Journal of Anatomy
ISSN: 0021-8782

Journal of Andrology
ISSN: 0196-3635

Journal of Antibiotics
ISSN: 0021-8820

Journal of Applied Physiology
ISSN: 8750-7587

Journal of Applied Toxicology
ISSN: 0260-437X

Journal of Artificial Organs
ISSN: 1434-7229

Journal of Assisted Reproduction and Genetics
ISSN: 1058-0468

Journal of Autism and Developmental Disorders
ISSN: 0162-3257

Journal of Autoimmunity
ISSN: 0896-8411

Journal of Automated Methods and Management in Chemistry
ISSN: 1463-9246

Journal of Back and Musculoskeletal Rehabilitation
ISSN: 1053-8127

Journal of Bacteriology
ISSN: 0021-9193

Journal of Biochemical and Biophysical Methods
ISSN: 0165-022X

Journal of Biological Chemistry
ISSN: 0021-9258

Journal of Biological Inorganic Chemistry
ISSN: 0949-8257

Journal of Biological Physics
ISSN: 0092-0606

Journal of Biological Rhythms
ISSN: 0748-7304

Journal of Biology
ISSN: 1475-4924

Journal of Biomaterials Science, Polymer Edition
ISSN: 0920-5063

Journal of Biomechanics
ISSN: 0021-9290

Journal of Biomedical Informatics
ISSN: 1532-0464

Journal of Biomedical Science
ISSN: 1021-7770

Journal of Biomolecular NMR
ISSN: 0925-2738

Journal of Biomolecular Screening
ISSN: 1087-0571

Journal of Biomolecular Structure and Dynamics
ISSN: 0739-1102

Journal of Biopharmaceutical Statistics
ISSN: 1054-3406

Journal of Biotechnology
ISSN: 0168-1656

Journal of Bone and Joint Surgery - Series A
ISSN: 0021-9355

Journal of Bone and Joint Surgery - Series B
ISSN: 0301-620X

Journal of Bone and Mineral Metabolism
ISSN: 0914-8779

Journal of Brief Therapy
ISSN: 1533-2098

Journal of Bronchology
ISSN: 1070-8030

Journal of Cancer Education
ISSN: 0885-8195

Journal of Cancer Research and Clinical Oncology
ISSN: 0171-5216

Journal of Cancer Research and Clinical Oncology, Supplement
ISSN: 0943-9382

Journal of Cardiac Failure
ISSN: 1071-9164

Journal of Cardiothoracic and Vascular Anesthesia
ISSN: 1053-0770

Journal of Cardiovascular Electrophysiology
ISSN: 1045-3873

Journal of Cardiovascular Magnetic Resonance
ISSN: 1097-6647

Journal of Cardiovascular Pharmacology
ISSN: 0160-2446

Journal of Cardiovascular Pharmacology and Therapeutics
ISSN: 1074-2484

Journal of Cataract and Refractive Surgery
ISSN: 0886-3350

Journal of Cell Biology
ISSN: 0021-9525

Journal of Cell Science
ISSN: 0021-9533

Journal of Cellular Biochemistry
ISSN: 0730-2312

Journal of Cellular Physiology
ISSN: 0021-9541

Journal of Cerebral Blood Flow and Metabolism
ISSN: 0271-678X

Journal of Chemical Neuroanatomy
ISSN: 0891-0618

Journal of Child Neurology
ISSN: 0883-0738

Journal of Chromatography A
ISSN: 0021-9673

Journal of Chromatography B: Analytical Technologies in the Biomedical and Life Sciences
ISSN: 1570-0232

Journal of Clinical and Basic Cardiology
ISSN: 1561-2775

Journal of Clinical and Experimental Neuropsychology
ISSN: 1380-3395

Journal of Clinical Anesthesia
ISSN: 0952-8180

Journal of Clinical Densitometry
ISSN: 1094-6950

Journal of Clinical Endocrinology and Metabolism
ISSN: 0021-972X

Journal of Clinical Epidemiology
ISSN: 0895-4356

Journal of Clinical Gastroenterology
ISSN: 0192-0790

Journal of Clinical Immunology
ISSN: 0271-9142

Journal of Clinical Investigation
ISSN: 0021-9738

Journal of Clinical Microbiology
ISSN: 0095-1137

Journal of Clinical Monitoring and Computing
ISSN: 1387-1307

Journal of Clinical Neuromuscular Disease
ISSN: 1522-0443

Journal of Clinical Neuroscience
ISSN: 0967-5868

Journal of Clinical Oncology
ISSN: 0732-183X

Journal of Clinical Pathology
ISSN: 0021-9746

Journal of Clinical Psychiatry
ISSN: 0160-6689

Journal of Clinical Psychology in Medical Settings
ISSN: 1068-9583

Journal of Clinical Psychopharmacology
ISSN: 0271-0749

Journal of Clinical Rheumatology
ISSN: 1076-1608

Journal of Clinical Virology
ISSN: 1386-6532

Journal of Cognitive Neuroscience
ISSN: 0898-929X

Journal of Colloid and Interface Science
ISSN: 0021-9797

Journal of Comparative Neurology
ISSN: 0021-9967

Journal of Computational Biology
ISSN: 1066-5277

Journal of Computational Neuroscience
ISSN: 0929-5313

Journal of Computer Assisted Tomography
ISSN: 0363-8715

Journal of Computer-Aided Molecular Design
ISSN: 0920-654X

Journal of Contaminant Hydrology
ISSN: 0169-7722

Journal of Controlled Release
ISSN: 0168-3659

Journal of Cosmetic and Laser Therapy
ISSN: 1462-883X

Journal of Cosmetic Dermatology
ISSN: 1473-2130

Journal of Craniofacial Surgery
ISSN: 1049-2275

Journal of Cutaneous Medicine and Surgery
ISSN: 1203-4754

Journal of Dermatological Science
ISSN: 0923-1811

Journal of Dermatological Treatment
ISSN: 0954-6634

Journal of Developmental and Behavioral Pediatrics
ISSN: 0196-206X

Journal of Diabetes and its Complications
ISSN: 1056-8727

Journal of Diagnostic Medical Sonography
ISSN: 8756-4793

Journal of Digital Imaging
ISSN: 0897-1889

Journal of Drug Targeting
ISSN: 1061-186X

Journal of Electrocardiology
ISSN: 0022-0736

Journal of Electromyography and Kinesiology
ISSN: 1050-6411

Journal of Emergency Medicine
ISSN: 0736-4679

Journal of Endocrinology
ISSN: 0022-0795

Journal of Endourology
ISSN: 0892-7790

Journal of Endovascular Therapy
ISSN: 1526-6028

Journal of Environmental Monitoring
ISSN: 1464-0325

Journal of Environmental Pathology, Toxicology and Oncology
ISSN: 0731-8898

Journal of Enzyme Inhibition and Medicinal Chemistry
ISSN: 8755-5093

Journal of Ethnicity in Substance Abuse
ISSN: 1533-2640

Journal of Evaluation in Clinical Practice
ISSN: 1356-1294

Journal of Experimental and Clinical Cancer Research
ISSN: 0392-9078

Journal of Experimental Medicine
ISSN: 0022-1007

Journal of Experimental Therapeutics and Oncology
ISSN: 1359-4117

Journal of Forensic Odonto-Stomatology
ISSN: 0258-414X

Journal of Forensic Psychiatry and Psychology
ISSN: 1478-9949

Journal of Forensic Sciences
ISSN: 0022-1198

Journal of Gastroenterology
ISSN: 0944-1174

Journal of Gastroenterology and Hepatology
ISSN: 0815-9319

Journal of Gene Medicine
ISSN: 1099-498X

Journal of General Virology
ISSN: 0022-1317

Journal of Geriatric Psychiatry and Neurology
ISSN: 0891-9887

Journal of Glaucoma
ISSN: 1057-0829

Journal of Gynecologic Surgery
ISSN: 1042-4067

Journal of Hand Surgery
ISSN: 0363-5023

Journal of Headache and Pain
ISSN: 1129-2369

Journal of Health Psychology
ISSN: 1359-1053

Journal of Health Services Research and Policy
ISSN: 1355-8196

Journal of Heart and Lung Transplantation
ISSN: 1053-2498

Journal of Heart Valve Disease
ISSN: 0966-8519

Journal of Hepatology
ISSN: 0168-8278

Journal of Herbal Pharmacotherapy
ISSN: 1522-8940

Journal of Heredity
ISSN: 0022-1503

Journal of Histochemistry and Cytochemistry
ISSN: 0022-1554

Journal of Hospital Marketing and Public Relations
ISSN: 1539-0942

Journal of Hypertension
ISSN: 0263-6352

Journal of Hypertension, Supplement
ISSN: 0952-1178

Journal of Immigrant and Minority Health
ISSN: 1557-1912

Journal of Immunoassay and Immunochemistry
ISSN: 1532-1819

Journal of Immunological Methods
ISSN: 0022-1759

Journal of Immunology
ISSN: 0022-1767

Journal of Immunotherapy
ISSN: 1524-9557

Journal of Immunotoxicology
ISSN: 1547-691X

Journal of Infectious Diseases
ISSN: 0022-1899

Journal of Infusion Nursing
ISSN: 1533-1458

Journal of Intellectual Disabilities
ISSN: 1744-6295

Journal of Intensive Care Medicine
ISSN: 0885-0666

Journal of Interferon and Cytokine Research
ISSN: 1079-9907

Journal of Internal Medicine
ISSN: 0954-6820

Journal of Interventional Cardiac Electrophysiology
ISSN: 1383-875X

Journal of Interventional Cardiology
ISSN: 0896-4327

Journal of Investigative Dermatology
ISSN: 0022-202X

Journal of Investigative Surgery
ISSN: 0894-1939

Journal of Laparoendoscopic and Advanced Surgical Techniques - Part A
ISSN: 1092-6429

Journal of Leukocyte Biology
ISSN: 0741-5400

Journal of Lipid Research
ISSN: 0022-2275

Journal of Liposome Research
ISSN: 0898-2104

Journal of Magnetic Resonance Imaging
ISSN: 1053-1807

Journal of Mass Spectrometry
ISSN: 1076-5174

Journal of Materials Science: Materials in Medicine
ISSN: 0957-4530

Journal of Maternal-Fetal and Neonatal Medicine
ISSN: 1476-7058

Journal of Medical Genetics
ISSN: 0022-2593

Journal of Medical Microbiology
ISSN: 0022-2615

Journal of Medical Primatology
ISSN: 0047-2565

Journal of Medical Speech-Language Pathology
ISSN: 1065-1438

Journal of Medicinal Food
ISSN: 1096-620X

Journal of Membrane Science
ISSN: 0376-7388

Journal of Men's Health and Gender
ISSN: 1571-8913

Journal of Microbiological Methods
ISSN: 0167-7012

Journal of Microscopy
ISSN: 0022-2720

Journal of Midwifery and Women's Health
ISSN: 1526-9523

Journal of Molecular and Cellular Cardiology
ISSN: 0022-2828

Journal of Molecular Biology
ISSN: 0022-2836

Journal of Molecular Endocrinology
ISSN: 0952-5041

Journal of Molecular Graphics and Modelling
ISSN: 1093-3263

Journal of Molecular Histology
ISSN: 1567-2379

Journal of Molecular Modeling
ISSN: 1610-2940

Journal of Muscle Research and Cell Motility
ISSN: 0142-4319

Journal of Neural Engineering
ISSN: 1741-2560

Journal of Neural Transmission
ISSN: 0300-9564

Journal of Neural Transmission, Supplement
ISSN: 0303-6995

Journal of Neuro-Ophthalmology
ISSN: 1070-8022

Journal of Neurobiology
ISSN: 0022-3034

Journal of Neurochemistry
ISSN: 0022-3042

Journal of Neurocytology
ISSN: 0300-4864

Journal of Neuroendocrinology
ISSN: 0953-8194

Journal of Neurogenetics
ISSN: 0167-7063

Journal of Neuroimmunology
ISSN: 0165-5728

Journal of Neurolinguistics
ISSN: 0911-6044

Journal of Neurology
ISSN: 0340-5354

Journal of Neurology, Neurosurgery and Psychiatry
ISSN: 0022-3050

Journal of Neurology, Supplement
ISSN: 0939-1517

Journal of Neuropathology and Experimental Neurology
ISSN: 0022-3069

Journal of Neurophysiology
ISSN: 0022-3077

Journal of Neuropsychiatry and Clinical Neurosciences
ISSN: 0895-0172

Journal of Neuroscience
ISSN: 0270-6474

Journal of Neuroscience Methods
ISSN: 0165-0270

Journal of Neuroscience Research
ISSN: 0360-4012

Journal of Neurosurgery
ISSN: 0022-3085

Journal of Neurosurgical Anesthesiology
ISSN: 0898-4921

Journal of Neurotherapy
ISSN: 1087-4208

Journal of NeuroVirology
ISSN: 1355-0284

Journal of Nuclear Cardiology
ISSN: 1071-3581

Journal of Nuclear Medicine
ISSN: 0161-5505

Journal of Nutrition, Health and Aging
ISSN: 1279-7707

Journal of Nutritional and Environmental Medicine
ISSN: 1359-0847

Journal of Obstetrics and Gynaecology
ISSN: 0144-3615

Journal of Occupational and Environmental Hygiene
ISSN: 1545-9624

Journal of Occupational Health Psychology
ISSN: 1076-8998

Journal of Oncology Pharmacy Practice
ISSN: 1078-1552

Journal of Oral Pathology and Medicine
ISSN: 0904-2512

Journal of Organ Dysfunction
ISSN: 1747-1060

Journal of Orofacial Orthopedics
ISSN: 1434-5293

Journal of Orthopaedic Research
ISSN: 0736-0266

Journal of Orthopaedic Trauma
ISSN: 0890-5339

Journal of Orthopaedics and Traumatology
ISSN: 1590-9921

Journal of Otolaryngology
ISSN: 0381-6605

Journal of Paediatrics and Child Health
ISSN: 1034-4810

Journal of Parenteral and Enteral Nutrition
ISSN: 0148-6071

Journal of Pathology
ISSN: 0022-3417

Journal of Pediatric and Adolescent Gynecology
ISSN: 1083-3188

Journal of Pediatric Gastroenterology and Nutrition
ISSN: 0277-2116

Journal of Pediatric Hematology/ Oncology
ISSN: 1077-4114

Journal of Pediatric Orthopaedics
ISSN: 0271-6798

Journal of Pediatric Orthopaedics Part B
ISSN: 1060-152X

Journal of Pediatric Surgery
ISSN: 0022-3468

Journal of Pediatric Urology
ISSN: 1477-5131

Journal of Pediatrics
ISSN: 0022-3476

Journal of Pelvic Medicine and Surgery
ISSN: 1542-5983

Journal of Peptide Science
ISSN: 1075-2617

Journal of Perinatal Medicine
ISSN: 0300-5577

Journal of Pharmaceutical and Biomedical Analysis
ISSN: 0731-7085

Journal of Pharmaceutical Finance, Economics and Policy
ISSN: 1538-5698

Journal of Pharmacokinetics and Pharmacodynamics
ISSN: 1567-567X

Journal of Pharmacology and Experimental Therapeutics
ISSN: 0022-3565

Journal of Photochemistry and Photobiology B: Biology
ISSN: 1011-1344

Journal of Physiology
ISSN: 0022-3751

Journal of Plastic, Reconstructive and Aesthetic Surgery
ISSN: 1748-6815

Journal of Proteome Research
ISSN: 1535-3893

Journal of Psychiatric Research
ISSN: 0022-3956

Journal of Psychopharmacology
ISSN: 0269-8811

Journal of Public Health
ISSN: 1741-3842

Journal of Public Health Management and Practice
ISSN: 1078-4659

Journal of Radiological Protection
ISSN: 0952-4746

Journal of Radiology Nursing
ISSN: 1546-0843

Journal of Reconstructive Microsurgery
ISSN: 0743-684X

Journal of Refractive Surgery
ISSN: 1081-597X

Journal of Rehabilitation Research and Development
ISSN: 0748-7711

Journal of Reproductive Immunology
ISSN: 0165-0378

Journal of Reproductive Medicine for the Obstetrician and Gynecologist
ISSN: 0024-7758

Journal of Rheumatology
ISSN: 0315-162X

Journal of Separation Science
ISSN: 1615-9306

Journal of Sexual Medicine
ISSN: 1743-6095

Journal of Shoulder and Elbow Surgery
ISSN: 1058-2746

Journal of Sleep Research
ISSN: 0962-1105

Journal of Sleep Research, Supplement
ISSN: 0966-6826

Journal of Spinal Disorders and Techniques
ISSN: 1536-0652

Journal of Spirochetal and Tick-borne Diseases
ISSN: 1060-0051

Journal of Stroke and Cerebrovascular Diseases
ISSN: 1052-3057

Journal of Structural and Functional Genomics
ISSN: 1345-711X

Journal of Structural Biology
ISSN: 1047-8477

Journal of Substance Abuse Treatment
ISSN: 0740-5472

Journal of Substance Use
ISSN: 1465-9891

Journal of Surgical Oncology
ISSN: 0022-4790

Journal of Surgical Research
ISSN: 0022-4804

Journal of the Acoustical Society of America
ISSN: 0001-4966

Journal of the Air and Waste Management Association
ISSN: 1047-3289

Journal of the American Academy of Audiology
ISSN: 1050-0545

Journal of the American Academy of Child and Adolescent Psychiatry
ISSN: 0890-8567

Journal of the American Academy of Dermatology
ISSN: 0190-9622

Journal of the American College of Cardiology
ISSN: 0735-1097

Journal of the American College of Surgeons
ISSN: 1072-7515

Journal of the American Medical Association
ISSN: 0098-7484

Journal of the American Society of Nephrology
ISSN: 1046-6673

Journal of the Chartered Institution of Water and Environmental Management
ISSN: 0951-7359

Journal of the European Academy of Dermatology and Venereology
ISSN: 0926-9959

Journal of the International Neuropsychological Society
ISSN: 1355-6177

Journal of the National Cancer Institute
ISSN: 0027-8874

Journal of the National Medical Association
ISSN: 0027-9684

Journal of the Neurological Sciences
ISSN: 0022-510X

Journal of the Peripheral Nervous System
ISSN: 1085-9489

Journal of the Society for Gynecologic Investigation
ISSN: 1071-5576

Journal of Theoretical Biology
ISSN: 0022-5193

Journal of Thoracic and Cardiovascular Surgery
ISSN: 0022-5223

Journal of Thrombosis and Haemostasis
ISSN: 1538-7933

Journal of Thrombosis and Thrombolysis
ISSN: 0929-5305

Journal of Toxicology and Environmental Health - Part A
ISSN: 1528-7394

Journal of Trace Elements in Medicine and Biology
ISSN: 0946-672X

Journal of Trauma - Injury, Infection and Critical Care
ISSN: 0022-5282

Journal of Trauma and Dissociation
ISSN: 1529-9732

Journal of Urban Health
ISSN: 1099-3460

Journal of Urology
ISSN: 0022-5347

Journal of Vascular and Interventional Radiology
ISSN: 1051-0443

Journal of Vascular Research
ISSN: 1018-1172

Journal of Vascular Surgery
ISSN: 0741-5214

Journal of Vestibular Research: Equilibrium and Orientation
ISSN: 0957-4271

Journal of Veterinary Cardiology
ISSN: 1760-2734

Journal of Veterinary Pharmacology and Therapeutics
ISSN: 0140-7783

Journal of Viral Hepatitis
ISSN: 1352-0504

Journal of Virological Methods
ISSN: 0166-0934

Journal of Virology
ISSN: 0022-538X

Journal of Whiplash and Related Disorders
ISSN: 1533-2888

Journal of Women's Health
ISSN: 1540-9996

Journal of Women's Imaging
ISSN: 1084-824X

Journals of Gerontology - Series A Biological Sciences and Medical Sciences
ISSN: 1079-5006

Journals of Gerontology - Series B Psychological Sciences and Social Sciences
ISSN: 1079-5014

Kidney and Blood Pressure Research
ISSN: 1420-4096

Kidney International
ISSN: 0085-2538

Knee
ISSN: 0968-0160

Lab on a Chip - Miniaturisation for Chemistry and Biology
ISSN: 1473-0197

Laboratory Investigation
ISSN: 0023-6837

Lancet
ISSN: 0140-6736

Lancet Infectious Diseases
ISSN: 1473-3099

Lancet Neurology
ISSN: 1474-4422

Lancet Oncology
ISSN: 1470-2045

Langenbeck's Archives of Surgery
ISSN: 1435-2443

Language and Cognitive Processes
ISSN: 0169-0965

Laryngo- Rhino- Otologie
ISSN: 0935-8943

Laryngoscope
ISSN: 0023-852X

Lasers in Medical Science
ISSN: 0268-8921

Lasers in Surgery and Medicine
ISSN: 0196-8092

Learning and Memory
ISSN: 1072-0502

Legal and Criminological Psychology
ISSN: 1355-3259

Letters in Drug Design and Discovery
ISSN: 1570-1808

Leukemia
ISSN: 0887-6924

Leukemia and Lymphoma
ISSN: 1042-8194

Leukemia Research
ISSN: 0145-2126

Liver Transplantation
ISSN: 1527-6465

Logopedics Phoniatrics Vocology
ISSN: 1401-5439

Logopedics Phoniatrics Vocology, Supplement
ISSN: 1401-5455

Lung
ISSN: 0341-2040

Lung Cancer
ISSN: 0169-5002

Lupus
ISSN: 0961-2033

Lymphatic Research and Biology
ISSN: 1539-6851

Lymphology
ISSN: 0024-7766

Magnetic Resonance Imaging
ISSN: 0730-725X

Magnetic Resonance Materials in Physics, Biology and Medicine
ISSN: 0968-5243

Manual Therapy
ISSN: 1356-689X

Maternal and Child Health Journal
ISSN: 1092-7875

Maternal and Child Nutrition
ISSN: 1740-8695

Matrix Biology
ISSN: 0945-053X

Mechanisms of Ageing and Development
ISSN: 0047-6374

Mechanisms of Development
ISSN: 0925-4773

Mediators of Inflammation
ISSN: 0962-9351

Medical Clinics of North America
ISSN: 0025-7125

Medical Dosimetry
ISSN: 0958-3947

Medical Electron Microscopy
ISSN: 0918-4287

Medical Engineering and Physics
ISSN: 1350-4533

Medical Hypotheses
ISSN: 0306-9877

Medical Image Analysis
ISSN: 1361-8415

Medical Microbiology and Immunology
ISSN: 0300-8584

Medical Oncology
ISSN: 1357-0560

Medical Physics
ISSN: 0094-2405

Medicinal Chemistry
ISSN: 1573-4064

Medicine
ISSN: 0025-7974

Melanoma Research
ISSN: 0960-8931

Memory
ISSN: 0965-8211

Mental Retardation and Developmental Disabilities Research Reviews
ISSN: 1080-4013

Metabolic Engineering
ISSN: 1096-7176

Metabolic Syndrome and Related Disorders
ISSN: 1540-4196

Metabolism: Clinical and Experimental
ISSN: 0026-0495

Methods
ISSN: 1046-2023

Methods in Enzymology
ISSN: 0076-6879

Methods of Information in Medicine
ISSN: 0026-1270

Microbes and Infection
ISSN: 1286-4579

Microbial Drug Resistance
ISSN: 1076-6294

Microbial Ecology in Health and Disease
ISSN: 0891-060X

Microbial Pathogenesis
ISSN: 0882-4010

Microbiology
ISSN: 1350-0872

Microcirculation
ISSN: 1073-9688

Microscopy Research and Technique
ISSN: 1059-910X

Microsurgery
ISSN: 0738-1085

Microvascular Research
ISSN: 0026-2862

Minimally Invasive Therapy and Allied Technologies
ISSN: 1364-5706

Mitochondrion
ISSN: 1567-7249

Modern Pathology
ISSN: 0893-3952

Modern Rheumatology
ISSN: 1439-7595

Molecular and Biochemical Parasitology
ISSN: 0166-6851

Molecular and Cellular Biology
ISSN: 0270-7306

Molecular and Cellular Endocrinology
ISSN: 0303-7207

Molecular and Cellular Neuroscience
ISSN: 1044-7431

Molecular and Cellular Probes
ISSN: 0890-8508

Molecular and Cellular Proteomics
ISSN: 1535-9476

Molecular Biology of the Cell
ISSN: 1059-1524

Molecular Brain Research
ISSN: 0169-328X

Molecular Cancer Research
ISSN: 1541-7786

Molecular Cancer Therapeutics
ISSN: 1535-7163

Molecular Carcinogenesis
ISSN: 0899-1987

Molecular Diagnosis and Therapy
ISSN: 1177-1062

Molecular Diversity
ISSN: 1381-1991

Molecular Endocrinology
ISSN: 0888-8809

Molecular Genetics and Genomics
ISSN: 1617-4615

Molecular Genetics and Metabolism
ISSN: 1096-7192

Molecular Human Reproduction
ISSN: 1360-9947

Molecular Imaging and Biology
ISSN: 1536-1632

Molecular Immunology
ISSN: 0161-5890

Molecular Medicine
ISSN: 1076-1551

Molecular Membrane Biology
ISSN: 0968-7688

Molecular Microbiology
ISSN: 0950-382X

Molecular Pain
ISSN: 1744-8069

Molecular Pharmaceutics
ISSN: 1543-8384

Molecular Pharmacology
ISSN: 0026-895X

Molecular Psychiatry
ISSN: 1359-4184

Molecular Reproduction and Development
ISSN: 1040-452X

Molecular Systems Biology
ISSN: 1744-4292

Molecular Vision
ISSN: 1090-0535

Movement Disorders
ISSN: 0885-3185

Muscle and Nerve
ISSN: 0148-639X

Mutagenesis
ISSN: 0267-8357

Mutation Research - Fundamental and Molecular Mechanisms of Mutagenesis
ISSN: 0027-5107

Mutation Research - Genetic Toxicology and Environmental Mutagenesis
ISSN: 1383-5718

Mutation Research - Reviews in Mutation Research
ISSN: 1383-5742

Mycoses
ISSN: 0933-7407

Mycoses, Supplement
ISSN: 0943-7312

Nanotoxicology
ISSN: 1743-5390

Nature
ISSN: 0028-0836

Nature Biotechnology
ISSN: 1087-0156

Nature Cell Biology
ISSN: 1465-7392

Nature Chemical Biology
ISSN: 1552-4450

Nature Clinical Practice Cardiovascular Medicine
ISSN: 1743-4297

Nature Clinical Practice Endocrinology and Metabolism
ISSN: 1745-8366

Nature Clinical Practice Gastroenterology and Hepatology
ISSN: 1743-4378

Nature Clinical Practice Nephrology
ISSN: 1745-8323

Nature Clinical Practice Neurology
ISSN: 1745-834X

Nature Clinical Practice Oncology
ISSN: 1743-4254

Nature Clinical Practice Rheumatology
ISSN: 1745-8382

Nature Clinical Practice Urology
ISSN: 1743-4270

Nature Genetics
ISSN: 1061-4036

Nature Immunology
ISSN: 1529-2908

Nature Medicine
ISSN: 1078-8956

Nature Methods
ISSN: 1548-7091

Nature Neuroscience
ISSN: 1097-6256

Nature Reviews Cancer
ISSN: 1474-175X

Nature Reviews Drug Discovery
ISSN: 1474-1776

Nature Reviews Genetics
ISSN: 1471-0056

Nature Reviews Immunology
ISSN: 1474-1733

Nature Reviews Microbiology
ISSN: 1740-1526

Nature Reviews Molecular Cell Biology
ISSN: 1471-0072

Nature Reviews Neuroscience
ISSN: 1471-003X

Nature Structural and Molecular Biology
ISSN: 1545-9993

Neoplasia
ISSN: 1522-8002

Nephrology
ISSN: 1320-5358

Nephrology Dialysis Transplantation
ISSN: 0931-0509

Nephron - Clinical Practice
ISSN: 1660-2110

Nephron - Experimental Nephrology
ISSN: 1660-2129

Nephron - Physiology
ISSN: 1660-2137

Neural Networks
ISSN: 0893-6080

Neurobiology of Aging
ISSN: 0197-4580

Neurobiology of Disease
ISSN: 0969-9961

Neurochemical Research
ISSN: 0364-3190

Neurochemistry International
ISSN: 0197-0186

Neurocomputing
ISSN: 0925-2312

Neurocritical Care
ISSN: 1541-6933

Neurodegenerative Diseases
ISSN: 1660-2854

Neuroembryology and Aging
ISSN: 1661-3406

Neuroendocrinology
ISSN: 0028-3835

Neurogastroenterology and Motility
ISSN: 1350-1925

Neurogenetics
ISSN: 1364-6745

NeuroImage
ISSN: 1053-8119

Neuroimaging Clinics of North America
ISSN: 1052-5149

NeuroImmunoModulation
ISSN: 1021-7401

Neuroinformatics
ISSN: 1539-2791

Neurologic Clinics
ISSN: 0733-8619

Neurologist
ISSN: 1074-7931

Neurology
ISSN: 0028-3878

Neurology and Clinical Neurophysiology
ISSN: 1526-8748

Neurology in Practice
ISSN: 1473-7086

Neurology Psychiatry and Brain Research
ISSN: 0941-9500

NeuroMolecular Medicine
ISSN: 1535-1084

Neuromuscular Disorders
ISSN: 0960-8966

Neuron
ISSN: 0896-6273

Neuron Glia Biology
ISSN: 1740-925X

Neuropathology
ISSN: 0919-6544

Neuropathology and Applied Neurobiology
ISSN: 0305-1846

Neuropediatrics
ISSN: 0174-304X

Neuropeptides
ISSN: 0143-4179

Neuropharmacology
ISSN: 0028-3908

Neuropsychobiology
ISSN: 0302-282X

Neuropsychology
ISSN: 0894-4105

Neuropsychopharmacology
ISSN: 0893-133X

Neuroradiology
ISSN: 0028-3940

NeuroReport
ISSN: 0959-4965

Neuroscience
ISSN: 0306-4522

Neuroscience and Biobehavioral Reviews
ISSN: 0149-7634

Neuroscience Letters
ISSN: 0304-3940

Neuroscience Research
ISSN: 0168-0102

Neuroscientist
ISSN: 1073-8584

Neurosurgery
ISSN: 0148-396X

Neurosurgery Clinics of North America
ISSN: 1042-3680

Neurosurgery Quarterly
ISSN: 1050-6438

Neurosurgical Review
ISSN: 0344-5607

NeuroToxicology
ISSN: 0161-813X

Neurotoxicology and Teratology
ISSN: 0892-0362

New England Journal of Medicine
ISSN: 0028-4793

New Genetics and Society
ISSN: 1463-6778

New Journal of Chemistry
ISSN: 1144-0546

Nicotine and Tobacco Research
ISSN: 1462-2203

Nitric Oxide - Biology and Chemistry
ISSN: 1089-8603

NMR in Biomedicine
ISSN: 0952-3480

Nucleic Acids Research
ISSN: 0305-1048

Nursing and Health Sciences
ISSN: 1441-0745

Nutrition
ISSN: 0899-9007

Nutrition Bulletin
ISSN: 1471-9827

Nutrition Research
ISSN: 0271-5317

Nutritional Neuroscience
ISSN: 1028-415X

Obesity Surgery
ISSN: 0960-8923

Obstetrics and Gynecology
ISSN: 0029-7844

Obstetrics and Gynecology Clinics of North America
ISSN: 0889-8545

Occupational and Environmental Medicine
ISSN: 1351-0711

Occupational Medicine
ISSN: 0962-7480

Oligonucleotides
ISSN: 1545-4576

OMICS A Journal of Integrative Biology
ISSN: 1536-2310

Oncogene
ISSN: 0950-9232

Oncologist
ISSN: 1083-7159

Oncology
ISSN: 0030-2414

Oncology Research
ISSN: 0965-0407

Operative Orthopadie und Traumatologie
ISSN: 0934-6694

Operative Techniques in General Surgery
ISSN: 1524-153X

Operative Techniques in Orthopaedics
ISSN: 1048-6666

Operative Techniques in Otolaryngology - Head and Neck Surgery
ISSN: 1043-1810

Operative Techniques in Plastic and Reconstructive Surgery
ISSN: 1071-0949

Operative Techniques in Sports Medicine
ISSN: 1060-1872

Operative Techniques in Thoracic and Cardiovascular Surgery
ISSN: 1522-2942

Ophthalmic and Physiological Optics
ISSN: 0275-5408

Ophthalmic Epidemiology
ISSN: 0928-6586

Ophthalmic Genetics
ISSN: 1381-6810

Ophthalmic Plastic and Reconstructive Surgery
ISSN: 0740-9303

Ophthalmic Research
ISSN: 0030-3747

Ophthalmic Surgery Lasers and Imaging
ISSN: 1542-8877

Ophthalmology
ISSN: 0161-6420

Ophthalmology Clinics of North America
ISSN: 0896-1549

Optometry and Vision Science
ISSN: 1040-5488

Oral Diseases
ISSN: 1354-523X

Oral Oncology
ISSN: 1368-8375

ORL
ISSN: 0301-1569

Orthopedic Clinics of North America
ISSN: 0030-5898

Osteoarthritis and Cartilage
ISSN: 1063-4584

Osteoporosis International
ISSN: 0937-941X

Otology and Neurotology
ISSN: 1531-7129

Paediatric Anaesthesia
ISSN: 1155-5645

Paediatric Respiratory Reviews
ISSN: 1526-0542

Pain
ISSN: 0304-3959

Pancreas
ISSN: 0885-3177

Pancreatology
ISSN: 1424-3903

Parasite Immunology
ISSN: 0141-9838

Parasitology
ISSN: 0031-1820

Parasitology International
ISSN: 1383-5769

Parasitology Research
ISSN: 0932-0113

Parkinsonism and Related Disorders
ISSN: 1353-8020

Particle and Fibre Toxicology
ISSN: 1743-8977

Particulate Science and Technology
ISSN: 0272-6351

Pathobiology
ISSN: 1015-2008

Pathology Case Reviews
ISSN: 1082-9784

Pathology International
ISSN: 1320-5463

Pathophysiology
ISSN: 0928-4680

Pathophysiology of Haemostasis and Thrombosis
ISSN: 1424-8832

Patient Education and Counseling
ISSN: 0738-3991

Pediatric Allergy and Immunology
ISSN: 0905-6157

Pediatric and Developmental Pathology
ISSN: 1093-5266

Pediatric Asthma, Allergy and Immunology
ISSN: 0883-1874

Pediatric Blood and Cancer
ISSN: 1545-5009

Pediatric Critical Care Medicine
ISSN: 1529-7535

Pediatric Dermatology
ISSN: 0736-8046

Pediatric Diabetes
ISSN: 1399-543X

Pediatric Diabetes, Supplement
ISSN: 1745-1426

Pediatric Drugs
ISSN: 1174-5878

Pediatric Endosurgery and Innovative Techniques
ISSN: 1092-6410

Pediatric Infectious Disease Journal
ISSN: 0891-3668

Pediatric Nephrology
ISSN: 0931-041X

Pediatric Neurology
ISSN: 0887-8994

Pediatric Neurosurgery
ISSN: 1016-2291

Pediatric Pulmonology
ISSN: 8755-6863

Pediatric Radiology
ISSN: 0301-0449

Pediatric Rehabilitation
ISSN: 1363-8491

Pediatric Research
ISSN: 0031-3998

Pediatric Surgery International
ISSN: 0179-0358

Pediatrics
ISSN: 0031-4005

Pediatrics International
ISSN: 1328-8067

Peptides
ISSN: 0196-9781

Perfusion
ISSN: 0267-6591

Peritoneal Dialysis International
ISSN: 0896-8608

Personalized Medicine
ISSN: 1741-0541

Perspectives in Biology and Medicine
ISSN: 0031-5982

Pflugers Archiv European Journal of Physiology
ISSN: 0031-6768

Pharmaceutical Development and Technology
ISSN: 1083-7450

Pharmaceutical Research
ISSN: 0724-8741

PharmacoEconomics
ISSN: 1170-7690

Pharmacoepidemiology and Drug Safety
ISSN: 1053-8569

Pharmacogenetics and Genomics
ISSN: 1744-6872

Pharmacogenomics Journal
ISSN: 1470-269X

Pharmacological Research
ISSN: 1043-6618

Pharmacological Reviews
ISSN: 0031-6997

Pharmacology
ISSN: 0031-7012

Pharmacology and Therapeutics
ISSN: 0163-7258

Pharmacology Biochemistry and Behavior
ISSN: 0091-3057

Pharmacopsychiatry
ISSN: 0176-3679

Pharmacy Education
ISSN: 1560-2214

Phlebology
ISSN: 0268-3555

Photochemical and Photobiological Sciences
ISSN: 1474-905X

Photodermatology Photoimmunology and Photomedicine
ISSN: 0905-4383

Photodiagnosis and Photodynamic Therapy
ISSN: 1572-1000

Physical Biology
ISSN: 1478-3975

Physical Medicine and Rehabilitation Clinics of North America
ISSN: 1047-9651

Physical Therapy in Sport
ISSN: 1466-853X

Physics in Medicine and Biology
ISSN: 0031-9155

Physiological Genomics
ISSN: 1094-8341

Physiological Measurement
ISSN: 0967-3334

Physiological Reviews
ISSN: 0031-9333

Physiology and Behavior
ISSN: 0031-9384

Physiotherapy
ISSN: 0031-9406

Phytochemistry Reviews
ISSN: 1568-7767

Phytomedicine
ISSN: 0944-7113

Pituitary
ISSN: 1386-341X

Placenta
ISSN: 0143-4004

Plastic and Reconstructive Surgery
ISSN: 0032-1052

Platelets
ISSN: 0953-7104

Pneumologie
ISSN: 0934-8387

PPmP Psychotherapie Psychosomatik Medizinische Psychologie
ISSN: 0937-2032

Prenatal Diagnosis
ISSN: 0197-3851

Prevention and Control
ISSN: 1573-2088

Preventive Medicine
ISSN: 0091-7435

Primary Care - Clinics in Office Practice
ISSN: 0095-4543

Primary Care and Community Psychiatry
ISSN: 1746-8841

Proceedings of the National Academy of Sciences of the United States of America
ISSN: 0027-8424

Proceedings of the Royal Society - Biological Sciences (Series B)
ISSN: 0962-8452

Progress in Brain Research
ISSN: 0079-6123

Progress in Drug Research
ISSN: 0071-786X

Progress in Lipid Research
ISSN: 0163-7827

Progress in Neurobiology
ISSN: 0301-0082

Progress in Pediatric Cardiology
ISSN: 1058-9813

Progress in Retinal and Eye Research
ISSN: 1350-9462

Prostaglandins Leukotrienes and Essential Fatty Acids
ISSN: 0952-3278

Prostate
ISSN: 0270-4137

Prostate Cancer and Prostatic Diseases
ISSN: 1365-7852

Protein Engineering, Design and Selection
ISSN: 1741-0126

Protein Science
ISSN: 0961-8368

Proteins: Structure, Function and Genetics
ISSN: 0887-3585

Proteomics
ISSN: 1615-9853

Psychiatric Clinics of North America
ISSN: 0193-953X

Psychiatric Genetics
ISSN: 0955-8829

Psychiatry Research
ISSN: 0165-1781

Psychiatry Research - Neuroimaging
ISSN: 0925-4927

Psychoanalytic Psychotherapy
ISSN: 0266-8734

Psychogeriatrics
ISSN: 1346-3500

Psychology, Health and Medicine
ISSN: 1354-8506

Psychoneuroendocrinology
ISSN: 0306-4530

Psychopharmacology
ISSN: 0033-3158

Psychophysiology
ISSN: 0048-5772

Psychosomatic Medicine
ISSN: 0033-3174

Psychotherapy and Psychosomatics
ISSN: 0033-3190

Psychotherapy Research
ISSN: 1050-3307

Pteridines
ISSN: 0933-4807

Public Health Medicine
ISSN: 1465-1505

Public Health Nutrition
ISSN: 1368-9800

Public Health Reports
ISSN: 0033-3549

Public Health Reviews
ISSN: 0301-0422

Pulmonary Pharmacology and Therapeutics
ISSN: 1094-5539

QJM - Monthly Journal of the Association of Physicians
ISSN: 1460-2725

QSAR and Combinatorial Science
ISSN: 1611-020X

Qualitative Health Research
ISSN: 1049-7323

Quality Assurance Journal
ISSN: 1087-8378

Quality of Life Research
ISSN: 0962-9343

Quarterly Reviews of Biophysics
ISSN: 0033-5835

Radiation Research
ISSN: 0033-7587

Radiography
ISSN: 1078-8174

Radiologic Clinics of North America
ISSN: 0033-8389

Radiology
ISSN: 0033-8419

Radiotherapy and Oncology
ISSN: 0167-8140

Redox Report
ISSN: 1351-0002

Regional Anesthesia and Pain Medicine
ISSN: 1098-7339

Regulatory Peptides
ISSN: 0167-0115

Regulatory Toxicology and Pharmacology
ISSN: 0273-2300

Rejuvenation Research
ISSN: 1549-1684

Renal Failure
ISSN: 0886-022X

Reproduction
ISSN: 1470-1626

Reproductive Health Matters
ISSN: 0968-8080

Reproductive Medicine and Biology
ISSN: 1445-5781

Reproductive Medicine Review
ISSN: 0962-2799

Research Communications in Biological Psychology and Psychiatry
ISSN: 1087-0695

Research Communications in Molecular Pathology and Pharmacology
ISSN: 1078-0297

Research in Microbiology
ISSN: 0923-2508

Respiration
ISSN: 0025-7931

Respiratory Medicine
ISSN: 0954-6111

Respiratory Medicine Extra
ISSN: 1744-9049

Respiratory Medicine: COPD Update
ISSN: 1745-0454

Respiratory Physiology and Neurobiology
ISSN: 1569-9048

Respirology
ISSN: 1323-7799

Restorative Neurology and Neuroscience
ISSN: 0922-6028

Resuscitation
ISSN: 0300-9572

Reviews in Gynaecological and Perinatal Practice
ISSN: 1871-2320

Reviews in Medical Microbiology
ISSN: 0954-139X

Reviews in the Neurosciences
ISSN: 0334-1763

Rheumatic Disease Clinics of North America
ISSN: 0889-857X

Rheumatology
ISSN: 1462-0324

Rheumatology International
ISSN: 0172-8172

RNA
ISSN: 1355-8382

RoFo Fortschritte auf dem Gebiet der Rontgenstrahlen und der Bildgebenden Verfahren
ISSN: 1438-9029

Safety Science
ISSN: 0925-7535

Santo Tomas Journal of Medicine
ISSN: 0115-1126

Sarcoidosis Vasculitis and Diffuse Lung Diseases
ISSN: 1124-0490

Sarcoma
ISSN: 1357-714X

Scandinavian Cardiovascular Journal
ISSN: 1401-7431

Scandinavian Cardiovascular Journal, Supplement
ISSN: 1401-7458

Scandinavian Journal of Clinical and Laboratory Investigation
ISSN: 0036-5513

Scandinavian Journal of Clinical and Laboratory Investigation, Supplement
ISSN: 0085-591X

Scandinavian Journal of Gastroenterology
ISSN: 0036-5521

Scandinavian Journal of Gastroenterology, Supplement
ISSN: 0085-5928

Scandinavian Journal of Immunology
ISSN: 0300-9475

Scandinavian Journal of Immunology, Supplement
ISSN: 0301-6323

Scandinavian Journal of Public Health
ISSN: 1403-4948

Scandinavian Journal of Public Health, Supplement
ISSN: 1403-4956

Scandinavian Journal of Rheumatology
ISSN: 0300-9742

Scandinavian Journal of Rheumatology, Supplement
ISSN: 0301-3847

Scandinavian Journal of Urology and Nephrology
ISSN: 0036-5599

Scandinavian Journal of Urology and Nephrology, Supplement
ISSN: 0300-8886

Scandinavian Journal of Work, Environment and Health
ISSN: 0355-3140

Scanning
ISSN: 0161-0457

Schizophrenia Bulletin
ISSN: 0586-7614

Schizophrenia Research
ISSN: 0920-9964

Science
ISSN: 0036-8075

Science and Sports
ISSN: 0765-1597

Science of the Total Environment
ISSN: 0048-9697

Seizure
ISSN: 1059-1311

Seminars in Anesthesia, Perioperative Medicine and Pain
ISSN: 0277-0326

Seminars in Arthritis and Rheumatism
ISSN: 0049-0172

Seminars in Breast Disease
ISSN: 1092-4450

Seminars in Cerebrovascular Diseases and Stroke
ISSN: 1528-9931

Seminars in Diagnostic Pathology
ISSN: 0740-2570

Seminars in Hematology
ISSN: 0037-1963

Seminars in Liver Disease
ISSN: 0272-8087

Seminars in Nephrology
ISSN: 0270-9295

Seminars in Neurology
ISSN: 0271-8235

Seminars in Oncology
ISSN: 0093-7754

Seminars in Ophthalmology
ISSN: 0882-0538

Seminars in Pain Medicine
ISSN: 1537-5897

Seminars in Pediatric Surgery
ISSN: 1055-8586

Seminars in Perinatology
ISSN: 0146-0005

Seminars in Radiation Oncology
ISSN: 1053-4296

Seminars in Respiratory and Critical Care Medicine
ISSN: 1069-3424

Seminars in Roentgenology
ISSN: 0037-198X

Seminars in Spine Surgery
ISSN: 1040-7383

Seminars in Surgical Oncology
ISSN: 8756-0437

Seminars in Thrombosis and Hemostasis
ISSN: 0094-6176

Seminars in Ultrasound, CT and MRI
ISSN: 0887-2171

Seminars in Vascular Surgery
ISSN: 0895-7967

Sexualities, Evolution and Gender
ISSN: 1479-2508

Sexuality, Reproduction and Menopause
ISSN: 1546-2501

Sexually Transmitted Infections
ISSN: 1368-4973

Signal Transduction
ISSN: 1615-4053

Skeletal Radiology
ISSN: 0364-2348

Skin Pharmacology and Physiology
ISSN: 1660-5527

Skull Base
ISSN: 1531-5010

Sleep
ISSN: 0161-8105

Sleep and Biological Rhythms
ISSN: 1446-9235

Sleep and Breathing
ISSN: 1520-9512

Sleep Medicine
ISSN: 1389-9457

Somatic Cell and Molecular Genetics
ISSN: 0740-7750

Somatosensory and Motor Research
ISSN: 0899-0220

Spinal Cord
ISSN: 1362-4393

Spine
ISSN: 0362-2436

Spine Journal
ISSN: 1529-9430

Sports Medicine and Arthroscopy Review
ISSN: 1062-8592

Springer Seminars in Immunopathology
ISSN: 0344-4325

Stem Cells and Development
ISSN: 1547-3287

Strabismus
ISSN: 0927-3972

Stress
ISSN: 1025-3890

Stroke
ISSN: 0039-2499

Structure
ISSN: 0969-2126

Substance Use and Misuse
ISSN: 1082-6084

Supportive Care in Cancer
ISSN: 0941-4355

Surgery
ISSN: 0039-6060

Surgery for Obesity and Related Diseases
ISSN: 1550-7289

Surgical and Radiologic Anatomy
ISSN: 0930-1038

Surgical Clinics of North America
ISSN: 0039-6109

Surgical Endoscopy
ISSN: 0930-2794

Surgical Infections
ISSN: 1096-2964

Surgical Laparoscopy, Endoscopy and Percutaneous Techniques
ISSN: 1530-4515

Surgical Oncology
ISSN: 0960-7404

Survey of Ophthalmology
ISSN: 0039-6257

Synapse
ISSN: 0887-4476

Systematic and Applied Microbiology
ISSN: 0723-2020

Techniques in Foot and Ankle Surgery
ISSN: 1536-0644

Techniques in Hand and Upper Extremity Surgery
ISSN: 1089-3393

Techniques in Knee Surgery
ISSN: 1536-0636

Techniques in Ophthalmology
ISSN: 1542-1929

Technology and Health Care
ISSN: 0928-7329

Telemedicine Journal and e-Health
ISSN: 1530-5627

Tetrahedron
ISSN: 0040-4020

Tetrahedron Asymmetry
ISSN: 0957-4166

Thalamus and Related Systems
ISSN: 1472-9288

Therapeutic Apheresis and Dialysis
ISSN: 1744-9979

Therapeutic Drug Monitoring
ISSN: 0163-4356

Therapie
ISSN: 0040-5957

Thoracic and Cardiovascular Surgeon
ISSN: 0171-6425

Thoracic and Cardiovascular Surgeon, Supplement
ISSN: 0946-4778

Thoracic Surgery Clinics
ISSN: 1547-4127

Thorax
ISSN: 0040-6376

Thrombosis and Haemostasis
ISSN: 0340-6245

Thrombosis Research
ISSN: 0049-3848

Thyroid
ISSN: 1050-7256

Tissue Antigens
ISSN: 0001-2815

Tissue Engineering
ISSN: 1076-3279

Topics in Clinical Nutrition
ISSN: 0883-5691

Topics in Emergency Medicine
ISSN: 0164-2340

Topics in Magnetic Resonance Imaging
ISSN: 0899-3459

Toxicologic Pathology
ISSN: 0192-6233

Toxicology
ISSN: 0300-483X

Toxicology Letters
ISSN: 0378-4274

Toxicology Mechanisms and Methods
ISSN: 1051-7235

Toxicon
ISSN: 0041-0101

TrAC - Trends in Analytical Chemistry
ISSN: 0165-9936

Trace Elements and Electrolytes
ISSN: 0946-2104

Traffic
ISSN: 1398-9219

Transactions of the American Ophthalmological Society
ISSN: 1545-6110

Transplant Immunology
ISSN: 0966-3274

Transplant Infectious Disease
ISSN: 1398-2273

Transplant International
ISSN: 0934-0874

Transplantation
ISSN: 0041-1337

Transplantation Proceedings
ISSN: 0041-1345

Travel Medicine and Infectious Disease
ISSN: 1477-8939

Treatments in Endocrinology
ISSN: 1175-6349

Treatments in Respiratory Medicine
ISSN: 1176-3450

Trends in Biochemical Sciences
ISSN: 0968-0004

Trends in Biotechnology
ISSN: 0167-7799

Trends in Cardiovascular Medicine
ISSN: 1050-1738

Trends in Cell Biology
ISSN: 0962-8924

Trends in Endocrinology and Metabolism
ISSN: 1043-2760

Trends in Genetics
ISSN: 0168-9525

Trends in Microbiology
ISSN: 0966-842X

Trends in Neurosciences
ISSN: 0166-2236

Trends in Pharmacological Sciences
ISSN: 0165-6147

Tuberculosis
ISSN: 1472-9792

Tumor Biology
ISSN: 1010-4283

Twin Research and Human Genetics
ISSN: 1832-4274

Ultrasonics Sonochemistry
ISSN: 1350-4177

Ultrasound in Medicine and Biology
ISSN: 0301-5629

Ultrasound in Obstetrics and Gynecology
ISSN: 0960-7692

Ultrasound International
ISSN: 0971-6874

Ultrasound Quarterly
ISSN: 0894-8771

Ultrastructural Pathology
ISSN: 0191-3123

Urologia Internationalis
ISSN: 0042-1138

Urologic Clinics of North America
ISSN: 0094-0143

Urologic Oncology: Seminars and Original Investigations
ISSN: 1078-1439

Urological Research
ISSN: 0300-5623

Urology
ISSN: 0090-4295

Vaccine
ISSN: 0264-410X

Vaccines: Children and Practice
ISSN: 1460-6887

Value in Health
ISSN: 1098-3015

Vascular
ISSN: 1708-5381

Vascular Disease Prevention
ISSN: 1567-2700

Vascular Medicine
ISSN: 1358-863X

Vascular Pharmacology
ISSN: 1537-1891

Vector-Borne and Zoonotic Diseases
ISSN: 1530-3667

Virchows Archiv
ISSN: 0945-6317

Virology
ISSN: 0042-6822

Virus Genes
ISSN: 0920-8569

Virus Research
ISSN: 0168-1702

Vision Research
ISSN: 0042-6989

Visual Cognition
ISSN: 1350-6285

Visual Impairment Research
ISSN: 1388-235X

Visual Neuroscience
ISSN: 0952-5238

Vox Sanguinis
ISSN: 0042-9007

Vox Sanguinis, Supplement
ISSN: 1741-6892

Waste Management
ISSN: 0956-053X

Waste Management and Research
ISSN: 0734-242X

Water Environment Research
ISSN: 1061-4303

Water Research
ISSN: 0043-1354

Wound Repair and Regeneration
ISSN: 1067-1927

Xenotransplantation
ISSN: 0908-665X

Yeast
ISSN: 0749-503X

NUMBER OF JOURNALS PER CLASSIFICATION

This list is alphabetically compiled.

Code	Classification	No.
E.4.1	Allergy and Clinical Immunology	68
E.7	Anatomy, Histology, Cytology and Anthropology	45
D.7	Anesthesiology	86
E.10.2	Applied Microbiology	16
C.1.7	Arthritis, Rheumatism and Connective Tissue Diseases	43
E.5.1	Bacteriology and Mycology	18
E.3.1	Biochemistry	144
E.3	Biochemistry and Clinical Chemistry	20
E.11.1	Biophysics	34
E.11	Biophysics, Medical Instrumentation and Biomedical Engineering	53
E.10	Biotechnology and Applied Science	69
C.1.1	Cardiology and Angiology	166
D.6	Cardiovascular and Thoracic Surgery	33
E.3.3	Cell Biology	88
C.1.2	Chest Diseases	75
E.3.2	Clinical Chemistry	27
C.1.4	Clinical Genetics	50
C.0	Clinical Medicine	421
E.5	Clinical Microbiology	95
C.1.3.1	Clinical Oncology	108
C.11	Clinical Pathology	46
A.1.1	Clinical Pharmacology	83
F.1	Dentistry and Stomatology	13
C.5	Dermatology and Venereology	93
E.8.1	Developmental Biology	47
E.8	Developmental Biology and Gerontology	3
A.1.2	Drug Dependence	32
A.0	Drugs and Chemicals	9
E.2	Endocrinology	133
B.2.1	Epidemiology	26
E.10.1	Experimental Microbiology	23
E.0	Experimental Science	100
E.12	Forensic Science	73
C.1.8	Gastroenterology	94
E.9.1	General Genetics	54
C.11.1	General Pathology	19
D.1	General and Abdominal Surgery	93
E.9	Genetics and Molecular Biology	55
C.3	Geriatrics	52
E.8.2	Gerontology	30
B.4.2	Health Care Management	31
B.4.1	Health Economics	17
B.4	Health Policy, Economics and Management	41
C.1.5	Hematology	80
C.1.8.1	Hepatology	19
E.4	Immunology and Serology	116
B.1	Industrial Medicine and Occupational Health	99
F.3	Informatics	27
D.7.1	Intensive Care	62
C.1	Internal Medicine	96
C.5.1	Leprosy	7
E.11.2	Medical Instrumentation	32
E.9.2	Molecular Biology	85
C.1.6	Nephrology	47
C.6.1	Neurology and Neurosurgery	196
C.6	Neurology and Psychiatry	84
E.1	Neuroscience	178
C.9.2	Nuclear Medicine	22
B.5	Nursing	27
E.13	Nutrition	64
C.4	Obstetrics and Gynecology	140
C.1.3	Oncology	117
C.7	Ophthalmology	84
D.3	Orthopedic Surgery	79
F.0	Other Specialities	1
C.8	Otorhinolaryngology	82
E.5.2	Parasitology	24
C.11.2	Pathological Anatomy	37
C.2	Pediatrics and Pediatric Surgery	206
A.3	Pharmaceutical and Medicinal Chemistry	151
A.1	Pharmacology and Drug Therapy	203
A.4	Pharmacy	127
C.10.2	Physical Medicine	25
E.6	Physiology	136
D.2	Plastic Surgery	44
B.3	Pollution Control	74
C.6.2	Psychiatry	344
B.2	Public Health	219
C.9.1	Radiology	99
C.9	Radiology and Nuclear Medicine	48
C.10.1	Rehabilitation	64
C.10	Rehabilitation and Physical Medicine	31
B.0	Social Medicine	2
D.0	Surgery and Anesthesiology	7
A.2	Toxicology	96
D.4	Transplantation	21
B.2.2	Tropical Medicine	14
D.5	Urological Surgery	52
C.1.1.1	Vascular Diseases	43
F.2	Veterinary Medicine	16
E.5.3	Virology	42

NUMBER OF JOURNALS PER CLASSIFICATION

This list is ordered by classification code.

Code	Classification	No.
A.0	Drugs and Chemicals	9
A.1	Pharmacology and Drug Therapy	203
A.1.1	Clinical Pharmacology	83
A.1.2	Drug Dependence	32
A.2	Toxicology	96
A.3	Pharmaceutical and Medicinal Chemistry	151
A.4	Pharmacy	127
B.0	Social Medicine	2
B.1	Industrial Medicine and Occupational Health	99
B.2	Public Health	219
B.2.1	Epidemiology	26
B.2.2	Tropical Medicine	14
B.3	Pollution Control	74
B.4	Health Policy, Economics and Management	41
B.4.1	Health Economics	17
B.4.2	Health Care Management	31
B.5	Nursing	27
C.0	Clinical Medicine	421
C.1	Internal Medicine	96
C.1.1	Cardiology and Angiology	166
C.1.1.1	Vascular Diseases	43
C.1.2	Chest Diseases	75
C.1.3	Oncology	117
C.1.3.1	Clinical Oncology	108
C.1.4	Clinical Genetics	50
C.1.5	Hematology	80
C.1.6	Nephrology	47
C.1.7	Arthritis, Rheumatism and Connective Tissue Diseases	43
C.1.8	Gastroenterology	94
C.1.8.1	Hepatology	19
C.2	Pediatrics and Pediatric Surgery	206
C.3	Geriatrics	52
C.4	Obstetrics and Gynecology	140
C.5	Dermatology and Venereology	93
C.5.1	Leprosy	7
C.6	Neurology and Psychiatry	84
C.6.1	Neurology and Neurosurgery	196
C.6.2	Psychiatry	344
C.7	Ophthalmology	84
C.8	Otorhinolaryngology	82
C.9	Radiology and Nuclear Medicine	48
C.9.1	Radiology	99
C.9.2	Nuclear Medicine	22
C.10	Rehabilitation and Physical Medicine	31
C.10.1	Rehabilitation	64
C.10.2	Physical Medicine	25
C.11	Clinical Pathology	46
C.11.1	General Pathology	19
C.11.2	Pathological Anatomy	37
D.0	Surgery and Anesthesiology	7
D.1	General and Abdominal Surgery	93
D.2	Plastic Surgery	44
D.3	Orthopedic Surgery	79
D.4	Transplantation	21
D.5	Urological Surgery	52
D.6	Cardiovascular and Thoracic Surgery	33
D.7	Anesthesiology	86
D.7.1	Intensive Care	62
E.0	Experimental Science	100
E.1	Neuroscience	178
E.2	Endocrinology	133
E.3	Biochemistry and Clinical Chemistry	20
E.3.1	Biochemistry	144
E.3.2	Clinical Chemistry	27
E.3.3	Cell Biology	88
E.4	Immunology and Serology	116
E.4.1	Allergy and Clinical Immunology	68
E.5	Clinical Microbiology	95
E.5.1	Bacteriology and Mycology	18
E.5.2	Parasitology	24
E.5.3	Virology	42
E.6	Physiology	136
E.7	Anatomy, Histology, Cytology and Anthropology	45
E.8	Developmental Biology and Gerontology	3
E.8.1	Developmental Biology	47
E.8.2	Gerontology	30
E.9	Genetics and Molecular Biology	55
E.9.1	General Genetics	54
E.9.2	Molecular Biology	85
E.10	Biotechnology and Applied Science	69
E.10.1	Experimental Microbiology	23
E.10.2	Applied Microbiology	16
E.11	Biophysics, Medical Instrumentation and Biomedical Engineering	53
E.11.1	Biophysics	34
E.11.2	Medical Instrumentation	32
E.12	Forensic Science	73
E.13	Nutrition	64
F.0	Other Specialities	1
F.1	Dentistry and Stomatology	13
F.2	Veterinary Medicine	16
F.3	Informatics	27

LIST OF JOURNALS PER SUBJECT CLASSIFICATION

A.0 Drugs and Chemicals

African Journal of Traditional, Complementary and Alternative Medicines
Alimentary Pharmacology and Therapeutics Symposium Series
Chinese Journal of Integrative Medicine
Current Computer-Aided Drug Design
Current Enzyme Inhibition
Expert Opinion on Drug Metabolism and Toxicology
Jordan Journal of Applied Sciences - Natural Sciences
Journal of Applied Biomedicine
Publications of Takeda Research Laboratories

A.1 Pharmacology and Drug Therapy

Acta Pharmacologica Sinica
Advanced Drug Delivery Reviews
Adverse Drug Reaction Bulletin
Agri
Alcohol
American Journal of Drug Delivery
American Journal of Therapeutics
Annexins
Annual Review of Pharmacology and Toxicology
Anti-Cancer Agents in Medicinal Chemistry
Anti-Cancer Drugs
Anti-Infective Agents in Medicinal Chemistry
Anti-Inflammatory and Anti-Allergy Agents in Medicinal Chemistry
Antibiotiques
Antiviral Chemistry and Chemotherapy
Antiviral Research
Antiviral Therapy
Applied Genomics and Proteomics
Arzneimittel-Forschung/Drug Research
Asia Pacific Journal of Pharmacology
Australian Prescriber
Autonomic and Autacoid Pharmacology
Basic and Clinical Pharmacology and Toxicology
Basic and Clinical Pharmacology and Toxicology, Supplement
Behavioural Pharmacology
Biochemical Pharmacology
Biogenic Amines
British Journal of Pharmacology
CNS Drug Reviews
CNS and Neurological Disorders - Drug Targets
Canadian Journal of Physiology and Pharmacology
Cancer Biotherapy and Radiopharmaceuticals
Cancer Chemotherapy and Pharmacology
Cancer Chemotherapy and Pharmacology, Supplement
Cardiovascular Drug Reviews
Cardiovascular and Hematological Agents in Medicinal Chemistry
Cardiovascular and Hematological Disorders - Drug Targets
Chemico-Biological Interactions
Chemotherapie Journal
Chemotherapie Journal, Supplement
Chinese Journal of Antibiotics
Chinese Journal of Infection and Chemotherapy
Chinese Journal of Pharmacology and Toxicology
Chinese Pharmacological Bulletin
Chronobiology International
Ciencia y Tecnologia Pharmaceutica
Clinical Therapeutics
Comparative Biochemistry and Physiology - C Toxicology and Pharmacology
Contemporary Clinical Trials
Core Evidence
Critical Reviews in Therapeutic Drug Carrier Systems
Current Drug Metabolism
Current Neuropharmacology
Current Opinion in Investigational Drugs
Current Pharmacogenomics
Current Therapeutic Research - Clinical and Experimental
Current Topics in Pharmacology
Current Vascular Pharmacology
Cytotherapy
Drug Delivery
Drug Development Research
Drug Development and Industrial Pharmacy
Drug Metabolism and Disposition
Drug News and Perspectives
Drugs and Aging
Drugs and Therapy Perspectives
Drugs in Context
Drugs in Context (US)
Drugs in R and D
EOS Rivista di Immunologia ed Immunofarmacologia
Endocrine, Metabolic and Immune Disorders - Drug Targets
Environmental Toxicology and Pharmacology
European Journal of Herbal Medicine
European Journal of Pharmacology
European Review for Medical and Pharmacological Sciences
Experimental and Clinical Psychopharmacology
Expert Opinion on Biological Therapy
Expert Opinion on Drug Safety
Expert Opinion on Emerging Drugs
Expert Opinion on Investigational Drugs
Expert Opinion on Therapeutic Patents
Expert Review of Anticancer Therapy
Expert Review of Cardiovascular Therapy
Expert Review of Pharmacoeconomics and Outcomes Research
Expert Review of Vaccines
Fitoterapia
Folia Pharmacologica Japonica
Frontiers in Drug Design and Discovery
Fundamental and Clinical Pharmacology
Geneesmiddelenbulletin
Homeopathy
Human Vaccines
IDrugs

Immunopharmacology and Immunotoxicology
Indian Journal of Pharmacology
Indian Journal of Physiology and Pharmacology
Infectious Disorders - Drug Targets
Inflammation and Allergy - Drug Targets
Inflammopharmacology
International Immunopharmacology
International Journal of Antimicrobial Agents
International Journal of Immunopathology and Pharmacology
International Journal of Neuroprotection and Neurogeneration
International Journal of Pharmaceutical Medicine
International Journal of Pharmacology
Investigational New Drugs
Iranian Journal of Pharmacology and Therapeutics
Japanese Journal of Antibiotics
Japanese Journal of Chemotherapy
Japanese Pharmacology and Therapeutics
Journal of Antibiotics
Journal of Antimicrobial Chemotherapy
Journal of Applied Research
Journal of Basic and Clinical Physiology and Pharmacology
Journal of Cardiovascular Pharmacology
Journal of Cardiovascular Pharmacology and Therapeutics
Journal of Controlled Release
Journal of Ethnopharmacology
Journal of Experimental Therapeutics and Oncology
Journal of Herbal Pharmacotherapy
Journal of Herbs, Spices and Medicinal Plants
Journal of Immunoassay and Immunochemistry
Journal of Immunotherapy
Journal of Liposome Research
Journal of Medicinal Plants
Journal of Natural Remedies
Journal of Oncology Pharmacy Practice
Journal of Pharmaceutical Finance, Economics and Policy
Journal of Pharmacological Sciences
Journal of Pharmacological and Toxicological Methods
Journal of Pharmacology and Experimental Therapeutics
Journal of Pharmacy Practice
Journal of Pharmacy and Pharmacology
Journal of Physiology and Pharmacology
Journal of Psychopharmacology
Journal of Veterinary Pharmacology and Therapeutics
Journal of the American Pharmacists Association
Klinik Psikofarmakoloji Bulteni
Korean Journal of Physiology and Pharmacology
Life Sciences
MeReC Bulletin
MeReC Extra
Medical Letter on Drugs and Therapeutics
Medicinal Research Reviews
Medicus
Microbial Drug Resistance
Molecular Cancer Therapeutics
Molecular Diagnosis and Therapy
Molecular Interventions
Molecular Pharmacology
Naunyn-Schmiedeberg's Archives of Pharmacology
Neuropharmacology
News in Medical Products
Oligonucleotides
P and T
Paediatric and Perinatal Drug Therapy
Pediatric Drugs
Personalized Medicine
Pharma-Kritik
Pharmaca
Pharmaceutical Biology
Pharmaceutical Care and Research
PharmacoEconomics
PharmacoEconomics - Italian Research Articles
Pharmacogenetics and Genomics
Pharmacogenomics
Pharmacogenomics Journal
Pharmacological Reports
Pharmacological Research
Pharmacological Reviews
Pharmacology
Pharmacology Biochemistry and Behavior
Pharmacometrics
Phytomedica
Phytomedicine
Phytotherapie Europeenne
Phytotherapy Research
Planta Medica
Prescrire International
Proceedings of the Western Pharmacology Society
Progress in Drug Research
Psychopharmacology
Psychopharmakotherapie
Psychopharmakotherapie, Supplement
Pulmonary Pharmacology and Therapeutics
QSAR and Combinatorial Science
Regulatory Toxicology and Pharmacology
Research Communications in Pharmacology and Toxicology
Reviews in Contemporary Pharmacotherapy
Revista Brasileira de Plantas Medicinais
Revista Espanola de Quimioterapia
Ricerca e Pratica
Supportive Cancer Therapy
Synapse
Toxicological Reviews
Toxicology and Applied Pharmacology
Treatments in Respiratory Medicine
Trends in Pharmacological Sciences
Value in Health
Vascular Pharmacology
WHO Drug Information
Xenobiotica
Zeitschrift fur Phytotherapie

A.1.1 Clinical Pharmacology

Alimentary Pharmacology and Therapeutics
Alimentary Pharmacology and Therapeutics Symposium Series
American Journal Geriatric Pharmacotherapy
American Journal of Cardiovascular Drugs
Annals of Pharmacotherapy
Antimicrobial Agents and Chemotherapy
BMC Clinical Pharmacology
BMC Pharmacology
BioDrugs
Biomedicine and Pharmacotherapy
Biopharmaceutics and Drug Disposition
Breast Care

British Journal of Clinical Pharmacology
British Journal of Clinical Pharmacology, Supplement
CNS Drugs
Canadian Journal of Clinical Pharmacology
Cardiovascular Drugs and Therapy
Chemotherapy
Chinese Journal of Infection and Chemotherapy
Chinese Traditional and Herbal Drugs
Clinical Drug Investigation
Clinical Neuropharmacology
Clinical Pharmacokinetics
Clinical Pharmacology and Therapeutics
Clinical Psychopharmacology and Neuroscience
Clinical and Experimental Pharmacology and Physiology
Current Drug Metabolism
Current Topics in Pharmacology
Drug Metabolism Reviews
Drug Metabolism and Drug Interactions
Drug Resistance Updates
Drug and Therapeutics Bulletin
Drugs
Drugs of Today
Drugs under Experimental and Clinical Research
Epitheorese Klinikes Farmakologias kai Farmakokinetikes
European Journal of Clinical Pharmacology
European Journal of Drug Metabolism and Pharmacokinetics
European Neuropsychopharmacology
Expert Opinion on Pharmacotherapy
Fundamental and Clinical Pharmacology
Giornale Italiano di Farmacia Clinica
Human Psychopharmacology
Inflammation Research
International Clinical Psychopharmacology
International Journal of Clinical Pharmacology Research
International Journal of Clinical Pharmacology and Therapeutics
International Journal of Neuropsychopharmacology
International Journal of Pharmaceutical Compounding
Japanese Journal of Clinical Pharmacology and Therapeutics
Japanese Journal of Neuropsychopharmacology
Japanese Pharmacology and Therapeutics
Journal of Aging and Pharmacotherapy
Journal of Anaesthesiology Clinical Pharmacology
Journal of Biopharmaceutical Statistics
Journal of Cancer Integrative Medicine
Journal of Chemotherapy
Journal of Child and Adolescent Psychopharmacology
Journal of Clinical Pharmacology
Journal of Clinical Pharmacy and Therapeutics
Journal of Clinical Psychopharmacology
Journal of Korean Society for Clinical Pharmacology and Therapeutics
Journal of Ocular Pharmacology and Therapeutics
Journal of Pharmacokinetics and Pharmacodynamics
Managed Care Pharmacy Update
Methods and Findings in Experimental and Clinical Pharmacology
NeuroRx
Neuropsychopharmacology
Oral Therapeutics and Pharmacology
Pharmacoepidemiology and Drug Safety
Pharmacology and Therapeutics
Pharmacopsychiatry
Pharmacotherapy
Progress in Neuro-Psychopharmacology and Biological Psychiatry
Review of Clinical Pharmacology and Pharmacokinetics, International Edition
Signal
Skin Pharmacology and Physiology
Therapeutic Drug Monitoring
Therapeutics and Clinical Risk Management
Therapie
Therapy
Trends in Medicine
Vascular Health and Risk Management

A.1.2 Drug Dependence

Addiction
Addiction Biology
Addictive Behaviors
Addictive Disorders and their Treatment
Adicciones
Alcohol
Alcohol and Alcoholism
Alcoholism
Alcoholism: Clinical and Experimental Research
American Journal of Drug and Alcohol Abuse
American Journal on Addictions
Drug and Alcohol Dependence
European Addiction Research
Harm Reduction Journal
Heroin Addiction and Related Clinical Problems
International Journal of Drug Policy
JAVA - Journal of the Association for Vascular Access
Journal of Addictive Diseases
Journal of Drug Education
Journal of Drug Issues
Journal of Ethnicity in Substance Abuse
Journal of Psychoactive Drugs
Journal of Studies on Alcohol
Journal of Substance Abuse Treatment
Journal of Substance Use
NAD Publication
NIDA Research Monograph Series
Research Communications in Alcohol and Substances of Abuse
Substance Abuse
Substance Use and Misuse
Sucht
Suchtmedizin in Forschung und Praxis

A.2 Toxicology

Acta Toxicologica
Annual Review of Pharmacology and Toxicology
Aquatic Toxicology
Archives of Environmental Contamination and Toxicology
Archives of Toxicology
Arhiv za Higijenu Rada i Toksikologiju
Basic and Clinical Pharmacology and Toxicology
Basic and Clinical Pharmacology and Toxicology, Supplement
Biomarkers

Birth Defects Research Part B - Developmental and Reproductive Toxicology
Bulletin of Environmental Contamination and Toxicology
Cardiovascular Toxicology
Cell Biology and Toxicology
Chemical Research in Toxicology
Chemico-Biological Interactions
Chinese Journal of Endemiology
Chinese Journal of Pharmacology and Toxicology
Clinical Toxicology
Comparative Biochemistry and Physiology - C Toxicology and Pharmacology
Contact Dermatitis
Critical Reviews in Toxicology
Current Topics in Toxicology
Cutaneous and Ocular Toxicology
DNA Repair
Dermatologie in Beruf und Umwelt
Drug Metabolism and Disposition
Drug Safety
Drug and Chemical Toxicology
Ecotoxicology
Ecotoxicology and Environmental Safety
Environmental Engineering Science
Environmental Toxicology
Environmental Toxicology and Chemistry
Environmental Toxicology and Pharmacology
Environmental and Molecular Mutagenesis
Experimental and Toxicologic Pathology
Expert Opinion on Drug Metabolism and Toxicology
Food Additives and Contaminants
Food Chemistry
Food and Chemical Toxicology
Frontiers in Medicinal Chemistry
Human and Experimental Toxicology
IARC Monographs on the Evaluation of Carcinogenic Risks to Humans
IPCS Concise International Chemical Assessment Documents
Industrial Health
Inflammation Research
Inhalation Toxicology
International Journal of Clinical Pharmacology and Therapeutics
International Journal of Medical Toxicology and Legal Medicine
International Journal of Toxicology
Japanese Journal of Forensic Toxicology
Journal of Analytical Toxicology
Journal of Applied Toxicology
Journal of Biochemical and Molecular Toxicology
Journal of Environmental Pathology, Toxicology and Oncology
Journal of Environmental Science and Health - Part B Pesticides, Food Contaminants, and Agricultural Wastes
Journal of Environmental Science and Health - Part C Environmental Carcinogenesis and Ecotoxicology Reviews
Journal of Forensic Medicine and Toxicology
Journal of Hazardous Materials
Journal of Immunotoxicology
Journal of Occupational Health
Journal of Pharmacological and Toxicological Methods
Journal of Toxicological Sciences
Journal of Toxicology and Environmental Health - Part A
Journal of Toxicology and Environmental Health - Part B: Critical Reviews
Mutagenesis
Mutation Research - Fundamental and Molecular Mechanisms of Mutagenesis
Mutation Research - Genetic Toxicology and Environmental Mutagenesis
Mutation Research - Reviews in Mutation Research
Nanotoxicology
NeuroToxicology
Neurotoxicology and Teratology
Particle and Fibre Toxicology
Regulatory Toxicology and Pharmacology
Reproductive Toxicology
Research Communications in Molecular Pathology and Pharmacology
Research Communications in Pharmacology and Toxicology
Reviews in Toxicology
Revista Brasileira de Toxicologia
Revista de Toxicologia
Scandinavian Journal of Work, Environment and Health
Toxicologic Pathology
Toxicological Reviews
Toxicological Sciences
Toxicology
Toxicology International
Toxicology Letters
Toxicology Mechanisms and Methods
Toxicology and Applied Pharmacology
Toxicology and Industrial Health
Toxicology in Vitro
Toxicon
Toxin Reviews
Umweltwissenschaften und Schadstoff-Forschung
Veterinary and Human Toxicology
Xenobiotica

A.3 Pharmaceutical and Medicinal Chemistry

AAPS PharmSciTech
AIDS Research and Therapy
Analyst
Analytica Chimica Acta
Analytical Chemistry
Annexins
Annual Report of Shionogi Research Laboratory
Anti-Cancer Agents in Medicinal Chemistry
Antimicrobial Agents and Chemotherapy
Archiv der Pharmazie
Arzneimittel-Forschung/Drug Research
Assay and Drug Development Technologies
Biological and Pharmaceutical Bulletin
Bioorganic Chemistry
Bioorganic and Medicinal Chemistry
Bioorganic and Medicinal Chemistry Letters
Biosciences Biotechnology Research Asia
Bollettino Chimico Farmaceutico
Bulletin of the Chemical Society of Japan
Canadian Journal of Chemistry
Carbohydrate Research
Cardiovascular Revascularization Medicine
Cardiovascular and Hematological Disorders - Drug Targets
Central Nervous System Agents in Medicinal Chemistry

Chemical Communications
Chemical and Pharmaceutical Bulletin
Chemistry and Biodiversity
Chemtracts
Chinese Journal of Antibiotics
Chinese Journal of Natural Medicines
Chinese Pharmaceutical Journal
Chirality
Chromatographia
Clinical Proteomics
Clinical Research and Regulatory Affairs
Colloid and Polymer Science
Combinatorial Chemistry and High Throughput Screening
Current Cancer Therapy Reviews
Current Computer-Aided Drug Design
Current Drug Delivery
Current Drug Discovery Technologies
Current Drug Targets
Current Enzyme Inhibition
Current Medicinal Chemistry
Current Opinion in Drug Discovery and Development
Current Organic Chemistry
Current Pharmaceutical Analysis
Current Pharmaceutical Biotechnology
Current Pharmaceutical Design
Current Topics in Medicinal Chemistry
Current Topics in Nutraceutical Research
Drug Design Reviews Online
Drug Development Research
Drug Development and Industrial Pharmacy
Drug Discovery Today: BIOSILICO
Drug Discovery Today: Disease Mechanisms
Drug Discovery Today: Disease Models
Drug Discovery Today: TARGETS
Drug Discovery Today: Technologies
Drug Discovery Today: Therapeutic Strategies
Drug Discovery World
Drugs
Drugs in Context
Drugs in Context (US)
Drugs of Today
Drugs of the Future
Drugs under Experimental and Clinical Research
EBR - European Biopharmaceutical Review
Enhancer - Biotherapy of Cancer
European Journal of Medicinal Chemistry
Expert Opinion on Therapeutic Targets
Flavour and Fragrance Journal
Frontiers in Drug Design and Discovery
Frontiers in Medicinal Chemistry
Future Cardiology
Helvetica Chimica Acta
IEE Proceedings: Systems Biology
Indian Drugs
Infectious Disorders - Drug Targets
International Journal of Cosmetic Science
International Journal of Medicinal Mushrooms
Investigational New Drugs
Japanese Journal of Antibiotics
Journal of Antibiotics
Journal of Asian Natural Products Research
Journal of Bioactive and Compatible Polymers
Journal of Biological Sciences
Journal of Biomolecular Screening
Journal of Chromatography A
Journal of Chromatography B: Analytical Technologies in the Biomedical and Life Sciences
Journal of Commercial Biotechnology
Journal of Computer-Aided Molecular Design
Journal of Enzyme Inhibition and Medicinal Chemistry
Journal of Food and Drug Analysis
Journal of Heterocyclic Chemistry
Journal of Inflammation
Journal of Labelled Compounds and Radiopharmaceuticals
Journal of Liquid Chromatography and Related Technologies
Journal of Mass Spectrometry
Journal of Medicinal Chemistry
Journal of Molecular Graphics and Modelling
Journal of Natural Products
Journal of Organic Chemistry
Journal of Pharmaceutical Sciences
Journal of Pharmaceutical and Biomedical Analysis
Journal of Planar Chromatography - Modern TLC
Journal of Radioanalytical and Nuclear Chemistry
Journal of Separation Science
Journal of the American Chemical Society
Journal of the Indian Chemical Society
Korean Journal of Pharmacognosy
Letters in Drug Design and Discovery
Marine Drugs
Medicinal Chemistry
Medicinal Chemistry Research
Mini-Reviews in Medicinal Chemistry
Molecular Cancer
Molecular Diagnosis and Therapy
Molecular Diversity
Molecular Pain
Molecular Pharmaceutics
Nanomedicine: Nanotechnology, Biology, and Medicine
Natural Medicines
Natural Product Research
Natural Product Sciences
Nature Reviews Drug Discovery
New Journal of Chemistry
Personalized Medicine
Pharmaceutical Biology
Pharmaceutical Chemistry Journal
Pharmaceutical Statistics
Pharmaceuticals
Pharmacogenomics Journal
Pharmazie
Phytochemical Analysis
Phytochemistry Reviews
Planta Medica
Progress in Drug Research
Pteridines
Signal
Structure
Synlett
Synthesis
Synthetic Communications
Tetrahedron
Tetrahedron Asymmetry
Tetrahedron Letters
TransPharma
Turkish Journal of Pharmaceutical Sciences
Virologica Sinica
Yakugaku Zasshi

A.4 Pharmacy

AAPS Journal
AAPS PharmSciTech
Acta Farmaceutica Bonaerense
Acta Pharmaceutica
Acta Pharmaceutica Hungarica
Acta Pharmaceutica Turcica
Acta Poloniae Pharmaceutica - Drug Research
Actualites Pharmaceutiques
Actualites Pharmaceutiques Hospitalieres
Advances in Pharmacy
American Journal of Drug Delivery
American Journal of Health-System Pharmacy
American Pharmaceutical Outsourcing
American Pharmaceutical Review
Anales de la Real Academia Nacional de Farmacia
Ankara Universitesi Eczacilik Fakultesi Dergisi
Annales Pharmaceutiques Francaises
Ars Pharmaceutica
Atencion Farmaceutica
BioPharm International
COSSMA
Canadian Journal of Hospital Pharmacy
Canadian Pharmacists Journal
Ceska a Slovenska Farmacie
Chemical and Pharmaceutical Bulletin
Chinese Pharmaceutical Journal
Chinese Traditional and Herbal Drugs
Circular Farmaceutica
Clinical Research and Regulatory Affairs
Colloids and Surfaces B: Biointerfaces
Consultant Pharmacist
Current Cancer Drug Targets
Current Drug Delivery
Current Drug Targets
Current Nanoscience
Current Opinion in Colloid and Interface Science
Current Pharmaceutical Analysis
Current Pharmaceutical Biotechnology
Deutsche Apotheker Zeitung
Drug Benefit Trends
Drug Design Reviews Online
Drug Development and Industrial Pharmacy
Drug Discovery Today
Drug Topics
East and Central African Journal of Pharmaceutical Sciences
European Journal of Pharmaceutical Sciences
European Journal of Pharmaceutics and Biopharmaceutics
European Pharmaceutical Contractor
Expert Opinion on Drug Delivery
Fabad Journal of Pharmaceutical Sciences
Farmaceuticky Obzor
Farmaceutico Hospitales
Farmaceutisch Tijdschrift voor Belgie
Farmaceutski Glasnik
Farmacevtski Vestnik
Farmacia Hospitalaria
Formulary
Frontiers in Drug Design and Discovery
Frontiers in Medicinal Chemistry
Gazi Universitesi Eczacilik Fakultesi Dergisi
Hospital Pharmacist
Hospital Pharmacy
Human Genomics
Indian Drugs
International Journal of Pharmaceutical Compounding
International Journal of Pharmaceutics
International Journal of Pharmacy Practice
Journal de Pharmacie Clinique
Journal de Pharmacie de Belgique
Journal of Aerosol Science
Journal of China Pharmaceutical University
Journal of Colloid and Interface Science
Journal of Drug Delivery Science and Technology
Journal of Drug Targeting
Journal of Generic Medicines
Journal of Microencapsulation
Journal of Oncology Pharmacy Practice
Journal of Pain and Palliative Care Pharmacotherapy
Journal of Pharmaceutical Finance, Economics and Policy
Journal of Pharmaceutical Sciences
Journal of Pharmaceutical and Biomedical Analysis
Journal of Pharmacy Practice and Research
Journal of Pharmacy Technology
Journal of Pharmacy and Pharmaceutical Sciences
Journal of Pharmacy and Pharmacology
Journal of Pharmacy of Istanbul University
Journal of Phytomedicine and Therapeutics
Krankenhauspharmazie
Letters in Drug Design and Discovery
Managed Care Pharmacy Update
Medizinische Monatsschrift fur Pharmazeuten
Molecular Pharmaceutics
PAT
PDA Journal of Pharmaceutical Science and Technology
PZ Prisma
Pharmaceutical Care Espana
Pharmaceutical Care and Research
Pharmaceutical Development and Technology
Pharmaceutical Engineering
Pharmaceutical Journal
Pharmaceutical Research
Pharmaceutical Technology
Pharmaceutical Technology Europe
Pharmaceutical Visions
Pharmaceuticals
Pharmaceutisch Weekblad
Pharmacia
Pharmacien Hospitalier
Pharmacy Education
Pharmacy World and Science
Pharmacy in Practice
Pharmazeutische Industrie
Pharmazeutische Zeitung
Pharmazie
Pharmazie in Unserer Zeit
Powder Technology
Proceedings of the Controlled Release Society
Quality Assurance Journal
Revista Brasileira de Ciencias Farmaceuticas/Brazilian Journal of Pharmaceutical Sciences
Revista Mexicana de Ciencias Farmaceuticas
Revista Portuguesa de Farmacia
Revista de Ciencias Farmaceuticas
S.T.P. Pharma Pratiques
Saudi Pharmaceutical Journal
Scientia Pharmaceutica
Seguimiento Farmacoterapeutico

Turkish Journal of Pharmaceutical Sciences

B.0 Social Medicine

Environnement, Risques et Sante
Men's Health Journal

B.1 Industrial Medicine and Occupational Health

Air Medical Journal
American Journal of Industrial Medicine
American Journal of Sports Medicine
Annals of Agricultural and Environmental Medicine
Annals of Occupational Hygiene
Applied Ergonomics
Arbeitsmedizin Sozialmedizin Umweltmedizin
Arbeitsmedizin Sozialmedizin Umweltmedizin, Supplement
Archives des Maladies Professionnelles et de l'Environnement
Archives of Environmental Health
Archivos de Medicina del Deporte
Arhiv za Higijenu Rada i Toksikologiju
Athletic Therapy Today
Aviation Space and Environmental Medicine
BMC Medicine
British Journal of Sports Medicine
Ceske Pracovni Lekarstvi
Clinical Journal of Sport Medicine
Clinics in Occupational and Environmental Medicine
Clinics in Sports Medicine
Dermatologie in Beruf und Umwelt
Diving and Hyperbaric Medicine
Ergonomics
European Journal of Applied Physiology
Exercise and Sport Sciences Reviews
Exogenous Dermatology
Fuss und Sprunggelenk
Geneeskunde en Sport
Giornale Italiano di Medicina del Lavoro ed Ergonomia
Hong Kong Journal of Occupational Therapy
Human Factors
Indian Journal of Occupational and Environmental Medicine
Indoor and Built Environment
Industrial Health
International Archives of Occupational and Environmental Health
International Journal of Environmental Health Research
International Journal of Hygiene and Environmental Health
International Journal of Industrial Ergonomics
International Journal of Low Radiation
International Journal of Men's Health
International Journal of Occupational Medicine and Environmental Health
International Journal of Occupational and Environmental Health
International Journal of Sport Nutrition and Exercise Metabolism
International Journal of Sports Medicine
International Journal of Sports Medicine, Supplement
Japanese Journal of Aerospace and Environmental Medicine
Journal of Applied Biomechanics
Journal of Athletic Training
Journal of Ecophysiology and Occupational Health
Journal of Exposure Science and Environmental Epidemiology
Journal of Occupational Health
Journal of Occupational Health Psychology
Journal of Occupational Health and Safety - Australia and New Zealand
Journal of Occupational Rehabilitation
Journal of Occupational and Environmental Hygiene
Journal of Occupational and Environmental Medicine
Journal of Radiological Protection
Journal of Science and Medicine in Sport
Journal of Sport Rehabilitation
Journal of Sports Medicine and Physical Fitness
Journal of Sports Sciences
Journal of Toxicology and Environmental Health - Part B: Critical Reviews
Journal of UOEH
Journal of Water and Health
Measurement in Physical Education and Exercise Science
Medical Problems of Performing Artists
Medicina del Lavoro
Medicina dello Sport
Medicine and Science in Sports and Exercise
Military Medicine
Noise and Health
Occupational Health
Occupational Medicine
Occupational and Environmental Medicine
Operative Techniques in Sports Medicine
Pediatric Exercise Science
Physical Therapy in Sport
Physician and Sportsmedicine
Pracovni Lekarstvi
Radiation Protection Dosimetry
Research in Sports Medicine
Revista Brasileira de Medicina do Esporte
Revista de la Asociacion Espanola de Especialistas en Medicina del Trabajo
Safety Science
Scandinavian Journal of Occupational Therapy
Scandinavian Journal of Work, Environment and Health
Schweizerische Zeitschrift fur Sportmedizin und Sporttraumatologie
Science and Sports
Sozial- und Praventivmedizin
Sports Medicine
Technology and Disability
Toxicology and Industrial Health
Umweltmedizin in Forschung und Praxis
VHSJ - Veterans Health System Journal
Work
Work and Stress
Zeitschrift fur Larmbekampfung
Zeitschrift fur Umweltmedizin
Zentralblatt fur Arbeitsmedizin, Arbeitsschutz und Ergonomie

B.2 Public Health

AIDS Care - Psychological and Socio-Medical Aspects of AIDS/HIV
AIDS Education and Prevention
AIDS Patient Care and STDs
AIDS and Behavior
AIDS and Public Policy Journal
Academic Medicine
Academic Psychiatry
African Journal of AIDS Research
Air Medical Journal
American Journal of Health Promotion
American Journal of Human Biology
American Journal of Preventive Medicine
American Journal of Public Health
Annali dell'Istituto Superiore di Sanita
Annals of Family Medicine
Annual Review of Public Health
Anthropology and Medicine
Arbeitsmedizin Sozialmedizin Umweltmedizin
Arbeitsmedizin Sozialmedizin Umweltmedizin, Supplement
Archives des Maladies Professionnelles et de l'Environnement
Archives of Public Health
Asia-Pacific Journal of Public Health
Australia and New Zealand Health Policy
Australian Journal of Primary Health
Australian Journal of Rural Health
Australian and New Zealand Journal of Public Health
BMC Health Services Research
BMC International Health and Human Rights
BMC Medical Education
BMC Medical Ethics
BMC Medicine
BMC Oral Health
BMC Public Health
Biometrics
Bulletin of the World Health Organization
Cahiers Sante
Canadian Journal of Community Mental Health
Canadian Journal of Public Health
Central European Journal of Public Health
Children's Health Care
Chinese Journal of Endemiology
Chronic Diseases in Canada
Community Mental Health Journal
Critical Public Health
Culture, Health and Sexuality
Current Nutrition and Food Science
Current Sports Medicine Reports
Developing World Bioethics
Drug Information Journal
Eastern Mediterranean Health Journal
Economics and Human Biology
Education for Health: Change in Learning and Practice
Education for Primary Care
Emerging Infectious Diseases
Enfermedades Emergentes
Environmental Health Criteria
Environmental Health Perspectives
Environmental Health and Preventive Medicine
Environmental Research
Epidemiologia e Psichiatria Sociale
Epidemiology and Infection
Ethnicity and Disease
Ethnicity and Health
European Journal of Allied Health
European Journal of Cardiovascular Prevention and Rehabilitation
European Journal of Health Law
European Journal of Public Health
Families, Systems and Health
Family Medicine
Family Practice
Fluoride
Food and Drug Law Journal
Future of Children
Gesundheitswesen
Gesundheitswesen, Supplement
Globalization and Health
HEC Forum
Health Care Analysis
Health Education Journal
Health Education Research
Health Expectations
Health Policy
Health Policy and Planning
Health Promotion International
Health Services Research
Health Services and Outcomes Research Methodology
Health and Place
Health and Population: Perspectives and Issues
Health and Quality of Life Outcomes
Health, Risk and Society
High Blood Pressure and Cardiovascular Prevention
Hygiena
Hygiene + Medizin
Inhalation Toxicology
International Journal for Equity in Health
International Journal for Quality in Health Care
International Journal of Adolescent Medicine and Health
International Journal of Disaster Medicine
International Journal of Environmental Research and Public Health
International Journal of Health Geographics
International Journal of Health Planning and Management
International Journal of Health Promotion and Education
International Journal of Health Services
International Journal of Low Radiation
International Journal of Men's Health
International Journal of Risk and Safety in Medicine
International Journal of STD and AIDS
Iranian Journal of Public Health
Journal of Adolescent Health
Journal of Biolaw and Business
Journal of Cancer Education
Journal of Community Health
Journal of Community Health Nursing
Journal of Comparative Family Studies
Journal of Environmental Health
Journal of Epidemiology and Community Health
Journal of Ethnicity in Substance Abuse
Journal of Evaluation in Clinical Practice
Journal of Family Practice
Journal of HIV/AIDS Prevention in Children and Youth
Journal of Health Politics, Policy and Law
Journal of Health Services Research and Policy
Journal of Health and Human Services Administration

Journal of Health and Social Policy
Journal of Human Nutrition and Dietetics
Journal of Immigrant and Minority Health
Journal of Infectious Diseases
Journal of Medical Screening
Journal of Medical Teachers Federation
Journal of Medicine and Philosophy
Journal of Midwifery and Women's Health
Journal of Prevention and Intervention in the Community
Journal of Preventive Medicine and Hygiene
Journal of Primary Prevention
Journal of Psychosocial Oncology
Journal of Public Health
Journal of Public Health
Journal of Public Health Management and Practice
Journal of Public Health Policy
Journal of Research in Crime and Delinquency
Journal of Rural Health
Journal of Substance Use
Journal of The Royal Society for the Promotion of Health
Journal of Trace Elements in Medicine and Biology
Journal of Travel Medicine
Journal of Tropical Pediatrics
Journal of Urban Health
Journal of Water and Health
Journal of the American Board of Family Practice
Journal of the American Health Information Management Association
Journal of the Food Hygienic Society of Japan
Maternal and Child Health Journal
Maternal and Child Nutrition
Maturitas
Medecine et Maladies Infectieuses
Medical Decision Making
Medical Education
Medical Education, Supplement
Medical Teacher
Medicina Preventiva
Medizinische Ausbildung
New Zealand Public Health Surveillance Report
Nicotine and Tobacco Research
Noise and Health
North and West London Journal of General Practice
Obesity Reviews
Obesity Reviews, Supplement
Paediatrics and Child Health
Pakistan Journal of Nutrition
Patient Education and Counseling
Perspectives on Sexual and Reproductive Health
Population Health Metrics
Preventive Medicine
Problemy HIV i AIDS
Psychiatric Bulletin
Psychology, Health and Medicine
Public Health
Public Health Medicine
Public Health Nutrition
Public Health Reports
Public Health Reviews
Qualitative Health Research
Quality and Safety in Health Care
Quality in Primary Care
Recht und Psychiatrie
Reproductive Health
Reproductive Health Matters
Research on Aging
Revista Espanola de Salud Publica
Revista Panamericana de Salud Publica/Pan American Journal of Public Health
Revista de Saude Publica
Revista do Instituto de Medicina Tropical de Sao Paulo
Revue d'Epidemiologie et de Sante Publique
Risk Analysis
Salud Publica de Mexico
Sante Publique
Scandinavian Journal of Primary Health Care
Scandinavian Journal of Primary Health Care, Supplement
Scandinavian Journal of Public Health
Scandinavian Journal of Public Health, Supplement
Sexual Health
Sexually Transmitted Diseases
Social Science and Medicine
Social Work in Health Care
Southern African Journal of HIV Medicine
Sozial- und Praventivmedizin
Taiwan Journal of Public Health
Theoretical Medicine and Bioethics
Traffic Injury Prevention
Tropical Medicine and International Health
Umweltmedizin in Forschung und Praxis
Vaccines: Children and Practice
Vascular Health and Risk Management
Water Science and Technology: Water Supply
Water, Air, and Soil Pollution
Women and Health
Work Based Learning in Primary Care
World Health Organization - Technical Report Series
Zeitschrift fur Umweltmedizin

B.2.1 Epidemiology

Alzheimer's and Dementia
American Journal of Epidemiology
Annals of Epidemiology
Cancer Causes and Control
Cancer Epidemiology Biomarkers and Prevention
Clinical Practice and Epidemiology in Mental Health
Emerging Themes in Epidemiology
Epidemiologic Perspectives and Innovations
Epidemiologic Reviews
Epidemiologie, Mikrobiologie, Imunologie
Epidemiology
European Journal of Epidemiology
Genetic Epidemiology
Infektoloski Glasnik
International Journal of Epidemiology
Journal of Clinical Epidemiology
Journal of Immigrant and Minority Health
Medicina Preventiva
Neuroepidemiology
Norsk Epidemiologi
Ophthalmic Epidemiology
Paediatric and Perinatal Epidemiology
Prevention and Control
Statistical Methods in Medical Research
Statistics in Medicine
Vascular Disease Prevention

B.2.2 Tropical Medicine

Acta Tropica
African Health Sciences
American Journal of Tropical Medicine and Hygiene
Annals of African Medicine
Annals of Tropical Medicine and Parasitology
Foodborne Pathogens and Disease
Giornale Italiano di Medicina Tropicale
Journal of Water and Health
Medecine Tropicale
Revista da Sociedade Brasileira de Medicina Tropical
Transactions of the Royal Society of Tropical Medicine and Hygiene
Travel Medicine and Infectious Disease
Tropical Doctor
Tropical Medicine and International Health

B.3 Pollution Control

Advances in Applied Microbiology
Advances in Environmental Research
Aerosol Science and Technology
Analyst
Annals of Agricultural and Environmental Medicine
Aquatic Toxicology
Archives of Environmental Contamination and Toxicology
Asia Pacific Journal of Environmental Law
Atmospheric Environment
Aviation Space and Environmental Medicine
Bioresource Technology
Bulletin of Environmental Contamination and Toxicology
Chemosphere
Combustion and Flame
Critical Reviews in Environmental Science and Technology
EcoHealth
Ecotoxicology
Ecotoxicology and Environmental Safety
Environment
Environment International
Environmental Conservation
Environmental Engineering Science
Environmental Health and Preventive Medicine
Environmental Health: A Global Access Science Source
Environmental Management
Environmental Monitoring and Assessment
Environmental Pollution
Environmental Research
Environmental Science and Policy
Environmental Science and Technology
Environmental Technology
Environmental Toxicology
Environmental Toxicology and Chemistry
Environmental Toxicology and Pharmacology
Gefahrstoffe Reinhaltung der Luft
International Journal of Environment and Pollution
International Journal of Environmental Health Research
International Journal of Environmental Research and Public Health
International Journal of Life Cycle Assessment
Journal of Atmospheric Chemistry
Journal of Chemical Technology and Biotechnology
Journal of Contaminant Hydrology
Journal of Ecophysiology and Occupational Health
Journal of Environmental Biology
Journal of Environmental Health
Journal of Environmental Management
Journal of Environmental Monitoring
Journal of Environmental Quality
Journal of Environmental Radioactivity
Journal of Environmental Science and Health - Part A Toxic/Hazardous Substances and Environmental Engineering
Journal of Environmental Science and Health - Part B Pesticides, Food Contaminants, and Agricultural Wastes
Journal of Environmental Science and Health - Part C Environmental Carcinogenesis and Ecotoxicology Reviews
Journal of Exposure Science and Environmental Epidemiology
Journal of Hazardous Materials
Journal of Toxicology and Environmental Health - Part A
Journal of the Air and Waste Management Association
Journal of the Chartered Institution of Water and Environmental Management
Marine Environmental Research
Marine Pollution Bulletin
Resources, Conservation and Recycling
Reviews on Environmental Health
Risk Analysis
Science of the Total Environment
Umweltmedizin in Forschung und Praxis
Umweltwissenschaften und Schadstoff-Forschung
Waste Management
Waste Management and Research
Water Environment Research
Water Research
Water Science and Technology
Water Science and Technology: Water Supply
Water, Air, and Soil Pollution
Zeitschrift fur Larmbekampfung
Zeitschrift fur Umweltmedizin

B.4 Health Policy, Economics and Management

Alzheimer's and Dementia
Applied Health Economics and Health Policy
Australia and New Zealand Health Policy
Chronic Illness
Cost Effectiveness and Resource Allocation
Disease Management
Disease Management and Health Outcomes
Evaluation and the Health Professions
Evidence-based Obstetrics and Gynecology
Gesundheitsokonomie und Qualitatsmanagement
Globalization and Health
Good Clinical Practice Journal
Health Policy
Health Policy and Planning
Health Research Policy and Systems
Health Services Research
Health Services and Outcomes Research Methodology
Health Technology Assessment
Human Resources for Health
Inquiry
International Journal of Electronic Healthcare

International Journal of Health Planning and Management
International Journal of Health Services
Journal of Health Politics, Policy and Law
Journal of Health Services Research and Policy
Journal of Health and Human Services Administration
Journal of Health and Social Policy
Journal of Hospital Marketing and Public Relations
Journal of Integrated Care Pathways
Journal of Pharmaceutical Finance, Economics and Policy
Journal of Public Health Management and Practice
Journal of the American Health Information Management Association
Managed Care Interface
Medical Care Research and Review
Medical Crossfire
Mental Health Services Research
Pain Physician
PharmacoEconomics - Italian Research Articles
Recht und Psychiatrie
Social Work in Health Care
TEN Trends in Evidence-Based Neuropsychiatry

B.4.1 Health Economics

Economics and Human Biology
European Journal of Health Economics
Expert Review of Pharmacoeconomics and Outcomes Research
Health Care Financing Review
Health Economics
International Journal of Technology Assessment in Health Care
Journal d'Economie Medicale
Journal of Health Care Finance
Journal of Health Economics
Journal of Medical Economics
Journal of Medical Practice Management
Journal of Mental Health Policy and Economics
Marketing Health Services
MeReC Briefing
PharmacoEconomics
Research in Healthcare Financial Management
Value in Health

B.4.2 Health Care Management

AIDS Patient Care and STDs
Acta Hospitalia
Administration and Policy in Mental Health
American Journal of Health-System Pharmacy
American Journal of Managed Care
Canadian Journal of Hospital Pharmacy
Care Management Journals
Clinical Governance
Clinician in Management
Communication and Medicine
Critical Pathways in Cardiology
Drug Benefit Trends
Evidence-based Cardiovascular Medicine
Health Care Management Review
Health Services Management Research
Healthcare Review Online
Hospitals and Health Networks
Informatics in Primary Care
International Journal of Healthcare Technology and Management
Journal of Ambulatory Care Management
Journal of Behavioral Health Services and Research
Journal of Healthcare Management
Journal of Nursing Administration
Journal of Pharmacy Practice and Research
Journal of Prevention and Intervention in the Community
Medical Group Management Journal
Qualitative Health Research
Quality Assurance Journal
Quality and Safety in Health Care
Quality in Primary Care
Salus

B.5 Nursing

AANA Journal
BMC Nursing
Cancer Nursing
Critical Care Nursing Quarterly
European Journal of Cardiovascular Nursing
Explore: The Journal of Science and Healing
Hispanic Healthcare International
International Journal of Nursing Education Scholarship
Journal of Community Health Nursing
Journal of Dementia Care
Journal of Infusion Nursing
Journal of Intravenous Nursing
Journal of Midwifery and Women's Health
Journal of Nursing Administration
Journal of Radiology Nursing
Matronas Profesion
Nursing and Health Sciences
Online Brazilian Journal of Nursing
Online Journal of Issues in Nursing
Online Journal of Knowledge Synthesis for Nursing
Online Journal of Nursing Informatics
Practical Cardiovascular Risk Management
Revue Francophone de Psycho-Oncologie
Scandinavian Journal of Caring Sciences
Topics in Advanced Practice Nursing
Work Based Learning in Primary Care
Worldviews on Evidence-Based Nursing

C.0 Clinical Medicine

AIDS Reader
ASGE Clinical Update
Academic Journal of Second Military Medical University
Academic Journal of Xi'an Jiaotong University
Acta Academiae Medicinae Hubei
Acta Academiae Medicinae Sinicae
Acta Biomedica de l'Ateneo Parmense
Acta Clinica Belgica
Acta Clinica Croatica
Acta Clinica Croatica, Supplement
Acta Endoscopica

Acta Facultatis Medicae Fluminensis
Acta Medica Austriaca
Acta Medica Auxologica
Acta Medica Bulgarica
Acta Medica Croatica
Acta Medica Iranica
Acta Medica Mediterranea
Acta Medica Nagasakiensia
Acta Medica Portuguesa
Acta Medica Romana
Acta Scientiarum - Biological Sciences
Acta Scientiarum - Health Sciences
Advanced Studies in Medicine
Advances in Therapy
African Health Sciences
African Journal of Traditional, Complementary and Alternative Medicines
Aging Health
Aktuelle Ernahrungsmedizin
Akupunktur und Traditionelle Chinesische Medizin
Alternative Medicine Review
Alternative Therapies in Health and Medicine
Alternative and Complementary Therapies
American Family Physician
American Journal of Chinese Medicine
American Journal of Medicine
American Journal of the Medical Sciences
Anales del Sistema Sanitario de Navarra
Annales Academiae Medicae Gedanensis
Annali Degli Ospedali San Camillo e Forlanini
Annals of African Medicine
Annals of Fundeni Hospital
Annals of Medical Sciences
Annals of Medicine
Annals of Saudi Medicine
Annals of the Academy of Medicine Singapore
Annual Review of Medicine
Archives Italiennes de Biologie
Archives of Hellenic Medicine
Archives of Iranian Medicine
Archives of Medical Research
Archives of Medical Science
Archives of the Balkan Medical Union
Archivos de Medicina
Arquivos Brasileiros de Medicina
Arquivos de Medicina
Artificial Cells, Blood Substitutes, and Biotechnology
Arztezeitschrift fur Naturheilverfahren und Regulationsmedizin
Atencion Primaria
Azerbaijan Medical Journal
BMC Complementary and Alternative Medicine
BMC Family Practice
BMC Medicine
Bahrain Medical Bulletin
Bangladesh Medical Research Council Bulletin
Biologische Medizin
British Journal of General Practice
British Journal of Hospital Medicine
British Medical Bulletin
British Medical Journal
Bulgarian Medicine
Bull RGKMC
Bulletin of Hunan Medical University
Bulletin of National Institute of Health Sciences
Bulletin, Postgraduate Institute of Medical Education and Research, Chandigarh
CME Cancer Medicine
Canadian Family Physician
Canadian Medical Association Journal
Case Reports and Clinical Practice Review
Casopis Lekaru Ceskych
Chang Gung Medical Journal
Chinese Journal of Emergency Medicine
Chinese Journal of Endemiology
Chinese Journal of Evidence-Based Medicine
Chinese Journal of Integrative Medicine
Chinese Medical Journal
Chinese Medical Sciences Journal
Chinesische Medizin
Chiropractic and Osteopathy
Chronic Illness
Cleveland Clinic Journal of Medicine
Clinica Terapeutica
Clinical Cornerstone
Clinical Medicine and Research
Clinical Medicine, Journal of the Royal College of Physicians of London
Clinical Science
Clinical Teacher
Clinical Trials
Clinical and Experimental Medicine
Clinical and Investigative Medicine
Clinics in Family Practice
Clinics in Laboratory Medicine
Cochrane Database of Systematic Reviews
Colombia Medica
Complementary Therapies in Medicine
Comprehensive Therapy
Connecticut Medicine
Consultant
Core Evidence
Croatian Medical Journal
Current Medical Research and Opinion
Current Medical Research and Opinion, Supplement
Current News in Medicine
Current Opinion in Clinical Experimental Research
Current Opinion in European Medicine
Danish Medical Bulletin
Daru
Deutsche Medizinische Wochenschrift
Deutsches Arzteblatt
Disease Markers
Disease-a-Month
Dokkyo Journal of Medical Sciences
Drug R and D Backgrounders
Eastern Journal of Medicine
Eastern Mediterranean Health Journal
Emirates Medical Journal
Endoskopie
Erciyes Tip Dergisi
Ethnicity and Disease
European Journal of Acupuncture
European Journal of Clinical Investigation
European Journal of Emergency Medicine
European Journal of General Medicine
European Journal of General Practice
European Journal of Inflammation
European Journal of Medical Research
European Journal of Oriental Medicine

Evidence-Based Healthcare and Public Health
Evidence-Based Integrative Medicine
Evidence-Based Medicine
Evidence-based Complementary and Alternative Medicine
Explore: The Journal of Science and Healing
FORUM - Trends in Experimental and Clinical Medicine
Family Medicine and Primary Care Review
Focus on Alternative and Complementary Therapies
Forschende Komplementarmedizin
Fudan University Journal of Medical Sciences
Gaceta Medica de Mexico
Gazi Medical Journal
Gazzetta Medica Italiana Archivio per le Scienze Mediche
Gender Medicine
General Medicine
Good Clinical Practice Journal
Goztepe Tip Dergisi
Gulhane Medical Journal
HIV Medicine
Harefuah
Herbals and Complementary Medicine
High Altitude Medicine and Biology
Hippokratia
Hirosaki Medical Journal
Hiroshima Journal of Medical Sciences
Homeopathy
Hong Kong Medical Journal
Hong Kong Practitioner
Huisarts en Wetenschap
IJCI - International Journal of Clinical Investigation
IRYO - Japanese Journal of National Medical Services
Iatreia
Impegno Ospedaliero, Sezione Scientifica
In Vivo
Indian Journal of Medical Sciences
Inflammation
Informatics in Primary Care
Informe Medico
Integrative Medicine
Intensivmedizin und Notfallmedizin
International Journal of Biostatistics
International Journal of Clinical Practice
International Journal of Clinical Practice, Supplement
International Journal of Medical Sciences
International Journal of Medicine
International Journal of Prisoner Health
International Journal of Transgenderism
International Medical Journal
Investigacion Clinica
Investigacion Medica Internacional
Iranian Journal of Medical Sciences
Irish Journal of Medical Science
Irish Medical Journal
Israel Medical Association Journal
JK Practitioner
JMS - Journal of Medical Society
Japan Medical Association Journal
Jordan Journal of Applied Sciences - Natural Sciences
Jordan Medical Journal
Journal International Medical Sciences Academy
Journal Medical Libanais
Journal of Alternative and Complementary Medicine
Journal of Applied Research
Journal of Cancer Integrative Medicine
Journal of Central South University (Medical Sciences)
Journal of Chinese Integrative Medicine
Journal of Chinese Medicine
Journal of Clinical Densitometry
Journal of Clinical Investigation
Journal of Clinical Outcomes Management
Journal of Clinical Research
Journal of Complementary and Integrative Medicine
Journal of Integrated Care Pathways
Journal of International Medical Research
Journal of Interprofessional Care
Journal of Liaquat University of Medical and Health Sciences
Journal of Medical Sciences
Journal of Medical Sciences
Journal of Medical Teachers Federation
Journal of Medicinal Plants
Journal of Medicine
Journal of Men's Health and Gender
Journal of Molecular Medicine
Journal of Nara Medical Association
Journal of Negative Results in BioMedicine
Journal of Nippon Medical School
Journal of Outcomes Research
Journal of Palliative Medicine
Journal of Postgraduate Medical Institute
Journal of Postgraduate Medicine
Journal of Research in Medical Sciences
Journal of Sichuan University (Medical Science Edition)
Journal of Terminal Oncology
Journal of Tokyo Medical University
Journal of Traditional Chinese Medicine
Journal of Translational Medicine
Journal of Women's Health
Journal of Xi'an Jiaotong University (Medical Sciences)
Journal of the American Medical Association
Journal of the American Medical Directors Association
Journal of the American Osteopathic Association
Journal of the Bahrain Medical Society
Journal of the Chinese Medical Association
Journal of the College of Physicians and Surgeons Pakistan
Journal of the Formosan Medical Association
Journal of the Indian Medical Association
Journal of the Medical Association of Thailand
Journal of the Medical Society of Toho University
Journal of the National Medical Association
Journal of the Nepal Medical Association
Journal of the Osaka City Medical Center
Journal of the Pakistan Medical Association
Journal of the Royal Society of Medicine
Journal of the Royal Society of Medicine, Supplement
Journal of the Showa Medical Association
Journal of the Wakayama Medical Society
Journal, Indian Academy of Clinical Medicine
Kaohsiung Journal of Medical Sciences
Kathmandu University Medical Journal
Keio Journal of Medicine
Kitakanto Medical Journal
Klinikarzt
Kobe Journal of Medical Sciences
Koomesh
Kumamoto Medical Journal
Kurume Medical Journal
Lancet
Lege Artis Medicine
Lekarsky Obzor

Lijecnicki Vjesnik
Lijecnicki Vjesnik, Supplement
Louvain Medical
MMW-Fortschritte der Medizin
Maghreb Medical
Malaysian Journal of Medical Sciences
Mapfre Medicina
Marmara Medical Journal
Mayo Clinic Proceedings
McGill Journal of Medicine
MedGenMed Medscape General Medicine
Medecine Therapeutique
Medecine et Hygiene
Medica Jadertina
Medical Clinics of North America
Medical Crossfire
Medical Forum Monthly
Medical Hypotheses
Medical Journal Armed Forces India
Medical Journal of Australia
Medical Journal of Kobe University
Medical Journal of Malaysia
Medical Journal of Minami Osaka Hospital
Medical Journal of the Islamic Republic of Iran
Medical Principles and Practice
Medical Quality
Medical Science Monitor
Medicina
Medicina
Medicina
Medicina Clinica
Medicina Paliativa
Medicine
Medicine Today
Medicinski Glasnik
Medicus
Medicus
Medizinische Klinik
Medizinische Welt
Men's Health Journal
Mid-Taiwan Journal of Medicine
Middle East Journal of Emergency Medicine
Mie Medical Journal
Minerva Medica
Modern Medicine
Molecular Cancer
Molecular Diagnosis and Therapy
Molecular Systems Biology
Mount Sinai Journal of Medicine
National Medical Journal of China
National Medical Journal of India
Nederlands Tijdschrift voor Geneeskunde
Netherlands Journal of Medicine
New England Journal of Medicine
New Medicine
New Zealand Medical Journal
Nobel Medicus
Nordisk Tidsskrift for Biologisk Medisin
North and West London Journal of General Practice
Ochsner Journal
Ondokuz Mayis Universitesi Tip Dergisi
Online Journal of Health and Allied Sciences
Orvosi Hetilap
Orvoskepzes
Ospedale Maggiore
Osteologicky Bulletin
Osteologie
Osteoporosis International
PLoS Medicine
Pakistan Journal of Medical Sciences
Palliative Medicine
Personalized Medicine
Photodiagnosis and Photodynamic Therapy
Phytotherapie
Policlinico - Sezione Medica
Polska Medycyna Paliatywna
Polski Merkuriusz Lekarski
Postgraduate Medical Journal
Postgraduate Medicine
Prakticky Lekar
Prensa Medica Argentina
Presse Medicale
Primary Care - Clinics in Office Practice
Primary Care Case Reviews
Progress in Palliative Care
QJM - Monthly Journal of the Association of Physicians
Qatar Medical Journal
Recenti Progressi in Medicina
Research and Clinical Forums
Revista Brasileira de Medicina
Revista Clinica Espanola
Revista Espanola de Enfermedades Metabolicas Oseas
Revista Medica de Aguascalientes
Revista Medica de Rosario
Revista de Calidad Asistencial
Revista de Investigacion Clinica
Revue Medicale de Bruxelles
Revue Medicale de Liege
Revue du Praticien
Rivista Medica
Rivista di Medicina di Laboratorio
Round Table Series - Royal Society of Medicine
SENDROM
Salud (i) Sciencia
Salus
Santo Tomas Journal of Medicine
Sao Paulo Medical Journal
Sapporo Medical Journal
Saudi Medical Journal
Schweizerische Rundschau fur Medizin - Praxis
Schweizerische Zeitschrift fur GanzheitsMedizin
Scientific Review of Alternative Medicine
Scottish Medical Journal
Scripta Medica Facultatis Medicae Universitatis Brunensis Masarykianae
Seminars in Integrative Medicine
Singapore General Hospital Proceedings
Singapore Medical Journal
South African Family Practice
South African Medical Journal
Southern Medical Journal
Swiss Medical Weekly
Tagliche Praxis
Teikyo Medical Journal
Telemedicine Journal and e-Health
TheScientificWorldJournal
Therapeutic Research
Therapeutics and Clinical Risk Management
Therapeutische Umschau
Tidsskrift for den Norske Laegeforening

Tijdschrift voor Geneeskunde
Today's Therapeutic Trends
Tohoku Journal of Experimental Medicine
Tokai Journal of Experimental and Clinical Medicine
Tokyo Jikeikai Medical Journal
Topics in Emergency Medicine
Tunisie Medicale
Turkish Journal of Medical Sciences
Turkiye Klinikleri Journal of Medical Sciences
Tzu Chi Medical Journal
Ugeskrift for Laeger
Ulster Medical Journal
VHSJ - Veterans Health System Journal
West African Journal of Medicine
West Indian Medical Journal
Wiener Klinische Wochenschrift
Wiener Klinische Wochenschrift, Supplement
Wiener Medizinische Wochenschrift
Wisconsin Medical Journal
Women's Health
Yakhteh
Yokohama Medical Journal
Yonago Acta Medica
Yonsei Medical Journal
Zeitschrift fur Allgemeinmedizin
Zeitschrift fur Arztliche Fortbildung und Qualitatssicherung

C.1 Internal Medicine

Acta Diabetologica
Acute Medicine
Adipocytes
Advances in Clinical and Experimental Medicine
American Journal of Hypertension
Anales de Medicina Interna
Annali Italiani di Medicina Interna
Annals of Internal Medicine
Anti-Infective Agents in Medicinal Chemistry
Archives of Internal Medicine
Archivos de la Facultad de Medicina de Zaragoza
Atlas of the Urologic Clinics of North America
Blood Pressure
Blood Pressure Monitoring
Blood Pressure, Supplement
Breast Journal
British Journal of Diabetes and Vascular Disease
Canadian Journal of Diabetes
Clinical Cases in Mineral and Bone Metabolism
Clinical and Experimental Hypertension
Contagion: Reports, Cases, and Commentaries in HIV and Infectious Disease Research
Current Diabetes Reviews
Current HIV/AIDS Reports
Current Hypertension Reports
Current Hypertension Reviews
Current Oncology Reports
Current Opinion in Endocrinology and Diabetes
Current Opinion in Nephrology and Hypertension
Current Osteoporosis Reports
Diabetes
Diabetes Care
Diabetes and Metabolism
Diabetes and Vascular Disease Research
Diabetes und Stoffwechsel
Diabetes, Nutrition and Metabolism - Clinical and Experimental
Diabetes, Obesity and Metabolism
Diabetes, Obesity and Metabolism, Supplement
Diabetic Medicine
Diabetologia
Diabetologia Croatica
Diabetologia Doswiadczalna i Kliniczna
Diabetologia Polska
Diabetologie Metabolismus Endokrinologie Vyziva
Electrolyte and Blood Pressure
European Journal of Internal Medicine
Evidence-based Cardiovascular Medicine
Ginekologia Onkologiczna
Giornale Italiano di Diabetologia e Metabolismo
Heart and Metabolism
Hypertension
Hypertension Research
Hypertension in Pregnancy
IM - Internal Medicine
Immunology, Endocrine and Metabolic Agents in Medicinal Chemistry
Insulin
Internal Medicine
Internal Medicine Clinical and Laboratory
Internal Medicine Journal
Internist
Internista
Internistische Praxis
JK Science
JRAAS - Journal of the Renin-Angiotensin-Aldosterone System
Journal fur Hypertonie
Journal of Diabetes and its Complications
Journal of Endocrinology, Metabolism and Diabetes of South Africa
Journal of General Internal Medicine
Journal of Human Hypertension
Journal of Hypertension
Journal of Hypertension, Supplement
Journal of Internal Medicine
Journal of Internal Medicine of India
Journal of Internal Medicine of Taiwan
Journal of Water and Health
Journal of the Diabetic Association of India
Journal of the Japan Diabetes Society
Korean Circulation Journal
Korean Journal of Internal Medicine
Medicina Interna de Mexico
Metabolic Syndrome and Related Disorders
Nature Clinical Practice Endocrinology and Metabolism
Nieren- und Hochdruckkrankheiten
Nutrition and Metabolic Disorders in HIV Infection
Nutrition and Metabolism
Obesity Management
Obesity in Practice
Pediatric Diabetes
Pediatric Diabetes, Supplement
Pediatric Endocrinology Reviews
Polskie Archiwum Medycyny Wewnetrznej
Practical Diabetes International
Presse Medicale
Reviews in Endocrine and Metabolic Disorders
Revue de Medecine Interne

Trends in Medicine
Vnitrni Lekarstvi

C.1.1 Cardiology and Angiology

ACC Cardiosource Review Journal
Acta Cardiologica
Acta Cardiologica Sinica
Acta Myologica
Acute Cardiac Care
American Heart Journal
American Journal of Cardiology
Anadolu Kardiyoloji Dergisi
Annales de Cardiologie et d'Angeiologie
Annals of Noninvasive Electrocardiology
Applied Cardiopulmonary Pathophysiology
Archives des Maladies du Coeur et des Vaisseaux
Archives des Maladies du Coeur et des Vaisseaux - Pratique
Archivos de Cardiologia de Mexico
Arquivos Brasileiros de Cardiologia
Arteriosclerosis, Thrombosis, and Vascular Biology
Atherosclerosis
Atherosclerosis Supplements
Autoimmunity Reviews
BMC Cardiovascular Disorders
Basic Research in Cardiology
Basic Research in Cardiology, Supplement
British Journal of Cardiology
Canadian Journal of Cardiology
CardioVascular and Interventional Radiology
Cardiology
Cardiology Clinics
Cardiology International
Cardiology Review
Cardiology in Review
Cardiology in the Young
Cardiovascular Diabetology
Cardiovascular Drug Reviews
Cardiovascular Drugs and Therapy
Cardiovascular Engineering
Cardiovascular Journal of South Africa
Cardiovascular Pathology
Cardiovascular Radiation Medicine
Cardiovascular Research
Cardiovascular Revascularization Medicine
Cardiovascular Toxicology
Cardiovascular Ultrasound
Cardiovascular and Hematological Agents in Medicinal Chemistry
Cardiovascular and Hematological Disorders - Drug Targets
Catheterization and Cardiovascular Interventions
Circulation
Circulation Journal
Circulation Research
Clinical Cardiology
Congenital Heart Disease
Continuing Medical Education - Cardiology
Cor et Vasa
Coronary Artery Disease
Critical Pathways in Cardiology
Current Cardiology Reports
Current Cardiology Reviews
Current Controlled Trials in Cardiovascular Medicine
Current Heart Failure Reports
Current Neurology and Neuroscience Reports
Current Opinion in Cardiology
Current Problems in Cardiology
Current Treatment Options in Cardiovascular Medicine
Current Vascular Pharmacology
Dialogues in Cardiovascular Medicine
Echocardiography
Electrolyte and Blood Pressure
Europace
European Heart Journal
European Heart Journal, Supplement
European Journal of Cardiovascular Nursing
European Journal of Cardiovascular Prevention and Rehabilitation
European Journal of Echocardiography
European Journal of Heart Failure
European Journal of Heart Failure, Supplement
European Journal of Lymphology and Related Problems
Evidence-based Cardiovascular Medicine
Experimental and Clinical Cardiology
Expert Review of Cardiovascular Therapy
Folia Cardiologica
Future Cardiology
Giornale Italiano di Cardiologia
Hart Bulletin
Heart
Heart Failure Clinics
Heart Failure Reviews
Heart Rhythm
Heart and Lung: Journal of Acute and Critical Care
Heart and Metabolism
Heart and Vessels
Hellenic Journal of Cardiology
Herz
Herz, Supplement
Herzschrittmachertherapie und Elektrophysiologie
High Blood Pressure and Cardiovascular Prevention
Indian Heart Journal
Indian Pacing and Electrophysiology Journal
International Heart Journal
International Journal of Cardiology
International Journal of Cardiovascular Imaging
International Journal of Sports Medicine, Supplement
Investigacion Cardiovascular
Italian Heart Journal
Italian Heart Journal Supplement
Italian Heart Journal, Supplement
Italian Journal of Vascular and Endovascular Surgery
Japanese Heart Journal
Japanese Journal of Interventional Cardiology
Journal for Vascular Ultrasound
Journal fur Kardiologie
Journal of Cardiac Failure
Journal of Cardiology
Journal of Cardiothoracic-Renal Research
Journal of Cardiovascular Electrophysiology
Journal of Cardiovascular Magnetic Resonance
Journal of Cardiovascular Medicine
Journal of Cardiovascular Pharmacology
Journal of Cardiovascular Pharmacology and Therapeutics
Journal of Clinical and Basic Cardiology
Journal of Electrocardiology
Journal of Geriatric Cardiology
Journal of Heart Valve Disease

Journal of Interventional Cardiac Electrophysiology
Journal of Interventional Cardiology
Journal of Invasive Cardiology
Journal of Molecular and Cellular Cardiology
Journal of Nuclear Cardiology
Journal of Thrombosis and Haemostasis
Journal of Thrombosis and Thrombolysis
Journal of the American College of Cardiology
Journal of the American Society of Echocardiography
Journal of the Hoffman Heart Institute of Connecticut
Kardiologia
Kardiologia Polska
Korean Circulation Journal
Korean Journal of Cardiovascular Diseases
Lymphologie in Forschung und Praxis
Medecine Therapeutique - Cardio
Mediterranean Journal of Pacing and Electrophysiology
Microcirculation
Minerva Cardioangiologica
Monaldi Archives for Chest Disease - Cardiac Series
Monaldi Archives for Chest Disease - Pulmonary Series
Nature Clinical Practice Cardiovascular Medicine
Netherlands Heart Journal
PACE - Pacing and Clinical Electrophysiology
Pediatric Cardiology
Pharma Fokus Herz-Kreislauf
Polski Przeglad Kardiologiczny
Practical Cardiovascular Risk Management
Prevention and Control
Progress in Cardiovascular Diseases
Progress in Pediatric Cardiology
Respiration and Circulation
Reviews in Cardiovascular Medicine
Revista Espanola de Cardiologia
Revista Latina de Cardiologia
Revista Portuguesa de Cardiologia
Seminars in Cerebrovascular Diseases and Stroke
Techniques in Vascular and Interventional Radiology
Texas Heart Institute Journal
Trends in Cardiovascular Medicine
Turk Kardiyoloji Dernegi Arsivi
Vascular Disease Management
Vascular Health and Risk Management
Vascular Pharmacology
Zeitschrift fur Gefassmedizin

C.1.1.1 Vascular Diseases

Acta Angiologica
Angeiologie
Angiologia
Angiology
British Journal of Diabetes and Vascular Disease
Cardiovascular Revascularization Medicine
Cardiovascular and Hematological Agents in Medicinal Chemistry
Cardiovascular and Hematological Disorders - Drug Targets
Chinese Journal of Cerebrovascular Diseases
Cor et Vasa
Current Cardiology Reviews
Diabetes and Vascular Disease Research
Electrolyte and Blood Pressure
Future Cardiology
Giornale Italiano di Cardiologia
Heart and Metabolism
International Angiology
International Journal of Angiology
International Journal of Stroke
Italian Heart Journal, Supplement
Jornal Vascular Brasileiro
Journal des Maladies Vasculaires
Journal for Vascular Ultrasound
Journal of Cardiovascular Medicine
Korean Circulation Journal
Microvascular Research
Nadcisnienie Tetnicze
Perfusion
Phlebologie
Phlebology
Phlebolymphology
Practical Cardiovascular Risk Management
Prevention and Control
Sang Thrombose Vaisseaux
Seminars in Vascular Medicine
Stroke
Vasa - Journal of Vascular Diseases
Vascular
Vascular Disease Management
Vascular Disease Prevention
Vascular Health and Risk Management
Vascular Medicine
Vasomed

C.1.2 Chest Diseases

American Journal of Physiology - Lung Cellular and Molecular Physiology
American Journal of Respiratory Cell and Molecular Biology
American Journal of Respiratory and Critical Care Medicine
Applied Cardiopulmonary Pathophysiology
Archivos de Bronconeumologia
Asthma Journal
Atemwegs- und Lungenkrankheiten
BMC Pulmonary Medicine
COPD: Journal of Chronic Obstructive Pulmonary Disease
Canadian Respiratory Journal
Chest
Children's Pulmonology
Chinese Journal of Lung Cancer
Chronic Respiratory Disease
Clinical Lung Cancer
Clinical Pulmonary Medicine
Clinics in Chest Medicine
Current Opinion in Pulmonary Medicine
Current Respiratory Medicine Reviews
European Respiratory Journal
European Respiratory Journal, Supplement
European Respiratory Review
Experimental Lung Research
GIMT - Giornale Italiano delle Malattie del Torace
Heart and Lung: Journal of Acute and Critical Care
International Journal of Respiratory Care
International Journal of Tuberculosis and Lung Disease
Japanese Journal of Chest Diseases
Japanese Journal of Lung Cancer

Journal of Aerosol Medicine: Deposition, Clearance, and Effects in the Lung
Journal of Asthma
Journal of Bronchology
Journal of Cardiothoracic-Renal Research
Journal of Clinical Sleep Medicine
Journal of Cystic Fibrosis
Journal of Respiratory Diseases
Journal of Thoracic Imaging
Kekkaku
Lotta Contro la Tuberculosi e le Malattie Polmonari Sociali
Lung
Lung Cancer
Minerva Pneumologica
Monaldi Archives for Chest Disease - Cardiac Series
Monaldi Archives for Chest Disease - Pulmonary Series
Paediatric Respiratory Reviews
Pediatric Pulmonology
Pneumologie
Pneumonologia i Alergologia Polska
Primary Care Respiratory Journal
Proceedings of the American Thoracic Society
Pulmonary Pharmacology and Therapeutics
Rassegna di Patologia dell'Apparato Respiratorio
Respiration
Respiration and Circulation
Respiratory Care
Respiratory Care Clinics of North America
Respiratory Medicine
Respiratory Medicine Extra
Respiratory Medicine: COPD Update
Respiratory Physiology and Neurobiology
Respiratory Research
Respirology
Revista Portuguesa de Pneumologia
Revista del Instituto Nacional de Enfermedades Respiratorias
Revue de Pneumologie Clinique
Revue des Maladies Respiratoires
Sarcoidosis Vasculitis and Diffuse Lung Diseases
Seminars in Respiratory and Critical Care Medicine
Studia Pneumologica et Phthiseologica
Thoracic Surgery Clinics
Thorax
Treatments in Respiratory Medicine
Tuberculosis
Tuberculosis and Respiratory Diseases
Tuberkuloz ve Toraks

C.1.3 Oncology

Acta Oncologica
Acta Oncologica, Supplement
Advances in Breast Cancer
Advances in Cancer Research
Advances in Gastrointestinal Cancers
Angiogenesis
Annals of Cancer Research and Therapy
Anti-Cancer Agents in Medicinal Chemistry
Anti-Cancer Drugs
Anticancer Research
Biochimica et Biophysica Acta - Reviews on Cancer
Biotherapy
Birth Defects Research Part B - Developmental and Reproductive Toxicology
Brachytherapy
Brain Tumor Pathology
Breast Cancer Online
Breast Cancer Research
British Journal of Cancer
Cancer Biology and Therapy
Cancer Biotherapy and Radiopharmaceuticals
Cancer Cell
Cancer Cell International
Cancer Chemotherapy and Pharmacology
Cancer Chemotherapy and Pharmacology, Supplement
Cancer Detection and Prevention
Cancer Epidemiology Biomarkers and Prevention
Cancer Forum
Cancer Gene Therapy
Cancer Genomics and Proteomics
Cancer Immunity
Cancer Immunology, Immunotherapy
Cancer Informatics
Cancer Investigation
Cancer Journal
Cancer Letters
Cancer Molecular Biology
Cancer Research
Cancer Science
Carcinogenesis
Chinese Journal of Cancer Research
Chinese Journal of Lung Cancer
Clinical Breast Cancer
Clinical Lung Cancer
Clinical Lymphoma and Myeloma
Clinical and Experimental Metastasis
Clinical and Translational Oncology
Current Cancer Drug Targets
Current Cancer Therapy Reviews
Current Oncology Reports
Deutsche Zeitschrift fur Onkologie
Enhancer - Biotherapy of Cancer
European Journal of Cancer
European Journal of Cancer Prevention
European Journal of Cancer, Supplement
Experimental Oncology
Familial Cancer
Future Oncology
Gann Monographs on Cancer Research
Genes Chromosomes and Cancer
Hereditary Cancer in Clinical Practice
IARC Monographs on the Evaluation of Carcinogenic Risks to Humans
Integrative Cancer Therapies
International Journal of Cancer
International Journal of Cancer Research
International Journal of Hyperthermia
International Journal of Radiation Oncology Biology Physics
International Seminars in Surgical Oncology
Journal of Biological Regulators and Homeostatic Agents
Journal of Cancer Education
Journal of Cancer Research and Clinical Oncology
Journal of Cancer Research and Clinical Oncology, Supplement
Journal of Carcinogenesis
Journal of Environmental Pathology, Toxicology and

Oncology
Journal of Environmental Science and Health - Part C Environmental Carcinogenesis and Ecotoxicology Reviews
Journal of Experimental Therapeutics and Oncology
Journal of Experimental and Clinical Cancer Research
Journal of Immunotherapy
Journal of JASTRO
Journal of Mammary Gland Biology and Neoplasia
Journal of Neuro-Oncology
Journal of Oncology Pharmacy Practice
Journal of Women's Cancer
Journal of the National Cancer Institute
Lancet Oncology
Leukemia Research
Melanoma Research
Molecular Cancer
Molecular Cancer Research
Molecular Carcinogenesis
Nature Clinical Practice Oncology
Nature Reviews Cancer
Neoplasia
Neoplasma
Neuro-Oncology
Nowotwory
Nutrition and Cancer
Oncogene
Oncologist
Oncology
Oncology Forum
Oncology Research
Onkologia Polska
Oral Oncology
PLoS Genetics
Pathology Oncology Research
Radiotherapy and Oncology
Reports of Practical Oncology and Radiotherapy
Seminars in Cancer Biology
Strahlentherapie und Onkologie
Supportive Cancer Therapy
Surgical Oncology
Technology in Cancer Research and Treatment
Tumor Biology
Tumor Research
Tumori
Urologic Oncology: Seminars and Original Investigations
World Journal of Surgical Oncology

C.1.3.1 Clinical Oncology

Advances in Breast Cancer
Advances in Gastrointestinal Cancers
American Journal of Cancer
American Journal of Clinical Oncology: Cancer Clinical Trials
American Journal of Oncology Review
Annals of Cancer Research and Therapy
Annals of Oncology
Annals of Surgical Oncology
Archive of Oncology
Asia-Pacific Journal of Clinical Oncology
BMC Cancer
Breast
Breast Cancer Online
Breast Cancer Research and Treatment
Breast Care
Breast Journal
Bulletin du Cancer
CME Cancer Medicine
CME Journal of Gynecologic Oncology
Ca-A Cancer Journal for Clinicians
Cancer
Cancer Biology and Therapy
Cancer Causes and Control
Cancer Control
Cancer Genetics and Cytogenetics
Cancer Informatics
Cancer Nursing
Cancer Treatment Reviews
Cancer and Metastasis Reviews
Cancer/Radiotherapie
Cellular Oncology
Chinese Journal of Clinical Oncology
Clinical Advances in Hematology and Oncology
Clinical Cancer Research
Clinical Colorectal Cancer
Clinical Dermatology: Retinoids and other Treatments
Clinical Genitourinary Cancer
Clinical Oncology
Clinical and Translational Oncology
Community Oncology
Critical Reviews in Oncology/Hematology
Current Cancer Therapy Reviews
Current Oncology
Current Opinion in Oncology
Current Problems in Cancer
Current Treatment Options in Oncology
CytoJournal
Endocrine-Related Cancer
Esophagus
European Journal of Gynaecological Oncology
European Journal of Oncology
European Journal of Surgical Oncology
Future Oncology
Gastric Cancer
Ginekologia Onkologiczna
Gynecologic Oncology
Hematological Oncology
Hematology/Oncology Clinics of North America
Hereditary Cancer in Clinical Practice
International Journal of Clinical Oncology
International Journal of Gynecological Cancer
JNCCN Journal of the National Comprehensive Cancer Network
Japanese Journal of Clinical Oncology
Japanese Journal of Lung Cancer
Journal of B.U.ON.
Journal of Cancer Integrative Medicine
Journal of Clinical Oncology
Journal of Pediatric Hematology/Oncology
Journal of Psychosocial Oncology
Journal of Supportive Oncology
Journal of Surgical Oncology
Journal of Terminal Oncology
Klinicka Onkologie
Leukemia
Leukemia and Lymphoma
Libri Oncologici

Lung Cancer
Medical Oncology
ONCOLOGY
Oncologia
Oncologie
Oncology Report
Onkologe
Onkologia Polska
Onkologie
Pediatric Blood and Cancer
Pediatric Hematology and Oncology
Prostate Cancer and Prostatic Diseases
Psycho-Oncology
Radiology and Oncology
Revisiones en Cancer
Revista da Sociedade Brasileira de Cancerologia
Revue Francophone de Psycho-Oncologie
Sarcoma
Seminars in Breast Disease
Seminars in Oncology
Seminars in Radiation Oncology
Seminars in Surgical Oncology
Signal
Skin Cancer
Supportive Cancer Therapy
Supportive Care in Cancer
Surgical Oncology Clinics of North America
Tumor Diagnostik und Therapie
Turkish Journal of Cancer
UHOD - Uluslararasi Hematoloji-Onkoloji Dergisi
Women's Oncology Review
Wspolczesna Onkologia

C.1.4 Clinical Genetics

Acta Geneticae Medicae et Gemellologiae
American Journal of Human Genetics
American Journal of Medical Genetics
American Journal of Medical Genetics - Neuropsychiatric Genetics
American Journal of Medical Genetics - Seminars in Medical Genetics
Annals of Human Genetics
Annual Review of Genomics and Human Genetics
BAG - Journal of Basic and Applied Genetics
Balkan Journal of Medical Genetics
Behavior Genetics
CONTINUUM Lifelong Learning in Neurology
Chinese Journal of Medical Genetics
Clinical Dysmorphology
Clinical Genetics
Community Genetics
Comparative and Functional Genomics
Current Gene Therapy
Current Opinion in Molecular Therapeutics
European Journal of Human Genetics
Genes Chromosomes and Cancer
Genes and Immunity
Genetic Counseling
Genetic Epidemiology
Genetic Testing
Genetics in Medicine
Genomics
Hereditary Cancer in Clinical Practice
Human Gene Therapy
Human Genetics
Human Heredity
Human Mutation
Indian Journal of Human Genetics
Journal of Endocrine Genetics
Journal of Genetic Counseling
Journal of Human Genetics
Journal of Inherited Metabolic Disease
Journal of Medical Genetics
Malignant Hyperthermia
Medizinische Genetik
Mental Retardation and Developmental Disabilities Research Reviews
Molecular Therapy
Nature Genetics
Neurogenetics
Ophthalmic Genetics
PLoS Genetics
Pharmacogenomics
Prenatal Diagnosis
Psychiatric Genetics
Twin Research and Human Genetics
Virology Journal

C.1.5 Hematology

Acta Haematologica
Acta Haematologica Polonica
American Journal of Hematology
Annals of Hematology
Annals of Hematology, Supplement
BMC Blood Disorders
Best Practice and Research in Clinical Haematology
Biology of Blood and Marrow Transplantation
Blood
Blood Cells, Molecules, and Diseases
Blood Coagulation and Fibrinolysis
Blood Reviews
Bone Marrow Transplantation
British Journal of Haematology
Cardiovascular and Hematological Agents in Medicinal Chemistry
Cardiovascular and Hematological Disorders - Drug Targets
Clinical Advances in Hematology and Oncology
Clinical Hemorheology and Microcirculation
Clinical and Applied Thrombosis/Hemostasis
Clinical and Laboratory Haematology
Clinical and Laboratory Haematology, Supplement
Clinical and Transfusion Haematology
Comparative Clinical Pathology
Critical Reviews in Oncology/Hematology
Current Hematology Reports
Current Opinion in Hematology
Current Studies in Hematology and Blood Transfusion
Cytometry Part B - Clinical Cytometry
European Journal of Haematology
European Journal of Haematology, Supplement
Experimental Hematology
HAEMA
Haematologica
Haematologica, Supplement

Haemophilia
Haemophilia, Supplement
Hamostaseologie
Hematological Oncology
Hematologie
Hematology
Hematology Journal
Hematology/Oncology Clinics of North America
Hemoglobin
Immunohematology
Indian Journal of Hematology and Blood Transfusion
International Journal of Hematology
Journal of Clinical Apheresis
Journal of Leukocyte Biology
Journal of Pediatric Hematology/Oncology
Journal of Thrombosis and Thrombolysis
Laboratory Hematology
Leukemia
Leukemia Research
Leukemia and Lymphoma
Oncologist
Pathophysiology of Haemostasis and Thrombosis
Pediatric Hematology and Oncology
Platelets
Reviews in Clinical and Experimental Hematology
Sangre
Seminars in Hematology
Seminars in Thrombosis and Hemostasis
Stem Cells and Development
Therapeutic Apheresis and Dialysis
Thrombosis Journal
Thrombosis Research
Thrombosis and Haemostasis
Transfusion
Transfusion Clinique et Biologique
Transfusion Medicine
Transfusion Medicine Reviews
Transfusion Medicine and Hemotherapy
Transfusion and Apheresis Science
Transfuze a Hematologie Dnes
Trasfusione del Sangue
Turkish Journal of Haematology
UHOD - Uluslararasi Hematoloji-Onkoloji Dergisi
Vascular Health and Risk Management
Vox Sanguinis
Vox Sanguinis, Supplement

C.1.6 Nephrology

Advances in Chronic Kidney Disease
Aktuality v Nefrologii
American Journal of Kidney Diseases
American Journal of Nephrology
Andrologia
Archivio Italiano di Urologia e Andrologia
Asian Journal of Andrology
Atlas of the Urologic Clinics of North America
BMC Nephrology
Bangladesh Renal Journal
Blood Purification
Clinical Nephrology
Clinical and Experimental Nephrology
Connective Tissue Research
Current Opinion in Nephrology and Hypertension
Dialysis and Transplantation
EDTNA-ERCA Journal
Hemodialysis International
Hong Kong Journal of Nephrology
International Braz J Urol
International Urology and Nephrology
Japanese Journal of Nephrology
Journal of Cardiothoracic-Renal Research
Journal of Nephrology
Journal of the American Society of Nephrology
Kidney International
Kidney and Blood Pressure Research
Minerva Urologica e Nefrologica
Nature Clinical Practice Nephrology
Nephrologie et Therapeutique
Nephrology
Nephrology Dialysis Transplantation
Nephrology, Dialysis and Transplantation
Nephron - Clinical Practice
Nephron - Experimental Nephrology
Nephron - Physiology
Neurourology and Urodynamics
Nieren- und Hochdruckkrankheiten
Pediatric Nephrology
Peritoneal Dialysis International
Renal Failure
Revista de Nefrologia, Dialisis y Transplante
Scandinavian Journal of Urology and Nephrology
Scandinavian Journal of Urology and Nephrology, Supplement
Seminars in Nephrology
Therapeutic Apheresis and Dialysis
Turk Uroloji Dergisi

C.1.7 Arthritis, Rheumatism and Connective Tissue Diseases

APLAR Journal of Rheumatology
Aktuelle Rheumatologie
Aktuelle Rheumatologie, Supplement
Annals of the Rheumatic Diseases
Arthritis Care and Research
Arthritis Research and Therapy
Arthritis and Rheumatism
BMC Musculoskeletal Disorders
Best Practice and Research in Clinical Rheumatology
Bulletin on the Rheumatic Diseases
Bulletin: Hospital for Joint Diseases
Ceska Revmatologie
Clinical Rheumatology
Clinical and Experimental Rheumatology
Current Opinion in Rheumatology
Current Rheumatology Reports
Current Rheumatology Reviews
Joint Bone Spine
Journal of Clinical Rheumatology
Journal of Musculoskeletal Pain
Journal of Rheumatology
Journal of Rheumatology and Medical Rehabilitation
Lupus
Matrix Biology

Modern Rheumatology
Nature Clinical Practice Rheumatology
Osteoarthritis and Cartilage
Progressi in Reumatologia
Reumatismo
Reumatologia
Reumatologia Clinica
Revista Brasileira de Reumatologia
Rheumatic Disease Clinics of North America
Rheumatologia
Rheumatology
Rheumatology
Rheumatology International
Scandinavian Journal of Rheumatology
Scandinavian Journal of Rheumatology, Supplement
Seminarios de la Fundacion Espanola de Reumatologia
Seminars in Arthritis and Rheumatism
Zeitschrift fur Rheumatologie
Zeitschrift fur Rheumatologie, Supplement

C.1.8 Gastroenterology

Abdominal Imaging
Acta Gastro-Enterologica Belgica
Advances in Gastrointestinal Cancers
American Journal of Gastroenterology
American Journal of Physiology - Gastrointestinal and Liver Physiology
Archives of Gastroenterohepatology
Argomenti di Gastroenterologia Clinica
Arquivos de Gastroenterologia
BMC Gastroenterology
Best Practice and Research in Clinical Gastroenterology
CME Journal Gastroenterology, Hepatology and Nutrition
Canadian Journal of Gastroenterology
Ceska a Slovenska Gastroenterologie a Hepatologie
Chinese Journal of Digestive Diseases
Chinese Journal of Gastroenterology
Chirurgische Gastroenterologie Interdisziplinar
Clinical Gastroenterology and Hepatology
Clinics in Colon and Rectal Surgery
Coloproctology
Colorectal Disease
Croatian Journal of Gastroenterology and Hepatology
Current Gastroenterology Reports
Current Issues in Intestinal Microbiology
Current Opinion in Gastroenterology
Current Treatment Options in Gastroenterology
Digestion
Digestive Diseases
Digestive Diseases and Sciences
Digestive Endoscopy
Digestive and Liver Disease
Diseases of the Colon and Rectum
Diseases of the Esophagus
Dysphagia
Endoscopic Forum for Digestive Disease
Esophagus
European Journal of Gastroenterology and Hepatology
Evidence-Based Gastroenterology
GED - Gastrenterologia Endoscopia Digestiva
Gastric Cancer
Gastroenterologia Polska
Gastroenterologia y Hepatologia
Gastroenterological Endoscopy
Gastroenterologie Clinique et Biologique
Gastroenterology
Gastroenterology Clinics of North America
Gastroenterology International
Gastroenterology and Hepatology
Gastrointestinal Endoscopy
Gastrointestinal Endoscopy Clinics of North America
Giornale Italiano di Endoscopia Digestiva
Gut
HPB
HPB, Supplement
Helicobacter
Helicobacter, Supplement
Hellenic Journal of Gastroenterology
Hepato-Gastro
Hepato-Gastroenterology
Hepatobiliary and Pancreatic Diseases International
Indian Journal of Gastroenterology
Inflammatory Bowel Diseases
International Journal of Colorectal Disease
International Journal of Gastrointestinal Cancer
Japanese Journal of Gastroenterological Surgery
Japanese Journal of Gastroenterology
Journal fur Gastroenterologische und Hepatologische Erkrankungen
Journal of Clinical Gastroenterology
Journal of Gastroenterology
Journal of Gastroenterology and Hepatology
Journal of Gastrointestinal and Liver Diseases
Journal of Health, Population and Nutrition
Journal of Pediatric Gastroenterology and Nutrition
Journal of the Pancreas
Minerva Gastroenterologica e Dietologica
Nature Clinical Practice Gastroenterology and Hepatology
Neurogastroenterology and Motility
Pancreas
Pancreatology
Pediatria Wspolczesna
Phlebologie
Practical Gastroenterology
Proktologia
Reviews in Gastroenterological Disorders
Revista Espanola de Enfermedades Digestivas
Scandinavian Journal of Gastroenterology
Scandinavian Journal of Gastroenterology, Supplement
Seminars in Colon and Rectal Surgery
Techniques in Coloproctology
Techniques in Gastrointestinal Endoscopy
Turkish Journal of Gastroenterology
Verdauungskrankheiten
World Chinese Journal of Digestology
World Journal of Gastroenterology
Zeitschrift fur Gastroenterologie

C.1.8.1 Hepatology

Acta Hepatologica Japonica
Annals of Hepatology
Ceska a Slovenska Gastroenterologie a Hepatologie
Clinics in Liver Disease
Comparative Hepatology

Current Hepatitis Reports
Gastroenterology and Hepatology
HPB, Supplement
Hepatologia Polska
Hepatology
Hepatology Research
Journal fur Gastroenterologische und Hepatologische Erkrankungen
Journal of Gastroenterology and Hepatology
Journal of Hepatology
Journal of Viral Hepatitis
Liver International
Liver Transplantation
Nature Clinical Practice Gastroenterology and Hepatology
Seminars in Liver Disease

C.2 Pediatrics and Pediatric Surgery

ANAE - Approche Neuropsychologique des Apprentissages chez l'Enfant
Acta Paediatrica Taiwanica
Acta Paediatrica, International Journal of Paediatrics
Acta Paediatrica, International Journal of Paediatrics, Supplement
Acta Pediatrica Espanola
Aggiornamento Pediatrico
Ambulatory Pediatrics
American Journal of Perinatology
Anales de Pediatria
Annales Nestle
Annales de Pediatrie
Annals of Tropical Paediatrics
Archives de Pediatrie
Archives of Disease in Childhood
Archives of Disease in Childhood: Education and Practice Edition
Archives of Disease in Childhood: Fetal and Neonatal Edition
Archives of Pediatrics and Adolescent Medicine
Autism
BMC Pediatrics
Binocular Vision and Strabismus Quarterly
Biology of the Neonate
Brain and Development
Cardiology in the Young
Cesko-Slovenska Pediatrie
Child Abuse and Neglect
Child Neuropsychology
Child Psychiatry and Human Development
Child Psychology and Psychiatry Review
Child and Adolescent Mental Health
Child and Adolescent Psychiatric Clinics of North America
Child's Nervous System
Child: Care, Health and Development
Children's Health Care
Children's Pulmonology
Chinese Journal of Contemporary Pediatrics
Clinical Child Psychology and Psychiatry
Clinical Dysmorphology
Clinical Pediatric Emergency Medicine
Clinical Pediatric Endocrinology
Clinical Pediatrics
Clinics in Perinatology
Cocuk Cerrahisi Dergisi
Cocuk Sagligi ve Hastaliklari Dergisi
Congenital Heart Disease
Current Opinion in Pediatrics
Current Paediatrics
Current Pediatric Research
Current Pediatric Reviews
Current Problems in Pediatric and Adolescent Health Care
Developmental Medicine and Child Neurology
Early Human Development
Endokrynologia, Diabetologia i Choroby Przemiany Materii Wieku Rozwojowego
Epidemiologia e Psichiatria Sociale
European Child and Adolescent Psychiatry, Supplement
European Journal of Paediatric Neurology
European Journal of Pediatric Dermatology
European Journal of Pediatric Surgery
European Journal of Pediatric Surgery, Supplement
European Journal of Pediatrics
European Journal of Pediatrics, Supplement
Fetal Diagnosis and Therapy
Fetal and Maternal Medicine Review
Fetal and Pediatric Pathology
Future of Children
Gynaecologia et Perinatologia, Supplement
Hong Kong Journal of Paediatrics
Imago
Indian Journal of Pediatrics
Indian Journal of Practical Pediatrics
Indian Pediatrics
Infancy
Infant Behavior and Development
Infant Mental Health Journal
Infanto
International Journal of Adolescent Medicine and Health
International Journal of Pediatric Otorhinolaryngology
International Pediatrics
Italian Journal of Pediatrics
Jornal de Pediatria
Journal de Pediatrie et de Puericulture
Journal of AAPOS
Journal of Abnormal Child Psychology
Journal of Adolescence
Journal of Adolescent Health
Journal of Autism and Developmental Disorders
Journal of Child Neurology
Journal of Child Psychology and Psychiatry and Allied Disciplines
Journal of Child Psychotherapy
Journal of Child and Adolescent Mental Health
Journal of Child and Adolescent Psychopharmacology
Journal of Developmental and Behavioral Pediatrics
Journal of HIV/AIDS Prevention in Children and Youth
Journal of Human Lactation
Journal of Indian Association for Child and Adolescent Mental Health
Journal of Maternal-Fetal and Neonatal Medicine
Journal of Neonatology
Journal of Paediatrics and Child Health
Journal of Pediatric Endocrinology and Metabolism
Journal of Pediatric Gastroenterology and Nutrition
Journal of Pediatric Hematology/Oncology
Journal of Pediatric Neurology
Journal of Pediatric Neurosciences
Journal of Pediatric Ophthalmology and Strabismus

Journal of Pediatric Orthopaedics
Journal of Pediatric Orthopaedics Part B
Journal of Pediatric Psychology
Journal of Pediatric Surgery
Journal of Pediatric Urology
Journal of Pediatric and Adolescent Gynecology
Journal of Pediatrics
Journal of Perinatal Medicine
Journal of Perinatology
Journal of Tropical Pediatrics
Journal of the American Academy of Child and Adolescent Psychiatry
Journal of the Canadian Academy of Child and Adolescent Psychiatry
Kinderanalyse
Klinische Padiatrie
Maternal and Child Nutrition
Medecine Therapeutique Pediatrie
Medico e Bambino
Mental Retardation and Developmental Disabilities Research Reviews
Middle East Paediatrics
Minerva Pediatrica
Monatsschrift fur Kinderheilkunde
Nascer e Crescer
Neuropediatrics
Neuropsychiatrie de l'Enfance et de l'Adolescence
No To Hattatsu
Ophthalmic Genetics
Otorinolaringologia Pediatrica
Padiatrische Praxis
Paediatria Croatica
Paediatric Anaesthesia
Paediatric Respiratory Reviews
Paediatric and Perinatal Drug Therapy
Paediatric and Perinatal Epidemiology
Paediatrics and Child Health
Pediatria Catalana
Pediatria Integral
Pediatria Medica e Chirurgica
Pediatria Oggi Medica e Chirurgica
Pediatria Polska
Pediatria Wspolczesna
Pediatria i Medycyna Rodzinna
Pediatric Allergy and Immunology
Pediatric Annals
Pediatric Asthma, Allergy and Immunology
Pediatric Blood and Cancer
Pediatric Cardiac Surgery Annual
Pediatric Cardiology
Pediatric Case Reviews
Pediatric Clinics of North America
Pediatric Critical Care Medicine
Pediatric Dermatology
Pediatric Diabetes
Pediatric Diabetes, Supplement
Pediatric Drugs
Pediatric Emergency Care
Pediatric Endocrinology Reviews
Pediatric Endosurgery and Innovative Techniques
Pediatric Exercise Science
Pediatric Hematology and Oncology
Pediatric Infectious Disease Journal
Pediatric Nephrology
Pediatric Neurology
Pediatric Neurosurgery
Pediatric Physical Therapy
Pediatric Pulmonology
Pediatric Radiology
Pediatric Rehabilitation
Pediatric Research
Pediatric Surgery International
Pediatric Transplantation
Pediatric and Developmental Pathology
Pediatrics
Pediatrics International
Pediatrics in Review
Pediatrika
Pediatriya
Perinatology
Physical and Occupational Therapy in Pediatrics
Practica Pediatrica
Progress in Pediatric Cardiology
Przeglad Pediatryczny
Quaderni ACP
Revista Chilena de Pediatria
Revista Espanola de Pediatria
Revue Internationale de Pediatrie
Rivista Italiana di Medicina dell'Adolescenza
SAGGI - Child Development and Disabilities
SD Revista Medica Internacional sobre el Sindrome de Down
Saludarte
Seminars in Fetal and Neonatal Medicine
Seminars in Pediatric Infectious Diseases
Seminars in Pediatric Surgery
Seminars in Perinatology
Singapore Paediatric Journal
Tijdschrift voor Kindergeneeskunde
Turk Pediatri Arsivi
Turkish Journal of Pediatrics
Twin Research and Human Genetics
Vaccines: Children and Practice
Zeitschrift fur Epileptologie
Zeitschrift fur Geburtshilfe und Neonatologie
Zeitschrift fur Geburtshilfe und Neonatologie, Supplement
Zeitschrift fur Kinder- und Jugendpsychiatrie und Psychotherapie

C.3 Geriatrics

Aging Health
Aging Male
Aging and Mental Health
Alzheimer Disease and Associated Disorders
Alzheimer's and Dementia
American Journal Geriatric Pharmacotherapy
American Journal of Alzheimer's Disease and other Dementias
American Journal of Geriatric Psychiatry
Annals of Long-Term Care
Arquivos de Geriatria e Gerontologia
BMC Geriatrics
Biogerontology
CME Journal Geriatric Medicine
Clinical Geriatrics
Clinics in Geriatric Medicine
Dementia and Geriatric Cognitive Disorders

Drugs and Aging
Geriatric and Medical Intelligence
Geriatrics
Geriatrics Today: Journal of the Canadian Geriatrics Society
Geriatrics and Aging
Geriatrika
Gerokomos
Giornale di Gerontologia
Hallym International Journal of Aging
International Journal of Geriatric Psychiatry
International Psychogeriatrics
International Urology and Nephrology
Journal of Aging and Health
Journal of Aging and Pharmacotherapy
Journal of Aging and Physical Activity
Journal of Applied Gerontology
Journal of Geriatric Cardiology
Journal of Geriatric Physical Therapy
Journal of Geriatric Psychiatry and Neurology
Journal of Mental Health and Aging
Journal of the American Geriatrics Society
Physical and Occupational Therapy in Geriatrics
Psychogeriatria Polska
Psychogeriatrics
Psychologie et NeuroPsychiatrie du Vieillissement
Psychology and Aging
Research and Practice in Alzheimer's Disease
Reviews in Clinical Gerontology
Revista Espanola de Geriatria y Gerontologia
Revue de Geriatrie
Tijdschrift voor Gerontologie en Geriatrie
Tijdschrift voor Verpleeghuisgeneeskunde
Topics in Geriatric Rehabilitation
Turk Geriatri Dergisi
Zeitschrift fur Gerontologie und Geriatrie
Zeitschrift fur Gerontopsychologie und -psychiatrie

C.4 Obstetrics and Gynecology

Acta Ginecologica
Acta Obstetricia et Gynecologica Scandinavica
Actualites Reproduction Humaine
American Journal of Obstetrics and Gynecology
American Journal of Perinatology
Archives of Gynecology and Obstetrics
Archivio di Ostetricia e Ginecologia
Australian and New Zealand Journal of Obstetrics and Gynaecology
BJOG: An International Journal of Obstetrics and Gynaecology
BMC Pregnancy and Childbirth
BMC Women's Health
Bangladesh Journal of Obstetrics and Gynecology
Best Practice and Research in Clinical Obstetrics and Gynaecology
Birth
Breast Care
CME Journal of Gynecologic Oncology
Ceska Gynekologie
Ciencia Ginecologika
Climacteric
Clinica e Investigacion en Ginecologia y Obstetricia
Clinical Obstetrics and Gynecology
Clinical and Experimental Obstetrics and Gynecology
Contraception
Cuadernos de Medicina Reproductiva
Current Obstetrics and Gynaecology
Current Opinion in Obstetrics and Gynecology
Current Sexual Health Reports
Current Women's Health Reviews
ESHRE Monographs
Electronic Journal of Human Sexuality
European Journal of Contraception and Reproductive Health Care
European Journal of Gynaecological Oncology
European Journal of Obstetrics Gynecology and Reproductive Biology
Evidence-based Obstetrics and Gynecology
Fertility and Sterility
Fetal Diagnosis and Therapy
Fetal and Maternal Medicine Review
Geburtshilfe und Frauenheilkunde
Ginecologia y Obstetricia Clinica
Ginecologia y Obstetricia de Mexico
Ginekologia Onkologiczna
Ginekologia Praktyczna
Giornale Italiano di Ostetricia e Ginecologia
Gynaecologia et Perinatologia
Gynaecologia et Perinatologia, Supplement
Gynakologe
Gynakologie fur Hausarzte
Gynakologisch-geburtshilfliche Rundschau
Gynakologische Endokrinologie
Gynakologische Praxis
Gynecologic Oncology
Gynecologic and Obstetric Investigation
Gynecological Endocrinology
Gynecological Surgery
Gynecologie Obstetrique Fertilite
Human Fertility
Human Reproduction Update
Hypertension in Pregnancy
Infectious Diseases in Obstetrics and Gynecology
International Journal of Fertility and Women's Medicine
International Journal of Gynecological Cancer
International Journal of Gynecological Pathology
International Journal of Gynecology and Obstetrics
International Journal of Obstetric Anesthesia
International Urogynecology Journal and Pelvic Floor Dysfunction
Italian Journal of Gynaecology and Obstetrics
Japanese Journal of Fertility and Sterility
Jinekoloji ve Obstetri Bulteni
Jinekoloji ve Obstetrik Dergisi
Jornal Brasileiro de Reproducao Assistida
Journal de Gynecologie Obstetrique et Biologie de la Reproduction
Journal fur Fertilitat und Reproduktion
Journal fur Menopause
Journal fur Reproduktionsmedizin und Endokrinologie
Journal of Assisted Reproduction and Genetics
Journal of Experimental and Clinical Assisted Reproduction
Journal of Family Planning and Reproductive Health Care
Journal of Gynecologic Surgery
Journal of Human Lactation
Journal of Lower Genital Tract Disease
Journal of Maternal-Fetal and Neonatal Medicine
Journal of Midwifery and Women's Health

Journal of Minimally Invasive Gynecology
Journal of Obstetrics and Gynaecology
Journal of Obstetrics and Gynaecology Research
Journal of Pediatric and Adolescent Gynecology
Journal of Perinatal Medicine
Journal of Perinatology
Journal of Psychosomatic Obstetrics and Gynecology
Journal of Reproductive Medicine for the Obstetrician and Gynecologist
Journal of Reproductive and Infant Psychology
Journal of Sexual Medicine
Journal of Women's Imaging
Journal of the British Menopause Society
Journal of the Society for Gynecologic Investigation
Journal of the Turkish German Gynecology Association Artemis
Magyar Noorvosok Lapja
Maternal and Child Nutrition
Menopause
Menopause Review
Middle East Fertility Society Journal
Minerva Ginecologica
Molecular Human Reproduction
Nederlands Tijdschrift voor Obstetrie en Gynaecologie
Obstetrical and Gynecological Survey
Obstetrics and Gynecology
Obstetrics and Gynecology Clinics of North America
Perinatology
Placenta
Polish Journal of Gynaecological Investigations
Prenatal Diagnosis
Przeglad Menopauzalny
References en Gynecologie Obstetrique
Reproducao e Climaterio
Reproduction
Reproduction Humaine et Hormones
Reproductive BioMedicine Online
Reproductive Health
Reproductive Health Matters
Reproductive Medicine Review
Reproductive Medicine and Biology
Reproductive Toxicology
Revista Iberoamericana de Fertilidad y Reproduccion Humana
Revista Iberoamericana de Revisiones en Menopausia
Revue du Praticien - Gynecologie et Obstetrique
Rivista di Ostetricia e Ginecologia
Seksuologia Polska
Sexuality, Reproduction and Menopause
Sexuologie
South African Journal of Obstetrics and Gynaecology
Taiwanese Journal of Obstetrics and Gynecology
Tijdschrift voor Fertiliteitsonderzoek
Twin Research and Human Genetics
Ultrasound Review of Obstetrics and Gynecology
Ultrasound in Obstetrics and Gynecology
Urogynaecologia International Journal
Women's Health Issues
Zeitschrift fur Geburtshilfe und Neonatologie
Zeitschrift fur Geburtshilfe und Neonatologie, Supplement
Zentralblatt fur Gynakologie

C.5 Dermatology and Venereology

Acta Dermatovenerologica Alpina, Pannonica et Adriatica
Acta Dermatovenerologica Croatica
Actas Dermo-Sifiliograficas
Aktuelle Dermatologie
American Journal of Clinical Dermatology
American Journal of Dermatopathology
Anais Brasileiros de Dermatologia
Annales de Dermatologie et de Venereologie
Annali Italiani di Dermatologia Allergologica Clinica e Sperimentale
Archives of Dermatological Research
Archives of Dermatological Research, Supplement
Archives of Dermatology
Australasian Journal of Dermatology
BMC Dermatology
British Journal of Dermatology
Cesko-Slovenska Dermatologie
Clinical Dermatology: Retinoids and other Treatments
Clinical and Experimental Dermatology
Clinical and Experimental Dermatology, Supplement
Clinics in Dermatology
Contact Dermatitis
Cosmetic Dermatology
Deri Hastaliklari ve Frengi Arsivi
Dermatitis
Dermatologia Clinica
Dermatologia Cosmetica, Medica y Quirurgica
Dermatologia Kliniczna
Dermatologia Revista Mexicana
Dermatologic Clinics
Dermatologic Surgery
Dermatologic Therapy
Dermatologie in Beruf und Umwelt
Dermatology
Dermatology Online Journal
European Journal of Dermatology
European Journal of Pediatric Dermatology
Exogenous Dermatology
Experimental Dermatology
Experimental Dermatology, Supplement
Forum for Nordic Dermato-Venerology
Giornale Italiano di Dermatologia e Venereologia
Haut
Hautarzt
Herpes
Hong Kong Dermatology and Venereology Bulletin
Hong Kong Journal of Dermatology and Venereology
Indian Journal of Dermatology, Venereology and Leprology
Infectious Diseases in Obstetrics and Gynecology
International Journal of Cosmetic Surgery and Aesthetic Dermatology
International Journal of Dermatology
International Journal of Lower Extremity Wounds
International Journal of STD and AIDS
JDDG - Journal of the German Society of Dermatology
Journal of Applied Cosmetology
Journal of Clinical Dermatology
Journal of Cosmetic Dermatology
Journal of Cosmetic and Laser Therapy
Journal of Cutaneous Medicine and Surgery
Journal of Cutaneous Pathology
Journal of Dermatological Science

Journal of Dermatological Treatment
Journal of Dermatology
Journal of Investigative Dermatology
Journal of Investigative Dermatology Symposium Proceedings
Journal of Pakistan Association of Dermatologists
Journal of Plastic Dermatology
Journal of Speech, Language, and Hearing Research
Journal of the American Academy of Dermatology
Journal of the European Academy of Dermatology and Venereology
Korean Journal of Dermatology
Kosmetische Medizin
Medicina Cutanea Ibero-Latino-Americana
Melanoma Research
Nederlands Tijdschrift voor Dermatologie en Venereologie
Nishinihon Journal of Dermatology
Nouvelles Dermatologiques
Pediatric Dermatology
Photodermatology Photoimmunology and Photomedicine
Photomedicine and Laser Surgery
Postepy Dermatologii I Alergologii
Przeglad Dermatologiczny
Revista Argentina de Dermatologia
Seminars in Cutaneous Medicine and Surgery
Sexual Health
Sexually Transmitted Diseases
Sexually Transmitted Infections
Skin Cancer
Skin Pharmacology and Physiology
Skin Research
Skin Research and Technology
World Wide Wounds
Wound Repair and Regeneration
Wounds UK

C.5.1 Leprosy

Bangladesh Journal of Dermatology, Venereology and Leprology
Dermatologica Sinica
Indian Journal of Dermatology, Venereology and Leprology
Indian Journal of Leprosy
International Journal of Leprosy and Other Mycobacterial Diseases
Japanese Journal of Leprosy
Leprosy Review

C.6 Neurology and Psychiatry

ANAE - Approche Neuropsychologique des Apprentissages chez l'Enfant
Aging and Mental Health
American Journal of Alzheimer's Disease and other Dementias
American Journal of Electroneurodiagnostic Technology
American Journal of Medical Genetics - Neuropsychiatric Genetics
Aphasiology
Applied Neuropsychology
Archives of Clinical Neuropsychology
Archivos de Psiquiatria
Arquivos de Neuro-Psiquiatria
Behavioural Neurology
Bipolar Disorders
Bipolar Disorders, Supplement
Brain and Cognition
Brain, Behavior, and Immunity
CNS Drug Reviews
CNS Drugs
CNS Spectrums
Child Neuropsychology
Clinical Case Studies
Clinical Neuroscience Research
Cognitive Behaviour Therapy
Cognitive Neuropsychology
Cognitive and Behavioral Neurology
Current Alzheimer Research
Current Neuropharmacology
Dementia and Geriatric Cognitive Disorders
Developmental Neuropsychology
Dialogues in Clinical Neuroscience
Emotion
Encephale
Epilepsy and Behavior
European Archives of Psychiatry and Clinical Neuroscience
European Child and Adolescent Psychiatry, Supplement
European Neuropsychopharmacology
Experimental Neurology
Fortschritte der Neurologie Psychiatrie
GOZE
Human Movement Science
Infanto
International Journal of Neuropsychopharmacology
Journal fur Neurologie, Neurochirurgie und Psychiatrie
Journal of Alzheimer's Disease
Journal of Biological Rhythms
Journal of Chronic Fatigue Syndrome
Journal of Forensic Neuropsychology
Journal of Geriatric Psychiatry and Neurology
Journal of Medical Speech-Language Pathology
Journal of Neurology, Neurosurgery and Psychiatry
Journal of Neuropsychiatry and Clinical Neurosciences
Journal of Neurotherapy
Journal of Psychotherapy Integration
Journal of the International Neuropsychological Society
Lancet Neurology
Mental Retardation and Developmental Disabilities Research Reviews
Nervenarzt
Neuro-Psy News
Neurocase
Neurology Psychiatry and Brain Research
Neurology in Practice
Neuropsychiatrie
Neuropsychiatrie de l'Enfance et de l'Adolescence
Neuropsychobiology
Neuropsychological Rehabilitation
Neuropsychology
Neuropsychology Review
Neurotoxicology and Teratology
Psychiatry and Clinical Neurosciences
PsychoNeuro
Psychologie et NeuroPsychiatrie du Vieillissement
Psychoneuroendocrinology
Research and Practice in Alzheimer's Disease

Restorative Neurology and Neuroscience
Revista Brasileira de Neurologia e Psiquiatria
SAGGI - Child Development and Disabilities
Schweizer Archiv fur Neurologie und Psychiatrie
Seizure
Sleep Medicine
Sleep Medicine Reviews
Sleep Research Online
Sleep and Biological Rhythms
Spine Journal
TEN Trends in Evidence-Based Neuropsychiatry
Thalamus and Related Systems

C.6.1 Neurology and Neurosurgery

Acta Myologica
Acta Neurochirurgica
Acta Neurochirurgica, Supplement
Acta Neurologica Belgica
Acta Neurologica Scandinavica
Acta Neurologica Taiwanica
Acta Neuropathologica
Acupuncture and Electro-Therapeutics Research
Acupuncture in Medicine
African Journal of Neurological Sciences
Aktualnosci Neurologiczne
Aktuelle Neurologie
Aktuelle Neurologie, Supplement
Akupunktur und Traditionelle Chinesische Medizin
Alzheimer's and Dementia
American Journal of Neuroradiology
Amyotrophic Lateral Sclerosis and Other Motor Neuron Disorders
Annali Italiani di Medicina Interna, Supplement
Annals of Neurology
Archives of Neurology
Autonomic Neuroscience: Basic and Clinical
BMC Neurology
Bolest
Bollettino - Lega Italiana contro l'Epilessia
Brain Injury
Brain Topography
Brain and Development
Brazilian Journal of Epilepsy and Clinical Neurophysiology
British Journal of Neurosurgery
CONTINUUM Lifelong Learning in Neurology
Canadian Journal of Neurological Sciences
Cephalalgia
Cephalalgia, Supplement
Cerebrovascular Diseases
Ceska a Slovenska Neurologie a Neurochirurgie
Child's Nervous System
Chinese Journal of Cerebrovascular Diseases
Chinese Journal of Contemporary Neurology and Neurosurgery
Chinese Journal of Neurology
Clinical Autonomic Research
Clinical EEG and Neuroscience
Clinical Journal of Pain
Clinical Neurology
Clinical Neurology and Neurosurgery
Clinical Neuropathology
Clinical Neuropharmacology
Clinical Neurophysiology
Confinia Cephalalgica
Cortex
Critical Reviews in Neurobiology
Current Alzheimer Research
Current Neurovascular Research
Current Opinion in Neurology
Current Pain and Headache Reports
Current Treatment Options in Neurology
Deutsche Zeitschrift fur Akupunktur
Developmental Medicine and Child Neurology
Electromyography and Clinical Neurophysiology
Epilepsia
Epilepsies
Epilepsy Research
Epileptic Disorders
European Journal of Neurology
European Journal of Neurology, Supplement
European Journal of Paediatric Neurology
European Neurology
Expert Review of Neurotherapeutics
Focus on Parkinson's Disease
Folia Neuropathologica
Functional Neurology
Headache
Headache Currents
Headache and Pain: Diagnostic Challenges, Current Therapy
Human Brain Mapping
International Journal of Language and Communication Disorders
International Journal of Neuroprotection and Neurogeneration
International Journal of Osteopathic Medicine
International Journal of Stroke
International MS Journal
Interventional Neuroradiology
Japanese Journal of Neurosurgery
Journal of Child Neurology
Journal of Clinical Neuromuscular Disease
Journal of Clinical Neurophysiology
Journal of Clinical Neuroscience
Journal of Clinical Sleep Medicine
Journal of Comparative Neurology
Journal of Head Trauma Rehabilitation
Journal of LGBT Health Research
Journal of Neuro-Oncology
Journal of Neuro-Ophthalmology
Journal of NeuroVirology
Journal of Neuroimaging
Journal of Neuroimmunology
Journal of Neurolinguistics
Journal of Neurology
Journal of Neurology, Supplement
Journal of Neuropathology and Experimental Neurology
Journal of Neuroradiology
Journal of Neurosurgery
Journal of Neurosurgical Anesthesiology
Journal of Neurosurgical Sciences
Journal of Neurotrauma
Journal of Pain
Journal of Pain and Symptom Management
Journal of Pediatric Neurology
Journal of Spinal Cord Medicine
Journal of Spinal Disorders and Techniques
Journal of Stroke and Cerebrovascular Diseases

Journal of the Japan Epilepsy Society
Journal of the Neurological Sciences
Journal of the Peripheral Nervous System
Klinische Neurophysiologie
Klinische Neuroradiologie
Metabolic Brain Disease
Minimally Invasive Neurosurgery
Movement Disorders
Multiple Sclerosis
Muscle and Nerve
Nature Clinical Practice Neurology
Nervenheilkunde
Neuro-Ophthalmology
Neuro-Ophthalmology Japan
Neuroanatomy
Neurochirurgie
Neurocirugia
Neurocritical Care
Neurodegenerative Diseases
Neuroepidemiology
Neuroimaging Clinics of North America
Neurologia
Neurologia Croatica
Neurologia Medico-Chirurgica
Neurologia Suplementos
Neurologia i Neurochirurgia Polska
Neurologic Clinics
Neurological Research
Neurological Sciences
Neurological Surgery
Neurologie und Rehabilitation
Neurologist
Neurology
Neurology India
Neurology and Clinical Neurophysiology
Neuromuscular Disorders
Neuropathology and Applied Neurobiology
Neuropediatrics
Neurophysiologie Clinique
Neurophysiologie-Labor
Neuroradiology
Neurorehabilitation and Neural Repair
Neurosurgery
Neurosurgery Clinics of North America
Neurosurgery Quarterly
Neurosurgical Review
Neurourology and Urodynamics
No To Hattatsu
Pain Clinic
Pain Medicine
Pain Medicine
Pain Practice
Pan Arab Journal of Neurosurgery
Parkinsonism and Related Disorders
Pediatric Neurology
Pediatric Neurosurgery
Practical Neurology
Psychologie et NeuroPsychiatrie du Vieillissement
Reviews in Neurological Diseases
Revista Brasileira de Neurologia
Revista Ecuatoriana de Neurologia
Revista Neurologica Argentina
Revista de Neurologia
Revue Neurologique
Rivista di Neuroradiologia
Schmerz
Seminars in Cerebrovascular Diseases and Stroke
Seminars in Neurology
Seminars in Pain Medicine
Seminars in Pediatric Neurology
Sen
Sinapse
Skull Base
Spinal Cord
Spine
Spine Journal
Stereotactic and Functional Neurosurgery
Stroke
Surgical Neurology
Topics in Spinal Cord Injury Rehabilitation
Topics in Stroke Rehabilitation
Turk Serebrovaskuler Hastaliklar Dergisi
Turkish Neurosurgery
Udar Mozgu - Problemy Interdyscyplinarne
Yeni Symposium
Zeitschrift fur Epileptologie
Zentralblatt fur Neurochirurgie

C.6.2 Psychiatry

AIDS and Behavior
Academic Psychiatry
Acta Neuropsychiatrica
Acta Psychiatrica Scandinavica
Acta Psychiatrica Scandinavica, Supplement
Addictive Behaviors
Administration and Policy in Mental Health
Advances in Mind-Body Medicine
Advances in Psychiatric Treatment
Aggression and Violent Behavior
Aging, Neuropsychology, and Cognition
Alcoholism Treatment Quarterly
Alzheimer Disease and Associated Disorders
Alzheimer's and Dementia
American Journal of Clinical Hypnosis
American Journal of Family Therapy
American Journal of Forensic Psychiatry
American Journal of Forensic Psychology
American Journal of Geriatric Psychiatry
American Journal of Orthopsychiatry
American Journal of Psychiatric Rehabilitation
American Journal of Psychiatry
American Journal of Psychoanalysis
American Journal of Psychotherapy
American Journal on Mental Retardation
Anadolu Psikiyatri Dergisi
Anales de Psiquiatria
Annales Medico-Psychologiques
Annals of Behavioral Medicine
Annals of Clinical Psychiatry
Annals of General Psychiatry
Annual Review of Clinical Psychology
Anuario de Psicologia
Applied Cognitive Psychology
Applied Psychophysiology Biofeedback
Archives of General Psychiatry
Archives of Psychiatry and Psychotherapy
Archives of Sexual Behavior

Archives of Women's Mental Health
Arts in Psychotherapy
Attachment and Human Development
Australasian Psychiatry
Australian Journal of Clinical Hypnotherapy and Hypnosis
Australian Journal of Clinical and Experimental Hypnosis
Australian and New Zealand Journal of Psychiatry
Autism
BMC Psychiatry
Behavior Modification
Behavior Therapy
Behavioral Interventions
Behavioral Medicine
Behavioral Neuroscience
Behavioral Sciences and the Law
Behaviour Change
Behaviour Research and Therapy
Behavioural and Cognitive Psychotherapy
Biological Psychiatry
Biological Psychology
Bridging Eastern and Western Psychiatry
British Journal of Clinical Psychology
British Journal of Developmental Disabilities
British Journal of Health Psychology
British Journal of Psychiatry
British Journal of Psychotherapy
Bulletin of the Menninger Clinic
CPD Bulletin Old Age Psychiatry
Canadian Journal of Community Mental Health
Canadian Journal of Human Sexuality
Canadian Journal of Psychiatry
Ceska a Slovenska Psychiatrie
Child Abuse and Neglect
Child Psychiatry and Human Development
Child Psychology and Psychiatry Review
Child and Adolescent Mental Health
Child and Adolescent Psychiatric Clinics of North America
Clinical Case Studies
Clinical Child Psychology and Psychiatry
Clinical Neuropsychiatry
Clinical Neuropsychologist
Clinical Practice and Epidemiology in Mental Health
Clinical Psychology Forum
Clinical Psychology Review
Clinical Psychology and Psychotherapy
Clinical Psychology: Science and Practice
Clinical Psychopharmacology and Neuroscience
Cognitive Neuropsychiatry
Cognitive Therapy and Research
Cognitive and Behavioral Practice
Community Mental Health Journal
Comprehensive Psychiatry
Confrontations Psychiatriques
Contemporary Hypnosis
Counselling Psychology Quarterly
Criminal Behaviour and Mental Health
Cultural Diversity and Ethnic Minority Psychology
Current Opinion in Psychiatry
Current Psychiatry Reports
Current Psychiatry Reviews
Current Psychosis and Therapeutics Reports
Cyberpsychology and Behavior
Depression and Anxiety
Development and Psychopathology
Dyslexia
ESHRE Monographs
Eating Behaviors
Eating Disorders
Eating and Weight Disorders
Electronic Journal of Communicative Psychoanalysis
Emotional and Behavioural Difficulties
Ethical Human Psychology and Psychiatry
Ethique et Sante
European Archives of Psychiatry and Clinical Neuroscience
European Child and Adolescent Psychiatry
European Child and Adolescent Psychiatry, Supplement
European Eating Disorders Review
European Journal of Psychiatry
European Journal of Special Needs Education
European Psychiatry
Evolution Psychiatrique
Familiendynamik
Forum der Psychoanalyse
Gender and Psychoanalysis
General Hospital Psychiatry
German Journal of Psychiatry
Group Analysis
Group Dynamics
Harvard Review of Psychiatry
Health Psychology
Homeostasis in Health and Disease
Hong Kong Journal of Psychiatry
Human Psychopharmacology
Imago
Infant Mental Health Journal
Informacao Psiquiatrica
Information Psychiatrique
International Clinical Psychopharmacology
International Forum of Psychoanalysis
International Journal of Aromatherapy
International Journal of Behavioral Medicine
International Journal of Clinical and Experimental Hypnosis
International Journal of Eating Disorders
International Journal of Geriatric Psychiatry
International Journal of Group Psychotherapy
International Journal of Law and Psychiatry
International Journal of Methods in Psychiatric Research
International Journal of Offender Therapy and Comparative Criminology
International Journal of Psychiatry in Clinical Practice
International Journal of Psychiatry in Medicine
International Journal of Psychoanalysis
International Journal of Psychophysiology
International Journal of Psychotherapy
International Journal of Social Psychiatry
International Psychogeriatrics
International Review of Psychiatry
Irish Journal of Psychological Medicine
Italian Journal of Psychiatry and Behavioural Sciences
Italian Journal of Psychopathology
Japanese Journal of Neuropsychopharmacology
Jornal Brasileiro de Psiquiatria
Journal of Abnormal Child Psychology
Journal of Abnormal Psychology
Journal of Adolescence
Journal of Affective Disorders
Journal of Aggression, Maltreatment and Trauma
Journal of Analytical Psychology
Journal of Anxiety Disorders
Journal of Autism and Developmental Disorders

Journal of Behavior Therapy and Experimental Psychiatry
Journal of Behavioral Health Services and Research
Journal of Behavioral Medicine
Journal of Brief Therapy
Journal of Child Psychology and Psychiatry and Allied Disciplines
Journal of Child Psychotherapy
Journal of Child Sexual Abuse
Journal of Child and Adolescent Mental Health
Journal of Child and Adolescent Psychopharmacology
Journal of Clinical Psychiatry
Journal of Clinical Psychoanalysis
Journal of Clinical Psychology
Journal of Clinical Psychology in Medical Settings
Journal of Clinical Psychopharmacology
Journal of Clinical and Experimental Neuropsychology
Journal of Cognition and Development
Journal of Cognitive Psychotherapy
Journal of Communication Disorders
Journal of Consulting and Clinical Psychology
Journal of Contemporary Psychotherapy
Journal of Counseling Psychology
Journal of Dementia Care
Journal of Dual Diagnosis
Journal of ECT
Journal of Emotional Abuse
Journal of Family Psychotherapy
Journal of Forensic Psychiatry and Psychology
Journal of Health Psychology
Journal of Homosexuality
Journal of Indian Association for Child and Adolescent Mental Health
Journal of Intellectual Disabilities
Journal of Intellectual Disability Research
Journal of Intellectual Disability Research, Supplement
Journal of Intellectual and Developmental Disability
Journal of Mental Health
Journal of Mental Health Policy and Economics
Journal of Mental Health and Aging
Journal of Mind and Behavior
Journal of Nervous and Mental Disease
Journal of Pediatric Psychology
Journal of Personality Assessment
Journal of Personality Disorders
Journal of Psychiatric Practice
Journal of Psychiatric Research
Journal of Psychiatry and Law
Journal of Psychiatry and Neuroscience
Journal of Psychoactive Drugs
Journal of Psychopathology and Behavioral Assessment
Journal of Psychopharmacology
Journal of Psychophysiology
Journal of Psychosomatic Obstetrics and Gynecology
Journal of Psychosomatic Research
Journal of Psychotherapy Integration
Journal of Reproductive and Infant Psychology
Journal of Sex Education and Therapy
Journal of Sex Research
Journal of Sex and Marital Therapy
Journal of Trauma and Dissociation
Journal of Traumatic Stress
Journal of the American Academy of Child and Adolescent Psychiatry
Journal of the American Academy of Psychiatry and the Law
Journal of the American Academy of Psychoanalysis and Dynamic Psychiatry
Journal of the Canadian Academy of Child and Adolescent Psychiatry
Kinderanalyse
Klinik Psikiyatri Dergisi
Klinik Psikofarmakoloji Bulteni
Krankenhauspsychiatrie
Krankenhauspsychiatrie, Supplement
Mental Health Aspects of Developmental Disabilities
Mental Health, Religion and Culture
Minerva Psichiatrica
Molecular Psychiatry
Neuropsychiatric Disease and Treatment
Neuropsychologia
Neuropsychopharmacology
Nordic Journal of Psychiatry
Nordic Journal of Psychiatry, Supplement
Occupational Therapy in Mental Health
PPmP Psychotherapie Psychosomatik Medizinische Psychologie
Pharmacopsychiatry
Primary Care Companion to the Journal of Clinical Psychiatry
Primary Care Mental Health
Primary Care and Community Psychiatry
Primary Psychiatry
Progress in Neuro-Psychopharmacology and Biological Psychiatry
Psicologia Conductual
Psicoterapia Cognitiva e Comportamentale
Psicoterapia e Scienze Umane
Psihoterapija
Psiquiatria Biologica
Psiquis
Psyche
Psychiatria Danubina
Psychiatria Polska
Psychiatria i Psychologia Kliniczna
Psychiatric Bulletin
Psychiatric Clinics of North America
Psychiatric Genetics
Psychiatric Quarterly
Psychiatric Rehabilitation Journal
Psychiatric Services
Psychiatrie
Psychiatrische Praxis
Psychiatrische Praxis, Supplement
Psychiatry
Psychiatry Investigation
Psychiatry Research
Psychiatry Research - Neuroimaging
Psychiatry, Psychology and Law
Psycho-Oncology
Psychoanalysis and Psychotherapy
Psychoanalytic Psychology
Psychoanalytic Psychotherapy
Psychoanalytic Quarterly
Psychodynamic Practice
Psychogeriatria Polska
Psychogeriatrics
Psychological Assessment
Psychological Medicine
Psychology and Aging
Psychology and Psychotherapy: Theory, Research and Practice

Psychology of Addictive Behaviors
Psychology, Health and Medicine
Psychopathology
Psychopharmakotherapie
Psychopharmakotherapie, Supplement
Psychophysiology
Psychosomatic Medicine
Psychosomatics
Psychoterapia
Psychotherapeut
Psychotherapies
Psychotherapy
Psychotherapy Research
Psychotherapy and Psychosomatics
Recht und Psychiatrie
Rehabilitation Psychology
Research Communications in Biological Psychology and Psychiatry
Research in Developmental Disabilities
Revista Brasileira de Psiquiatria
Revista da Associacao Brasileira de Medicina Psicossomatica
Revista de Psiquiatria Clinica
Revista de Psiquiatria Consiliar e de Ligacao
Revista de Psiquiatria de la Facultad de Medicina de Barcelona
Revista de Psiquiatria do Rio Grande do Sul
Revista del Hospital Psiquiatrico de la Habana
Revue Francophone de Psycho-Oncologie
Ricerca in Psicoterapia
Ricerche in Psichiatria
Rivista di Psichiatria
Salud Mental
Scandinavian Psychoanalytic Review
Schizophrenia Bulletin
Schizophrenia Research
Sexual Addiction and Compulsivity
Sexual and Relationship Therapy
Sexualities, Evolution and Gender
Sexuologie
Social Psychiatry and Psychiatric Epidemiology
Socijalna Psihijatrija
South African Journal of Psychiatry
South African Psychiatry Review
Stress and Health
Sucht
Suicide and Life-Threatening Behavior
Therapeutic Communities: the International Journal for Therapeutic and Supportive Organizations
Tijdschrift voor Psychiatrie
Transcultural Psychiatry
Verhaltenstherapie
Violence and Victims
World Journal of Biological Psychiatry
Yeni Symposium
Zeitschrift fur Gerontopsychologie und -psychiatrie
Zeitschrift fur Kinder- und Jugendpsychiatrie und Psychotherapie
Zeitschrift fur Klinische Psychologie und Psychotherapie
Zeitschrift fur Psychiatrie, Psychologie und Psychotherapie
Zeitschrift fur Psychosomatische Medizin und Psychotherapie

C.7 Ophthalmology

Acta Ophthalmologica Scandinavica
Acta Ophthalmologica Scandinavica, Supplement
American Journal of Ophthalmology
Annali di Ottalmologia e Clinica Oculistica
Annals of Ophthalmology
Archives of Ophthalmology
Arquivos Brasileiros de Oftalmologia
Asian Journal of Ophthalmology
BMC Ophthalmology
Binocular Vision and Strabismus Quarterly
British Journal of Ophthalmology
CME Journal Ophthalmology
Canadian Journal of Ophthalmology
Ceska a Slovenska Oftalmologie
Chinese Journal of Ophthalmology
Chinese Ophthalmic Research
Clinical and Experimental Ophthalmology
Clinical and Refractive Optometry
Clinical and Surgical Ophthalmology
Community Eye Health Journal
Comprehensive Ophthalmology Update
Contact Lens and Anterior Eye
Cornea
Current Eye Research
Current Opinion in Ophthalmology
Cutaneous and Ocular Toxicology
Documenta Ophthalmologica
EVRS Educational Electronic Journal
European Journal of Ophthalmology
Evidence-Based Ophthalmology
Experimental Eye Research
Eye
Eye and Contact Lens
Folia Ophthalmologica Japonica
Graefe's Archive for Clinical and Experimental Ophthalmology
Indian Journal of Ophthalmology
International Journal of Ophthalmology
International Ophthalmology
International Ophthalmology Clinics
Investigative Ophthalmology and Visual Science
Japanese Journal of Clinical Ophthalmology
Japanese Journal of Ophthalmology
Journal Francais d'Ophtalmologie
Journal of AAPOS
Journal of Cataract and Refractive Surgery
Journal of Glaucoma
Journal of Neuro-Ophthalmology
Journal of Ocular Pharmacology and Therapeutics
Journal of Pediatric Ophthalmology and Strabismus
Journal of Refractive Surgery
Journal of Vision
Journal of Visual Impairment and Blindness
Klinika Oczna
Klinische Monatsblatter fur Augenheilkunde
Molecular Vision
Neuro-Ophthalmology
Neuro-Ophthalmology Japan
Ocular Immunology and Inflammation
Ocular Surface
Ophthalmic Epidemiology
Ophthalmic Genetics

Ophthalmic Plastic and Reconstructive Surgery
Ophthalmic Research
Ophthalmic Surgery Lasers and Imaging
Ophthalmic and Physiological Optics
Ophthalmologe
Ophthalmologica
Ophthalmology
Ophthalmology Clinics of North America
Ophthalmology in China
Optometry and Vision Science
Orbit
Progress in Retinal and Eye Research
Retina
Revista Mexicana de Oftalmologia
Rivista Italiana di Ossigeno-Ozonoterapia
Seminars in Ophthalmology
Spektrum der Augenheilkunde
Strabismus
Survey of Ophthalmology
Techniques in Ophthalmology
Transactions of the American Ophthalmological Society
Vision Research
Visual Impairment Research

C.8 Otorhinolaryngology

Acta Oto-Laryngologica
Acta Oto-Laryngologica, Supplement
Acta Oto-Rhino-Laryngologica Belgica
American Journal of Audiology
American Journal of Otolaryngology - Head and Neck Medicine and Surgery
American Journal of Rhinology
Annals of Otology, Rhinology and Laryngology
Archives of Otolaryngology - Head and Neck Surgery
Atlas of the Oral and Maxillofacial Surgery Clinics of North America
Audiological Medicine
Audiology and Neurotology
Auris Nasus Larynx
Australian Journal of Otolaryngology
Australian and New Zealand Journal of Audiology
B-ENT
BMC Ear, Nose and Throat Disorders
CME Bulletin Otorhinolaryngology Head and Neck Surgery
Clinical Linguistics and Phonetics
Clinical Otolaryngology
Cochlear Implants International
Current Opinion in Otolaryngology and Head and Neck Surgery
Dentomaxillofacial Radiology
Dysphagia
Ear and Hearing
Ear, Nose and Throat Journal
Equilibrium Research
Equilibrium Research, Supplement
European Archives of Oto-Rhino-Laryngology
Folia Phoniatrica et Logopaedica
Forum Logopadie
HNO
Head and Neck
Hearing Journal
Hearing Research
Indian Journal of Otolaryngology and Head and Neck Surgery
Indian Journal of Otology
International Journal of Audiology
International Journal of Oral and Maxillofacial Surgery
International Journal of Pediatric Otorhinolaryngology
International Tinnitus Journal
JARO - Journal of the Association for Research in Otolaryngology
Japan Journal of Logopedics and Phoniatrics
Journal of Fluency Disorders
Journal of Laryngology and Otology
Journal of Medical Speech-Language Pathology
Journal of Otolaryngology
Journal of Otolaryngology of Japan
Journal of Taiwan Otolaryngology - Head and Neck Surgery
Journal of Vestibular Research: Equilibrium and Orientation
Journal of Voice
Journal of the Acoustical Society of America
Journal of the American Academy of Audiology
Laryngo- Rhino- Otologie
Laryngoscope
Logopedics Phoniatrics Vocology
Logopedics Phoniatrics Vocology, Supplement
Nederlands Tijdschrift voor Keel- Neus- Oorheelkunde
Noise and Health
ORL
ORL-DIPS
Operative Techniques in Otolaryngology - Head and Neck Surgery
Oto-Rhino-Laryngology Tokyo
Otolaryngologic Clinics of North America
Otolaryngology - Head and Neck Surgery
Otolaryngology - Head and Neck Surgery (Tokyo)
Otology and Neurotology
Otorinolaringologia
Otorinolaringologia Pediatrica
Otorinolaryngologie a Foniatrie
Practica Oto-Rhino-Laryngologica
Practica Otologica, Supplement
Revista Brasileira de Otorrinolaringologia
Revista Chilena de Neuro-Psiquiatria
Revista de Logopedia, Foniatria y Audiologia
Revue de Laryngologie Otologie Rhinologie
Rhinology
Rivista Italiana di Otorinolaringologia Audiologia e Foniatria
Seminars in Hearing
Seminars in Speech and Language
Sleep and Breathing
Sprache Stimme Gehor
Trends in Amplification

C.9 Radiology and Nuclear Medicine

Academic Radiology
Annals of the ICRP
Applied Radiation and Isotopes
Chinese Journal of Cerebrovascular Diseases
Chinese Journal of Interventional Imaging and Therapy
Chinese Journal of Medical Imaging Technology
Clinical Oncology
Clinical Radiology Extra

Computer Aided Surgery
Congenital Heart Disease
Critical Reviews in Computed Tomography
Current Medical Imaging Reviews
Diagnostic and Interventional Radiology
European Journal of Radiology
Health Physics
Human Brain Mapping
Imagerie de la Femme
Imaging
International Journal of Cardiovascular Imaging
International Journal of Radiation Biology
International Journal of Radiation Oncology Biology Physics
JACR Journal of the American College of Radiology
Journal for Vascular Ultrasound
Journal of Environmental Radioactivity
Journal of JASTRO
Journal of Neuroimaging
Journal of the ICRU
Klinische Neuroradiologie
Medical Dosimetry
Medical Image Analysis
Molecular Imaging and Biology
NCRP Report
NeuroImage
Neuroimaging Clinics of North America
Quarterly Journal of Nuclear Medicine and Molecular Imaging
RAS - Radioloski Arhiv Srbije
Radiation Medicine - Medical Imaging and Radiation Oncology
Radiation Physics and Chemistry
Radiation Protection Dosimetry
Radiation Research
Radiation and Environmental Biophysics
Radiography
Radiology and Oncology
Rays - International Journal of Radiological Sciences
Rentgenologiya i Radiologiya
Reports of Practical Oncology and Radiotherapy
RoFo Fortschritte auf dem Gebiet der Rontgenstrahlen und der Bildgebenden Verfahren
Seminars in Radiation Oncology

C.9.1 Radiology

Abdominal Imaging
Acta Oncologica
Acta Oncologica, Supplement
Acute Cardiac Care
American Journal of Neuroradiology
American Journal of Roentgenology
Applied Radiology
Asian Oceanian Journal of Radiology
Australasian Radiology
BMC Medical Genetics
BMC Medical Imaging
British Journal of Radiology
CME Journal Radiology
Canadian Association of Radiologists Journal
Cancer/Radiotherapie
CardioVascular and Interventional Radiology
Cardiovascular Radiation Medicine
Cardiovascular Ultrasound
Catheterization and Cardiovascular Interventions
Ceska Radiologie
Chinese Journal of Radiology
Chinese Journal of Radiology
Clinical Imaging
Clinical MRI
Clinical Radiology
Computerized Medical Imaging and Graphics
Current Problems in Diagnostic Radiology
Dentomaxillofacial Radiology
Echocardiography
Emergency Radiology
European Radiology
European Radiology, Supplement
Feuillets de Radiologie
Imaging Decisions MRI
Indian Journal of Radiology and Imaging
Interventional Neuroradiology
Investigative Radiology
Japanese Journal of Clinical Radiology
Japanese Journal of Interventional Cardiology
Journal Belge de Radiologie
Journal de Radiologie
Journal of Cardiovascular Magnetic Resonance
Journal of Clinical Densitometry
Journal of Clinical Ultrasound
Journal of Computer Assisted Tomography
Journal of Diagnostic Medical Sonography
Journal of Diagnostic Radiography and Imaging
Journal of Digital Imaging
Journal of Interventional Radiology
Journal of Magnetic Resonance Imaging
Journal of Medical Ultrasonics
Journal of Medical Ultrasound
Journal of Neuroradiology
Journal of Radiological Protection
Journal of Radiotherapy in Practice
Journal of Thoracic Imaging
Journal of Ultrasound in Medicine
Journal of Vascular and Interventional Radiology
Journal of Women's Imaging
Journal of the American Society of Echocardiography
Journal of the Hong Kong College of Radiologists
Magnetic Resonance Imaging
Magnetic Resonance Imaging Clinics of North America
Magnetic Resonance Materials in Physics, Biology and Medicine
Magnetic Resonance in Medicine
MedicaMundi
Minimally Invasive Therapy and Allied Technologies
Neuroradiology
Oral Radiology
Pediatric Radiology
Polish Journal of Radiology
Radiologe
Radiologia
Radiologia Medica
Radiologic Clinics of North America
Radiology
Radiology Now
Radiotherapy and Oncology
Rivista Italiana di Ossigeno-Ozonoterapia
Rivista di Neuroradiologia

Rontgenpraxis
Seminars in Interventional Radiology
Seminars in Musculoskeletal Radiology
Seminars in Roentgenology
Seminars in Ultrasound, CT and MRI
Skeletal Radiology
Strahlentherapie und Onkologie
Surgical and Radiologic Anatomy
Techniques in Vascular and Interventional Radiology
Technology in Cancer Research and Treatment
Topics in Magnetic Resonance Imaging
Ultraschall in der Medizin
Ultraschall in der Medizin, Supplement
Ultrasonic Imaging
Ultrasound International
Ultrasound Quarterly
Ultrasound Review of Obstetrics and Gynecology
Ultrasound in Medicine and Biology
Ultrasound in Obstetrics and Gynecology

C.9.2 Nuclear Medicine

ANZ Nuclear Medicine
Annals of Nuclear Medicine
BMC Nuclear Medicine
Brachytherapy
Clinical Nuclear Medicine
European Journal of Nuclear Medicine and Molecular Imaging
Hellenic Journal of Nuclear Medicine
Iranian Journal of Nuclear Medicine
Iranian Journal of Radiation Research
Journal of Labelled Compounds and Radiopharmaceuticals
Journal of Nuclear Cardiology
Journal of Nuclear Medicine
Journal of Nuclear Medicine Technology
Journal of Radioanalytical and Nuclear Chemistry
Kakuigaku
Medecine Nucleaire
Nuclear Medicine Communications
Nuclear Medicine Review
Nuclear Medicine and Biology
NuklearMedizin
Revista Espanola de Medicina Nuclear
Seminars in Nuclear Medicine

C.10 Rehabilitation and Physical Medicine

Advances in Physiotherapy
American Journal of Physical Medicine and Rehabilitation
Annales de Readaptation et de Medecine Physique
Archives of Physical Medicine and Rehabilitation
Athletic Therapy Today
Clinical Kinesiology
Critical Reviews in Physical and Rehabilitation Medicine
Europa Medicophysica
Fizjoterapia
Fizyoterapi Rehabilitasyon
Gait and Posture
Human Movement Science
International Journal of Osteopathic Medicine
Isokinetics and Exercise Science
Journal of Aging and Physical Activity
Journal of Bodywork and Movement Therapies
Journal of Fluency Disorders
Journal of Hand Therapy
Journal of Whiplash and Related Disorders
Measurement in Physical Education and Exercise Science
Medicina Clinica e Termale
Motor Control
Ortopedia Traumatologia Rehabilitacja
Physical Medicine and Rehabilitation Clinics of North America
Physical and Occupational Therapy in Geriatrics
Physical and Occupational Therapy in Pediatrics
Physikalische Medizin Rehabilitationsmedizin Kurortmedizin
Rehabilitace a Fyzikalni Lekarstvi
Research in Developmental Disabilities
Technology and Disability
Turkiye Fiziksel Tip ve Rehabilitasyon Dergisi

C.10.1 Rehabilitation

American Journal of Psychiatric Rehabilitation
Assistive Technology
Australian Occupational Therapy Journal
British Journal of Occupational Therapy
Canadian Journal of Occupational Therapy
Canadian Journal of Respiratory Therapy
Chinese Journal of Clinical Rehabilitation
Chinese Journal of Rehabilitation Medicine
Clinical Rehabilitation
Disability and Rehabilitation
Ergotherapie und Rehabilitation
European Journal of Cardiovascular Prevention and Rehabilitation
European Journal of Special Needs Education
Eurorehab
Giornale Italiano di Medicina del Lavoro ed Ergonomia
Hrvatska Revija Za Rehabilitacijska Istrazivanja
IEEE Transactions on Neural Systems and Rehabilitation Engineering
International Journal of Language and Communication Disorders
International Journal of Rehabilitation Research
Japan Journal of Logopedics and Phoniatrics
Journal de Traumatologie du Sport
Journal of Back and Musculoskeletal Rehabilitation
Journal of Burn Care and Rehabilitation
Journal of Cardiopulmonary Rehabilitation
Journal of Head Trauma Rehabilitation
Journal of Intellectual and Developmental Disability
Journal of Medical Speech-Language Pathology
Journal of NeuroEngineering and Rehabilitation
Journal of Occupational Rehabilitation
Journal of Rehabilitation Medicine
Journal of Rehabilitation Medicine, Supplement
Journal of Rehabilitation Research and Development
Journal of Rheumatology and Medical Rehabilitation
Journal of Sport Rehabilitation
Journal of Visual Impairment and Blindness
Journal of Vocational Rehabilitation
NeuroRehabilitation

Neurologie und Rehabilitation
Neuropsychological Rehabilitation
Neurorehabilitation and Neural Repair
OTJR Occupation, Participation and Health
Occupational Therapy International
Occupational Therapy in Health Care
Pediatric Rehabilitation
Postepy Rehabilitacji
Pravention und Rehabilitation
Prosthetics and Orthotics International
Psychiatric Rehabilitation Journal
Quality of Life Research
Rehabilitacia
Rehabilitacja Medyczna
Rehabilitation
Rehabilitation Psychology
Research in Sports Medicine
Revista de Logopedia, Foniatria y Audiologia
SAGGI - Child Development and Disabilities
Seleccion
Sexuality and Disability
Tijdschrift voor Verpleeghuisgeneeskunde
Topics in Geriatric Rehabilitation
Topics in Spinal Cord Injury Rehabilitation
Topics in Stroke Rehabilitation
Trends in Amplification
Work

C.10.2 Physical Medicine

Adapted Physical Activity Quarterly
Australian Journal of Physiotherapy
Current Sports Medicine Reports
Fizjoterapia Polska
Hong Kong Physiotherapy Journal
Human Movement
International Journal of Behavioral Nutrition and Physical Activity
International Journal of Lower Extremity Wounds
Journal of Athletic Training
Journal of Exercise Physiology Online
Journal of Geriatric Physical Therapy
Journal of Manipulative and Physiological Therapeutics
Journal of Manual and Manipulative Therapy
Journal of Orthopaedic and Sports Physical Therapy
Manual Therapy
Operative Techniques in Sports Medicine
Osteopathische Medizin
Patologia Del Aparato Locomotor
Pediatric Physical Therapy
Physical Therapy
Physiotherapy
Physiotherapy Singapore
Physiotherapy Theory and Practice
Seleccion
Thermology International

C.11 Clinical Pathology

Acta Neuropathologica
Advances in Anatomic Pathology
American Journal of Clinical Pathology
American Journal of Dermatopathology
American Journal of Pathology
Annales de Pathologie
Annual Review of Pathology
Archives of Pathology and Laboratory Medicine
BMC Clinical Pathology
Brain Tumor Pathology
Breast Care
CPD Bulletin Cellular Pathology
Cardiovascular Pathology
Cellular Oncology
Clinical Immunology
Clinical Neuropathology
Comparative Clinical Pathology
Current Atherosclerosis Reports
Diagnostic Molecular Pathology
Fetal and Pediatric Pathology
Folia Neuropathologica
Human Pathology
Indian Journal of Pathology and Microbiology
International Journal of Gynecological Pathology
Journal of Cellular Pathology
Journal of Clinical Pathology
Journal of Cutaneous Pathology
Journal of Molecular Diagnostics
Journal of Neuropathology and Experimental Neurology
Journal of Oral Pathology and Medicine
Journal of Pathology
Laboratory Medicine
Modern Pathology
Neuropathology and Applied Neurobiology
Pathologe
Pathologica
Pathology
Pathology Case Reviews
Pathology International
Pathology Research and Practice
Patologia Del Aparato Locomotor
Patoloji Bulteni
Pediatric and Developmental Pathology
Polish Journal of Pathology
Toxicologic Pathology
Wspolczesna Onkologia

C.11.1 General Pathology

APMIS
APMIS, Supplement
Amyloid
Angiogenesis
Brain Pathology
Chinese Journal of Pathology
Differentiation
Experimental and Molecular Pathology
Experimental and Toxicologic Pathology
International Journal of Experimental Pathology
Journal of Comparative Pathology
Laboratory Investigation
Neuropathology
Pathobiology
Pathologie Biologie
Pathology Oncology Research

Pathophysiology
Rassegna di Patologia dell'Apparato Respiratorio
Virchows Archiv

C.11.2 Pathological Anatomy

Acta Anatomica Sinica
Acta Cytologica
Acta Histochemica
American Journal of Forensic Medicine and Pathology
American Journal of Surgical Pathology
Analytical and Quantitative Cytology and Histology
Annals of Diagnostic Pathology
Applied Immunohistochemistry and Molecular Morphology
Archivos de la Facultad de Medicina de Zaragoza
Biotechnic and Histochemistry
Cells and Materials
Chinese Journal of Andrology
Clinical Dysmorphology
Current Diagnostic Pathology
CytoJournal
Cytometry Part A
Cytometry Part B - Clinical Cytometry
Cytopathology
Cytopathology, Supplement
Diagnostic Cytopathology
Endocrine Pathology
European Journal of Histochemistry
Folia Histochemica et Cytobiologica
Forensic Science International
Histochemistry and Cell Biology
Histology and Histopathology
Histopathology
International Journal of Legal Medicine
International Journal of Surgical Pathology
International Review of Cytology
Journal of Histochemistry and Cytochemistry
Journal of Histotechnology
Medical Electron Microscopy
Morphologie
Neuroanatomy
Seminars in Diagnostic Pathology
Ultrastructural Pathology

D.0 Surgery and Anesthesiology

Acta Cirurgica Brasileira
Cardiovascular Revascularization Medicine
International Journal of Medical Robotics and Computer Assisted Surgery
International Journal of Oral and Maxillofacial Surgery
Surgery for Obesity and Related Diseases
Surgical Innovation
Surgical Practice

D.1 General and Abdominal Surgery

ANZ Journal of Surgery
Acta Chirurgica Austriaca
Ambulatory Surgery
American Journal of Surgery
American Surgeon
Anales de Cirugia Cardiaca y Vascular
Annales de Chirurgie
Annals of Surgery
Annals of Surgical Oncology
Annals of the Royal College of Surgeons of England
Archives of Surgery
Asian Journal of Surgery
Atlas of the Oral and Maxillofacial Surgery Clinics of North America
BMC Surgery
Breast
Breast Disease
British Journal of Surgery
Canadian Journal of Surgery
Chinese Journal of Traumatology - English Edition
Chirurgia
Chirurgia Polska
Chirurgische Gastroenterologie Interdisziplinar
Chirurgische Praxis
Clinics in Colon and Rectal Surgery
Coloproctology
Computer Aided Surgery
Current Problems in Surgery
Current Surgery
Digestive Surgery
European Journal of Surgical Oncology
European Surgery - Acta Chirurgica Austriaca
European Surgical Research
Gastrointestinal Endoscopy Clinics of North America
Hernia
Imagerie de la Femme
Injury
International Journal of Lower Extremity Wounds
International Journal of Surgery
International Seminars in Surgical Oncology
International Wound Journal
Japanese Journal of Gastroenterological Surgery
Journal of Hepato-Biliary-Pancreatic Surgery
Journal of Investigative Surgery
Journal of Laparoendoscopic and Advanced Surgical Techniques - Part A
Journal of Pelvic Medicine and Surgery
Journal of Surgical Oncology
Journal of Surgical Research
Journal of Trauma - Injury, Infection and Critical Care
Journal of Vascular Access
Journal of the American College of Surgeons
Langenbeck's Archives of Surgery
Lasers in Medical Science
Lasers in Surgery and Medicine
Liver Transplantation
Microsurgery
Minerva Chirurgica
Minimal Invasive Chirurgie
Minimally Invasive Therapy and Allied Technologies
Nigerian Journal of Surgical Research
Obesity Surgery
Operative Techniques in General Surgery
Operative Techniques in Otolaryngology - Head and Neck Surgery
Oral and Maxillofacial Surgery Clinics of North America
Photomedicine and Laser Surgery

Polski Przeglad Chirurgiczny
Proktologia
Scandinavian Journal of Surgery
Seminars in Colon and Rectal Surgery
Seminars in Surgical Oncology
South African Journal of Surgery
Surgeon
Surgery
Surgery Today
Surgery for Obesity and Related Diseases
Surgical Chronicles
Surgical Clinics of North America
Surgical Endoscopy
Surgical Infections
Surgical Laparoscopy, Endoscopy and Percutaneous Techniques
Surgical Oncology
Surgical Oncology Clinics of North America
Techniques in Coloproctology
Techniques in Gastrointestinal Endoscopy
Tecniche Chirurgiche in Ortopedia e Traumatologia
Turkish Journal of Surgery
Viszeralchirurgie
World Journal of Surgery
World Journal of Surgical Oncology
World Wide Wounds
Wound Repair and Regeneration
Zeitschrift fur Wundheilung
Zentralblatt fur Chirurgie
Zentralblatt fur Chirurgie, Supplement

D.2 Plastic Surgery

Acta Chirurgiae Plasticae
Aesthetic Plastic Surgery
Aesthetic Surgery Journal
Annales de Chirurgie Plastique et Esthetique
Annals of Plastic Surgery
Asian Journal of Oral and Maxillofacial Surgery
Atlas of the Oral and Maxillofacial Surgery Clinics of North America
British Journal of Oral and Maxillofacial Surgery
Burns
Chirurgie de la Main
Cirugia Plastica Ibero-Latinoamericana
Cleft Palate-Craniofacial Journal
Clinics in Plastic Surgery
Current Opinion in Otolaryngology and Head and Neck Surgery
Dermatologic Surgery
European Journal of Plastic Surgery
Facial Plastic Surgery
Facial Plastic Surgery Clinics of North America
Hand Clinics
Handchirurgie Mikrochirurgie Plastische Chirurgie
International Journal of Cosmetic Surgery and Aesthetic Dermatology
Japanese Journal of Plastic and Reconstructive Surgery
Journal of Burn Care and Rehabilitation
Journal of Cosmetic and Laser Therapy
Journal of Cranio-Maxillofacial Surgery
Journal of Craniofacial Surgery
Journal of Cutaneous Medicine and Surgery
Journal of Facial and Somato Prosthetics
Journal of Hand Surgery
Journal of Hand Surgery
Journal of Oral and Maxillofacial Surgery
Journal of Plastic, Reconstructive and Aesthetic Surgery
Journal of Reconstructive Microsurgery
Kosmetische Medizin
Microsurgery
Operative Techniques in Plastic and Reconstructive Surgery
Ophthalmic Plastic and Reconstructive Surgery
Oral and Maxillofacial Surgery Clinics of North America
Otolaryngology - Head and Neck Surgery
Otolaryngology - Head and Neck Surgery (Tokyo)
Plastic and Reconstructive Surgery
Rivista Italiana di Chirurgia Plastica
Scandinavian Journal of Plastic and Reconstructive Surgery and Hand Surgery
Scandinavian Journal of Plastic and Reconstructive Surgery and Hand Surgery, Supplement

D.3 Orthopedic Surgery

Acta Chirurgiae Orthopaedicae et Traumatologiae Cechoslovaca
Acta Orthopaedica
Acta Orthopaedica Belgica
Acta Orthopaedica, Supplement
Aktuelle Traumatologie
Aktuelle Traumatologie, Supplement
Archives of Orthopaedic and Trauma Surgery
Arthroscopy - Journal of Arthroscopic and Related Surgery
Artroscopia
Atlas of Hand Clinics
Bone
Bulletin: Hospital for Joint Diseases
CME Orthopaedics
Chinese Journal of Traumatology - English Edition
Chirurgia del Piede
Clinical Orthopaedics and Related Research
Clinics in Podiatric Medicine and Surgery
Current Opinion in Orthopaedics
Current Orthopaedics
Emergency Medicine Journal
European Journal of Orthopaedic Surgery and Traumatology
European Journal of Trauma
European Spine Journal
Foot
Foot and Ankle International
Foot and Ankle Surgery
Fuss und Sprunggelenk
HIP International
Hokkaido Journal of Orthopaedics and Traumatology
International Orthopaedics
Journal de Traumatologie du Sport
Journal of Arthroplasty
Journal of Bone and Joint Surgery - Series A
Journal of Bone and Joint Surgery - Series B
Journal of Bone and Mineral Research
Journal of Foot and Ankle Surgery
Journal of Hand Surgery
Journal of Musculoskeletal Neuronal Interactions
Journal of Musculoskeletal Research
Journal of Orofacial Orthopedics

Journal of Orthopaedic Research
Journal of Orthopaedic Science
Journal of Orthopaedic Trauma
Journal of Orthopaedics and Traumatology
Journal of Pediatric Orthopaedics
Journal of Pediatric Orthopaedics Part B
Journal of Shoulder and Elbow Surgery
Journal of Spinal Disorders and Techniques
Journal of Whiplash and Related Disorders
Knee
Medecine et Chirurgie du Pied
Minerva Ortopedica e Traumatologica
Operative Orthopadie und Traumatologie
Operative Techniques in Orthopaedics
Operative Techniques in Sports Medicine
Orthopade
Orthopedic Clinics of North America
Orthopedics
Ortopedia Traumatologia Rehabilitacja
Osteologicky Bulletin
Osteologie
Osteopathische Medizin
Patologia Del Aparato Locomotor
Revista de Ortopedia y Traumatologia
Revue de Chirurgie Orthopedique et Reparatrice de l'Appareil Moteur
Seminars in Arthroplasty
Seminars in Musculoskeletal Radiology
Seminars in Spine Surgery
Spine Journal
Sports Medicine and Arthroscopy Review
Techniques in Foot and Ankle Surgery
Techniques in Hand and Upper Extremity Surgery
Techniques in Knee Surgery
Techniques in Orthopaedics
Techniques in Shoulder and Elbow Surgery
Tecniche Chirurgiche in Ortopedia e Traumatologia
Trauma
Unfallchirurg
Zeitschrift fur Orthopadie und Ihre Grenzgebiete

D.4 Transplantation

American Journal of Transplantation
Bone Marrow Transplantation
Breast Care
Cell and Tissue Banking
Clinical Transplantation
Clinical Transplantation, Supplement
Dialysis and Transplantation
Experimental and Clinical Transplantation
Journal of Heart and Lung Transplantation
Nephrology Dialysis Transplantation
Organogenesis
Pediatric Transplantation
Progress in Transplantation
Revista de Nefrologia, Dialisis y Transplante
Transplant Infectious Disease
Transplant International
TransplantLinc
Transplantation
Transplantation Proceedings
Transplantation Reviews
Transplantationsmedizin: Organ der Deutschen Transplantationsgesellschaft

D.5 Urological Surgery

Abdominal Imaging
Acta Urologica Japonica
American Journal of Urology Review
Andrologie
Annales d'Urologie
Archivio Italiano di Urologia e Andrologia
Atlas of the Urologic Clinics of North America
BJU International
BMC Urology
Chinese Journal of Andrology
Clinical Genitourinary Cancer
Current Opinion in Urology
Current Prostate Reports
Current Sexual Health Reports
Current Urology Reports
EAU Update Series
Electronic Journal of Human Sexuality
European Urology
European Urology, Supplements
Fertility and Sterility
International Journal of Impotence Research
International Journal of Urology
International Urogynecology Journal and Pelvic Floor Dysfunction
International Urology and Nephrology
Japanese Journal of Clinical Urology
Japanese Journal of Urology
Journal fur Urologie und Urogynakologie
Journal of Endourology
Journal of Men's Health and Gender
Journal of Nephrology, Urology and Transplantation
Journal of Pediatric Urology
Journal of Sexual Medicine
Journal of Sexual and Reproductive Medicine
Journal of Urology
Korean Journal of Urology
Magyar Urologia
Minerva Urologica e Nefrologica
Nature Clinical Practice Urology
Nederlands Tijdschrift voor Urologie
Nishinihon Journal of Urology
Prostate
Prostate Cancer and Prostatic Diseases
Scandinavian Journal of Urology and Nephrology
Scandinavian Journal of Urology and Nephrology, Supplement
Seksuologia Polska
Urodinamica
Urologe - Ausgabe A
Urologia Internationalis
Urologic Clinics of North America
Urologic Oncology: Seminars and Original Investigations
Urological Research
Urology

D.6 Cardiovascular and Thoracic Surgery

Angiologia
Annals of Thoracic Surgery
Annals of Vascular Surgery
Asian Cardiovascular and Thoracic Annals
Congenital Heart Disease
European Journal of Cardio-thoracic Surgery
European Journal of Vascular and Endovascular Surgery
Gefasschirurgie
Heart Lung and Circulation
Heart Surgery Forum
Interactive Cardiovascular and Thoracic Surgery
Japanese Journal of Thoracic and Cardiovascular Surgery
Jornal Vascular Brasileiro
Journal of Cardiac Surgery
Journal of Cardiothoracic and Vascular Anesthesia
Journal of Cardiothoracic-Renal Research
Journal of Cardiovascular Surgery
Journal of Endovascular Therapy
Journal of Heart Valve Disease
Journal of Thoracic and Cardiovascular Surgery
Journal of Vascular Surgery
Kardiotechnik
Operative Techniques in Thoracic and Cardiovascular Surgery
Pediatric Cardiac Surgery Annual
Phlebology
Scandinavian Cardiovascular Journal
Scandinavian Cardiovascular Journal, Supplement
Seminars in Thoracic and Cardiovascular Surgery
Seminars in Vascular Surgery
Thoracic and Cardiovascular Surgeon
Thoracic and Cardiovascular Surgeon, Supplement
Vascular and Endovascular Surgery
Zeitschrift fur Herz-, Thorax- und Gefasschirurgie

D.7 Anesthesiology

AANA Journal
Acta Anaesthesiologica Belgica
Acta Anaesthesiologica Italica / Anaesthesia and Intensive Care in Italy
Acta Anaesthesiologica Scandinavica
Acta Anaesthesiologica Scandinavica, Supplement
Acta Anaesthesiologica Taiwanica
Actualizaciones en Anestesiologia y Reanimacion
Acute Pain
Agri
Anaesthesia
Anaesthesia and Intensive Care
Anaesthesia, Supplement
Anaesthesiology and Intensive Care
Anaesthesist
Anasthesiologie Intensivmedizin Notfallmedizin Schmerztherapie
Anasthesiologie Intensivmedizin Notfallmedizin Schmerztherapie, Supplement
Anasthesiologie und Intensivmedizin
Anestesia en Mexico
Anestezi Dergisi
Anesteziologie a Intenzivni Medicina
Anestezjologia Intensywna Terapia
Anesthesia and Analgesia
Anesthesia and Resuscitation
Anesthesiology
Anesthesiology Clinics of North America
Annales Francaises d'Anesthesie et de Reanimation
BMC Anesthesiology
BMC Palliative Care
Best Practice and Research in Clinical Anaesthesiology
British Journal of Anaesthesia
CPD Anaesthesia
Canadian Journal of Anesthesia
Canadian Journal of Emergency Medicine
Clinical Journal of Pain
Clinical Pediatric Emergency Medicine
Current Anaesthesia and Critical Care
Current Opinion in Anaesthesiology
DOLOR
Difficult Airways
Douleur et Analgesie
Douleurs
Egyptian Journal of Anaesthesia
European Journal of Anaesthesiology
European Journal of Anaesthesiology, Supplement
European Journal of Pain
Gogus-Kalp-Damar Anestezi ve Yogun Bakim Dernegi Dergisi
Hokuriku Journal of Anesthesiology
International Anesthesiology Clinics
International Journal of Obstetric Anesthesia
JAVA - Journal of the Association for Vascular Access
Japanese Journal of Anesthesiology
Journal fur Anasthesie und Intensivbehandlung
Journal of Anaesthesiology Clinical Pharmacology
Journal of Anesthesia
Journal of Cardiothoracic and Vascular Anesthesia
Journal of Clinical Anesthesia
Journal of Headache and Pain
Journal of Japanese Dental Society of Anesthesiology
Journal of Neurosurgical Anesthesiology
Journal of Pain
Journal of Pain and Palliative Care Pharmacotherapy
Journal of Pain and Symptom Management
Journal of Terminal Oncology
Malignant Hyperthermia
Minerva Anestesiologica
Molecular Pain
Nederlands Tijdschrift voor Anesthesiologie
Neuromodulation
Paediatric Anaesthesia
Pain
Pain Clinic
Pain Digest
Pain Physician
Pain Research and Management
Regional Anesthesia and Pain Medicine
Reviews in Analgesia
Revista Brasileira de Anestesiologia
Revista Mexicana de Anestesiologia
Revista de la Sociedad Espanola del Dolor
Schmerz
Seminars in Anesthesia, Perioperative Medicine and Pain
Seminars in Cardiothoracic and Vascular Anesthesia
Techniques in Regional Anesthesia and Pain Management
Turk Anesteziyoloji ve Reanimasyon Dernegi Dergisi

Ulusal Travma ve Acil Cerrahi Dergisi
Update in Anaesthesia

D.7.1 Intensive Care

Academic Emergency Medicine
Acute Medicine
American Journal of Emergency Medicine
American Journal of Respiratory and Critical Care Medicine
Annals of Emergency Medicine
Applied Cardiopulmonary Pathophysiology
BMC Emergency Medicine
British Journal of Intensive Care
Canadian Journal of Respiratory Therapy
Care of the Critically Ill
Chinese Critical Care Medicine
Clinical Intensive Care
Clinical Nutrition, Supplement
Critical Care
Critical Care Clinics
Critical Care Medicine
Critical Care Nursing Quarterly
Critical Care and Shock
Current Opinion in Clinical Nutrition and Metabolic Care
Current Opinion in Critical Care
EMA - Emergency Medicine Australasia
Emergency Medicine Clinics of North America
Emergency Medicine Journal
Heart and Lung: Journal of Acute and Critical Care
Hong Kong Journal of Emergency Medicine
Indian Journal of Critical Care Medicine
Intensiv- und Notfallbehandlung
Intensive Care Medicine
Intensive Care Medicine, Supplement
Intensivmedizin und Notfallmedizin
International Journal of Emergency Management
International Journal of Intensive Care
JAVA - Journal of the Association for Vascular Access
JEUR
Journal of Clinical Monitoring and Computing
Journal of Critical Care
Journal of Critical Illness
Journal of Emergency Medicine
Journal of Emergency Primary Health Care
Journal of Intensive Care Medicine
Journal of Intravenous Nursing
Journal of Organ Dysfunction
Journal of Parenteral and Enteral Nutrition
Medicina Intensiva
Modern Aspects in Sepsis
Neurocritical Care
Notarzt
Notfall und Hausarztmedizin
Nutricion Hospitalaria
Pediatric Critical Care Medicine
Prehospital Emergency Care
Puesta al Dia en Urgencias, Emergencias y Catastrofes
Reanimation
Resuscitation
Rivista Italiana di Nutrizione Parenterale ed Enterale
Seminars in Respiratory and Critical Care Medicine
Shock
Southern African Journal of Critical Care
Transfusion Medicine and Hemotherapy
Tuberkuloz ve Toraks
Ulusal Travma ve Acil Cerrahi Dergisi
Wilderness and Environmental Medicine

E.0 Experimental Science

ATLA Alternatives to Laboratory Animals
AWIC Bulletin
Academic Journal of Xi'an Jiaotong University
Acta Biologica Szegediensis
Acta Cientifica Venezolana
Acta Medica et Biologica
Advances in Clinical and Experimental Medicine
Advances in Experimental Medicine and Biology
Altex
Animal Welfare
Annals of Clinical and Laboratory Science
Annals of the New York Academy of Sciences
Annual Review of Biomedical Sciences
Apoptosis
Australian Journal of Medical Science
Azerbaijan Medical Journal
BMC Medical Research Methodology
BioSystems
Biological Procedures Online
Biological Research
Biological Reviews of the Cambridge Philosophical Society
Biologicals
Biomaterials
Biomedical Research
Biomedical Research
Biomedical Reviews
Biomedicine
Bioscience Reports
Brazilian Journal of Medical and Biological Research
British Journal of Biomedical Science
Bulletin of Experimental Biology and Medicine
Cell Preservation Technology
Cell Transplantation
Cellular and Molecular Life Sciences
Clinical Laboratory
Clinical and Experimental Medicine
Clinics in Laboratory Medicine
Comparative Medicine
Comptes Rendus - Biologies
Contemporary Topics in Laboratory Animal Science
Critical Reviews in Clinical Laboratory Sciences
Cryobiology
Current News in Medicine
Current Opinion in Clinical Experimental Research
Current Opinion in European Medicine
Current Opinion in Lipidology
Dynamic Medicine
Experimental Animals
Experimental Biology and Medicine
Expert Opinion on Biological Therapy
Expert Opinion on Therapeutic Targets
FASEB Journal
Folia Biologica
Folia Biologica
ILAR Journal
Indian Journal of Biotechnology

Indian Journal of Medical Research
Indian Journal of Medical Research, Supplement
International Journal of Medicine, Biology and the Environment
Iranian Biomedical Journal
Jordan Journal of Applied Sciences - Natural Sciences
Journal of Applied Animal Welfare Science
Journal of Applied Biomedicine
Journal of Biology
Journal of Biomedical Science
Journal of Biomedicine and Biotechnology
Journal of Biosciences
Journal of Biosocial Science
Journal of Clinical Laboratory Analysis
Journal of Experimental Animal Science
Journal of Health Science
Journal of Investigative Medicine
Journal of Medical Investigation
Journal of Medical Primatology
Journal of Negative Results in BioMedicine
Journal of Photochemistry and Photobiology B: Biology
Journal of Theoretical Biology
Journal of Theoretical Medicine
Journal of Translational Medicine
LaboratoriumsMedizin
Laboratory Animals
Mathematical Biosciences
Mathematical Medicine and Biology
Medecine/Sciences
Molecular Biotechnology
Nature
Nature Medicine
Naturwissenschaften
PLoS Biology
Periodicum Biologorum
Perspectives in Biology and Medicine
Placebo
Proceedings of the Royal Society - Biological Sciences (Series B)
Publications of Takeda Research Laboratories
Reproduction Nutrition Development
Scandinavian Journal of Laboratory Animal Science
Science
Theory in Biosciences
Today's Life Science
Yale Journal of Biology and Medicine

E.1 Neuroscience

ANAE - Approche Neuropsychologique des Apprentissages chez l'Enfant
Acta Neurobiologiae Experimentalis
Acta Neuropsychologica
Alzheimer's and Dementia
Annual Review of Neuroscience
Archivos de Neurociencias
Audiology and Neurotology
Autonomic Neuroscience: Basic and Clinical
Autonomic and Autacoid Pharmacology
BMC Neuroscience
Behavioral Neuroscience
Behavioral and Brain Functions
Behavioral and Brain Sciences
Behavioral and Cognitive Neuroscience Reviews
Behavioural Brain Research
Biogenic Amines
Bipolar Disorders
Bipolar Disorders, Supplement
Bolest
Brain
Brain Pathology
Brain Research
Brain Research Bulletin
Brain Research Reviews
Brain Topography
Brain and Cognition
Brain and Language
Brain and Nerve
Brain, Behavior and Evolution
CNS and Neurological Disorders - Drug Targets
Cellular and Molecular Neurobiology
Central Nervous System Agents in Medicinal Chemistry
Cerebellum
Cerebral Cortex
Cerebrospinal Fluid Research
Chemical Senses
Chinese Journal of Contemporary Neurology and Neurosurgery
Clinical Autonomic Research
Clinical Neurophysiology
Clinical Psychopharmacology and Neuroscience
Cognition
Cognition and Emotion
Cognitive Neuropsychology
Cognitive Processing
Cognitive Systems Research
Cognitive, Affective and Behavioral Neuroscience
Consciousness and Cognition
Current Neurovascular Research
Current Opinion in Neurobiology
Current Topics in Neurochemistry
Developmental Neuroscience
Developmental Psychobiology
Dialogues in Clinical Neuroscience
Dyslexia
European Journal of Cognitive Psychology
European Journal of Neurology, Supplement
European Journal of Neuroscience
European Journal of Pain
European Neuropsychopharmacology
Evidence-Based Ophthalmology
Experimental Brain Research
Experimental Neurology
Folia Neuropathologica
Frontiers in Neuroendocrinology
GLIA
Gene Expression Patterns
Genes, Brain and Behavior
Hippocampus
Homeostasis in Health and Disease
International Journal of Developmental Neuroscience
International Journal of Neuroscience
International Journal of Stroke
Invertebrate Neuroscience
Journal of Applied Biomedicine
Journal of Biological Rhythms
Journal of Brain Science
Journal of Clinical Sleep Medicine

Journal of Cognitive Neuroscience
Journal of Computational Neuroscience
Journal of Headache and Pain
Journal of Integrative Neuroscience
Journal of Molecular Neuroscience
Journal of Motor Behavior
Journal of Neural Engineering
Journal of Neural Transmission
Journal of Neural Transmission, Supplement
Journal of Neurobiology
Journal of Neurochemistry
Journal of Neurocytology
Journal of Neuroendocrinology
Journal of Neurogenetics
Journal of Neuroinflammation
Journal of Neurophysiology
Journal of Neuropsychiatry and Clinical Neurosciences
Journal of Neuroscience
Journal of Neuroscience Methods
Journal of Neuroscience Research
Journal of Pain
Journal of Pediatric Neurosciences
Journal of Sleep Research
Journal of Sleep Research, Supplement
Journal of Vestibular Research: Equilibrium and Orientation
Journal of the History of the Neurosciences
Journal of the International Neuropsychological Society
Journal of the Neurological Sciences
Journal of the Peripheral Nervous System
Language and Cognitive Processes
Laterality
Learning and Behavior
Learning and Memory
Memory
Molecular Brain Research
Molecular Neurobiology
Molecular and Cellular Neuroscience
Motor Control
Nature Clinical Practice Neurology
Nature Neuroscience
Nature Reviews Neuroscience
Neural Networks
Neural Plasticity
NeuroImmunoModulation
NeuroMolecular Medicine
NeuroReport
NeuroSignals
NeuroToxicology
Neurobiology of Disease
Neurochemical Research
Neurochemistry International
Neurocomputing
Neurodegenerative Diseases
Neuroembryology and Aging
Neuroendocrinology
Neuroforum
Neuroinformatics
Neurological Sciences
Neurology Psychiatry and Brain Research
Neuromodulation
Neuron
Neuron Glia Biology
Neuropathology
Neuropathology and Applied Neurobiology
Neuropeptides
Neuropharmacology
Neurophysiology
Neuropsychiatric Disease and Treatment
Neuropsychobiology
Neuropsychology
Neuroscience
Neuroscience Letters
Neuroscience Research
Neuroscience and Behavioral Physiology
Neuroscience and Biobehavioral Reviews
Neurosciences
Neuroscientist
Neurotoxicology and Teratology
Nutritional Neuroscience
Pain
Peptides
Pharmacology Biochemistry and Behavior
Placebo
Progress in Brain Research
Progress in Neurobiology
Psychiatry Investigation
Restorative Neurology and Neuroscience
Reviews in the Neurosciences
Rivista di Neurobiologia
Sleep and Biological Rhythms
Sleep and Hypnosis
Somnologie
Stress
Synapse
Thalamus and Related Systems
Trends in Cognitive Sciences
Trends in Neurosciences
Visual Cognition
Visual Neuroscience
World Journal of Biological Psychiatry
Yeni Symposium

E.2 Endocrinology

Acta Diabetologica
Adipocytes
American Journal of Physiology - Endocrinology and Metabolism
Andrologie
Annales d'Endocrinologie
Archives of Andrology
Asian Journal of Andrology
Autonomic and Autacoid Pharmacology
Azerbaijan Journal of Metabolism
BMC Endocrine Disorders
Best Practice and Research in Clinical Endocrinology and Metabolism
CME Bulletin Endocrinology and Diabetes
Calcified Tissue International
Canadian Journal of Diabetes
Cardiovascular Diabetology
Clinical Cases in Mineral and Bone Metabolism
Clinical Diabetes
Clinical Endocrinology
Clinical Pediatric Endocrinology
Clinical Reviews in Bone and Mineral Metabolism
Comparative Biochemistry and Physiology - C Toxicology and Pharmacology

Current Diabetes Reports
Current Diabetes Reviews
Current Opinion in Endocrinology and Diabetes
Diabetes
Diabetes Care
Diabetes Research and Clinical Practice
Diabetes Spectrum
Diabetes Technology and Therapeutics
Diabetes and Metabolism
Diabetes und Stoffwechsel
Diabetes, Nutrition and Metabolism - Clinical and Experimental
Diabetes, Obesity and Metabolism
Diabetes, Obesity and Metabolism, Supplement
Diabetes/Metabolism Research and Reviews
Diabetic Medicine
Diabetologia
Diabetologia Croatica
Diabetologia Doswiadczalna i Kliniczna
Diabetologie Metabolismus Endokrinologie Vyziva
Endocrine
Endocrine Journal
Endocrine Pathology
Endocrine Practice
Endocrine Regulations
Endocrine Research
Endocrine Reviews
Endocrine-Related Cancer
Endocrinologia
Endocrinologia y Nutricion
Endocrinologist
Endocrinology
Endocrinology and Metabolism Clinics of North America
Endokrynologia, Diabetologia i Choroby Przemiany Materii Wieku Rozwojowego
European Journal of Endocrinology
European Journal of Endocrinology, Supplement
Experimental Diabesity Research
Experimental and Clinical Endocrinology and Diabetes
Fish Physiology and Biochemistry
Folia Morphologica
Frontiers in Neuroendocrinology
General and Comparative Endocrinology
Giornale Italiano di Diabetologia e Metabolismo
Growth Hormone and IGF Research
Gynakologische Endokrinologie
Gynecological Endocrinology
Heart and Metabolism
Hormone Research
Hormone and Metabolic Research
Hormone and Metabolic Research, Supplement
Hormones and Behavior
Insulin
International Journal of Andrology
International Journal of Andrology, Supplement
International Journal of Diabetes and Metabolism
International Journal of Gastrointestinal Cancer
International Journal of Obesity
Iranian Journal of Diabetes and Lipid Disorders
JRAAS - Journal of the Renin-Angiotensin-Aldosterone System
Journal fur Mineralstoffwechsel
Journal fur Reproduktionsmedizin und Endokrinologie
Journal of Andrology
Journal of Bone and Mineral Metabolism
Journal of Bone and Mineral Research
Journal of Clinical Endocrinology and Metabolism
Journal of Diabetes and its Complications
Journal of Endocrine Genetics
Journal of Endocrinological Investigation
Journal of Endocrinology
Journal of Endocrinology, Metabolism and Diabetes of South Africa
Journal of Molecular Endocrinology
Journal of Musculoskeletal Neuronal Interactions
Journal of Neuroendocrinology
Journal of Pediatric Endocrinology and Metabolism
Journal of Pineal Research
Journal of Steroid Biochemistry and Molecular Biology
Journal of the Diabetic Association of India
Journal of the Japan Diabetes Society
Journal of the Pancreas
Medecine Therapeutique Endocrinologie et Reproduction
Menopause Review
Metabolic Syndrome and Related Disorders
Metabolism: Clinical and Experimental
Molecular Endocrinology
Molecular and Cellular Endocrinology
Nature Clinical Practice Endocrinology and Metabolism
NeuroImmunoModulation
Neuroendocrinology
Neuroendocrinology Letters
Obesity Reviews, Supplement
Organogenesis
Pancreas
Pancreatology
Pediatric Endocrinology Reviews
Peptides
Pituitary
Practical Diabetes International
Przeglad Menopauzalny
Psychoneuroendocrinology
Regulatory Peptides
Reproductive Biology and Endocrinology
Reviews in Endocrine and Metabolic Disorders
Revista Argentina de Endocrinologia y Metabolismo
Revista Espanola de Obesidad
Romanian Journal of Endocrinology
Seksuologia Polska
Seminars in Reproductive Medicine
Sexuality, Reproduction and Menopause
Steroids
Thyroid
Thyroidology Clinical and Experimental
Treatments in Endocrinology
Trends in Endocrinology and Metabolism

E.3 Biochemistry and Clinical Chemistry

Acta Scientiarum - Biological Sciences
Acta Scientiarum - Health Sciences
Amino Acids
BioMetals
CLB Chemie in Labor und Biotechnik
Cell Stress and Chaperones
Clinical and Vaccine Immunology
Critical Reviews in Biochemistry and Molecular Biology
International Journal of Biochemistry and Cell Biology

Journal of Clinical Biochemistry and Nutrition
Journal of Laboratory and Clinical Medicine
Journal of Receptor and Signal Transduction
Journal of Separation Science
Jugoslovenska Medicinska Biohemija
Lab on a Chip - Miniaturisation for Chemistry and Biology
Molecular Systems Biology
NMR in Biomedicine
Research Communications in Biochemistry and Cell and Molecular Biology
Spectroscopy
TheScientificWorldJournal

E.3.1 Biochemistry

Acta Biomaterialia
Advances in Biophysics
Advances in Carbohydrate Chemistry and Biochemistry
Advances in Protein Chemistry
American Journal of Physiology - Endocrinology and Metabolism
Analytical Biochemistry
Annals of Nutrition and Metabolism
Annexins
Annual Review of Biochemistry
Annual Review of Biophysics and Biomolecular Structure
Antioxidants and Redox Signaling
Applied Biochemistry and Biotechnology
Archives of Biochemistry and Biophysics
Archives of Physiology and Biochemistry
BMC Biochemistry
BMC Structural Biology
Best Practice and Research in Clinical Endocrinology and Metabolism
BioFactors
Biochemia Medica
Biochemical Genetics
Biochemical Journal
Biochemical Pharmacology
Biochemical Society Transactions
Biochemical and Biophysical Research Communications
Biochemistry
Biochemistry and Molecular Biology Education
Biochimica et Biophysica Acta - Bioenergetics
Biochimica et Biophysica Acta - Biomembranes
Biochimica et Biophysica Acta - General Subjects
Biochimica et Biophysica Acta - Molecular Basis of Disease
Biochimica et Biophysica Acta - Molecular Cell Research
Biochimica et Biophysica Acta - Molecular and Cell Biology of Lipids
Biochimica et Biophysica Acta - Proteins and Proteomics
Biochimie
Bioconjugate Chemistry
Bioelectrochemistry
Biological Chemistry
Biological Trace Element Research
Biomacromolecules
Biomarkers
Biophysical Chemistry
Biopolymers
Biopolymers - Peptide Science Section
Briefings in Functional Genomics and Proteomics
Cancer Genomics and Proteomics
Cell Cycle
ChemBioChem
Chemical Biology and Drug Design
Chemistry and Biodiversity
Chemistry and Physics of Lipids
Clinical Proteomics
Clinical Psychology Forum
Colloids and Surfaces A: Physicochemical and Engineering Aspects
Colloids and Surfaces B: Biointerfaces
Comparative Biochemistry and Physiology - B Biochemistry and Molecular Biology
Comparative Biochemistry and Physiology - Part D: Genomics and Proteomics
Current Enzyme Inhibition
Current Molecular Medicine
Current Opinion in Chemical Biology
Current Opinion in Colloid and Interface Science
Current Opinion in Structural Biology
Current Pharmaceutical Analysis
Current Protein and Peptide Science
Current Topics in Neurochemistry
Current Topics in Peptide and Protein Research
Diabetes/Metabolism Research and Reviews
Endocrinology and Metabolism Clinics of North America
European Journal of Nutrition
Experimental and Molecular Medicine
FEBS Journal
FEBS Letters
Food Chemistry
Free Radical Biology and Medicine
Free Radical Research
Frontiers in Medicinal Chemistry
Genomics, Proteomics and Bioinformatics
Glycobiology
Glycoconjugate Journal
Hormone and Metabolic Research
Hormone and Metabolic Research, Supplement
Human Genomics
IUBMB Life
International Journal of Biological Macromolecules
International Journal of Biological Markers
International Journal of Peptide Research and Therapeutics
Journal of Biochemical and Biophysical Methods
Journal of Biochemistry
Journal of Biological Chemistry
Journal of Biological Inorganic Chemistry
Journal of Biomolecular NMR
Journal of Cellular Biochemistry
Journal of Clinical Endocrinology and Metabolism
Journal of Enzyme Inhibition and Medicinal Chemistry
Journal of Inorganic Biochemistry
Journal of Lipid Research
Journal of Mass Spectrometry
Journal of Molecular Catalysis B: Enzymatic
Journal of Molecular Graphics and Modelling
Journal of Molecular Modeling
Journal of Molecular Neuroscience
Journal of Neurochemistry
Journal of Nutritional Biochemistry
Journal of Nutritional Science and Vitaminology
Journal of Peptide Science
Journal of Physiology and Biochemistry
Journal of Proteome Research
Journal of Steroid Biochemistry and Molecular Biology

Journal of Trace Elements in Medicine and Biology
Journal of the American Society for Mass Spectrometry
Journal of the Royal Society Interface
Lipids
Lipids in Health and Disease
Metabolic Brain Disease
Metabolism: Clinical and Experimental
Methods
Methods in Enzymology
Molecular Biology Reports
Molecular Biotechnology
Molecular and Cellular Proteomics
Nature Structural and Molecular Biology
Neurochemical Research
Neurochemistry International
Nitric Oxide - Biology and Chemistry
Nucleosides, Nucleotides and Nucleic Acids
Nutrition Clinique et Metabolisme
Nutrition Research
Photochemical and Photobiological Sciences
Progress in Lipid Research
Prostaglandins Leukotrienes and Essential Fatty Acids
Prostaglandins and Other Lipid Mediators
Protein Journal
Protein Science
Proteins: Structure, Function and Genetics
Proteome Science
Proteomics
Pteridines
Redox Report
Steroids
Structure
TrAC - Trends in Analytical Chemistry
Trace Elements and Electrolytes
Traffic
Trends in Biochemical Sciences
Ultrasonics Sonochemistry

E.3.2 Clinical Chemistry

Annals of Clinical Biochemistry
Biochemia Medica
Biomedical Chromatography
CPD Bulletin Clinical Biochemistry
Chromatographia
Clinica Chimica Acta
Clinical Biochemistry
Clinical Chemistry
Clinical Chemistry and Laboratory Medicine
Clinical Psychology Forum
Electrophoresis
Immuno-Analyse et Biologie Specialisee
Indian Journal of Clinical Biochemistry
International Journal of Bioinformatics Research and Applications
Japanese Journal of Clinical Chemistry
Journal of Automated Methods and Management in Chemistry
Journal of Chromatography A
Journal of Chromatography B: Analytical Technologies in the Biomedical and Life Sciences
Journal of Clinical Ligand Assay
Journal of Laboratory and Clinical Medicine
Klinicka Biochemie a Metabolismus
Laboratory Medicine
Nederlands Tijdschrift voor Klinische Chemie en Laboratoriumgeneeskunde
New Zealand Journal of Medical Laboratory Science
Quimica Clinica
Scandinavian Journal of Clinical and Laboratory Investigation
Scandinavian Journal of Clinical and Laboratory Investigation, Supplement

E.3.3 Cell Biology

Aging Cell
American Journal of Physiology - Cell Physiology
Angiogenesis
Annual Review of Cell and Developmental Biology
Autophagy
BMC Biology
BMC Cell Biology
BMC Chemical Biology
BioEssays
Biocell
Biochemistry and Cell Biology
Biochimica et Biophysica Acta - Molecular Cell Research
Biology of the Cell
Cancer Cell
Cell
Cell Biochemistry and Function
Cell Biology Education
Cell Biology International
Cell Biology and Toxicology
Cell Calcium
Cell Communication and Adhesion
Cell Communication and Signaling
Cell Cycle
Cell Death and Differentiation
Cell Metabolism
Cell Motility and the Cytoskeleton
Cell Proliferation
Cell Proliferation, Supplement
Cell Structure and Function
Cell and Tissue Research
Cellular Microbiology
Cellular Oncology
Cellular Physiology and Biochemistry
Cellular Signalling
Cellular and Molecular Biology
Cellular and Molecular Biology Letters
Cellular and Molecular Neurobiology
Chemotaxis and Migration
Current Opinion in Cell Biology
Cytokine and Growth Factor Reviews
Cytotechnology
DNA and Cell Biology
Endothelium: Journal of Endothelial Cell Research
European Journal of Cell Biology
Experimental Cell Research
Growth Factors
Histochemistry and Cell Biology
IEE Proceedings: Systems Biology
Immunology and Cell Biology
In Vitro Cellular and Developmental Biology - Animal

International Journal of Biological Sciences
Journal of Bioenergetics and Biomembranes
Journal of Bioinformatics and Computational Biology
Journal of Biological Sciences
Journal of Cell Biology
Journal of Cell Science
Journal of Cellular Physiology
Journal of Inflammation
Journal of Membrane Biology
Journal of Membrane Science
Journal of Molecular Histology
Journal of Muscle Research and Cell Motility
Journal of Receptor and Signal Transduction
Methods in Cell Science
Mitochondrion
Molecular Biology of the Cell
Molecular Cancer Research
Molecular Cell
Molecular Membrane Biology
Molecular and Cellular Biochemistry
Molecular and Cellular Biology
Nature Cell Biology
Nature Chemical Biology
Nature Methods
Neuron Glia Biology
Nuclear Receptor
Pathobiology
Photochemical and Photobiological Sciences
Physical Biology
Pigment Cell Research
Proceedings of the American Thoracic Society
Progress in Histochemistry and Cytochemistry
RNA Biology
Seminars in Cell and Developmental Biology
Signal Transduction
Stem Cells
Traffic
Trends in Cell Biology

E.4 Immunology and Serology

AIDS Research and Human Retroviruses
APMIS
APMIS, Supplement
Advances in Immunology
American Journal of Reproductive Immunology
Annual Review of Immunology
Anti-Inflammatory and Anti-Allergy Agents in Medicinal Chemistry
Archivum Immunologiae et Therapiae Experimentalis
Autoimmunity Reviews
BMC Immunology
Bioconjugate Chemistry
Biology of Blood and Marrow Transplantation
Brain, Behavior, and Immunity
Cancer Immunity
Cancer Immunology, Immunotherapy
Cellular Immunology
Central-European Journal of Immunology
Chinese Journal of Endemiology
Chinese Journal of Microbiology and Immunology
Clinical Application of Immunology
Clinical Transplantation, Supplement
Clinical and Applied Immunology Reviews
Clinical and Developmental Immunology
Clinical and Experimental Immunology
Clinical and Vaccine Immunology
Comparative Immunology, Microbiology and Infectious Diseases
Critical Reviews in Immunology
Current Immunology Reviews
Current Opinion in Immunology
Current Trends in Immunology
Cytokine
Cytokine and Growth Factor Reviews
Cytotherapy
Developmental and Comparative Immunology
EOS Rivista di Immunologia ed Immunofarmacologia
Epidemiologie, Mikrobiologie, Imunologie
European Cytokine Network
European Journal of Immunology
European Journal of Inflammation
Expert Review of Anti-Infective Therapy
Expert Review of Clinical Immunology
Expert Review of Vaccines
FEMS Immunology and Medical Microbiology
Genes and Immunity
Genetic Vaccines and Therapy
Human Antibodies
Human Vaccines
Hybridoma
Immunity
Immunity and Ageing
Immuno-Analyse et Biologie Specialisee
Immunobiology
Immunogenetics
Immunohematology
Immunologic Research
Immunological Investigations
Immunological Reviews
Immunologiya
Immunology
Immunology Letters
Immunology and Cell Biology
Immunology, Supplement
Immunopharmacology and Immunotoxicology
Inflammation and Allergy - Drug Targets
Inmunologia
International Immunology
International Immunopharmacology
International Journal of Biological Markers
International Journal of Immunogenetics
International Journal of Immunopathology and Pharmacology
International Reviews of Immunology
Iranian Journal of Immunology
Journal of Autoimmunity
Journal of Biological Regulators and Homeostatic Agents
Journal of Clinical Ligand Assay
Journal of Endotoxin Research
Journal of Experimental Medicine
Journal of Immune Based Therapies and Vaccines
Journal of Immunoassay and Immunochemistry
Journal of Immunological Methods
Journal of Immunology
Journal of Immunotherapy
Journal of Immunotoxicology
Journal of Interferon and Cytokine Research

Journal of Neuroimmunology
Journal of Reproductive Immunology
Lymphatic Research and Biology
Lymphology
Mediators of Inflammation
Microbes and Infection
Microbiology and Immunology
Modern Aspects of Immunobiology
Molecular Immunology
Molecular Systems Biology
Nature Immunology
Nature Reviews Immunology
NeuroImmunoModulation
Oral Microbiology and Immunology
PLoS Pathogens
Parasite Immunology
Russian Journal of Immunology
Scandinavian Journal of Immunology
Scandinavian Journal of Immunology, Supplement
Seminars in Immunology
Springer Seminars in Immunopathology
Stem Cells and Development
Tissue Antigens
Transplant Immunology
Travel Medicine and Infectious Disease
Trends in Immunology
Turkish Journal of Immunology
Vaccine
Vector-Borne and Zoonotic Diseases
Veterinary Immunology and Immunopathology
Viral Immunology
Xenotransplantation

E.4.1 Allergy and Clinical Immunology

AIDS
Aerobiologia
Alergia Astma Immunologia
Alergie
Alergologia e Inmunologia Clinica
Allergo Journal
Allergologia et Immunopathologia
Allergologie
Allergology International
Allergy and Clinical Immunology International
Allergy: European Journal of Allergy and Clinical Immunology
Allergy: European Journal of Allergy and Clinical Immunology, Supplement
Annals of Allergy, Asthma and Immunology
Anti-Inflammatory and Anti-Allergy Agents in Medicinal Chemistry
Asian Pacific Journal of Allergy and Immunology
Autoimmunity
BioDrugs
Biotherapy
CPD Bulletin Immunology and Allergy
Chemical Immunology and Allergy
Clinical Immunology
Clinical Reviews in Allergy and Immunology
Clinical and Diagnostic Laboratory Immunology
Clinical and Experimental Allergy
Clinical and Experimental Allergy Reviews
Clinical and Molecular Allergy
Clinical and Vaccine Immunology
Current Allergy and Asthma Reports
Current Immunology Reviews
Current Opinion in Allergy and Clinical Immunology
Current Opinion in Organ Transplantation
Current Topics in Microbiology and Immunology
Current Trends in Immunology
Cytotherapy
Endocrine, Metabolic and Immune Disorders - Drug Targets
Enhancer - Biotherapy of Cancer
European Annals of Allergy and Clinical Immunology
Expert Review of Clinical Immunology
Human Immunology
Immunology and Allergy Clinics of North America
Immunology, Endocrine and Metabolic Agents in Medicinal Chemistry
Inflammation and Allergy - Drug Targets
International Archives of Allergy and Immunology
International Review of Allergology and Clinical Immunology
Iranian Journal of Allergy, Asthma and Immunology
Iranian Journal of Immunology
Italian Journal of Allergy and Clinical Immunology
Japanese Journal of Allergology
Journal of Acquired Immune Deficiency Syndromes
Journal of Allergy and Clinical Immunology
Journal of Autoimmune Diseases
Journal of Clinical Immunology
Journal of Immunotoxicology
Journal of Investigational Allergology and Clinical Immunology
Journal of Microbiology, Immunology and Infection
Klinicka Imunologia a Alergologia
Medical Immunology
Medical Microbiology and Immunology
Ocular Immunology and Inflammation
Pathobiology
Pediatric Allergy and Immunology
Pediatric Asthma, Allergy and Immunology
Postepy Dermatologii I Alergologii
Revue Francaise d'Allergologie et d'Immunologie Clinique
Transfusion Medicine and Hemotherapy
Transplantationsmedizin: Organ der Deutschen Transplantationsgesellschaft
Trasfusione del Sangue
Vaccines: Children and Practice

E.5 Clinical Microbiology

APMIS
APMIS, Supplement
Acta Microbiologica Hellenica
American Journal of Infection Control
Annals of Clinical Microbiology and Antimicrobials
BMC Infectious Diseases
BMC Microbiology
Brazilian Journal of Infectious Diseases
CPD Infection
Canadian Journal of Infectious Diseases and Medical Microbiology
Chemotherapie Journal
Chemotherapie Journal, Supplement

Chinese Journal of Infection and Chemotherapy
Chinese Journal of Microbiology and Immunology
Clinical Infectious Diseases
Clinical Microbiology Newsletter
Clinical Microbiology Reviews
Clinical Microbiology and Infection
Clinical Microbiology and Infection, Supplement
Clinical and Diagnostic Laboratory Immunology
Clinical and Vaccine Immunology
Comparative Immunology, Microbiology and Infectious Diseases
Current Infectious Disease Reports
Current Issues in Intestinal Microbiology
Current Opinion in Infectious Diseases
Current Opinion in Microbiology
Current Topics in Microbiology and Immunology
Diagnostic Microbiology and Infectious Disease
Emerging Infectious Diseases
Enfermedades Emergentes
Enfermedades Infecciosas y Microbiologia
Enfermedades Infecciosas y Microbiologia Clinica
Epidemiologie, Mikrobiologie, Imunologie
European Journal of Clinical Microbiology and Infectious Diseases
FEMS Immunology and Medical Microbiology
Indian Journal of Medical Microbiology
Indian Journal of Pathology and Microbiology
Infection
Infection Control and Hospital Epidemiology
Infection, Genetics and Evolution
Infection, Supplement
Infections in Medicine
Infectious Disease Clinics of North America
Infectious Diseases in Clinical Practice
Infectious Disorders - Drug Targets
Infektoloski Glasnik
Infezioni in Medicina
International Journal of Antimicrobial Agents
International Journal of Infectious Diseases
International Journal of Medical Microbiology
International Journal of STD and AIDS
Japanese Journal of Infectious Diseases
Journal of Chemotherapy
Journal of Clinical Microbiology
Journal of HIV Therapy
Journal of Hospital Infection
Journal of Infection
Journal of Infection and Chemotherapy
Journal of Infectious Diseases
Journal of Medical Microbiology
Journal of Microbiology, Immunology and Infection
Journal of Water and Health
Klinicka Mikrobiologie a Infekcni Lekarstvi
Krankenhaushygiene und Infektionsverhutung
Lancet Infectious Diseases
Medecine et Maladies Infectieuses
Medical Microbiology and Immunology
Microbes and Infection
Microbial Drug Resistance
Microbial Ecology in Health and Disease
Microbial Pathogenesis
Microbiology and Immunology
Molecular Systems Biology
Nature Reviews Microbiology
Oral Microbiology and Immunology
PLoS Pathogens
Pediatric Infectious Disease Journal
Postepy Mikrobiologii
Problems of Infectious and Parasitic Diseases
Remedia - Klinicka Mikrobiologie
Reviews in Medical Microbiology
Revista Argentina de Microbiologia
Revista Espanola de Quimioterapia
Revista Latinoamericana de Microbiologia
Scandinavian Journal of Infectious Diseases
Scandinavian Journal of Infectious Diseases, Supplement
Seminars in Pediatric Infectious Diseases
Sexually Transmitted Diseases
Sexually Transmitted Infections
Surgical Infections
Transplant Infectious Disease
Trends in Microbiology
Vaccine
Vector-Borne and Zoonotic Diseases
Veterinary Microbiology

E.5.1 Bacteriology and Mycology

Anaerobe
Clinical Microbiology and Infection, Supplement
Critical Reviews in Microbiology
Foodborne Pathogens and Disease
Helicobacter
International Journal of Systematic and Evolutionary Microbiology
Japanese Journal of Medical Mycology
Journal de Mycologie Medicale
Journal of Bacteriology
Journal of Bacteriology and Virology
Journal of Endotoxin Research
Journal of Spirochetal and Tick-borne Diseases
Korean Journal of Medical Mycology
Medical Mycology
Mikologia Lekarska
Mycoses
Mycoses, Supplement
Revista Iberoamericana de Micologia

E.5.2 Parasitology

Acta Tropica
Advances in Parasitology
American Journal of Tropical Medicine and Hygiene
Annals of Tropical Medicine and Parasitology
Experimental Parasitology
Filaria Journal
International Journal for Parasitology
Journal of Helminthology
Journal of Parasitology
Journal of Vector Borne Diseases
Kinetoplastid Biology and Disease
Malaria Journal
Medecine Tropicale
Molecular and Biochemical Parasitology
PLoS Pathogens
Parasite Immunology

Parasitology
Parasitology International
Parasitology Research
Revista da Sociedade Brasileira de Medicina Tropical
Trends in Parasitology
Tropical Doctor
Tropical Medicine and International Health
Veterinary Parasitology

E.5.3 Virology

AIDS
AIDS Reader
AIDS Research and Human Retroviruses
AIDS Research and Therapy
AIDS Reviews
Acta Virologica
African Journal of AIDS Research
Antiviral Chemistry and Chemotherapy
Antiviral Research
Current HIV Research
Current HIV/AIDS Reports
Current Hepatitis Reports
Dengue Bulletin
HIV Clinical Trials
HIV Medicine
HIV and AIDS Review
Herpes
International Journal of Virology
Intervirology
Journal of Acquired Immune Deficiency Syndromes
Journal of Bacteriology and Virology
Journal of Clinical Virology
Journal of General Virology
Journal of HIV Therapy
Journal of Interferon and Cytokine Research
Journal of Medical Virology
Journal of NeuroVirology
Journal of Viral Hepatitis
Journal of Virological Methods
Journal of Virology
PLoS Pathogens
Problemy HIV i AIDS
Retrovirology
Reviews in Medical Virology
Southern African Journal of HIV Medicine
Viral Immunology
Virologica Sinica
Virologie
Virology
Virology Journal
Virus Genes
Virus Research

E.6 Physiology

Acta Physiologica
Acta Physiologica Hungarica
Acta Physiologica Scandinavica, Supplement
Adipocytes
American Journal of Human Biology
American Journal of Physiology - Cell Physiology
American Journal of Physiology - Endocrinology and Metabolism
American Journal of Physiology - Gastrointestinal and Liver Physiology
American Journal of Physiology - Heart and Circulatory Physiology
American Journal of Physiology - Lung Cellular and Molecular Physiology
American Journal of Physiology - Regulatory Integrative and Comparative Physiology
American Journal of Physiology - Renal Physiology
Annals of Behavioral Medicine
Annual Review of Physiology
Appetite
Applied Psychophysiology Biofeedback
Archives of Physiology and Biochemistry
Archives of Sexual Behavior
Audiology and Neurotology
Autonomic Neuroscience: Basic and Clinical
BMC Physiology
Behavioral and Brain Sciences
Behavioural Brain Research
Behavioural Processes
Bioelectromagnetics
Biological Psychology
Biological Rhythm Research
Brain and Cognition
Bulletin of the Physical Fitness Research Institute
Canadian Journal of Physiology and Pharmacology
Cardiovascular Research
Cell Calcium
Cell Metabolism
Cell Structure and Function
Cellular Physiology and Biochemistry
Chinese Journal of Andrology
Chinese Journal of Neuroscience
Chinese Journal of Physiology
Chronobiology International
Circulation
Circulation Journal
Circulation Research
Clinical Hemorheology and Microcirculation
Clinical Physiology and Functional Imaging
Clinical and Experimental Pharmacology and Physiology
Cognition
Comparative Biochemistry and Physiology - A Molecular and Integrative Physiology
Comparative Biochemistry and Physiology - Part D: Genomics and Proteomics
Critical Reviews in Neurobiology
Current Heart Failure Reports
Developmental Psychobiology
Electrolyte and Blood Pressure
Electromyography and Clinical Neurophysiology
Endothelium: Journal of Endothelial Cell Research
European Journal of Applied Physiology
Experimental Physiology
Fish Physiology and Biochemistry
General Physiology and Biophysics
High Altitude Medicine and Biology
Hormones and Behavior
Human Reproduction
Indian Journal of Physiology and Pharmacology
International Journal of Psychophysiology

International Journal of Sports Medicine, Supplement
International Journal of Tissue Reactions
JARO - Journal of the Association for Research in Otolaryngology
Japanese Journal of Physiology
Journal of Applied Physiology
Journal of Basic and Clinical Physiology and Pharmacology
Journal of Bioenergetics and Biomembranes
Journal of Cardiovascular Electrophysiology
Journal of Cellular Physiology
Journal of Cerebral Blood Flow and Metabolism
Journal of Circadian Rhythms
Journal of Clinical Neurophysiology
Journal of Electromyography and Kinesiology
Journal of General Physiology
Journal of Jilin University Medicine Edition
Journal of Laboratory and Clinical Medicine
Journal of Medical Sciences
Journal of Membrane Biology
Journal of Motor Behavior
Journal of Muscle Research and Cell Motility
Journal of Neurophysiology
Journal of Organ Dysfunction
Journal of Physiology
Journal of Physiology Paris
Journal of Physiology and Biochemistry
Journal of Physiology and Pharmacology
Journal of Psychophysiology
Journal of Sleep Research
Journal of Sleep Research, Supplement
Journal of Smooth Muscle Research
Journal of Thermal Biology
Journal of Thrombosis and Haemostasis
Journal of Vascular Research
Journal of Veterinary Cardiology
Kidney and Blood Pressure Research
Korean Journal of Physiology and Pharmacology
Laterality
Lung
Mediterranean Journal of Pacing and Electrophysiology
Memory and Cognition
Metabolic Engineering
Microcirculation
Molecular Neurobiology
Neural Networks
Neurobiology of Learning and Memory
Neurogastroenterology and Motility
Neurophysiologie Clinique
Neurophysiology
Neuroscience and Behavioral Physiology
Ophthalmic and Physiological Optics
Pflugers Archiv European Journal of Physiology
Physiological Genomics
Physiological Research
Physiological Reviews
Physiology and Behavior
Proceedings of the American Thoracic Society
Progress in Neurobiology
Prostaglandins Leukotrienes and Essential Fatty Acids
Prostaglandins and Other Lipid Mediators
Psychophysiology
Respiration
Respiratory Physiology and Neurobiology
Seminars in Reproductive Medicine
Shock
Sleep
Sleep and Hypnosis
Somatosensory and Motor Research
Somnologie
Spine
Stress
Synapse
Tissue and Cell
Vision Research

E.7 Anatomy, Histology, Cytology and Anthropology

Acta Anatomica Sinica
Acta Histochemica
Acta Histochemica et Cytochemica
Advances in Anatomic Pathology
American Journal of Physical Anthropology
Analytical and Quantitative Cytology and Histology
Anatomical Record - Part A Discoveries in Molecular, Cellular, and Evolutionary Biology
Anatomical Record - Part B New Anatomist
Anatomical Science International
Anatomy and Embryology
Annals of Anatomy
Archives of Histology and Cytology
Archivos de la Facultad de Medicina de Zaragoza
Biocell
Biotechnic and Histochemistry
Cell and Tissue Research
Cells Tissues Organs
Chinese Journal of Andrology
Clinical Anatomy
Cytometry Part A
Development
Development Growth and Differentiation
Developmental Dynamics
European Journal of Anatomy
European Journal of Cell Biology
European Journal of Histochemistry
European Journal of Morphology
Folia Histochemica et Cytobiologica
Folia Morphologica
Histochemistry and Cell Biology
Histology and Histopathology
International Review of Cytology
Journal of Anatomy
Journal of Chemical Neuroanatomy
Journal of Histochemistry and Cytochemistry
Journal of Histotechnology
Journal of Molecular Histology
Journal of Neurocytology
Journal of Structural Biology
Microscopy Research and Technique
Morphologie
Neuroanatomy
Scanning
Surgical and Radiologic Anatomy
Ultramicroscopy

E.8 Developmental Biology and Gerontology

Complexus
Growth, Development and Aging
International Journal of Behavioral Development

E.8.1 Developmental Biology

Acta Anatomica Sinica
American Journal of Human Biology
Anatomy and Embryology
Arthropod Structure and Development
BMC Developmental Biology
BioEssays
Biology of Reproduction
Biology of the Neonate
Birth Defects Research Part A - Clinical and Molecular Teratology
Birth Defects Research Part C - Embryo Today: Reviews
Clinical and Developmental Immunology
Cloning and Stem Cells
Congenital Anomalies
Current Opinion in Genetics and Development
Current Opinion in Pharmacology
Development
Development Genes and Evolution
Development Growth and Differentiation
Developmental Biology
Developmental Cell
Developmental Dynamics
Differentiation
Early Human Development
European Journal of Histochemistry
Genes and Development
Genesis
Human Development
Human Reproduction
In Vitro Cellular and Developmental Biology - Animal
Infancy
International Journal of Biological Sciences
International Journal of Developmental Biology
International Journal of Developmental Neuroscience
Journal of Assisted Reproduction and Genetics
Journal of Cognition and Development
Journal of Neurogenetics
Mechanisms of Development
Molecular Human Reproduction
Molecular Reproduction and Development
Neuroembryology and Aging
Organogenesis
Pigment Cell Research
Placenta
Reproduction, Fertility and Development
Reproductive Biology and Endocrinology
Reproductive Toxicology
Seminars in Cell and Developmental Biology

E.8.2 Gerontology

Age and Ageing
Ageing Research Reviews
Aging - Clinical and Experimental Research
Aging Cell
Aging, Neuropsychology, and Cognition
Archives of Gerontology and Geriatrics
Arquivos de Geriatria e Gerontologia
CPD Bulletin Old Age Psychiatry
Canadian Journal on Aging
Clinical Gerontologist
Experimental Aging Research
Experimental Gerontology
Gerontologist
Gerontology
Giornale di Gerontologia
Hallym International Journal of Aging
Immunity and Ageing
International Journal of Aging and Human Development
Journal of Nutrition, Health and Aging
Journal of the American Aging Association
Journals of Gerontology - Series A Biological Sciences and Medical Sciences
Journals of Gerontology - Series B Psychological Sciences and Social Sciences
Mechanisms of Ageing and Development
Neurobiology of Aging
Rejuvenation Research
Research on Aging
Revista Espanola de Geriatria y Gerontologia
Revista Multidisciplinar de Gerontologia
Tijdschrift voor Gerontologie en Geriatrie
Zeitschrift fur Gerontologie und Geriatrie

E.9 Genetics and Molecular Biology

Acta Scientiarum - Biological Sciences
Acta Scientiarum - Health Sciences
Ageing Research Reviews
Annual Review of Genomics and Human Genetics
BMC Genomics
BioEssays
Biomolecular Engineering
Cancer Gene Therapy
Cancer Genomics and Proteomics
Chromosome Research
Clinical Proteomics
Current Gene Therapy
Current Genomics
Current Opinion in Genetics and Development
Current Opinion in Molecular Therapeutics
DNA Sequence - Journal of DNA Sequencing and Mapping
Drug Discovery Today: TARGETS
Experimental and Molecular Medicine
Expert Review of Molecular Diagnostics
Expert Review of Proteomics
Familial Cancer
Gene Expression
Gene Expression Patterns
Gene Therapy
Gene Therapy and Regulation
Genes, Brain and Behavior
Genetics and Molecular Biology
Genetics and Molecular Research
Genome Letters

Genome Research
Genomics, Proteomics and Bioinformatics
Infection, Genetics and Evolution
Journal of Gene Medicine
Journal of Proteome Research
Journal of Structural and Functional Genomics
Microcirculation
Molecular Biology Reports
Molecular Cancer Therapeutics
Molecular Cell
Molecular Human Reproduction
Molecular Medicine
Molecular Reproduction and Development
Molecular Systems Biology
Molecular and Cellular Proteomics
Nature Reviews Genetics
NeuroMolecular Medicine
OMICS A Journal of Integrative Biology
Oligonucleotides
Pharmacogenomics Journal
Physiological Genomics
Primary Psychiatry
RNA
Somatic Cell and Molecular Genetics
Statistical Applications in Genetics and Molecular Biology
TheScientificWorldJournal

E.9.1 General Genetics

Acta Genetica Sinica
Annual Review of Genetics
Applied Genomics and Proteomics
BAG - Journal of Basic and Applied Genetics
BMC Genetics
Biochemical Genetics
Biochimica et Biophysica Acta - Gene Structure and Expression
Birth Defects Research Part B - Developmental and Reproductive Toxicology
Cancer Genetics and Cytogenetics
Cell and Chromosome
Chromosoma
Comparative Biochemistry and Physiology - Part D: Genomics and Proteomics
Current Genetics
Cytogenetic and Genome Research
DNA Repair
Environmental and Molecular Mutagenesis
European Journal of Medical Genetics
Functional and Integrative Genomics
Gene
Genes and Development
Genes and Genetic Systems
Genes and Immunity
Genes to Cells
Genesis
Genetical Research
Genetics
Genome Biology
Genomics
Hereditas
Human Biology
Human Molecular Genetics
Hybridoma
IEE Proceedings: Systems Biology
Immunogenetics
International Journal of Immunogenetics
Journal of Assisted Reproduction and Genetics
Journal of Heredity
Journal of Molecular Neuroscience
Journal of Neurogenetics
Mammalian Genome
Molecular Diagnosis and Therapy
Molecular Genetics and Genomics
Molecular Genetics and Metabolism
Mutagenesis
Mutation Research - Fundamental and Molecular Mechanisms of Mutagenesis
Mutation Research - Genetic Toxicology and Environmental Mutagenesis
Mutation Research - Reviews in Mutation Research
New Genetics and Society
PLoS Computational Biology
PLoS Genetics
PLoS Pathogens
Pharmacogenetics and Genomics
Proteins: Structure, Function and Genetics
Trends in Genetics

E.9.2 Molecular Biology

American Journal of Respiratory Cell and Molecular Biology
Autophagy
BMC Molecular Biology
Biochimica et Biophysica Acta - Gene Structure and Expression
Biochimica et Biophysica Acta - Molecular Cell Research
Briefings in Functional Genomics and Proteomics
Cancer Molecular Biology
Cell Communication and Signaling
Cell Cycle
Cell Metabolism
Cellular and Molecular Biology
Cellular and Molecular Biology Letters
Chemistry and Biodiversity
Cold Spring Harbor Symposia on Quantitative Biology
Comparative Biochemistry and Physiology - B Biochemistry and Molecular Biology
Comparative Biochemistry and Physiology - Part D: Genomics and Proteomics
Critical Reviews in Biochemistry and Molecular Biology
Critical Reviews in Eukaryotic Gene Expression
Current Issues in Molecular Biology
Current Topics in Peptide and Protein Research
DNA Research
DNA and Cell Biology
Drug Discovery Today: Disease Mechanisms
Drug Discovery Today: Disease Models
Drug Discovery Today: Technologies
Drug Discovery Today: Therapeutic Strategies
EMBO Journal
EMBO Reports
Expert Reviews in Molecular Medicine
Genome Letters
Human Genomics

IEE Proceedings: Systems Biology
IUBMB Life
In Silico Biology
International Journal of Biological Sciences
International Journal of Immunogenetics
International Journal of Molecular Sciences
Journal of Biomolecular NMR
Journal of Biomolecular Structure and Dynamics
Journal of Computational Biology
Journal of Molecular Biology
Journal of Molecular Diagnostics
Journal of Molecular Endocrinology
Journal of Molecular Evolution
Journal of Molecular Modeling
Journal of Molecular and Cellular Cardiology
Journal of Organ Dysfunction
Journal of Receptor and Signal Transduction
Journal of Steroid Biochemistry and Molecular Biology
Journal of Structural Biology
Microbiology and Molecular Biology Reviews
Molecular Aspects of Medicine
Molecular Biology and Evolution
Molecular Biology of the Cell
Molecular Brain Research
Molecular Cancer Research
Molecular Endocrinology
Molecular Microbiology
Molecular Psychiatry
Molecular Therapy
Molecular and Cellular Biochemistry
Molecular and Cellular Biology
Molecular and Cellular Endocrinology
Molecular and Cellular Probes
Molecules and Cells
Nanomedicine: Nanotechnology, Biology, and Medicine
Nature Chemical Biology
Nature Methods
Nature Reviews Molecular Cell Biology
Nature Structural and Molecular Biology
Nuclear Receptor
Nucleic Acids Research
Oncogene
PLoS Computational Biology
PLoS Genetics
PLoS Pathogens
Pharmacogenetics and Genomics
Physical Biology
Proceedings of the National Academy of Sciences of the United States of America
Protein Engineering, Design and Selection
Proteomics
RNA Biology
Research Communications in Biochemistry and Cell and Molecular Biology
Trends in Molecular Medicine
Virus Genes

E.10 Biotechnology and Applied Science

Acta Biomaterialia
African Journal of Biotechnology
Annexins
Applied Biochemistry and Biotechnology
Applied Biotechnology, Food Science and Policy
Applied Microbiology and Biotechnology
Asian Journal of Microbiology, Biotechnology and Environmental Sciences
Australasian Biotechnology
BMC Biotechnology
BioTechniques
Biocatalysis and Biotransformation
Biochemical Engineering Journal
Biocontrol Science
Biofilms
Biofutur
Biomedical Materials
Bioprocess and Biosystems Engineering
Biosciences Biotechnology Research Asia
Biotechnology Annual Review
Biotechnology and Applied Biochemistry
Biotechnology and Bioengineering
Biotechnology and Bioprocess Engineering
Biotechnology and Genetic Engineering Reviews
Biotecnologia Aplicada
CLB Chemie in Labor und Biotechnik
Cell Preservation Technology
Cloning and Stem Cells
Comparative and Functional Genomics
Critical Reviews in Biotechnology
Current Nanoscience
Current Opinion in Biotechnology
Current Pharmaceutical Biotechnology
Cytotechnology
Electronic Journal of Biotechnology
Expert Review of Proteomics
FEMS Microbiology Letters
Genetic Vaccines and Therapy
Genome Letters
IEE Proceedings: Systems Biology
International Journal of Biological Sciences
International Journal of Biotechnology
International Journal of Dairy Technology
Journal of Biolaw and Business
Journal of Bioscience and Bioengineering
Journal of Biotechnology
Journal of Chemical Technology and Biotechnology
Journal of Commercial Biotechnology
Journal of Industrial Microbiology and Biotechnology
Journal of Microbiology and Biotechnology
Journal of Molecular Microbiology and Biotechnology
Journal of Nanobiotechnology
Journal of Neural Engineering
Journal of Proteome Research
Journal of the Royal Society Interface
Korean Journal of Microbiology and Biotechnology
Macromolecular Bioscience
Microbial Cell Factories
Minerva Biotecnologica
Molecular Biotechnology
Nano Today
Nanomedicine: Nanotechnology, Biology, and Medicine
Nature Biotechnology
Nature Methods
OMICS A Journal of Integrative Biology
Organogenesis
Phytochemistry Reviews
TransPharma
Trends in Biotechnology

Yeast

E.10.1 Experimental Microbiology

Advances in Microbial Physiology
Annual Review of Microbiology
Antonie van Leeuwenhoek, International Journal of General and Molecular Microbiology
Archives of Microbiology
Canadian Journal of Microbiology
Cellular Microbiology
Current Microbiology
FEMS Microbiology Letters
FEMS Microbiology Reviews
FEMS Yeast Research
Fungal Genetics and Biology
Infection and Immunity
International Journal of Systematic and Evolutionary Microbiology
Journal of Bacteriology
Journal of Microbiological Methods
Microbiology
Microbiology and Molecular Biology Reviews
Molecular Biotechnology
Molecular Microbiology
Nature Reviews Microbiology
Plasmid
Postepy Mikrobiologii
Research in Microbiology

E.10.2 Applied Microbiology

Advances in Applied Microbiology
Applied Microbiology and Biotechnology
Applied and Environmental Microbiology
Biofilms
Engineering in Life Sciences
Enzyme and Microbial Technology
FEMS Microbiology Ecology
Foodborne Pathogens and Disease
International Journal of Biological Sciences
International Journal of Food Microbiology
Journal of Applied Microbiology
Journal of General and Applied Microbiology
Letters in Applied Microbiology
Molecular Biotechnology
Polish Journal of Microbiology
Systematic and Applied Microbiology

E.11 Biophysics, Medical Instrumentation and Biomedical Engineering

ASAIO Journal
Acta of Bioengineering and Biomechanics
Annual Review of Biomedical Engineering
Applied Bioinformatics
Applied Radiology
Artificial Organs
Australasian Physical and Engineering Sciences in Medicine
BMC Bioinformatics
Bio-Medical Materials and Engineering
BioMedical Engineering Online
Bioinformatics
Biomedical Engineering
Biomedical Engineering - Applications, Basis and Communications
Biomedical Materials
Biosensors and Bioelectronics
Cells and Materials
Chemometrics and Intelligent Laboratory Systems
Chinese Journal of Biomedical Engineering
Clinical Biomechanics
Colloids and Surfaces A: Physicochemical and Engineering Aspects
Critical Reviews in Biomedical Engineering
European Cells and Materials
European Radiology, Supplement
IEEE Engineering in Medicine and Biology Magazine
IEEE Transactions on Biomedical Engineering
IEEE Transactions on Neural Systems and Rehabilitation Engineering
ITBM-RBM
ITBM-RBM News
International Journal of Artificial Organs
Japanese Journal of Medical Electronics and Biological Engineering
Journal of Applied Biomaterials and Biomechanics
Journal of Artificial Organs
Journal of Biomaterials Science, Polymer Edition
Journal of Biomechanical Engineering
Journal of Biomedical Materials Research - Part A
Journal of Biomedical Materials Research - Part B Applied Biomaterials
Journal of Clinical Engineering
Journal of Facial and Somato Prosthetics
Journal of Materials Science: Materials in Medicine
Journal of Medical Engineering and Technology
Journal of Medical Systems
Journal of Medical and Biological Engineering
Journal of Nanobiotechnology
Journal on Information Technology in Healthcare
Kardiotechnik
Medical Engineering and Physics
Medical and Biological Engineering and Computing
Natural Computing
Pharmaceutical Engineering
Pharmaceutical Technology
Physiological Measurement
Technology and Health Care
Tissue Engineering

E.11.1 Biophysics

Advances in Biophysics
Archives of Biochemistry and Biophysics
Bioelectromagnetics
Biomedical Materials
Biophysical Journal
Biorheology
Current Pharmaceutical Analysis
European Biophysics Journal

International Journal of Medical Robotics and Computer Assisted Surgery
International Journal of Radiation Oncology Biology Physics
Journal of Applied Biomechanics
Journal of Biochemical and Biophysical Methods
Journal of Biological Physics
Journal of Biomechanics
Journal of Membrane Biology
Journal of the Acoustical Society of America
Journal of the ICRU
Journal of the Royal Society Interface
Magnetic Resonance Materials in Physics, Biology and Medicine
Medecine Nucleaire
Medical Physics
Nanomedicine: Nanotechnology, Biology, and Medicine
Neuroanatomy
Nitric Oxide - Biology and Chemistry
Particulate Science and Technology
Photochemical and Photobiological Sciences
Photodiagnosis and Photodynamic Therapy
Photomedicine and Laser Surgery
Physica Medica
Physical Biology
Physics in Medicine and Biology
Quarterly Reviews of Biophysics
Radiation and Environmental Biophysics
Zeitschrift fur Medizinische Physik

E.11.2 Medical Instrumentation

American Journal of Electroneurodiagnostic Technology
Biomedical Instrumentation and Technology
Biomedical Microdevices
Biomedical Sciences Instrumentation
Biomedizinische Technik
Cardiovascular Engineering
Computer Aided Surgery
Computer Methods and Programs in Biomedicine
Computerized Medical Imaging and Graphics
Computers in Biology and Medicine
Diabetes Technology and Therapeutics
Endoscopy
European Journal of Echocardiography
Expert Review of Medical Devices
International Journal of Medical Informatics
JALA - Journal of the Association for Laboratory Automation
Journal of Bioinformatics and Computational Biology
Journal of Biomedical Informatics
Journal of Clinical Monitoring and Computing
Journal of Electrocardiology
Journal of Extra-Corporeal Technology
Journal of Long-Term Effects of Medical Implants
Journal of Microscopy
Journal of the American Medical Informatics Association
Lasers in Medical Science
Magnetic Resonance Imaging
Medical Laser Application
Medizintechnik
News in Medical Products
Scanning
Ultramicroscopy
Zentralsterilisation - Central Service

E.12 Forensic Science

Acta Criminologiae et Medicinae Legalis Japonica
Aggression and Violent Behavior
Altex
American Journal of Forensic Medicine and Pathology
American Journal of Forensic Psychiatry
American Journal of Forensic Psychology
American Journal of Law and Medicine
Anil Aggrawal's Internet Journal of Forensic Medicine and Toxicology
Archiv fur Kriminologie
Australian Journal of Forensic Sciences
Behavioral Sciences and the Law
Bioethics
Blutalkohol
Child Abuse and Neglect
Chinese Journal of Forensic Medicine
Clinical Risk
Criminal Behaviour and Mental Health
Death Studies
European Journal of Health Law
Forensic Science International
HEC Forum
International Journal of Drug Policy
International Journal of Law and Psychiatry
International Journal of Legal Medicine
International Journal of Medical Toxicology and Legal Medicine
International Journal of Offender Therapy and Comparative Criminology
Japanese Journal of Forensic Toxicology
Japanese Journal of Legal Medicine
Journal International de Bioethique
Journal de Medecine Legale Droit Medical
Journal of Aggression, Maltreatment and Trauma
Journal of Child Sexual Abuse
Journal of Clinical Forensic Medicine
Journal of Drug Education
Journal of Drug Issues
Journal of Emotional Abuse
Journal of Forensic Identification
Journal of Forensic Medicine and Toxicology
Journal of Forensic Neuropsychology
Journal of Forensic Odonto-Stomatology
Journal of Forensic Psychiatry and Psychology
Journal of Forensic Psychology Practice
Journal of Forensic Sciences
Journal of Legal Medicine
Journal of Medical Ethics
Journal of Medicine and Philosophy
Journal of Psychiatry and Law
Journal of Research in Crime and Delinquency
Journal of Studies on Alcohol
Journal of Substance Use
Journal of Whiplash and Related Disorders
Journal of the American Academy of Psychiatry and the Law
Journal of the Canadian Society of Forensic Science
Law and Human Behavior
Legal Medicine

Legal and Criminological Psychology
Medecine et Droit
Medical Humanities
Medical Law International
Medicine and Law
Medico-Legal Update
Psychiatry, Psychology and Law
Recht und Psychiatrie
Rechtsmedizin
Romanian Journal of Legal Medicine
Science and Justice - Journal of the Forensic Science Society
Suicide and Life-Threatening Behavior
Theoretical Medicine and Bioethics
Traffic Injury Prevention
Trauma
Violence and Victims
Wound Ballistics Review
Z Zagadnien Nauk Sadowych

E.13 Nutrition

Aktuelle Ernahrungsmedizin
American Journal of Clinical Nutrition
Annals of Nutrition and Metabolism
Annual Review of Nutrition
Appetite
British Journal of Nutrition
Cahiers de Nutrition et de Dietetique
Chinese Journal of Clinical Nutrition
Clinical Nutrition
Clinical Nutrition, Supplement
Current Nutrition and Food Science
Current Opinion in Clinical Nutrition and Metabolic Care
Current Topics in Nutraceutical Research
Diabetes, Nutrition and Metabolism - Clinical and Experimental
Ernahrungs Umschau
European Journal of Clinical Nutrition
European Journal of Nutrition
Food Chemistry
International Journal for Vitamin and Nutrition Research
International Journal of Behavioral Nutrition and Physical Activity
International Journal of Food Sciences and Nutrition
International Journal of Sport Nutrition and Exercise Metabolism
Journal fur Ernahrungsmedizin
Journal of Clinical Biochemistry and Nutrition
Journal of Human Nutrition and Dietetics
Journal of Medicinal Food
Journal of Nutraceuticals, Functional and Medical Foods
Journal of Nutrition
Journal of Nutrition, Health and Aging
Journal of Nutritional Biochemistry
Journal of Nutritional Science and Vitaminology
Journal of Nutritional and Environmental Medicine
Journal of Orthomolecular Medicine
Journal of Parenteral and Enteral Nutrition
Journal of the American College of Nutrition
Maternal and Child Nutrition
Nutricion Clinica y Dietetica Hospitalaria
Nutricion Hospitalaria
Nutrition
Nutrition Bulletin
Nutrition Clinique et Metabolisme
Nutrition Journal
Nutrition Research
Nutrition Research Reviews
Nutrition Reviews
Nutrition and Cancer
Nutrition and Metabolic Disorders in HIV Infection
Nutrition and Metabolism
Nutrition in Clinical Care
Nutrition in Clinical Practice
Nutritional Neuroscience
Nutritional Sciences Journal
Obesity Management
Pakistan Journal of Nutrition
Proceedings of the Nutrition Society
Progress in Nutrition
Public Health Nutrition
Revista Espanola de Nutricion Comunitaria
Revista de Nutricao
Rivista Italiana di Nutrizione Parenterale ed Enterale
Scandinavian Journal of Food and Nutrition
Scandinavian Journal of Nutrition/Naringsforskning
South African Journal of Clinical Nutrition
Topics in Clinical Nutrition

F.0 Other Specialities

Biology Letters

F.1 Dentistry and Stomatology

Acta Odontologica Scandinavica
Atlas of the Oral and Maxillofacial Surgery Clinics of North America
BMC Oral Health
Critical Reviews in Oral Biology and Medicine
International Journal of Oral and Maxillofacial Surgery
Journal of Contemporary Dental Practice
Journal of Oral Pathology and Medicine
Journal of Periodontology
Medicina Oral, Patologia Oral y Cirugia Bucal
Oral Diseases
Oral Microbiology and Immunology
Oral and Maxillofacial Surgery Clinics of North America
Salus

F.2 Veterinary Medicine

American Journal of Veterinary Research
Canadian Journal of Veterinary Research
Canadian Veterinary Journal
In Practice
Japanese Journal of Veterinary Research
Journal of Veterinary Cardiology
Journal of the American Animal Hospital Association
Journal of the American Veterinary Medical Association
Journal of the South African Veterinary Association
Research in Veterinary Science

Veterinary Clinics of North America - Small Animal Practice
Veterinary Journal
Veterinary Quarterly
Veterinary Record
Veterinary Research
Veterinary Research Communications

F.3 Informatics

Applied Bioinformatics
Artificial Intelligence in Medicine
BMC Medical Informatics and Decision Making
Biomedical Digital Libraries
Biomedical Signal Processing and Control
Briefings in Bioinformatics
Drug Information Journal
Health Informatics Journal
Healthcare Review Online
Informatics in Primary Care
International Journal of Bioinformatics Research and Applications
International Journal of Electronic Healthcare
International Journal of Healthcare Technology and Management
International Journal of Technology Assessment in Health Care
Journal of Medical Internet Research
Journal of the American Medical Informatics Association
Journal of the Medical Library Association
Journal on Information Technology in Healthcare
Medical Decision Making
Medical Informatics and the Internet in Medicine
Medical and Biological Engineering and Computing
Methods of Information in Medicine
Natural Computing
Neuroinformatics
Online Journal of Nursing Informatics
Online Publishing in Medicine
Theoretical Biology and Medical Modelling

LIST OF JOURNALS PER COUNTRY OF PUBLICATION

Country	Country Code	Number of Journals
Argentina	ARG	12
Australia	AUS	64
Austria	AUT	43
Azerbaijan	AZE	2
Bahrain	BHR	2
Bangladesh	BGD	6
Belgium	BEL	19
Bosnia and Herzegovina	BIH	1
Brazil	BRA	48
Bulgaria	BGR	14
Canada	CAN	58
Chile	CHL	4
China	CHN	66
Colombia	COL	3
Croatia	HRV	30
Cuba	CUB	2
Cyprus	CYP	1
Czech Republic	CZE	42
Denmark	DNK	22
Ecuador	ECU	1
Egypt	EGY	4
Finland	FIN	3
France	FRA	138
Germany	DEU	339
Greece	GRC	16
Hong Kong	HKG	17
Hungary	HUN	11
India	IND	72
Indonesia	IDN	1
Iran	IRN	17
Ireland	IRL	46
Israel	ISR	12
Italy	ITA	165
Jamaica	JAM	1
Japan	JPN	166
Jordan	JOR	2
Kenya	KEN	2
Lebanon	LBN	1
Macedonia	MKD	1
Malaysia	MYS	3
Mexico	MEX	21
Nepal	NPL	2
Netherlands	NLD	310
New Zealand	NZL	39
Nigeria	NGA	5
Norway	NOR	39
Pakistan	PAK	14
Poland	POL	85
Portugal	PRT	9
Qatar	QAT	2
Romania	ROM	5
Russian Federation	RUS	2
Saudi Arabia	SAU	5
Serbia and Montenegro	SCG	6
Singapore	SGP	11
Slovakia	SVK	11
Slovenia	SVN	3
South Africa	ZAF	14
South Korea	KOR	21
Spain	ESP	113
Sweden	SWE	4
Switzerland	CHE	122
Taiwan	TWN	21
Thailand	THA	2
Tunisia	TUN	2
Turkey	TUR	58
Uganda	UGA	1
Ukraine	UKR	2
United Arab Emirates	ARE	2
United Kingdom	GBR	1196
United States	USA	1489
Venezuela	VEN	4

LIST OF JOURNALS PER COUNTRY OF PUBLICATION

Argentina ARG

Acta Farmaceutica Bonaerense
BAG - Journal of Basic and Applied Genetics
Biocell
Medicina
Prensa Medica Argentina
Revista Argentina de Dermatologia
Revista Argentina de Endocrinologia y Metabolismo
Revista Argentina de Microbiologia
Revista Medica de Rosario
Revista Neurologica Argentina
Revista de Nefrologia, Dialisis y Transplante
Salud (i) Sciencia

Australia AUS

ANZ Journal of Surgery
ANZ Nuclear Medicine
APLAR Journal of Rheumatology
Anaesthesia and Intensive Care
Anatomical Science International
Asia Pacific Journal of Environmental Law
Australasian Biotechnology
Australasian Journal of Dermatology
Australasian Physical and Engineering Sciences in Medicine
Australasian Psychiatry
Australasian Radiology
Australian Journal of Clinical Hypnotherapy and Hypnosis
Australian Journal of Clinical and Experimental Hypnosis
Australian Journal of Forensic Sciences
Australian Journal of Medical Science
Australian Journal of Otolaryngology
Australian Journal of Physiotherapy
Australian Journal of Primary Health
Australian Journal of Rural Health
Australian Occupational Therapy Journal
Australian Prescriber
Australian and New Zealand Journal of Audiology
Australian and New Zealand Journal of Obstetrics and Gynaecology
Australian and New Zealand Journal of Psychiatry
Australian and New Zealand Journal of Public Health
Behaviour Change
Cancer Forum
Chinese Journal of Digestive Diseases
Clinical and Experimental Ophthalmology
Clinical and Experimental Pharmacology and Physiology
Congenital Anomalies
Development Growth and Differentiation
Diseases of the Esophagus
Diving and Hyperbaric Medicine
EMA - Emergency Medicine Australasia
Heart Lung and Circulation
Hong Kong Journal of Emergency Medicine
Immunology and Cell Biology
Internal Medicine Journal
International Journal of Biological Sciences
International Journal of Medical Sciences
International Journal of Urology
Journal of Emergency Primary Health Care
Journal of Forensic Odonto-Stomatology
Journal of Gastroenterology and Hepatology
Journal of Occupational Health and Safety - Australia and New Zealand
Journal of Paediatrics and Child Health
Journal of Pharmacy Practice and Research
Medical Journal of Australia
Medicine Today
Nephrology
Neuropathology
Nursing and Health Sciences
Pathology International
Pediatrics International
Psychiatry, Psychology and Law
Psychogeriatrics
Reproduction, Fertility and Development
Reproductive Medicine and Biology
Respirology
Sexual Health
Sleep and Biological Rhythms
Today's Life Science
Twin Research and Human Genetics

Austria AUT

Acta Chirurgica Austriaca
Acta Medica Austriaca
Acta Neurochirurgica
Acta Neurochirurgica, Supplement
Amino Acids
Archives of Women's Mental Health
Chemotaxis and Migration
Current News in Medicine
Current Opinion in Clinical Experimental Research
Current Opinion in European Medicine
Difficult Airways
European Journal of Acupuncture
European Journal of Allied Health
European Surgery - Acta Chirurgica Austriaca
Herbals and Complementary Medicine
International Journal of Psychotherapy
Journal fur Ernahrungsmedizin
Journal fur Fertilitat und Reproduktion
Journal fur Gastroenterologische und Hepatologische Erkrankungen
Journal fur Hypertonie
Journal fur Kardiologie
Journal fur Menopause
Journal fur Mineralstoffwechsel
Journal fur Neurologie, Neurochirurgie und Psychiatrie
Journal fur Reproduktionsmedizin und Endokrinologie
Journal fur Urologie und Urogynakologie
Journal of Clinical and Basic Cardiology
Journal of Neural Transmission

Journal of Neural Transmission, Supplement
Malignant Hyperthermia
Medical Quality
Modern Aspects in Sepsis
News in Medical Products
Online Publishing in Medicine
Pain Medicine
Placebo
Scientia Pharmaceutica
Spektrum der Augenheilkunde
Thermology International
Wiener Klinische Wochenschrift
Wiener Klinische Wochenschrift, Supplement
Wiener Medizinische Wochenschrift
Zeitschrift fur Gefassmedizin

Azerbaijan AZE

Azerbaijan Journal of Metabolism
Azerbaijan Medical Journal

Bahrain BHR

Bahrain Medical Bulletin
Journal of the Bahrain Medical Society

Bangladesh BGD

Bangladesh Journal of Dermatology, Venereology and Leprology
Bangladesh Journal of Obstetrics and Gynecology
Bangladesh Medical Research Council Bulletin
Bangladesh Renal Journal
Journal of Health, Population and Nutrition
Journal of Medical Teachers Federation

Belgium BEL

Acta Anaesthesiologica Belgica
Acta Cardiologica
Acta Clinica Belgica
Acta Gastro-Enterologica Belgica
Acta Hospitalia
Acta Neurologica Belgica
Acta Orthopaedica Belgica
Acta Oto-Rhino-Laryngologica Belgica
Archives of Public Health
B-ENT
Electromyography and Clinical Neurophysiology
European Journal of Lymphology and Related Problems
Farmaceutisch Tijdschrift voor Belgie
Journal Belge de Radiologie
Journal de Pharmacie de Belgique
Louvain Medical
Revue Medicale de Bruxelles
Revue Medicale de Liege
Tijdschrift voor Geneeskunde

Bosnia and Herzegovina BIH

Medicinski Glasnik

Brazil BRA

Acta Cirurgica Brasileira
Acta Scientiarum - Biological Sciences
Acta Scientiarum - Health Sciences
Anais Brasileiros de Dermatologia
Annual Review of Biomedical Sciences
Arquivos Brasileiros de Cardiologia
Arquivos Brasileiros de Medicina
Arquivos Brasileiros de Oftalmologia
Arquivos de Gastroenterologia
Arquivos de Geriatria e Gerontologia
Arquivos de Neuro-Psiquiatria
Brazilian Journal of Epilepsy and Clinical Neurophysiology
Brazilian Journal of Infectious Diseases
Brazilian Journal of Medical and Biological Research
GED - Gastrenterologia Endoscopia Digestiva
Genetics and Molecular Biology
Genetics and Molecular Research
Infanto
Informacao Psiquiatrica
International Braz J Urol
Jornal Brasileiro de Psiquiatria
Jornal Brasileiro de Reproducao Assistida
Jornal Vascular Brasileiro
Jornal de Pediatria
Medicina
Online Brazilian Journal of Nursing
Reproducao e Climaterio
Revista Brasileira de Anestesiologia
Revista Brasileira de Ciencias Farmaceuticas/Brazilian Journal of Pharmaceutical Sciences
Revista Brasileira de Medicina
Revista Brasileira de Medicina do Esporte
Revista Brasileira de Neurologia
Revista Brasileira de Neurologia e Psiquiatria
Revista Brasileira de Otorrinolaringologia
Revista Brasileira de Plantas Medicinais
Revista Brasileira de Psiquiatria
Revista Brasileira de Reumatologia
Revista Brasileira de Toxicologia
Revista da Associacao Brasileira de Medicina Psicossomatica
Revista da Sociedade Brasileira de Cancerologia
Revista da Sociedade Brasileira de Medicina Tropical
Revista de Ciencias Farmaceuticas
Revista de Nutricao
Revista de Psiquiatria Clinica
Revista de Psiquiatria do Rio Grande do Sul
Revista de Saude Publica
Revista do Instituto de Medicina Tropical de Sao Paulo
Sao Paulo Medical Journal

Bulgaria BGR

Acta Medica Bulgarica
Anaesthesiology and Intensive Care

Biomedical Reviews
Bulgarian Medicine
Clinical Application of Immunology
Clinical and Transfusion Haematology
Endocrinologia
General Medicine
Nephrology, Dialysis and Transplantation
Pediatriya
Pharmacia
Problems of Infectious and Parasitic Diseases
Rentgenologiya i Radiologiya
Rheumatology

Canada CAN

Biochemistry and Cell Biology
Biological Procedures Online
Canadian Association of Radiologists Journal
Canadian Family Physician
Canadian Journal of Anesthesia
Canadian Journal of Cardiology
Canadian Journal of Chemistry
Canadian Journal of Clinical Pharmacology
Canadian Journal of Community Mental Health
Canadian Journal of Diabetes
Canadian Journal of Emergency Medicine
Canadian Journal of Gastroenterology
Canadian Journal of Hospital Pharmacy
Canadian Journal of Human Sexuality
Canadian Journal of Infectious Diseases and Medical Microbiology
Canadian Journal of Microbiology
Canadian Journal of Neurological Sciences
Canadian Journal of Occupational Therapy
Canadian Journal of Ophthalmology
Canadian Journal of Physiology and Pharmacology
Canadian Journal of Psychiatry
Canadian Journal of Public Health
Canadian Journal of Respiratory Therapy
Canadian Journal of Surgery
Canadian Journal of Veterinary Research
Canadian Journal on Aging
Canadian Medical Association Journal
Canadian Pharmacists Journal
Canadian Respiratory Journal
Canadian Veterinary Journal
Chronic Diseases in Canada
Clinical and Experimental Obstetrics and Gynecology
Clinical and Investigative Medicine
Clinical and Refractive Optometry
Clinical and Surgical Ophthalmology
Current Oncology
Dermatitis
Electronic Journal of Communicative Psychoanalysis
European Journal of Gynaecological Oncology
Experimental and Clinical Cardiology
Geriatrics Today: Journal of the Canadian Geriatrics Society
Geriatrics and Aging
Journal of Child Neurology
Journal of Comparative Family Studies
Journal of Medical Internet Research
Journal of Orthomolecular Medicine
Journal of Otolaryngology
Journal of Pharmacy and Pharmaceutical Sciences
Journal of Psychiatry and Neuroscience
Journal of Rheumatology
Journal of Sexual and Reproductive Medicine
Journal of the Canadian Academy of Child and Adolescent Psychiatry
Journal of the Canadian Society of Forensic Science
McGill Journal of Medicine
Paediatrics and Child Health
Pain Research and Management
Peritoneal Dialysis International
Vascular

Chile CHL

Biological Research
Electronic Journal of Biotechnology
Revista Chilena de Neuro-Psiquiatria
Revista Chilena de Pediatria

China CHN

Academic Journal of Second Military Medical University
Academic Journal of Xi'an Jiaotong University
Acta Academiae Medicinae Hubei
Acta Academiae Medicinae Sinicae
Acta Anatomica Sinica
Acta Genetica Sinica
Bulletin of Hunan Medical University
Chinese Critical Care Medicine
Chinese Journal of Andrology
Chinese Journal of Antibiotics
Chinese Journal of Biomedical Engineering
Chinese Journal of Cancer Research
Chinese Journal of Cerebrovascular Diseases
Chinese Journal of Clinical Nutrition
Chinese Journal of Clinical Oncology
Chinese Journal of Clinical Rehabilitation
Chinese Journal of Contemporary Neurology and Neurosurgery
Chinese Journal of Contemporary Pediatrics
Chinese Journal of Emergency Medicine
Chinese Journal of Endemiology
Chinese Journal of Evidence-Based Medicine
Chinese Journal of Forensic Medicine
Chinese Journal of Gastroenterology
Chinese Journal of Infection and Chemotherapy
Chinese Journal of Integrative Medicine
Chinese Journal of Interventional Imaging and Therapy
Chinese Journal of Lung Cancer
Chinese Journal of Medical Genetics
Chinese Journal of Medical Imaging Technology
Chinese Journal of Microbiology and Immunology
Chinese Journal of Natural Medicines
Chinese Journal of Neurology
Chinese Journal of Neuroscience
Chinese Journal of Ophthalmology
Chinese Journal of Pathology
Chinese Journal of Pharmacology and Toxicology
Chinese Journal of Radiology
Chinese Journal of Rehabilitation Medicine
Chinese Journal of Traumatology - English Edition

Chinese Medical Journal
Chinese Medical Sciences Journal
Chinese Ophthalmic Research
Chinese Pharmaceutical Journal
Chinese Pharmacological Bulletin
Chinese Traditional and Herbal Drugs
Fudan University Journal of Medical Sciences
Genomics, Proteomics and Bioinformatics
Hepatobiliary and Pancreatic Diseases International
International Journal of Ophthalmology
Journal of Central South University (Medical Sciences)
Journal of China Pharmaceutical University
Journal of Chinese Integrative Medicine
Journal of Clinical Dermatology
Journal of Geriatric Cardiology
Journal of Interventional Radiology
Journal of Jilin University Medicine Edition
Journal of Medical Ultrasound
Journal of Sichuan University (Medical Science Edition)
Journal of Traditional Chinese Medicine
Journal of Xi'an Jiaotong University (Medical Sciences)
National Medical Journal of China
Ophthalmology in China
Pharmaceutical Care and Research
Virologica Sinica
World Chinese Journal of Digestology
World Journal of Gastroenterology

Colombia COL

Colombia Medica
Iatreia
Saludarte

Croatia HRV

Acta Clinica Croatica
Acta Clinica Croatica, Supplement
Acta Dermatovenerologica Croatica
Acta Facultatis Medicae Fluminensis
Acta Medica Croatica
Acta Pharmaceutica
Alcoholism
Arhiv za Higijenu Rada i Toksikologiju
Biochemia Medica
Croatian Journal of Gastroenterology and Hepatology
Croatian Medical Journal
Diabetologia Croatica
Farmaceutski Glasnik
Gynaecologia et Perinatologia
Gynaecologia et Perinatologia, Supplement
Hrvatska Revija Za Rehabilitacijska Istrazivanja
Infektoloski Glasnik
Libri Oncologici
Lijecnicki Vjesnik
Lijecnicki Vjesnik, Supplement
Medica Jadertina
Medicina
Medicus
Neurologia Croatica
Paediatria Croatica
Periodicum Biologorum
Pharmaca
Psihoterapija
Psychiatria Danubina
Socijalna Psihijatrija

Cuba CUB

Biotecnologia Aplicada
Revista del Hospital Psiquiatrico de la Habana

Cyprus CYP

Modern Medicine

Czech Republic CZE

Acta Chirurgiae Orthopaedicae et Traumatologiae Cechoslovaca
Acta Chirurgiae Plasticae
Aktuality v Nefrologii
Alergie
Anesteziologie a Intenzivni Medicina
Bolest
Casopis Lekaru Ceskych
Central European Journal of Public Health
Ceska Gynekologie
Ceska Radiologie
Ceska Revmatologie
Ceska a Slovenska Farmacie
Ceska a Slovenska Gastroenterologie a Hepatologie
Ceska a Slovenska Neurologie a Neurochirurgie
Ceska a Slovenska Oftalmologie
Ceska a Slovenska Psychiatrie
Ceske Pracovni Lekarstvi
Cesko-Slovenska Dermatologie
Cesko-Slovenska Pediatrie
Cor et Vasa
Diabetologie Metabolismus Endokrinologie Vyziva
Endoskopie
Epidemiologie, Mikrobiologie, Imunologie
Folia Biologica
Homeostasis in Health and Disease
Hygiena
Journal of Applied Biomedicine
Klinicka Biochemie a Metabolismus
Klinicka Mikrobiologie a Infekcni Lekarstvi
Klinicka Onkologie
Osteologicky Bulletin
Otorinolaryngologie a Foniatrie
Physiological Research
Pracovni Lekarstvi
Prakticky Lekar
Psychiatrie
Rehabilitace a Fyzikalni Lekarstvi
Remedia - Klinicka Mikrobiologie
Scripta Medica Facultatis Medicae Universitatis Brunensis Masarykianae
Studia Pneumologica et Phthiseologica
Transfuze a Hematologie Dnes
Vnitrni Lekarstvi

Denmark DNK

APMIS
APMIS, Supplement
Acta Anaesthesiologica Scandinavica
Acta Anaesthesiologica Scandinavica, Supplement
Acta Ophthalmologica Scandinavica, Supplement
Acta Psychiatrica Scandinavica, Supplement
Allergy: European Journal of Allergy and Clinical Immunology, Supplement
Bipolar Disorders
Bipolar Disorders, Supplement
Danish Medical Bulletin
European Journal of Haematology, Supplement
European Respiratory Journal, Supplement
Journal of Medical Primatology
Oral Diseases
Pediatric Allergy and Immunology
Pediatric Diabetes
Pediatric Diabetes, Supplement
Photodermatology Photoimmunology and Photomedicine
Scandinavian Journal of Laboratory Animal Science
Scandinavian Psychoanalytic Review
Traffic
Ugeskrift for Laeger

Ecuador ECU

Revista Ecuatoriana de Neurologia

Egypt EGY

Cancer Molecular Biology
Eastern Mediterranean Health Journal
Egyptian Journal of Anaesthesia
Middle East Fertility Society Journal

Finland FIN

NAD Publication
Scandinavian Journal of Surgery
Scandinavian Journal of Work, Environment and Health

France FRA

ANAE - Approche Neuropsychologique des Apprentissages chez l'Enfant
Acta Endoscopica
Actualites Pharmaceutiques
Actualites Pharmaceutiques Hospitalieres
Actualites Reproduction Humaine
African Journal of Neurological Sciences
Andrologie
Angeiologie
Annales Francaises d'Anesthesie et de Reanimation
Annales Medico-Psychologiques
Annales Pharmaceutiques Francaises
Annales d'Endocrinologie
Annales d'Urologie
Annales de Cardiologie et d'Angeiologie
Annales de Chirurgie
Annales de Chirurgie Plastique et Esthetique
Annales de Dermatologie et de Venereologie
Annales de Pathologie
Annales de Pediatrie
Annales de Readaptation et de Medecine Physique
Antibiotiques
Archives de Pediatrie
Archives des Maladies Professionnelles et de l'Environnement
Archives des Maladies du Coeur et des Vaisseaux
Archives des Maladies du Coeur et des Vaisseaux - Pratique
Biofutur
Biomedicine and Pharmacotherapy
Bulletin du Cancer
Cahiers Sante
Cahiers de Nutrition et de Dietetique
Cancer/Radiotherapie
Cellular and Molecular Biology
Chirurgie de la Main
Comptes Rendus - Biologies
Confrontations Psychiatriques
Diabetes and Metabolism
Dialogues in Cardiovascular Medicine
Dialogues in Clinical Neuroscience
Douleurs
EVRS Educational Electronic Journal
Encephale
Environnement, Risques et Sante
Epilepsies
Epileptic Disorders
Ethique et Sante
European Annals of Allergy and Clinical Immunology
European Cytokine Network
European Journal of Dermatology
European Journal of Medical Genetics
European Journal of Medicinal Chemistry
European Journal of Orthopaedic Surgery and Traumatology
European Psychiatry
Evolution Psychiatrique
Feuillets de Radiologie
Gastroenterologie Clinique et Biologique
Gynecologie Obstetrique Fertilite
Hematologie
Hepato-Gastro
Hernia
IARC Monographs on the Evaluation of Carcinogenic Risks to Humans
ITBM-RBM
ITBM-RBM News
Imagerie de la Femme
Immuno-Analyse et Biologie Specialisee
Information Psychiatrique
International Journal of Tuberculosis and Lung Disease
JEUR
Joint Bone Spine
Journal Francais d'Ophtalmologie
Journal International de Bioethique
Journal d'Economie Medicale
Journal de Gynecologie Obstetrique et Biologie de la Reproduction
Journal de Medecine Legale Droit Medical
Journal de Pediatrie et de Puericulture

Journal de Pharmacie Clinique
Journal de Radiologie
Journal de Traumatologie du Sport
Journal des Maladies Vasculaires
Journal of Drug Delivery Science and Technology
Journal of Neuroradiology
Journal of Nutrition, Health and Aging
Journal of Physiology Paris
Medecine Nucleaire
Medecine Therapeutique
Medecine Therapeutique - Cardio
Medecine Therapeutique Endocrinologie et Reproduction
Medecine Therapeutique Pediatrie
Medecine Tropicale
Medecine et Chirurgie du Pied
Medecine et Droit
Medecine et Maladies Infectieuses
Medecine/Sciences
Menopause Review
Microbes and Infection
Morphologie
Nephrologie et Therapeutique
Neuro-Psy News
Neurochirurgie
Neurophysiologie Clinique
Neuropsychiatrie de l'Enfance et de l'Adolescence
Nouvelles Dermatologiques
Nutrition Clinique et Metabolisme
Oncologie
Pathologie Biologie
Pharmacien Hospitalier
Phlebolymphology
Phytotherapie
Phytotherapie Europeenne
Prescrire International
Presse Medicale
Psychologie et NeuroPsychiatrie du Vieillissement
Reanimation
References en Gynecologie Obstetrique
Reproduction Humaine et Hormones
Reproduction Nutrition Development
Research and Practice in Alzheimer's Disease
Research in Microbiology
Revue Francaise d'Allergologie et d'Immunologie Clinique
Revue Francophone de Psycho-Oncologie
Revue Internationale de Pediatrie
Revue Neurologique
Revue d'Epidemiologie et de Sante Publique
Revue de Chirurgie Orthopedique et Reparatrice de l'Appareil Moteur
Revue de Geriatrie
Revue de Laryngologie Otologie Rhinologie
Revue de Medecine Interne
Revue de Pneumologie Clinique
Revue des Maladies Respiratoires
Revue du Praticien
Revue du Praticien - Gynecologie et Obstetrique
S.T.P. Pharma Pratiques
Sang Thrombose Vaisseaux
Sante Publique
Science and Sports
Surgical and Radiologic Anatomy
Transfusion Clinique et Biologique
Veterinary Research
Virologie

Germany DEU

Acta Diabetologica
Acta Histochemica
Acta Neuropathologica
Aktuelle Dermatologie
Aktuelle Ernahrungsmedizin
Aktuelle Neurologie
Aktuelle Neurologie, Supplement
Aktuelle Rheumatologie
Aktuelle Rheumatologie, Supplement
Aktuelle Traumatologie
Aktuelle Traumatologie, Supplement
Akupunktur und Traditionelle Chinesische Medizin
Allergo Journal
Allergologie
Allergy and Clinical Immunology International
Altex
Anaesthesist
Anasthesiologie Intensivmedizin Notfallmedizin Schmerztherapie
Anasthesiologie Intensivmedizin Notfallmedizin Schmerztherapie, Supplement
Anasthesiologie und Intensivmedizin
Anatomy and Embryology
Annals of Anatomy
Annals of Hematology
Annals of Hematology, Supplement
Applied Cardiopulmonary Pathophysiology
Applied Microbiology and Biotechnology
Arbeitsmedizin Sozialmedizin Umweltmedizin
Arbeitsmedizin Sozialmedizin Umweltmedizin, Supplement
Archiv der Pharmazie
Archiv fur Kriminologie
Archives of Dermatological Research
Archives of Dermatological Research, Supplement
Archives of Gynecology and Obstetrics
Archives of Microbiology
Archives of Orthopaedic and Trauma Surgery
Archives of Toxicology
Arzneimittel-Forschung/Drug Research
Arztezeitschrift fur Naturheilverfahren und Regulationsmedizin
Atemwegs- und Lungenkrankheiten
Basic Research in Cardiology
Basic Research in Cardiology, Supplement
Biological Chemistry
Biologische Medizin
Biomedizinische Technik
Bioprocess and Biosystems Engineering
Blutalkohol
CLB Chemie in Labor und Biotechnik
Cancer Chemotherapy and Pharmacology
Cancer Chemotherapy and Pharmacology, Supplement
Cancer Immunology, Immunotherapy
Cell and Tissue Research
ChemBioChem
Chemistry and Biodiversity
Chemotherapie Journal
Chemotherapie Journal, Supplement
Child's Nervous System
Chinesische Medizin
Chirurgische Praxis
Chromatographia

Chromosoma
Clinical Autonomic Research
Clinical Chemistry and Laboratory Medicine
Clinical Laboratory
Clinical Nephrology
Clinical Neuropathology
Clinical Research in Cardiology
Clinical and Experimental Medicine
Cognitive Processing
Colloid and Polymer Science
Coloproctology
Communication and Medicine
Current Genetics
Current Topics in Microbiology and Immunology
Dermatologie in Beruf und Umwelt
Deutsche Apotheker Zeitung
Deutsche Medizinische Wochenschrift
Deutsche Zeitschrift fur Akupunktur
Deutsche Zeitschrift fur Onkologie
Deutsches Arzteblatt
Development Genes and Evolution
Diabetes und Stoffwechsel
Diabetologia
Electrophoresis
Endoscopy
Engineering in Life Sciences
Ergotherapie und Rehabilitation
Ernahrungs Umschau
European Archives of Oto-Rhino-Laryngology
European Archives of Psychiatry and Clinical Neuroscience
European Biophysics Journal
European Child and Adolescent Psychiatry
European Child and Adolescent Psychiatry, Supplement
European Journal of Applied Physiology
European Journal of Cell Biology
European Journal of Clinical Microbiology and Infectious Diseases
European Journal of Clinical Pharmacology
European Journal of Health Economics
European Journal of Immunology
European Journal of Medical Research
European Journal of Nuclear Medicine and Molecular Imaging
European Journal of Nutrition
European Journal of Pediatric Surgery
European Journal of Pediatric Surgery, Supplement
European Journal of Pediatrics
European Journal of Pediatrics, Supplement
European Journal of Plastic Surgery
European Journal of Trauma
European Radiology
European Radiology, Supplement
European Spine Journal
Experimental Brain Research
Experimental and Clinical Endocrinology and Diabetes
Experimental and Toxicologic Pathology
Familiendynamik
Fortschritte der Neurologie Psychiatrie
Forum Logopadie
Forum der Psychoanalyse
Functional and Integrative Genomics
Fuss und Sprunggelenk
Geburtshilfe und Frauenheilkunde
Gefahrstoffe Reinhaltung der Luft
Gefasschirurgie
German Journal of Psychiatry
Gesundheitsokonomie und Qualitatsmanagement
Gesundheitswesen
Gesundheitswesen, Supplement
Graefe's Archive for Clinical and Experimental Ophthalmology
Gynakologe
Gynakologische Endokrinologie
Gynakologische Praxis
Gynecological Surgery
HNO
Hamostaseologie
Handchirurgie Mikrochirurgie Plastische Chirurgie
Haut
Hautarzt
Herz
Herz, Supplement
Herzschrittmachertherapie und Elektrophysiologie
Histochemistry and Cell Biology
Hormone and Metabolic Research
Hormone and Metabolic Research, Supplement
Human Genetics
Hygiene + Medizin
Immunobiology
Immunogenetics
Infection
Infection, Supplement
Intensiv- und Notfallbehandlung
Intensive Care Medicine
Intensive Care Medicine, Supplement
Intensivmedizin und Notfallmedizin
International Archives of Occupational and Environmental Health
International Journal of Clinical Pharmacology and Therapeutics
International Journal of Colorectal Disease
International Journal of Hygiene and Environmental Health
International Journal of Legal Medicine
International Journal of Life Cycle Assessment
International Journal of Medical Microbiology
International Journal of Sports Medicine
International Journal of Sports Medicine, Supplement
International Journal of Transgenderism
International Orthopaedics
Internist
Internistische Praxis
Invertebrate Neuroscience
Journal fur Anasthesie und Intensivbehandlung
Journal of Biological Inorganic Chemistry
Journal of Cancer Research and Clinical Oncology
Journal of Cancer Research and Clinical Oncology, Supplement
Journal of Experimental Animal Science
Journal of Industrial Microbiology and Biotechnology
Journal of Molecular Medicine
Journal of Molecular Modeling
Journal of Neurology
Journal of Neurology, Supplement
Journal of Orofacial Orthopedics
Journal of Perinatal Medicine
Journal of Psychophysiology
Journal of Public Health
Journal of Separation Science
Journal of Trace Elements in Medicine and Biology
Journal of Urban Health

Kardiotechnik
Kinderanalyse
Klinikarzt
Klinische Monatsblatter fur Augenheilkunde
Klinische Neurophysiologie
Klinische Neuroradiologie
Klinische Padiatrie
Kosmetische Medizin
Krankenhaushygiene und Infektionsverhutung
Krankenhauspharmazie
Krankenhauspsychiatrie
Krankenhauspsychiatrie, Supplement
LaboratoriumsMedizin
Langenbeck's Archives of Surgery
Laryngo- Rhino- Otologie
Lymphologie in Forschung und Praxis
MMW-Fortschritte der Medizin
Macromolecular Bioscience
Magnetic Resonance Materials in Physics, Biology and Medicine
Medical Laser Application
Medical Microbiology and Immunology
Medical and Biological Engineering and Computing
Medizinische Ausbildung
Medizinische Genetik
Medizinische Klinik
Medizinische Monatsschrift fur Pharmazeuten
Medizinische Welt
Medizintechnik
Methods of Information in Medicine
Minimal Invasive Chirurgie
Minimally Invasive Neurosurgery
Modern Aspects of Immunobiology
Molecular Genetics and Genomics
Monatsschrift fur Kinderheilkunde
Naturwissenschaften
Naunyn-Schmiedeberg's Archives of Pharmacology
Nervenarzt
Nervenheilkunde
Neuroforum
Neurogenetics
Neurologie und Rehabilitation
Neurology Psychiatry and Brain Research
Neuropediatrics
Neurophysiologie-Labor
Neuropsychiatrie
Neuroradiology
Neurosurgical Review
Nieren- und Hochdruckkrankheiten
Notarzt
Notfall und Hausarztmedizin
NuklearMedizin
Onkologe
Onkologie
Operative Orthopadie und Traumatologie
Ophthalmologe
Orthopade
Osteopathische Medizin
PPmP Psychotherapie Psychosomatik Medizinische Psychologie
PZ Prisma
Padiatrische Praxis
Parasitology Research
Pathologe
Pathology Research and Practice
Pediatric Nephrology
Pediatric Radiology
Pediatric Surgery International
Pflugers Archiv European Journal of Physiology
Pharma Fokus Herz-Kreislauf
Pharmacopsychiatry
Pharmazeutische Industrie
Pharmazeutische Zeitung
Pharmazie
Pharmazie in Unserer Zeit
Phlebologie
Physikalische Medizin Rehabilitationsmedizin Kurortmedizin
Phytomedicine
Planta Medica
Pneumologie
Pravention und Rehabilitation
Progress in Histochemistry and Cytochemistry
Proteomics
Psyche
Psychiatrische Praxis
Psychiatrische Praxis, Supplement
PsychoNeuro
Psychopharmacology
Psychopharmakotherapie
Psychopharmakotherapie, Supplement
Psychotherapeut
QSAR and Combinatorial Science
Radiologe
Recht und Psychiatrie
Rechtsmedizin
Rehabilitation
Rheumatology International
RoFo Fortschritte auf dem Gebiet der Rontgenstrahlen und der Bildgebenden Verfahren
Rontgenpraxis
Schmerz
Sexuologie
Signal Transduction
Skeletal Radiology
Sleep and Breathing
Social Psychiatry and Psychiatric Epidemiology
Sprache Stimme Gehor
Springer Seminars in Immunopathology
Strahlentherapie und Onkologie
Sucht
Suchtmedizin in Forschung und Praxis
Supportive Care in Cancer
Synlett
Synthesis
Systematic and Applied Microbiology
Tagliche Praxis
Theory in Biosciences
Thoracic and Cardiovascular Surgeon
Thoracic and Cardiovascular Surgeon, Supplement
Thrombosis and Haemostasis
Trace Elements and Electrolytes
TransplantLinc
Transplantationsmedizin: Organ der Deutschen Transplantationsgesellschaft
Tumor Diagnostik und Therapie
Ultraschall in der Medizin
Ultraschall in der Medizin, Supplement
Umweltmedizin in Forschung und Praxis
Umweltwissenschaften und Schadstoff-Forschung

Unfallchirurg
Urologe - Ausgabe A
Urological Research
Vasomed
Verdauungskrankheiten
Virchows Archiv
Viszeralchirurgie
Zeitschrift fur Allgemeinmedizin
Zeitschrift fur Arztliche Fortbildung und Qualitatssicherung
Zeitschrift fur Epileptologie
Zeitschrift fur Gastroenterologie
Zeitschrift fur Geburtshilfe und Neonatologie
Zeitschrift fur Geburtshilfe und Neonatologie, Supplement
Zeitschrift fur Gerontologie und Geriatrie
Zeitschrift fur Herz-, Thorax- und Gefasschirurgie
Zeitschrift fur Klinische Psychologie und Psychotherapie
Zeitschrift fur Larmbekampfung
Zeitschrift fur Medizinische Physik
Zeitschrift fur Orthopadie und Ihre Grenzgebiete
Zeitschrift fur Phytotherapie
Zeitschrift fur Psychosomatische Medizin und Psychotherapie
Zeitschrift fur Rheumatologie
Zeitschrift fur Rheumatologie, Supplement
Zeitschrift fur Umweltmedizin
Zeitschrift fur Wundheilung
Zentralblatt fur Arbeitsmedizin, Arbeitsschutz und Ergonomie
Zentralblatt fur Chirurgie
Zentralblatt fur Chirurgie, Supplement
Zentralblatt fur Gynakologie
Zentralblatt fur Neurochirurgie
Zentralsterilisation - Central Service

Greece GRC

Acta Microbiologica Hellenica
Anticancer Research
Archives of Hellenic Medicine
Cancer Genomics and Proteomics
Epitheorese Klinikes Farmakologias kai Farmakokinetikes
HAEMA
Hellenic Journal of Cardiology
Hellenic Journal of Gastroenterology
Hellenic Journal of Nuclear Medicine
Hepato-Gastroenterology
Hippokratia
In Vivo
Journal of B.U.ON.
Journal of Musculoskeletal Neuronal Interactions
Review of Clinical Pharmacology and Pharmacokinetics, International Edition
Surgical Chronicles

Hong Kong HKG

Asian Journal of Ophthalmology
Asian Journal of Surgery
Hong Kong Dermatology and Venereology Bulletin
Hong Kong Journal of Dermatology and Venereology
Hong Kong Journal of Nephrology
Hong Kong Journal of Occupational Therapy
Hong Kong Journal of Paediatrics
Hong Kong Journal of Psychiatry
Hong Kong Medical Journal
Hong Kong Physiotherapy Journal
Hong Kong Practitioner
Journal of the Formosan Medical Association
Journal of the Hong Kong College of Radiologists
Kaohsiung Journal of Medical Sciences
Oncology Forum
Santo Tomas Journal of Medicine
Taiwanese Journal of Obstetrics and Gynecology

Hungary HUN

Acta Biologica Szegediensis
Acta Pharmaceutica Hungarica
Acta Physiologica Hungarica
CME Journal of Gynecologic Oncology
Journal of Planar Chromatography - Modern TLC
Lege Artis Medicine
Magyar Noorvosok Lapja
Magyar Urologia
Orvosi Hetilap
Orvoskepzes
Pathology Oncology Research

India IND

Anil Aggrawal's Internet Journal of Forensic Medicine and Toxicology
Asian Journal of Microbiology, Biotechnology and Environmental Sciences
Asian Oceanian Journal of Radiology
Biomedical Research
Biomedicine
Biosciences Biotechnology Research Asia
Bull RGKMC
Bulletin, Postgraduate Institute of Medical Education and Research, Chandigarh
Current Pediatric Research
Current Topics in Neurochemistry
Current Topics in Peptide and Protein Research
Current Topics in Pharmacology
Current Topics in Toxicology
Current Trends in Immunology
Dengue Bulletin
Health and Population: Perspectives and Issues
Indian Drugs
Indian Heart Journal
Indian Journal of Biotechnology
Indian Journal of Clinical Biochemistry
Indian Journal of Critical Care Medicine
Indian Journal of Dermatology, Venereology and Leprology
Indian Journal of Gastroenterology
Indian Journal of Hematology and Blood Transfusion
Indian Journal of Human Genetics
Indian Journal of Leprosy
Indian Journal of Medical Microbiology
Indian Journal of Medical Research
Indian Journal of Medical Research, Supplement
Indian Journal of Medical Sciences
Indian Journal of Occupational and Environmental Medicine

Indian Journal of Ophthalmology
Indian Journal of Otolaryngology and Head and Neck Surgery
Indian Journal of Otology
Indian Journal of Pathology and Microbiology
Indian Journal of Pediatrics
Indian Journal of Pharmacology
Indian Journal of Physiology and Pharmacology
Indian Journal of Practical Pediatrics
Indian Journal of Radiology and Imaging
Indian Pacing and Electrophysiology Journal
Indian Pediatrics
International Journal of Medical Toxicology and Legal Medicine
JK Practitioner
JK Science
JMS - Journal of Medical Society
Journal International Medical Sciences Academy
Journal of Anaesthesiology Clinical Pharmacology
Journal of Biosciences
Journal of Ecophysiology and Occupational Health
Journal of Environmental Biology
Journal of Forensic Medicine and Toxicology
Journal of Indian Association for Child and Adolescent Mental Health
Journal of Internal Medicine of India
Journal of Natural Remedies
Journal of Neonatology
Journal of Pediatric Neurosciences
Journal of Postgraduate Medicine
Journal of Vector Borne Diseases
Journal of the Diabetic Association of India
Journal of the Indian Chemical Society
Journal of the Indian Medical Association
Journal, Indian Academy of Clinical Medicine
Medical Journal Armed Forces India
Medico-Legal Update
National Medical Journal of India
Neurology India
Online Journal of Health and Allied Sciences
Perinatology
Phytomedica
Toxicology International
Ultrasound International

Indonesia IDN

Critical Care and Shock

Iran IRN

Acta Medica Iranica
Archives of Iranian Medicine
Daru
Iranian Biomedical Journal
Iranian Journal of Allergy, Asthma and Immunology
Iranian Journal of Diabetes and Lipid Disorders
Iranian Journal of Immunology
Iranian Journal of Medical Sciences
Iranian Journal of Nuclear Medicine
Iranian Journal of Pharmacology and Therapeutics
Iranian Journal of Public Health
Iranian Journal of Radiation Research
Journal of Medicinal Plants
Journal of Research in Medical Sciences
Koomesh
Medical Journal of the Islamic Republic of Iran
Yakhteh

Ireland IRL

Ageing Research Reviews
Archives of Gerontology and Geriatrics
Atherosclerosis
Atherosclerosis Supplements
Auris Nasus Larynx
BioSystems
Cancer Letters
Chemico-Biological Interactions
Chemistry and Physics of Lipids
Clinical Neurophysiology
Computer Methods and Programs in Biomedicine
Critical Reviews in Oncology/Hematology
Diabetes Research and Clinical Practice
Drug and Alcohol Dependence
Early Human Development
European Journal of Obstetrics Gynecology and Reproductive Biology
European Journal of Radiology
Forensic Science International
Health Policy
Hepatology Research
International Journal of Cardiology
International Journal of Gynecology and Obstetrics
International Journal of Medical Informatics
International Journal of Pediatric Otorhinolaryngology
Irish Journal of Medical Science
Irish Journal of Psychological Medicine
Irish Medical Journal
Journal of Dermatological Science
Journal of Ethnopharmacology
Journal of Reproductive Immunology
Lung Cancer
Maturitas
Mechanisms of Ageing and Development
Mechanisms of Development
Molecular and Cellular Endocrinology
Neuroscience Letters
Neuroscience Research
Parasitology International
Patient Education and Counseling
Progress in Pediatric Cardiology
Psychiatry Research
Psychiatry Research - Neuroimaging
Radiotherapy and Oncology
Resuscitation
Toxicology
Toxicology Letters

Israel ISR

Drug Metabolism and Drug Interactions
Harefuah
International Journal of Adolescent Medicine and Health

Israel Medical Association Journal
Journal of Basic and Clinical Physiology and Pharmacology
Journal of Endocrine Genetics
Journal of Pediatric Endocrinology and Metabolism
Medicine and Law
Pediatric Endocrinology Reviews
Public Health Reviews
Reviews in the Neurosciences
Reviews on Environmental Health

Italy ITA

Acta Anaesthesiologica Italica / Anaesthesia and Intensive Care in Italy
Acta Biomedica de l'Ateneo Parmense
Acta Geneticae Medicae et Gemellologiae
Acta Medica Auxologica
Acta Medica Mediterranea
Acta Medica Romana
Acta Myologica
Aggiornamento Pediatrico
Aging - Clinical and Experimental Research
Annali Degli Ospedali San Camillo e Forlanini
Annali Italiani di Dermatologia Allergologica Clinica e Sperimentale
Annali Italiani di Medicina Interna
Annali Italiani di Medicina Interna, Supplement
Annali dell'Istituto Superiore di Sanita
Annali di Ottalmologia e Clinica Oculistica
Archives Italiennes de Biologie
Archivio Italiano di Urologia e Andrologia
Archivio di Ostetricia e Ginecologia
Argomenti di Gastroenterologia Clinica
Artroscopia
Bollettino - Lega Italiana contro l'Epilessia
Bollettino Chimico Farmaceutico
Chirurgia
Chirurgia del Piede
Clinica Terapeutica
Clinical Cases in Mineral and Bone Metabolism
Clinical Neuropsychiatry
Clinical and Experimental Rheumatology
Confinia Cephalalgica
Cortex
Dermatologia Clinica
Diabetes, Nutrition and Metabolism - Clinical and Experimental
EOS Rivista di Immunologia ed Immunofarmacologia
Eating and Weight Disorders
Epidemiologia e Psichiatria Sociale
Europa Medicophysica
European Journal of Histochemistry
European Journal of Inflammation
European Journal of Oncology
European Journal of Ophthalmology
European Journal of Pediatric Dermatology
European Review for Medical and Pharmacological Sciences
FORUM - Trends in Experimental and Clinical Medicine
Functional Neurology
GIMT - Giornale Italiano delle Malattie del Torace
Gastroenterology International
Gazzetta Medica Italiana Archivio per le Scienze Mediche
Geriatric and Medical Intelligence
Giornale Italiano di Cardiologia
Giornale Italiano di Dermatologia e Venereologia
Giornale Italiano di Diabetologia e Metabolismo
Giornale Italiano di Endoscopia Digestiva
Giornale Italiano di Farmacia Clinica
Giornale Italiano di Medicina Tropicale
Giornale Italiano di Medicina del Lavoro ed Ergonomia
Giornale Italiano di Ostetricia e Ginecologia
Giornale di Gerontologia
HIP International
Haematologica
Haematologica, Supplement
Heroin Addiction and Related Clinical Problems
IJCI - International Journal of Clinical Investigation
Imago
Impegno Ospedaliero, Sezione Scientifica
Infezioni in Medicina
Internal Medicine Clinical and Laboratory
International Angiology
International Journal of Artificial Organs
International Journal of Biological Markers
International Journal of Immunopathology and Pharmacology
International Journal of Medicine, Biology and the Environment
Internista
Interventional Neuroradiology
Italian Heart Journal
Italian Heart Journal Supplement
Italian Heart Journal, Supplement
Italian Journal of Allergy and Clinical Immunology
Italian Journal of Gynaecology and Obstetrics
Italian Journal of Pediatrics
Italian Journal of Psychiatry and Behavioural Sciences
Italian Journal of Psychopathology
Italian Journal of Vascular and Endovascular Surgery
Journal of Applied Biomaterials and Biomechanics
Journal of Applied Cosmetology
Journal of Biological Regulators and Homeostatic Agents
Journal of Cardiovascular Surgery
Journal of Chemotherapy
Journal of Endocrinological Investigation
Journal of Experimental and Clinical Cancer Research
Journal of Headache and Pain
Journal of Mental Health Policy and Economics
Journal of Nephrology
Journal of Neurosurgical Sciences
Journal of Orthopaedics and Traumatology
Journal of Plastic Dermatology
Journal of Preventive Medicine and Hygiene
Journal of Sports Medicine and Physical Fitness
Journal of Vascular Access
Journal of the Pancreas
Lotta Contro la Tuberculosi e le Malattie Polmonari Sociali
Medicina Clinica e Termale
Medicina del Lavoro
Medicina dello Sport
Medico e Bambino
Mediterranean Journal of Pacing and Electrophysiology
Minerva Anestesiologica
Minerva Biotecnologica
Minerva Cardioangiologica
Minerva Chirurgica
Minerva Gastroenterologica e Dietologica

Minerva Ginecologica
Minerva Medica
Minerva Ortopedica e Traumatologica
Minerva Pediatrica
Minerva Pneumologica
Minerva Psichiatrica
Minerva Urologica e Nefrologica
Monaldi Archives for Chest Disease - Cardiac Series
Monaldi Archives for Chest Disease - Pulmonary Series
Neurological Sciences
Ospedale Maggiore
Otorinolaringologia
Otorinolaringologia Pediatrica
Pathologica
Pediatria Medica e Chirurgica
Pediatria Oggi Medica e Chirurgica
PharmacoEconomics - Italian Research Articles
Physica Medica
Policlinico - Sezione Medica
Progress in Nutrition
Progressi in Reumatologia
Psicoterapia Cognitiva e Comportamentale
Psicoterapia e Scienze Umane
Quaderni ACP
Quarterly Journal of Nuclear Medicine and Molecular Imaging
Radiologia Medica
Rassegna di Patologia dell'Apparato Respiratorio
Rays - International Journal of Radiological Sciences
Recenti Progressi in Medicina
Reumatismo
Reviews in Clinical and Experimental Hematology
Ricerca e Pratica
Ricerca in Psicoterapia
Ricerche in Psichiatria
Rivista Italiana di Chirurgia Plastica
Rivista Italiana di Medicina dell'Adolescenza
Rivista Italiana di Nutrizione Parenterale ed Enterale
Rivista Italiana di Ossigeno-Ozonoterapia
Rivista Italiana di Otorinolaringologia Audiologia e Foniatria
Rivista Medica
Rivista di Medicina di Laboratorio
Rivista di Neurobiologia
Rivista di Neuroradiologia
Rivista di Ostetricia e Ginecologia
Rivista di Psichiatria
SAGGI - Child Development and Disabilities
Sarcoidosis Vasculitis and Diffuse Lung Diseases
Techniques in Coloproctology
Tecniche Chirurgiche in Ortopedia e Traumatologia
Thyroidology Clinical and Experimental
Trasfusione del Sangue
Trends in Medicine
Tumori
Urodinamica
Urogynaecologia International Journal

Jamaica JAM

West Indian Medical Journal

Japan JPN

Acta Criminologiae et Medicinae Legalis Japonica
Acta Hepatologica Japonica
Acta Histochemica et Cytochemica
Acta Medica Nagasakiensia
Acta Medica et Biologica
Acta Urologica Japonica
Advances in Biophysics
Allergology International
Anesthesia and Resuscitation
Annals of Nuclear Medicine
Annual Report of Shionogi Research Laboratory
Archives of Histology and Cytology
Asian Journal of Oral and Maxillofacial Surgery
Biocontrol Science
Biological and Pharmaceutical Bulletin
Biomedical Research
Biotherapy
Brain Tumor Pathology
Brain and Nerve
Bulletin of National Institute of Health Sciences
Bulletin of the Chemical Society of Japan
Bulletin of the Physical Fitness Research Institute
Cell Structure and Function
Chemical and Pharmaceutical Bulletin
Circulation Journal
Clinical Neurology
Clinical Pediatric Endocrinology
Clinical and Experimental Nephrology
Dokkyo Journal of Medical Sciences
Endocrine Journal
Endoscopic Forum for Digestive Disease
Environmental Health and Preventive Medicine
Equilibrium Research
Equilibrium Research, Supplement
Esophagus
Experimental Animals
Folia Ophthalmologica Japonica
Folia Pharmacologica Japonica
Gastric Cancer
Gastroenterological Endoscopy
Genes and Genetic Systems
Heart and Vessels
Hirosaki Medical Journal
Hiroshima Journal of Medical Sciences
Hokkaido Journal of Orthopaedics and Traumatology
Hokuriku Journal of Anesthesiology
Hypertension Research
IRYO - Japanese Journal of National Medical Services
Industrial Health
Internal Medicine
International Heart Journal
International Journal of Clinical Oncology
International Medical Journal
Japan Journal of Logopedics and Phoniatrics
Japan Medical Association Journal
Japanese Heart Journal
Japanese Journal of Aerospace and Environmental Medicine
Japanese Journal of Allergology
Japanese Journal of Anesthesiology
Japanese Journal of Antibiotics
Japanese Journal of Chemotherapy
Japanese Journal of Chest Diseases

Japanese Journal of Clinical Chemistry
Japanese Journal of Clinical Ophthalmology
Japanese Journal of Clinical Pharmacology and Therapeutics
Japanese Journal of Clinical Radiology
Japanese Journal of Clinical Urology
Japanese Journal of Fertility and Sterility
Japanese Journal of Forensic Toxicology
Japanese Journal of Gastroenterological Surgery
Japanese Journal of Gastroenterology
Japanese Journal of Infectious Diseases
Japanese Journal of Interventional Cardiology
Japanese Journal of Legal Medicine
Japanese Journal of Leprosy
Japanese Journal of Lung Cancer
Japanese Journal of Medical Electronics and Biological Engineering
Japanese Journal of Medical Mycology
Japanese Journal of Nephrology
Japanese Journal of Neuropsychopharmacology
Japanese Journal of Neurosurgery
Japanese Journal of Ophthalmology
Japanese Journal of Physiology
Japanese Journal of Plastic and Reconstructive Surgery
Japanese Journal of Thoracic and Cardiovascular Surgery
Japanese Journal of Urology
Japanese Journal of Veterinary Research
Japanese Pharmacology and Therapeutics
Journal of Anesthesia
Journal of Antibiotics
Journal of Artificial Organs
Journal of Biochemistry
Journal of Bone and Mineral Metabolism
Journal of Cardiology
Journal of Clinical Biochemistry and Nutrition
Journal of Gastroenterology
Journal of General and Applied Microbiology
Journal of Health Science
Journal of Hepato-Biliary-Pancreatic Surgery
Journal of Human Genetics
Journal of Infection and Chemotherapy
Journal of JASTRO
Journal of Japanese Dental Society of Anesthesiology
Journal of Medical Investigation
Journal of Medical Ultrasonics
Journal of Nara Medical Association
Journal of Nippon Medical School
Journal of Nutritional Science and Vitaminology
Journal of Obstetrics and Gynaecology Research
Journal of Occupational Health
Journal of Orthopaedic Science
Journal of Otolaryngology of Japan
Journal of Pharmacological Sciences
Journal of Smooth Muscle Research
Journal of Tokyo Medical University
Journal of Toxicological Sciences
Journal of UOEH
Journal of the Food Hygienic Society of Japan
Journal of the Japan Diabetes Society
Journal of the Japan Epilepsy Society
Journal of the Medical Society of Toho University
Journal of the Osaka City Medical Center
Journal of the Showa Medical Association
Journal of the Wakayama Medical Society
Kakuigaku
Keio Journal of Medicine
Kekkaku
Kitakanto Medical Journal
Kobe Journal of Medical Sciences
Kumamoto Medical Journal
Kurume Medical Journal
Medical Electron Microscopy
Medical Journal of Kobe University
Medical Journal of Minami Osaka Hospital
Microbiology and Immunology
Mie Medical Journal
Modern Rheumatology
Natural Medicines
Neuro-Ophthalmology Japan
Neurologia Medico-Chirurgica
Neurological Surgery
Nishinihon Journal of Dermatology
Nishinihon Journal of Urology
No To Hattatsu
Oral Radiology
Oral Therapeutics and Pharmacology
Oto-Rhino-Laryngology Tokyo
Otolaryngology - Head and Neck Surgery (Tokyo)
Pharmacometrics
Practica Oto-Rhino-Laryngologica
Practica Otologica, Supplement
Publications of Takeda Research Laboratories
Radiation Medicine - Medical Imaging and Radiation Oncology
Respiration and Circulation
Sapporo Medical Journal
Skin Research
Surgery Today
Teikyo Medical Journal
Therapeutic Research
Tohoku Journal of Experimental Medicine
Tokai Journal of Experimental and Clinical Medicine
Tokyo Jikeikai Medical Journal
Tumor Research
Yakugaku Zasshi
Yokohama Medical Journal
Yonago Acta Medica

Jordan JOR

Jordan Journal of Applied Sciences - Natural Sciences
Jordan Medical Journal

Kenya KEN

African Journal of Biotechnology
East and Central African Journal of Pharmaceutical Sciences

Lebanon LBN

Journal Medical Libanais

Macedonia MKD

Balkan Journal of Medical Genetics

Malaysia MYS

Asia-Pacific Journal of Public Health
Malaysian Journal of Medical Sciences
Medical Journal of Malaysia

Mexico MEX

Anestesia en Mexico
Annals of Hepatology
Archivos de Cardiologia de Mexico
Archivos de Neurociencias
Dermatologia Cosmetica, Medica y Quirurgica
Dermatologia Revista Mexicana
Enfermedades Infecciosas y Microbiologia
Gaceta Medica de Mexico
Ginecologia y Obstetricia de Mexico
Investigacion Medica Internacional
Medicina Interna de Mexico
Practica Pediatrica
Revista Latinoamericana de Microbiologia
Revista Medica de Aguascalientes
Revista Mexicana de Anestesiologia
Revista Mexicana de Ciencias Farmaceuticas
Revista Mexicana de Oftalmologia
Revista de Investigacion Clinica
Revista del Instituto Nacional de Enfermedades Respiratorias
Salud Mental
Salud Publica de Mexico

Nepal NPL

Journal of the Nepal Medical Association
Kathmandu University Medical Journal

Netherlands NLD

Acta Tropica
Advanced Drug Delivery Reviews
Aerobiologia
American Journal of Psychoanalysis
Analytica Chimica Acta
Angiogenesis
Anti-Cancer Agents in Medicinal Chemistry
Anti-Infective Agents in Medicinal Chemistry
Anti-Inflammatory and Anti-Allergy Agents in Medicinal Chemistry
Antiviral Research
Antonie van Leeuwenhoek, International Journal of General and Molecular Microbiology
Apoptosis
Aquatic Toxicology
Artificial Intelligence in Medicine
Autoimmunity Reviews
Autonomic Neuroscience: Basic and Clinical
Behavioural Brain Research
Behavioural Neurology
Behavioural Processes
Bio-Medical Materials and Engineering
BioFactors
BioMetals
Biochemical Engineering Journal
Biochimica et Biophysica Acta - Bioenergetics
Biochimica et Biophysica Acta - Biomembranes
Biochimica et Biophysica Acta - Gene Structure and Expression
Biochimica et Biophysica Acta - General Subjects
Biochimica et Biophysica Acta - Molecular Basis of Disease
Biochimica et Biophysica Acta - Molecular Cell Research
Biochimica et Biophysica Acta - Molecular and Cell Biology of Lipids
Biochimica et Biophysica Acta - Proteins and Proteomics
Biochimica et Biophysica Acta - Reviews on Cancer
Biochimie
Bioelectrochemistry
Biogenic Amines
Biogerontology
Biological Psychology
Biomolecular Engineering
Biophysical Chemistry
Biorheology
Biotechnology Annual Review
Brain Research
Brain Research Reviews
Brain and Development
Breast Disease
CNS and Neurological Disorders - Drug Targets
COSSMA
Cancer Causes and Control
Cancer and Metastasis Reviews
Cardiovascular Engineering
Cardiovascular Research
Cardiovascular and Hematological Agents in Medicinal Chemistry
Cardiovascular and Hematological Disorders - Drug Targets
Cell Biology and Toxicology
Cell and Tissue Banking
Cellular Oncology
Central Nervous System Agents in Medicinal Chemistry
Chemometrics and Intelligent Laboratory Systems
Chromosome Research
Clinica Chimica Acta
Clinical Hemorheology and Microcirculation
Clinical Neurology and Neurosurgery
Clinical and Experimental Metastasis
Cognition
Cognitive Systems Research
Colloids and Surfaces A: Physicochemical and Engineering Aspects
Colloids and Surfaces B: Biointerfaces
Combinatorial Chemistry and High Throughput Screening
Community Mental Health Journal
Contact Lens and Anterior Eye
Current Alzheimer Research
Current Cancer Drug Targets
Current Cancer Therapy Reviews
Current Cardiology Reviews
Current Computer-Aided Drug Design
Current Diabetes Reviews
Current Drug Delivery
Current Drug Discovery Technologies
Current Drug Metabolism
Current Drug Targets
Current Enzyme Inhibition

Current Gene Therapy
Current Genomics
Current HIV Research
Current Hypertension Reviews
Current Immunology Reviews
Current Medical Imaging Reviews
Current Medicinal Chemistry
Current Molecular Medicine
Current Nanoscience
Current Neuropharmacology
Current Neurovascular Research
Current Nutrition and Food Science
Current Organic Chemistry
Current Pediatric Reviews
Current Pharmaceutical Analysis
Current Pharmaceutical Biotechnology
Current Pharmaceutical Design
Current Pharmacogenomics
Current Protein and Peptide Science
Current Psychiatry Reviews
Current Respiratory Medicine Reviews
Current Rheumatology Reviews
Current Topics in Medicinal Chemistry
Current Vascular Pharmacology
Current Women's Health Reviews
Cytotechnology
DNA Repair
Disease Markers
Documenta Ophthalmologica
Drug Design Reviews Online
EAU Update Series
Economics and Human Biology
Ecotoxicology
Endocrine, Metabolic and Immune Disorders - Drug Targets
Enhancer - Biotherapy of Cancer
Environmental Monitoring and Assessment
Environmental Toxicology and Pharmacology
Epilepsy Research
European Journal of Cardio-thoracic Surgery
European Journal of Cardiovascular Nursing
European Journal of Epidemiology
European Journal of General Practice
European Journal of Health Law
European Journal of Heart Failure
European Journal of Heart Failure, Supplement
European Journal of Internal Medicine
European Journal of Pharmaceutical Sciences
European Journal of Pharmaceutics and Biopharmaceutics
European Journal of Pharmacology
European Neuropsychopharmacology
European Urology
European Urology, Supplements
FEBS Letters
Familial Cancer
Fish Physiology and Biochemistry
Fitoterapia
Focus on Parkinson's Disease
Frontiers in Drug Design and Discovery
Frontiers in Medicinal Chemistry
Gait and Posture
Gene
Gene Expression Patterns
Gene Therapy and Regulation
Geneeskunde en Sport
Geneesmiddelenbulletin
Glycoconjugate Journal
HEC Forum
Hart Bulletin
Health Care Analysis
Health Services and Outcomes Research Methodology
Hearing Research
Heart Failure Reviews
Heart Rhythm
Huisarts en Wetenschap
Human Antibodies
Human Movement Science
Immunology Letters
Immunology, Endocrine and Metabolic Agents in Medicinal Chemistry
In Silico Biology
Infection, Genetics and Evolution
Infectious Disorders - Drug Targets
Inflammation and Allergy - Drug Targets
Inflammopharmacology
International Immunopharmacology
International Journal of Antimicrobial Agents
International Journal of Biological Macromolecules
International Journal of Cardiovascular Imaging
International Journal of Drug Policy
International Journal of Food Microbiology
International Journal of Industrial Ergonomics
International Journal of Infectious Diseases
International Journal of Pharmaceutics
International Journal of Psychophysiology
International Journal of Risk and Safety in Medicine
International Ophthalmology
International Urology and Nephrology
Isokinetics and Exercise Science
JACR Journal of the American College of Radiology
Journal de Mycologie Medicale
Journal of Affective Disorders
Journal of Alzheimer's Disease
Journal of Atmospheric Chemistry
Journal of Back and Musculoskeletal Rehabilitation
Journal of Biochemical and Biophysical Methods
Journal of Biological Physics
Journal of Biomaterials Science, Polymer Edition
Journal of Biomolecular NMR
Journal of Bioscience and Bioengineering
Journal of Biotechnology
Journal of Cardiothoracic-Renal Research
Journal of Chemical Neuroanatomy
Journal of Chromatography A
Journal of Chromatography B: Analytical Technologies in the Biomedical and Life Sciences
Journal of Clinical Monitoring and Computing
Journal of Clinical Virology
Journal of Community Health
Journal of Computational Neuroscience
Journal of Computer-Aided Molecular Design
Journal of Contaminant Hydrology
Journal of Controlled Release
Journal of Cystic Fibrosis
Journal of Hazardous Materials
Journal of Health Economics
Journal of Hepatology
Journal of Immunological Methods
Journal of Inherited Metabolic Disease
Journal of Interventional Cardiac Electrophysiology
Journal of Materials Science: Materials in Medicine

Journal of Membrane Science
Journal of Men's Health and Gender
Journal of Microbiological Methods
Journal of Minimally Invasive Gynecology
Journal of Molecular Catalysis B: Enzymatic
Journal of Molecular Histology
Journal of Muscle Research and Cell Motility
Journal of Neurocytology
Journal of Neuroimmunology
Journal of Neuroscience Methods
Journal of Pharmaceutical and Biomedical Analysis
Journal of Photochemistry and Photobiology B: Biology
Journal of Radioanalytical and Nuclear Chemistry
Journal of Structural and Functional Genomics
Journal of Thrombosis and Thrombolysis
Journal of Vestibular Research: Equilibrium and Orientation
Journal of Veterinary Cardiology
Journal of Virological Methods
Journal of Vocational Rehabilitation
Journal of the Neurological Sciences
Knee
Legal Medicine
Letters in Drug Design and Discovery
Matrix Biology
MedicaMundi
Medical Image Analysis
Medicinal Chemistry
Mental Health Services Research
Methods in Cell Science
Mini-Reviews in Medicinal Chemistry
Mitochondrion
Molecular Biology Reports
Molecular Brain Research
Molecular Diversity
Molecular and Biochemical Parasitology
Molecular and Cellular Biochemistry
Mutation Research - Fundamental and Molecular Mechanisms of Mutagenesis
Mutation Research - Genetic Toxicology and Environmental Mutagenesis
Mutation Research - Reviews in Mutation Research
Nano Today
Nederlands Tijdschrift voor Anesthesiologie
Nederlands Tijdschrift voor Dermatologie en Venereologie
Nederlands Tijdschrift voor Geneeskunde
Nederlands Tijdschrift voor Keel- Neus- Oorheelkunde
Nederlands Tijdschrift voor Klinische Chemie en Laboratoriumgeneeskunde
Nederlands Tijdschrift voor Obstetrie en Gynaecologie
Nederlands Tijdschrift voor Urologie
Netherlands Heart Journal
Netherlands Journal of Medicine
NeuroRehabilitation
NeuroToxicology
Neurocomputing
Pain
Pain Clinic
Pathophysiology
Pharmaceutisch Weekblad
Pharmacy World and Science
Photodiagnosis and Photodynamic Therapy
Phytochemistry Reviews
Powder Technology
Prevention and Control
Primary Care Respiratory Journal
Progress in Brain Research
Public Health
Quality of Life Research
Regulatory Peptides
Resources, Conservation and Recycling
Respiratory Physiology and Neurobiology
Restorative Neurology and Neuroscience
Reviews in Endocrine and Metabolic Disorders
Reviews in Gynaecological and Perinatal Practice
Reviews in Toxicology
Rhinology
Safety Science
Schizophrenia Research
Science of the Total Environment
Sleep Medicine
Spectroscopy
Technology and Disability
Technology and Health Care
Theoretical Medicine and Bioethics
Tijdschrift voor Fertiliteitsonderzoek
Tijdschrift voor Gerontologie en Geriatrie
Tijdschrift voor Kindergeneeskunde
Tijdschrift voor Psychiatrie
Tijdschrift voor Verpleeghuisgeneeskunde
TrAC - Trends in Analytical Chemistry
Transactions of the Royal Society of Tropical Medicine and Hygiene
Transplant Immunology
Ultramicroscopy
Ultrasonics Sonochemistry
Vascular Disease Prevention
Veterinary Immunology and Immunopathology
Veterinary Microbiology
Veterinary Parasitology
Veterinary Quarterly
Veterinary Research Communications
Virus Genes
Virus Research
Water, Air, and Soil Pollution
Work

New Zealand NZL

American Journal of Cancer
American Journal of Cardiovascular Drugs
American Journal of Clinical Dermatology
American Journal of Drug Delivery
Applied Bioinformatics
Applied Biotechnology, Food Science and Policy
Applied Genomics and Proteomics
Applied Health Economics and Health Policy
BioDrugs
CNS Drugs
Cancer Informatics
Clinical Drug Investigation
Clinical Pharmacokinetics
Disease Management and Health Outcomes
Drug Safety
Drugs
Drugs and Aging
Drugs and Therapy Perspectives
Drugs in R and D
Evidence-Based Integrative Medicine

Fluoride
Healthcare Review Online
High Blood Pressure and Cardiovascular Prevention
International Journal of Pharmaceutical Medicine
Molecular Diagnosis and Therapy
Neuropsychiatric Disease and Treatment
New Zealand Journal of Medical Laboratory Science
New Zealand Medical Journal
New Zealand Public Health Surveillance Report
Pediatric Drugs
PharmacoEconomics
Sports Medicine
Therapeutics and Clinical Risk Management
Therapie
Toxicological Reviews
Treatments in Endocrinology
Treatments in Respiratory Medicine
Vaccines: Children and Practice
Vascular Health and Risk Management

Nigeria NGA

African Journal of Traditional, Complementary and Alternative Medicines
Annals of African Medicine
Journal of Phytomedicine and Therapeutics
Nigerian Journal of Surgical Research
West African Journal of Medicine

Norway NOR

Acta Odontologica Scandinavica
Acta Oncologica
Acta Oncologica, Supplement
Acta Oto-Laryngologica
Acta Oto-Laryngologica, Supplement
Acta Paediatrica, International Journal of Paediatrics
Acta Paediatrica, International Journal of Paediatrics, Supplement
Advances in Physiotherapy
Amyotrophic Lateral Sclerosis and Other Motor Neuron Disorders
Annals of Medicine
Blood Pressure
Blood Pressure, Supplement
Cognitive Behaviour Therapy
International Journal of Disaster Medicine
Nordic Journal of Psychiatry
Nordic Journal of Psychiatry, Supplement
Nordisk Tidsskrift for Biologisk Medisin
Norsk Epidemiologi
Scandinavian Cardiovascular Journal
Scandinavian Cardiovascular Journal, Supplement
Scandinavian Journal of Clinical and Laboratory Investigation
Scandinavian Journal of Clinical and Laboratory Investigation, Supplement
Scandinavian Journal of Gastroenterology
Scandinavian Journal of Gastroenterology, Supplement
Scandinavian Journal of Infectious Diseases
Scandinavian Journal of Infectious Diseases, Supplement
Scandinavian Journal of Occupational Therapy
Scandinavian Journal of Plastic and Reconstructive Surgery and Hand Surgery
Scandinavian Journal of Plastic and Reconstructive Surgery and Hand Surgery, Supplement
Scandinavian Journal of Primary Health Care
Scandinavian Journal of Primary Health Care, Supplement
Scandinavian Journal of Public Health
Scandinavian Journal of Public Health, Supplement
Scandinavian Journal of Rheumatology
Scandinavian Journal of Rheumatology, Supplement
Scandinavian Journal of Urology and Nephrology
Scandinavian Journal of Urology and Nephrology, Supplement
Tidsskrift for den Norske Laegeforening
World Journal of Biological Psychiatry

Pakistan PAK

International Journal of Cancer Research
International Journal of Pharmacology
International Journal of Virology
Journal of Biological Sciences
Journal of Liaquat University of Medical and Health Sciences
Journal of Medical Sciences
Journal of Nephrology, Urology and Transplantation
Journal of Pakistan Association of Dermatologists
Journal of Postgraduate Medical Institute
Journal of the College of Physicians and Surgeons Pakistan
Journal of the Pakistan Medical Association
Medical Forum Monthly
Pakistan Journal of Medical Sciences
Pakistan Journal of Nutrition

Poland POL

Acta Angiologica
Acta Haematologica Polonica
Acta Neurobiologiae Experimentalis
Acta Neuropsychologica
Acta Poloniae Pharmaceutica - Drug Research
Acta Toxicologica
Acta of Bioengineering and Biomechanics
Advances in Clinical and Experimental Medicine
Aktualnosci Neurologiczne
Alergia Astma Immunologia
Anestezjologia Intensywna Terapia
Annales Academiae Medicae Gedanensis
Annals of Agricultural and Environmental Medicine
Archives of Medical Science
Archives of Psychiatry and Psychotherapy
Case Reports and Clinical Practice Review
Cellular and Molecular Biology Letters
Central-European Journal of Immunology
Chirurgia Polska
Dermatologia Kliniczna
Diabetologia Doswiadczalna i Kliniczna
Diabetologia Polska
Endokrynologia, Diabetologia i Choroby Przemiany Materii Wieku Rozwojowego
Family Medicine and Primary Care Review
Fizjoterapia

Fizjoterapia Polska
Folia Biologica
Folia Cardiologica
Folia Histochemica et Cytobiologica
Folia Morphologica
Folia Neuropathologica
Gastroenterologia Polska
Ginekologia Onkologiczna
Ginekologia Praktyczna
HIV and AIDS Review
Hepatologia Polska
Hereditary Cancer in Clinical Practice
Human Movement
International Journal of Occupational Medicine and Environmental Health
International Review of Allergology and Clinical Immunology
Journal of Physiology and Pharmacology
Kardiologia Polska
Klinika Oczna
Mikologia Lekarska
Nadcisnienie Tetnicze
Neurologia i Neurochirurgia Polska
New Medicine
Nowotwory
Nuclear Medicine Review
Onkologia Polska
Ortopedia Traumatologia Rehabilitacja
Pediatria Polska
Pediatria Wspolczesna
Pediatria i Medycyna Rodzinna
Pharmacological Reports
Pneumonologia i Alergologia Polska
Polish Journal of Gynaecological Investigations
Polish Journal of Microbiology
Polish Journal of Pathology
Polish Journal of Radiology
Polska Medycyna Paliatywna
Polski Merkuriusz Lekarski
Polski Przeglad Chirurgiczny
Polski Przeglad Kardiologiczny
Polskie Archiwum Medycyny Wewnetrznej
Postepy Dermatologii I Alergologii
Postepy Mikrobiologii
Postepy Rehabilitacji
Problemy HIV i AIDS
Proktologia
Przeglad Dermatologiczny
Przeglad Menopauzalny
Przeglad Pediatryczny
Psychiatria Polska
Psychiatria i Psychologia Kliniczna
Psychogeriatria Polska
Psychoterapia
Rehabilitacja Medyczna
Reports of Practical Oncology and Radiotherapy
Reumatologia
Seksuologia Polska
Sen
Udar Mozgu - Problemy Interdyscyplinarne
Wspolczesna Onkologia
Z Zagadnien Nauk Sadowych

Portugal PRT

Acta Medica Portuguesa
Arquivos de Medicina
Nascer e Crescer
Revista Portuguesa de Cardiologia
Revista Portuguesa de Farmacia
Revista Portuguesa de Pneumologia
Revista de Psiquiatria Consiliar e de Ligacao
Sinapse
Skin Cancer

Qatar QAT

Middle East Journal of Emergency Medicine
Qatar Medical Journal

Romania ROM

Annals of Fundeni Hospital
Archives of the Balkan Medical Union
Journal of Gastrointestinal and Liver Diseases
Romanian Journal of Endocrinology
Romanian Journal of Legal Medicine

Russian Federation RUS

Immunologiya
Russian Journal of Immunology

Saudi Arabia SAU

Annals of Saudi Medicine
Neurosciences
Pan Arab Journal of Neurosurgery
Saudi Medical Journal
Saudi Pharmaceutical Journal

Serbia and Montenegro SCG

Archive of Oncology
Archives of Gastroenterohepatology
Children's Pulmonology
Jugoslovenska Medicinska Biohemija
Medicus
RAS - Radioloski Arhiv Srbije

Singapore SGP

American Journal of Chinese Medicine
Annals of the Academy of Medicine Singapore
Asia Pacific Journal of Pharmacology
Asian Cardiovascular and Thoracic Annals
Journal of Bioinformatics and Computational Biology
Journal of Integrative Neuroscience

Journal of Musculoskeletal Research
Physiotherapy Singapore
Singapore General Hospital Proceedings
Singapore Medical Journal
Singapore Paediatric Journal

Slovakia SVK

Acta Virologica
Endocrine Regulations
Eurorehab
Farmaceuticky Obzor
General Physiology and Biophysics
Kardiologia
Klinicka Imunologia a Alergologia
Lekarsky Obzor
Neoplasma
Rehabilitacia
Rheumatologia

Slovenia SVN

Acta Dermatovenerologica Alpina, Pannonica et Adriatica
Farmacevtski Vestnik
Radiology and Oncology

South Africa ZAF

African Journal of AIDS Research
Cardiovascular Journal of South Africa
Journal of Child and Adolescent Mental Health
Journal of Endocrinology, Metabolism and Diabetes of South Africa
Journal of the South African Veterinary Association
South African Family Practice
South African Journal of Clinical Nutrition
South African Journal of Obstetrics and Gynaecology
South African Journal of Psychiatry
South African Journal of Surgery
South African Medical Journal
South African Psychiatry Review
Southern African Journal of Critical Care
Southern African Journal of HIV Medicine

South Korea KOR

Biotechnology and Bioprocess Engineering
Clinical Psychopharmacology and Neuroscience
Electrolyte and Blood Pressure
Experimental and Molecular Medicine
Journal of Bacteriology and Virology
Journal of Korean Society for Clinical Pharmacology and Therapeutics
Journal of Microbiology and Biotechnology
Korean Circulation Journal
Korean Journal of Cardiovascular Diseases
Korean Journal of Dermatology
Korean Journal of Internal Medicine
Korean Journal of Medical Mycology
Korean Journal of Microbiology and Biotechnology
Korean Journal of Pharmacognosy
Korean Journal of Physiology and Pharmacology
Korean Journal of Urology
Molecules and Cells
Natural Product Sciences
Psychiatry Investigation
Tuberculosis and Respiratory Diseases
Yonsei Medical Journal

Spain ESP

AIDS Reviews
Acta Ginecologica
Acta Pediatrica Espanola
Actas Dermo-Sifiliograficas
Actualizaciones en Anestesiologia y Reanimacion
Adicciones
Alergologia e Inmunologia Clinica
Allergologia et Immunopathologia
Anales de Cirugia Cardiaca y Vascular
Anales de Medicina Interna
Anales de Pediatria
Anales de Psiquiatria
Anales de la Real Academia Nacional de Farmacia
Anales del Sistema Sanitario de Navarra
Angiologia
Anuario de Psicologia
Archivos de Bronconeumologia
Archivos de Medicina
Archivos de Medicina del Deporte
Archivos de Psiquiatria
Archivos de la Facultad de Medicina de Zaragoza
Ars Pharmaceutica
Atencion Farmaceutica
Atencion Primaria
Ciencia Ginecologika
Ciencia y Tecnologia Pharmaceutica
Circular Farmaceutica
Cirugia Plastica Ibero-Latinoamericana
Clinica e Investigacion en Ginecologia y Obstetricia
Clinical and Translational Oncology
Cuadernos de Medicina Reproductiva
DOLOR
Drug News and Perspectives
Drug R and D Backgrounders
Drugs of Today
Drugs of the Future
Endocrinologia y Nutricion
Enfermedades Emergentes
Enfermedades Infecciosas y Microbiologia Clinica
European Journal of Anatomy
European Journal of Psychiatry
Farmaceutico Hospitales
Farmacia Hospitalaria
GOZE
Gastroenterologia y Hepatologia
Geriatrika
Gerokomos
Ginecologia y Obstetricia Clinica
Histology and Histopathology
Inmunologia
International Journal of Developmental Biology

Investigacion Cardiovascular
Journal of Investigational Allergology and Clinical Immunology
Journal of Physiology and Biochemistry
Mapfre Medicina
Matronas Profesion
Medicina Clinica
Medicina Cutanea Ibero-Latino-Americana
Medicina Intensiva
Medicina Oral, Patologia Oral y Cirugia Bucal
Medicina Paliativa
Medicina Preventiva
Methods and Findings in Experimental and Clinical Pharmacology
Neurocirugia
Neurologia
Neurologia Suplementos
Nutricion Clinica y Dietetica Hospitalaria
Nutricion Hospitalaria
Nutrition and Metabolic Disorders in HIV Infection
ORL-DIPS
Oncologia
Patologia Del Aparato Locomotor
Pediatria Catalana
Pediatria Integral
Pediatrika
Pharmaceutical Care Espana
Psicologia Conductual
Psiquiatria Biologica
Psiquis
Puesta al Dia en Urgencias, Emergencias y Catastrofes
Quimica Clinica
Radiologia
Reumatologia Clinica
Revisiones en Cancer
Revista Clinica Espanola
Revista Espanola de Cardiologia
Revista Espanola de Enfermedades Digestivas
Revista Espanola de Enfermedades Metabolicas Oseas
Revista Espanola de Geriatria y Gerontologia
Revista Espanola de Medicina Nuclear
Revista Espanola de Nutricion Comunitaria
Revista Espanola de Obesidad
Revista Espanola de Pediatria
Revista Espanola de Quimioterapia
Revista Espanola de Salud Publica
Revista Iberoamericana de Fertilidad y Reproduccion Humana
Revista Iberoamericana de Micologia
Revista Iberoamericana de Revisiones en Menopausia
Revista Latina de Cardiologia
Revista Multidisciplinar de Gerontologia
Revista de Calidad Asistencial
Revista de Logopedia, Foniatria y Audiologia
Revista de Neurologia
Revista de Ortopedia y Traumatologia
Revista de Psiquiatria de la Facultad de Medicina de Barcelona
Revista de Toxicologia
Revista de la Asociacion Espanola de Especialistas en Medicina del Trabajo
Revista de la Sociedad Espanola del Dolor
SD Revista Medica Internacional sobre el Sindrome de Down
Sangre
Seguimiento Farmacoterapeutico
Seleccion
Seminarios de la Fundacion Espanola de Reumatologia

Sweden SWE

Forum for Nordic Dermato-Venerology
Neuroendocrinology Letters
Scandinavian Journal of Food and Nutrition
Scandinavian Journal of Nutrition/Naringsforskning

Switzerland CHE

Acta Haematologica
American Journal of Nephrology
Annales Nestle
Annals of Nutrition and Metabolism
Archivum Immunologiae et Therapiae Experimentalis
Audiology and Neurotology
Biology of the Neonate
Blood Purification
Brain, Behavior and Evolution
Breast Care
Bulletin of the World Health Organization
Cancer Immunity
Cardiology
Cells Tissues Organs
Cellular Physiology and Biochemistry
Cellular and Molecular Life Sciences
Cerebrovascular Diseases
Chemical Immunology and Allergy
Chemotherapy
Chirurgische Gastroenterologie Interdisziplinar
Community Genetics
Complexus
Current Studies in Hematology and Blood Transfusion
Cytogenetic and Genome Research
Dementia and Geriatric Cognitive Disorders
Dermatology
Developmental Neuroscience
Digestion
Digestive Diseases
Digestive Surgery
Douleur et Analgesie
Drugs under Experimental and Clinical Research
EDTNA-ERCA Journal
Environmental Health Criteria
European Addiction Research
European Journal of Drug Metabolism and Pharmacokinetics
European Neurology
European Respiratory Journal
European Respiratory Review
European Surgical Research
Exogenous Dermatology
Fetal Diagnosis and Therapy
Folia Phoniatrica et Logopaedica
Forschende Komplementarmedizin
Gann Monographs on Cancer Research
Genetic Counseling
Gerontology
Gynakologie fur Hausarzte

Gynakologisch-geburtshilfliche Rundschau
Gynecologic and Obstetric Investigation
Helvetica Chimica Acta
Hormone Research
Human Development
Human Heredity
IPCS Concise International Chemical Assessment Documents
Inflammation Research
International Archives of Allergy and Immunology
International Journal for Vitamin and Nutrition Research
International Journal of Bioinformatics Research and Applications
International Journal of Biotechnology
International Journal of Clinical Pharmacology Research
International Journal of Electronic Healthcare
International Journal of Emergency Management
International Journal of Environment and Pollution
International Journal of Environmental Research and Public Health
International Journal of Low Radiation
International Journal of Molecular Sciences
International Journal of Tissue Reactions
Intervirology
Journal of Biomedical Science
Journal of Molecular Microbiology and Biotechnology
Journal of Vascular Research
Kidney and Blood Pressure Research
Marine Drugs
Medecine et Hygiene
Medical Principles and Practice
Nephron - Clinical Practice
Nephron - Experimental Nephrology
Nephron - Physiology
NeuroImmunoModulation
NeuroSignals
Neurodegenerative Diseases
Neuroembryology and Aging
Neuroendocrinology
Neuroepidemiology
Neuropsychobiology
ORL
Oncology
Ophthalmic Research
Ophthalmologica
Osteologie
Pancreatology
Pathobiology
Pathophysiology of Haemostasis and Thrombosis
Pediatric Neurosurgery
Pharma-Kritik
Pharmaceuticals
Pharmacology
Progress in Drug Research
Psychopathology
Psychotherapies
Psychotherapy and Psychosomatics
Respiration
Schweizer Archiv fur Neurologie und Psychiatrie
Schweizerische Rundschau fur Medizin - Praxis
Schweizerische Zeitschrift fur GanzheitsMedizin
Schweizerische Zeitschrift fur Sportmedizin und Sporttraumatologie
Skin Pharmacology and Physiology
Sozial- und Praventivmedizin
Stereotactic and Functional Neurosurgery
Swiss Medical Weekly
Therapeutische Umschau
Transfusion Medicine and Hemotherapy
Tumor Biology
Urologia Internationalis
Vasa - Journal of Vascular Diseases
Verhaltenstherapie
WHO Drug Information
World Health Organization - Technical Report Series
Zeitschrift fur Gerontopsychologie und -psychiatrie
Zeitschrift fur Kinder- und Jugendpsychiatrie und Psychotherapie
Zeitschrift fur Psychiatrie, Psychologie und Psychotherapie

Taiwan TWN

Acta Anaesthesiologica Taiwanica
Acta Cardiologica Sinica
Acta Neurologica Taiwanica
Acta Paediatrica Taiwanica
Biomedical Engineering - Applications, Basis and Communications
Chang Gung Medical Journal
Chinese Journal of Physiology
Chinese Journal of Radiology
Chinese Pharmaceutical Journal
Dermatologica Sinica
Journal of Food and Drug Analysis
Journal of Internal Medicine of Taiwan
Journal of Medical Sciences
Journal of Medical and Biological Engineering
Journal of Microbiology, Immunology and Infection
Journal of Taiwan Otolaryngology - Head and Neck Surgery
Journal of the Chinese Medical Association
Mid-Taiwan Journal of Medicine
Nutritional Sciences Journal
Taiwan Journal of Public Health
Tzu Chi Medical Journal

Thailand THA

Asian Pacific Journal of Allergy and Immunology
Journal of the Medical Association of Thailand

Tunisia TUN

Maghreb Medical
Tunisie Medicale

Turkey TUR

Acta Pharmaceutica Turcica
Agri
Anadolu Kardiyoloji Dergisi
Anadolu Psikiyatri Dergisi
Anestezi Dergisi
Ankara Universitesi Eczacilik Fakultesi Dergisi
Annals of Medical Sciences

Cocuk Cerrahisi Dergisi
Cocuk Sagligi ve Hastaliklari Dergisi
Deri Hastaliklari ve Frengi Arsivi
Diagnostic and Interventional Radiology
Eastern Journal of Medicine
Erciyes Tip Dergisi
European Journal of General Medicine
Experimental and Clinical Transplantation
Fabad Journal of Pharmaceutical Sciences
Fizyoterapi Rehabilitasyon
Gazi Medical Journal
Gazi Universitesi Eczacilik Fakultesi Dergisi
Gogus-Kalp-Damar Anestezi ve Yogun Bakim Dernegi Dergisi
Goztepe Tip Dergisi
Gulhane Medical Journal
Jinekoloji ve Obstetri Bulteni
Jinekoloji ve Obstetrik Dergisi
Journal of Pediatric Neurology
Journal of Pharmacy of Istanbul University
Journal of Rheumatology and Medical Rehabilitation
Journal of the Turkish German Gynecology Association Artemis
Klinik Psikiyatri Dergisi
Klinik Psikofarmakoloji Bulteni
Marmara Medical Journal
Neuroanatomy
Nobel Medicus
Ondokuz Mayis Universitesi Tip Dergisi
Patoloji Bulteni
SENDROM
Sleep and Hypnosis
Tuberkuloz ve Toraks
Turk Anesteziyoloji ve Reanimasyon Dernegi Dergisi
Turk Geriatri Dergisi
Turk Kardiyoloji Dernegi Arsivi
Turk Pediatri Arsivi
Turk Serebrovaskuler Hastaliklar Dergisi
Turk Uroloji Dergisi
Turkish Journal of Cancer
Turkish Journal of Gastroenterology
Turkish Journal of Haematology
Turkish Journal of Immunology
Turkish Journal of Medical Sciences
Turkish Journal of Pediatrics
Turkish Journal of Pharmaceutical Sciences
Turkish Journal of Surgery
Turkish Neurosurgery
Turkiye Fiziksel Tip ve Rehabilitasyon Dergisi
Turkiye Klinikleri Journal of Medical Sciences
UHOD - Uluslararasi Hematoloji-Onkoloji Dergisi
Ulusal Travma ve Acil Cerrahi Dergisi
Yeni Symposium

Uganda UGA

African Health Sciences

Ukraine UKR

Bridging Eastern and Western Psychiatry
Experimental Oncology

United Arab Emirates ARE

Emirates Medical Journal
International Journal of Diabetes and Metabolism

United Kingdom GBR

AIDS
AIDS Care - Psychological and Socio-Medical Aspects of AIDS/HIV
AIDS Research and Therapy
ATLA Alternatives to Laboratory Animals
Acta Biomaterialia
Acta Neurologica Scandinavica
Acta Neuropsychiatrica
Acta Obstetricia et Gynecologica Scandinavica
Acta Ophthalmologica Scandinavica
Acta Orthopaedica
Acta Orthopaedica, Supplement
Acta Pharmacologica Sinica
Acta Physiologica
Acta Physiologica Scandinavica, Supplement
Acta Psychiatrica Scandinavica
Acupuncture in Medicine
Acute Cardiac Care
Acute Medicine
Acute Pain
Addiction
Addiction Biology
Addictive Behaviors
Advances in Breast Cancer
Advances in Environmental Research
Advances in Gastrointestinal Cancers
Advances in Microbial Physiology
Advances in Parasitology
Advances in Psychiatric Treatment
Adverse Drug Reaction Bulletin
Age and Ageing
Aggression and Violent Behavior
Aging Cell
Aging Health
Aging Male
Aging and Mental Health
Aging, Neuropsychology, and Cognition
Alcohol and Alcoholism
Alcoholism: Clinical and Experimental Research
Alimentary Pharmacology and Therapeutics
Alimentary Pharmacology and Therapeutics Symposium Series
Allergy: European Journal of Allergy and Clinical Immunology
Ambulatory Pediatrics
Ambulatory Surgery
American Journal of Family Therapy
American Journal of Reproductive Immunology
American Journal of Transplantation
American Journal on Addictions
Amyloid
Anaerobe
Anaesthesia
Anaesthesia, Supplement
Analyst
Andrologia

Animal Welfare
Annals of Clinical Biochemistry
Annals of Clinical Microbiology and Antimicrobials
Annals of Clinical Psychiatry
Annals of General Psychiatry
Annals of Human Genetics
Annals of Oncology
Annals of Tropical Medicine and Parasitology
Annals of Tropical Paediatrics
Annals of the ICRP
Annals of the Rheumatic Diseases
Annals of the Royal College of Surgeons of England
Anthropology and Medicine
Anti-Cancer Drugs
Antiviral Chemistry and Chemotherapy
Antiviral Therapy
Aphasiology
Appetite
Applied Cognitive Psychology
Applied Ergonomics
Applied Radiation and Isotopes
Archives of Andrology
Archives of Clinical Neuropsychology
Archives of Disease in Childhood
Archives of Disease in Childhood: Education and Practice Edition
Archives of Disease in Childhood: Fetal and Neonatal Edition
Archives of Physiology and Biochemistry
Arthritis Research and Therapy
Arthropod Structure and Development
Arts in Psychotherapy
Asia-Pacific Journal of Clinical Oncology
Asian Journal of Andrology
Asthma Journal
Atmospheric Environment
Attachment and Human Development
Audiological Medicine
Australia and New Zealand Health Policy
Autism
Autoimmunity
Autonomic and Autacoid Pharmacology
BJOG: An International Journal of Obstetrics and Gynaecology
BJU International
BMC Anesthesiology
BMC Biochemistry
BMC Bioinformatics
BMC Biology
BMC Biotechnology
BMC Blood Disorders
BMC Cancer
BMC Cardiovascular Disorders
BMC Cell Biology
BMC Chemical Biology
BMC Clinical Pathology
BMC Clinical Pharmacology
BMC Complementary and Alternative Medicine
BMC Dermatology
BMC Developmental Biology
BMC Ear, Nose and Throat Disorders
BMC Emergency Medicine
BMC Endocrine Disorders
BMC Evolutionary Biology
BMC Family Practice
BMC Gastroenterology
BMC Genetics
BMC Genomics
BMC Geriatrics
BMC Health Services Research
BMC Immunology
BMC Infectious Diseases
BMC International Health and Human Rights
BMC Medical Education
BMC Medical Ethics
BMC Medical Genetics
BMC Medical Imaging
BMC Medical Informatics and Decision Making
BMC Medical Research Methodology
BMC Medicine
BMC Microbiology
BMC Molecular Biology
BMC Musculoskeletal Disorders
BMC Nephrology
BMC Neurology
BMC Neuroscience
BMC Nuclear Medicine
BMC Nursing
BMC Ophthalmology
BMC Oral Health
BMC Palliative Care
BMC Pediatrics
BMC Pharmacology
BMC Physiology
BMC Pregnancy and Childbirth
BMC Psychiatry
BMC Public Health
BMC Pulmonary Medicine
BMC Structural Biology
BMC Surgery
BMC Urology
BMC Women's Health
Basic and Clinical Pharmacology and Toxicology
Basic and Clinical Pharmacology and Toxicology, Supplement
Behavioral Interventions
Behavioral Sciences and the Law
Behavioral and Brain Functions
Behavioral and Brain Sciences
Behaviour Research and Therapy
Behavioural Pharmacology
Behavioural and Cognitive Psychotherapy
Best Practice and Research in Clinical Anaesthesiology
Best Practice and Research in Clinical Endocrinology and Metabolism
Best Practice and Research in Clinical Gastroenterology
Best Practice and Research in Clinical Haematology
Best Practice and Research in Clinical Obstetrics and Gynaecology
Best Practice and Research in Clinical Rheumatology
BioMedical Engineering Online
Biocatalysis and Biotransformation
Biochemical Journal
Biochemical Society Transactions
Bioethics
Biofilms
Bioinformatics
Biological Reviews of the Cambridge Philosophical Society
Biological Rhythm Research
Biologicals

Biology Letters
Biology of the Cell
Biomarkers
Biomaterials
Biomedical Chromatography
Biomedical Digital Libraries
Biomedical Materials
Biomedical Signal Processing and Control
Bioorganic and Medicinal Chemistry
Bioorganic and Medicinal Chemistry Letters
Biopharmaceutics and Drug Disposition
Bioresource Technology
Biosensors and Bioelectronics
Biotechnic and Histochemistry
Biotechnology and Applied Biochemistry
Biotechnology and Genetic Engineering Reviews
Blood Coagulation and Fibrinolysis
Blood Pressure Monitoring
Blood Reviews
Bone Marrow Transplantation
Brain
Brain Injury
Breast
Breast Cancer Online
Breast Cancer Research
Briefings in Bioinformatics
Briefings in Functional Genomics and Proteomics
British Journal of Anaesthesia
British Journal of Biomedical Science
British Journal of Cancer
British Journal of Cardiology
British Journal of Clinical Pharmacology
British Journal of Clinical Pharmacology, Supplement
British Journal of Clinical Psychology
British Journal of Dermatology
British Journal of Developmental Disabilities
British Journal of Diabetes and Vascular Disease
British Journal of General Practice
British Journal of Haematology
British Journal of Health Psychology
British Journal of Hospital Medicine
British Journal of Intensive Care
British Journal of Neurosurgery
British Journal of Nutrition
British Journal of Occupational Therapy
British Journal of Ophthalmology
British Journal of Oral and Maxillofacial Surgery
British Journal of Pharmacology
British Journal of Psychiatry
British Journal of Psychotherapy
British Journal of Radiology
British Journal of Sports Medicine
British Journal of Surgery
British Medical Bulletin
British Medical Journal
Burns
CME Bulletin Endocrinology and Diabetes
CME Bulletin Otorhinolaryngology Head and Neck Surgery
CME Cancer Medicine
CME Journal Gastroenterology, Hepatology and Nutrition
CME Journal Geriatric Medicine
CME Journal Ophthalmology
CME Journal Radiology
CME Orthopaedics
CNS Drug Reviews
CPD Anaesthesia
CPD Bulletin Cellular Pathology
CPD Bulletin Clinical Biochemistry
CPD Bulletin Immunology and Allergy
CPD Bulletin Old Age Psychiatry
CPD Infection
Cancer Cell International
Cancer Detection and Prevention
Cancer Science
Cancer Treatment Reviews
Carbohydrate Research
Carcinogenesis
Cardiology International
Cardiology in the Young
Cardiovascular Diabetology
Cardiovascular Drug Reviews
Cardiovascular Ultrasound
Care of the Critically Ill
Cell Biochemistry and Function
Cell Biology International
Cell Calcium
Cell Communication and Adhesion
Cell Communication and Signaling
Cell Death and Differentiation
Cell Proliferation
Cell Proliferation, Supplement
Cell and Chromosome
Cellular Microbiology
Cephalalgia
Cephalalgia, Supplement
Cerebellum
Cerebrospinal Fluid Research
Chemical Biology and Drug Design
Chemical Communications
Chemical Senses
Chemosphere
Child Abuse and Neglect
Child Neuropsychology
Child Psychology and Psychiatry Review
Child and Adolescent Mental Health
Child: Care, Health and Development
Chiropractic and Osteopathy
Chronic Respiratory Disease
Chronic Illness
Climacteric
Clinical Biomechanics
Clinical Dermatology: Retinoids and other Treatments
Clinical Dysmorphology
Clinical Endocrinology
Clinical Genetics
Clinical Governance
Clinical Intensive Care
Clinical Linguistics and Phonetics
Clinical MRI
Clinical Medicine, Journal of the Royal College of Physicians of London
Clinical Microbiology and Infection
Clinical Microbiology and Infection, Supplement
Clinical Neuropsychologist
Clinical Neuroscience Research
Clinical Nutrition
Clinical Nutrition, Supplement
Clinical Oncology
Clinical Otolaryngology
Clinical Pediatric Emergency Medicine

Clinical Physiology and Functional Imaging
Clinical Practice and Epidemiology in Mental Health
Clinical Psychology Forum
Clinical Psychology and Psychotherapy
Clinical Radiology
Clinical Radiology Extra
Clinical Rehabilitation
Clinical Rheumatology
Clinical Risk
Clinical Science
Clinical Teacher
Clinical Transplantation
Clinical Transplantation, Supplement
Clinical Trials
Clinical and Developmental Immunology
Clinical and Experimental Allergy
Clinical and Experimental Allergy Reviews
Clinical and Experimental Dermatology
Clinical and Experimental Dermatology, Supplement
Clinical and Experimental Immunology
Clinical and Laboratory Haematology
Clinical and Laboratory Haematology, Supplement
Clinical and Molecular Allergy
Clinician in Management
Cochlear Implants International
Cochrane Database of Systematic Reviews
Cognition and Emotion
Cognitive Neuropsychiatry
Cognitive Neuropsychology
Colorectal Disease
Community Eye Health Journal
Comparative Clinical Pathology
Comparative Hepatology
Comparative Immunology, Microbiology and Infectious Diseases
Complementary Therapies in Medicine
Computer Aided Surgery
Computerized Medical Imaging and Graphics
Computers in Biology and Medicine
Congenital Heart Disease
Connective Tissue Research
Contact Dermatitis
Contemporary Hypnosis
Continuing Medical Education - Cardiology
Coronary Artery Disease
Cost Effectiveness and Resource Allocation
Counselling Psychology Quarterly
Criminal Behaviour and Mental Health
Critical Care
Critical Public Health
Culture, Health and Sexuality
Current Allergy and Asthma Reports
Current Anaesthesia and Critical Care
Current Atherosclerosis Reports
Current Cardiology Reports
Current Controlled Trials in Cardiovascular Medicine
Current Diabetes Reports
Current Diagnostic Pathology
Current Eye Research
Current Gastroenterology Reports
Current Hypertension Reports
Current Infectious Disease Reports
Current Issues in Intestinal Microbiology
Current Issues in Molecular Biology
Current Medical Research and Opinion
Current Medical Research and Opinion, Supplement
Current Neurology and Neuroscience Reports
Current Obstetrics and Gynaecology
Current Oncology Reports
Current Opinion in Anaesthesiology
Current Opinion in Biotechnology
Current Opinion in Cell Biology
Current Opinion in Chemical Biology
Current Opinion in Colloid and Interface Science
Current Opinion in Drug Discovery and Development
Current Opinion in Genetics and Development
Current Opinion in Immunology
Current Opinion in Infectious Diseases
Current Opinion in Investigational Drugs
Current Opinion in Lipidology
Current Opinion in Microbiology
Current Opinion in Molecular Therapeutics
Current Opinion in Nephrology and Hypertension
Current Opinion in Neurobiology
Current Opinion in Neurology
Current Opinion in Obstetrics and Gynecology
Current Opinion in Pharmacology
Current Opinion in Psychiatry
Current Opinion in Structural Biology
Current Opinion in Urology
Current Orthopaedics
Current Paediatrics
Current Pain and Headache Reports
Current Psychiatry Reports
Current Treatment Options in Cardiovascular Medicine
Current Treatment Options in Gastroenterology
Current Treatment Options in Neurology
Current Treatment Options in Oncology
CytoJournal
Cytokine
Cytokine and Growth Factor Reviews
Cytopathology
Cytopathology, Supplement
Cytotherapy
DNA Research
DNA Sequence - Journal of DNA Sequencing and Mapping
Dentomaxillofacial Radiology
Dermatologic Surgery
Developing World Bioethics
Development
Development and Psychopathology
Developmental Medicine and Child Neurology
Developmental and Comparative Immunology
Diabetes and Vascular Disease Research
Diabetes, Obesity and Metabolism
Diabetes, Obesity and Metabolism, Supplement
Diabetes/Metabolism Research and Reviews
Diabetic Medicine
Differentiation
Digestive Endoscopy
Disability and Rehabilitation
Drug Delivery
Drug Discovery Today
Drug Discovery Today: BIOSILICO
Drug Discovery Today: Disease Mechanisms
Drug Discovery Today: Disease Models
Drug Discovery Today: TARGETS
Drug Discovery Today: Technologies
Drug Discovery Today: Therapeutic Strategies
Drug Discovery World

Drug Resistance Updates
Drug and Therapeutics Bulletin
Drugs in Context
Drugs in Context (US)
Dynamic Medicine
Dyslexia
EBR - European Biopharmaceutical Review
EMBO Journal
EMBO Reports
ESHRE Monographs
Eating Behaviors
Eating Disorders
Ecotoxicology and Environmental Safety
Education for Health: Change in Learning and Practice
Education for Primary Care
Emergency Medicine Journal
Emerging Themes in Epidemiology
Endocrine-Related Cancer
Endothelium: Journal of Endothelial Cell Research
Environment International
Environmental Conservation
Environmental Health: A Global Access Science Source
Environmental Pollution
Environmental Science and Policy
Environmental Technology
Epidemiologic Perspectives and Innovations
Epidemiologic Reviews
Epidemiology and Infection
Ergonomics
Ethnicity and Health
Europace
European Cells and Materials
European Eating Disorders Review
European Heart Journal
European Heart Journal, Supplement
European Journal of Anaesthesiology
European Journal of Anaesthesiology, Supplement
European Journal of Cancer
European Journal of Cancer Prevention
European Journal of Cancer, Supplement
European Journal of Cardiovascular Prevention and Rehabilitation
European Journal of Clinical Investigation
European Journal of Clinical Nutrition
European Journal of Cognitive Psychology
European Journal of Contraception and Reproductive Health Care
European Journal of Echocardiography
European Journal of Endocrinology
European Journal of Endocrinology, Supplement
European Journal of Gastroenterology and Hepatology
European Journal of Haematology
European Journal of Herbal Medicine
European Journal of Human Genetics
European Journal of Morphology
European Journal of Neurology
European Journal of Neurology, Supplement
European Journal of Neuroscience
European Journal of Oriental Medicine
European Journal of Paediatric Neurology
European Journal of Pain
European Journal of Public Health
European Journal of Special Needs Education
European Journal of Surgical Oncology
European Journal of Vascular and Endovascular Surgery
European Pharmaceutical Contractor
Evidence-Based Healthcare and Public Health
Evidence-Based Medicine
Evidence-based Cardiovascular Medicine
Evidence-based Complementary and Alternative Medicine
Evidence-based Obstetrics and Gynecology
Experimental Dermatology
Experimental Dermatology, Supplement
Experimental Eye Research
Experimental Physiology
Expert Opinion on Biological Therapy
Expert Opinion on Drug Delivery
Expert Opinion on Drug Metabolism and Toxicology
Expert Opinion on Drug Safety
Expert Opinion on Emerging Drugs
Expert Opinion on Investigational Drugs
Expert Opinion on Pharmacotherapy
Expert Opinion on Therapeutic Patents
Expert Opinion on Therapeutic Targets
Expert Review of Anti-Infective Therapy
Expert Review of Anticancer Therapy
Expert Review of Cardiovascular Therapy
Expert Review of Clinical Immunology
Expert Review of Medical Devices
Expert Review of Molecular Diagnostics
Expert Review of Neurotherapeutics
Expert Review of Pharmacoeconomics and Outcomes Research
Expert Review of Proteomics
Expert Review of Vaccines
Expert Reviews in Molecular Medicine
Eye
FEBS Journal
FEMS Immunology and Medical Microbiology
FEMS Microbiology Ecology
FEMS Microbiology Letters
FEMS Microbiology Reviews
FEMS Yeast Research
Family Practice
Fetal and Maternal Medicine Review
Fetal and Pediatric Pathology
Filaria Journal
Flavour and Fragrance Journal
Focus on Alternative and Complementary Therapies
Food Additives and Contaminants
Food Chemistry
Food and Chemical Toxicology
Foot
Foot and Ankle Surgery
Free Radical Research
Fundamental and Clinical Pharmacology
Future Cardiology
Future Oncology
Gene Therapy
Genes and Immunity
Genes to Cells
Genes, Brain and Behavior
Genetic Vaccines and Therapy
Genetical Research
Genome Biology
Globalization and Health
Glycobiology
Good Clinical Practice Journal
Group Analysis
Growth Factors

Growth Hormone and IGF Research
Gut
Gynecological Endocrinology
HIV Medicine
HPB
HPB, Supplement
Haemophilia
Haemophilia, Supplement
Harm Reduction Journal
Harvard Review of Psychiatry
Headache Currents
Health Economics
Health Education Journal
Health Education Research
Health Expectations
Health Informatics Journal
Health Policy and Planning
Health Promotion International
Health Research Policy and Systems
Health Services Management Research
Health Technology Assessment
Health and Place
Health and Quality of Life Outcomes
Health, Risk and Society
Heart
Heart and Metabolism
Helicobacter
Helicobacter, Supplement
Hematological Oncology
Hematology
Hematology Journal
Hereditas
Herpes
Histopathology
Homeopathy
Hospital Pharmacist
Human Fertility
Human Genomics
Human Molecular Genetics
Human Psychopharmacology
Human Reproduction
Human Reproduction Update
Human Resources for Health
Human and Experimental Toxicology
IDrugs
IEE Proceedings: Systems Biology
IUBMB Life
Imaging
Imaging Decisions MRI
Immunity and Ageing
Immunological Reviews
Immunology
Immunology, Supplement
In Practice
Indoor and Built Environment
Infant Behavior and Development
Informatics in Primary Care
Inhalation Toxicology
Injury
Interactive Cardiovascular and Thoracic Surgery
International Clinical Psychopharmacology
International Forum of Psychoanalysis
International Immunology
International Journal for Equity in Health
International Journal for Parasitology
International Journal for Quality in Health Care
International Journal of Andrology
International Journal of Andrology, Supplement
International Journal of Aromatherapy
International Journal of Audiology
International Journal of Behavioral Development
International Journal of Behavioral Nutrition and Physical Activity
International Journal of Biochemistry and Cell Biology
International Journal of Clinical Practice
International Journal of Clinical Practice, Supplement
International Journal of Clinical and Experimental Hypnosis
International Journal of Cosmetic Science
International Journal of Dairy Technology
International Journal of Dermatology
International Journal of Developmental Neuroscience
International Journal of Environmental Health Research
International Journal of Epidemiology
International Journal of Experimental Pathology
International Journal of Food Sciences and Nutrition
International Journal of Geriatric Psychiatry
International Journal of Health Geographics
International Journal of Health Planning and Management
International Journal of Health Promotion and Education
International Journal of Healthcare Technology and Management
International Journal of Hyperthermia
International Journal of Immunogenetics
International Journal of Impotence Research
International Journal of Intensive Care
International Journal of Language and Communication Disorders
International Journal of Law and Psychiatry
International Journal of Medical Robotics and Computer Assisted Surgery
International Journal of Medicine
International Journal of Methods in Psychiatric Research
International Journal of Neuroprotection and Neurogeneration
International Journal of Neuropsychopharmacology
International Journal of Neuroscience
International Journal of Obesity
International Journal of Obstetric Anesthesia
International Journal of Oral and Maxillofacial Surgery
International Journal of Osteopathic Medicine
International Journal of Pharmacy Practice
International Journal of Prisoner Health
International Journal of Psychiatry in Clinical Practice
International Journal of Psychoanalysis
International Journal of Radiation Biology
International Journal of Rehabilitation Research
International Journal of Respiratory Care
International Journal of STD and AIDS
International Journal of Social Psychiatry
International Journal of Stroke
International Journal of Surgery
International Journal of Systematic and Evolutionary Microbiology
International Journal of Toxicology
International MS Journal
International Review of Psychiatry
International Reviews of Immunology
International Seminars in Surgical Oncology
International Urogynecology Journal and Pelvic Floor Dysfunction

International Wound Journal
JDDG - Journal of the German Society of Dermatology
JRAAS - Journal of the Renin-Angiotensin-Aldosterone System
Japanese Journal of Clinical Oncology
Journal of Adolescence
Journal of Aerosol Science
Journal of Analytical Psychology
Journal of Anatomy
Journal of Antimicrobial Chemotherapy
Journal of Anxiety Disorders
Journal of Applied Microbiology
Journal of Applied Toxicology
Journal of Asian Natural Products Research
Journal of Autoimmune Diseases
Journal of Autoimmunity
Journal of Behavior Therapy and Experimental Psychiatry
Journal of Bioactive and Compatible Polymers
Journal of Biology
Journal of Biomechanics
Journal of Biosocial Science
Journal of Bodywork and Movement Therapies
Journal of Bone and Joint Surgery - Series B
Journal of Carcinogenesis
Journal of Cell Science
Journal of Cellular Pathology
Journal of Chemical Technology and Biotechnology
Journal of Child Psychology and Psychiatry and Allied Disciplines
Journal of Child Psychotherapy
Journal of Chinese Medicine
Journal of Circadian Rhythms
Journal of Clinical Forensic Medicine
Journal of Clinical Neuroscience
Journal of Clinical Pathology
Journal of Clinical Pharmacy and Therapeutics
Journal of Clinical Research
Journal of Clinical and Experimental Neuropsychology
Journal of Commercial Biotechnology
Journal of Comparative Pathology
Journal of Cosmetic Dermatology
Journal of Cosmetic and Laser Therapy
Journal of Cranio-Maxillofacial Surgery
Journal of Cutaneous Pathology
Journal of Dementia Care
Journal of Dermatological Treatment
Journal of Dermatology
Journal of Diagnostic Radiography and Imaging
Journal of Drug Targeting
Journal of Electromyography and Kinesiology
Journal of Endocrinology
Journal of Endotoxin Research
Journal of Environmental Management
Journal of Environmental Monitoring
Journal of Environmental Radioactivity
Journal of Enzyme Inhibition and Medicinal Chemistry
Journal of Epidemiology and Community Health
Journal of Evaluation in Clinical Practice
Journal of Experimental Therapeutics and Oncology
Journal of Experimental and Clinical Assisted Reproduction
Journal of Family Planning and Reproductive Health Care
Journal of Forensic Psychiatry and Psychology
Journal of Gene Medicine
Journal of General Virology
Journal of Generic Medicines
Journal of HIV Therapy
Journal of Hand Surgery
Journal of Health Psychology
Journal of Health Services Research and Policy
Journal of Heart Valve Disease
Journal of Helminthology
Journal of Hospital Infection
Journal of Human Hypertension
Journal of Human Nutrition and Dietetics
Journal of Hypertension, Supplement
Journal of Immune Based Therapies and Vaccines
Journal of Infection
Journal of Inflammation
Journal of Integrated Care Pathways
Journal of Intellectual Disabilities
Journal of Intellectual Disability Research
Journal of Intellectual Disability Research, Supplement
Journal of Intellectual and Developmental Disability
Journal of Internal Medicine
Journal of International Medical Research
Journal of Interprofessional Care
Journal of Investigative Dermatology Symposium Proceedings
Journal of Investigative Surgery
Journal of Labelled Compounds and Radiopharmaceuticals
Journal of Laryngology and Otology
Journal of Legal Medicine
Journal of Mass Spectrometry
Journal of Maternal-Fetal and Neonatal Medicine
Journal of Medical Economics
Journal of Medical Engineering and Technology
Journal of Medical Ethics
Journal of Medical Genetics
Journal of Medical Microbiology
Journal of Medical Screening
Journal of Medicine and Philosophy
Journal of Mental Health
Journal of Microencapsulation
Journal of Microscopy
Journal of Molecular Biology
Journal of Molecular Endocrinology
Journal of Molecular and Cellular Cardiology
Journal of Nanobiotechnology
Journal of Negative Results in BioMedicine
Journal of Neural Engineering
Journal of NeuroEngineering and Rehabilitation
Journal of Neurochemistry
Journal of Neuroendocrinology
Journal of Neurogenetics
Journal of Neuroinflammation
Journal of Neurolinguistics
Journal of Neurology, Neurosurgery and Psychiatry
Journal of Nutritional and Environmental Medicine
Journal of Obstetrics and Gynaecology
Journal of Occupational and Environmental Hygiene
Journal of Oral Pathology and Medicine
Journal of Organ Dysfunction
Journal of Outcomes Research
Journal of Pathology
Journal of Pediatric Urology
Journal of Peptide Science
Journal of Perinatology
Journal of Pharmacy and Pharmacology
Journal of Physiology
Journal of Pineal Research

Journal of Plastic, Reconstructive and Aesthetic Surgery
Journal of Psychiatric Research
Journal of Psychopharmacology
Journal of Psychosomatic Obstetrics and Gynecology
Journal of Public Health
Journal of Radiological Protection
Journal of Radiotherapy in Practice
Journal of Rehabilitation Medicine
Journal of Rehabilitation Medicine, Supplement
Journal of Reproductive and Infant Psychology
Journal of Rural Health
Journal of Science and Medicine in Sport
Journal of Sex and Marital Therapy
Journal of Sexual Medicine
Journal of Sleep Research
Journal of Sleep Research, Supplement
Journal of Sports Sciences
Journal of Steroid Biochemistry and Molecular Biology
Journal of Substance Use
Journal of The Royal Society for the Promotion of Health
Journal of Theoretical Biology
Journal of Theoretical Medicine
Journal of Thermal Biology
Journal of Thrombosis and Haemostasis
Journal of Toxicology and Environmental Health - Part A
Journal of Translational Medicine
Journal of Travel Medicine
Journal of Tropical Pediatrics
Journal of Veterinary Pharmacology and Therapeutics
Journal of Viral Hepatitis
Journal of Water and Health
Journal of the British Menopause Society
Journal of the Chartered Institution of Water and Environmental Management
Journal of the European Academy of Dermatology and Venereology
Journal of the History of the Neurosciences
Journal of the International Neuropsychological Society
Journal of the National Cancer Institute
Journal of the Royal Society Interface
Journal of the Royal Society of Medicine
Journal of the Royal Society of Medicine, Supplement
Journal on Information Technology in Healthcare
Kidney International
Kinetoplastid Biology and Disease
Lab on a Chip - Miniaturisation for Chemistry and Biology
Laboratory Animals
Laboratory Investigation
Lancet
Lancet Infectious Diseases
Lancet Neurology
Language and Cognitive Processes
Lasers in Medical Science
Laterality
Legal and Criminological Psychology
Leprosy Review
Letters in Applied Microbiology
Leukemia
Leukemia Research
Leukemia and Lymphoma
Lipids in Health and Disease
Liver International
Liver Transplantation
Logopedics Phoniatrics Vocology
Logopedics Phoniatrics Vocology, Supplement
Lupus
Malaria Journal
Manual Therapy
Marine Environmental Research
Marine Pollution Bulletin
Maternal and Child Nutrition
Mathematical Medicine and Biology
MeReC Briefing
MeReC Bulletin
MeReC Extra
Medical Education
Medical Education, Supplement
Medical Engineering and Physics
Medical Humanities
Medical Hypotheses
Medical Immunology
Medical Informatics and the Internet in Medicine
Medical Law International
Medical Mycology
Medical Teacher
Melanoma Research
Memory
Men's Health Journal
Mental Health, Religion and Culture
Microbial Cell Factories
Microbial Ecology in Health and Disease
Microbial Pathogenesis
Microbiology
Middle East Paediatrics
Minimally Invasive Therapy and Allied Technologies
Modern Pathology
Molecular Aspects of Medicine
Molecular Cancer
Molecular Human Reproduction
Molecular Immunology
Molecular Membrane Biology
Molecular Microbiology
Molecular Pain
Molecular Psychiatry
Molecular Systems Biology
Molecular and Cellular Probes
Multiple Sclerosis
Mutagenesis
Mycoses
Mycoses, Supplement
NMR in Biomedicine
Nanotoxicology
Natural Product Research
Nature
Nature Cell Biology
Nature Chemical Biology
Nature Clinical Practice Cardiovascular Medicine
Nature Clinical Practice Endocrinology and Metabolism
Nature Clinical Practice Gastroenterology and Hepatology
Nature Clinical Practice Nephrology
Nature Clinical Practice Neurology
Nature Clinical Practice Oncology
Nature Clinical Practice Rheumatology
Nature Clinical Practice Urology
Nature Immunology
Nature Medicine
Nature Methods
Nature Reviews Cancer
Nature Reviews Drug Discovery
Nature Reviews Genetics

Nature Reviews Immunology
Nature Reviews Microbiology
Nature Reviews Molecular Cell Biology
Nature Reviews Neuroscience
Nephrology Dialysis Transplantation
Neural Networks
Neuro-Ophthalmology
NeuroReport
Neurocase
Neurochemistry International
Neurogastroenterology and Motility
Neurology in Practice
Neuromuscular Disorders
Neuron Glia Biology
Neuropathology and Applied Neurobiology
Neuropeptides
Neuropharmacology
Neuropsychologia
Neuropsychological Rehabilitation
Neuropsychopharmacology
Neuroscience
Neuroscience and Biobehavioral Reviews
New Genetics and Society
New Journal of Chemistry
Nicotine and Tobacco Research
Noise and Health
North and West London Journal of General Practice
Nuclear Medicine Communications
Nuclear Receptor
Nucleic Acids Research
Nutrition Bulletin
Nutrition Journal
Nutrition Research Reviews
Nutrition and Metabolism
Nutritional Neuroscience
Obesity Reviews
Obesity Reviews, Supplement
Obesity in Practice
Occupational Health
Occupational Medicine
Occupational Therapy International
Occupational and Environmental Medicine
Ocular Immunology and Inflammation
Oncogene
Ophthalmic Epidemiology
Ophthalmic Genetics
Ophthalmic and Physiological Optics
Oral Microbiology and Immunology
Oral Oncology
Orbit
Osteoarthritis and Cartilage
Osteoporosis International
Paediatric Anaesthesia
Paediatric Respiratory Reviews
Paediatric and Perinatal Drug Therapy
Paediatric and Perinatal Epidemiology
Palliative Medicine
Parasite Immunology
Parasitology
Parkinsonism and Related Disorders
Particle and Fibre Toxicology
Pathology
Pediatric Hematology and Oncology
Pediatric Rehabilitation
Pediatric Transplantation
Perfusion
Personalized Medicine
Pharmaceutical Biology
Pharmaceutical Journal
Pharmaceutical Statistics
Pharmaceutical Technology Europe
Pharmaceutical Visions
Pharmacoepidemiology and Drug Safety
Pharmacogenetics and Genomics
Pharmacogenomics
Pharmacogenomics Journal
Pharmacological Research
Pharmacy Education
Pharmacy in Practice
Phlebology
Photochemical and Photobiological Sciences
Physical Biology
Physical Therapy in Sport
Physics in Medicine and Biology
Physiological Measurement
Physiotherapy
Physiotherapy Theory and Practice
Phytochemical Analysis
Phytotherapy Research
Pigment Cell Research
Placenta
Platelets
Population Health Metrics
Postgraduate Medical Journal
Practical Cardiovascular Risk Management
Practical Diabetes International
Practical Neurology
Prenatal Diagnosis
Primary Care Mental Health
Primary Care and Community Psychiatry
Proceedings of the Nutrition Society
Proceedings of the Royal Society - Biological Sciences (Series B)
Progress in Lipid Research
Progress in Neurobiology
Progress in Palliative Care
Progress in Retinal and Eye Research
Prostaglandins Leukotrienes and Essential Fatty Acids
Prostate Cancer and Prostatic Diseases
Prosthetics and Orthotics International
Protein Engineering, Design and Selection
Proteome Science
Psychiatric Bulletin
Psychiatric Genetics
Psychiatry and Clinical Neurosciences
Psycho-Oncology
Psychoanalytic Psychotherapy
Psychodynamic Practice
Psychological Medicine
Psychology and Psychotherapy: Theory, Research and Practice
Psychology, Health and Medicine
Psychoneuroendocrinology
Psychotherapy Research
Pteridines
Public Health Medicine
Public Health Nutrition
Pulmonary Pharmacology and Therapeutics
QJM - Monthly Journal of the Association of Physicians
Quality Assurance Journal

Quality and Safety in Health Care
Quality in Primary Care
Quarterly Reviews of Biophysics
Radiation Physics and Chemistry
Radiation Protection Dosimetry
Radiography
Radiology Now
Redox Report
Reproduction
Reproductive BioMedicine Online
Reproductive Biology and Endocrinology
Reproductive Health
Reproductive Health Matters
Reproductive Medicine Review
Research and Clinical Forums
Research in Sports Medicine
Research in Veterinary Science
Respiratory Medicine
Respiratory Medicine Extra
Respiratory Medicine: COPD Update
Respiratory Research
Retrovirology
Reviews in Clinical Gerontology
Reviews in Contemporary Pharmacotherapy
Reviews in Medical Microbiology
Reviews in Medical Virology
Rheumatology
Round Table Series - Royal Society of Medicine
Scandinavian Journal of Caring Sciences
Scandinavian Journal of Immunology
Scandinavian Journal of Immunology, Supplement
Schizophrenia Bulletin
Science and Justice - Journal of the Forensic Science Society
Scottish Medical Journal
Seizure
Seminars in Cancer Biology
Seminars in Cell and Developmental Biology
Seminars in Fetal and Neonatal Medicine
Seminars in Immunology
Seminars in Ophthalmology
Sexual Addiction and Compulsivity
Sexual and Relationship Therapy
Sexualities, Evolution and Gender
Sexually Transmitted Infections
Signal
Skin Research and Technology
Sleep Medicine Reviews
Social Science and Medicine
Somatosensory and Motor Research
Somnologie
Spinal Cord
Statistical Methods in Medical Research
Statistics in Medicine
Strabismus
Stress
Stress and Health
Surgeon
Surgical Oncology
Tetrahedron
Tetrahedron Asymmetry
Tetrahedron Letters
Thalamus and Related Systems
TheScientificWorldJournal
Theoretical Biology and Medical Modelling
Therapeutic Communities: the International Journal for Therapeutic and Supportive Organizations
Therapy
Thorax
Thrombosis Journal
Thrombosis Research
Tissue Antigens
Tissue and Cell
Toxicologic Pathology
Toxicology Mechanisms and Methods
Toxicology and Industrial Health
Toxicology in Vitro
Toxicon
TransPharma
Transcultural Psychiatry
Transfusion Medicine
Transfusion and Apheresis Science
Transplant Infectious Disease
Transplant International
Trauma
Trends in Biochemical Sciences
Trends in Biotechnology
Trends in Cell Biology
Trends in Cognitive Sciences
Trends in Genetics
Trends in Immunology
Trends in Microbiology
Trends in Molecular Medicine
Trends in Neurosciences
Trends in Parasitology
Trends in Pharmacological Sciences
Tropical Doctor
Tropical Medicine and International Health
Tuberculosis
Ulster Medical Journal
Ultrasound Review of Obstetrics and Gynecology
Ultrasound in Obstetrics and Gynecology
Ultrastructural Pathology
Update in Anaesthesia
Vaccine
Vascular Medicine
Veterinary Journal
Veterinary Record
Virology Journal
Vision Research
Visual Cognition
Visual Impairment Research
Vox Sanguinis
Vox Sanguinis, Supplement
Waste Management
Waste Management and Research
Water Research
Water Science and Technology
Water Science and Technology: Water Supply
Women's Health
Women's Oncology Review
Work Based Learning in Primary Care
Work and Stress
World Journal of Surgical Oncology
World Wide Wounds
Worldviews on Evidence-Based Nursing
Wounds UK
Xenobiotica
Xenotransplantation
Yeast

United States USA

AANA Journal
AAPS Journal
AAPS PharmSciTech
ACC Cardiosource Review Journal
AIDS Education and Prevention
AIDS Patient Care and STDs
AIDS Reader
AIDS Research and Human Retroviruses
AIDS and Behavior
AIDS and Public Policy Journal
ASAIO Journal
ASGE Clinical Update
AWIC Bulletin
Abdominal Imaging
Academic Emergency Medicine
Academic Medicine
Academic Psychiatry
Academic Radiology
Acta Cytologica
Acupuncture and Electro-Therapeutics Research
Adapted Physical Activity Quarterly
Addictive Disorders and their Treatment
Adipocytes
Administration and Policy in Mental Health
Advanced Studies in Medicine
Advances in Anatomic Pathology
Advances in Applied Microbiology
Advances in Cancer Research
Advances in Carbohydrate Chemistry and Biochemistry
Advances in Chronic Kidney Disease
Advances in Experimental Medicine and Biology
Advances in Immunology
Advances in Mind-Body Medicine
Advances in Pharmacy
Advances in Protein Chemistry
Advances in Therapy
Aerosol Science and Technology
Aesthetic Plastic Surgery
Aesthetic Surgery Journal
Air Medical Journal
Alcohol
Alcoholism Treatment Quarterly
Alternative Medicine Review
Alternative Therapies in Health and Medicine
Alternative and Complementary Therapies
Alzheimer Disease and Associated Disorders
Alzheimer's and Dementia
American Family Physician
American Heart Journal
American Journal Geriatric Pharmacotherapy
American Journal of Alzheimer's Disease and other Dementias
American Journal of Audiology
American Journal of Cardiology
American Journal of Clinical Hypnosis
American Journal of Clinical Nutrition
American Journal of Clinical Oncology: Cancer Clinical Trials
American Journal of Clinical Pathology
American Journal of Dermatopathology
American Journal of Drug and Alcohol Abuse
American Journal of Electroneurodiagnostic Technology
American Journal of Emergency Medicine
American Journal of Epidemiology
American Journal of Forensic Medicine and Pathology
American Journal of Forensic Psychiatry
American Journal of Forensic Psychology
American Journal of Gastroenterology
American Journal of Geriatric Psychiatry
American Journal of Health Promotion
American Journal of Health-System Pharmacy
American Journal of Hematology
American Journal of Human Biology
American Journal of Human Genetics
American Journal of Hypertension
American Journal of Industrial Medicine
American Journal of Infection Control
American Journal of Kidney Diseases
American Journal of Law and Medicine
American Journal of Managed Care
American Journal of Medical Genetics
American Journal of Medical Genetics - Neuropsychiatric Genetics
American Journal of Medical Genetics - Seminars in Medical Genetics
American Journal of Medicine
American Journal of Neuroradiology
American Journal of Obstetrics and Gynecology
American Journal of Oncology Review
American Journal of Ophthalmology
American Journal of Orthopsychiatry
American Journal of Otolaryngology - Head and Neck Medicine and Surgery
American Journal of Pathology
American Journal of Perinatology
American Journal of Physical Anthropology
American Journal of Physical Medicine and Rehabilitation
American Journal of Physiology - Cell Physiology
American Journal of Physiology - Endocrinology and Metabolism
American Journal of Physiology - Gastrointestinal and Liver Physiology
American Journal of Physiology - Heart and Circulatory Physiology
American Journal of Physiology - Lung Cellular and Molecular Physiology
American Journal of Physiology - Regulatory Integrative and Comparative Physiology
American Journal of Physiology - Renal Physiology
American Journal of Preventive Medicine
American Journal of Psychiatric Rehabilitation
American Journal of Psychiatry
American Journal of Psychotherapy
American Journal of Public Health
American Journal of Respiratory Cell and Molecular Biology
American Journal of Respiratory and Critical Care Medicine
American Journal of Rhinology
American Journal of Roentgenology
American Journal of Sports Medicine
American Journal of Surgery
American Journal of Surgical Pathology
American Journal of Therapeutics
American Journal of Tropical Medicine and Hygiene
American Journal of Urology Review
American Journal of Veterinary Research
American Journal of the Medical Sciences

American Journal on Mental Retardation
American Pharmaceutical Outsourcing
American Pharmaceutical Review
American Surgeon
Analytical Biochemistry
Analytical Chemistry
Analytical and Quantitative Cytology and Histology
Anatomical Record - Part A Discoveries in Molecular, Cellular, and Evolutionary Biology
Anatomical Record - Part B New Anatomist
Anesthesia and Analgesia
Anesthesiology
Anesthesiology Clinics of North America
Angiology
Annals of Allergy, Asthma and Immunology
Annals of Behavioral Medicine
Annals of Cancer Research and Therapy
Annals of Clinical and Laboratory Science
Annals of Diagnostic Pathology
Annals of Emergency Medicine
Annals of Epidemiology
Annals of Family Medicine
Annals of Internal Medicine
Annals of Long-Term Care
Annals of Neurology
Annals of Noninvasive Electrocardiology
Annals of Occupational Hygiene
Annals of Ophthalmology
Annals of Otology, Rhinology and Laryngology
Annals of Pharmacotherapy
Annals of Plastic Surgery
Annals of Surgery
Annals of Surgical Oncology
Annals of Thoracic Surgery
Annals of Vascular Surgery
Annals of the New York Academy of Sciences
Annexins
Annual Review of Biochemistry
Annual Review of Biomedical Engineering
Annual Review of Biophysics and Biomolecular Structure
Annual Review of Cell and Developmental Biology
Annual Review of Clinical Psychology
Annual Review of Genetics
Annual Review of Genomics and Human Genetics
Annual Review of Immunology
Annual Review of Medicine
Annual Review of Microbiology
Annual Review of Neuroscience
Annual Review of Nutrition
Annual Review of Pathology
Annual Review of Pharmacology and Toxicology
Annual Review of Physiology
Annual Review of Public Health
Antimicrobial Agents and Chemotherapy
Antioxidants and Redox Signaling
Applied Biochemistry and Biotechnology
Applied Immunohistochemistry and Molecular Morphology
Applied Neuropsychology
Applied Psychophysiology Biofeedback
Applied Radiology
Applied and Environmental Microbiology
Archives of Biochemistry and Biophysics
Archives of Dermatology
Archives of Environmental Contamination and Toxicology
Archives of Environmental Health
Archives of General Psychiatry
Archives of Internal Medicine
Archives of Medical Research
Archives of Neurology
Archives of Ophthalmology
Archives of Otolaryngology - Head and Neck Surgery
Archives of Pathology and Laboratory Medicine
Archives of Pediatrics and Adolescent Medicine
Archives of Physical Medicine and Rehabilitation
Archives of Sexual Behavior
Archives of Surgery
Arteriosclerosis, Thrombosis, and Vascular Biology
Arthritis Care and Research
Arthritis and Rheumatism
Arthroscopy - Journal of Arthroscopic and Related Surgery
Artificial Cells, Blood Substitutes, and Biotechnology
Artificial Organs
Assay and Drug Development Technologies
Assistive Technology
Athletic Therapy Today
Atlas of Hand Clinics
Atlas of the Oral and Maxillofacial Surgery Clinics of North America
Atlas of the Urologic Clinics of North America
Autophagy
Aviation Space and Environmental Medicine
Behavior Genetics
Behavior Modification
Behavior Therapy
Behavioral Medicine
Behavioral Neuroscience
Behavioral and Cognitive Neuroscience Reviews
Binocular Vision and Strabismus Quarterly
BioEssays
BioPharm International
BioTechniques
Biochemical Genetics
Biochemical Pharmacology
Biochemical and Biophysical Research Communications
Biochemistry
Biochemistry and Molecular Biology Education
Bioconjugate Chemistry
Bioelectromagnetics
Biological Psychiatry
Biological Trace Element Research
Biology of Blood and Marrow Transplantation
Biology of Reproduction
Biomacromolecules
Biomedical Engineering
Biomedical Instrumentation and Technology
Biomedical Microdevices
Biomedical Sciences Instrumentation
Biometrics
Bioorganic Chemistry
Biophysical Journal
Biopolymers
Biopolymers - Peptide Science Section
Bioscience Reports
Biotechnology and Bioengineering
Birth
Birth Defects Research Part A - Clinical and Molecular Teratology
Birth Defects Research Part B - Developmental and Reproductive Toxicology
Birth Defects Research Part C - Embryo Today: Reviews

Blood
Blood Cells, Molecules, and Diseases
Bone
Brachytherapy
Brain Pathology
Brain Research Bulletin
Brain Topography
Brain and Cognition
Brain and Language
Brain, Behavior, and Immunity
Breast Cancer Research and Treatment
Breast Journal
Bulletin of Environmental Contamination and Toxicology
Bulletin of Experimental Biology and Medicine
Bulletin of the Menninger Clinic
Bulletin on the Rheumatic Diseases
Bulletin: Hospital for Joint Diseases
CNS Spectrums
CONTINUUM Lifelong Learning in Neurology
COPD: Journal of Chronic Obstructive Pulmonary Disease
Ca-A Cancer Journal for Clinicians
Calcified Tissue International
Cancer
Cancer Biology and Therapy
Cancer Biotherapy and Radiopharmaceuticals
Cancer Cell
Cancer Control
Cancer Epidemiology Biomarkers and Prevention
Cancer Gene Therapy
Cancer Genetics and Cytogenetics
Cancer Investigation
Cancer Journal
Cancer Nursing
Cancer Research
CardioVascular and Interventional Radiology
Cardiology Clinics
Cardiology Review
Cardiology in Review
Cardiovascular Drugs and Therapy
Cardiovascular Pathology
Cardiovascular Radiation Medicine
Cardiovascular Revascularization Medicine
Cardiovascular Toxicology
Care Management Journals
Catheterization and Cardiovascular Interventions
Cell
Cell Biology Education
Cell Cycle
Cell Metabolism
Cell Motility and the Cytoskeleton
Cell Preservation Technology
Cell Stress and Chaperones
Cell Transplantation
Cells and Materials
Cellular Immunology
Cellular Signalling
Cellular and Molecular Neurobiology
Cerebral Cortex
Chemical Research in Toxicology
Chemtracts
Chest
Child Psychiatry and Human Development
Child and Adolescent Psychiatric Clinics of North America
Children's Health Care
Chirality
Chronobiology International
Circulation
Circulation Research
Cleft Palate-Craniofacial Journal
Cleveland Clinic Journal of Medicine
Clinical Advances in Hematology and Oncology
Clinical Anatomy
Clinical Biochemistry
Clinical Breast Cancer
Clinical Cancer Research
Clinical Cardiology
Clinical Case Studies
Clinical Chemistry
Clinical Child Psychology and Psychiatry
Clinical Colorectal Cancer
Clinical Cornerstone
Clinical Diabetes
Clinical EEG and Neuroscience
Clinical Gastroenterology and Hepatology
Clinical Genitourinary Cancer
Clinical Geriatrics
Clinical Gerontologist
Clinical Imaging
Clinical Immunology
Clinical Infectious Diseases
Clinical Journal of Pain
Clinical Journal of Sport Medicine
Clinical Kinesiology
Clinical Lung Cancer
Clinical Lymphoma and Myeloma
Clinical Medicine and Research
Clinical Microbiology Newsletter
Clinical Microbiology Reviews
Clinical Neuropharmacology
Clinical Nuclear Medicine
Clinical Obstetrics and Gynecology
Clinical Orthopaedics and Related Research
Clinical Pediatrics
Clinical Pharmacology and Therapeutics
Clinical Proteomics
Clinical Psychology Review
Clinical Psychology: Science and Practice
Clinical Pulmonary Medicine
Clinical Research and Regulatory Affairs
Clinical Reviews in Allergy and Immunology
Clinical Reviews in Bone and Mineral Metabolism
Clinical Therapeutics
Clinical Toxicology
Clinical and Applied Immunology Reviews
Clinical and Applied Thrombosis/Hemostasis
Clinical and Diagnostic Laboratory Immunology
Clinical and Experimental Hypertension
Clinical and Vaccine Immunology
Clinics in Chest Medicine
Clinics in Colon and Rectal Surgery
Clinics in Dermatology
Clinics in Family Practice
Clinics in Geriatric Medicine
Clinics in Laboratory Medicine
Clinics in Liver Disease
Clinics in Occupational and Environmental Medicine
Clinics in Perinatology
Clinics in Plastic Surgery
Clinics in Podiatric Medicine and Surgery
Clinics in Sports Medicine

Cloning and Stem Cells
Cognitive Therapy and Research
Cognitive and Behavioral Neurology
Cognitive and Behavioral Practice
Cognitive, Affective and Behavioral Neuroscience
Cold Spring Harbor Symposia on Quantitative Biology
Combustion and Flame
Community Oncology
Comparative Biochemistry and Physiology - A Molecular and Integrative Physiology
Comparative Biochemistry and Physiology - B Biochemistry and Molecular Biology
Comparative Biochemistry and Physiology - C Toxicology and Pharmacology
Comparative Biochemistry and Physiology - Part D: Genomics and Proteomics
Comparative Medicine
Comparative and Functional Genomics
Comprehensive Ophthalmology Update
Comprehensive Psychiatry
Comprehensive Therapy
Connecticut Medicine
Consciousness and Cognition
Consultant
Consultant Pharmacist
Contagion: Reports, Cases, and Commentaries in HIV and Infectious Disease Research
Contemporary Clinical Trials
Contemporary Topics in Laboratory Animal Science
Contraception
Core Evidence
Cornea
Cosmetic Dermatology
Critical Care Clinics
Critical Care Medicine
Critical Care Nursing Quarterly
Critical Pathways in Cardiology
Critical Reviews in Biochemistry and Molecular Biology
Critical Reviews in Biomedical Engineering
Critical Reviews in Biotechnology
Critical Reviews in Clinical Laboratory Sciences
Critical Reviews in Computed Tomography
Critical Reviews in Environmental Science and Technology
Critical Reviews in Eukaryotic Gene Expression
Critical Reviews in Immunology
Critical Reviews in Microbiology
Critical Reviews in Neurobiology
Critical Reviews in Oral Biology and Medicine
Critical Reviews in Physical and Rehabilitation Medicine
Critical Reviews in Therapeutic Drug Carrier Systems
Critical Reviews in Toxicology
Cryobiology
Cultural Diversity and Ethnic Minority Psychology
Current HIV/AIDS Reports
Current Heart Failure Reports
Current Hematology Reports
Current Hepatitis Reports
Current Microbiology
Current Opinion in Allergy and Clinical Immunology
Current Opinion in Cardiology
Current Opinion in Clinical Nutrition and Metabolic Care
Current Opinion in Critical Care
Current Opinion in Endocrinology and Diabetes
Current Opinion in Gastroenterology
Current Opinion in Hematology
Current Opinion in Oncology
Current Opinion in Ophthalmology
Current Opinion in Organ Transplantation
Current Opinion in Orthopaedics
Current Opinion in Otolaryngology and Head and Neck Surgery
Current Opinion in Pediatrics
Current Opinion in Pulmonary Medicine
Current Opinion in Rheumatology
Current Osteoporosis Reports
Current Problems in Cancer
Current Problems in Cardiology
Current Problems in Diagnostic Radiology
Current Problems in Pediatric and Adolescent Health Care
Current Problems in Surgery
Current Prostate Reports
Current Psychosis and Therapeutics Reports
Current Rheumatology Reports
Current Sexual Health Reports
Current Sports Medicine Reports
Current Surgery
Current Therapeutic Research - Clinical and Experimental
Current Topics in Nutraceutical Research
Current Urology Reports
Cutaneous and Ocular Toxicology
Cyberpsychology and Behavior
Cytometry Part A
Cytometry Part B - Clinical Cytometry
DNA and Cell Biology
Death Studies
Depression and Anxiety
Dermatologic Clinics
Dermatologic Therapy
Dermatology Online Journal
Developmental Biology
Developmental Cell
Developmental Dynamics
Developmental Neuropsychology
Developmental Psychobiology
Diabetes
Diabetes Care
Diabetes Spectrum
Diabetes Technology and Therapeutics
Diagnostic Cytopathology
Diagnostic Microbiology and Infectious Disease
Diagnostic Molecular Pathology
Dialysis and Transplantation
Digestive Diseases and Sciences
Digestive and Liver Disease
Disease Management
Disease-a-Month
Diseases of the Colon and Rectum
Drug Benefit Trends
Drug Development Research
Drug Development and Industrial Pharmacy
Drug Information Journal
Drug Metabolism Reviews
Drug Metabolism and Disposition
Drug Topics
Drug and Chemical Toxicology
Dysphagia
Ear and Hearing
Ear, Nose and Throat Journal
Echocardiography
EcoHealth

Electronic Journal of Human Sexuality
Emergency Medicine Clinics of North America
Emergency Radiology
Emerging Infectious Diseases
Emotion
Emotional and Behavioural Difficulties
Endocrine
Endocrine Pathology
Endocrine Practice
Endocrine Research
Endocrine Reviews
Endocrinologist
Endocrinology
Endocrinology and Metabolism Clinics of North America
Environment
Environmental Engineering Science
Environmental Health Perspectives
Environmental Management
Environmental Research
Environmental Science and Technology
Environmental Toxicology
Environmental Toxicology and Chemistry
Environmental and Molecular Mutagenesis
Enzyme and Microbial Technology
Epidemiology
Epilepsia
Epilepsy and Behavior
Ethical Human Psychology and Psychiatry
Ethnicity and Disease
European Journal of Emergency Medicine
Evaluation and the Health Professions
Evidence-Based Gastroenterology
Evidence-Based Ophthalmology
Exercise and Sport Sciences Reviews
Experimental Aging Research
Experimental Biology and Medicine
Experimental Cell Research
Experimental Diabesity Research
Experimental Gerontology
Experimental Hematology
Experimental Lung Research
Experimental Neurology
Experimental Parasitology
Experimental and Clinical Psychopharmacology
Experimental and Molecular Pathology
Explore: The Journal of Science and Healing
Eye and Contact Lens
FASEB Journal
Facial Plastic Surgery
Facial Plastic Surgery Clinics of North America
Families, Systems and Health
Family Medicine
Fertility and Sterility
Food and Drug Law Journal
Foodborne Pathogens and Disease
Foot and Ankle International
Formulary
Free Radical Biology and Medicine
Frontiers in Neuroendocrinology
Fungal Genetics and Biology
Future of Children
GLIA
Gastroenterology
Gastroenterology Clinics of North America
Gastroenterology and Hepatology
Gastrointestinal Endoscopy
Gastrointestinal Endoscopy Clinics of North America
Gender Medicine
Gender and Psychoanalysis
Gene Expression
General Hospital Psychiatry
General and Comparative Endocrinology
Genes Chromosomes and Cancer
Genes and Development
Genesis
Genetic Epidemiology
Genetic Testing
Genetics
Genetics in Medicine
Genome Letters
Genome Research
Genomics
Geriatrics
Gerontologist
Group Dynamics
Growth, Development and Aging
Gynecologic Oncology
HIV Clinical Trials
Hallym International Journal of Aging
Hand Clinics
Head and Neck
Headache
Headache and Pain: Diagnostic Challenges, Current Therapy
Health Care Financing Review
Health Care Management Review
Health Physics
Health Psychology
Health Services Research
Hearing Journal
Heart Failure Clinics
Heart Surgery Forum
Heart and Lung: Journal of Acute and Critical Care
Hematology/Oncology Clinics of North America
Hemodialysis International
Hemoglobin
Hepatology
High Altitude Medicine and Biology
Hippocampus
Hispanic Healthcare International
Hormones and Behavior
Hospital Pharmacy
Hospitals and Health Networks
Human Biology
Human Brain Mapping
Human Factors
Human Gene Therapy
Human Immunology
Human Mutation
Human Pathology
Human Vaccines
Hybridoma
Hypertension
Hypertension in Pregnancy
IEEE Engineering in Medicine and Biology Magazine
IEEE Transactions on Biomedical Engineering
IEEE Transactions on Neural Systems and Rehabilitation Engineering
ILAR Journal
IM - Internal Medicine
Immunity

Immunohematology
Immunologic Research
Immunological Investigations
Immunology and Allergy Clinics of North America
Immunopharmacology and Immunotoxicology
In Vitro Cellular and Developmental Biology - Animal
Infancy
Infant Mental Health Journal
Infection Control and Hospital Epidemiology
Infection and Immunity
Infections in Medicine
Infectious Disease Clinics of North America
Infectious Diseases in Clinical Practice
Infectious Diseases in Obstetrics and Gynecology
Inflammation
Inflammatory Bowel Diseases
Inquiry
Insulin
Integrative Cancer Therapies
Integrative Medicine
International Anesthesiology Clinics
International Journal of Aging and Human Development
International Journal of Angiology
International Journal of Behavioral Medicine
International Journal of Biostatistics
International Journal of Cancer
International Journal of Cosmetic Surgery and Aesthetic Dermatology
International Journal of Eating Disorders
International Journal of Fertility and Women's Medicine
International Journal of Gastrointestinal Cancer
International Journal of Group Psychotherapy
International Journal of Gynecological Cancer
International Journal of Gynecological Pathology
International Journal of Health Services
International Journal of Hematology
International Journal of Leprosy and Other Mycobacterial Diseases
International Journal of Lower Extremity Wounds
International Journal of Medicinal Mushrooms
International Journal of Men's Health
International Journal of Nursing Education Scholarship
International Journal of Occupational and Environmental Health
International Journal of Offender Therapy and Comparative Criminology
International Journal of Peptide Research and Therapeutics
International Journal of Pharmaceutical Compounding
International Journal of Psychiatry in Medicine
International Journal of Radiation Oncology Biology Physics
International Journal of Sport Nutrition and Exercise Metabolism
International Journal of Surgical Pathology
International Journal of Technology Assessment in Health Care
International Ophthalmology Clinics
International Pediatrics
International Psychogeriatrics
International Review of Cytology
International Tinnitus Journal
Investigational New Drugs
Investigative Ophthalmology and Visual Science
Investigative Radiology
JALA - Journal of the Association for Laboratory Automation
JARO - Journal of the Association for Research in Otolaryngology
JAVA - Journal of the Association for Vascular Access
JNCCN Journal of the National Comprehensive Cancer Network
Journal for Vascular Ultrasound
Journal of AAPOS
Journal of Abnormal Child Psychology
Journal of Abnormal Psychology
Journal of Acquired Immune Deficiency Syndromes
Journal of Addictive Diseases
Journal of Adolescent Health
Journal of Aerosol Medicine: Deposition, Clearance, and Effects in the Lung
Journal of Aggression, Maltreatment and Trauma
Journal of Aging and Health
Journal of Aging and Pharmacotherapy
Journal of Aging and Physical Activity
Journal of Allergy and Clinical Immunology
Journal of Alternative and Complementary Medicine
Journal of Ambulatory Care Management
Journal of Analytical Toxicology
Journal of Andrology
Journal of Applied Animal Welfare Science
Journal of Applied Biomechanics
Journal of Applied Gerontology
Journal of Applied Physiology
Journal of Applied Research
Journal of Arthroplasty
Journal of Assisted Reproduction and Genetics
Journal of Asthma
Journal of Athletic Training
Journal of Autism and Developmental Disorders
Journal of Automated Methods and Management in Chemistry
Journal of Bacteriology
Journal of Behavioral Health Services and Research
Journal of Behavioral Medicine
Journal of Biochemical and Molecular Toxicology
Journal of Bioenergetics and Biomembranes
Journal of Biolaw and Business
Journal of Biological Chemistry
Journal of Biological Rhythms
Journal of Biomechanical Engineering
Journal of Biomedical Informatics
Journal of Biomedical Materials Research - Part A
Journal of Biomedical Materials Research - Part B Applied Biomaterials
Journal of Biomedicine and Biotechnology
Journal of Biomolecular Screening
Journal of Biomolecular Structure and Dynamics
Journal of Biopharmaceutical Statistics
Journal of Bone and Joint Surgery - Series A
Journal of Bone and Mineral Research
Journal of Brain Science
Journal of Brief Therapy
Journal of Bronchology
Journal of Burn Care and Rehabilitation
Journal of Cancer Education
Journal of Cancer Integrative Medicine
Journal of Cardiac Failure
Journal of Cardiac Surgery
Journal of Cardiopulmonary Rehabilitation
Journal of Cardiothoracic and Vascular Anesthesia

Journal of Cardiovascular Electrophysiology
Journal of Cardiovascular Magnetic Resonance
Journal of Cardiovascular Medicine
Journal of Cardiovascular Pharmacology
Journal of Cardiovascular Pharmacology and Therapeutics
Journal of Cataract and Refractive Surgery
Journal of Cell Biology
Journal of Cellular Biochemistry
Journal of Cellular Physiology
Journal of Cerebral Blood Flow and Metabolism
Journal of Child Sexual Abuse
Journal of Child and Adolescent Psychopharmacology
Journal of Chronic Fatigue Syndrome
Journal of Clinical Anesthesia
Journal of Clinical Apheresis
Journal of Clinical Densitometry
Journal of Clinical Endocrinology and Metabolism
Journal of Clinical Engineering
Journal of Clinical Epidemiology
Journal of Clinical Gastroenterology
Journal of Clinical Immunology
Journal of Clinical Investigation
Journal of Clinical Laboratory Analysis
Journal of Clinical Ligand Assay
Journal of Clinical Microbiology
Journal of Clinical Neuromuscular Disease
Journal of Clinical Neurophysiology
Journal of Clinical Oncology
Journal of Clinical Outcomes Management
Journal of Clinical Pharmacology
Journal of Clinical Psychiatry
Journal of Clinical Psychoanalysis
Journal of Clinical Psychology
Journal of Clinical Psychology in Medical Settings
Journal of Clinical Psychopharmacology
Journal of Clinical Rheumatology
Journal of Clinical Sleep Medicine
Journal of Clinical Ultrasound
Journal of Cognition and Development
Journal of Cognitive Neuroscience
Journal of Cognitive Psychotherapy
Journal of Colloid and Interface Science
Journal of Communication Disorders
Journal of Community Health Nursing
Journal of Comparative Neurology
Journal of Complementary and Integrative Medicine
Journal of Computational Biology
Journal of Computer Assisted Tomography
Journal of Consulting and Clinical Psychology
Journal of Contemporary Dental Practice
Journal of Contemporary Psychotherapy
Journal of Counseling Psychology
Journal of Craniofacial Surgery
Journal of Critical Care
Journal of Critical Illness
Journal of Cutaneous Medicine and Surgery
Journal of Developmental and Behavioral Pediatrics
Journal of Diabetes and its Complications
Journal of Diagnostic Medical Sonography
Journal of Digital Imaging
Journal of Drug Education
Journal of Drug Issues
Journal of Dual Diagnosis
Journal of ECT
Journal of Electrocardiology
Journal of Emergency Medicine
Journal of Emotional Abuse
Journal of Endourology
Journal of Endovascular Therapy
Journal of Environmental Health
Journal of Environmental Pathology, Toxicology and Oncology
Journal of Environmental Quality
Journal of Environmental Science and Health - Part A Toxic/Hazardous Substances and Environmental Engineering
Journal of Environmental Science and Health - Part B Pesticides, Food Contaminants, and Agricultural Wastes
Journal of Environmental Science and Health - Part C Environmental Carcinogenesis and Ecotoxicology Reviews
Journal of Ethnicity in Substance Abuse
Journal of Exercise Physiology Online
Journal of Experimental Medicine
Journal of Exposure Science and Environmental Epidemiology
Journal of Extra-Corporeal Technology
Journal of Facial and Somato Prosthetics
Journal of Family Practice
Journal of Family Psychotherapy
Journal of Fluency Disorders
Journal of Foot and Ankle Surgery
Journal of Forensic Identification
Journal of Forensic Neuropsychology
Journal of Forensic Psychology Practice
Journal of Forensic Sciences
Journal of General Internal Medicine
Journal of General Physiology
Journal of Genetic Counseling
Journal of Geriatric Physical Therapy
Journal of Geriatric Psychiatry and Neurology
Journal of Glaucoma
Journal of Gynecologic Surgery
Journal of HIV/AIDS Prevention in Children and Youth
Journal of Hand Surgery
Journal of Hand Therapy
Journal of Head Trauma Rehabilitation
Journal of Health Care Finance
Journal of Health Politics, Policy and Law
Journal of Health and Human Services Administration
Journal of Health and Social Policy
Journal of Healthcare Management
Journal of Heart and Lung Transplantation
Journal of Herbal Pharmacotherapy
Journal of Herbs, Spices and Medicinal Plants
Journal of Heredity
Journal of Heterocyclic Chemistry
Journal of Histochemistry and Cytochemistry
Journal of Histotechnology
Journal of Homosexuality
Journal of Hospital Marketing and Public Relations
Journal of Human Lactation
Journal of Hypertension
Journal of Immigrant and Minority Health
Journal of Immunoassay and Immunochemistry
Journal of Immunology
Journal of Immunotherapy
Journal of Immunotoxicology
Journal of Infectious Diseases
Journal of Infusion Nursing
Journal of Inorganic Biochemistry

Journal of Intensive Care Medicine
Journal of Interferon and Cytokine Research
Journal of Interventional Cardiology
Journal of Intravenous Nursing
Journal of Invasive Cardiology
Journal of Investigative Dermatology
Journal of Investigative Medicine
Journal of LGBT Health Research
Journal of Laboratory and Clinical Medicine
Journal of Laparoendoscopic and Advanced Surgical Techniques - Part A
Journal of Leukocyte Biology
Journal of Lipid Research
Journal of Liposome Research
Journal of Liquid Chromatography and Related Technologies
Journal of Long-Term Effects of Medical Implants
Journal of Lower Genital Tract Disease
Journal of Magnetic Resonance Imaging
Journal of Mammary Gland Biology and Neoplasia
Journal of Manipulative and Physiological Therapeutics
Journal of Manual and Manipulative Therapy
Journal of Medical Practice Management
Journal of Medical Speech-Language Pathology
Journal of Medical Systems
Journal of Medical Virology
Journal of Medicinal Chemistry
Journal of Medicinal Food
Journal of Medicine
Journal of Membrane Biology
Journal of Mental Health and Aging
Journal of Midwifery and Women's Health
Journal of Mind and Behavior
Journal of Molecular Diagnostics
Journal of Molecular Evolution
Journal of Molecular Graphics and Modelling
Journal of Molecular Neuroscience
Journal of Motor Behavior
Journal of Musculoskeletal Pain
Journal of Natural Products
Journal of Nervous and Mental Disease
Journal of Neuro-Oncology
Journal of Neuro-Ophthalmology
Journal of NeuroVirology
Journal of Neurobiology
Journal of Neuroimaging
Journal of Neuropathology and Experimental Neurology
Journal of Neurophysiology
Journal of Neuropsychiatry and Clinical Neurosciences
Journal of Neuroscience
Journal of Neuroscience Research
Journal of Neurosurgery
Journal of Neurosurgical Anesthesiology
Journal of Neurotherapy
Journal of Neurotrauma
Journal of Nuclear Cardiology
Journal of Nuclear Medicine
Journal of Nuclear Medicine Technology
Journal of Nursing Administration
Journal of Nutraceuticals, Functional and Medical Foods
Journal of Nutrition
Journal of Nutritional Biochemistry
Journal of Occupational Health Psychology
Journal of Occupational Rehabilitation
Journal of Occupational and Environmental Medicine
Journal of Ocular Pharmacology and Therapeutics
Journal of Oncology Pharmacy Practice
Journal of Oral and Maxillofacial Surgery
Journal of Organic Chemistry
Journal of Orthopaedic Research
Journal of Orthopaedic Trauma
Journal of Orthopaedic and Sports Physical Therapy
Journal of Pain
Journal of Pain and Palliative Care Pharmacotherapy
Journal of Pain and Symptom Management
Journal of Palliative Medicine
Journal of Parasitology
Journal of Parenteral and Enteral Nutrition
Journal of Pediatric Gastroenterology and Nutrition
Journal of Pediatric Hematology/Oncology
Journal of Pediatric Ophthalmology and Strabismus
Journal of Pediatric Orthopaedics
Journal of Pediatric Orthopaedics Part B
Journal of Pediatric Psychology
Journal of Pediatric Surgery
Journal of Pediatric and Adolescent Gynecology
Journal of Pediatrics
Journal of Pelvic Medicine and Surgery
Journal of Periodontology
Journal of Personality Assessment
Journal of Personality Disorders
Journal of Pharmaceutical Finance, Economics and Policy
Journal of Pharmaceutical Sciences
Journal of Pharmacokinetics and Pharmacodynamics
Journal of Pharmacological and Toxicological Methods
Journal of Pharmacology and Experimental Therapeutics
Journal of Pharmacy Practice
Journal of Pharmacy Technology
Journal of Prevention and Intervention in the Community
Journal of Primary Prevention
Journal of Proteome Research
Journal of Psychiatric Practice
Journal of Psychiatry and Law
Journal of Psychoactive Drugs
Journal of Psychopathology and Behavioral Assessment
Journal of Psychosocial Oncology
Journal of Psychosomatic Research
Journal of Psychotherapy Integration
Journal of Public Health Management and Practice
Journal of Public Health Policy
Journal of Radiology Nursing
Journal of Receptor and Signal Transduction
Journal of Reconstructive Microsurgery
Journal of Refractive Surgery
Journal of Rehabilitation Research and Development
Journal of Reproductive Medicine for the Obstetrician and Gynecologist
Journal of Research in Crime and Delinquency
Journal of Respiratory Diseases
Journal of Sex Education and Therapy
Journal of Sex Research
Journal of Shoulder and Elbow Surgery
Journal of Speech, Language, and Hearing Research
Journal of Spinal Cord Medicine
Journal of Spinal Disorders and Techniques
Journal of Spirochetal and Tick-borne Diseases
Journal of Sport Rehabilitation
Journal of Stroke and Cerebrovascular Diseases
Journal of Structural Biology
Journal of Studies on Alcohol

Journal of Substance Abuse Treatment
Journal of Supportive Oncology
Journal of Surgical Oncology
Journal of Surgical Research
Journal of Terminal Oncology
Journal of Thoracic Imaging
Journal of Thoracic and Cardiovascular Surgery
Journal of Toxicology and Environmental Health - Part B: Critical Reviews
Journal of Trauma - Injury, Infection and Critical Care
Journal of Trauma and Dissociation
Journal of Traumatic Stress
Journal of Ultrasound in Medicine
Journal of Urology
Journal of Vascular Surgery
Journal of Vascular and Interventional Radiology
Journal of Virology
Journal of Vision
Journal of Visual Impairment and Blindness
Journal of Voice
Journal of Whiplash and Related Disorders
Journal of Women's Cancer
Journal of Women's Health
Journal of Women's Imaging
Journal of the Acoustical Society of America
Journal of the Air and Waste Management Association
Journal of the American Academy of Audiology
Journal of the American Academy of Child and Adolescent Psychiatry
Journal of the American Academy of Dermatology
Journal of the American Academy of Psychiatry and the Law
Journal of the American Academy of Psychoanalysis and Dynamic Psychiatry
Journal of the American Aging Association
Journal of the American Animal Hospital Association
Journal of the American Board of Family Practice
Journal of the American Chemical Society
Journal of the American College of Cardiology
Journal of the American College of Nutrition
Journal of the American College of Surgeons
Journal of the American Geriatrics Society
Journal of the American Health Information Management Association
Journal of the American Medical Association
Journal of the American Medical Directors Association
Journal of the American Medical Informatics Association
Journal of the American Osteopathic Association
Journal of the American Pharmacists Association
Journal of the American Society for Mass Spectrometry
Journal of the American Society of Echocardiography
Journal of the American Society of Nephrology
Journal of the American Veterinary Medical Association
Journal of the Hoffman Heart Institute of Connecticut
Journal of the ICRU
Journal of the Medical Library Association
Journal of the National Medical Association
Journal of the Peripheral Nervous System
Journal of the Society for Gynecologic Investigation
Journals of Gerontology - Series A Biological Sciences and Medical Sciences
Journals of Gerontology - Series B Psychological Sciences and Social Sciences
Laboratory Hematology
Laboratory Medicine
Lancet Oncology
Laryngoscope
Lasers in Surgery and Medicine
Law and Human Behavior
Learning and Behavior
Learning and Memory
Life Sciences
Lipids
Lung
Lymphatic Research and Biology
Lymphology
Magnetic Resonance Imaging
Magnetic Resonance Imaging Clinics of North America
Magnetic Resonance in Medicine
Mammalian Genome
Managed Care Interface
Managed Care Pharmacy Update
Marketing Health Services
Maternal and Child Health Journal
Mathematical Biosciences
Mayo Clinic Proceedings
Measurement in Physical Education and Exercise Science
MedGenMed Medscape General Medicine
Mediators of Inflammation
Medical Care Research and Review
Medical Clinics of North America
Medical Crossfire
Medical Decision Making
Medical Dosimetry
Medical Group Management Journal
Medical Letter on Drugs and Therapeutics
Medical Oncology
Medical Physics
Medical Problems of Performing Artists
Medical Science Monitor
Medicinal Chemistry Research
Medicinal Research Reviews
Medicine
Medicine and Science in Sports and Exercise
Memory and Cognition
Menopause
Mental Health Aspects of Developmental Disabilities
Mental Retardation and Developmental Disabilities Research Reviews
Metabolic Brain Disease
Metabolic Engineering
Metabolic Syndrome and Related Disorders
Metabolism: Clinical and Experimental
Methods
Methods in Enzymology
Microbial Drug Resistance
Microbiology and Molecular Biology Reviews
Microcirculation
Microscopy Research and Technique
Microsurgery
Microvascular Research
Military Medicine
Molecular Biology and Evolution
Molecular Biology of the Cell
Molecular Biotechnology
Molecular Cancer Research
Molecular Cancer Therapeutics
Molecular Carcinogenesis
Molecular Cell
Molecular Endocrinology
Molecular Genetics and Metabolism

Molecular Imaging and Biology
Molecular Interventions
Molecular Medicine
Molecular Neurobiology
Molecular Pharmaceutics
Molecular Pharmacology
Molecular Reproduction and Development
Molecular Therapy
Molecular Vision
Molecular and Cellular Biology
Molecular and Cellular Neuroscience
Molecular and Cellular Proteomics
Motor Control
Mount Sinai Journal of Medicine
Movement Disorders
Muscle and Nerve
NCRP Report
NIDA Research Monograph Series
Nanomedicine: Nanotechnology, Biology, and Medicine
Natural Computing
Nature Biotechnology
Nature Genetics
Nature Neuroscience
Nature Structural and Molecular Biology
Neoplasia
Neural Plasticity
Neuro-Oncology
NeuroImage
NeuroMolecular Medicine
NeuroRx
Neurobiology of Aging
Neurobiology of Disease
Neurobiology of Learning and Memory
Neurochemical Research
Neurocritical Care
Neuroimaging Clinics of North America
Neuroinformatics
Neurologic Clinics
Neurological Research
Neurologist
Neurology
Neurology and Clinical Neurophysiology
Neuromodulation
Neuron
Neurophysiology
Neuropsychology
Neuropsychology Review
Neurorehabilitation and Neural Repair
Neuroscience and Behavioral Physiology
Neuroscientist
Neurosurgery
Neurosurgery Clinics of North America
Neurosurgery Quarterly
Neurotoxicology and Teratology
Neurourology and Urodynamics
New England Journal of Medicine
Nitric Oxide - Biology and Chemistry
Nuclear Medicine and Biology
Nucleosides, Nucleotides and Nucleic Acids
Nutrition
Nutrition Research
Nutrition Reviews
Nutrition and Cancer
Nutrition in Clinical Care
Nutrition in Clinical Practice
OMICS A Journal of Integrative Biology
ONCOLOGY
OTJR Occupation, Participation and Health
Obesity Management
Obesity Surgery
Obstetrical and Gynecological Survey
Obstetrics and Gynecology
Obstetrics and Gynecology Clinics of North America
Occupational Therapy in Health Care
Occupational Therapy in Mental Health
Ochsner Journal
Ocular Surface
Oligonucleotides
Oncologist
Oncology Report
Oncology Research
Online Journal of Issues in Nursing
Online Journal of Knowledge Synthesis for Nursing
Online Journal of Nursing Informatics
Operative Techniques in General Surgery
Operative Techniques in Orthopaedics
Operative Techniques in Otolaryngology - Head and Neck Surgery
Operative Techniques in Plastic and Reconstructive Surgery
Operative Techniques in Sports Medicine
Operative Techniques in Thoracic and Cardiovascular Surgery
Ophthalmic Plastic and Reconstructive Surgery
Ophthalmic Surgery Lasers and Imaging
Ophthalmology
Ophthalmology Clinics of North America
Optometry and Vision Science
Oral and Maxillofacial Surgery Clinics of North America
Organogenesis
Orthopedic Clinics of North America
Orthopedics
Otolaryngologic Clinics of North America
Otolaryngology - Head and Neck Surgery
Otology and Neurotology
P and T
PACE - Pacing and Clinical Electrophysiology
PAT
PDA Journal of Pharmaceutical Science and Technology
PLoS Biology
PLoS Computational Biology
PLoS Genetics
PLoS Medicine
PLoS Pathogens
Pain Digest
Pain Medicine
Pain Physician
Pain Practice
Pancreas
Particulate Science and Technology
Pathology Case Reviews
Pediatric Annals
Pediatric Asthma, Allergy and Immunology
Pediatric Blood and Cancer
Pediatric Cardiac Surgery Annual
Pediatric Cardiology
Pediatric Case Reviews
Pediatric Clinics of North America
Pediatric Critical Care Medicine
Pediatric Dermatology
Pediatric Emergency Care

tric Endosurgery and Innovative Techniques
diatric Exercise Science
Pediatric Infectious Disease Journal
Pediatric Neurology
Pediatric Physical Therapy
Pediatric Pulmonology
Pediatric Research
Pediatric and Developmental Pathology
Pediatrics
Pediatrics in Review
Peptides
Perspectives in Biology and Medicine
Perspectives on Sexual and Reproductive Health
Pharmaceutical Chemistry Journal
Pharmaceutical Development and Technology
Pharmaceutical Engineering
Pharmaceutical Research
Pharmaceutical Technology
Pharmacological Reviews
Pharmacology Biochemistry and Behavior
Pharmacology and Therapeutics
Pharmacotherapy
Photomedicine and Laser Surgery
Physical Medicine and Rehabilitation Clinics of North America
Physical Therapy
Physical and Occupational Therapy in Geriatrics
Physical and Occupational Therapy in Pediatrics
Physician and Sportsmedicine
Physiological Genomics
Physiological Reviews
Physiology and Behavior
Pituitary
Plasmid
Plastic and Reconstructive Surgery
Postgraduate Medicine
Practical Gastroenterology
Prehospital Emergency Care
Preventive Medicine
Primary Care - Clinics in Office Practice
Primary Care Case Reviews
Primary Care Companion to the Journal of Clinical Psychiatry
Primary Psychiatry
Proceedings of the American Thoracic Society
Proceedings of the Controlled Release Society
Proceedings of the National Academy of Sciences of the United States of America
Proceedings of the Western Pharmacology Society
Progress in Cardiovascular Diseases
Progress in Neuro-Psychopharmacology and Biological Psychiatry
Progress in Transplantation
Prostaglandins and Other Lipid Mediators
Prostate
Protein Journal
Protein Science
Proteins: Structure, Function and Genetics
Psychiatric Clinics of North America
Psychiatric Quarterly
Psychiatric Rehabilitation Journal
Psychiatric Services
Psychiatry
Psychoanalysis and Psychotherapy
Psychoanalytic Psychology
Psychoanalytic Quarterly
Psychological Assessment
Psychology and Aging
Psychology of Addictive Behaviors
Psychophysiology
Psychosomatic Medicine
Psychosomatics
Psychotherapy
Public Health Reports
Qualitative Health Research
RNA
RNA Biology
Radiation Research
Radiation and Environmental Biophysics
Radiologic Clinics of North America
Radiology
Regional Anesthesia and Pain Medicine
Regulatory Toxicology and Pharmacology
Rehabilitation Psychology
Rejuvenation Research
Renal Failure
Reproductive Toxicology
Research Communications in Alcohol and Substances of Abuse
Research Communications in Biochemistry and Cell and Molecular Biology
Research Communications in Biological Psychology and Psychiatry
Research Communications in Molecular Pathology and Pharmacology
Research Communications in Pharmacology and Toxicology
Research in Developmental Disabilities
Research in Healthcare Financial Management
Research on Aging
Respiratory Care
Respiratory Care Clinics of North America
Retina
Reviews in Analgesia
Reviews in Cardiovascular Medicine
Reviews in Gastroenterological Disorders
Reviews in Neurological Diseases
Revista Panamericana de Salud Publica/Pan American Journal of Public Health
Rheumatic Disease Clinics of North America
Risk Analysis
Sarcoma
Scanning
Science
Scientific Review of Alternative Medicine
Seminars in Anesthesia, Perioperative Medicine and Pain
Seminars in Arthritis and Rheumatism
Seminars in Arthroplasty
Seminars in Breast Disease
Seminars in Cardiothoracic and Vascular Anesthesia
Seminars in Cerebrovascular Diseases and Stroke
Seminars in Colon and Rectal Surgery
Seminars in Cutaneous Medicine and Surgery
Seminars in Diagnostic Pathology
Seminars in Hearing
Seminars in Hematology
Seminars in Integrative Medicine
Seminars in Interventional Radiology
Seminars in Liver Disease
Seminars in Musculoskeletal Radiology
Seminars in Nephrology

Seminars in Neurology
Seminars in Nuclear Medicine
Seminars in Oncology
Seminars in Pain Medicine
Seminars in Pediatric Infectious Diseases
Seminars in Pediatric Neurology
Seminars in Pediatric Surgery
Seminars in Perinatology
Seminars in Radiation Oncology
Seminars in Reproductive Medicine
Seminars in Respiratory and Critical Care Medicine
Seminars in Roentgenology
Seminars in Speech and Language
Seminars in Spine Surgery
Seminars in Surgical Oncology
Seminars in Thoracic and Cardiovascular Surgery
Seminars in Thrombosis and Hemostasis
Seminars in Ultrasound, CT and MRI
Seminars in Vascular Medicine
Seminars in Vascular Surgery
Sexuality and Disability
Sexuality, Reproduction and Menopause
Sexually Transmitted Diseases
Shock
Skull Base
Sleep
Sleep Research Online
Social Work in Health Care
Somatic Cell and Molecular Genetics
Southern Medical Journal
Spine
Spine Journal
Sports Medicine and Arthroscopy Review
Statistical Applications in Genetics and Molecular Biology
Stem Cells
Stem Cells and Development
Steroids
Stroke
Structure
Substance Abuse
Substance Use and Misuse
Suicide and Life-Threatening Behavior
Supportive Cancer Therapy
Surgery
Surgery for Obesity and Related Diseases
Surgical Clinics of North America
Surgical Endoscopy
Surgical Infections
Surgical Innovation
Surgical Laparoscopy, Endoscopy and Percutaneous Techniques
Surgical Neurology
Surgical Oncology Clinics of North America
Surgical Practice
Survey of Ophthalmology
Synapse
Synthetic Communications
TEN Trends in Evidence-Based Neuropsychiatry
Techniques in Foot and Ankle Surgery
Techniques in Gastrointestinal Endoscopy
Techniques in Hand and Upper Extremity Surgery
Techniques in Knee Surgery
Techniques in Ophthalmology
Techniques in Orthopaedics
Techniques in Regional Anesthesia and Pain Management
Techniques in Shoulder and Elbow Surgery
Techniques in Vascular and Interventional Radiology
Technology in Cancer Research and Treatment
Telemedicine Journal and e-Health
Texas Heart Institute Journal
Therapeutic Apheresis and Dialysis
Therapeutic Drug Monitoring
Thoracic Surgery Clinics
Thyroid
Tissue Engineering
Today's Therapeutic Trends
Topics in Advanced Practice Nursing
Topics in Clinical Nutrition
Topics in Emergency Medicine
Topics in Geriatric Rehabilitation
Topics in Magnetic Resonance Imaging
Topics in Spinal Cord Injury Rehabilitation
Topics in Stroke Rehabilitation
Toxicological Sciences
Toxicology and Applied Pharmacology
Toxin Reviews
Traffic Injury Prevention
Transactions of the American Ophthalmological Society
Transfusion
Transfusion Medicine Reviews
Transplantation
Transplantation Proceedings
Transplantation Reviews
Trauma
Travel Medicine and Infectious Disease
Trends in Amplification
Trends in Cardiovascular Medicine
Trends in Endocrinology and Metabolism
Ultrasonic Imaging
Ultrasound Quarterly
Ultrasound in Medicine and Biology
Urologic Clinics of North America
Urologic Oncology: Seminars and Original Investigations
Urology
VHSJ - Veterans Health System Journal
Value in Health
Vascular Disease Management
Vascular Pharmacology
Vascular and Endovascular Surgery
Vector-Borne and Zoonotic Diseases
Veterinary Clinics of North America - Small Animal Practice
Veterinary and Human Toxicology
Violence and Victims
Viral Immunology
Virology
Visual Neuroscience
Water Environment Research
Wilderness and Environmental Medicine
Wisconsin Medical Journal
Women and Health
Women's Health Issues
World Journal of Surgery
Wound Ballistics Review
Wound Repair and Regeneration
Yale Journal of Biology and Medicine

VEN

...ıfica Venezolana
...e Medico
...stigacion Clinica
...alus

List per Country